THE COMPLEX VARIABLES PROBLEM SOLVER®

REGISTERED TRADEMARK

A Complete Solution Guide to Any Textbook

Emil G. Milewski, Ph.D.

Research and Education Association
61 Ethel Road West
Piscataway, New Jersey 08854

THE COMPLEX VARIABLES PROBLEM SOLVER ®

Copyright © 1987 by Research and Education Association. All rights reserved. No part of this book may be reproduced in any form without permission of the publisher.

Printed in the United States of America

Library of Congress Card Catalog Number 87-60365

International Standard Book Number 0-87891-604-0

Revised Printing, 1992

PROBLEM SOLVER is a registered trademark of
Research and Education Association, Piscataway, New Jersey

WHAT THIS BOOK IS FOR

Students have generally found complex variables a difficult subject to understand and learn. Despite the publication of hundreds of textbooks in this field, each one intended to provide an improvement over previous textbooks, students continue to remain perplexed as a result of the numerous conditions that often must be remembered and correlated in solving a problem. Various possible interpretations of terms used in complex variables have also contributed to much of the difficulties experienced by students.

In a study of the problem, REA found the following basic reasons underlying students' difficulties with complex variables taught in schools:

(a) No systematic rules of analysis have been developed which students may follow in a step-by-step manner to solve the usual problems encountered. This results from the fact that the numerous different conditions and principles which may be involved in a problem lead to many possible different methods of solution. To prescribe a set of rules to be followed for each of the possible variations, would involve an enormous number of rules and steps to be searched through by students, and this task would perhaps be more burdensome than solving the problem directly with some accompanying trial and error to find the correct solution route.

(b) Textbooks currently available will usually explain a given principle in a few pages written by a professional who has an insight in the subject matter that is not shared by students. The explanations are often written in an abstract manner which leaves the students confused as to the application of the principle. The explanations given are not sufficiently detailed and extensive to make the student aware of the wide range of applications and different aspects of the principle being studied. The numerous possible variations of principles and their applications are usually not discussed, and it is left for the students to discover these for themselves while doing exercises. Accordingly, the average student is expected to rediscover that which has been long known and practiced, but not published or explained extensively.

(c) The examples usually following the explanation of a topic are too few in number and too simple to enable the student to obtain a thorough grasp of the principles involved. The explanations do not provide sufficient basis to enable a student to solve problems that may be subsequently assigned for homework or given on examinations.

The examples are presented in abbreviated form which leaves out much material between steps, and requires that students derive the omitted material themselves. As a result, students find the examples difficult to understand--contrary to the purpose of the examples.

Examples are, furthermore, often worded in a confusing manner. They do not state the problem and then present the solution. Instead, they pass through a general discussion, never revealing what is to be solved for.

Examples, also, do not always include diagrams/graphs, wherever appropriate, and students do not obtain the training to draw diagrams or graphs to simplify and organize their thinking.

(d) Students can learn the subject only by doing the exercises themselves and reviewing them in class, to obtain experience in applying the principles with their different ramifications.

In doing the exercises by themselves, students find that they are required to devote considerably more time to complex variables than to other subjects of comparable credits, because they are uncertain with regard to the selection and application of the theorems and principles involved. It is also often necessary for students to discover those "tricks" not revealed in their texts (or review books), that make it possible to solve problems easily. Students must usually resort to methods of trial-and-error to discover these "tricks", and as a result they find that they may sometimes spend several hours to solve a single problem.

(e) When reviewing the exercises in classrooms, instructors usually request students to take turns in writing solutions on the boards and explaining them to the class. Students often find it difficult to explain in a manner that holds the interest of the class, and enables the remaining students to follow the material written on the boards. The remaining students seated in the class are, furthermore, too occupied with copying the material from the boards, to listen to the oral explanations and concentrate on the methods of solution.

This book is intended to aid students in complex variables in overcoming the difficulties described, by supplying detailed illustrations of the solution methods which are usually not apparent to students. The solution methods are illustrated by problems selected from those that are most often assigned for class work and given on examinations. The problems are arranged in order of complexity to enable students to learn and understand a particular topic by reviewing the problems in sequence. The problems are illustrated with detailed step-by-step explanations, to save the students the large amount of time that is often needed to fill in the gaps that are usually found between steps of illustrations in textbooks or review/outline books.

The staff of REA considers complex variables a subject that is best learned by allowing students to view the methods of analysis and solution techniques themselves. This approach to learning the subject matter is similar to that practiced in various scientific laboratories, particularly in the medical fields.

In using this book, students may review and study the illustrated problems at their own pace; they are not limited to the time allowed for explaining problems on the board in class.

When students want to look up a particular type of problem and solution, they can readily locate it in the book by referring to the index which has been extensively prepared. It is also possible to locate a particular type of problem by glancing at just the material within the boxed portions. To facilitate rapid scanning of the problems, each problem has a heavy border around it. Furthermore, each problem is identified with a number immediately above the problem at the right-hand margin.

To obtain maximum benefit from the book, students should familiarize themselves with the section, "How To Use This Book," located in the front pages.

To meet the objectives of this book, staff members of REA have selected problems usually encountered in assignments and examinations, and have solved each problem meticulously to illustrate the steps which are difficult for students to comprehend. Special gratitude is expressed to them for their efforts in this area, as well as to the numerous contributors who devoted brief periods of time to this work.

Gratitude is also expressed to the many persons involved in the difficult task of typing the manuscript with its endless changes, and to the REA art staff who prepared the numerous detailed illustrations together with the layout and physical features of the book.

The difficult task of coordinating the efforts of all persons was carried out by Carl Fuchs. His conscientious work deserves much appreciation. He also trained and supervised art and production personnel in the preparation of the book for printing.

Finally, special thanks are due to Helen Kaufmann for her unique talents in rendering those difficult border-line decisions and in making constructive suggestions related to the design and organization of the book.

Max Fogiel, Ph.D.
Program Director

HOW TO USE THIS BOOK

This book can be an invaluable aid to students in complex variables as a supplement to their textbooks. The book is subdivided into 24 chapters, each dealing with a separate topic. The subject matter is developed beginning with complex numbers, geometric representations, De Moivre's Theorem, sequences and series, continuous functions, limits, Cauchy's Theorem, power series, and extending through radius of convergence, residues, Taylor series, and Laurent's series. Also included are problems on special kinds of integrals, conformal mappings, and symmetry principle. An extensive number of applications have been included, since these appear to be more troublesome to students.

TO LEARN AND UNDERSTAND A TOPIC THOROUGHLY

1. Refer to your class text and read the section pertaining to the topic. You should become acquainted with the principles discussed there. These principles, however, may not be clear to you at that time.

2. Then locate the topic you are looking for by referring to the "Table of Contents" in front of this book, "The Complex Variables Problem Solver."

3. Turn to the page where the topic begins and review the problems under each topic, in the order given. For each topic, the problems are arranged in order of complexity, from the simplest to the more difficult. Some problems may appear similar to others, but each problem has been selected to illustrate a different point or solution method.

To learn and understand a topic thoroughly and retain its contents, it will be generally necessary for students to review the problems several times. Repeated review is essential in order to gain experience in recognizing the principles that should be applied, and in selecting the best solution technique.

TO FIND A PARTICULAR PROBLEM

To locate one or more problems related to a particular subject matter, refer to the index. In using the index, be certain to note that the numbers given there refer to problem numbers, not page numbers. This arrangement of the index is intended to facilitate finding a problem more rapidily, since two or more problems may appear on a page.

If a particular type of problem cannot be found readily, it is recommended that the student refer to the "Table of Contents" in the front pages, and then turn to the chapter which is applicable to the problem being sought. By scanning or glancing at the material that is boxed, it will generally be possible to find problems related to the one being sought, without consuming considerable time. After the problems have been located, the solutions can be reviewed and studied in detail. For this purpose of locating problems rapidly, students should acquaint themselves with the organization of the book as found in the "Table of Contents".

In preparing for an exam, locate the topics to be covered on the exam in the "Table of Contents," and then review the problems under those topics **several times**. This should equip the student with what might be **needed for** the exam.

CONTENTS

Chapter No.   **Page No.**

1 COMPLEX NUMBERS, INTRODUCTORY REMARKS   1

       Integers, Rational and Real Numbers   1
       Complex Numbers, Imaginary Unit "i"   6
       Fundamental Operations with Complex Numbers   8
       Field of Complex Numbers   16
       Complex Conjugate   19
       Absolute Value   21
       Inequalities and Identities   26
       Polynomials   30

2 GEOMETRIC REPRESENTATION OF COMPLEX NUMBERS   35

       Representations of Complex Numbers   35
       Applications of Complex Numbers in Geometry   39
       Applications in Physics   65

3 DE MOIVRE'S THEOREM   67

       Polar Form of Complex Numbers   68
       Euler's Formula   75
       Applications in Geometry   79
       Dot and Cross Products   89
       De Moivre's Theorem   94
       Applications in Trigonometry, Identities   108
       Series Polynomials   123

4 COMPLEX NUMBERS AS A METRIC SPACE   135

Elements of Set Theory   136
Metric Spaces   139
Open Sets   145
Topological Spaces   148
Closure of a Set   154
Regions, Domains   157
Sequences   161
Complete and Compact Spaces   165
Theorems and Properties   169

5 SEQUENCES AND SERIES OF COMPLEX NUMBERS   176

Sequences   177
Series   191
Test for Convergence   197

6 CONTINUOUS MAPPINGS, CONTINUOUS CURVES, STEREOGRAPHIC PROJECTION   217

Mappings   218
Continuous Mappings, Homomorphisms   221
Properties of Topological Spaces   226
Curves   235
Convex Sets, Domains   240
Continuous Curves   244
Stereographic Projection, Riemann Sphere   246

7 FUNCTIONS, LIMITS AND CONTINUITY   261

Complex Functions   261
Limits   266
Continuity   270
Uniform Continuity   275
Functions Continuous or Connected, Compact, and Bounded Sets   281

8 DIFFERENTIATION, ANALYTIC FUNCTIONS, CAUCHY-RIEMANN CONDITIONS 285

Derivatives 285
Analytic Functions, Cauchy-Riemann Equations 296
Properties of Analytic Functions 301
Some Applications 313

9 DIFFERENTIAL OPERATORS HARMONIC FUNCTIONS 317

Differential Operators, Gradient, Divergence, and Curl 317
Harmonic Functions 322
Differential Equations 328

10 ELEMENTARY FUNCTIONS, MULTIPLE-VALUED FUNCTIONS 332

Exponential Functions 332
Trigonometric Functions 335
Hyperbolic Functions 337
Logarithmic Functions 340
Multiple-Valued Functions 346

11 COMPLEX INTEGRALS, CAUCHY'S THEOREM 356

Complex Line Integrals 356
Green's Theorem 364
Cauchy's Theorem, The Cauchy-Gorsat Theorem 369
Integrals Independent of Path 374
Applications 377
Indefinite Integrals 384

12 CAUCHY'S INTEGRAL FORMULA AND RELATED THEOREMS 388

Cauchy's Integral Formula 389
Applications 396
Theorems 401
Functions with Poles and Zeros 408

13 POWER SERIES 420

    Cauchy Convergence Criterion   420
    Some Useful Theorems   425
    Uniform Convergence   437
    Power Series of Analytic Functions   442

14 TAYLOR SERIES 447

    Taylor Theorem   447
    Mac Laurin Series   451
    Taylor Expansion   456
    Applications   458
    Taylor Series for Rational Functions   461
    More Applications   468

15 LAURENT EXPANSION 474

    Laurent Series Expansion   474
    Examples   482
    Applications   490
    Lagrange's Expansion   498
    Analytic Continuation   502

16 SINGULARITIES, RATIONAL AND MEROMORPHIC FUNCTIONS 506

    Singularities   506
    Poles   512
    Properties of Singularities   517
    Removable Singularities, Singularities at Infinity   525
    Theorems   533
    Rational and Meromorphic Functions   536

17 RESIDUE CALCULUS 542

    Residues   542
    Residues at Poles   545
    Residues of Functions $\frac{p(z)}{q(z)}$ 553
    Residue Theorem   555
    Applications   560

18 APPLICATIONS OF RESIDUE CALCULUS, INTEGRATION OF REAL FUNCTIONS 574

19 APPLICATIONS OF RESIDUE CALCULUS PART II 611

20 MAPPING BY ELEMENTARY FUNCTIONS AND LINEAR FRACTIONAL TRANSFORMATIONS 662

Linear Transformation 662
Rotation, Translation 664
Function $\frac{1}{z}$ 666
Linear Fractional Transformation 672
Fixed Points 675
Transformations of Regions 676
Some Special Functions 683

21 CONFORMAL MAPPINGS, BOUNDARY VALUE PROBLEM 686

Critical and Non-Critical Points 687
Conformal Mapping 689
Theorems, Jacobian of the Transformation 699
Harmonic Conjugate, Dirichlet Problem 706
Applications 713
Boundary Value Problem, Neumann Problem, Poisson's Formula 722

22 APPLICATIONS IN PHYSICS 727

Heat Flow 727
Elasticity 746
Hydrodynamics and Aerodynamics 749
Flow Around an Object 766
Electrostatic Fields 776

23 APPLICATIONS OF CONFORMAL MAPPINGS : THE SCHWARZ-CHRISTOFFEL TRANSFORMATION   801

Mapping of Polygons onto the Real Axis   801
Triangles, Rectangles and Degenerate Polygons   804
Various Polygons   821
Applications in Physics   824

24 SPECIAL TOPICS OF COMPLEX ANALYSIS   828

Analytic Continuation   828
Principle of Reflection   837
Gamma Function   839
Beta Function   849
Infinite Products   856
Differential Equations   863
Bessel Function   871
Legendre Polynomials   875
Hypergeometric Function   878
Zeta Function   879
Doubly Periodic Functions, Elliptic Functions   881
Asymptotic Expansions   890

APPENDIX   893

INDEX   900

CHAPTER 1

COMPLEX NUMBERS, INTRODUCTORY REMARKS

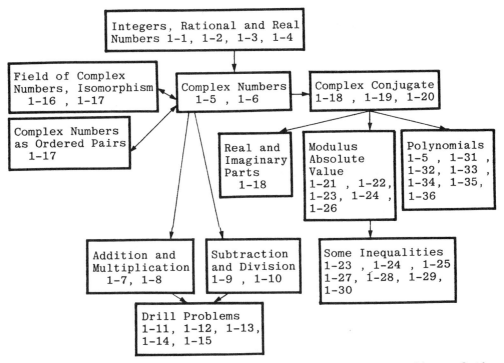

This chart is provided to facilitate rapid understanding of the interrelationships of the topics and subject matter in this chapter. Also shown are the problem numbers associated with the subject matter.

INTEGERS, RATIONAL AND REAL NUMBERS

● PROBLEM 1-1

Describe the following sets:

1. the set N of positive integers
2. the set Z of all integers and zero
3. the set E of positive even integers
4. the set Q of rational numbers
5. the set R of real numbers

Using a diagram, illustrate the relationships between the sets of the type set - subset.

Solution: In this book we shall use the commonly accepted logical notation. A collection of things is called a set, a family, or a class. For our purposes, all three terms mean the same thing. Sets are designated by upper-case letters, and their elements by the corresponding lower-case letters: $a \in A$ is read as "a belongs to the set A" or "a is an element of A". For example: Texas $\in$ USA. On the other hand, the notation $b \notin A$ is read as "b is not an element of A".

A set of objects having a particular property is denoted by $\{x : P\}$. This statement is read "the set of all x which possess the property P". Sometimes, we simply list all of the elements of a set. In such cases, we explicitly show the contents of the set. For example, $\{1,7,4\}$ is the set consisting of the three elements 1, 4, and 7.

1. The set N of positive integers (natural numbers) is

$$N \equiv \{1,2,3,\ldots\} \tag{1}$$

Note that "$\equiv$" indicates "equal by definition". N is closed under the operation of addition and multiplication. That is, if a and b are natural numbers, then the sum $a+b$ and the product ab are also natural numbers.

2. The set Z consists of

$$Z \equiv \{\ldots-4,-3,-2,-1,0,1,2,3,\ldots\} \tag{2}$$

It is easy to see that Z is closed under the operation of addition, multiplication and subtraction. However, it is not closed under the operation of division because $4 \in Z$ & $3 \in Z$, but $\frac{4}{3} \notin Z$.

3. The set of positive even integers is given by

$$E \equiv \{e : e = 2n, \; n \in N\} \tag{3}$$

4. The set of rational numbers is

$$Q \equiv \{q : q = \frac{m}{n}, \; m \in Z, \; n \in Z, \; n \neq 0\} \tag{4}$$

Q is closed under the four basic arithmetic operations: addition, subtraction, multiplication and division.

5. The set of real numbers R consists of all rational numbers and all irrational numbers. An irrational number is a number that cannot be expressed as a ratio $\frac{a}{b}$, where a and b are integers and $b \neq 0$. An example of an irrational number is $\sqrt{2} = 1.4142\ldots$.

A one-to-one correspondence can be established between the set of real numbers and a line. Each number is represented by a point of the line called the real axis.

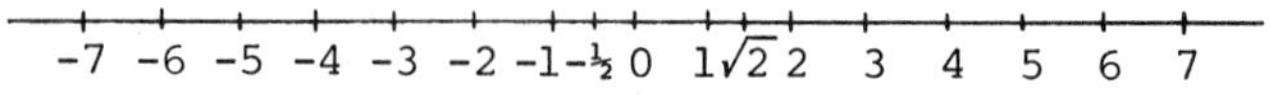

$-7 \quad -6 \quad -5 \quad -4 \quad -3 \quad -2 \; -1 -\tfrac{1}{2} \; 0 \quad 1 \sqrt{2} \; 2 \quad 3 \quad 4 \quad 5 \quad 6 \quad 7$

Fig. 1

A set A is a subset of B, i.e. is included in B, when

$$x \in A \Rightarrow x \in B \qquad (5)$$

The notation $\Rightarrow$ is read as: then. We write $A \subset B$ to indicate that A is a subset of B. Let us note that $A \subset B$ does not exclude the possibility that $A = B$.

Any positive even integer is a positive integer, therefore

$$E \subset N \qquad (6)$$

Any positive integer is an integer, thus

$$N \subset Z \qquad (7)$$

Any integer or zero can be expressed in the form $\frac{a}{1}$, thus

$$Z \subset Q \qquad (8)$$

By definition, any rational number is a real number, thus

$$Q \subset R \qquad (9)$$

We have

$$E \subset N \subset Z \subset Q \subset R \qquad (10)$$

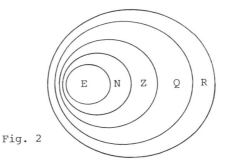

Fig. 2

● **PROBLEM 1-2**

Consider a linear polynomial equation

$$ax + b = 0 \qquad (1)$$

Using the definitions given in Problem 1-1, find the smallest set containing the solutions of eq.(1) when

1. $a = 1$ and $b \in E$
2. $a \in Z$ and $b \in N$
3. $a \in Q$ and $b \in Q$
4. $a \in R$ and $b \in Q$.

Solution: The solution of eq.(1) is

$$x = -\frac{b}{a} \qquad (2)$$

1. If a = 1 and b ∈ E, then x = -b. Of course x ∉ E because E is the set of positive even integers. The smallest set containing x is Z.

2. Since a ∈ Z and b ∈ N, $-\frac{b}{a} \in Q$.

3. Since both a and b are rational numbers, $-\frac{b}{a}$ is a rational number, hence, $-\frac{b}{a} \in Q$.

4. Here, a ∈ R and b ∈ Q. Let us note that $-\frac{b}{a}$ does not have to be a rational number. For example, when a = √2 and b = 1, $-\frac{b}{a}$ is not a rational number.

Hence, $-\frac{b}{a}$ is a real number, $-\frac{b}{a} \in R$.

● **PROBLEM 1-3**

Imagine that due to some perfidious law, all irrational numbers have been banned from private, public and scientific life. As a law-abiding citizen and a good mathematician, what would you do to make this quandary bearable?

Solution: Certainly you know that there are several theories on real numbers. Axiomatic theories, even though convenient, are not very instructive. The theory of Cantor* (1845-1918), the founder of the theory of sets, is particularly handy. It takes as a starting point the set of rational numbers Q and considers sequences of rational numbers. These so-called Cauchy sequences play a key role. In the space α of Cauchy sequences, Cantor introduces the equivalence relation ~. The set of real numbers R is defined as a quotient space

$$R \equiv \frac{\alpha}{\sim} \quad (1)$$

The algebraic operations are carried over from the set of rational numbers Q to the space R. The reader will find a variety of books on Cantor's method, and it is not within our scope to present it here. An important conclusion of this theory (noted by Hausdorff) is that the space R is the complement of the set of rational numbers,

$$R = \overline{Q} \quad (2)$$

A practical conclusion from this is that if we take any real number r ∈ R and any positive rational number e ∈ Q, regardless of how small, there exists a rational number q ∈ Q such that

$$|r - q| < e \quad (3)$$

Each real number can be approximated with any required accuracy by rational numbers. Computers, electronics, space technology etc. use rational numbers.

At some point in your life, you were √19 feet tall. Chances are that when measuring your height, you used the approximation 4 ft. 4 inches, and everybody was satisfied with the accuracy. The answer to the question posed in the problem is to approximate all irrational numbers with rational numbers.

*Georg Cantor was born in 1845 in Petrograd and he died in 1918 in Halle. Cantor, the son of a Danish merchant, was one of the most prominent German mathematicians. Cantor studied at Zürich, Gottingen, and Berlin. In 1872, he became a professor in Halle. Cantor introduced many original ideas into the theory of sets (theory of aggregates). He also worked on the foundations of the theory of numbers, these findings were published in the "Annalen" in 1879.

● **PROBLEM 1-4**

1. A linear equation is given

$$ax + b = 0 \qquad (1)$$

where a and b are rational numbers. Is the solution of this equation always a rational number?

2. Is the solution of the equation

$$ax^2 + bx + c = 0 \qquad (2)$$

where $a, b, c \in R$ (i.e. all coefficients are real numbers) always a real number?

<u>Solution</u>: 1. The solution of eq.(1) is

$$x = -\frac{b}{a}, \quad a \neq 0 \qquad (3)$$

Since a and b are rational numbers, $-\frac{b}{a}$ is always a rational number.

2. Dividing eq.(2) by $a \neq 0$, we obtain the equation

$$x^2 + \frac{b}{a}x + \frac{c}{a} = 0 \qquad (4)$$

which can be transformed to

$$x^2 + \frac{b}{a}x + \frac{c}{a} = \left(x^2 + \frac{b}{a}x + \frac{b^2}{4a^2}\right) + \frac{c}{a} - \frac{b^2}{4a^2}$$

$$= \left(x + \frac{b}{2a}\right)^2 + \frac{-b^2 + 4ac}{4a^2} = 0 \qquad (5)$$

so that

$$\left(x + \frac{b}{2a}\right)^2 = \frac{b^2 - 4ac}{4a^2} \qquad (6)$$

When $b^2 - 4ac > 0$ eq.(6) has two solutions

$$x_1 = \sqrt{\frac{b^2 - 4ac}{4a^2}} - \frac{b}{2a} = \frac{-b + \sqrt{b^2 - 4ac}}{2a} \qquad (7)$$

$$x_2 = -\sqrt{\frac{b^2 - 4ac}{4a^2}} - \frac{b}{2a} = \frac{-b - \sqrt{b^2 - 4ac}}{2a} \qquad (8)$$

When $b^2 - 4ac = 0$, one solution exists

$$x = -\frac{b}{2a} \qquad (9)$$

The situation becomes more complicated when $b^2 - 4ac < 0$. The left-hand side of eq.(6) is then positive, while the right-hand side is negative. Such equations cannot be solved in terms of real numbers.

COMPLEX NUMBERS, IMAGINARY UNIT "i"

● **PROBLEM** 1-5

A complex number z can be expressed in the form

$$z = a + bi \qquad (1)$$

where a and b are real numbers and i has the property

$$i^2 = -1 \qquad (2)$$

Find the solutions of the following equations using complex numbers

1. $x^2 + 1 = 0$ \qquad (3)

2. $2x^2 + x + 3 = 0$ \qquad (4)

Solution: The representation of complex numbers in the form $z = a + bi$ is probably the most popular. The mathematical operations of addition, subtraction and multiplication of real numbers carry on into the complex number system. However, one must remember that $i^2 = -1$. Thus,

1. $x^2 + 1 = 0$ \qquad (5)

$$x^2 = -1 \qquad (6)$$

$$x_1 = \sqrt{-1}, \quad x_2 = -\sqrt{-1} \qquad (7)$$

We can write, using the "i" symbol,

$$x_1 = i, \quad x_2 = -i \qquad (8)$$

Let us now verify our results. Substituting $x_1 = i$ into eq.(3) we find

$$x^2 + 1 = i^2 + 1 = -1 + 1 = 0 \qquad (9)$$

Substituting $x_2 = -i$ into eq.(3) results in

$$x^2 + 1 = (-i)^2 + 1 = i^2 + 1 = 0 \qquad (10)$$

2. To solve eq.(4), we use the quadratic formula to obtain the two solutions.

$$x_1 = \frac{-1 + \sqrt{1-4(2)(3)}}{4} = -\frac{1}{4} + i\frac{\sqrt{23}}{4} \qquad (11)$$

$$x_2 = \frac{-1 - \sqrt{1-4(2)(3)}}{4} = -\frac{1}{4} - i\frac{\sqrt{23}}{4} \qquad (12)$$

Let us verify that x_1 and x_2 are indeed the roots of eq.(4).

Substituting x_1 into eq.(4) results in

$$2\left(-\frac{1}{4} + i\frac{\sqrt{23}}{4}\right)^2 + \left(-\frac{1}{4} + i\frac{\sqrt{23}}{4}\right) + 3$$
$$= 2\left(\frac{1}{16} - i\frac{\sqrt{23}}{8} - \frac{23}{16}\right) - \frac{1}{4} + i\frac{\sqrt{23}}{4} + 3 = 0 \qquad (13)$$

Substituting x_2 into eq.(4) gives

$$2\left(-\frac{1}{4} - i\frac{\sqrt{23}}{4}\right)^2 + \left(-\frac{1}{4} - i\frac{\sqrt{23}}{4}\right) + 3$$
$$= 2\left(\frac{1}{16} + i\frac{\sqrt{23}}{8} - \frac{23}{16}\right) + \left(-\frac{1}{4} - i\frac{\sqrt{23}}{4}\right) + 3 = 0 \qquad (14)$$

Complex numbers were introduced into mathematics in connection with the solution of algebraic equations. Since it is impossible to solve the polynomial equation

$$x^2 + 1 = 0 \qquad (15)$$

in the domain of real numbers, the imaginary unit i, defined by the equation

$$i^2 = -1 \qquad (16)$$

was introduced by Euler in 1777 and fully incorporated into the notation of complex numbers by Gauss.

One should execute caution in handling i, $\sqrt{-1}$ etc. The following fallacy indicates why.

$$-1 = i^2 = i \cdot i = \sqrt{-1}\sqrt{-1} = \sqrt{(-1)(-1)} = \sqrt{1} = 1$$

thus $-1 = 1$. Explain what went wrong.

• **PROBLEM 1-6**

Define the system of complex numbers. When are two numbers equal?

Solution: The system of complex numbers is the set of all numbers a+ib where a and b are real, with two binary operations, addition and multiplication, defined as follows:

$$(a+ib) + (c+id) \equiv a+c+i(b+d) \qquad (1)$$

$$(a+ib) \cdot (c+id) \equiv (ac-bd) + i(ad+bc) \qquad (2)$$

The system of complex numbers, denoted $(C,+,\cdot)$ is a field whose multiplicative neutral element is 1,

$$1 = 1 + i0 \qquad (3)$$

whose additive neutral element is 0,
$$0 = 0 + i0 \qquad (4)$$
and in which $i^2 = -1$.

The field of real numbers R is a subset of C. Strictly speaking, the elements of C of the form $a = a + i0$ form a subfield of C, which is isomorphic as a field to the field of real numbers. The function $a \to a \cdot 1 + i0$ from R into C maps R one-to-one onto this subfield, preserving products and sums, so it is an isomorphism.

Two complex numbers
$$\begin{aligned} z_1 &= a_1 + ib_1 \\ z_2 &= a_2 + ib_2 \end{aligned} \qquad (5)$$
are equal if, and only if,
$$a_1 = a_2$$
and $\qquad (6)$
$$b_1 = b_2$$

The student should keep in mind that the statement $z_1 > z_2$ is meaningless unless both z_1 and z_2 are real.

FUNDAMENTAL OPERATIONS WITH COMPLEX NUMBERS

• **PROBLEM 1-7**

---

1. Prove the theorem:

    For all complex numbers, addition is commutative
    $$z_1 + z_2 = z_2 + z_1 \qquad (1)$$
    and associative
    $$z_1 + (z_2 + z_3) = (z_1 + z_2) + z_3 \qquad (2)$$

2. Reduce the following numbers to the form $a + ib$.

    $(a + ib) + (-a + ib)$

    $(4 + 2i) + (3 - 3i)$

    $(2 + 4i) - (6 - 3i) + (4 + 7i)$

Solution: 1. Let
$$\begin{aligned} z_1 &= a_1 + ib_1 & (3) \\ z_2 &= a_2 + ib_2 & (4) \end{aligned}$$
Then
$$\begin{aligned} z_1 + z_2 &= (a_1 + ib_1) + (a_2 + ib_2) \\ &= (a_1+a_2) + i(b_1+b_2) = (a_2+a_1) + i(b_2+b_1) \\ &= (a_2+ib_2) + (a_1+ib_1) = z_2 + z_1 \end{aligned} \qquad (5)$$

Here we used the fact that the addition of real numbers is commutative. We have,

$$z_1 + (z_2+z_3) = (a_1+ib_1) + [(a_2+ib_2) + (a_3+ib_3)]$$
$$= (a_1+ib_1) + [(a_2+a_3) + i(b_2+b_3)]$$
$$= [a_1 + (a_2+a_3)] + i[b_1 + (b_2+b_3)] \quad (6)$$
$$= [(a_1+a_2) + a_3] + i[(b_1+b_2) + b_3]$$
$$= [(a_1+a_2) + i(b_1+b_2)] + a_3 + ib_3$$
$$= (z_1+z_2) + z_3$$

We used the fact that addition of real numbers is associative.

2. $\quad (a+ib) + (-a+ib) = ib + ib = 2bi \quad (7)$

$\quad (4+2i) + (3-3i) = 7 - i \quad (8)$

$\quad (2+4i) - (6-3i) + (4+7i) \quad (9)$

$= 2 + 4i - 6 + 3i + 4 + 7i = 14i \quad (10)$

● **PROBLEM 1-8**

1. Prove that:

   (a) Multiplication of complex numbers is commutative, i.e.
   $$z_1 \cdot z_2 = z_2 \cdot z_1 \quad (1)$$

   (b) Multiplication is associative
   $$z_1 \cdot (z_2 \cdot z_3) = (z_1 \cdot z_2) \cdot z_3 \quad (2)$$

   (c) Multiplication is distributive with respect to addition
   $$z_1 \cdot (z_2 + z_3) = z_1 \cdot z_2 + z_1 \cdot z_3 \quad (3)$$

2. Reduce the following numbers to the form $a + ib$.

   (a) $(3-2i) \cdot (4+5i) + (3-2i) \cdot (4-5i) \quad (4)$

   (b) $(2-i) \cdot (4+3i) \cdot (5+2i) \quad (5)$

Solution: Let
$$z_1 = a_1 + ib_1$$
$$z_2 = a_2 + ib_2 \quad (7)$$

(a) Since multiplication of real numbers is commutative, we obtain
$$z_1 \cdot z_2 = (a_1+ib_1) \cdot (a_2+ib_2) = a_1a_2 + i(a_1b_2+b_1a_2)$$
$$+ i^2b_1b_2 = a_1a_2 - b_1b_2 + i(a_1b_2+a_2b_1) \quad (8)$$
$$= a_2a_1 - b_2b_1 + i(b_2a_1+a_2b_1) = (a_2+ib_2)(a_1+ib_1)$$
$$= z_2 \cdot z_1$$

(b) Multiplication of real numbers is associative, thus
$$z_1 \cdot (z_2 \cdot z_3) = (a_1+ib_1) \cdot [(a_2+ib_2) \cdot (a_3+ib_3)]$$

$$\begin{aligned}
&= (a_1+ib_1) \cdot [(a_2a_3-b_2b_3) + i(b_2a_3+a_2b_3)] \\
&= a_1(a_2a_3-b_2b_3) - b_1(b_2a_3+a_2b_3) \\
&\quad + i[b_1(a_2a_3-b_2b_3) + a_1(b_2a_3+a_2b_3)] \quad &(9) \\
&= (a_1a_2-b_1b_2)a_3 - (a_1b_2+a_2b_1)b_3 \\
&\quad + i[(a_1b_2+b_1a_2)a_3 + (a_1a_2-b_1b_2)b_3] \\
&= [(a_1+ib_1) \cdot (a_2+ib_2)] \cdot (a_3+ib_3) \\
&= (z_1 \cdot z_2) \cdot z_3
\end{aligned}$$

(c) Now, to prove eq.(3), we write

$$\begin{aligned}
z_1 \cdot (z_2+z_3) &= (a_1+ib_1) \cdot [(a_2+ib_2) + (a_3+ib_3)] \\
&= (a_1+ib_1) \cdot [(a_2+a_3) + i(b_2+b_3)] \\
&= a_1(a_2+a_3) - b_1(b_2+b_3) + ib_1(a_2+a_3) + ia_1(b_2+b_3) \\
&= (a_1a_2-b_1b_2) + i(b_1a_2+a_1b_2) \quad &(10) \\
&\quad + (a_1a_3-b_1b_3) + i(a_1b_3+b_1a_3) \\
&= z_1 \cdot z_2 + z_1 \cdot z_3
\end{aligned}$$

Instead of writing $z_1 \cdot (z_2 \cdot z_3)$, we skip the brackets and write $z_1 \cdot z_2 \cdot z_3$.

2.(a) $(3-2i) \cdot (4+5i) + (3-2i) \cdot (4-5i)$
$= (3-2i) \cdot [(4+5i) + (4-5i)]$ (11)
$= (3-2i) \cdot 8 = 24 - 16i$

(b) $(2-i) \cdot (4+3i) \cdot (5+2i) = (11+2i)(5+2i)$
$= 51 + 32i$ (12)

• **PROBLEM 1-9**

> 1. Prove that if $z_1 = a_1 + ib_1$ and $z_2 = a_2 + ib_2$ are complex numbers, then there exists a unique complex number z such that
>
> $$z_1 + z = z_2 \quad (1)$$
>
> 2. Given two complex numbers z_1 and z_2, we define the difference z, denoted by $z_2 - z_1$, to be a complex number such that
>
> $$z_1 + z = z_2 \quad (2)$$
>
> Find
>
> $(3+4i) - (2-3i)$ (3)
>
> $(7-7i) - (7+6i)$ (4)

Solution: 1. Let $z = a + ib.$ (5)

The equation $z_1 + z = z_2$ is true if and only if

$$a_1 + a = a_2 \quad \text{and} \quad (6)$$
$$b_1 + b = b_2 \quad (7)$$

or equivalently, if and only if

$$a = a_2 - a_1 \quad \text{and} \quad (8)$$

$$b = b_2 - b_1 \quad (9)$$

Thus z exists and is unique.

2.
$$(3+4i) - (2-3i) = 1 + 7i \quad (10)$$

$$(7-7i) - (7+6i) = -13i \quad (11)$$

It immediately follows from the above theorem that the difference of two complex numbers exists and is unique.

● **PROBLEM** 1-10

1. Prove the following theorem:

   Given two complex numbers $z_1 = a_1 + ib_1$ and $z_2 = a_2 + ib_2$, $z_2 \neq 0$, there exists a unique complex number z, such that

   $$zz_2 = z_1 \quad (1)$$

   The quotient of z_1 and z_2 is denoted by $\frac{z_1}{z_2}$.

2. Reduce the following numbers to the form $a + ib$.

   (a) $\dfrac{3 - 2i}{4 + 3i}$ \quad (2)

   (b) $\dfrac{2 - 2i}{7 + i} + \dfrac{3 + 4i}{2 - 3i}$ \quad (3)

Solution: 1. Let $\quad z = a + ib.$ \quad (4)

We shall show that eq.(1) determines uniquely a and b. We have

$$zz_2 = (a+ib)(a_2+ib_2) \quad (5)$$

$$= (aa_2 - bb_2) + i(ba_2 + ab_2) = a_1 + ib_1$$

Eq.(5) is true if, and only if,

$$aa_2 - bb_2 = a_1 \quad (6)$$

$$ba_2 + ab_2 = b_1 \quad (7)$$

Solving for a and b, we find

$$a = \frac{a_1 a_2 + b_1 b_2}{a_2^2 + b_2^2} \quad (8)$$

$$b = \frac{b_1 a_2 - a_1 b_2}{a_2^2 + b_2^2} \quad (9)$$

Hence, $z = a + ib$ is uniquely determined.

2.(a) $$\frac{3-2i}{4+3i} = \frac{(3-2i)(4-3i)}{(4+3i)(4-3i)}$$

$$= \frac{12-9i-8i+6i^2}{16-12i+12i-9i^2} = \frac{6-17i}{16+9} \qquad (10)$$

$$= \frac{6}{25} - \frac{17}{25}i$$

(b) $$\frac{2-2i}{7+i} + \frac{3+4i}{2-3i} = \frac{(2-2i)(7-i)}{(7+i)(7-i)} + \frac{(3+4i)(2+3i)}{(2-3i)(2+3i)}$$

$$= \frac{6}{25} - \frac{8}{25}i - \frac{6}{13} + \frac{17}{13}i \qquad (11)$$

$$= \left(\frac{6}{25} - \frac{6}{13}\right) + \left(\frac{17}{13} - \frac{8}{25}\right)i$$

● **PROBLEM 1-11**

1. Reduce the following numbers to the form $a + ib$

   (a) $i^4 - 3i^3 + 4i^2 + 2i - 6$ (1)

   (b) $\left(\dfrac{2i}{1+i}\right)^4$ (2)

2. Prove that $z_1 z_2 = 0$ if and only if at least one of the numbers z_1, z_2 is zero.

<u>Solution</u>: 1.(a) $i^4 - 3i^3 + 4i^2 + 2i - 6$

$$= (-1)(-1) - 3\cdot(-1)i + 4\cdot(-1) + 2i - 6 \qquad (3)$$

$$= 1 + 3i - 4 + 2i - 6 = -9 + 5i$$

(b) $$\left(\frac{2i}{1+i}\right)^4 = \left[\left(\frac{2i}{1+i}\right)^2\right]^2 = \left(\frac{-4}{2i}\right)^2 = -4 \qquad (4)$$

2. Let $z_1 = a_1 + ib_1$ and $z_2 = a_2 + ib_2$, then

$$z_1 z_2 = (a_1+ib_1)(a_2+ib_2)$$

$$= (a_1 a_2 - b_1 b_2) + i(a_1 b_2 + a_2 b_1) = 0 \qquad (5)$$

Eq.(5) implies that

$$a_1 a_2 - b_1 b_2 = 0 \qquad (6)$$

$$a_1 b_2 + a_2 b_1 = 0 \qquad (7)$$

Therefore

$$0 = (a_1 a_2 - b_1 b_2)^2 + (a_1 b_2 + a_2 b_1)^2$$
$$= (a_1^2 + b_1^2)(a_2^2 + b_2^2) \qquad (8)$$

If both $a_1^2 + b_1^2$ and $a_2^2 + b_2^2$ are different from zero, eq.(8) is not true.

Hence, $z_1 z_2 = 0$ if and only if at least one of the numbers z_1, z_2 is zero.

Let us note that if z is a complex number and n is a positive integer, then

$$z^n = \underbrace{z \cdot z \cdot \ldots \cdot z}_{n \text{ times}}$$

• PROBLEM 1-12

1. Show that if

$$z = \cos\alpha + i\sin\alpha, \quad 0 < \alpha < 2\pi \tag{1}$$

then

$$\frac{1+z}{1-z} = i\cot\frac{\alpha}{2} \tag{2}$$

2. If a and b are real numbers, show that

$$(a^2 - iab - b^2)(a+ib) = a^3 - ib^3 \tag{3}$$

Derive the equation

$$[(a^2-b^2)^2 + a^2 b^2](a^2+b^2) = a^6 + b^6 \tag{4}$$

from eq.(3).

Solution: 1. Substituting

$$\cot\frac{\alpha}{2} = \frac{1+\cos\alpha}{\sin\alpha} \tag{5}$$

and eq.(1) into eq.(2), we find

$$\frac{1+\cos\alpha + i\sin\alpha}{1-\cos\alpha - i\sin\alpha} = \frac{i(1+\cos\alpha)}{\sin\alpha} \tag{6}$$

Eq.(6) can be transformed to

$$\sin\alpha + \sin\alpha\cos\alpha + i\sin^2\alpha = (i+i\cos\alpha)(1-\cos\alpha-i\sin\alpha)$$
$$= i - i\cos\alpha + \sin\alpha + i\cos\alpha - i\cos^2\alpha + \cos\alpha\sin\alpha \tag{7}$$

Eq.(7) leads to

$$i\sin^2\alpha = i - i\cos^2\alpha \tag{8}$$

$$i(\sin^2\alpha + \cos^2\alpha) = i \tag{9}$$

Eqs.(8) and (9) are obviously true, since $\sin^2\alpha + \cos^2\alpha = 1$. Thus, we have prove eq.(2).

2. The left-hand side of eq.(3) yields

$$(a^2-iab-b^2)(a+ib) = a^3 + ia^2 b - ia^2 b + ab^2 - ab^2 - ib^3$$
$$= a^3 - ib^3 \tag{10}$$

Let us replace, in eq.(10), a by a^2 and b by $-ib^2$.

$$a \to a^2 \tag{11}$$
$$b \to -ib^2 \tag{12}$$

We find

$$[a^4 - ia^2(-ib^2) - (-ib^2)^2][a^2 + i(-ib^2)]$$
$$= [a^4 - a^2b^2 + b^4][a^2 + b^2]$$
$$= [(a^2-b^2)^2 + a^2b^2](a^2+b^2) \tag{13}$$
$$= (a^2)^3 - i(-ib^2)^3 = a^6 + b^6$$

• **PROBLEM 1-13**

Show that if
$$i^n = -i \tag{1}$$
then
$$n = 4m - 1, \quad \text{where} \tag{2}$$
n and m are integers.

Solution: Substituting $n = 4m - 1$ into eq.(1), we get

$$i^{4m-1} = \frac{i^{4m}}{i} = -i \tag{3}$$

Thus,

$$i^{4m} = (i^4)^m = 1^m \tag{4}$$
$$= 1 = i \cdot (-i) = 1$$

Therefore, if $n = 4m - 1$ then $i^n = -i = \frac{1}{i}$. Multiplying by i, we get

$$i^n i = i^{n+1} = 1 \tag{5}$$

From
$$i^1 = i, \; i^2 = -1, \; i^3 = -i, \; i^4 = 1, \; i^5 = i, \; i^6 = -1 \tag{6}$$

etc. we see that every fourth element equals one.

In general
$$i^{4m} = 1 \tag{7}$$

Combining eqs.(5) and (7), we find

$$i^{n+1} = i^{4m} \tag{8}$$

Hence,

$$n = 4m - 1 \tag{9}$$

• **PROBLEM 1-14**

Find real numbers x and y such that
$$4x + 2ixy - iy = 5 + 3i \tag{1}$$

Solution: Two complex numbers $z_1 = a_1 + ib_1$ and $z_2 = a_2 + ib_2$ are equal if and only if

$$a_1 = a_2 \quad \text{and} \quad b_1 = b_2 \tag{2}$$

From eq.(1)

$$4x + i(2xy - y) = 5 + 3i \tag{3}$$

we obtain

$$\begin{aligned} 4x &= 5 \\ 2xy - y &= 3 \end{aligned} \tag{4}$$

Hence,

$$x = \frac{5}{4}, \quad y = 2 \tag{5}$$

● **PROBLEM 1-15**

In 1702, Leibnitz showed that

$$x^4 + a^4 = (x+a\sqrt{-\sqrt{-1}})(x-a\sqrt{-\sqrt{-1}})(x+a\sqrt{\sqrt{-1}})(x-a\sqrt{\sqrt{-1}}) \tag{1}$$

Prove that he was right.

Solution: Multiplying the first two terms of eq.(1), we find

$$\begin{aligned} (x+a\sqrt{-\sqrt{-1}})(x-a\sqrt{-\sqrt{-1}}) &= x^2 - a^2(-\sqrt{-1}) \\ &= x^2 + a^2\sqrt{-1} \end{aligned} \tag{2}$$

Multiplying the last two terms results in

$$(x+a\sqrt{\sqrt{-1}})(x-a\sqrt{\sqrt{-1}}) = x^2 - a^2\sqrt{-1} \tag{3}$$

Thus,

$$\begin{aligned} (x^2+a^2\sqrt{-1})(x^2-a^2\sqrt{-1}) &= (x^2)^2 - (a^2\sqrt{-1})^2 \\ &= x^4 - a^4(-1) = x^4 + a^4 \end{aligned} \tag{4}$$

You would certainly appreciate the clarity and simplicity of the modern algebraic notation if you would know that eq.(1) was written in the form

"Ex irrationalibus oriuntur quantitates impossibiles seu imaginariae, quarum mira est natura, et tamen non contemnenda utilitas; etsi enim ipsae per se aliquid impossibile significent, tamen non tantum ostendunt fontem impossibilitatis, et quomoolo quaestio corrigi potuerit, ne esset impossibilis, sed etiam interventu ipsarum exprimi possunt quantitates reales." See Werke, Gerhardt ed. (Berlin, 1850), VII, 69.

FIELD OF COMPLEX NUMBERS

● **PROBLEM 1-16**

An important concept in mathematics is the notion of a field. A set S on which two operations of addition and multiplication are defined is called a field if the operations satisfy the following axioms:

I. If $a \in S$, $b \in S$, then $a+b \in S$ and $ab \in S$.

II. Commutative Laws.

  If $a \in S$, $b \in S$, then

  $$a + b = b + a$$

  $$ab = ba$$

III. Associative Laws.

  If $a, b, c \in S$, then

  $$(a+b)+c = a+(b+c)$$

  $$a(bc) = (ab)c$$

IV. Distributive Law.

  If $a \in S$, $b \in S$, $c \in S$, then

  $$a(b+c) = ab + ac$$

V. There exists an element $0 \in S$ such that for every $a \in S$

  $$a + 0 = 0 + a = a$$

VI. There exists an element $1 \in S$ such that for every $a \in S$

  $$a \cdot 1 = 1 \cdot a = a$$

Show that the zero element, $0 \in S$, and the unit element, $1 \in S$ are unique.

Show that the set of all complex numbers, that is numbers $a + ib$, where a and b are real numbers, with addition and multiplication, as defined in Problem 1-6, forms a field.

Solution: To show that $0 \in S$ is unique, let us assume that it isn't, that is that there exist two zeros 0 and 0*. Then by axiom V

$$0^* + 0 = 0^* \tag{1}$$

$$0^* + 0 = 0 \tag{2}$$

and hence,

$$0 = 0^* \tag{3}$$

Now, if there are two unit elements 1 and 1*, then by axiom VI

$$1* \cdot 1 = 1* \tag{4}$$

$$1* \cdot 1 = 1 \tag{5}$$

and hence

$$1 = 1* \tag{6}$$

The set of complex numbers is usually denoted by C. The operations of addition and multiplication were defined in Problem 1-6. Since C is the set of all numbers $\alpha + i\beta$, then if $a \in C$ and $b \in C$, then $a + b \in C$ and $ab \in C$. In Problems 1-7 and 1-8, we showed that multiplication and addition are commutative, associative and distributive. The zero element is $0 + i0 = 0$, because

$$(\alpha+i\beta) + (0+i0) = \alpha + i\beta \tag{7}$$

For all $\alpha + i\beta \in C$. The unit element is $1 + i0$. Indeed if $\alpha + i\beta$ is any complex number, then

$$(\alpha+i\beta)(1+i0) = \alpha + i\beta \tag{8}$$

Thus, the set C of all complex numbers with addition and multiplication as defined in Problem 1-6 is a field. It has all the properties and satisfies all the theorems concerning fields.

● **PROBLEM 1-17**

Consider the set of all ordered pairs (a,b) of real numbers, with two binary operations, addition, +, and multiplication, ·, defined as follows:

$$(a,b) + (c,d) = (a+c, b+d) \tag{1}$$

$$(a,b) \cdot (c,d) = (ac-bd, ad+bc) \tag{2}$$

It is very easy to show that this set is a field, with the zero element (0,0), indeed

$$(a,b) + (0,0) = (a,b) \tag{3}$$

and the unit element (1,0)

$$(a,b) \cdot (1,0) = (a-0, 0+b) = (a,b). \tag{4}$$

Two pairs (a,b) and (c,d) are equal if, and only if, a = c and b = d. A one-to-one correspondence can be established between all the pairs of the form (a,0) and the real numbers a

$$(a,0) \leftrightarrow a \tag{5}$$

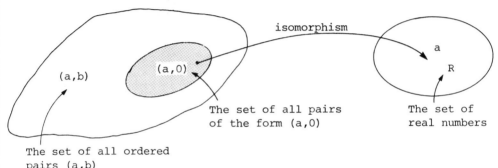

Fig. 1

From eqs.(1) and (2), we can deduce the following properties of the one-to-one correspondence

$$(a,0) \leftrightarrow a$$

$$(a,0) + (b,0) = (a+b, 0) \leftrightarrow a + b \qquad (6)$$

$$(a,0) \cdot (b,0) = (ab, 0) \leftrightarrow ab \qquad (7)$$

A one-to-one function which satisfies eqs.(6), (7) and (5) is called an isomorphism. We say that the set of all pairs of the type (a,0) is isomorphic with the set of real numbers R. In practice, it is convenient to write

$$(a,0) = a \qquad (8)$$

Now comes the most interesting part.

Denote (0,1) by i,
$$(0,1) := i \qquad (9)$$

It is not without a reason that we use the same "i" as in the complex number a + ib. Compute

$$i^2 \quad \text{and} \qquad (10)$$

$$(a,0) + (0,1)(b,0) \qquad (11)$$

Can you draw any conclusions based on the statement (a,0) = a?

Solution: Since (0,1) = i we get

$$i^2 = (0,1) \cdot (0,1) = (-1,0) = -1 \qquad (12)$$

$$(a,b) = (a,0) + (0,b) = (a,0) + (0,1)(b,0) \qquad (13)$$

Substituting $(a,0) = a$

$$(b,0) = b \qquad (14)$$

$$(0,1) = i$$

into eq.(13), we find

$$(a,b) = a + ib \qquad (15)$$

That leads us to the conclusion that a complex number can be represented by a + ib, or by an ordered pair (a,b).

The set of real numbers R can be regarded as a subset of the set of complex numbers C.

Fig. 2 of Problem 1-1 can be modified to include the set of complex numbers, as shown.

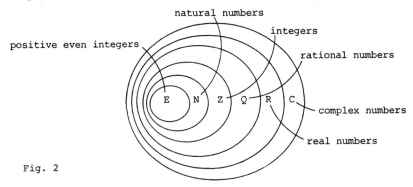

Fig. 2

In this book we will use the representation a + ib rather than (a,b).

COMPLEX CONJUGATE

● **PROBLEM 1-18**

I. Definition

The conjugate $\bar{z}$ of a complex number

$$z = a + ib \qquad (1)$$

is defined by

$$\bar{z} \equiv a - ib \qquad (2)$$

Show that

$$\overline{z_1 + z_2} = \bar{z}_1 + \bar{z}_2 \qquad (3)$$

$$\overline{z_1 z_2} = \bar{z}_1 \bar{z}_2 \qquad (4)$$

$$\bar{\bar{z}} = z \qquad (5)$$

II. Definition

The real part of the complex number z = a + ib is denoted by Re z, or Re(z), hence

$$Re(z) = a \qquad (6)$$

The imaginary part of z is denoted by Im(z), hence,

and
$$\text{Im}(z) = b \qquad (7)$$
$$z = a + ib = \text{Re}(z) + i\,\text{Im}(z) \qquad (8)$$

Show that
$$\text{Re}(z) = \tfrac{1}{2}(z+\bar{z}) \qquad (9)$$
$$\text{Im}(z) = \frac{1}{2i}(z-\bar{z}) \qquad (10)$$

<u>Solution</u>: I. If $z_1 = a_1 + ib_1$ and $z_2 = a_2 + ib_2$, then

$$\overline{z_1+z_2} = \overline{a_1+ib_1+a_2+ib_2} = \overline{(a_1+a_2)+i(b_1+b_2)}$$
$$= (a_1+a_2) - i(b_1+b_2) = (a_1-ib_1) + (a_2-ib_2) \qquad (11)$$
$$= \bar{z}_1 + \bar{z}_2$$

$$\overline{z_1 z_2} = \overline{(a_1+ib_1)(a_2+ib_2)} = \overline{(a_1a_2-b_1b_2)+i(a_1b_2+a_2b_1)}$$
$$= (a_1a_2-b_1b_2) - i(a_1b_2+a_2b_1) = \bar{z}_1\bar{z}_2 \qquad (12)$$

$$\bar{\bar{z}} = \overline{\overline{(a+ib)}} = \overline{(a-ib)} = a + ib = z \qquad (13)$$

II. $\tfrac{1}{2}(z+\bar{z}) = \tfrac{1}{2}(a+ib+a-ib) = a = \text{Re}(z) \qquad (14)$

$\dfrac{1}{2i}(z-\bar{z}) = \dfrac{1}{2i}(a+ib-a+ib) = \dfrac{2ib}{2i} = b = \text{Im}(z) \qquad (15)$

From eqs.(3), (4) and (5), it follows that if $R(z_1, z_2, z_3, \ldots)$ is a rational expression of complex numbers $z_1, z_2, z_3, \ldots$, then

$$\overline{R(z_1, z_2, z_3, \ldots)} = R(\bar{z}_1, \bar{z}_2, \bar{z}_3, \ldots) \qquad (16)$$

• **PROBLEM 1-19**

Prove that if the product of two complex numbers z_1 and z_2 is real and different from zero, then there exists a real number α such that
$$z_1 = \alpha \bar{z}_2 \qquad (1)$$

<u>Solution</u>: Let
$$z_1 = a_1 + ib_1 \qquad (2)$$
$$z_2 = a_2 + ib_2 \qquad (3)$$

The product of z_1 and z_2 is real and different from zero
$$z_1 z_2 \ne 0 \qquad (4)$$

Hence, $z_2 \ne 0$ and $\bar{z}_2 \ne 0 \qquad (5)$

and we can write

$$\frac{z_1}{\bar{z}_2} = \frac{z_1}{a_2 - ib_2} = \frac{z_1(a_2+ib_2)}{(a_2-ib_2)(a_2+ib_2)} = \frac{z_1 z_2}{a_2^2 + b_2^2} = \alpha \qquad (6)$$

Because z_1z_2 is real and different from zero and $a_2^2 + b_2^2$ is real and different from zero, we obtain, from eq.(6)

$$z_1 = \alpha \bar{z}_2 \tag{7}$$

● **PROBLEM 1-20**

Find two numbers whose sum is 5 and whose product is 9.

<u>Solution</u>: Let us denote the numbers by z_1 and z_2. Then

$$z_1 + z_2 = 5 \tag{1}$$

and

$$z_1 z_2 = 9$$

At this point, we don't know if z_1 and z_2 are real or complex.

$$z_1 = 5 - z_2 \tag{2}$$

$$(5-z_2)z_2 = 9 \tag{3}$$

$$z_2^2 - 5z_2 + 9 = 0 \tag{4}$$

Solving eq.(4), we find

$$\begin{aligned} z_{2,1} &= \frac{5}{2} + i\frac{\sqrt{11}}{2} \\ z_{2,2} &= \frac{5}{2} - i\frac{\sqrt{11}}{2} \end{aligned} \tag{5}$$

Thus,

$$z_{1,1} = 5 - z_{2,1} = \frac{5}{2} - i\frac{\sqrt{11}}{2} \tag{6}$$

$$z_{1,2} = 5 - z_{2,2} = \frac{5}{2} + i\frac{\sqrt{11}}{2} \tag{7}$$

There is only one set of numbers $z_1 = \frac{5}{2} - i\frac{\sqrt{11}}{2}$ and $z_2 = \frac{5}{2} + i\frac{\sqrt{11}}{2}$ satisfying eq.(1). It is clear that

$$z_1 = \bar{z}_2 \tag{8}$$

ABSOLUTE VALUE

● **PROBLEM 1-21**

Definition

The modulus or absolute value of a complex number $z = a + ib$, written $|z|$, is given by

$$|z| \equiv \sqrt{a^2+b^2} \tag{1}$$

The absolute value of a complex number is a nonnegative real number.

1. Evaluate the absolute value of the following numbers

   $3 + 2i$

   $5i$

   $(2-2i)(4+i)$

   -6

2. Show that if z_1 and z_2 are complex numbers, then

   $|z| = |\bar{z}|$ \hfill (2)

   $|z|^2 = z\bar{z}$ \hfill (3)

   $|Re(z)| \leq |z|$, $|Im(z)| \leq |z|$ \hfill (4)

   $|z_1||z_2| = |z_1 z_2|$ \hfill (5)

Solution: 1. Applying eq.(1), we find

$$|3+2i| = \sqrt{9+4} = \sqrt{13} \tag{6}$$

$$|5i| = \sqrt{25} = 5 \tag{7}$$

$$|(2-2i)(4+i)| = |10-6i| = \sqrt{100+36} = \sqrt{136} \tag{8}$$

$$|-6| = \sqrt{36} = 6 \tag{9}$$

2.
$$|z| = \sqrt{a^2+b^2} = \sqrt{a^2+(-b)^2} = |\bar{z}| \tag{10}$$

$$z\bar{z} = (a+ib)(a-ib) = a^2 + b^2 = |z|^2 \tag{11}$$

Since

$$|z|^2 = |Re(z)|^2 + |Im(z)|^2 \tag{12}$$

it follows that

$$|z| \geq |Re(z)|^2 \quad \text{and} \tag{13}$$

$$|z| \geq |Im(z)|^2 \tag{14}$$

Now, we shall prove the last identity $|z_1 z_2| = |z_1||z_2|$.

$$|z_1 z_2|^2 = (z_1 z_2)(\overline{z_1 z_2}) = (z_1 z_2)(\bar{z}_1 \bar{z}_2) \tag{15}$$

$$= (z_1 \bar{z}_1)(z_2 \bar{z}_2) = |z_1|^2 |z_2|^2$$

Note that to prove eq.(5), we applied eq.(11). Eq.(15) leads to

$$|z_1 z_2 \ldots z_{n-1} z_n| = |z_1||z_2|\ldots|z_{n-1}||z_n| \tag{16}$$

Thus,

$$|z^n| = |z|^n \tag{17}$$

● **PROBLEM 1-22**

Show that the only solution of the equation
$$2 - \frac{1}{z} = \bar{z} \tag{1}$$
is $z = 1$.

Solution: Multiplying eq.(1) by z, we find
$$2z - 1 = \bar{z}z = |z|^2 \tag{2}$$
or
$$2z = |z|^2 + 1 \tag{3}$$

Note that the right-hand side of eq.(3) is a real number, therefore the left-hand side also has to be a real number, $z \in R$. Thus, we can drop the absolute value symbol and write

$$2z = z^2 + 1$$

or

$$z^2 - 2z + 1 = (z - 1)^2 = 0 \tag{4}$$

The only solution is
$$z = 1 \tag{5}$$

● **PROBLEM 1-23**

Prove the theorem known as the triangle inequality. The absolute value of the sum of two complex numbers cannot exceed the sum of their absolute values, i.e.
$$|z_1 + z_2| \leq |z_1| + |z_2| \tag{1}$$

Solution: Since $|z|^2 = z\bar{z}$, we have

$$|z_1+z_2|^2 = (z_1+z_2)(\overline{z_1+z_2}) = (z_1+z_2)(\bar{z}_1+\bar{z}_2)$$
$$= z_1\bar{z}_1 + z_1\bar{z}_2 + \bar{z}_1 z_2 + z_2\bar{z}_2 \tag{2}$$

Since $2\operatorname{Re}(z) = z + \bar{z}$, eq.(2) yields

$$|z_1 + z_2|^2 = |z_1|^2 + 2\operatorname{Re}(z_1\bar{z}_2) + |z_2|^2$$
$$\leq |z_1|^2 + 2|z_1\bar{z}_2| + |z_2|^2$$
$$= |z_1|^2 + 2|z_1||z_2| + |z_2|^2$$
$$= (|z_1|+|z_2|)^2 \tag{3}$$

Hence,
$$|z_1+z_2|^2 \leq (|z_1|+|z_2|)^2 \tag{4}$$
and
$$|z_1+z_2| \leq |z_1| + |z_2| \tag{5}$$

Using mathematical induction, we can extend eq.(5) to any finite number of complex numbers

$$\left|\sum_{K=1}^{n} z_K\right| \leq \sum_{K=1}^{n} |z_K| \tag{6}$$

• **PROBLEM 1-24**

Prove that if z_1 and z_2 are complex numbers, then

$$\left||z_1| - |z_2|\right| \leq |z_1 - z_2| \tag{1}$$

Solution: Applying

$$|z|^2 = z\bar{z} \quad \text{and} \tag{2}$$

$$2\text{Re}(z) = z + \bar{z} \tag{3}$$

we find

$$|z_1-z_2|^2 = (z_1-z_2)(\overline{z_1-z_2}) = (z_1-z_2)(\bar{z}_1-\bar{z}_2)$$

$$= z_1\bar{z}_1 - z_1\bar{z}_2 - \bar{z}_1 z_2 + z_2\bar{z}_2 \tag{4}$$

$$= |z_1|^2 - 2\text{Re}(z_1\bar{z}_2) + |z_2|^2$$

Since,

$$|\text{Re}(z)| \leq |z| \tag{5}$$

we have

$$-|z| \leq \text{Re}(z) \leq |z| \tag{6}$$

and

$$|z| \geq -\text{Re}(z) \geq -|z| \tag{7}$$

Therefore, eq.(4) leads to

$$|z_1-z_2|^2 = |z_1|^2 - 2\text{Re}(z_1\bar{z}_2) + |z_2|^2 \tag{8}$$

$$\geq |z_1|^2 - 2|z_1\bar{z}_2| + |z_2|^2 = (|z_1| - |z_2|)^2$$

Hence,

$$|z_1-z_2| \geq |z_1| - |z_2| \tag{9}$$

• **PROBLEM 1-25**

Prove that if $z = a + ib$, then

$$|a| + |b| \leq \sqrt{2}|a+ib| \leq \sqrt{2}(|a|+|b|) \tag{1}$$

Solution: Note that $|a|$ indicates the absolute value of a, while $|a+ib|$ is the modulus of a complex number.

From eq.(1), we obtain

$$|a| + |b| \leq \sqrt{2}\sqrt{a^2+b^2} \tag{2}$$

$$a^2 + 2|a||b| + b^2 \leq 2(a^2+b^2) \tag{3}$$

$$2|a||b| \le a^2 + b^2 \tag{4}$$

$$0 \le a^2 - 2|a||b| + b^2 \tag{5}$$

Hence,

$$0 \le (|a| - |b|)^2 \tag{6}$$

Eq.(6), and therefore eq.(1), becomes an equality when $|a| = |b|$.

To show the other part i.e.

$$\sqrt{2}|a+ib| \le \sqrt{2}(|a| + |b|) \tag{7}$$

we observe that

$$|a+ib| = \sqrt{a^2+b^2} \le |a| + |b| \tag{8}$$

and

$$a^2 + b^2 \le a^2 + 2|ab| + b^2 \tag{9}$$

$$0 \le 2|ab| \tag{10}$$

The similarity between the notation $|a+ib|$ denoting the modulus of a complex number and $|a|$ denoting the absolute value of a is not accidental. If z is a complex number such that $z = a + ib$, then when $b = 0$, we obtain

$$|z| = |a+ib| = \sqrt{a^2+b^2} = \sqrt{a^2} = |a| \tag{11}$$

Thus, the modulus reduces to the absolute value when the complex number has no imaginary part. The absolute value or modulus notation should not cause any confusion.

• **PROBLEM 1-26**

Show that

$$|z_1+z_2|^2 + |z_1+\overline{z}_2|^2 = 2(|z_1|^2+|z_2|^2) + 4\mathrm{Re}z_1\mathrm{Re}z_2 \tag{1}$$

Solution: By applying

$$|z|^2 = z\overline{z} \quad \text{and} \quad (\overline{z_1+z_2}) = \overline{z}_1 + \overline{z}_2 \tag{2}$$

we shall transform the left-hand side of eq.(1).

$$|z_1+z_2|^2 + |z_1+\overline{z}_2|^2 = (z_1+z_2)(\overline{z}_1+\overline{z}_2)$$

$$+ (z_1+\overline{z}_2)(\overline{z}_1+z_2) = z_1\overline{z}_1 + z_1\overline{z}_2 + \overline{z}_1z_2$$

$$+ z_2\overline{z}_2 + z_1\overline{z}_1 + z_1z_2 + \overline{z}_1\overline{z}_2 + z_2\overline{z}_2$$

$$= 2|z_1|^2 + 2|z_2|^2 + z_1(z_2+\overline{z}_2) + \overline{z}_1(z_2+\overline{z}_2) \tag{3}$$

$$= 2(|z_1|^2+|z_2|^2) + (z_1+\overline{z}_1)(z_2+\overline{z}_2)$$

Since $z_1 = a_1 + ib_1$ and $\overline{z}_1 = a_1 - ib_1$, we obtain

$$z_1 + \overline{z}_1 = 2a_1 = 2\mathrm{Re}z_1 \tag{4}$$

$$z_2 + \bar{z}_2 = 2a_2 = 2\mathrm{Re}z_2 \tag{5}$$

Substituting eqs.(4) and (5) into eq.(3) results in

$$|z_1+z_2|^2 + |z_1+\bar{z}_2|^2 = 2(|z_1|^2+|z_2|^2) + 4\mathrm{Re}z_1\mathrm{Re}z_2 \tag{6}$$

INEQUALITIES AND IDENTITIES

• **PROBLEM 1-27**

Show that if $|z| \leq 1$, then
$$|z-1| + |z+1| \leq 2\sqrt{2} \tag{1}$$

Solution: Since both sides of eq.(1) are positive, we can evaluate the second power of eq.(1).

$$|z-1|^2 + |z+1|^2 + 2|z-1||z+1| \leq 8 \tag{2}$$

Hence,

$$(z-1)(\bar{z}-1) + (z+1)(\bar{z}+1) + 2|(z-1)(z+1)| \leq 8 \tag{3}$$

and

$$|z|^2 - z - \bar{z} + 1 + |z|^2 + z + \bar{z} + 1 + 2|z^2-1| \leq 8 \tag{4}$$

or

$$2|z|^2 + 2 + 2|z^2-1| \leq 8$$

$$|z|^2 + |z^2-1| \leq 3 \tag{5}$$

$$|z|^2 + |z^2+(-1)| \leq |z|^2 + |z^2| + |-1|$$
$$= |z|^2 + |z|^2 + 1 \tag{6}$$

Note that

$$|z^2| = |zz| = |z||z| = |z|^2 \tag{7}$$

Since $|z| \leq 1$, also $|z|^2 \leq 1$, and thus

$$|z|^2 + |z|^2 + 1 \leq 1 + 1 + 1 = 3 \tag{8}$$

and

$$|z|^2 + |z^2-1| \leq 3 \tag{9}$$

That completes the proof.

• **PROBLEM 1-28**

Prove Lagrange's* identity for the complex numbers

$$\left|\sum_{j=1}^{n} z_j w_j\right|^2 = \sum_{j=1}^{n} |z_j|^2 \sum_{j=1}^{n} |w_j|^2 - \sum_{1\leq j<k\leq n} |z_j\bar{w}_k - z_k\bar{w}_j|^2 \tag{1}$$

where z_j, w_j are complex numbers.

Solution: We shall utilize the identity
$$|z|^2 = z\bar{z} \qquad (2)$$

The right-hand side of eq.(1) is equal to

$$(z_1\bar{z}_1 + z_2\bar{z}_2 + \ldots + z_n\bar{z}_n)(w_1\bar{w}_1 + w_2\bar{w}_2 + \ldots + w_n\bar{w}_n) - \sum_{1 \leq j < k \leq n} (z_j\bar{w}_k - z_k\bar{w}_j)(\bar{z}_j w_k - \bar{z}_k w_j)$$

$$= (z_1\bar{z}_1 + \ldots + z_n\bar{z}_n)(w_1\bar{w}_1 + \ldots + w_n\bar{w}_n) -$$

$$\sum_{1 \leq j < k \leq n} z_j\bar{z}_j w_k\bar{w}_k - \sum_{1 \leq j < k \leq n} z_k\bar{z}_k w_j\bar{w}_j + \qquad (3)$$

$$\sum_{1 \leq j < k \leq n} z_j\bar{z}_k w_j\bar{w}_k + \sum_{1 \leq j < k \leq n} z_k\bar{z}_j w_k\bar{w}_j$$

At this point let us observe that

$$\sum_{1 \leq j < k \leq n} a_j b_k + \sum_{1 \leq j < k \leq n} a_k b_j = \sum_{\substack{j,k=1 \\ j \neq k}}^{n} a_j b_k \qquad (4)$$

Hence, eq.(3) leads to

$$= \sum_{j,k=1}^{n} z_j\bar{z}_j w_k\bar{w}_k - \sum_{\substack{j,k=1 \\ j \neq k}}^{n} z_j\bar{z}_j w_k\bar{w}_k + \sum_{\substack{j,k=1 \\ j \neq k}}^{n} z_j\bar{z}_k w_j\bar{w}_k$$

$$= \sum_{j=1}^{n} z_j\bar{z}_j w_k\bar{w}_k + \sum_{\substack{j,k=1 \\ j \neq k}}^{n} z_j\bar{z}_k w_j\bar{w}_k \qquad (5)$$

$$= \sum_{j,k=1}^{n} z_j\bar{z}_k w_j\bar{w}_k = \sum_{j,k=1}^{n} (z_j w_j)\overline{(z_k w_k)}$$

$$= (z_1 w_1 + z_2 w_2 + \ldots + z_n w_n)(\overline{z_1 w_1} + \overline{z_2 w_2} + \ldots + \overline{z_n w_n})$$

$$= (\sum_{j=1}^{n} z_j w_j)(\sum_{j=1}^{n} \overline{z_j w_j}) = \left|\sum_{j=1}^{n} z_j w_j\right|^2$$

That completes the proof.

*Joseph Louis Lagrange was born in Turin in 1736 and died in Paris in 1813. Lagrange was one of the greatest mathematician of the eighteenth century. He was also a physicist and an astronomer.

• **PROBLEM 1-29**

> Prove Cauchy's inequality for the complex numbers.

Solution: This is a well known inequality and appears in different forms and in various branches of mathematics. For the complex numbers, it can be stated as follows:

$$\left| \sum_{j=1}^{n} z_j w_j \right|^2 \leq \left(\sum_{j=1}^{n} |z_j|^2 \right) \left(\sum_{j=1}^{n} |w_j|^2 \right) \quad (1)$$

For any complex number, $|z|^2$ is a real number and $|z|^2 \geq 0$. Thus, in Lagrange's identity (see Problem 1-28, eq.(1))

$$\sum_{1 \leq j < \kappa \leq n} |z_j \bar{w}_\kappa - z_\kappa \bar{w}_j|^2 \geq 0 \quad (2)$$

We obtain

$$\left| \sum_{j=1}^{n} z_j w_j \right|^2 = \sum_{j=1}^{n} |z_j|^2 \sum_{j=1}^{n} |w_j|^2 - \sum_{1 \leq j < \kappa \leq n} |z_j \bar{w}_\kappa - z_\kappa \bar{w}_j|$$

$$\leq \sum_{j=1}^{n} |z_j|^2 \sum_{j=1}^{n} |w_j|^2 \quad (3)$$

• **PROBLEM 1-30**

> Prove the triangle or Minkowski* inequality for the complex numbers:
>
> $$\sqrt{\sum_{j=1}^{n} |z_j + w_j|^2} \leq \sqrt{\sum_{j=1}^{n} |z_j|^2} + \sqrt{\sum_{j=1}^{n} |w_j|^2} \quad (1)$$

Solution: First, let us consider the left-hand side of eq.(1) and utilize inequality

$$|z_j + w_j| \leq |z_j| + |w_j| \quad (2)$$

$$\sum_{j=1}^{n} |z_j + w_j|^2 = \sum_{j=1}^{n} |z_j + w_j||z_j + w_j|$$

$$\leq \sum_{j=1}^{n} |z_j + w_j|(|z_j| + |w_j|)$$

$$= \sum_{j=1}^{n} |z_j + w_j||z_j| + \sum_{j=1}^{n} |z_j + w_j||w_j| \quad (3)$$

Applying Cauchy's inequality to the last two sums in eq.(3), we find

$$\sum_{j=1}^{n} |z_j + w_j||z_j| \leq \sqrt{\left(\sum_{j=1}^{n} |z_j+w_j|^2\right)\left(\sum_{j=1}^{n} |z_j|^2\right)} \qquad (4)$$

$$\sum_{j=1}^{n} |z_j + w_j||w_j| \leq \sqrt{\left(\sum_{j=1}^{n} |z_j+w_j|^2\right)\left(\sum_{j=1}^{n} |w_j|^2\right)} \qquad (5)$$

For those of you who are somewhat confused, let us repeat the reasoning in obtaining eqs.(4) and (5).

Cauchy's inequality can be written in the form

$$\left|\sum_{j=1}^{n} z_j v_j\right| \leq \sum_{j=1}^{n} |z_j v_j| \leq \sqrt{\left(\sum_{j=1}^{n} |z_j|^2\right)\left(\sum_{j=1}^{n} |v_j|^2\right)} \qquad (6)$$

Hence,

$$\sum_{j=1}^{n} |z_j + w_j||z_j| = \sum_{j=1}^{n} |(z_j+w_j)z_j|$$
$$\leq \sqrt{\left(\sum_{j=1}^{n} |z_j+w_j|^2\right)\left(\sum_{j=1}^{n} |z_j|^2\right)} \qquad (7)$$

From eqs.(3), (4) and (5) we conclude that

$$\sum_{j=1}^{n} |z_j + w_j|^2 \leq \sum_{j=1}^{n} |z_j + w_j||z_j| + \sum_{j=1}^{n} |z_j + w_j||w_j|$$
$$\leq \sqrt{\sum_{j=1}^{n} |z_j + w_j|^2}\left(\sqrt{\sum_{j=1}^{n} |z_j|^2} + \sqrt{\sum_{j=1}^{n} |w_j|^2}\right) \qquad (8)$$

From eq.(8), we obtain

$$\sqrt{\sum_{j=1}^{n} |z_j + w_j|^2} \leq \sqrt{\sum_{j=1}^{n} |z_j|^2} + \sqrt{\sum_{j=1}^{n} |w_j|^2} \qquad (9)$$

*Hermann Minkowski (1864-1909), a German mathematician, laid the foundation for the theory of convex bodies. Minkowski investigated the problem of regular lattices, which is closely connected with the theory of numbers and geometric crystallography. In 1905, after Einstein developed the special theory of relativity, Minkowski showed the mutual link of space and time. The separation of space from time is to a great degree relative depending on the system of reference. There is a single absolute form of existence of matter: space-time. These ideas were developed in his book "Raum and Zeit", Leipzig 1909.

POLYNOMIALS

● **PROBLEM 1-31**

Let $a_n, a_{n-1}, \ldots, a_1, a_0$ be real numbers. If α is a complex root of the polynomial equation

$$a_n z^n + a_{n-1} z^{n-1} + \ldots + a_1 z + a_0 = 0 \qquad (1)$$

then show that $\bar{\alpha}$ is also a root of the equation.

Solution: Since α is a root of eq.(1), we get

$$a_n \alpha^n + a_{n-1} \alpha^{n-1} + \ldots + a_1 \alpha + a_0 = 0 \qquad (2)$$

Taking the conjugate of eq.(2) and using the property

$$\overline{z_1 + z_2} = \bar{z}_1 + \bar{z}_2 \qquad (3)$$

we find

$$\overline{a_n \alpha^n} + \overline{a_{n-1} \alpha^{n-1}} + \ldots + \overline{a_1 \alpha} + \overline{a_0} = 0 \qquad (4)$$

Using the property

$$\overline{z_1 z_2} = \bar{z}_1 \bar{z}_2 \qquad (5)$$

and the fact that coefficients are real numbers

$$\bar{a}_n = a_n, \; \bar{a}_{n-1} = a_{n-1}, \ldots, \bar{a}_0 = a_0 \qquad (6)$$

we find

$$a_n \bar{\alpha}^n + a_{n-1} \bar{\alpha}^{n-1} + \ldots + a_1 \bar{\alpha} + a_0 = 0 \qquad (7)$$

Hence, if α is a complex root of a polynomial equation with real coefficients, then $\bar{\alpha}$ is also a root of the equation.

● **PROBLEM 1-32**

Verify that $2+i$ is a root of the equation

$$z^4 - 5z^3 + 3z^2 + 19z - 30 = 0 \qquad (1)$$

and find the other three roots.

Solution: Since $2+i$ is one of the roots of the equation, $\overline{2+i} = 2-i$ is also a root of the equation (see Problem 1-31). Let us denote the roots of eq.(1) by z_1, z_2, z_3, z_4, then

$$(z-z_1)(z-z_2)(z-z_3)(z-z_4) = z^4 - 5z^3 + 3z^2 + 19z - 30 \quad (2)$$

For $z_1 = 2+i$, $z_2 = 2-i$, we have

$$(z-2-i)(z-2+i) = z^2 - 4z + 5 \quad (3)$$

To find $(z-z_3)(z-z_4)$, we have to divide $z^4-5z^3+3z^2+19z-30$ by z^2-4z+5.

$$\begin{array}{r}
z^2-z-6 \\
z^2-4z+5 \overline{\smash{\big)} z^4-5z^3+3z^2+19z-30} \\
-(z^4-4z^3+5z^2) \\
\overline{-z^3-2z^2+19z } \\
-(-z^3+4z^2-5z) \\
\overline{-6z^2+24z-30} \\
-(-6z^2+24z-30) \\
\end{array} \quad (4)$$

Thus,

$$(z-z_3)(z-z_4) = z^2 - z - 6 \quad (5)$$

Solving the equation

$$z^2 - z - 6 = 0 \quad (6)$$

$$(z-3)(z+2) = 0$$

Thus, $z_3 = 3$ and $z_4 = -2$.

Therefore the four roots of eq.(1) are

$$z_1 = 2+i, \quad z_2 = 2-i, \quad z_3 = 3, \quad z_4 = -2 \quad (7)$$

• **PROBLEM 1-33**

Show that if the rational number $\frac{\alpha}{\beta}$ (where α and β have no common factor except ± 1) satisfies the polynomial equation

$$a_n z^n + a_{n-1} z^{n-1} + \ldots + a_1 z + a_0 = 0 \quad (1)$$

where $a_n, a_{n-1}, \ldots, a_0$ are integers, then α and β must be factors of a_0 and a_n, respectively.

Solution: Substituting $\frac{\alpha}{\beta}$ into eq.(1) and multiplying by β^n, we find

$$a_n \alpha^n + a_{n-1} \alpha^{n-1} \beta + \ldots + a_1 \alpha \beta^{n-1} + a_0 \beta^n = 0 \quad (2)$$

Dividing by α, we get

$$a_n \alpha^{n-1} + a_{n-1} \alpha^{n-2} \beta + a_1 \beta^{n-1} = -\frac{a_0 \beta^n}{\alpha} \quad (3)$$

Since $a_n, \ldots, a_1$ and α and β are integers, the left-hand side of eq.(3) is an integer. Hence, the right-hand side is also an integer. Let us note that α has no common factor with β.

It cannot divide β^n, therefore α divides a_0.

$\frac{a_0}{\alpha}$ is an integer.

Dividing eq.(2) by β we get

$$a_{n-1}\alpha^{n-1} + \ldots + a_1\alpha\beta^{n-2} + a_0\beta^{n-1} = \frac{a_n\alpha^n}{\beta} \qquad (4)$$

The left-hand side is, of course, an integer, thus the right-hand side is an integer. The real numbers α and β have no common factor. Therefore, β divides a_n.

$\frac{a_n}{\beta}$ is an integer.

• **PROBLEM 1-34**

Solve the equation

$$6z^4 - 47z^3 + 148z^2 - 167z + 52 = 0 \qquad (1)$$

Solution: The integer factors of 6 and 52 are, respectively (see Problem 1-33),

for 6 $\pm 1, \pm 2, \pm 3, \pm 6$

for 52 $\pm 1, \pm 2, \pm 4, \pm 13$ (2)

Hence, the possible solutions of eq.(1) are

$$\pm 1, \pm\frac{1}{2}, \pm\frac{1}{3}, \pm\frac{1}{6}, \pm\frac{2}{3}, \pm 4, \pm 2, \pm\frac{4}{3}, \pm 13, \pm\frac{13}{2}, \pm\frac{13}{3}, \pm\frac{13}{6} \qquad (3)$$

Given the proper time, one could easily verify that $\frac{1}{2}$ and $\frac{4}{3}$ are the solutions. Thus the polynomial

$$(z - \frac{1}{2})(z - \frac{4}{3}) = z^2 - \frac{1}{2}z - \frac{4}{3}z + \frac{2}{3} \qquad (4)$$

is a factor of $6z^4-47z^3+148z^2-167z+52$. Multiplying eq.(4) by 6 to obtain integer coefficients, we find

$$6z^2 - 11z + 4 \qquad (5)$$

By dividing, we find the other factor

$$6z^4 - 47z^3 + 148z^2 - 167z + 52 = (6z^2-11z+4)(z^2-6z+13) \qquad (6)$$

Solving the equation

$$z^2 - 6z + 13 = 0 \qquad (7)$$

we find

$$z_3 = \frac{6 + \sqrt{36-52}}{2} = 3 + 2i \qquad (8)$$

$$z_4 = \bar{z}_3 = \overline{(3+2i)} = 3 - 2i \qquad (9)$$

Hence, $\frac{1}{2}, \frac{4}{3}, 3+2i, 3-2i$ are the solutions of eq.(1).

● **PROBLEM 1-35**

Prove that the sum and the product of all the roots of

$$a_n z^n + a_{n-1} z^{n-1} + \ldots + a_1 z + a_0 = 0, \quad a_n \neq 0 \qquad (1)$$

is equal to

$$\sum_{j=1}^{n} z_j = \frac{-a_{n-1}}{a_n} \qquad (2)$$

$$\prod_{j=1}^{n} z_j = (-1)^n \frac{a_0}{a_n} \qquad (3)$$

Solution: The product symbol "$\prod$" is defined as

$$\prod_{j=1}^{k} b_j \equiv b_1 \cdot b_2 \cdot \ldots \cdot b_K$$

If $z_1, z_2, \ldots, z_n$ are all the roots, then eq.(1) can be written in the factored form

$$a_n (z-z_1)(z-z_2)\ldots(z-z_n)$$

$$= a_n \prod_{j=1}^{n} (z-z_j) = 0 \qquad (4)$$

Multiplying, we obtain

$$a_n \left[z^n - (z_1 + z_2 + \ldots + z_n) z^{n-1} + \ldots + (-1)^n z_1 z_2 \ldots z_n \right] = 0 \qquad (5)$$

Comparing the coefficients in eqs.(1) and eq.(5) results in

$$z_1 + z_2 + \ldots + z_n = \frac{-a_{n-1}}{a_n} \qquad (6)$$

$$z_1 z_2 \ldots z_n = (-1)^n \frac{a_0}{a_n} \qquad (7)$$

● **PROBLEM 1-36**

A number is called an algebraic number if it is a solution of a polynomial equation with integer coefficients.

$$a_n z^n + a_{n-1} z^{n-1} + \ldots + a_1 z + a_0 = 0 \qquad (1)$$

$a_n, a_{n-1}, \ldots, a_1, a_0$ are integers. Numbers which are not algebraic are called transcendental numbers.

1. Show that all rational numbers are algebraic.

2. Show that $\sqrt{2}+1$ and $\sqrt[3]{5}+2i$ are algebraic numbers.

Solution: 1. Any rational number can be represented in the form $\frac{a}{b}$ where a and b are integers. The solution of the equation

$$bz - a = 0 \tag{2}$$

is $z = \frac{a}{b}$. Hence, all rational numbers are algebraic numbers.

2. Let $z = \sqrt{2} + 1$, then

$$z - 1 = \sqrt{2} \tag{3}$$

The second power of eq.(3) is

$$z^2 - 2z + 1 = 2 \tag{4}$$

The number $\sqrt{2}+1$ is an algebraic number, since it is a solution of

$$z^2 - 2z - 1 = 0 \tag{5}$$

Now, let $z = \sqrt[3]{5} + 2i$, then

$$z - 2i = \sqrt[3]{5} \tag{6}$$

The third power of eq.(6) is

$$z^3 - 6iz^2 - 12z + 8i = 5 \tag{7}$$

or

$$z^3 - 12z - 5 = i(6z^2 - 8) \tag{8}$$

Taking the second power of eq.(8), we find

$$z^6 + 12z^4 - 10z^3 + 48z^2 + 120z + 89 = 0 \tag{9}$$

Thus, $\sqrt[3]{5} + 2i$ is an algebraic number. It has been proven that $\pi = 3.1415...$ and $e = 2.7182...$ are transcendental numbers. It is still known whether or not numbers such as πe or $\pi + e$ are transcendental. The algebraic theory of numbers was developed by Lagrange, Gauss, Dedekind, and Kummer among others.

CHAPTER 2

GEOMETRIC REPRESENTATION OF COMPLEX NUMBERS

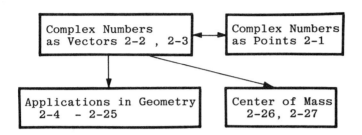

This chart is provided to facilitate rapid understanding of the interrelationships of the topics and subject matter in this chapter. Also shown are the problem numbers associated with the subject matter.

REPRESENTATIONS OF COMPLEX NUMBERS

● **PROBLEM 2-1**

1) Discuss the geometrical interpretation of the complex number and describe the complex plane.

2) Sketch the following numbers in the complex plane

$3 + 4i$, $2 - i$, $6i$, 5, $-1 - i$, $2i$

3) What is the relationship between the modulus of a complex number z and the distance of the point representing the number from the origin of the coordinate system?

Solution: 1) We will now briefly outline the geometrical interpretation of the complex numbers.

Any complex number can be represented by a point in a two-dimensional plane endowed with a rectangular coordinate system. The number $z = x + iy$ corresponds to the point $P(x,y)$, where the real part x of the number z is the abscissa and the imaginary part y is the ordinate. This plane is called the complex plane, also referred to as the z plane, or the Argand* diagram.

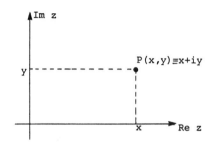
Fig. 1

In this plane, the x axis is the real axis and the y axis is the imaginary axis. This procedure establishes a one-to-one and onto correspondence between the field of complex numbers and the two-dimensional plane.

2) Since x is the real part and y is the imaginary part, the six numbers
$$3 + 4i,\ 2 - i,\ 6i,\ 5,\ -1 - i,\ 2i$$
are located as shown in Fig. 2. We use the terms "complex number" and "point" interchangeably.

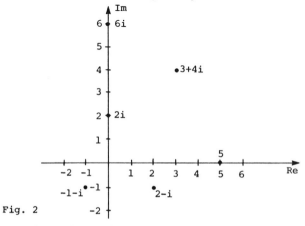
Fig. 2

*Jean Robert Argand was born in Geneva in 1768, and died in Paris in 1822. Argand was the first to give the geometrical representation of a complex number and applied it to show that every algebraic equation has a root. In 1806 Argand published his major book "Essai".

• **PROBLEM 2-2**

A complex number $z = x + iy$ can be represented by a point $P(x,y)$. The number $x + iy$ can also be considered as a vector $\overline{OP}$ whose initial point is the origin 0 and whose terminal point is $P(x,y)$. The vector $\overline{OP}$ is called the position vector of point P.

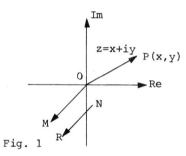
Fig. 1

Two vectors $\overline{OM}$ and $\overline{NR}$ having the same magnitude (length) and direction are considered equal

$$\overline{OM} = \overline{NR} \qquad (1)$$

even though they may have different initial points. We write

$$\overline{OP} = x + iy \qquad (2)$$

1. Based on vector representation of complex numbers, find graphically

   a) $z_1 + z_2$ (z_1, z_2 are any two arbitrary numbers)

   b) $(2-2i) + (1+5i)$

   c) $(1+2i) + (2-3i) + (-5+4i)$

2. Find graphically the difference

$$z_2 - z_1.$$

Solution: 1. Let P_1 and P_2 be the points z_1 and z_2.

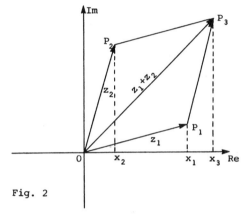

Fig. 2

Through P_1, draw $\overline{P_1P_3}$ equal and parallel to $\overline{OP_2}$. The coordinates of P_3 are (x_1+x_2, y_1+y_2). Hence, P_3 represents the point z_1+z_2. In vector notation,

$$\overline{OP}_1 + \overline{OP}_2 = \overline{OP}_3 \qquad (3)$$

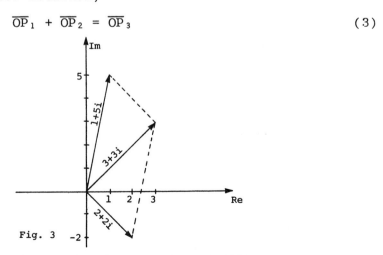

Fig. 3

b) Graphically the sum of the numbers (2-2i) and (1+5i) is found to be (3+3i), as shown in Fig. 3. This result can be easily verified analytically.

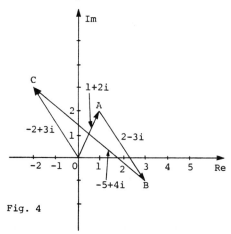

Fig. 4

To find the sum (1+2i) + (2-3i) + (-5+4i), we first draw vector $\overline{OA}$ = (1+2i). Then, from the point A we draw the vector $\overline{AB}$ = 2 - 3i. From point B, we then draw $\overline{BC}$ = -5 + 4i. Vector $\overline{OC}$ is the sum

$$\overline{OC} = \overline{OA} + \overline{AB} + \overline{BC} \tag{4}$$

2. To find the difference $z_2 - z_1$,

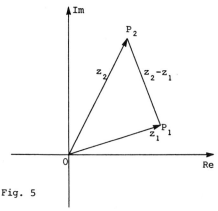

Fig. 5

we draw a vector from the point z_1 to the point z_2, $z = z_2 - z_1$. Therefore,

$$z_1 + (z_2-z_1) = \overline{OP_1} + \overline{P_1P_2} = \overline{OP_2} = z_2 \tag{5}$$

● **PROBLEM 2-3**

Give the graphical interpretation of the following inequalities:

1. $|z_1+z_2| \leq |z_1| + |z_2|$ (1)

2. $|z_1+z_2+z_3| \leq |z_1| + |z_2| + |z_3|$ (2)

3. $|z_1| - |z_2| \leq |z_1-z_2|$ (3)

Solution: 1. Note that $|z_1|$, $|z_2|$ and $|z_1+z_2|$ represent the lengths of the sides of a triangle (see Fig. 1). The sum of the lengths of two sides of a triangle is greater than or equal to the length of the third side.

Thus, $\quad |z_1+z_2| \leq |z_1| + |z_2|$

This property is known as the triangle inequality.

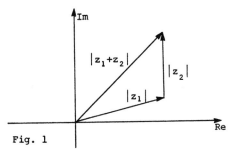

Fig. 1

2. $|z_1|$, $|z_2|$, $|z_3|$ and $|z_1+z_2+z_3|$ are shown in Fig. 2.

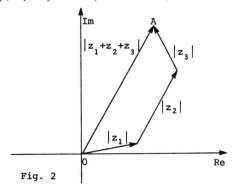

Fig. 2

The geometric interpretation of inequality (2) is the statement that in a plane a straight line is the shortest distance between two points O and A.

3. We have $\quad |z_1| = |z_1-z_2+z_2| \leq |z_1-z_2| + |z_2| \quad$ (4)

Thus,
$$|z_1| - |z_2| \leq |z_1-z_2| \quad (5)$$

The geometrical interpretation of inequality (5) is that a side of a triangle has a length greater than or equal to the difference in lengths of the other two sides.

APPLICATIONS OF COMPLEX NUMBERS IN GEOMETRY

• **PROBLEM 2-4**

Find the equation of

1. a circle of radius r and center at z_0.

2. an ellipse with major axis of length 2a and foci at (c,0) and (-c,0).

Solution: 1. A circle of radius r and center at z_0 is the collection of all points whose distance from z_0 is r, as shown in Fig. 1.

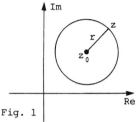

Fig. 1

Therefore, the equation of a circle in the complex plane is

$$|z - z_0| = r \tag{1}$$

The unit circle is the circle with radius r = 1 and center at the origin. A disk of radius r and center z_0 is the set of all points z satisfying the inequality,

$$|z - z_0| < r \tag{2}$$

2. An ellipse is the set of all points z such that the sum of the distances from z to the foci must be equal to the major axis of the ellipse (see Fig. 2).

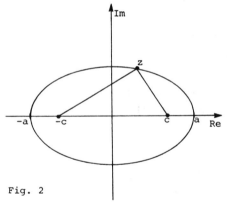

Fig. 2

Thus, the equation of an ellipse in the complex plane is

$$|z+c| + |z-c| = 2a \tag{3}$$

● **PROBLEM 2-5**

Find the loci of points z such that

1. $\text{Re}(z) \geq 3$

2. $\text{Re}(z) \geq 1$ and $\text{Im}(z) \leq 1$

3. $|z| < 4$

4. $|z| \geq 4$
5. $Re(z^2) = p$
6. $Im(z^2) = p$
7. $|z^2-1| = s > 0$
8. $|z^2-z| = 2$

Solution: 1. The region $Re(z) \geq 3$ is the half-plane shown in Fig. 1.

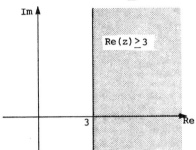

Fig. 1

2. The common part of the half-plane $Re(z) \geq 1$ and the half-plane $Im(z) \leq 1$ is shown in Fig. 2.

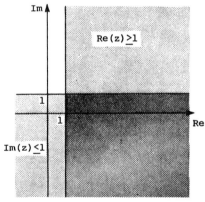

Fig. 2

3. The region $|z| < 4$ is a disk of radius 4 with center at the

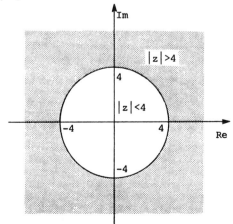

Fig. 3

origin as shown in Fig. 3.

4. The region $|z| \geq 4$ is also shown in Fig. 3. The sets $|z| < 4$ and $|z| \geq 4$ have no common elements.

5. Let $z = x + iy$, then

$$\text{Re}(z^2) = x^2 - y^2 = p \tag{1}$$

This equation represents a hyperbola (see Fig. 4). For $p = 0$, it becomes a pair of straight lines.

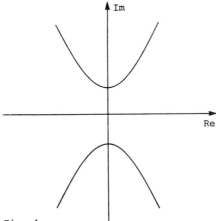

Fig. 4

6. We have $\text{Im}(z^2) = 2xy = p$. $\tag{2}$

The equation $xy = \frac{p}{2}$ represents a rectangular (equilateral) hyperbola, as shown in Fig. 5.

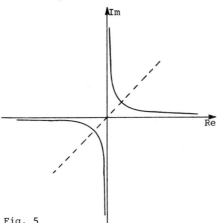

Fig. 5

7. Let $z = x + iy$, then

$$|z^2-1|^2 = |(x^2-y^2-1) + 2xyi|^2 = (x^2-y^2-1)^2 + (2xy)^2$$
$$= (x^2+y^2)^2 - 2x^2 + 2y^2 + 1 = s^2 \tag{3}$$

The equation

$$(x^2+y^2)^2 = 2(x^2-y^2) + (s^2-1) \tag{4}$$

represents the lemniscate of Bernoulli, as shown in Fig. 6.

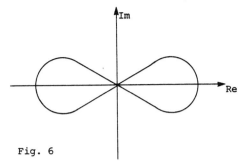

Fig. 6

This conclusion can be reached without long calculations. It follows that

$$|z^2-1| = |(z+1)(z-1)| = |z+1||z-1| = s \qquad (5)$$

$|z^2-1|$ is the product of the distances of z from 1 and -1.

8. Let $|z^2-z| = 2$. Then

$$|z^2-z| = |z(z-1)| = |z||z-1| = 2 \qquad (6)$$

is the product of the distances of z from 0 and 1. Since it is constant (equal to 2), eq.(6) represents the lemniscate.

● **PROBLEM 2-6**

Represent graphically all points z such that

$$\left|\frac{z+1}{z-1}\right| = 2 \qquad (1)$$

<u>Solution</u>: From eq.(1), we obtain

$$\left|\frac{z+1}{z-1}\right|^2 = \left(\frac{z+1}{z-1}\right)\left(\frac{\overline{z}+1}{\overline{z}-1}\right) = 4 \qquad (2)$$

or

$$(z+1)(\overline{z}+1) = 4(z-1)(\overline{z}-1) \qquad (3)$$

so that

$$z\overline{z} - \frac{5}{3}z - \frac{5}{3}\overline{z} + 1 = 0. \qquad (4)$$

Eq.(4) can be written in the form

$$(z - \frac{5}{3})(\overline{z} - \frac{5}{3}) - \frac{16}{9} = 0 \qquad (5)$$

or

$$\left|z - \frac{5}{3}\right|^2 = \frac{16}{9} \qquad (6)$$

and

$$\left|z - \frac{5}{3}\right| = \frac{4}{3} \qquad (7)$$

Eq.(7) represents a circle of radius $\frac{4}{3}$ with center at $(\frac{5}{3},0)$, as shown in Fig. 1. Thus, all the points z lying on the circle satisfy eq.(1).

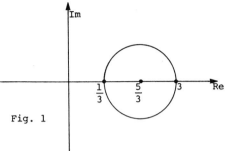

Fig. 1

The solution to this problem can also be found if we represent z by

$$z = x + iy$$

Then, from eq.(1), we obtain

$$|x+1 + iy| = 2|x-1 + iy| \qquad (8)$$

or

$$(x+1)^2 + y^2 = 4[(x-1)^2 + y^2] \qquad (9)$$

$$(x - \tfrac{5}{3})^2 + y^2 = \tfrac{16}{9} \qquad (10)$$

As expected, eq.(10) represents a circle of radius $\tfrac{4}{3}$ with center at $(\tfrac{5}{3}, 0)$.

• **PROBLEM 2-7**

Find the equation of the straight line passing through two given points $A(x_1, y_1)$ and $B(x_2, y_2)$.

Solution: Points A and B can be represented by the complex numbers

$$z_1 = x_1 + iy_1 \quad \text{and} \quad z_2 = x_2 + iy_2, \qquad (1)$$

as shown in Fig. 1.

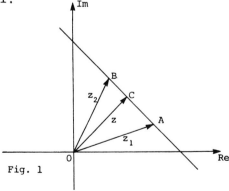

Fig. 1

Let z be the position vector of any point C on the line passing through A and B.

From Fig. 1, we obtain

$$\overline{OB} + \overline{BC} = \overline{OC} \quad \text{or} \quad z_2 + \overline{BC} = z \tag{2}$$

Hence,
$$\overline{BC} = z - z_2 \tag{3}$$

and
$$\overline{OB} + \overline{BA} = \overline{OA} \quad \text{or} \quad z_2 + \overline{BA} = z_1 \tag{4}$$

Hence,
$$\overline{BA} = z_1 - z_2 \tag{5}$$

Vectors $\overline{BC}$ and $\overline{BA}$ are collinear, therefore, we can write

$$\overline{BC} = t\,\overline{BA} \tag{6}$$

or
$$z - z_2 = t(z_1 - z_2) \tag{7}$$

where t is a real parameter.

The sought equation is

$$z = z_2 + t(z_1 - z_2) = tz_1 + (1-t)z_2 \tag{8}$$

or
$$x = x_2 + t(x_1 - x_2)$$
$$y = y_2 + t(y_1 - y_2) \tag{9}$$

Eq.(9) can be written in the form

$$\frac{x - x_2}{x_1 - x_2} = \frac{y - y_2}{y_1 - y_2} \tag{10}$$

Eq.(10) is the standard equation of the straight line. Eq.(9) is the parametric equation of the line.

Another way of describing a straight line is the symmetric equation. Since $\overline{BC}$ and $\overline{CA}$ are collinear, we get

$$\alpha\,\overline{BC} = \beta\,\overline{CA} \tag{11}$$

or
$$\alpha(z - z_2) = \beta(z_1 - z) \tag{12}$$

Solving eq.(12) for z, we find

$$z = \frac{\beta z_1 + \alpha z_2}{\alpha + \beta} \tag{13}$$

or
$$x = \frac{\beta x_1 + \alpha x_2}{\alpha + \beta} \quad;\quad y = \frac{\beta y_1 + \alpha y_2}{\alpha + \beta} \tag{14}$$

• **PROBLEM 2-8**

What requirement must be fulfilled for the points z_1, z_2 and z_3 to be collinear?

Solution: Let us assume that three complex numbers z_1, z_2 and

z_3 are collinear, that is, that all three numbers lie on one line, as shown in Fig. 1.

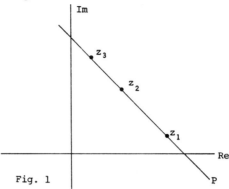

Fig. 1

In such case, the vectors $z_1 - z_3$ and $z_2 - z_3$ are parallel and we can write

$$z_1 - z_3 = t(z_2 - z_3) \tag{1}$$

where t is a real parameter. From eq.(1), we conclude that z_1, z_2, z_3 are collinear if and only if

$$\frac{z_1 - z_3}{z_2 - z_3} \tag{2}$$

is a real number.

● **PROBLEM 2-9**

Show that if the points z_1, z_2, z_3 are collinear, then there exist real numbers α, β, $\gamma \in R$, such that

$$|\alpha| + |\beta| + |\gamma| > 0 \tag{1}$$

$$\alpha + \beta + \gamma = 0 \tag{2}$$

$$\alpha z_1 + \beta z_2 + \gamma z_3 = 0 \tag{3}$$

Solution: As established in the preceeding problem (Problem 2-8), three complex numbers z_1, z_2, z_3 are collinear if and only if the ratio

$$\frac{z_1 - z_3}{z_2 - z_3} \tag{4}$$

is a real number.

Thus,

$$\frac{z_1 - z_3}{z_2 - z_3} = t \tag{5}$$

and

$$z_1 - tz_2 + tz_3 - z_3 = 0 \tag{6}$$

Let us denote

$$\alpha = 1, \quad \beta = -t, \quad \gamma = t-1 \tag{7}$$

We have

$$|\alpha| + |\beta| + |\gamma| > 0 \tag{8}$$

$$\alpha + \beta + \gamma = 0 \tag{9}$$

and
$$z_1 - tz_2 + (t-1)z_3 = \alpha z_1 + \beta z_2 + \gamma z_3 = 0 \tag{10}$$

● **PROBLEM 2-10**

Find the point z which divides the line segment (z_1, z_2) in the ratio $\frac{\alpha_1}{\alpha_2}$, $(\alpha_1 + \alpha_2) \neq 0$.

Solution: The points z_1, z_2 and z are collinear. The point z divides the segment (z_1, z_2) in the ratio $\frac{\alpha_1}{\alpha_2}$, as shown in Fig.1.

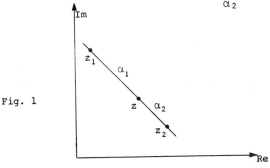

Fig. 1

Thus

$$\frac{z_1 - z}{z - z_2} = \frac{\alpha_1}{\alpha_2} \tag{1}$$

Solving eq.(1) for z, we obtain

$$\alpha_2 z_1 + \alpha_1 z_2 = \alpha_1 z + \alpha_2 z \tag{2}$$

and

$$z = \frac{\alpha_2 z_1 + \alpha_1 z_2}{\alpha_1 + \alpha_2} \tag{3}$$

● **PROBLEM 2-11**

Find the equation of the straight line perpendicular to a complex vector z_0.

Solution: Fig. 1 illustrates the situation.

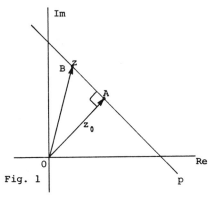

Fig. 1

Let z be an arbitrary point on the line p. We have

$$|OB|^2 = |OA|^2 + |AB|^2 \qquad (1)$$

Substituting

$$\overline{OB} = z, \quad \overline{OA} = z_0, \quad \overline{AB} = z - z_0 \qquad (2)$$

into eq.(1), we find

$$|z|^2 = |z_0|^2 + |z - z_0|^2 \qquad (3)$$

Hence,
$$\qquad (4)$$
$$z\bar{z} = z_0\bar{z}_0 + (z-z_0)(\bar{z}-\bar{z}_0) = z_0\bar{z}_0 + z\bar{z} - z\bar{z}_0 - z_0\bar{z} + z_0\bar{z}_0$$

From eq.(4), we find

$$z\bar{z}_0 + z_0\bar{z} = 2z_0\bar{z}_0 \qquad (5)$$

Thus, the equation of a straight line perpendicular to z_0 is

$$\bar{z}_0 z + z_0 \bar{z} = 2|z_0|^2 \qquad (6)$$

• **PROBLEM 2-12**

Prove that two vectors z_1 and z_2 are perpendicular if, and only if

$$z_1\bar{z}_2 + \bar{z}_1 z_2 = 0 \qquad (1)$$

Solution: The statement α if and only if β, means that α implies β and that β implies α. We shall accomodate the following notation:

α => β   reads α implies β

α<=> β   reads α if, and only if β

We have to prove that

$$\begin{pmatrix} z_1 \text{ and } z_2 \text{ are} \\ \text{perpendicular} \end{pmatrix} <=> \left(z_1\bar{z}_2 + \bar{z}_1 z_2 = 0 \right)$$

First we shall prove =>

We assume that z_1 and z_2 are perpendicular as shown in Fig. 1.

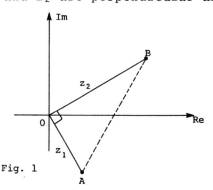

Fig. 1

Then

$$|OA|^2 + |OB|^2 = |AB|^2$$
$$|z_1|^2 + |z_2|^2 = |z_2-z_1|^2 \quad \text{or} \tag{2}$$

Eq.(2) leads to

$$|z_2-z_1|^2 = (z_2-z_1)(\bar{z}_2-\bar{z}_1) = z_2\bar{z}_2 - \bar{z}_1z_2 - z_1\bar{z}_2 + z_1\bar{z}_1$$
$$= |z_1|^2 + |z_2|^2 - (z_1\bar{z}_2 + \bar{z}_1z_2) \tag{3}$$

Therefore,

$$z_1\bar{z}_2 + \bar{z}_1z_2 = 0 \tag{4}$$

Now we shall prove $\Leftarrow$
If

$$z_1\bar{z}_2 + \bar{z}_1z_2 = 0$$

Then

$$|z_1|^2 + |z_2|^2 = z_1\bar{z}_1 + z_2\bar{z}_2 = z_1\bar{z}_1 + z_2\bar{z}_2 - z_1\bar{z}_2 - \bar{z}_1z_2$$
$$= (z_1-z_2)(\bar{z}_1-\bar{z}_2) = |z_1-z_2|^2 \tag{5}$$

Therefore,

$$|OA|^2 + |OB|^2 = |AB|^2, \tag{6}$$

which implies that z_1 and z_2 are perpendicular.

q.e.d.*

*q.e.d. [Latin quod erat demonstrandum] which was to be proved.

● **PROBLEM 2-13**

Using complex numbers, find the length of the median of a triangle whose vertices are $A(x_1,y_1)$, $B(x_2,y_2)$ and $C(x_3,y_3)$.

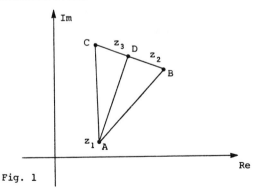

Fig. 1

Solution: Let z_1, z_2, z_3 be complex numbers representing vertices A, B and C, respectively.

From Fig. 1, we find

$$\overline{AB} = z_2 - z_1 \tag{1}$$

$$\overline{BC} = z_3 - z_2 \tag{2}$$

Since AD is the median,

$$|\overline{CD}| = |\overline{BD}| \tag{3}$$

and

$$\overline{BD} = \frac{z_3 - z_2}{2} \tag{4}$$

Therefore,

$$\overline{AD} = \overline{AB} + \overline{BD} \tag{5}$$

and

$$|\overline{AD}| = |\overline{AB} + \overline{BD}| = \left| z_2 - z_1 + \frac{z_3 - z_2}{2} \right|$$
$$= \left| \frac{z_2 + z_3 - 2z_1}{2} \right| \tag{6}$$

Remembering that

$$|z| \equiv |x+iy| \equiv \sqrt{x^2+y^2} \tag{7}$$

and substituting z_1, z_2, and z_3 into eq.(6), we find

$$|\overline{AD}| = \left| \frac{(x_2+x_3-2x_1) + i(y_2+y_3-2y_1)}{2} \right|$$
$$= \tfrac{1}{2}\sqrt{(x_2+x_3-2x_1)^2 + (y_2+y_3-2y_1)^2} \tag{8}$$

• **PROBLEM 2-14**

> Let z_1, z_2, z_3 be three complex numbers such that
> $$z_1 + z_2 + z_3 = 0 \tag{1}$$
> $$|z_1| = |z_2| = |z_3| = 1 \tag{2}$$
> Show that these numbers are the vertices of an equilateral triangle inscribed in the unit circle.

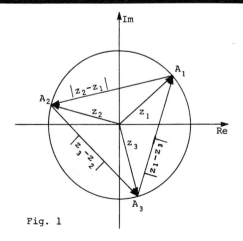

Fig. 1

Solution: From eq.(2), we see that z_1, z_2, z_3 are positioned on the unit circle. Let

$$z_1 = OA_1, \quad z_2 = OA_2, \quad z_3 = OA_3 \tag{3}$$

as shown in Fig. 1.
Then

$$\overline{A_1A_2} = z_2-z_1, \quad \overline{A_2A_3} = z_3-z_2, \quad \overline{A_3A_1} = z_1-z_3 \tag{4}$$

To show that the triangle is equilateral we have to prove that

$$|z_2-z_1| = |z_3-z_2| = |z_1-z_3| \tag{5}$$

First, let us show that

$$|z_2-z_1| = |z_3-z_2| \tag{6}$$

Since

$$z_1 + z_2 + z_3 = 0, \tag{7}$$

then

$$z_2 = -z_1 - z_3 \tag{8}$$

Eq.(6) yields

$$|2z_1+z_3| = |2z_3+z_1| \tag{9}$$

which is equivalent to

$$(2z_1+z_3)(2\bar{z}_1+\bar{z}_3) - (2z_3+z_1)(2\bar{z}_3+\bar{z}_1) = 0 \tag{10}$$

Multiplying and substituting $z_1\bar{z}_1 = z_3\bar{z}_3 = 1$ we find

$$4z_1\bar{z}_1 + 2z_1\bar{z}_3 + 2\bar{z}_1z_3 + z_3\bar{z}_3 - 4z_3\bar{z}_3 - 2\bar{z}_1z_3 - 2z_1\bar{z}_3 - z_1\bar{z}_1$$
$$= 4 + 1 - 4 - 1 = 0 \tag{11}$$

In a similar manner, we verify that

$$|z_3-z_2| = |z_1-z_3| \tag{12}$$

Hence, the triangle $A_1A_2A_3$ is equilateral and inscribed in the unit circle.

• **PROBLEM 2-15**

Let z_1, z_2, z_3 be the vertices of an equilateral triangle. Prove that

$$|z_2|^2 + z_1\bar{z}_3 + z_3\bar{z}_1 = |z_3|^2 + z_1\bar{z}_2 + z_2\bar{z}_1 \tag{1}$$

Solution: Since the triangle is equilateral

$$|z_2-z_1| = |z_3-z_2| = |z_1-z_3| \tag{2}$$

or

$$|z_2-z_1|^2 = |z_3-z_2|^2 = |z_1-z_3|^2 \tag{3}$$

$$(z_2-z_1)(\bar{z}_2-\bar{z}_1) = (z_3-z_2)(\bar{z}_3-\bar{z}_2) = (z_1-z_3)(\bar{z}_1-\bar{z}_3) \qquad (4)$$

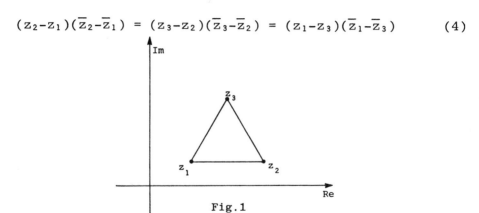

Fig.1

Eq.(4) leads to

$$|z_1|^2 + |z_2|^2 - \bar{z}_1 z_2 - z_1 \bar{z}_2 = |z_2|^2 + |z_3|^2 - z_3 \bar{z}_2 - z_2 \bar{z}_3 \qquad (5)$$
$$= |z_1|^2 + |z_3|^2 - z_1 \bar{z}_3 - z_3 \bar{z}_1$$

Eq.(5) is equivalent to

$$|z_1|^2 - z_2 \bar{z}_1 - z_1 \bar{z}_2 = |z_3|^2 - z_3 \bar{z}_2 - z_2 \bar{z}_3 \qquad (6)$$

$$|z_2|^2 - z_3 \bar{z}_2 - z_2 \bar{z}_3 = |z_1|^2 - z_1 \bar{z}_3 - z_3 \bar{z}_1 \qquad (7)$$

Subtracting eq.(7) from eq.(6) we find

$$|z_2|^2 + z_1 \bar{z}_3 + z_3 \bar{z}_1 = |z_3|^2 + z_1 \bar{z}_2 + z_2 \bar{z}_1 \qquad (8)$$

q.e.d.

• **PROBLEM 2-16**

1. Let z_1 and z_2 be two non-parallel vectors. Prove that if α and β are real numbers and

$$\alpha z_1 + \beta z_2 = 0, \qquad (1)$$

then $\qquad \alpha = \beta = 0 \qquad (2)$

2. Prove that the diagonals of a parallelogram bisect each other.

Solution: 1. Let $z_1 = a_1 + ib_1$ and $z_2 = a_2 + ib_2$, then

$$\alpha z_1 + \beta z_2 = (\alpha a_1 + \beta a_2) + i(\alpha b_1 + \beta b_2) = 0 \qquad (3)$$

Eq.(3) is equivalent to

$$\alpha a_1 + \beta a_2 = 0 \qquad (4)$$
$$\alpha b_1 + \beta b_2 = 0 \qquad (5)$$

The solution is
$$\alpha = \beta = 0 \qquad (6)$$

if $\frac{b_1}{a_1} \neq \frac{b_2}{a_2}$, i.e. if z_1 and z_2 are non-parallel.

2. The quadrilateral OABC shown in Fig. 1 is a parallelogram.

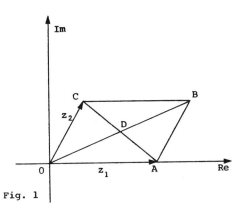

Fig. 1

We have

$$z_2 + \overline{CA} = z_1 \quad \text{and} \quad \overline{CA} = z_1 - z_2 \qquad (7)$$

A real parameter α exists such that

$$\overline{CD} = \alpha(z_1 - z_2) \qquad 0 \leq \alpha \leq 1 \qquad (8)$$

$$\overline{OB} = z_1 + z_2$$

A real parameter β exists such that

$$\overline{OD} = \beta(z_1 + z_2) \qquad 0 \leq \beta \leq 1 \qquad (9)$$

From the figure, we get

$$\overline{OC} + \overline{CD} = \overline{OD} \quad \text{and} \qquad (10)$$

$$z_2 + \alpha(z_1 - z_2) = \beta(z_1 + z_2) \quad \text{or}$$

$$z_1(\alpha - \beta) + z_2(1 - \alpha - \beta) = 0 \qquad (11)$$

Since z_1 and z_2 are non-parallel, by part I of this problem, we obtain

$$\begin{aligned} \alpha - \beta &= 0 \\ 1 - \alpha - \beta &= 0 \end{aligned} \qquad (12)$$

Hence,

$$\alpha = \beta = \frac{1}{2} \qquad (13)$$

and the diagonals bisect each other.

● **PROBLEM 2-17**

Find the point of intersection of the medians of a triangle.

Hint: Represent the vertices of the triangle by complex numbers.

Solution: The triangle is shown in Fig. 1.

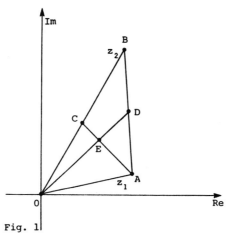

Fig. 1

The medians of a triangle intersect at one point. Therefore, it is enough to take into account two medians, say AC and OD. From the figure, we obtain

$$\overline{OD} = \overline{OA} + \overline{AD} \tag{1}$$

$$\overline{AD} = \tfrac{1}{2}\overline{AB} = \tfrac{1}{2}(z_2 - z_1) = \frac{z_2 - z_1}{2} \tag{2}$$

$$\overline{OD} = z_1 + \frac{z_2 - z_1}{2} = \frac{z_1 + z_2}{2} \tag{3}$$

$$\overline{OC} + \overline{CA} = \overline{OA} \tag{4}$$

$$\overline{CA} = \overline{OA} - \overline{OC} = z_1 - \frac{z_2}{2} \tag{5}$$

We can write

$$\overline{OE} = \alpha \, \overline{OD} = \alpha \, \frac{z_1 + z_2}{2} \tag{6}$$

$$\overline{CE} = \beta \, \overline{CA} = \beta \left(z_1 - \frac{z_2}{2} \right) \tag{7}$$

where α and β are positive real parameters. Again, from the figure

$$\overline{OC} + \overline{CE} = \overline{OE} \tag{8}$$

$$\frac{z_2}{2} + \beta \left(z_1 - \frac{z_2}{2} \right) = \alpha \, \frac{z_1 + z_2}{2} \tag{9}$$

or

$$z_1 \left(\beta - \frac{\alpha}{2} \right) + z_2 \left(\frac{1}{2} - \frac{\beta}{2} - \frac{\alpha}{2} \right) = 0 \tag{10}$$

Applying the results of Problem 2-16 (eqs.(1) and (2)), we get

$$\beta - \frac{\alpha}{2} = 0 \tag{11}$$

$$\frac{1}{2} - \frac{\beta}{2} - \frac{\alpha}{2} = 0 \tag{12}$$

Solving for α and β, we find

$$\alpha = \frac{2}{3}, \quad \beta = \frac{1}{3} \tag{13}$$

Therefore, the medians of a triangle trisect each other.

● **PROBLEM 2-18**

ABCD is a quadrilateral whose vertices are represented by the complex vectors z_1, z_2, z_3, z_4 as shown in Fig. 1. Prove that ABCD is a parallelogram if, and only if,

$$z_1 - z_2 + z_3 - z_4 = 0 \tag{1}$$

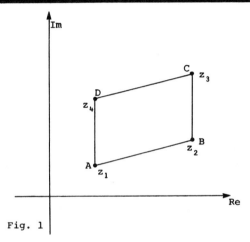

Fig. 1

Solution: We have to prove that

$$\begin{pmatrix} \text{ABCD is a} \\ \text{parallelogram} \end{pmatrix} \iff \left(z_1 - z_2 + z_3 - z_4 = 0 \right)$$

First we shall show that ⇒ is true. If ABCD is a parallelogram, then its sides AB and DC must be parallel and of equal length. Using vector notation, we can write

$$\overline{AB} = \overline{DC} \tag{2}$$

From Fig. 1, we have

$$\overline{AB} = z_2 - z_1 \tag{3}$$

$$\overline{DC} = z_3 - z_4 \tag{4}$$

Substituting eqs.(3) and (4) into eq.(2) results in

$$z_2 - z_1 = z_3 - z_4 \tag{5}$$

or

$$z_1 - z_2 + z_3 - z_4 = 0 \tag{6}$$

Now we will show that ⇐ is true. If $z_1 - z_2 + z_3 - z_4 = 0$, then ABCD is a parallelogram.

$$z_2 - z_1 = z_3 - z_4 \tag{7}$$

Since

$$\overline{AB} = z_2 - z_1 \quad \text{and} \quad \overline{DC} = z_3 - z_4, \tag{8}$$

we conclude that

$$\overline{AB} = \overline{DC} \tag{9}$$

Thus, ABCD is a parallelogram.

q.e.d.

From eq.(1) we conclude that the remaining two sides of ABCD, namely AD and BC are also parallel and of equal length. Therefore,

$$z_4 - z_1 = z_3 - z_2 \tag{10}$$

But

$$\overline{AD} = z_4 - z_1 \quad \text{and} \quad \overline{BC} = z_3 - z_2 \tag{11}$$

then

$$\overline{AD} = \overline{BC} \tag{12}$$

● **PROBLEM 2-19**

Prove that if the diagonals of a quadrilateral bisect each other, then the quadrilateral is a parallelogram.

<u>Solution</u>: Let z_1, z_2, z_3, z_4 represent vertices A, B, C, D as shown in Fig. 1.

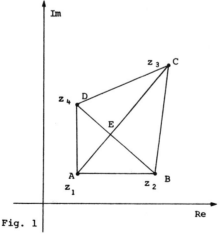

Fig. 1

Since the diagonals bisect each other, we have

$$\overline{AE} = \overline{EC}$$
$$\text{and} \tag{1}$$
$$\overline{DE} = \overline{EB}$$

on the other hand

$$\overline{AE} = \frac{\overline{AC}}{2} = \frac{z_3 - z_1}{2} \tag{2}$$

$$\overline{DE} = \frac{\overline{DB}}{2} = \frac{z_2 - z_4}{2} = \overline{EB} \tag{3}$$

From Fig. 1 we have

$$\overline{AB} = \overline{AE} + \overline{EB} \tag{4}$$

where

$$\overline{AB} = z_2 - z_1 \tag{5}$$

Substituting eqs.(2), (3) and (5) into eq.(4), we obtain

$$z_2 - z_1 = \frac{z_3 - z_1}{2} + \frac{z_2 - z_4}{2} \tag{6}$$

or

$$z_1 - z_2 + z_3 - z_4 = 0 \tag{7}$$

It was shown in Problem 2-18 that if z_1, z_2, z_3 and z_4 represent the vertices of a quadrilateral, then the quadrilateral is a parallelogram if and only if

$$z_1 - z_2 + z_3 - z_4 = 0$$

Thus, we conclude that ABCD is a parallelogram.

• **PROBLEM 2-20**

ABCD is a quadrilateral and E, F, G, H are midpoints of its sides. Prove that EFGH is a parallelogram.

Solution: Quadrilateral ABCD is shown in Fig. 1.

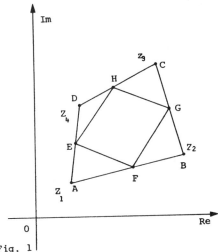

Fig. 1

Let z_1, z_2, z_3, z_4 be the position vectors of A, B, C, D, respectively.

To find the position vector of F we compute

$$\overline{OA} + \overline{AB} = \overline{OB} \tag{1}$$

$$\overline{AB} = \overline{OB} - \overline{OA} = z_2 - z_1 \qquad (2)$$

$$\overline{AF} = \frac{\overline{AB}}{2} = \frac{z_2 - z_1}{2} \qquad (3)$$

Thus,

$$\overline{OA} + \overline{AF} = \overline{OF} \qquad (4)$$

$$\overline{OF} = z_1 + \frac{z_2 - z_1}{2} = \frac{z_1 + z_2}{2} \qquad (5)$$

The position vector of F is $\frac{z_1 + z_2}{2}$.

In the same way, we compute the position vector of G as $\frac{z_2 + z_3}{2}$. Respectively, the position vector of H is $\frac{z_3 + z_4}{2}$, and the position vector of E is $\frac{z_1 + z_4}{2}$.

A quadrilateral ABCD is a parallelogram, if and only if,

$$z_A - z_B + z_C - z_D = 0 \qquad (6)$$

Substituting the position vectors of E, F, G, H we find

$$z_E - z_F + z_G - z_H = \frac{z_1+z_4}{2} - \frac{z_1+z_2}{2} + \frac{z_2+z_3}{2} - \frac{z_3+z_4}{2} = 0 \qquad (7)$$

Therefore EFGH is a parallelogram.

● **PROBLEM 2-21**

Prove the theorem:

If OABC is a parallelogram and AE bisects side OC, then AE trisects the diagonal OB (see Fig. 1).

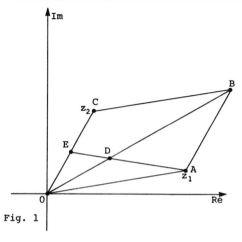

Fig. 1

Solution: Let z_1 be the position vector of A and z_2 be the position vector of C.

We have
$$\overline{OB} = z_1 + z_2 \tag{1}$$

A positive real number α exists such that
$$\overline{OD} = \alpha\,\overline{OB} = \alpha(z_1+z_2) \tag{2}$$

From Fig. 1, we obtain
$$\overline{OE} = \tfrac{1}{2}\,\overline{OC} = \tfrac{z_2}{2} \tag{3}$$

$$\overline{EA} = \overline{OA} - \overline{OE} = z_1 - \tfrac{z_2}{2} \tag{4}$$

A positive real number β exists such that
$$\overline{ED} = \beta\,\overline{EA} = \beta\left(z_1 - \tfrac{z_2}{2}\right) \tag{5}$$

Again, from the figure
$$\overline{OE} + \overline{ED} = \overline{OD} \tag{6}$$

Substituting eqs.(2), (3) and (5) into eq.(6) results in
$$\tfrac{z_2}{2} + \beta\left(z_1 - \tfrac{z_2}{2}\right) = \alpha(z_1+z_2) \tag{7}$$

or
$$z_1(\beta-\alpha) + z_2\left(\tfrac{1}{2} - \tfrac{\beta}{2} - \alpha\right) = 0 \tag{8}$$

Applying the results of Problem 2-16, part I, we obtain
$$\beta - \alpha = 0 \tag{9}$$

$$\tfrac{1}{2} - \tfrac{\beta}{2} - \alpha = 0 \tag{10}$$

Solving for α and β,
$$\alpha = \beta = \tfrac{2}{3} \tag{11}$$

and
$$\overline{OD} = \alpha\,\overline{OB} = \tfrac{2}{3}\,\overline{OB} \tag{12}$$

Indeed, AE trisects OB.

• **PROBLEM 2-22**

Prove that
$$|z_1+z_2|^2 + |z_1-z_2|^2 = 2|z_1|^2 + 2|z_2|^2 \tag{1}$$

What is the relationship between this identity and the following theorem?

The sum of the squares of the sides of a parallelogram is equal to the sum of the squares of the diagonals.

Solution: The left-hand side of eq.(1) can be transformed to

$$|z_1+z_2|^2 + |z_1-z_2|^2 = (z_1+z_2)(\bar{z}_1+\bar{z}_2) + (z_1-z_2)(\bar{z}_1-\bar{z}_2)$$

$$= z_1\bar{z}_1 + z_1\bar{z}_2 + \bar{z}_1 z_2 + z_2\bar{z}_2 + z_1\bar{z}_1 - z_1\bar{z}_2 - \bar{z}_1 z_2 + z_2\bar{z}_2 \qquad (2)$$

$$= 2z_1\bar{z}_1 + 2z_2\bar{z}_2 = 2|z_1|^2 + 2|z_2|^2$$

q.e.d.

Let OABC be a parallelogram, as shown in Fig. 1.

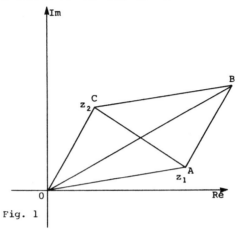

Fig. 1

We have

$$|\overline{OA}| = |z_1| = |\overline{CB}| \qquad (3)$$

$$|\overline{OC}| = |z_2| = |\overline{AB}| \qquad (4)$$

$$\overline{OB} = z_1 + z_2, \quad |\overline{OB}| = |z_1+z_2| \qquad (5)$$

$$\overline{CA} = z_2 - z_1, \quad |\overline{CA}| = |z_2-z_1| \qquad (6)$$

The left-hand side of eq.(1) is

$$|z_1+z_2|^2 + |z_1-z_2|^2 = |\overline{OB}|^2 + |\overline{CA}|^2 \qquad (7)$$

which is the sum of the squares of the diagonals. The right-hand side of eq.(1) is

$$2|z_1|^2 + 2|z_2|^2 = |OA|^2 + |CB|^2 + |OC|^2 + |AB|^2 \qquad (8)$$

which is the sum of the squares of the sides of a parallelogram.

By proving identity (1), we have proven the theorem.

• **PROBLEM 2-23**

Prove that the equation of a circle in the z plane can be written in the form

$$z\bar{z} + az + \overline{az} + b = 0 \qquad (1)$$

where a is a complex constant and b is a real constant.

Solution: The equation of a circle of radius r and center at z_0 is

$$|z - z_0| = r \qquad (2)$$

(see Problem 2-4).

Eq.(2) can be transformed as follows

$$|z - z_0|^2 = r^2 \qquad (3)$$

$$|z-z_0|^2 = (z-z_0)(\bar{z}-\bar{z}_0) = z\bar{z} - z\bar{z}_0 - z_0\bar{z} + z_0\bar{z}_0 = r^2 \qquad (4)$$

Let

$$-\bar{z}_0 = a \qquad (5)$$

then

$$-z_0 = \bar{a} \qquad (6)$$

Eq.(4) becomes

$$z\bar{z} + az + \bar{a}\bar{z} + z_0\bar{z}_0 - r^2 = 0 \qquad (7)$$

Denoting

$$b = z_0\bar{z}_0 - r^2 \qquad (8)$$

where b is of course real, we obtain

$$z\bar{z} + az + \bar{a}\bar{z} + b = 0 \qquad (9)$$

• **PROBLEM 2-24**

The complex numbers z_1, z_2, z_3, z_4 are such that

$$z_1 + z_2 + z_3 + z_4 = 0 \qquad (1)$$

and

$$|z_1| = |z_2| = |z_3| = |z_4| = 1 \qquad (2)$$

Show that z_1, z_2, z_3, z_4 are the vertices of a rectangle.

Solution: All vertices are located on the unit circle, as shown in Fig. 1.

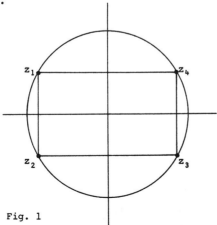

Fig. 1

To show that $z_1 z_2 z_3 z_4$ is a rectangle, we have to prove that

$$|z_4 - z_1| = |z_3 - z_2| \tag{3}$$

and

$$|z_1 - z_2| = |z_4 - z_3| \tag{4}$$

From eq.(1), we obtain

$$z_2 + z_3 = -z_4 - z_1 \tag{5}$$

or

$$|z_3 + z_2|^2 = |z_4 + z_1|^2 \tag{6}$$

Thus

$$(z_3 + z_2)(\bar{z}_3 + \bar{z}_2) = (z_4 + z_1)(\bar{z}_4 + \bar{z}_1) \tag{7}$$

and

$$z_3 \bar{z}_3 + z_3 \bar{z}_2 + z_2 \bar{z}_3 + z_2 \bar{z}_2 = z_4 \bar{z}_4 + z_4 \bar{z}_1 + z_1 \bar{z}_4 + z_1 \bar{z}_1 \tag{8}$$

Since

$$z_3 \bar{z}_3 = z_2 \bar{z}_2 = z_4 \bar{z}_4 = z_1 \bar{z}_1 = 1 \tag{9}$$

eq.(8) reduces to

$$z_3 \bar{z}_2 + z_2 \bar{z}_3 = z_4 \bar{z}_1 + z_1 \bar{z}_4 \tag{10}$$

Let us now prove eq.(3). We have

$$(z_4 - z_1)(\bar{z}_4 - \bar{z}_1) = (z_3 - z_2)(\bar{z}_3 - \bar{z}_2) \tag{11}$$

$$z_4 \bar{z}_4 - z_4 \bar{z}_1 - z_1 \bar{z}_4 + z_1 \bar{z}_1 = z_3 \bar{z}_3 - z_3 \bar{z}_2 - z_2 \bar{z}_3 + z_2 \bar{z}_2 \tag{12}$$

$$-z_4 \bar{z}_1 - z_1 \bar{z}_4 = -z_3 \bar{z}_2 - z_2 \bar{z}_3 \tag{13}$$

which is eq.(10).

To prove eq.(4), observe that

$$z_1 + z_2 = -z_3 - z_4 \tag{14}$$

and

$$(z_1 + z_2)(\bar{z}_1 + \bar{z}_2) = (-z_3 - z_4)(-\bar{z}_3 - \bar{z}_4) \tag{15}$$

$$z_1 \bar{z}_2 + z_2 \bar{z}_1 = z_3 \bar{z}_4 + z_4 \bar{z}_3 \tag{16}$$

Utilizing eqs.(9) and (16), we transform eq.(4) to

$$(z_1 - z_2)(\bar{z}_1 - \bar{z}_2) = (z_4 - z_3)(\bar{z}_4 - \bar{z}_3) \tag{17}$$

$$z_1 \bar{z}_1 - z_1 \bar{z}_2 - z_2 \bar{z}_1 + z_2 \bar{z}_2 = z_4 \bar{z}_4 - z_4 \bar{z}_3 - z_3 \bar{z}_4 + z_3 \bar{z}_3 \tag{18}$$

$$-z_1 \bar{z}_2 - z_2 \bar{z}_1 = -z_4 \bar{z}_3 - z_3 \bar{z}_4 \tag{19}$$

which is eq.(16).

That completes the proof of eqs.(3) and (4). Since z_1, z_2, z_3, z_4 are located on a circle, and since they satisfy eqs.(3) and (4), we conclude that $z_1 z_2 z_3 z_4$ is a rectangle.

• **PROBLEM 2-25**

Let K be a circle passing through the point $z = \pm a$ with the center at $z_0 = ib$.

Show that for any point z, the value of $z\bar{z} - 2b\,\text{Im}(z)$ determines the location of z with respect to K. That is,

for $z\bar{z} - 2b\,\text{Im}(z) \Rightarrow \begin{cases} > a^2 & z \text{ is exterior to K} \\ = a^2 & z \text{ is situated on K} \\ < a^2 & z \text{ is interior to K} \end{cases}$

Furthermore, show that z is interior to the circle K if, and only if, the point $\dfrac{a^2}{\bar{z}}$ is exterior to K.

Solution: The circle K is shown in Fig. 1.

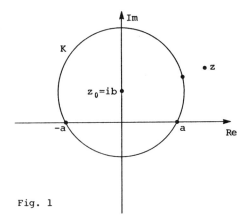

Fig. 1

The radius of circle k is

$$r = \sqrt{a^2 + b^2} \qquad (1)$$

Consider a point z exterior to K. Then

$$|z - z_0| > r \qquad (2)$$

Eq.(2) can be transformed to

$$|z - z_0|^2 = (z - z_0)(\bar{z} - \bar{z}_0)$$
$$= (z - ib)(\bar{z} + ib) = z\bar{z} + zib - \bar{z}ib + b^2 \qquad (3)$$
$$= z\bar{z} + ib(z - \bar{z}) + b^2 = z\bar{z} - 2b\,\text{Im}(z) + b^2 > a^2 + b^2$$

Thus, for a point z exterior to K,

$$z\bar{z} - 2b\,\text{Im}(z) > a^2 \qquad (4)$$

If z is on the circle, then

$$|z - z_0| = r \qquad (5)$$

and
$$|z-z_0|^2 = r^2 \tag{6}$$

Utilizing eq.(3), we obtain
$$|z-z_0|^2 = z\bar{z} - 2b\,\text{Im}(z) + b^2 = a^2 + b^2 \tag{7}$$

Thus, if z is on the circle K, then
$$z\bar{z} - 2b\,\text{Im}(z) = a^2 \tag{8}$$

In the same manner, we can show that if z is interior to K, then
$$z\bar{z} - 2b\,\text{Im}(z) < a^2 \tag{9}$$

Now we will prove
$$\begin{pmatrix} z \text{ is interior} \\ \text{to } K \end{pmatrix} \iff \begin{pmatrix} \dfrac{a^2}{z} \text{ is exterior to } K \end{pmatrix}$$

We shall utilize eqs.(4) and (9).

$\Rightarrow$

Since z is interior to K and $z = x + iy$,
$$z\bar{z} - 2b\,\text{Im}(z) = (x+iy)(x-iy) - 2by = x^2 + y^2 - 2by < a^2 \tag{10}$$

To show that $\dfrac{a^2}{z}$ is exterior to K, observe that

$$\left(\frac{a^2}{z}\right)\overline{\left(\frac{a^2}{z}\right)} - 2b\,\text{Im}\left(\frac{a^2}{z}\right)$$

$$= \frac{a^4}{(x+iy)(x-iy)} - 2b\,\text{Im}\left(\frac{a^2\bar{z}}{2\bar{z}}\right) = \frac{a^4}{x^2+y^2} - 2b\left(\frac{-a^2 y}{x^2+y^2}\right) \tag{11}$$

$$= \frac{a^4 + 2ba^2 y}{x^2 + y^2} > a^2$$

The last inequality can be derived from eq.(10), indeed
$$x^2 + y^2 < a^2 + 2by$$
$$a^2(x^2+y^2) < a^4 + 2ba^2 y \tag{12}$$
$$a^2 < \frac{a^4 + 2ba^2 y}{x^2 + y^2}$$

$\Leftarrow$

$\dfrac{a^2}{z}$ is exterior to K, therefore,

$$\left(\frac{a^2}{z}\right)\overline{\left(\frac{a^2}{z}\right)} - 2b\,\text{Im}\left(\frac{a^2}{z}\right) = a^2\,\frac{a^2 + 2by}{x^2+y^2} > a^2 \tag{13}$$

and
$$a^2 + 2by > x^2 + y^2 \tag{14}$$
$$x^2 + y^2 - 2by < a^2 \tag{15}$$
$$z\bar{z} - 2b\,\text{Im}(z) < a^2 \tag{16}$$

Therefore z is interior to K.

APPLICATIONS IN PHYSICS

● PROBLEM 2-26

Find the mass center of the system of three masses m_1, m_2, m_3, whose position vectors are z_1, z_2, z_3 respectively.

Solution: To find the mass center of the system shown in Fig. 1, consider the part of the system shown in Fig. 2.

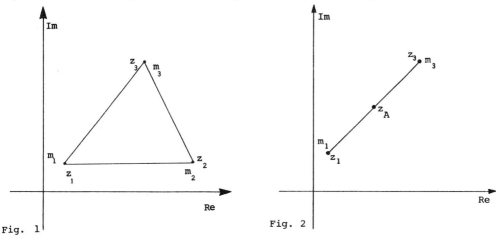

Fig. 1   Fig. 2

If z_A is the mass center of this system, then

$$m_1(z_A - z_1) = m_3(z_3 - z_A) \tag{1}$$

Solving for z_A we find

$$z_A = \frac{m_1 z_1 + m_3 z_3}{m_1 + m_3} \tag{2}$$

Now, the system consisting of m_1 at z_1 and m_3 at z_3 can be replaced by one mass equal $m_1 + m_3$ located at

$$z_A = \frac{m_1 z_1 + m_3 z_3}{m_1 + m_3},$$

as shown in Fig. 3.

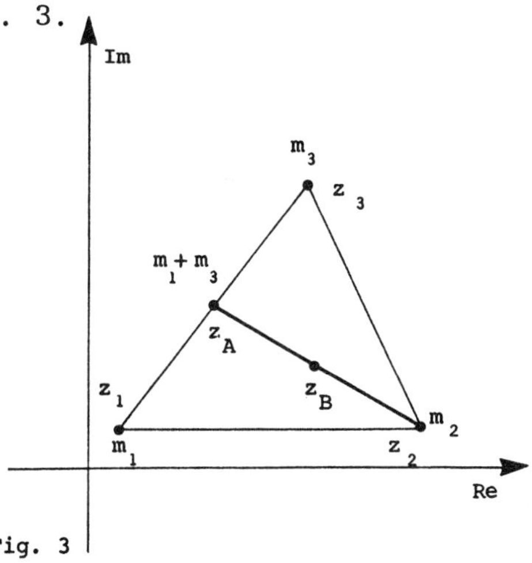

Fig. 3

The whole problem reduces to finding the mass center of the system consisting of mass m_2 at z_2 and $m_1 + m_3$ at z_A.

Modifying eq.(2), we find

$$z_B = \frac{(m_1+m_3)z_A + m_2 z_2}{(m_1+m_3) + m_2} \quad (3)$$

$$= \frac{m_1+m_3 \frac{m_1 z_1+m_3 z_3}{m_1+m_3} + m_2 z_2}{m_1 + m_2 + m_3} = \frac{m_1 z_1 + m_2 z_2 + m_3 z_3}{m_1 + m_2 + m_3}$$

Eq.(3) gives the position of the mass center of the system shown in Fig. 1.

• **PROBLEM 2-27**

Find the mass center of the system of masses $m_1, m_2, \ldots, m_n$ situated at $z_1, z_2, \ldots, z_n$.

Solution: First consider mass m_1 and mass m_2 situated at z_1 and z_2. The mass center is

$$z_1' = \frac{m_1 z_1 + m_2 z_2}{m_1 + m_2} \quad (1)$$

Thus m_1 and m_2 can be replaced by one mass $m_1 + m_2$ located at z_1' given by eq.(1).

Now, consider mass $m_1 + m_2$ located at z_1' and m_3 located at z_3. The mass center is

$$z_2' = \frac{z_1'(m_1+m_2) + m_3 z_3}{(m_1+m_2) + m_3}$$

$$= \frac{m_1 z_1 + m_2 z_2 + m_3 z_3}{m_1 + m_2 + m_3} \quad (2)$$

We see that masses m_1, m_2, m_3 at z_1, z_2, z_3 can be replaced by one mass $m_1 + m_2 + m_3$ located at z_2' given by eq.(2).

For $k < n$ the system $m_1, m_2, \ldots, m_k$ with position vectors $z_1, z_2, \ldots, z_k$ can be replaced by one mass $m_1 + m_2 + \ldots + m_k$ located at

$$z_{k-1}' = \frac{m_1 z_1 + \ldots + m_k z_k}{m_1 + \ldots + m_k} \quad (3)$$

For the next mass $k+1$, we obtain the total mass $m_1 + \ldots + m_k + m_{k+1}$ located at

$$z_k' = \frac{(m_1+\ldots+m_k)z_{k-1}' + m_{k+1} z_{k+1}}{(m_1+\ldots+m_k) + m_{k+1}} \quad (4)$$

$$= \frac{m_1 z_1 + \ldots + m_k z_k + m_{k+1} z_{k+1}}{m_1 + \ldots + m_k + m_{k+1}}$$

Thus, for a system of n masses $m_1, \ldots, m_n$ situated at $z_1, z_2, \ldots, z_n$, the mass center is

$$\frac{m_1 z_1 + m_2 z_2 + \ldots + m_n z_n}{m_1 + m_2 + \ldots + m_n} \quad (5)$$

CHAPTER 3

DE MOIVRE'S THEOREM

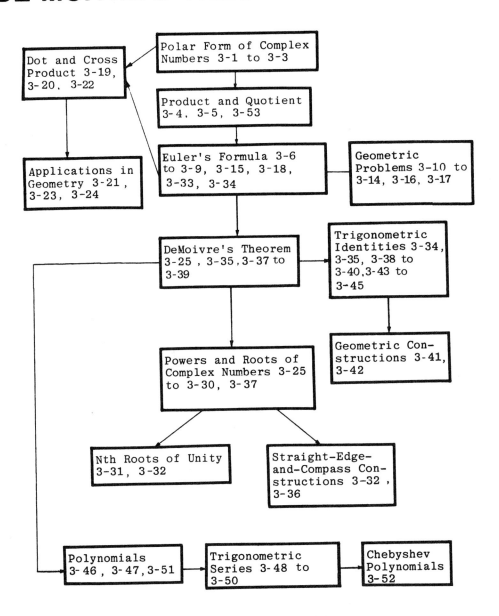

This chart is provided to facilitate rapid understanding of the interrelationships of the topics and subject matter in this chapter. Also shown are the problem numbers associated with the subject matter.

POLAR FORM OF COMPLEX NUMBERS

• **PROBLEM 3-1**

1. Describe the system of polar coordinates.
2. Let z be a complex number $z = x + iy$. Represent z in polar coordinates.
3. Express the following complex numbers in polar form

$$z_1 = 4 + 4\sqrt{3}\, i$$

$$z_2 = -2 + 2i$$

$$z_3 = -5i$$

Solution: 1. We shall introduce in the z plane the polar coordinates. The location of each point P of the plane is uniquely determined by two polar coordinates r and θ, as shown in Fig. 1.

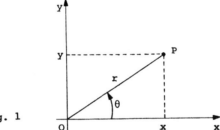

Fig. 1

Here, r is the length of the segment $\overline{OP}$, and θ is the angle between the x-axis and $\overline{OP}$. The relationship between the Cartesian coordinates (x,y) and the polar coordinates (r,θ) is

$$x = r\cos\theta, \quad y = r\sin\theta \tag{1}$$

2. A complex number $z = x + iy$ can be represented by a point in the z plane, as shown in Fig. 1. We have

$$r = |z| = \sqrt{x^2+y^2}, \quad \tan\theta = \frac{y}{x} \tag{2}$$

Here, θ is called the argument or amplitude of z. We write

$$\theta = \arg z \tag{3}$$

A complex number $z = x + iy$ can be written in the polar form

$$z = r(\cos\theta + i\sin\theta) \tag{4}$$

Note that any integral multiple of 2π can be added to θ, without changing the value of z. The value of arg z which satisfies

$$-\pi < \arg z \leq \pi \tag{5}$$

is called the principal value of arg z, and it is sometimes denoted by Arg z.

3. $z_1 = 4 + 4\sqrt{3}\, i$

$$r = \sqrt{4^2 + (4^2 \cdot 3)} = 8 \tag{6}$$

$x = r \cos\theta$ and

$$\cos\theta = \frac{x}{r} = \frac{4}{8} = \frac{1}{2} \tag{7}$$

Therefore $\theta = \frac{\pi}{3}$ and

$$4 + 4\sqrt{3}\, i = 8\left(\cos\frac{\pi}{3} + i \sin\frac{\pi}{3}\right) \tag{8}$$

$$\tag{9}$$

For $z_2 = -2 + 2i$

$$r = \sqrt{4+4} = 2\sqrt{2}$$

$$\cos\theta = \frac{x}{r} = \frac{-2}{2\sqrt{2}} = -\frac{\sqrt{2}}{2}$$

$$\theta = \frac{3\pi}{4}$$

thus $\quad -2 + 2i = 2\sqrt{2}\left(\cos\frac{3\pi}{4} + i \sin\frac{3\pi}{4}\right) \tag{10}$

For $z_3 = -5i$,

$$r = \sqrt{25} = 5$$

$$\cos\theta = \frac{x}{r} = 0$$

$$\theta = \frac{\pi}{2} \quad \text{or} \quad -\frac{\pi}{2}$$

Since $\sin\theta = \frac{y}{r} = -1$, we choose

$$\theta = -\frac{\pi}{2}$$

Thus
$$-5i = 5\left[\cos\left(-\frac{\pi}{2}\right) + i \sin\left(-\frac{\pi}{2}\right)\right]$$
$$= 5\left(\cos\frac{3\pi}{2} + i \sin\frac{3\pi}{2}\right) \tag{11}$$

● **PROBLEM 3-2**

Represent graphically the following complex numbers

1. $z_1 = 3(\cos 75° + i \sin 75°)$
2. $z_2 = 5(\cos 210° + i \sin 210°)$

3. $z_3 = 2(\cos \pi + i \sin \pi)$

4. the conjugate of $z_4 = 4(\cos 30° + i \sin 30°)$.

Solution: In general, a complex number can be written in the polar form

$$z = r(\cos \theta + i \sin \theta)$$

the graphical representation of which is shown in Fig. 1.

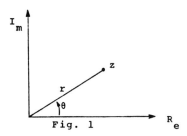

Fig. 1

1. In this case,

$$z_1 = 3(\cos 75° + i \sin 75°)$$

the distance r is 3, and the angle θ is $75°$, thus we can represent this number graphically as shown in Fig. 2.

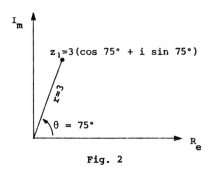

Fig. 2

2. In this case, $r = 5$ and $\theta = 210°$; the number z_2 is shown in Fig. 3.

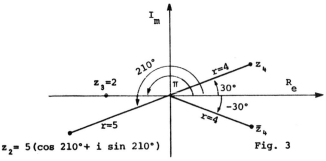

Fig. 3

3. Here, the angle is $\theta = \pi$, and $r = 2$. The number z_3 lies on the negative side of the x-axis, as shown in Fig. 3.

4. The conjugate of a number, or point

$$z = a + bi$$

is the reflection of the point about the real axis, and is given by

$$\bar{z} = a - bi$$

We have

$$|\bar{z}| = |z| = r$$

and

$$\arg z = -\arg \bar{z} = -\theta$$

The polar form of the conjugate is therefore

$$\bar{z} = r(\cos(-\theta) + i \sin(-\theta))$$

$$= r(\cos\theta - i \sin\theta)$$

The conjugate of z_4 is thus

$$\bar{z}_4 = 4(\cos 30° - i \sin 30°)$$

$$r = 4 \quad \theta = -30°$$

• **PROBLEM 3-3**

A car travels 15 miles 60° north of east, 10 miles northwest, and then 6 miles 30° south of west. Determine analytically and graphically the final location of the car.

Solution: First we shall use the graphical method. Let UN be a unit which represents one mile. Each part of the journey is represented by a vector.

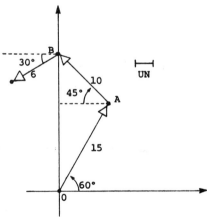

Fig. 1

From Fig 1, we can evaluate the distance of the car from 0.

$$OC \approx 17 \text{ miles}$$

Analytically we have

$$\overline{OC} = \overline{OA} + \overline{AB} + \overline{BC} \qquad (1)$$

$$\overline{OA} = 15(\cos 60° + i \sin 60°) = 15\left(\cos \frac{\pi}{3} + i \sin \frac{\pi}{3}\right) \tag{2}$$

$$= 15\left(\frac{1}{2} + i \frac{\sqrt{3}}{2}\right)$$

$$\overline{AB} = 10\left[\cos(90°+45°) + i \sin(90°+45°)\right] \tag{3}$$

$$= 10\left[\cos \frac{3\pi}{4} + i \sin \frac{3\pi}{4}\right] = 10\left[-\frac{\sqrt{2}}{2} + i \frac{\sqrt{2}}{2}\right]$$

$$\overline{BC} = 6\left[\cos(180°+30°) + i \sin(180°+30°)\right] \tag{4}$$

$$= 6\left[\cos\left(\pi + \frac{\pi}{6}\right) + i \sin\left(\pi + \frac{\pi}{6}\right)\right] = 6\left[-\frac{\sqrt{3}}{2} - i \frac{1}{2}\right]$$

Substituting into eq.(1) we find

$$\overline{OC} = 15\left(\frac{1}{2} + i \frac{\sqrt{3}}{2}\right) + 10\left(-\frac{\sqrt{2}}{2} + i \frac{\sqrt{2}}{2}\right) + 6\left(-\frac{\sqrt{3}}{2} - i \frac{1}{2}\right) \tag{5}$$

$$= (7.5 - 5\sqrt{2} - 3\sqrt{3}) + i(7.5\sqrt{3} + 5\sqrt{2} - 3)$$

$$= -4.78 + 17.07i$$

$$|\overline{OC}| = \sqrt{4.78^2 + 17.07^2} = 17.72 \tag{6}$$

• **PROBLEM 3-4**

1. Derive the formula for the product of two complex numbers given in polar form. Evaluate $z_1 \cdot z_2$, where

$$z_1 = 6\left(\cos \frac{3\pi}{2} + i \sin \frac{3\pi}{2}\right) \tag{1}$$

$$z_2 = 3\left(\cos \frac{\pi}{5} + i \sin \frac{\pi}{5}\right) \tag{2}$$

2. Derive the formula for the quotient of two complex numbers and evaluate $\frac{z_1}{z_2}$, where z_1 and z_2 are given above.

Solution: 1. Let
$$z_1 = r_1(\cos\theta_1 + i \sin\theta_1) \tag{3}$$
$$z_2 = r_2(\cos\theta_2 + i \sin\theta_2) \tag{4}$$

Then

$$z_1 \cdot z_2 = r_1(\cos\theta_1 + i \sin\theta_1) \cdot r_2(\cos\theta_2 + i \sin\theta_2)$$

$$= r_1 r_2 \left[(\cos\theta_1\cos\theta_2 - \sin\theta_1\sin\theta_2) + i(\sin\theta_1\cos\theta_2 + \cos\theta_1\sin\theta_2)\right]$$

$$= r_1 r_2 \left[\cos(\theta_1+\theta_2) + i \sin(\theta_1+\theta_2)\right] \tag{5}$$

From eq.(5) we conclude that
$$|z_1 \cdot z_2| = |z_1| \cdot |z_2| \qquad (6)$$
and
$$\arg(z_1 \cdot z_2) = \arg z_1 + \arg z_2 + 2n\pi \qquad (7)$$
$$n = 0, \pm 1, \pm 2, \ldots$$
Eqs.(6) and (7) can be graphically illustrated, as shown in Fig. 1.

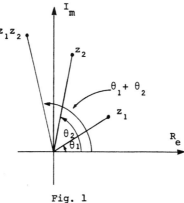

Fig. 1

For z_1 and z_2 given by eqs.(1) and (2), we have
$$|z_1 \cdot z_2| = |z_1| \cdot |z_2| = 6 \cdot 3 = 18 \qquad (8)$$
$$\arg(z_1 \cdot z_2) = \arg z_1 + \arg z_2 = \frac{3\pi}{2} + \frac{\pi}{5} = \frac{15\pi + 2\pi}{10} = \frac{17}{10}\pi \qquad (9)$$
Thus
$$z_1 \cdot z_2 = 18(\cos \frac{17}{10}\pi + i \sin \frac{17}{10}\pi) \qquad (10)$$

2. We have
$$\frac{z_1}{z_2} = z_1 \cdot z_2^{-1} \qquad (11)$$

By definition
$$z_2^{-1} \cdot z_2 = 1 \quad \text{and} \qquad (12)$$
$$1 = |z_2^{-1}| \cdot |z_2| = |z_2^{-1}| \cdot r_2$$
$$|z_2^{-1}| = \frac{1}{r_2} \qquad (13)$$

$$0 = \arg 1 = \arg(z_2^{-1} \cdot z_2) = \arg(z_2^{-1}) + \arg z_2$$
$$= \arg(z_2^{-1}) + \theta_2 \qquad (14)$$

Therefore
$$\arg z_2^{-1} = -\theta_2 \qquad (15)$$

From eqs.(11), (13) and (15), we obtain

$$\frac{z_1}{z_2} = \frac{r_1}{r_2}\left[\cos(\theta_1-\theta_2) + i\sin(\theta_1-\theta_2)\right] \tag{16}$$

Substituting eqs.(1) and (2) into eq.(16), we find

$$\frac{z_1}{z_2} = \frac{6}{3}\left[\cos\left(\frac{3\pi}{2}-\frac{\pi}{5}\right) + i\sin\left(\frac{3\pi}{2}-\frac{\pi}{5}\right)\right]$$
$$= 2\left[\cos\frac{13\pi}{10} + i\sin\frac{13\pi}{10}\right] \tag{17}$$

For the quotient of two complex numbers we obtain

$$\left|\frac{z_1}{z_2}\right| = \frac{|z_1|}{|z_2|} \tag{18}$$

$$\arg\left(\frac{z_1}{z_2}\right) = \arg z_1 - \arg z_2 \tag{19}$$

• **PROBLEM 3-5**

Evaluate the modulus of

$$z = \frac{1 + \cos\theta + i\sin\theta}{1 + \cos\phi + i\sin\phi} \tag{1}$$

Solution: Let us transform the numerator and denominator into polar form. Let

$$z_1 = 1 + \cos\theta + i\sin\theta = r_1(\cos\theta + i\sin\theta)$$

$$z_2 = 1 + \cos\phi + i\sin\phi = r_2(\cos\phi + i\sin\phi)$$

Since r_1 is the modulus of z_1, we obtain

$$r_1 = \sqrt{(1+\cos\theta)^2 + \sin^2\theta} = \sqrt{2(1+\cos\theta)}$$

Likewise

$$r_2 = \sqrt{(1+\cos\phi)^2 + \sin^2\phi} = \sqrt{2(1+\cos\phi)}$$

Now, since

$$z = \frac{z_1}{z_2}$$

then

$$|z| = \left|\frac{z_1}{z_2}\right| = \frac{|z_1|}{|z_2|} = \frac{r_1}{r_2}$$

Thus, the modulus of $z = \frac{z_1}{z_2}$ is

$$|z| = \frac{r_1}{r_2} = \sqrt{\frac{1+\cos\theta}{1+\cos\phi}}$$

EULER'S FORMULA

• **PROBLEM 3-6**

For the complex number $z = x + iy$, the function e^z is defined as follows.

$$e^z = e^{x+iy} = e^x(\cos y + i \sin y) \qquad (1)$$

When $x = 0$, $z = iy$ and we can write, by replacing y by θ,

$$e^{i\theta} = \cos\theta + i \sin\theta \qquad (2)$$

Verify the following properties of $e^{i\theta}$

1. $e^{i\theta_1} \cdot e^{i\theta_2} = e^{i(\theta_1+\theta_2)}$

2. $(e^{i\theta})^{-1} = e^{-i\theta}$

3. $\dfrac{e^{i\theta_1}}{e^{i\theta_2}} = e^{i(\theta_1-\theta_2)}$

4. $e^{i(\theta+2k\pi)} = e^{i\theta}$, $k = 0, \pm1, \pm2, \ldots$

Solution: 1. Using the representation of complex numbers given by eq.(2), we find

$$e^{i\theta_1} \cdot e^{i\theta_2} = (\cos\theta_1 + i \sin\theta_1)\cdot(\cos\theta_2 + i \sin\theta_2)$$

$$= (\cos\theta_1\cos\theta_2 - \sin\theta_1\sin\theta_2) + i(\sin\theta_1\cos\theta_2 + \sin\theta_2\cos\theta_1) \qquad (3)$$

$$= \cos(\theta_1+\theta_2) + i \sin(\theta_1+\theta_2) = e^{i(\theta_1+\theta_2)}$$

2. We have

$$(e^{i\theta})^{-1} = \frac{1}{e^{i\theta}} = \frac{1}{\cos\theta + i \sin\theta}$$

$$= \frac{\cos\theta - i \sin\theta}{\cos^2\theta + \sin^2\theta} = \cos(-\theta) + i \sin(-\theta) = e^{-i\theta} \qquad (4)$$

3. Utilizing eqs.(3) and (4), we obtain

$$\frac{e^{i\theta_1}}{e^{i\theta_2}} = e^{i\theta_1} \cdot e^{-i\theta_2} = e^{i[\theta_1+(-\theta_2)]} = e^{i(\theta_1-\theta_2)} \qquad (5)$$

4. Since

$$\cos\theta = \cos(\theta + 2k\pi)$$
$$\sin\theta = \sin(\theta + 2k\pi), \text{ where } k=0, \pm1, \pm2, \ldots \qquad (6)$$

the following is true

$$e^{i\theta} = \cos\theta + i \sin\theta$$

$$= \cos(\theta+2k\pi) + i\sin(\theta+2k\pi) \qquad (7)$$

$$= e^{i(\theta+2k\pi)}$$

The equation $e^{i\theta} = \cos\theta + i\sin\theta$ is known as Euler's formula.

Leonard Euler, born in Bâle (Basel) Switzerland in 1707, died in Petrograd (Leningrad) in 1783. Euler is widely considered as one of the greatest mathematicians of all time. John Bernoulli was the first professor of young Euler. At the recommendation of Bernoulli's sons, Daniel and Nicholas, Euler was offered the chair of mathematics in Petrograd. Euler completed the work of his predecessors on pure mathematical analysis in his "Introductro in Analysin Infinitorum". Among other things, he introduced the current abbreviations for the trigonometric functions and showed that the trigonometric and exponential functions are related by $e^{i\theta} = \cos\theta + i\sin\theta$. All in all, Euler's contributions to mathematics were enormous despite the fact that his life abounded in calamities. He became blind in 1768, and in 1770-71, he published three volumes of Dioptrica - a profound and detailed treatise on optics. In 1771, his house, together with many of his papers, burned. Euler died of apoplexy.

• **PROBLEM 3-7**

Using the representation of complex numbers

$$z = r e^{i\theta} \qquad (1)$$

prove that

1. the product of two complex numbers is a complex number whose modulus is the product of the two moduli and whose argument is the sum of the two arguments.

2. the quotient of two complex numbers $\frac{z_1}{z_2}$, $z_2 \neq 0$, is a complex number whose modulus is the quotient of the two moduli and whose argument is the difference of the two arguments of the two complex numbers.

<u>Solution</u>: 1. Let
$$z_1 = r_1 e^{i\theta_1} \qquad (2)$$
$$z_2 = r_2 e^{i\theta_2} \qquad (3)$$

The product of z_1 and z_2 is

$$z_1 z_2 = r_1 e^{i\theta_1} \cdot r_2 e^{i\theta_2} = r_1 r_2 e^{i(\theta_1+\theta_2)} \qquad (4)$$

Thus
$$|z_1 z_2| = r_1 r_2 \qquad (5)$$

and we can write
$$|z_1 z_2| = |z_1||z_2| \qquad (6)$$

From eq.(4), we conclude that

$$\arg(z_1 z_2) = \theta_1 + \theta_2 \tag{7}$$

Thus

$$\arg(z_1 z_2) = \arg z_1 + \arg z_2 + 2k\pi \tag{8}$$

$$k = 0, \pm 1, \pm 2, \ldots$$

2. $$\frac{z_1}{z_2} = \frac{r_1 e^{i\theta_1}}{r_2 e^{i\theta_2}} = \frac{r_1}{r_2} e^{i(\theta_1 - \theta_2)} \tag{9}$$

Thus

$$\left|\frac{z_1}{z_2}\right| = \frac{r_1}{r_2} = \frac{|z_1|}{|z_2|} \tag{10}$$

and

$$\arg\left(\frac{z_1}{z_2}\right) = \theta_1 - \theta_2 = \arg z_1 - \arg z_2 + 2k\pi \tag{11}$$

● **PROBLEM 3-8**

Compute

$$\frac{\left(\frac{3}{2}\sqrt{3} + \frac{3}{2} i\right)^6}{\left(\sqrt{\frac{5}{2}} + i\sqrt{\frac{5}{2}}\right)^3} \tag{1}$$

<u>Solution:</u>   Let us denote

$$z_1 = \frac{3}{2}\sqrt{3} + \frac{3}{2} i \tag{2}$$

$$z_2 = \sqrt{\frac{5}{2}} + \sqrt{\frac{5}{2}} i \tag{3}$$

Then

$$|z_1| = \sqrt{\frac{27}{4} + \frac{9}{4}} = 3 \quad \text{and}$$

$$\cos\theta_1 = \frac{x_1}{|z_1|} = \frac{3\sqrt{3}}{2 \cdot 3} = \frac{\sqrt{3}}{2} \tag{4}$$

Thus

$$\theta_1 = \frac{\pi}{6} \tag{5}$$

and

$$z_1 = 3\left(\cos\frac{\pi}{6} + i\sin\frac{\pi}{6}\right) = 3e^{i\frac{\pi}{6}} \tag{6}$$

For z_2, we have

$$|z_2| = \sqrt{\frac{5}{2} + \frac{5}{2}} = \sqrt{5} \quad \text{and}$$

$$\cos\theta_2 = \frac{\sqrt{5}}{\sqrt{2} \cdot \sqrt{5}} = \frac{1}{\sqrt{2}} \tag{7}$$

Hence

$$\theta_2 = \frac{\pi}{4} \quad \text{and}$$

$$z_2 = \sqrt{5}\left(\cos\frac{\pi}{4} + i\sin\frac{\pi}{4}\right) = \sqrt{5}\, e^{\pi i/4} \tag{8}$$

Substituting eqs.(6) and (8) into eq.(1), we find

$$\frac{z_1^6}{z_2^3} = \frac{\left(3e^{\pi i/6}\right)^6}{\left(\sqrt{5}\, e^{\pi i/4}\right)^3} = \frac{729}{5\sqrt{5}}\, e^{i\left(\pi - \frac{3\pi}{4}\right)}$$

$$= \frac{729}{5\sqrt{5}}\, e^{\pi i/4} = \frac{729}{5\sqrt{5}}\left(\cos\frac{\pi}{4} + i\sin\frac{\pi}{4}\right) \tag{9}$$

$$= \frac{729}{5\sqrt{5}}\left(\frac{1}{\sqrt{2}} + i\frac{1}{\sqrt{2}}\right) = \frac{729}{5\sqrt{5}}(1 + i)$$

• **PROBLEM 3-9**

Let $z = re^{i\theta}$ be a complex number and α a real parameter. What is the geometric interpretation of $z' = ze^{i\alpha}$? (1)

Solution: The number $z = re^{i\theta}$ is represented by the segment OA shown in Fig. 1.

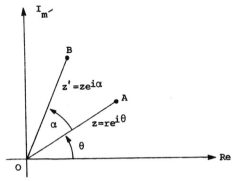

Fig. 1

Now, we have

$$z' = ze^{i\alpha} = re^{i\theta}e^{i\alpha} = re^{i(\theta+\alpha)} \tag{2}$$

From eq.(2) it is clear the numbers z and z' have the same modulus r (both points are at equal distance from the origin).

Notice however, that z' has an argument of $\theta + \alpha$. Thus, z' may be obtained by rotating z about the origin, through an angle α, in the counterclockwise direction. This indicates that $e^{i\alpha}$ can be considered as an operator that acts on a complex number z to produce a new number z' of the same absolute value as z but rotated through an angle α.

APPLICATIONS IN GEOMETRY

● **PROBLEM 3-10**

Prove that the triangles whose vertices are z_1, z_2, z_3 and z_1', z_2', z_3' are similar if, and only if,

$$\begin{vmatrix} 1 & z_1 & z_1' \\ 1 & z_2 & z_2' \\ 1 & z_3 & z_3' \end{vmatrix} = 0 \tag{1}$$

Solution: Since both triangles are similar, their corresponding angles are equal, as shown in Fig. 1.

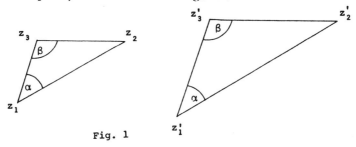

Fig. 1

We have (see Problem 3-9)

$$e^{i\alpha}(z_2-z_1) = z_3-z_1 \tag{2}$$

$$e^{i\alpha}(z_2'-z_1') = z_3'-z_1' \tag{3}$$

and

$$e^{i\beta}(z_1-z_3) = z_2-z_3 \tag{4}$$

$$e^{i\beta}(z_1'-z_3') = z_2'-z_3' \tag{5}$$

Dividing eq.(2) by eq.(3) results in

$$\frac{z_2 - z_1}{z_2' - z_1'} = \frac{z_3 - z_1}{z_3' - z_1'} \tag{6}$$

or

$$z_2 z_3' + z_1 z_2' + z_3 z_1' - z_2 z_1' - z_3 z_2' - z_3' z_1 = 0 \tag{7}$$

which is equivalent to

$$\begin{vmatrix} 1 & z_1 & z_1' \\ 1 & z_2 & z_2' \\ 1 & z_3 & z_3' \end{vmatrix} = 0 \qquad (8)$$

Note that the same equation can be obtained from eqs.(4) and (5).

● **PROBLEM 3-11**

Prove that two lines joining the points z_1, z_2 and z_3, z_4 are perpendicular if

$$\text{Re}\left(\frac{z_1 - z_2}{z_3 - z_4}\right) = 0 \qquad (1)$$

Here, Re denotes the real part of $\left(\frac{z_1 - z_2}{z_3 - z_4}\right)$.

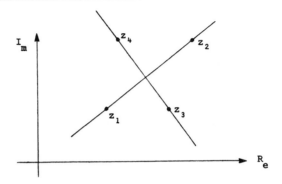

Fig. 1

Solution: The points and the lines are shown in Fig. 1.

If the vectors $z_2 - z_1$ and $z_4 - z_3$ are perpendicular, then

$$e^{\pm i \frac{\pi}{2}}(z_2 - z_1) = z_4 - z_3 \qquad (2)$$

The sign, + or −, in $e^{\pm i \frac{\pi}{2}}$ depends on how we denote the points. From eq.(2)

$$\frac{z_1 - z_2}{z_3 - z_4} = e^{\pm i \frac{\pi}{2}} \qquad (3)$$

Therefore

$$\arg \frac{z_1 - z_2}{z_3 - z_4} = \pm \frac{\pi}{2} \qquad (4)$$

The numbers whose argument is $\pm \frac{\pi}{2}$ are situated on the imaginary axis.

Hence

$$\text{Re}\left(\frac{z_1 - z_2}{z_3 - z_4}\right) = 0 \qquad (5)$$

• **PROBLEM 3-12**

Let $\triangle ABC$ be an isosceles triangle such that the angle at the vertex A is α.

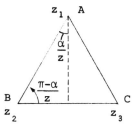

Prove that if the complex numbers z_1, z_2, z_3 represent the points A, B, C, respectively, then

$$(z_3-z_2)^2 = 4\sin^2 \frac{\alpha}{2}(z_1-z_2)(z_3-z_1) \tag{1}$$

<u>Solution</u>: Notice that from the figure, the vector z_3-z_1 may be obtained by rotating z_2-z_1, through an angle α, about the point z_1 (point A).

Thus

$$e^{i\alpha}(z_2-z_1) = z_3-z_1 \tag{2}$$

Let

$$z_3-z_2 = re^{i\theta} \tag{3}$$

and let

$$|z_1-z_2| = R \tag{4}$$

Then we have

$$z_1-z_2 = Re^{i\theta}e^{i\frac{\pi-\alpha}{2}} \tag{5}$$

Finally

$$(z_3-z_2)^2 = re^{i\theta} \cdot re^{i\theta} \tag{6}$$

and, since

$$\sin\frac{\alpha}{2} = \frac{r}{2R} \tag{7}$$

eq.(6) becomes

$$(z_3-z_2)^2 = 4R^2\sin^2\frac{\alpha}{2} e^{i\theta}e^{i\theta}$$

$$= 4\sin^2\frac{\alpha}{2} \cdot R^2 \left[\frac{z_1-z_2}{Re^{i\frac{\pi-\alpha}{2}}}\right]^2 \tag{8}$$

$$= 4 \frac{(z_1-z_2)(z_1-z_2)}{e^{i(\pi-\alpha)}} \sin^2\frac{\alpha}{2}$$

Substituting

$$z_1 - z_2 = \frac{-(z_3 - z_1)}{e^{i\alpha}} \qquad (9)$$

we obtain

$$(z_3 - z_2)^2 = -4 \sin^2 \frac{\alpha}{2} \frac{(z_1 - z_2)}{e^{i(\pi - \alpha)}} \cdot \frac{(z_3 - z_1)}{e^{i\alpha}}$$

$$= -4 \sin^2 \frac{\alpha}{2} (z_1 - z_2)(z_3 - z_1) e^{-i\pi} \qquad (10)$$

$$= 4 \sin^2 \frac{\alpha}{2} (z_1 - z_2)(z_3 - z_1)$$

● **PROBLEM 3-13**

Let z_1, z_2, z_3 be the vertices of an equilateral triangle. Prove that

$$z_1^2 + z_2^2 + z_3^2 = z_1 z_2 + z_3 z_1 \qquad (1)$$

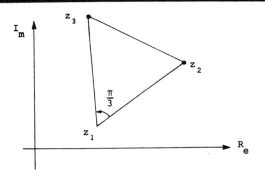

Fig. 1

Solution: The equilateral triangle with vertices z_1, z_2, z_3 is shown in Fig. 1.

Note that vectors $z_2 - z_1$ and $z_3 - z_1$ are of the same modulus, and $z_3 - z_1$ is obtained by rotating $z_2 - z_1$ through the angle $\frac{\pi}{3}$. Therefore

$$z_3 - z_1 = e^{i\frac{\pi}{3}} (z_2 - z_1) \qquad (2)$$

In the same manner, we find

$$z_1 - z_2 = e^{i\frac{\pi}{3}} (z_3 - z_2) \qquad (3)$$

Dividing eq.(2) by eq.(3) results in

$$\frac{z_3 - z_1}{z_1 - z_2} = \frac{z_2 - z_1}{z_3 - z_2} \qquad (4)$$

From eq.(4) we find

$$(z_3 - z_1)(z_3 - z_2) = (z_1 - z_2)(z_2 - z_1) \qquad (5)$$

or

$$z_1^2 + z_2^2 + z_3^2 = z_1 z_2 + z_1 z_3 + z_2 z_3 \qquad (6)$$

● **PROBLEM 3-14**

Prove that if four points z_1, z_2, z_3, z_4 lie on a circle, in that order, then

$$|(z_1-z_2)(z_3-z_4)| + |(z_2-z_3)(z_1-z_4)| = |(z_3-z_1)(z_4-z_2)| \qquad (1)$$

This theorem is known as Ptolemy's* theorem.

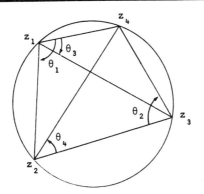

Fig. 1

Solution: Four arbitrary points located on a circle are shown in Fig. 1, with the angles θ_1, θ_2, θ_3, θ_4 as indicated.

The following relationships are true

$$|\theta_1 + \theta_2| = \pi \qquad (2)$$
$$|\theta_3| = |\theta_4| \qquad (3)$$

Note that since one is clockwise and the other counterclockwise, in eq.(3), we have

$$\theta_3 + \theta_4 = 0 \qquad (4)$$

To verify eqs.(2) and (3), we shall use Fig. 2, where z_0 is the center of the circle.

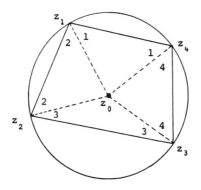

Fig. 2

Since the sum of all angles in the quadrilateral $z_1 z_2 z_3 z_4$ is 2π, we have

$$\hat{1} + \hat{1} + \hat{2} + \hat{2} + \hat{3} + \hat{3} + \hat{4} + \hat{4} = 2\pi$$

or

$$\hat{1} + \hat{2} + \hat{3} + \hat{4} = \pi \tag{5}$$

Therefore, since $1 + 2 = \theta_1$ and $3 + 4 = \theta_2$, we have

$$|\theta_1 + \theta_2| = \pi \tag{6}$$

In a similar way we can verify eq.(3). We will now write eq.(1) in the form

$$\left| -\frac{z_2 - z_1}{z_4 - z_1} \cdot \frac{z_4 - z_3}{z_2 - z_3} \right| + 1 = \left| \frac{z_3 - z_1}{z_4 - z_1} \cdot \frac{z_4 - z_2}{z_3 - z_2} \right| \tag{7}$$

Observe that eq.(7) without the absolute value sign, i.e,

$$\left(-\frac{z_2 - z_1}{z_4 - z_1} \cdot \frac{z_4 - z_3}{z_2 - z_3} \right) + 1 = \frac{z_3 - z_1}{z_4 - z_1} \cdot \frac{z_4 - z_2}{z_3 - z_2} \tag{8}$$

becomes an identity.

θ_1 is the smallest angle through which $z_4 - z_1$ must be rotated to coincide with the vector $z_2 - z_1$. Hence

$$\text{Arg } \frac{z_2 - z_1}{z_4 - z_1} = \theta_1 \tag{9}$$

Similarly

$$\text{Arg } \frac{z_4 - z_3}{z_2 - z_3} = \theta_2 \tag{10}$$

$$\text{Arg } \frac{z_3 - z_1}{z_4 - z_1} = \theta_3 \tag{11}$$

$$\text{Arg } \frac{z_4 - z_2}{z_3 - z_2} = \theta_4 \tag{12}$$

Then

$$\text{Arg } \left[\frac{z_2 - z_1}{z_4 - z_1} \cdot \frac{z_4 - z_3}{z_2 - z_3} \right] = \theta_1 + \theta_2 \tag{13}$$

and

$$|\theta_1 + \theta_2| = \pi$$

$$\text{Arg } \left[\frac{z_3 - z_1}{z_4 - z_1} \cdot \frac{z_4 - z_2}{z_3 - z_2} \right] = \theta_3 + \theta_4 = 0 \tag{14}$$

Thus both numbers $\frac{z_3 - z_1}{z_4 - z_1} \cdot \frac{z_4 - z_2}{z_3 - z_2}$ and $\frac{z_2 - z_1}{z_4 - z_1} \cdot \frac{z_4 - z_3}{z_2 - z_3}$ are located on the real axis. Therefore, in eq.(8), which is an identity, we can replace each of these numbers by it's modulus.

That completes the proof of eq.(7) and, hence, eq.(1). One can show that if four points satisfy eq.(1), then they lie on a circle.

*Claudius Ptolemaeus, an astronomer, geographer, and mathematician, lived in the second century A.D. Ptolemaeus is mostly known for his detailed description of the geocentric universe, a revolutionary but erroneous idea that dominated astronomy for over 1,300 years.

• PROBLEM 3-15

Let
$$z_1 = r_1 e^{i\theta_1} \quad \text{and} \quad z_2 = r_2 e^{i\theta_2} \tag{1}$$

and
$$z_3 = z_1 + z_2 = r_3 e^{i\theta_3} \tag{2}$$

Evaluate r_3 and θ_3.

Solution: We have

$$z_1 = r_1 e^{i\theta_1} = x_1 + iy_1 = r_1(\cos\theta_1 + i\sin\theta_1) \tag{3}$$

$$z_2 = r_2 e^{i\theta_2} = x_2 + iy_2 = r_2(\cos\theta_2 + i\sin\theta_2) \tag{4}$$

Therefore

$$\begin{aligned} z_3 = x_3 + iy_3 = r_3 e^{i\theta_3} &= (x_1 + x_2) + i(y_1 + y_2) \\ &= (r_1\cos\theta_1 + r_2\cos\theta_2) + i(r_1\sin\theta_1 + r_2\sin\theta_2) \end{aligned} \tag{5}$$

Thus

$$\begin{aligned} r_3 &= \sqrt{x_3^2 + y_3^2} = \sqrt{(r_1\cos\theta_1 + r_2\cos\theta_2)^2 + (r_1\sin\theta_1 + r_2\sin\theta_2)^2} \\ &= \left[r_1^2(\cos^2\theta_1 + \sin^2\theta_1) + r_2^2(\cos^2\theta_2 + \sin^2\theta_2) \right. \\ &\quad \left. + 2r_1 r_2(\cos\theta_1\cos\theta_2 + \sin\theta_1\sin\theta_2) \right]^{\frac{1}{2}} \\ &= \sqrt{r_1^2 + r_2^2 + 2r_1 r_2 \cos(\theta_1 - \theta_2)} \end{aligned} \tag{6}$$

Since
$$z_3 = x_3 + iy_3 = r_3(\cos\theta_3 + i\sin\theta_3) \tag{7}$$

we have
$$\tan\theta_3 = \frac{y_3}{x_3} \tag{8}$$

Substituting x_3 and y_3 into eq.(8) results in

$$\theta_3 = \tan^{-1}\left(\frac{r_1\sin\theta_1 + r_2\sin\theta_2}{r_1\cos\theta_1 + r_2\cos\theta_2}\right) \tag{9}$$

• PROBLEM 3-16

Show that on the circle
$$z = re^{i\theta}, \quad 0 \le \theta < 2\pi \tag{1}$$

the following holds

$$|e^{iz}| = e^{-r\sin\theta} \tag{2}$$

Solution: From Euler's formula, we have

$$z = re^{i\theta} = r(\cos\theta + i\sin\theta) \tag{3}$$

Hence

$$\begin{aligned}
|e^{iz}| &= |e^{ir(\cos\theta + i\sin\theta)}| \\
&= |e^{ir\cos\theta - r\sin\theta}| \\
&= |e^{-r\sin\theta} e^{ir\cos\theta}| \\
&= e^{-r\sin\theta} |e^{ir\cos\theta}| \\
&= e^{-r\sin\theta}
\end{aligned} \tag{4}$$

Notice that the modulus of $e^{ir\cos\theta}$ is 1.

● **PROBLEM 3-17**

Let z_1, z_2 be two complex numbers such that $|z_1| < 1$ and $|z_2| < 1$. Let T be the triangle having vertices $z_1, z_2, 1$. Show that there exists a real number α depending only on z_1 and z_2, $\alpha = \alpha(z_1, z_2)$, such that for every z, $z \neq 1$, belonging to T,

$$\frac{|1-z|}{1-|z|} \leq \alpha \tag{1}$$

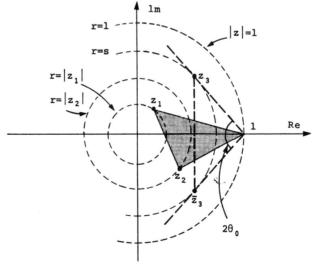

Fig. 1

Solution: The triangle T is shown in Fig. 1.

Let us choose a number $s < 1$ such that $|z_1| \leq s$ and $|z_2| \leq s$. (2)
The circle of radius s divides triangle T into two parts.

Let $|z| \leq s$, then

$$\frac{|1-z|}{1-|z|} \leq \frac{1+s}{1-s} \tag{3}$$

We shall denote

$$\frac{1+s}{1-s} = \alpha_1 = \alpha_1(s) = \alpha_1(z_1, z_2) \tag{4}$$

Hence, for any $z \in T$ such that $|z| \leq s$, we have

$$\frac{|1-z|}{1-|z|} \leq \alpha_1 \tag{5}$$

Now we have to establish a similar inequality for the remaining part of T. Let z_3 be the point on the circle $r = s$, such that the segment $[1, z_3]$ is tangent to this circle.

We shall denote the triangle $1, z_3, \bar{z}_3$ by T'.

Note that points $z \in T$, such that $|z| \geq s$, lie within triangle T'. Let

$$s_1 = \sqrt{1-s^2} \tag{6}$$

then

$$s_1 = |1-z_3| \tag{7}$$

Let the angle between $[1, z_3]$ and $[1, \bar{z}_3]$ be $2\theta_0$. Thus

$$0 < \theta_0 < \frac{\pi}{2}$$
$$s_1 = \cos\theta_0 \tag{8}$$

The points of T' can be represented in the form

$$z = 1 - r(\cos\theta + i\sin\theta) \tag{9}$$

$$0 < r < \cos\theta_0 \quad |\theta| \leq \theta_0 \tag{10}$$

We have

$$-2r\cos\theta + r^2 \leq -r\cos\theta_0 + \frac{1}{4}r^2\cos^2\theta_0 \tag{11}$$

Hence

$$\frac{r}{1-\sqrt{1-2r\cos\theta+r^2}} \leq \frac{2}{\cos\theta_0} \tag{12}$$

or

$$\frac{|1-z|}{1-|z|} \leq \frac{2}{\cos\theta_0} = \alpha_2 \tag{13}$$

Let
$$\alpha = \max(\alpha_1, \alpha_2) \qquad (14)$$

Then
$$\frac{|1-z|}{1-|z|} \leq \alpha \qquad (15)$$

for every $z \in T$, $z \neq 1$.

● **PROBLEM 3-18**

In this problem, we shall offer some kind of justification of Euler's formula

$$e^{i\theta} = \cos\theta + i\sin\theta \qquad (1)$$

The function e^x can be represented by the Taylor series expansion

$$e^x = 1 + \frac{x}{1!} + \frac{x^2}{2!} + \ldots + \frac{x^n}{n!} + \ldots = \sum_{n=0}^{\infty} \frac{x^n}{n!} \qquad (2)$$

Similarly, the $\sin\theta$ and $\cos\theta$ functions can be expanded in a Taylor series as

$$\sin\theta = \frac{\theta}{1!} - \frac{\theta^3}{3!} + \frac{\theta^5}{5!} + \ldots + \frac{(-1)^n \theta^{2n+1}}{(2n+1)!} + \ldots$$
$$= \sum_{n=0}^{\infty} \frac{(-1)^n \theta^{2n+1}}{(2n+1)!} \qquad (3)$$

$$\cos\theta = 1 - \frac{\theta^2}{2!} + \frac{\theta^4}{4!} + \ldots + \frac{(-1)^n \theta^{2n}}{(2n)!} + \ldots$$
$$= \sum_{n=0}^{\infty} \frac{(-1)^n \theta^{2n}}{(2n)!} \qquad (4)$$

Verify Euler's formula using Taylor series.

Solution: In eq.(2), let us substitute $x = i\theta$.

$$e^{i\theta} = \sum_{n=0}^{\infty} \frac{(i\theta)^n}{n!} = \sum_{n=0}^{\infty} \frac{(i\theta)^{2n}}{(2n)!} + \sum_{n=0}^{\infty} \frac{(i\theta)^{2n+1}}{(2n+1)!}$$

$$= \sum_{n=0}^{\infty} \frac{(-1)^n \theta^{2n}}{(2n)!} + i \sum_{n=0}^{\infty} \frac{(-1)^n \theta^{2n+1}}{(2n+1)!} \qquad (5)$$

$$= \cos\theta + i\sin\theta$$

This problem does not introduce the concept of a complex infinite series. It merely offers some justification of Euler's formula.

DOT AND CROSS PRODUCTS

● **PROBLEM 3-19**

1. Compute the scalar product (also called the dot product) of the two complex vectors

$$z_1 = 3 - 2i$$
$$z_2 = -4 - 6i \qquad (1)$$

2. Compute the cross product of z_1 and z_2 as given above.

Solution: 1. Let $z_1 = x_1 + iy_1$ and $z_2 = x_2 + iy_2$ be two complex numbers. The scalar product of z_1 and z_2 is defined by

$$z_1 \cdot z_2 \equiv x_1 x_2 + y_1 y_2 = |z_1||z_2|\cos\theta = \text{Re}(\bar{z}_1 z_2)$$
$$= \frac{1}{2}(\bar{z}_1 z_2 + z_1 \bar{z}_2) \qquad (2)$$

where θ is the angle between z_1 and z_2, $0 \leq \theta \leq \pi$.

It is easy to show that all parts of eq.(2) are indeed equal. For example

$$\text{Re}(\bar{z}_1 z_2) = \text{Re}\left[(x_1 - iy_1)(x_2 + iy_2)\right]$$
$$= \text{Re}\left[x_1 x_2 + y_1 y_2 + i x_1 y_2 - i x_2 y_1\right] = x_1 x_2 + y_1 y_2 \qquad (3)$$

For z_1 and z_2 given by eq.(1), we obtain

$$z_1 \cdot z_2 = x_1 x_2 + y_1 y_2 = (3-2i) \cdot (-4-6i)$$
$$= (3)(-4) + (-2)(-6) = 0 \qquad (4)$$

2. The cross product of z_1 and z_2 is defined as

$$z_1 \times z_2 \equiv |z_1||z_2|\sin\theta = x_1 y_2 - y_1 x_2$$
$$= \text{Im}(\bar{z}_1 z_2) = \frac{1}{2i}(\bar{z}_1 z_2 - z_1 \bar{z}_2) \qquad (5)$$

We have

$$z_1 \times z_2 = (3-2i) \times (-4-6i) = x_1 y_2 - y_1 x_2$$
$$= (3)(-6) - (-2)(-4) = -18 - 8 = -26 \qquad (6)$$

● **PROBLEM 3-20**

Let z_1 and z_2 be two non-zero complex numbers.

Prove the following properties of the dot and cross product.

1. z_1 and z_2 are perpendicular, if and only if $z_1 \cdot z_2 = 0$.

2. z_1 and z_2 are parallel if and only if $z_1 \times z_2 = 0$.

3. If z is the projection of z_1 on z_2, then
$$|z| = \frac{|z_1 \cdot z_2|}{|z_2|}$$
4. $z_1 \cdot z_2 = z_2 \cdot z_1$

$z_1 \times z_2 = (-z_2 \times z_1)$

Solution: 1. If z_1 and z_2 are perpendicular, then $\theta = \frac{\pi}{2}$ and
$$z_1 \cdot z_2 = |z_1||z_2|\cos\theta = |z_1||z_2|\cos\frac{\pi}{2} = 0 \qquad (1)$$
On the other hand, if $z_1 \cdot z_2 = 0$ and $z_1 \neq 0$, $z_2 \neq 0$, then
$$0 = z_1 \cdot z_2 = |z_1||z_2|\cos\theta \qquad (2)$$
Hence
$$\cos\theta = 0 \qquad (3)$$
and
$$\theta = \frac{\pi}{2}, \text{ and the vectors } z_1, z_2 \text{ are perpendicular.}$$

2. $\begin{pmatrix} z_1, z_2 \text{ are} \\ \text{parallel} \end{pmatrix} \iff (z_1 \times z_2 = 0)$

$\Rightarrow$

z_1 and z_2 are parallel, thus $\theta = 0$ and
$$z_1 \times z_2 = |z_1||z_2|\sin\theta = |z_1||z_2| \cdot 0 = 0 \qquad (4)$$

$\Leftarrow$

If $z_1 \times z_2 = 0$ then
$$0 = z_1 \times z_2 = |z_1||z_2|\sin\theta \qquad (5)$$
Thus
$$\sin\theta = 0 \quad \text{and} \quad \theta = 0.$$
Therefore, z_1 and z_2 are parallel.

3.

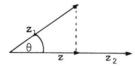

Fig. 1

From Fig. 1,
$$\cos\theta = \frac{|z|}{|z_1|} \qquad (6)$$
On the other hand
$$z_1 \cdot z_2 = |z_1||z_2|\cos\theta \qquad (7)$$
and
$$\cos\theta = \frac{|z_1 \cdot z_2|}{|z_1||z_2|} \qquad (8)$$

Substituting eq.(8) into eq.(6) we find

$$|z| = |z_1|\cos\theta = |z_1|\frac{|z_1 \cdot z_2|}{|z_1||z_2|} \qquad (9)$$

$$= \frac{|z_1 \cdot z_2|}{|z_2|}$$

4. $z_1 \cdot z_2 = |z_1||z_2|\cos\theta = |z_2||z_1|\cos\theta = z_2 \cdot z_1 \qquad (10)$

$z_1 \times z_2 = |z_1||z_2|\sin\theta = |z_2||z_1|\sin(-\theta)$
$= -|z_2||z_1|\sin\theta = -(z_2 \times z_1) \qquad (11)$

● **PROBLEM 3-21**

Prove that

1. $\bar{z}_1 z_2 = (z_1 \cdot z_2) + i(z_1 \times z_2) = |z_1||z_2|e^{i\theta} \qquad (1)$

2. The area of a parallelogram whose sides are z_1 and z_2 is

$$S = |z_1 \times z_2| \qquad (2)$$

<u>Solution</u>: 1. From the definition of the dot and cross product, we obtain, for $z_1 = x_1 + iy_1$ and $z_2 = x_2 + iy_2$,

$$\bar{z}_1 z_2 = (x_1 - iy_1) \cdot (x_2 + iy_2)$$
$$= (x_1 x_2 + y_1 y_2) + i(x_1 y_2 - x_2 y_1) \qquad (3)$$
$$= (z_1 \cdot z_2) + i(z_1 \times z_2) = |z_1||z_2|\cos\theta$$
$$+ i|z_1||z_2|\sin\theta$$
$$= |z_1||z_2|(\cos\theta + i\sin\theta) = |z_1||z_2|e^{i\theta}$$

2. The parallelogram is shown in Fig. 1.

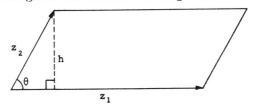

Fig. 1

The area of the parallelogram is

$$S = |z_1| \cdot h \qquad (4)$$

From the figure

$$\sin\theta = \frac{h}{|z_2|} \qquad (5)$$

Hence

$$S = |z_1| \cdot h = |z_1||z_2|\sin\theta = |z_1 \times z_2| \qquad (6)$$

The absolute value of $z_1 \times z_2$ was used in order to assure that S is positive.

• **PROBLEM 3-22**

Prove that if

$$z_1 = r_1 e^{i\theta_1} \quad \text{and} \quad z_2 = r_2 e^{i\theta_2} \qquad (1)$$

then

1. $\quad z_1 \cdot z_2 = r_1 r_2 \cos(\theta_2 - \theta_1)$ \hfill (2)

   $\quad z_1 \times z_2 = r_1 r_2 \sin(\theta_2 - \theta_1)$ \hfill (3)

2. $\quad z_1 \cdot (z_2 + z_3) = \left(z_1 \cdot z_2\right) + \left(z_1 \cdot z_3\right)$ \hfill (4)

   $\quad z_1 \times (z_2 + z_3) = \left(z_1 \times z_2\right) + \left(z_1 \times z_3\right)$ \hfill (5)

<u>Solution</u>: 1. From the definition of the dot and cross product, we obtain

$$z_1 \cdot z_2 = |z_1||z_2|\cos\theta = r_1 r_2 \cos(\theta_2 - \theta_1) \qquad (6)$$

$$z_1 \times z_2 = |z_1||z_2|\sin\theta = r_1 r_2 \sin(\theta_2 - \theta_1) \qquad (7)$$

2. Let $z_1 = x_1 + iy_1$, $z_2 = x_2 + iy_2$ and $z_3 = x_3 + iy_3$. We have

$$\begin{aligned}
z_1 \cdot (z_2 + z_3) &= (x_1 + iy_1) \cdot \left[(x_2 + iy_2) + (x_3 + iy_3)\right] \\
&= (x_1 + iy_1) \cdot \left[(x_2 + x_3) + i(y_2 + y_3)\right] \\
&= x_1(x_2 + x_3) + y_1(y_2 + y_3) = x_1 x_2 + x_1 x_3 + y_1 y_2 + y_1 y_3 \\
&= (x_1 x_2 + y_1 y_2) + (x_1 x_3 + y_1 y_3) \\
&= \left(z_1 \cdot z_2\right) + \left(z_1 \cdot z_3\right)
\end{aligned} \qquad (8)$$

$$\begin{aligned}
z_1 \times (z_2 + z_3) &= (x_1 + iy_1) \times \left[(x_2 + x_3) + i(y_2 + y_3)\right] \\
&= x_1(y_2 + y_3) - y_1(x_2 + x_3) \\
&= (x_1 y_2 - y_1 x_2) + (x_1 y_3 - y_1 x_3) \\
&= \left(z_1 \times z_2\right) + \left(z_1 \times z_3\right)
\end{aligned} \qquad (9)$$

• **PROBLEM 3-23**

Find the area of the triangle with vertices $A(z_1 = 3 + 2i)$, $B(z_2 = 4 + i)$ and $C(z_3 = 6 + 3i)$.

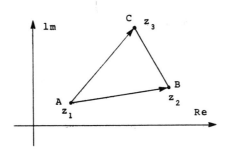

Fig. 1

Solution: The triangle ABC is shown in Fig. 1.
Notice that z_3-z_1 and z_2-z_1 are two complex vectors. The area of the parallelogram whose sides are (z_3-z_1) and (z_2-z_1) is

$$|(z_3-z_1) \times (z_2-z_1)| \tag{1}$$

Since the area of the triangle is half the area of the corresponding parallelogram, we have

$$\begin{aligned} S &= \tfrac{1}{2}\left|(z_3-z_1) \times (z_2-z_1)\right| \\ &= \tfrac{1}{2}\left|[(x_3+iy_3)-(x_1+iy_1)] \times [(x_2+iy_2)-(x_1+iy_1)]\right| \\ &= \tfrac{1}{2}\left|[(x_3-x_1)+i(y_3-y_1)] \times [(x_2-x_1)+i(y_2-y_1)]\right| \\ &= \tfrac{1}{2}\left|(x_3-x_1)(y_2-y_1)-(y_3-y_1)(x_2-x_1)\right| \\ &= \tfrac{1}{2}\left|x_3y_2-x_3y_1-x_1y_2+x_1y_1-x_2y_3+x_1y_3+x_2y_1-x_1y_1\right| \tag{2} \\ &= \tfrac{1}{2}\left|x_1y_2+x_2y_3+x_3y_1-x_1y_3-x_2y_1-x_3y_2\right| \\ &= \tfrac{1}{2}\left|\begin{matrix} 1 & 1 & 1 \\ x_1 & x_2 & x_3 \\ y_1 & y_2 & y_3 \end{matrix}\right| \end{aligned}$$

Substituting $A(3,2)$, $B(4,1)$ and $C(6,3)$ into eq.(2), we obtain

$$S = \tfrac{1}{2}\left|\begin{matrix} 1 & 1 & 1 \\ 3 & 4 & 6 \\ 2 & 1 & 3 \end{matrix}\right| = 7 \tag{3}$$

● **PROBLEM 3-24**

Find the area of the quadrilateral having vertices $z_1 = -4-2i$, $z_2 = 3-i$, $z_3 = -1+i$, $z_4 = 4+4i$.

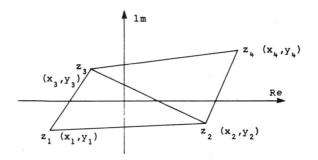

Fig. 1

<u>Solution</u>: The quadrilateral shown in Fig. 1 has vertices z_1, z_2, z_3, z_4.

The area of the quadrilateral is equal to the sum of the areas of the two triangles $z_1 z_2 z_3$ and $z_2 z_3 z_4$. Therefore

$$S = \frac{1}{2} \left| \begin{pmatrix} 1 & 1 & 1 \\ x_1 & x_2 & x_3 \\ y_1 & y_2 & y_3 \end{pmatrix} \right| + \frac{1}{2} \left| \begin{pmatrix} 1 & 1 & 1 \\ x_2 & x_4 & x_3 \\ y_2 & y_4 & y_3 \end{pmatrix} \right|$$

$$= \frac{1}{2} \left| \begin{pmatrix} 1 & 1 & 1 \\ -4 & 3 & -1 \\ -2 & -1 & 1 \end{pmatrix} \right| + \frac{1}{2} \left| \begin{pmatrix} 1 & 1 & 1 \\ 3 & 4 & -1 \\ -1 & 4 & 1 \end{pmatrix} \right|$$

$$= \frac{1}{2} |3 + 2 + 4 + 6 - 1 + 4| + \frac{1}{2} |4 + 1 + 12 + 4 + 4 - 3|$$

$$= \frac{1}{2}(18) + \frac{1}{2}(22) = 20$$

It should be noted that the above method can be applied when computing the area of any general polygon having n vertices. We simply divide the polygon into n-2 triangles and compute the area of each triangle. The area of the polygon is then the sum of the areas of the individual triangles.

DE MOIVRE'S THEOREM

● **PROBLEM 3-25**

Prove DeMoivre's* theorem

$$(\cos \theta + i \sin \theta)^n = \cos n\theta + i \sin n\theta \tag{1}$$

Find the formula to compute the nth roots of a complex number z.

<u>Solution</u>: We have shown that for

$$z_1 = r_1 e^{i \theta_1} \quad \text{and} \quad z_2 = r_2 e^{i \theta_2} \tag{2}$$

their product is

$$z_1 z_2 = r_1 r_2 \, e^{i(\theta_1 + \theta_2)} \tag{3}$$

Repeated application of eq.(3) for $z_1 = z_2 = z$ leads to the formula for the nth power of z

$$z^n = (re^{i\theta})^n = r^n e^{in\theta} \tag{4}$$

$$\text{for } n = 1, 2, 3, \ldots$$

In particular, if $r = 1$, then

$$z = e^{i\theta} = \cos\theta + i\sin\theta \tag{5}$$

and eq.(4) yields

$$(e^{i\theta})^n = e^{in\theta} \tag{6}$$

or

$$(\cos\theta + i\sin\theta)^n = \cos n\theta + i\sin n\theta \tag{7}$$

To find the nth roots of z, let us denote

$$z_0^n = z, \quad \text{where } z_0 = r_0 e^{i\theta_0} \tag{8}$$

$$\text{and } z = re^{i\theta}$$

Then

$$r_0^n \, e^{in\theta_0} = re^{i\theta} \tag{9}$$

Thus

$$r_0 = \sqrt[n]{r} \tag{10}$$

is the real positive nth root of r, and

$$n\theta_0 = \theta + 2k\pi, \quad k = 0, \pm 1, \pm 2, \ldots \tag{11}$$

Combining all the equations, we can write

$$\sqrt[n]{z} = z^{\frac{1}{n}} = \sqrt[n]{r} \, e^{i\left[\frac{\theta + 2k\pi}{n}\right]} \tag{12}$$

where $k = 0, 1, \ldots n-1$ \hfill (13)

From eq.(11), since $k = 0, \pm 1, \pm 2, \ldots$ one might guess that there are infinitely many complex numbers z_0 corresponding to the infinitely many values of k. It is not so. We will show later that when $k = n+p$, we obtain the same complex number as when $k = p$. Thus, we have to take n consecutive values of k to obtain all the nth roots of z, as indicated in eq.(13).

*Abraham DeMoivre (1667-1754) was one of the first mathematicians that used complex numbers in trigonometry. A French Huguenot, DeMoivre was jailed as a protestant as a result of the revocation of the Edict of Nantes in 1685. Upon release, he fled to England. DeMoivre's greatest contributions were in the fields of analytic trigonometry and the theory of probability. He discovered Stirling's formula, incorrectly attributed to James Stirling. The formula states that for a large number n, n! is approximately given by

$$n! \approx \frac{\sqrt{2\pi n}\; n^n}{e^n}$$

where $\quad n! = n(n-1)(n-2) \cdot \ldots \cdot 3 \cdot 2 \cdot 1$

DeMoivre, who never held a permanent position, supported himself by working as a tutor and a consultant on insurance and gambling. He was a close friend of Sir Isaac Newton.

● **PROBLEM 3-26**

Find the location of the n nth roots of a complex number z in the z plane.

Solution: If $z = re^{i\theta}$ is a complex number, then it has n nth roots given by

$$\sqrt[n]{z} = \sqrt[n]{r}\; \cos^{i\left[\frac{\theta}{n} + \frac{2\pi k}{n}\right]} \tag{1}$$

where $\quad k = 0, 1, 2, \ldots, n-1$

All the nth roots of $z = re^{i\theta}$ lie on a circle centered at the origin 0 and having radius equal to

$$z_0 = \sqrt[n]{r}$$

One of the roots (for k=0) has an argument $\frac{\theta}{n}$, and the others are uniformly spaced around the circumference of the circle. The roots are separated by an angle $\frac{2\pi}{n}$ as shown in Fig. 1.

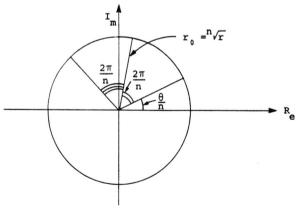

Fig. 1

There are n nth roots of any complex number $z \neq 0$. They have the same modulus and their arguments are separated by the same number $\frac{2\pi}{n}$. It is easy to verify that eq.(1) also holds for negative integers. In general, if m and n have no common factors, that is, if m and n are relatively prime integers, then

$$z^{\frac{m}{n}} = \sqrt[n]{r^m}\, e^{i\left[\frac{m\theta}{n} + \frac{2\pi m}{n}k\right]} \qquad (2)$$

$$k = 0, 1, 2, \ldots, n-1$$

• **PROBLEM 3-27**

Compute the fifth roots of -23, and then show them graphically in the z plane.

Solution: We have to find all values of z such that

$$z^5 = -23 \qquad (1)$$

The number -23 can be represented in polar form as

$$-23 = 23\left[\cos(\pi + 2k\pi) + i\,\sin(\pi + 2k\pi)\right] \qquad (2)$$

where $k = 0, \pm 1, \pm 2, \ldots$.

By DeMoivre's theorem

$$z = r(\cos\theta + i\,\sin\theta) \qquad (3)$$

$$z^5 = r^5(\cos 5\theta + i\,\sin 5\theta) = 23\left[\cos(\pi + 2k\pi) + i\,\sin(\pi + 2k\pi)\right] \qquad (4)$$

From eq.(4), we obtain

$$r^5 = 23 \quad \text{and} \quad 5\theta = \pi + 2k\pi \qquad (5)$$

Hence

$$z = \sqrt[5]{23}\left[\cos\left(\frac{\pi + 2k\pi}{5}\right) + i\,\sin\left(\frac{\pi + 2k\pi}{5}\right)\right] \qquad (6)$$

There are five fifth roots for k=0,1,2,3,4. For any other k, for example k=5,6,7.... or negative k, we obtain repetitions of the above five values.

We have

$$k = 0 \qquad z_1 = \sqrt[5]{23}\left(\cos\frac{\pi}{5} + i\,\sin\frac{\pi}{5}\right)$$

$$k = 1 \qquad z_2 = \sqrt[5]{23}\left(\cos\frac{3\pi}{5} + i\,\sin\frac{3\pi}{5}\right)$$

$$k = 2 \qquad z_3 = \sqrt[5]{23}\left(\cos\frac{5\pi}{5} + i\,\sin\frac{5\pi}{5}\right) = -\sqrt[5]{23} \qquad (7)$$

$$k = 3 \qquad z_4 = \sqrt[5]{23}\left(\cos\frac{7\pi}{5} + i\,\sin\frac{7\pi}{5}\right)$$

$$k = 4 \qquad z_5 = \sqrt[5]{23}\left(\cos\frac{9\pi}{5} + i\,\sin\frac{9\pi}{5}\right)$$

The roots are called the fifth roots of -23 and are collectively denoted $\sqrt[5]{-23}$.

The values of $\sqrt[5]{-23}$ are indicated in Fig. 1. They are equally spaced on the circumference of a circle with center at the origin and radius $\sqrt[5]{23}$.

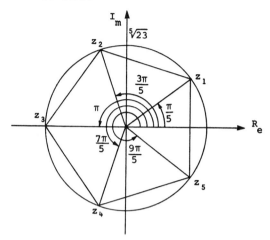

Fig. 1

Note that the roots represent the vertices of a regular polygon. In this case, a regular pentagon.

● **PROBLEM 3-28**

Find the fourth roots of $-\frac{3}{2} - i\frac{3\sqrt{3}}{2}$. Check the results graphically.

<u>Solution</u>: In the polar form

$$-\frac{3}{2} - i\frac{3\sqrt{3}}{2} = r(\cos\theta + i\sin\theta) \tag{1}$$

$$r = \sqrt{\left(-\frac{3}{2}\right)^2 + \left(\frac{3\sqrt{3}}{2}\right)^2} = 3$$

$$\cos\theta = -\frac{1}{2}, \quad \sin\theta = -\frac{\sqrt{3}}{2} \tag{2}$$

Hence

$$\theta = -\frac{2\pi}{3} \tag{3}$$

We have to find the fourth roots of

$$z = re^{i\theta} = 3e^{-2\pi i/3} \tag{4}$$

Thus

$$\sqrt[4]{z} = \sqrt[4]{3}\, e^{i\left[\frac{-\frac{2\pi}{3} + 2k\pi}{4}\right]} = \sqrt[4]{3}\, e^{i\frac{\pi(3k-1)}{6}} \tag{5}$$

For k = 0,1,2,3 we obtain

$$k = 0 \qquad \sqrt[4]{3}\, e^{-i\frac{\pi}{6}} = \sqrt[4]{3}\left[\cos\left(-\frac{\pi}{6}\right) + i\sin\left(-\frac{\pi}{6}\right)\right]$$

$$= \sqrt[4]{3}\left(\frac{\sqrt{3}}{2} - i\frac{1}{2}\right)$$

$$k = 1 \qquad \sqrt[4]{3}\, e^{i\frac{\pi}{3}} = \sqrt[4]{3}\left(\cos\frac{\pi}{3} + i\sin\frac{\pi}{3}\right) = \sqrt[4]{3}\left(\frac{1}{2} + i\frac{\sqrt{3}}{2}\right)$$

$$k = 2 \qquad \sqrt[4]{3}\, e^{i\frac{5\pi}{6}} = \sqrt[4]{3}\left(\cos\frac{5\pi}{6} + i\sin\frac{5\pi}{6}\right) = \sqrt[4]{3}\left(-\frac{\sqrt{3}}{2} + i\frac{1}{2}\right)$$

$$k = 3 \qquad \sqrt[4]{3}\, e^{i\frac{4\pi}{3}} = \sqrt[4]{3}\left(\cos\frac{4\pi}{3} + i\sin\frac{4\pi}{3}\right) = \sqrt[4]{3}\left(-\frac{1}{2} - i\frac{\sqrt{3}}{2}\right)$$

We can easily verify that all the roots are located on the circumference of a circle of radius $\sqrt[4]{3}$.

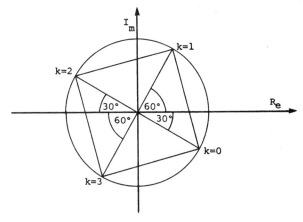

Fig. 1

In this case, the polygon formed by the roots is a square.

● **PROBLEM 3-29**

Evaluate

$$\frac{(2+i)^{60}}{(3+4i)^{50}} \qquad (1)$$

i.e., express it in the form $x + iy$.

Solution: Let $z_1 = 2 + i = r_1 e^{i\theta_1}$ and $\qquad (2)$

$$z_2 = 3 + 4i = r_2 e^{i\theta_2} \qquad (3)$$

Then

$$r_1 = \sqrt{4+1} = \sqrt{5}$$
$$r_2 = \sqrt{9+16} = 5 \qquad (4)$$

and

$$\cos\theta_1 = \frac{2}{\sqrt{5}} \qquad \sin\theta_1 = \frac{1}{\sqrt{5}}$$

$$\cos\theta_2 = \frac{3}{5} \qquad \sin\theta_2 = \frac{4}{5}$$
(5)

From eq.(5) we evaluate the angles

$$\theta_1 = 27°$$
$$\theta_2 = 53°$$
(6)

Eq.(1) can be written

$$\frac{(2+i)^{60}}{(3+4i)^{50}} = \frac{[\sqrt{5}\, e^{i\,27°}]^{60}}{[5\, e^{i\,53°}]^{50}}$$

$$= \frac{(5)^{30}}{(5)^{50}} \frac{e^{i\,1620°}}{e^{i\,2650°}} = \frac{1}{(5)^{20}} e^{i(1620° - 2650°)}$$

$$= (5)^{-20}\, e^{-i\,1030°} = (5)^{-20}\, e^{i(-310°)}$$

$$= 1.0486 \times 10^{-14}\,(\cos 50° + i\,\sin 50°)$$

$$= 1.0486 \times 10^{-14}\,(0.6427 + i\,0.7660)$$

$$= 6.74 \times 10^{-15} + i\,8.032 \times 10^{-15}$$

● **PROBLEM 3-30**

Find the square roots of
$$z = -4 - 3i \qquad (1)$$

Solution: In the polar form

$$-4 - 3i = 5\Big[\cos(\theta + 2k\pi) + i\,\sin(\theta + 2k\pi)\Big] \qquad (2)$$

Thus

$$\cos\theta = -\frac{4}{5} \quad \text{and} \quad \sin\theta = -\frac{3}{5}$$

The square roots of z are

$$\sqrt{5}\left[\cos\left(\frac{\theta + 2k\pi}{n}\right) + i\,\sin\left(\frac{\theta + 2k\pi}{n}\right)\right] \qquad (3)$$

Therefore

for $k = 0$ $\qquad \sqrt{5}\left(\cos\frac{\theta}{2} + i\,\sin\frac{\theta}{2}\right)$ (4)

for $k = 1$ $\qquad \sqrt{5}\left[\cos\left(\frac{\theta}{2} + \pi\right) + i\,\sin\left(\frac{\theta}{2} + \pi\right)\right]$

$$= -\sqrt{5}\left(\cos \frac{\theta}{2} + i \sin \frac{\theta}{2}\right) \quad (5)$$

Having $\cos\theta$ and $\sin\theta$, we find

$$\cos \frac{\theta}{2} = \pm\sqrt{\frac{1+\cos\theta}{2}} = \pm\sqrt{\frac{1}{10}} = \pm\frac{1}{\sqrt{10}} \quad (6)$$

$$\sin \frac{\theta}{2} = \pm\sqrt{\frac{1-\cos\theta}{2}} = \pm\sqrt{\frac{9}{10}} = \pm\frac{3}{\sqrt{10}} \quad (7)$$

Since both $\cos\theta$ and $\sin\theta$ are negative, θ is an angle in the third quadrant. Therefore, $\frac{\theta}{2}$ is an angle in the second quadrant. In such case, $\cos \frac{\theta}{2} < 0$ and $\sin \frac{\theta}{2} > 0$ and we obtain

$$\cos \frac{\theta}{2} = -\frac{1}{\sqrt{10}}, \quad \sin \frac{\theta}{2} = \frac{3}{\sqrt{10}} \quad (8)$$

From eqs. (4) and (5), the two roots are found to be

$$\sqrt{5}\left(-\frac{1}{\sqrt{10}} + i \frac{3}{\sqrt{10}}\right) = -\frac{1}{\sqrt{2}} + i \frac{3}{\sqrt{2}} \quad (9)$$

$$-\sqrt{5}\left(-\frac{1}{\sqrt{10}} + i \frac{3}{\sqrt{10}}\right) = \frac{1}{\sqrt{2}} - i \frac{3}{\sqrt{2}} \quad (10)$$

There is another method of solving this problem. Let $\alpha + i\beta$ be the solution. Then

$$(\alpha+i\beta)^2 = \alpha^2 - \beta^2 + 2\alpha\beta i = -4 - 3i \quad (11)$$

or

$$\alpha^2 - \beta^2 = -4$$
$$2\alpha\beta = -3 \quad (12)$$

This system of equations has two solutions

$$\alpha_1 = -\frac{1}{\sqrt{2}} \quad \beta_1 = \frac{3}{\sqrt{2}}$$
$$\alpha_2 = \frac{1}{\sqrt{2}} \quad \beta_2 = -\frac{3}{\sqrt{2}} \quad (13)$$

Thus, the corresponding square roots of 2 are

$$-\frac{1}{\sqrt{2}} + i \frac{3}{\sqrt{2}}$$
$$\frac{1}{\sqrt{2}} - i \frac{3}{\sqrt{2}} \quad (14)$$

• **PROBLEM 3-31**

Establish the general formula for the nth roots of unity. Find the three cube roots of unity.

Solution: The number 1 can be represented by

$$1 = \cos 0 + i \sin 0 = e^0 \tag{1}$$

Thus, the nth roots of unity are

$$e^{i\frac{2k\pi}{n}}, \quad k = 0, 1, \ldots n-1 \tag{2}$$

The three cube roots of unity are obtained for n=3

$$e^{i\frac{2k\pi}{3}} \quad k = 0, 1, 2 \tag{3}$$

Thus, the three roots are

for $k = 0$ $\quad e^0 = 1$

for $k = 1$ $\quad e^{i\frac{2\pi}{3}} = \frac{1}{2}(-1 + i\sqrt{3}) \tag{4}$

for $k = 2$ $\quad e^{i\frac{4\pi}{3}} = \frac{1}{2}(-1 - i\sqrt{3})$

The complex cube roots of unity are denoted by $1, \omega, \omega^2$. Note that since $\omega^3 = 1$, we have

$$(\omega-1)(\omega^2+\omega+1) = 0 \tag{5}$$

so that

$$\omega^2 + \omega + 1 = 0 \tag{6}$$

In the general case, we denote by ω_n the root corresponding to k=1 in eq.(2), i.e.

$$\omega_n = e^{i\frac{2\pi}{n}} \tag{7}$$

We can now write the n nth roots of unity as

$$1, \omega_n, \omega_n^2, \omega_n^3, \ldots \omega_n^{n-1} \tag{8}$$

These numbers are the vertices of a regular n-sided polygon inscribed in the circle $|z| = 1$.

Observe that for any z,

$$(1-z)(1+z+z^2+\ldots+z^{n-1}) = 1 - z^n \tag{9}$$

Since $\omega_n^n = 1$, we get

102

$$(1-\omega_n)(1+\omega_n+\omega_n^2+\ldots+\omega_n^{n-1}) = 0 \tag{10}$$

Because $\omega_n \neq 1$, we have

$$\omega_n^{n-1} + \ldots + \omega_n^2 + \omega_n + 1 = 0 \tag{11}$$

Finally, we observe that if z_0 is any nth root of z, then

$$z_0, \; z_0\omega_n, \; z_0\omega_n^2, \; \ldots, \; z_0\omega_n^{n-1} \tag{12}$$

are the nth roots of z.

● **PROBLEM 3-32**

Using only a straight edge and a compass, can you geometrically construct the fifth roots of unity?

Solution: The fifth roots of unity are the vertices of a regular pentagon inscribed in a circle of radius one, centered at the origin and having one vertex at the point $z = 1$. The question is how to construct a regular pentagon using a compass and a straight edge. We shall find the solution in three steps.

Step 1

Let AB be a unit segment. We will find point C such that

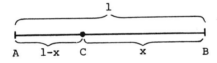

Fig. 1

$$\frac{1}{x} = \frac{x}{1-x} \tag{1}$$

or

$$x^2 + x - 1 = 0 \tag{2}$$

Thus, we have to find the solution of eq.(2). The following construction leads to the solution:

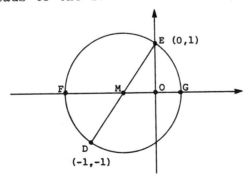

Fig. 2

Construct a circle having a diameter with endpoints $E(0,1)$, $D(-1,-1)$. It will now be shown that points F and G on the x-axis are the solutions of $x^2 + x - 1 = 0$. The coordinates of the center M of the circle are

$$x_0 = -\frac{1}{2}, \quad y_0 = 0 \tag{3}$$

and the radius is

$$|ME| = r = \sqrt{MO^2 + OE^2} = \sqrt{\tfrac{1}{4}+1} = \frac{\sqrt{5}}{2} \tag{4}$$

Therefore, the equation of the circle is

$$(x+\tfrac{1}{2})^2 + y^2 = \tfrac{5}{4} \tag{5}$$

To find the coordinates of F and G, let $y = 0$. Then

$$x^2 + x - 1 = 0 \tag{6}$$

which is eq.(2).

Eq.(6) has two solutions

$$x = \frac{-1-\sqrt{5}}{2} \tag{7}$$

and

$$x = \frac{-1+\sqrt{5}}{2} \tag{8}$$

The value of x in eq.(7) represents the distance of point F from the origin. Likewise, the positive root in eq.(8) denotes the distance of point G from the origin. We are only interested in the positive root

$$\frac{-1+\sqrt{5}}{2}$$

Fig. 2 shows how to construct $\frac{-1+\sqrt{5}}{2}$, since, as stated before, $OG = \frac{-1+\sqrt{5}}{2}$.

Step 2

Draw a circle of unit radius. Construct $x = \frac{-1+\sqrt{5}}{2}$ as described in Step 1. Starting from the point $x = 1$, $y = 0$, construct a regular decagon, whose side is $x = \frac{-1+\sqrt{5}}{2}$, as shown in Fig. 3.

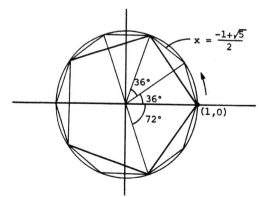

Fig. 3

A regular pentagon is formed by joining the alternate vertices.

Step 3

We will show that $\frac{-1+\sqrt{5}}{2}$ is indeed a side of a regular decagon. Let's take one slice of a decagon, as shown in Fig. 4.

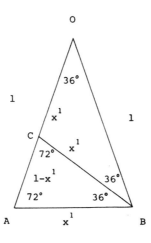

Fig. 4

Angle O is $36°$ and angles A and B are $72°$. The segment BC bisects angle B. We denote AB by x' (temporarily !). Then BC = x' and CO = x'. Therefore, AC = $1 - x'$.

Now, since the bisector BC of the angle of a triangle divides the opposite side into two segments that are proportional to the adjacent sides, we may write

$$\frac{x'}{1-x'} = \frac{1}{x'} \tag{9}$$

or

$$x'^2 + x' - 1 = 0 \tag{10}$$

which is eq.(2), with solution $\frac{-1+\sqrt{5}}{2}$.

q.e.d.

The problem of constructing regular polygons was known to the Greeks. They were able to construct a regular polygon of 2^m sides, $m > 1$. Gauss made a remarkable discovery in this field. At the age of nineteen, he found the construction of a regular 17-gon.

In Göttingen, after his death, a bronze statue of him was erected with a pedestal in the form of a regular 17-gon.

● **PROBLEM 3-33**

Show that

1. $\sin\theta = \dfrac{e^{i\theta} - e^{-i\theta}}{2i}$ \hfill (1)

2. $\cos\theta = \dfrac{e^{i\theta} + e^{-i\theta}}{2}$ \hfill (2)

Solution: Euler's formula for $e^{i\theta}$ and $e^{-i\theta}$ is

$$e^{i\theta} = \cos\theta + i\sin\theta \qquad (3)$$

$$e^{-i\theta} = \cos\theta - i\sin\theta \qquad (4)$$

Adding eqs.(3) and (4), we obtain

$$e^{i\theta} + e^{-i\theta} = 2\cos\theta \qquad (5)$$

or

$$\cos\theta = \dfrac{e^{i\theta} + e^{-i\theta}}{2} \qquad (6)$$

Subtracting eq.(4) from eq.(3) results in

$$e^{i\theta} - e^{-i\theta} = 2i\sin\theta$$

or

$$\sin\theta = \dfrac{e^{i\theta} - e^{-i\theta}}{2i} \qquad (7)$$

● **PROBLEM 3-34**

Prove the following formulas by applying Euler's formula.

1. $\cos^4\theta = \dfrac{1}{8}\cos 4\theta + \dfrac{1}{2}\cos 2\theta + \dfrac{3}{8}$ \hfill (1)

2. $\sin^3\theta = \dfrac{3}{4}\sin\theta - \dfrac{1}{4}\sin 3\theta$ \hfill (2)

Solution: 1. $\cos^4\theta = \left[\dfrac{e^{i\theta} + e^{-i\theta}}{2}\right]^4$

$= \dfrac{1}{16}\left[(e^{i\theta})^4 + 4(e^{i\theta})^3(e^{-i\theta}) + 6(e^{i\theta})^2(e^{-i\theta})^2 + 4(e^{i\theta})(e^{-i\theta})^3 + (e^{-i\theta})^4\right]$

$= \dfrac{1}{16}\left[e^{4i\theta} + 4e^{2i\theta} + 6 + 4e^{-2i\theta} + e^{-4i\theta}\right]$ \hfill (3)

$= \dfrac{1}{8}\left(\dfrac{e^{4i\theta} + e^{-4i\theta}}{2}\right) + \dfrac{1}{2}\left(\dfrac{e^{2i\theta} + e^{-2i\theta}}{2}\right) + \dfrac{3}{8}$

$$= \frac{1}{8}\cos 4\theta + \frac{1}{2}\cos 2\theta + \frac{3}{8}$$

2. $\sin^3\theta = \left(\dfrac{e^{i\theta} - e^{-i\theta}}{2i}\right)^3$

$$= -\frac{1}{8i}\left[(e^{i\theta})^3 - 3(e^{i\theta})^2(e^{-i\theta}) + 3(e^{i\theta})(e^{-i\theta})^2 - (e^{-i\theta})^3\right]$$

$$= -\frac{1}{8i}\left[e^{3i\theta} - 3e^{i\theta} + 3e^{-i\theta} - e^{-3i\theta}\right] \tag{4}$$

$$= \frac{3}{4}\left(\frac{e^{i\theta} - e^{-i\theta}}{2i}\right) - \frac{1}{4}\left(\frac{e^{3i\theta} - e^{-3i\theta}}{2i}\right)$$

$$= \frac{3}{4}\sin\theta - \frac{1}{4}\sin 3\theta$$

• **PROBLEM 3-35**

Apply DeMoivre's formula to prove

1. $\cos 2\theta = \cos^2\theta - \sin^2\theta$ \hfill (1)

   $\sin 2\theta = 2\sin\theta\cos\theta$ \hfill (2)

2. $\cos 3\theta = 4\cos^3\theta - 3\cos\theta$ \hfill (3)

   $\sin 3\theta = 3\sin\theta - 4\sin^3\theta$ \hfill (4)

Note that eq.(4) is the same as eq.(2) of Problem 3-34.

Solution: 1. From DeMoivre's formula

$$(\cos\theta + i\sin\theta)^n = \cos n\theta + i\sin n\theta$$

we find

$$(\cos\theta + i\sin\theta)^2 = \cos 2\theta + i\sin 2\theta \tag{5}$$

On the other hand

$$(\cos\theta + i\sin\theta)^2 = (\cos^2\theta - \sin^2\theta) + 2i\sin\theta\cos\theta \tag{6}$$

Eqs.(5) and (6) yield

$$\cos 2\theta + i\sin 2\theta = (\cos^2\theta - \sin^2\theta) + 2i\sin\theta\cos\theta \tag{7}$$

Comparing the real and imaginary parts, we find

$$\cos 2\theta = \cos^2\theta - \sin^2\theta \tag{8}$$

$$\sin 2\theta = 2\sin\theta\cos\theta \tag{9}$$

2. Again, from DeMoivre's formula

$$(\cos\theta + i\sin\theta)^3 = \cos 3\theta + i\sin 3\theta \tag{10}$$

But we also have

$$(\cos\theta + i\sin\theta)^3 = \cos^3\theta + 3i\cos^2\theta\sin\theta - 3\cos\theta\sin^2\theta$$
$$- i\sin^3\theta \qquad (11)$$
$$= \cos^3\theta - 3\cos\theta\sin^2\theta + i(3\cos^2\theta\sin\theta - \sin^3\theta)$$

Hence

$$\cos 3\theta = \cos^3\theta - 3\cos\theta\sin^2\theta = \cos\theta(\cos^2\theta - 3\sin^2\theta)$$
$$= \cos\theta(\cos^2\theta - 3 + 3\cos^2\theta) = 4\cos^3\theta - 3\cos\theta \qquad (12)$$

and

$$\sin 3\theta = 3\cos^2\theta\sin\theta - \sin^3\theta$$
$$= \sin\theta(3\cos^2\theta - \sin^2\theta) = \sin\theta(3 - 3\sin^2\theta - \sin^2\theta) \qquad (13)$$
$$= 3\sin\theta - 4\sin^3\theta$$

The derivation of the above identities gives a clear indication of how useful formulas from the real number system can be obtained from a theorem dealing with complex numbers.

APPLICATIONS IN TRIGONOMETRY, IDENTITIES

• PROBLEM 3-36

Solving mathematical problems was one of the highly regarded activities of the ancient Greeks. In many cases, the solution of a particular problem was a matter of life and death. Such was the case in 430 B.C., when a plague took a heavy toll of lives in the city of Athens. As a remedy, the oracle at Delos advised the Athenians to double the size (volume) of the altar of Apollo, which had the shape of a cube. This task had to be carried out by utilizing only a straight edge and a compass, the only available tools. It took over two millennia to prove that a cube could not, using only geometric constructions, be doubled in size.

Other famous problems that are impossible to solve with only a straight edge and compass are the squaring of a circle and the trisecting of an angle.

Certain angles can be trisected without difficulty as is the case with the right angle, since a 30° angle can be constructed.

Show that it is impossible to trisect any angle using a compass and an unmarked straight edge.

Hint:

Show that a 60° angle cannot be trisected. Use eq.(3) of Problem 3-35. Apply the theorem:

An angle can be constructed if and only if its cosine can be constructed.

Solution: Substituting $3\theta = 60°$ into eq.(3) of Problem 3-35, we obtain

$$\cos 60° = 4\cos^3 20° - 3\cos 20° \tag{1}$$

Let

$$x = 2\cos 20° \tag{2}$$

then

$$\frac{1}{2} = \frac{x^3}{2} - \frac{3x}{2} \tag{3}$$

or

$$x^3 - 3x - 1 = 0 \tag{4}$$

Eq.(4) is irreducible. Why? From a theorem from algebra:

If the equation $a_0 x^n + a_1 x^{n-1} + \ldots + a_n = 0$, (5)

where $a_0, a_1, \ldots, a_n$ are integers, has a rational root $\frac{\alpha}{\beta}$, then α is a factor of a_n and β is a factor of a_0. We conclude that the only possible rational roots are ± 1. Obviously $+1$ and -1 are not the roots and eq.(4) has no rational roots.

Hence, $x^3 - 3x - 1 = 0$ is an irreducible cubic equation. Its roots cannot be constructed with a straight edge and compass.

Thus, $\cos 20° = \frac{x}{2}$ cannot be constructed with a straight edge and compass. Therefore, since an angle of $20°$ cannot be constructed, a $60°$ angle cannot be trisected. We have proven that there is no procedure (using only an unmarked straight edge and a compass) to trisect an arbitrary angle.

● **PROBLEM 3-37**

Let z_k be the roots of the equation

$$(z+1)^7 + z^7 = 0 \tag{1}$$

Show that

$$\text{Re}(z_k) = -\frac{1}{2}, \quad k = 0, 1, \ldots, 6 \tag{2}$$

Solution: Let us rewrite eq.(1) as

$$(z+1)^7 = -z^7 \tag{3}$$

or

$$\left[-\frac{(z+1)}{z}\right]^7 = 1 \tag{4}$$

If we make the substitution

$$z' = -\frac{z+1}{z} \tag{5}$$

then eq.(5) is equivalent to

$$(z')^7 = 1 = \cos 2\pi + i \sin 2\pi \tag{6}$$

to which the roots of z' are

$$z'_k = \cos \frac{2k\pi}{7} + i \sin \frac{2k\pi}{7}$$

$$= \cos \frac{2k\pi}{7} = e^{i\frac{2k\pi}{7}}, \quad k = 0,1,2,3,4,5,6 \tag{7}$$

However, we wish to find the roots of z, not z'. Thus, going back to eq.(5), we obtain

$$z'_k = -\frac{z_k + 1}{z_k} = e^{i\frac{2k\pi}{7}} \tag{8}$$

so that the roots of z are

$$z_k = \frac{-1}{1 + e^{i\frac{2k\pi}{7}}}, \quad k = 0,1,2,3,4,5 \tag{9}$$

Notice that there are seven roots of z. Now, since we are interested in the real part of the roots z_k, we must manipulate eq.(9).

We have

$$z_k = \frac{-1}{1 + e^{i\frac{2k\pi}{7}}} = \frac{-1}{1 + \cos \frac{2k\pi}{7} + i \sin \frac{2k\pi}{7}}$$

$$= \frac{-1}{1 + \cos \frac{2k\pi}{7} + i \sin \frac{2k\pi}{7}} \cdot \frac{(1 + \cos \frac{2k\pi}{7} - i \sin \frac{2k\pi}{7})}{(1 + \cos \frac{2k\pi}{7} - i \sin \frac{2k\pi}{7})}$$

$$= \frac{-(1 + \cos \frac{2k\pi}{7}) + i \sin \frac{2k\pi}{7}}{2(1 + \cos \frac{2k\pi}{7})}$$

$$= -\frac{1}{2} + i \frac{\sin \frac{2k\pi}{7}}{\cos \frac{2k\pi}{7}} \tag{10}$$

Thus, the real part of z_k is

$$\text{Re}(z_k) = -\frac{1}{2}$$

All the roots are located on the line $x = -\frac{1}{2}$, as shown in Fig. 1.

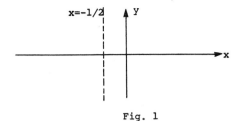

Fig. 1

• **PROBLEM 3-38**

Using the binomial formula and DeMoivre's theorem, prove the identities

1. $\cos 5\theta = 16\cos^5\theta - 20\cos^3\theta + 5\cos\theta$ \hfill (1)

2. $\dfrac{\sin 5\theta}{\sin\theta} = 16\cos^4\theta - 12\cos^2\theta + 1,\ \theta \neq 0, \pm\pi, \pm 2\pi, \ldots$ \hfill (2)

Solution: The binomial formula is
$$(a+b)^n = \binom{n}{0}a^n + \binom{n}{1}a^{n-1}b + \binom{n}{2}a^{n-2}b^2 + \ldots$$
$$\ldots + \binom{n}{r}a^{n-r}b^r + \ldots + \binom{n}{n}b^n \qquad (3)$$

Here, the coefficients $\binom{n}{r}$, sometimes denoted C_r^n or $_nC_r$, are given by

$$\binom{n}{r} = \frac{n!}{r!(n-r)!} \qquad (4)$$

We shall apply the binomial formula and DeMoivre's theorem as follows

$\cos 5\theta + i\sin 5\theta = (\cos\theta + i\sin\theta)^5$

$= \cos^5\theta + \binom{5}{1}\cos^4\theta(i\sin\theta) + \binom{5}{2}\cos^3\theta(i\sin\theta)^2$

$\qquad + \binom{5}{3}\cos^2\theta(i\sin\theta)^3 + \binom{5}{4}\cos\theta(i\sin\theta)^4 + (i\sin\theta)^5$

$= \cos^5\theta + 5i\cos^4\theta\sin\theta - 10\cos^3\theta\sin^2\theta - 10i\cos^2\theta\sin^3\theta \qquad (5)$

$\qquad + 5\cos\theta\sin^4\theta + i\sin^5\theta$

$= \cos^5\theta - 10\cos^3\theta\sin^2\theta + 5\cos\theta\sin^4\theta$

$\qquad + i(5\cos^4\theta\sin\theta - 10\cos^2\theta\sin^3\theta + \sin^5\theta)$

Comparing the real and imaginary parts in eq.(5), we obtain

$\cos 5\theta = \cos^5\theta - 10\cos^3\theta\sin^2\theta + 5\cos\theta\sin^4\theta$

$\qquad = \cos^5\theta - 10\cos^3\theta(1-\cos^2\theta) + 5\cos\theta(1-\cos^2\theta)^2 \qquad (6)$

$\qquad = 16\cos^5\theta - 20\cos^3\theta + 5\cos\theta$

and

$$\sin 5\theta = 5\cos^4\theta \sin\theta - 10\cos^2\theta \sin^3\theta + \sin^5\theta \qquad (7)$$

Thus

$$\begin{aligned}\frac{\sin 5\theta}{\sin\theta} &= 5\cos^4\theta - 10\cos^2\theta \sin^2\theta + \sin^4\theta \\ &= 5\cos^4\theta - 10\cos^2\theta(1-\cos^2\theta) + (1-\cos^2\theta)^2 \qquad (8) \\ &= 16\cos^4\theta - 12\cos^2\theta + 1\end{aligned}$$

Note that this holds for $\sin\theta \neq 0$, that is for

$$\theta \neq 0, \pm\pi, \pm 2\pi, \ldots$$

● **PROBLEM 3-39**

For $n = 2, 3, \ldots$ prove that

$$\cos\frac{2\pi}{n} + \cos\frac{4\pi}{n} + \cos\frac{6\pi}{n} + \ldots + \cos\frac{2(n-1)\pi}{n} = -1 \qquad (1)$$

$$\sin\frac{2\pi}{n} + \sin\frac{4\pi}{n} + \sin\frac{6\pi}{n} + \ldots + \sin\frac{2(n-1)\pi}{n} = 0 \qquad (2)$$

Solution: Observe that the angles in eqs.(1) and (2) resemble the angles in the formula for the nth roots of unity. Therefore, if we consider the equation

$$z^n = 1 \qquad (3)$$

or

$$z^n - 1 = 0 \qquad (4)$$

then the roots z_k, the nth roots of unity, are given by

$$z_k = e^{\frac{2k\pi}{n}}, \quad k = 0, 1, 2, \ldots, n-1$$

or

$$z_1 = 1, \quad z_2 = e^{\frac{2\pi}{n}}, \quad z_3 = e^{\frac{4\pi}{n}}, \ldots, z_n = e^{\frac{2(n-1)\pi}{n}} \qquad (5)$$

Now let us consider the following theorem:

The sum of all the roots of the equation

$$a_0 z^n + a_1 z^{n-1} + \ldots + a_{n-1} z + a_n = 0 \qquad (6)$$

is given by

$$\sum_{k=0}^{n-1} z_k = -\frac{a_1}{a_0} \qquad (7)$$

Now, in our case, $a_0 = 1$ and $a_1 = 0$, as obtained from eq.(4).

Thus, the sum of the roots of unity is equal to zero.

$$\sum_{k=0}^{n-1} z_k = 0 \tag{8}$$

Therefore

$$z_1 + z_2 + \ldots + z_n = 0 \tag{9}$$

or

$$1 + e^{i\frac{2\pi}{n}} + e^{i\frac{4\pi}{n}} + \ldots + e^{i\frac{2(n-1)}{n}\pi} = 0 \tag{10}$$

Applying Euler's formula

$$1 + \left(\cos\frac{2\pi}{n} + i\sin\frac{2\pi}{n}\right) + \left(\cos\frac{4\pi}{n} + i\sin\frac{4\pi}{n}\right)$$
$$+ \ldots + \left(\cos\frac{2(n-1)}{n}\pi + i\sin\frac{2(n-1)}{n}\pi\right) = 0$$

or

$$\left[1 + \cos\frac{2\pi}{n} + \cos\frac{4\pi}{n} + \ldots + \cos\frac{2(n-1)}{n}\pi\right]$$
$$+ i\left[\sin\frac{2\pi}{n} + \sin\frac{4\pi}{n} + \ldots + \sin\frac{2(n-1)}{n}\pi\right] = 0$$

Thus, since both the real and imaginary parts must equal zero, we obtain eqs.(1) and (2).

• **PROBLEM 3-40**

Prove that

$$\sin\frac{\pi}{n} \sin\frac{2\pi}{n} \sin\frac{3\pi}{n} \ldots \sin\frac{(n-1)\pi}{n} = \frac{n}{2^{n-1}} \tag{1}$$

for $n = 2, 3, \ldots$.

<u>Solution</u>: Consider the nth roots of unity

$$1, e^{i\frac{2\pi}{n}}, e^{i\frac{4\pi}{n}}, \ldots, e^{i\frac{2(n-1)\pi}{n}} \tag{2}$$

which are the solutions of the equation

$$z^n - 1 = 0 \tag{3}$$

We may write

$$z^n - 1 = (z-1)(z - e^{i\frac{2\pi}{n}})(z - e^{i\frac{4\pi}{n}}) \ldots (z - e^{i\frac{2(n-1)\pi}{n}}) \tag{4}$$

From Problem 3-31, eq.(9), we obtain

$$\frac{z^n - 1}{z - 1} = 1 + z + z^2 + \ldots + z^{n-1} \tag{5}$$

Dividing eq.(4) by z-1 and substituting eq.(5) results in

$$1 + z + \ldots + z^{n-1} = (z - e^{i\frac{2\pi}{n}})(z - e^{i\frac{4\pi}{n}}) \ldots (z - e^{i\frac{2(n-1)\pi}{n}}) \tag{6}$$

Setting z = 1 gives us

$$n = (1 - e^{i\frac{2\pi}{n}})(1 - e^{i\frac{4\pi}{n}}) \ldots (1 - e^{i\frac{2(n-1)\pi}{n}}) \tag{7}$$

The complex conjugate of eq.(7) is

$$n_c = (1 - e^{-i\frac{2\pi}{n}})(1 - e^{-i\frac{4\pi}{n}}) \ldots (1 - e^{-i\frac{2(n-1)\pi}{n}}) \tag{8}$$

Multiplying eq.(7) by eq.(8) and utilizing the identity

$$(1 - e^{i\frac{2k\pi}{n}})(1 - e^{-i\frac{2k\pi}{n}}) = 2 - 2\cos\frac{2k\pi}{n} \tag{9}$$

we obtain

$$n^2 = 2^{n-1}(1 - \cos\frac{2\pi}{n})(1 - \cos\frac{4\pi}{n}) \ldots (1 - \cos\frac{2(n-1)\pi}{n}) \tag{10}$$

Substituting the identity

$$1 - \cos\frac{2k\pi}{n} = 2\sin^2\frac{k\pi}{n} \tag{11}$$

into eq.(10) results in

$$n^2 = 2^{2(n-1)} \sin^2\frac{\pi}{n} \sin^2\frac{2\pi}{n} \ldots \sin^2\frac{(n-1)\pi}{n} \tag{12}$$

Taking the positive square root of eq.(12) finally gives

$$\frac{n}{2^{n-1}} = \sin\frac{\pi}{n} \sin\frac{2\pi}{n} \ldots \sin\frac{(n-1)\pi}{n} \tag{13}$$

• **PROBLEM 3-41**

A regular polygon of n sides inscribed in a circle of unit radius is given. Show that the product of the lengths of all diagonals from any vertex is $\frac{1}{4}\frac{n}{\sin^2\frac{\pi}{n}}$.

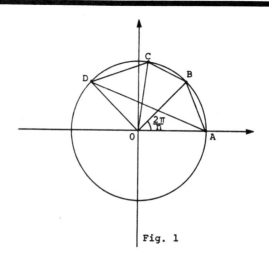

Fig. 1

Solution: Part of a regular polygon of n sides is shown in Fig. 1.

Note that n − 3 diagonals can be drawn from vertex A (or any other vertex). We denote

$$\alpha = \frac{2\pi}{n} \qquad (1)$$

From the figure, we have

$$AC = 2 \sin \frac{2\alpha}{2}$$

$$AD = 2 \sin \frac{3\alpha}{2} \qquad (2)$$

$$AE = 2 \sin \frac{4\alpha}{2}$$

Thus, the lengths of the diagonals are

$$2 \sin 2 \frac{\pi}{n}$$

$$2 \sin \frac{3\pi}{n} \qquad (3)$$

$$2 \sin \frac{4\pi}{n}$$

$$\vdots$$

$$2 \sin \frac{(n-2)\pi}{n}$$

Thus, the product of the lengths is

$$2 \sin \frac{2\pi}{n} \cdot 2 \sin \frac{3\pi}{n} \cdot 2 \sin \frac{4\pi}{n} \cdot \ldots \cdot 2 \sin \frac{(n-2)\pi}{n} \qquad (4)$$

Utilizing the results of Problem 3-40, we obtain

$$\text{Product} = 2^{n-3} \frac{\sin \frac{\pi}{n} \sin \frac{2\pi}{n} \sin \frac{3\pi}{n} \ldots \sin \frac{(n-2)\pi}{n} \sin \frac{(n-1)\pi}{n}}{\sin \frac{\pi}{n} \sin \frac{(n-1)\pi}{n}}$$

$$= \frac{2^{n-3} \cdot n}{2^{n-1} \sin \frac{\pi}{n} \sin \frac{(n-1)\pi}{n}} = \frac{n}{4} \cdot \frac{1}{\sin \frac{\pi}{n}} \cdot \frac{1}{\sin(\pi - \frac{\pi}{n})} \qquad (5)$$

$$= \frac{n}{4} \cdot \frac{1}{\sin^2 \frac{\pi}{n}}$$

● **PROBLEM 3-42**

A unit circle with center C located at the point z = 1 is shown in Fig. 1.

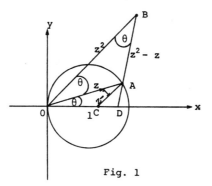

Fig. 1

Let A be any point on the circle represented by z, and let B represent z^2. Prove geometrically that

$$|z^2 - z| = |z| \tag{1}$$

$$3 \arg(z-1) = 3\arg z^2 = 2\arg(z^2-z) \tag{2}$$

Solution: From the figure

$$OA = |z|, \quad OB = |z^2| \tag{3}$$

Hence

$$\frac{OB}{OA} = \frac{OA}{1} = \frac{OA}{OC} \tag{4}$$

Let

$$\arg z = \theta \tag{5}$$

then ∢ DOB = 2θ and thus

$$\angle AOB = \theta \tag{6}$$

Triangles OCA and OAB are similar. Since OC = CA, we have

$$OA = AB \tag{7}$$

Therefore

$$OA = |z| = AB = |z^2-z| \tag{8}$$

From the figure, we obtain the following relationships

$$\angle BOD = 2\theta, \quad \angle OBD = \theta \quad \text{thus} \quad \angle ODB = \pi - 3\theta \tag{9}$$

and ∢ BDx = 3θ \quad(10)

From eqs.(9) and (10), and from the figure, we find

$$\arg z^2 = \arg(z-1) = 2\theta \tag{11}$$

$$\arg(z^2-z) = 3\theta \tag{12}$$

We conclude that

$$3 \arg z^2 = 3 \arg(z-1) = 2 \arg(z^2-z) \tag{13}$$

• **PROBLEM 3-43**

Prove the general formula

$$2^n \cos^n \theta = 2\cos n\theta + \binom{n}{1} 2\cos(n-2)\theta$$
$$+ \binom{n}{2} 2\cos(n-4)\theta + \cdots \tag{1}$$

Show that

$$\cos^5 \theta = \frac{1}{16}(\cos 5\theta + 5\cos 3\theta + 10\cos \theta) \tag{2}$$

Solution: Let

$$z = \cos\theta + i\sin\theta \tag{3}$$

Then

$$z^k = \cos k\theta + i\sin k\theta \tag{4}$$

and

$$z^{-k} = \cos(-k\theta) + i\sin(-k\theta) = \cos k\theta - \sin k\theta \tag{5}$$

From eqs. (4) and (5), we compute

$$2\cos k\theta = z^k + z^{-k} \tag{6}$$

and

$$2i\sin k\theta = z^k - z^{-k} \tag{7}$$

for all integers k.

The nth power of eq. (6) is

$$(2\cos k\theta)^n = 2^n \cos^n k\theta = (z^k + z^{-k})^n \tag{8}$$

Applying the binomial expression and setting k = 1, we find

$$2^n \cos^n \theta = (z + z^{-1})^n$$
$$= \binom{n}{0} z^n + \binom{n}{1} z^{n-1}(z^{-1}) + \binom{n}{2} z^{n-2}(z^{-1})^2 + \cdots$$
$$\cdots + \binom{n}{\ell} z^{n-\ell}(z^{-1})^\ell + \cdots + \binom{n}{n-1} z\,(z^{-1})^{n-1} + \binom{n}{n}(z^{-1})^n$$
$$= z^n + \binom{n}{1} z^{n-2} + \cdots + \binom{n}{\ell} z^{n-2\ell} + \cdots + z^{-n} \tag{9}$$

We collect terms in such a way that each pair consists of two terms equally distant from the two ends. That is, the first term with the last, the second with the one next to last, and so on. Each pair has terms with equal numerical coefficients. Therefore

$$\binom{n}{\ell} = \frac{n!}{(n-\ell)!\,\ell!} = \binom{n}{n-\ell} \tag{10}$$

so that we may write

117

$$2^n \cos^n \theta = (z^n + z^{-n}) + \binom{n}{1}(z^{n-2} + z^{-n+2})$$
$$+ \binom{n}{2}(z^{n-4} + z^{-n+4}) + \ldots \quad (11)$$

Applying eq.(6), we find

$$2^n \cos^n \theta = 2\cos n\theta + \binom{n}{1} 2\cos(n-2)\theta + \binom{n}{2} 2\cos(n-4)\theta + \ldots \quad (12)$$

In particular, for $n = 5$, we obtain

$$2^5 \cos^5 \theta = 2\cos 5\theta + \binom{5}{1} 2\cos 3\theta + \binom{5}{2} 2\cos \theta$$
$$= 2\cos 5\theta + 10\cos 3\theta + 20\cos \theta \quad (13)$$

and

$$\cos^5 \theta = \frac{1}{16}(\cos 5\theta + 5\cos 3\theta + 10\cos \theta) \quad (14)$$

Eq.(12) is helpful in evaluating certain integrals. For example, instead of evaluating the integral

$$\int \cos^5 x \, dx$$

we can evaluate the much simpler integral

$$\int \cos^5 x \, dx = \frac{1}{16} \int (\cos 5x + 5\cos 3x + 10\cos x) dx \quad (15)$$

• **PROBLEM 3-44**

Using the method described in Problem 3-43, derive a formula expressing $\sin^n \theta$ in terms of $\sin n\theta$, $\sin(n-2)\theta, \ldots \sin\theta$ or $\cos n\theta$, $\cos(n-2)\theta, \ldots, \cos 2\theta$.

<u>Solution</u>: If $\quad z = \cos \theta + i \sin \theta \quad (1)$

then
$$2i \sin \theta = z - z^{-1} \quad (2)$$

The nth power of eq.(2) is

$$(2i \sin \theta)^n = (z - z^{-1})^n \quad (3)$$

Applying the binomial expansion,

$$2^n i^n \sin^n \theta = z^n - \binom{n}{1} z^{n-2} + \binom{n}{2} z^{n-4} + \ldots$$
$$+ (-1)^k \binom{n}{k} z^{n-2k} + \ldots + (-1)^n z^{-n} \quad (4)$$

For odd n, the last term in the above expansion is $-z^{-n}$, while for even n, the last term is z^{-n}. It is most convenient to consider these two cases separately.

1. For even n, the terms in eq.(4) can be assembled in pairs in such a way that each pair consists of terms with the same distance from the ends of the sum. Thus, the first pair is $z^n + z^{-n}$, while the second is

$$-\binom{n}{1}z^{n-2} - \binom{n}{n-1}z^{-n+2} = -\binom{n}{1}(z^{n-2} + z^{-n+2})$$

etc.

Each pair consists of terms with either a plus sign, or a minus sign. Eq.(4) can be written

$$2^n i^n \sin^n \theta = (z^n + z^{-n}) - \binom{n}{1}(z^{n-2} + z^{-n+2})$$
$$+ \binom{n}{2}(z^{n-4} + z^{-n+4}) - \ldots \quad (5)$$
$$= 2\cos n\theta - \binom{n}{1}2\cos(n-2)\theta + \binom{n}{2}2\cos(n-4)\theta + \ldots$$

Since n is even,

$$i^n = (-1)^{\frac{n}{2}} \quad (6)$$

and

$$(-1)^{\frac{n}{2}} 2^n \sin^n \theta = 2\cos n\theta - \binom{n}{1}2\cos(n-2)\theta$$
$$+ \binom{n}{2}2\cos(n-4)\theta + \ldots \quad (7)$$

2. The situation is different for odd n. Collecting terms in eq.(4) in the same manner as in part 1, we obtain the first pair

$$z^n - z^{-n}$$

The second pair is

$$-\binom{n}{1}z^{n-2} + (-1)^{n-1}\binom{n}{n-1}z^{-n+2} = -\binom{n}{1}z^{n-2} + \binom{n}{1}z^{-n+2}$$

Eq.(4) becomes

$$2^n i^n \sin^n \theta = (z^n - z^{-n}) - \binom{n}{1}(z^{n-2} - z^{-n+2})$$
$$+ \binom{n}{2}(z^{n-4} - z^{-n+4}) - \ldots \quad (8)$$

Since

$$z^k - z^{-k} = 2i \sin k\theta \quad (9)$$

we get

$$2^n i^n \sin^n \theta = 2i \sin n\theta - \binom{n}{1}2i \sin(n-2)\theta$$
$$+ \binom{n}{2}2i \sin(n-4)\theta - \ldots \quad (10)$$

• **PROBLEM 3-45**

Prove

$$\cos^8\theta + \sin^8\theta = \frac{1}{64}(\cos 8\theta + 28\cos 4\theta + 35) \tag{1}$$

<u>Solution</u>: Applying eq.(1) of Problem 3-43 we find

$$2^8 \cos^8\theta = 2\cos 8\theta + \binom{8}{1}2\cos 6\theta + \binom{8}{2}2\cos 4\theta \\ + \binom{8}{3}2\cos 2\theta + \binom{8}{4} \tag{2}$$

From eq.(7) of Problem 3-44, $\sin^8\theta$ can be evaluated

$$(-1)^4 2^8 \sin^8\theta = 2\cos 8\theta - \binom{8}{1}2\cos 6\theta \\ + \binom{8}{2}2\cos 4\theta - \binom{8}{3}2\cos 2\theta + \binom{8}{4} \tag{3}$$

Adding eqs.(2) and (3),

$$2^8(\cos^8\theta + \sin^8\theta) = 4\cos 8\theta + 4\binom{8}{2}\cos 4\theta + 2\binom{8}{4} \tag{4}$$

Since

$$\binom{8}{2} = \frac{8!}{6!2!} = \frac{7 \cdot 8}{2} = 28$$

and $\tag{5}$

$$\binom{8}{4} = \frac{8!}{4!4!} = 70$$

eq.(4) can be written

$$\cos^8\theta + \sin^8\theta = \frac{1}{64}(\cos 8\theta + 28\cos 4\theta + 35) \tag{6}$$

• **PROBLEM 3-46**

Express $\cos n\theta$ and $\sin n\theta$ in terms of powers of $\cos\theta$ and $\sin\theta$.

<u>Solution</u>: DeMoivre's theorem states that for all positive integers

$$\cos n\theta + i\sin n\theta = (\cos\theta + i\sin\theta)^n \tag{1}$$

Applying the binomial expansion to eq.(1), we find

$$\cos n\theta + i\sin n\theta = (\cos\theta + i\sin\theta)^n$$

$$= \binom{n}{0}\cos^n\theta + \binom{n}{1}\cos^{n-1}\theta(i\sin\theta) + \binom{n}{2}\cos^{n-2}\theta(i\sin\theta)^2$$

120

$$+ \ldots + \binom{n}{k}\cos^{n-k}\theta(i\sin\theta)^k + \ldots + \binom{n}{n-1}\cos\theta(i\sin\theta)^{n-1}$$
$$+ \binom{n}{n}(i\sin\theta)^n \tag{2}$$

If we equate the real and imaginary parts, we obtain expressions for $\cos n\theta$ and $\sin n\theta$

$$\cos n\theta = \cos^n\theta - \binom{n}{2}\cos^{n-2}\theta \sin^2\theta + \binom{n}{4}\cos^{n-4}\theta \sin^4\theta - \ldots \tag{3}$$

$$\sin n\theta = \binom{n}{1}\cos^{n-1}\theta \sin\theta - \binom{n}{3}\cos^{n-3}\theta \sin^3\theta + \ldots \tag{4}$$

In a more compact form, eqs.(3) and (4) can be written

$$\cos n\theta = \sum_{k=0}^{\ell} (-1)^k \binom{n}{2k} \cos^{n-2k}\theta \sin^{2k}\theta \tag{5}$$

$$\sin n\theta = \sum_{k=0}^{m} (-1)^k \binom{n}{2k+1} \cos^{n-2k-1}\theta \sin^{2k+1}\theta \tag{6}$$

where

$$\binom{n}{2k} = \frac{n(n-1)(n-2)\ldots(n-2k+1)}{(2k)!} \quad \text{for } k > 0 \tag{7}$$

$$\ell = \frac{(2n-1)+(-1)^n}{4} = \left[\frac{n}{2}\right] \tag{8}$$

$$m = \left[\frac{n-1}{2}\right] \tag{9}$$

The notation $[x]$ means the greatest integer less than or equal to the real number x. For example

$[0.46] = 0$

$[1.002] = 1$

$[-1.27] = -2$

The function $f(x) = [x]$ is called the bracket function.

● **PROBLEM 3-47**

Express $\tan n\theta$ in terms of powers of $\tan\theta$. Compute $\tan n\theta$ for $n = 5$. Assume that $5\theta = k\pi$, where k is an integer. Can you draw any conclusions?

Solution: We have

$$\tan n\theta = \frac{\sin n\theta}{\cos n\theta} \tag{1}$$

$\cos n\theta \neq 0$

Substituting $\sin n\theta$ and $\cos n\theta$ from Problem 3-46, we obtain

$$\tan n\theta = \frac{\binom{n}{1}\cos^{n-1}\theta \sin\theta - \binom{n}{3}\cos^{n-3}\theta \sin^3\theta + \binom{n}{5}\cos^{n-5}\theta \sin^5\theta - \ldots}{\cos^n\theta - \binom{n}{2}\cos^{n-2}\theta \sin^2\theta + \binom{n}{4}\cos^{n-4}\theta \sin^4\theta - \ldots} \quad (2)$$

Dividing numerator and denominator by $\cos^n\theta$, we obtain (provided $\cos\theta \neq 0$)

$$\tan n\theta = \frac{\binom{n}{1}\tan\theta - \binom{n}{3}\tan^3\theta + \binom{n}{5}\tan^5\theta - \ldots}{1 - \binom{n}{2}\tan^2\theta + \binom{n}{4}\tan^4\theta - \ldots} \quad (3)$$

For $n = 5$, eq.(3) becomes

$$\tan 5\theta = \frac{\binom{5}{1}\tan\theta - \binom{5}{3}\tan^3\theta + \binom{5}{5}\tan^5\theta}{1 - \binom{5}{2}\tan^2\theta + \binom{5}{4}\tan^4\theta} \quad (4)$$

If
$$5\theta = k\pi, \quad \text{then}$$
$$\tan 5\theta = 0 \quad (5)$$

and from eq.(4), we get

$$\tan^5\theta - 10\tan^3\theta + 5\tan\theta = 0 \quad (6)$$

If $5\theta = k\pi$, where k is an integer, then

$$\theta = \frac{k\pi}{5} \quad (7)$$

and

$$\tan\theta = \tan\frac{k\pi}{5} \quad (8)$$

$\tan\alpha$ is a periodic function with period π. Hence

$$\tan\frac{\pi}{5} = \tan\frac{6\pi}{5} = \tan\frac{11\pi}{5} = \ldots \quad (9)$$

There are only five distinct values of $\tan\frac{k\pi}{5}$, i.e., for $k = 0, 1, 2, 3, 4$.

Each of these values must satisfy eq.(6). That is, the solutions of eq.(6) are

$$\tan 0, \quad \tan\frac{\pi}{5}, \quad \tan\frac{2\pi}{5}, \quad \tan\frac{3\pi}{5}, \quad \tan\frac{4\pi}{5} \quad (10)$$

Eq.(6) can be written

$$\tan\theta(\tan^4\theta - 10\tan^2\theta + 5) = 0 \quad (11)$$

Either $\tan\theta = 0$ or $\tan^4\theta - 10\tan^2\theta + 5 = 0$. For $k = 0$, $\tan 0 = 0$. We can therefore conclude that

$$\tan^4\theta - 10\tan^2\theta + 5 = 0 \tag{12}$$

has four solutions

$$\tan\frac{\pi}{5}, \quad \tan\frac{2\pi}{5}, \quad \tan\frac{3\pi}{5}, \quad \tan\frac{4\pi}{5} \tag{13}$$

SERIES POLYNOMIALS

• **PROBLEM 3-48**

Compute the sum of the series of the form

1. $a_0\cos\theta + a_1\cos(\theta+\alpha) + a_2\cos(\theta+2\alpha) + \ldots + a_{n-1}\cos\left[\theta+(n-1)\alpha\right]$ (1)

2. $a_0\sin\theta + a_1\sin(\theta+\alpha) + a_2\sin(\theta+2\alpha) + \ldots + a_{n-1}\sin\left[\theta+(n-1)\alpha\right]$ (2)

Consider the case where $a_0 = a_1 = \ldots = a_{n-1} = 1$.

Solution: If we denote the first series by C_n, and the second by S_n, then

$$C_n + i S_n = a_0(\cos\theta + i\sin\theta) + a_1\left[\cos(\theta+\alpha) + i\sin(\theta+\alpha)\right]$$
$$+ \ldots + a_{n-1}\left[\cos[\theta+(n-1)\alpha] + i\sin[\theta+(n-1)\alpha]\right] \tag{3}$$

Let
$$z = \cos\theta + i\sin\theta \tag{4}$$
and
$$w = \cos\alpha + i\sin\alpha \tag{5}$$

For all positive integers k (because $e^{i\alpha_1} \cdot e^{i\alpha_2} = e^{i(\alpha_1+\alpha_2)}$) we have

$$\cos(\theta+k\alpha) + i\sin(\theta+k\alpha)$$
$$= (\cos\theta + i\sin\theta)(\cos k\alpha + i\sin k\alpha) \tag{6}$$
$$= (\cos\theta + i\sin\theta)(\cos\alpha + i\sin\alpha)^k$$

Hence
$$\cos(\theta+k\alpha) + i\sin(\theta+k\alpha) = zw^k \tag{7}$$

and eq.(3) becomes

$$C_n + i S_n = a_0 z + a_1 zw + a_2 zw^2 + \ldots + a_{n-1} zw^{n-1}$$
$$= z(a_0 + a_1 w + a_2 w^2 + \ldots + a_{n-1} w^{n-1}) \tag{8}$$

Finding the value of $a_0 + a_1 w + a_2 w^2 + \ldots + a_{n-1} w^{n-1}$, multiplying it by z, and finally comparing real and imaginary parts, we can compute C_n and S_n from eq.(8).

For $a_0 = a_1 = \ldots = a_{n-1} = 1$, eq.(8) becomes

$$C_n + i S_n = z(1 + w + w^2 + \ldots + w^{n-1}) \quad (9)$$

If
$$A = 1 + w + w^2 + \ldots + w^{n-1} \quad (10)$$

then
$$wA = w + w^2 + \ldots + w^n \quad (11)$$

Subtracting eq.(11) from eq.(10) results in
$$A(1-w) = 1 - w^n \quad (12)$$

If $w \neq 1$, then
$$A = \frac{1 - w^n}{1 - w} \quad (13)$$

For $w = 1$, $A = n$.

Eq.(9) yields
$$C_n + i S_n = \begin{cases} zn & \text{for } w = 1 \\ \frac{z(1-w^n)}{1-w} & \text{for } w \neq 1 \end{cases} \quad (14)$$

If $w = \cos\alpha + i\sin\alpha = 1$, then $\sin\alpha = 0$, $\cos\alpha = 1$. Thus, $\alpha = 2k\pi$, where k is an integer. We obtain

$$\sin\frac{\alpha}{2} = 0 \quad (15)$$

We shall consider separately cases when $\sin\frac{\alpha}{2} = 0$ and $\sin\frac{\alpha}{2} \neq 0$.

I. For $\sin\frac{\alpha}{2} = 0$:

This condition implies that
$$w = \cos\alpha + i\sin\alpha = 1 \quad (16)$$

and
$$C_n + i S_n = zn = n(\cos\theta + i\sin\theta) \quad (17)$$

Equating real and imaginary parts yields
$$C_n = n\cos\theta \quad (18)$$
$$S_n = n\sin\theta \quad (19)$$

II. For $\sin\frac{\alpha}{2} \neq 0$, we have

$$C_n + i S_n = \frac{z(1-w^n)}{1-w} = \frac{(z-zw^n)(1-\bar{w})}{(1-w)(1-\bar{w})}$$

$$= \frac{z - z\bar{w} - zw^n + z\bar{w}w^n}{1 - \bar{w} - w + w\bar{w}} \tag{20}$$

Note that

$$w + \bar{w} = 2\cos\alpha$$
$$w\bar{w} = 1 \tag{21}$$

and

$$C_n + iS_n = \frac{z - z\bar{w} - zw^n + zw^{n-1}}{2 - 2\cos\alpha} \tag{22}$$

Furthermore,

$$z\bar{w} = (\cos\theta + i\sin\theta)(\cos\alpha - i\sin\alpha)$$
$$= \cos(\theta-\alpha) + i\sin(\theta-\alpha) \tag{23}$$

and eq.(22) becomes

$$C_n + iS_n = \frac{(\cos\theta + i\sin\theta) - \cos(\theta-\alpha) - i\sin(\theta-\alpha) - \cos(\theta+n\alpha)}{2 - 2\cos\alpha}$$
$$+ \frac{-i\sin(\theta+n\alpha) + \cos[\theta+(n-1)\alpha] + i\sin[\theta+(n-1)\alpha]}{2 - 2\cos\alpha} \tag{24}$$

Equating real and imaginary parts, we find

$$C_n = \frac{\cos\theta - \cos(\theta-\alpha) - \cos(\theta+n\alpha) + \cos[\theta+(n-1)\alpha]}{2 - 2\cos\alpha} \tag{25}$$

$$S_n = \frac{\sin\theta - \sin(\theta-\alpha) - \sin(\theta+n\alpha) + \sin[\theta+(n-1)\alpha]}{2 - 2\cos\alpha} \tag{26}$$

Utilizing the identities

$$2 - 2\cos\alpha = 4\sin^2\frac{\alpha}{2} \tag{27}$$

$$\cos\theta - \cos(\theta-\alpha) = -2\sin(\theta - \frac{\alpha}{2})\sin\frac{\alpha}{2} \tag{28}$$

$$\cos(\theta+n\alpha) - \cos[\theta+(n-1)\alpha] = -2\sin[\theta+(n-\frac{1}{2})\alpha]\sin\frac{\alpha}{2} \tag{29}$$

we reduce eq.(25) to

$$C_n = \frac{-2\sin(\theta - \frac{\alpha}{2}) + 2\sin[\theta+(n-\frac{1}{2})\alpha]}{4\sin\frac{\alpha}{2}}$$
$$= \frac{\cos[\theta+(n-1)\frac{\alpha}{2}]\sin\frac{n\alpha}{2}}{\sin\frac{\alpha}{2}} \tag{30}$$

Similarly, S_n can be simplified to

$$S_n = \frac{\sin[\theta+(n-1)\frac{\alpha}{2}]\sin\frac{n\alpha}{2}}{\sin\frac{\alpha}{2}} \tag{31}$$

• **PROBLEM 3-49**

Evaluate the sum

$$A = 1 + x\cos\theta + x^2\cos 2\theta + \ldots + x^n \cos n\theta \qquad (1)$$

Hint: Use the method described in Problem 3-48.

Solution: We have to find C_n and S_n such that $C_n + i\,S_n$ gives a series which we can sum. Let

$$C_n = x\cos\theta + x^2\cos 2\theta + \ldots + x^n \cos n\theta \qquad (2)$$

$$S_n = x\sin\theta + x^2\sin 2\theta + \ldots + x^n \sin n\theta \qquad (3)$$

Thus

$$C_n + 1 = A$$

From eqs.(2) and (3), we obtain

$$C_n + i\,S_n = x(\cos\theta + i\sin\theta) + x^2(\cos 2\theta + i\sin 2\theta)$$
$$+ \ldots + x^n(\cos n\theta + i\sin n\theta) \qquad (4)$$

$$= x(\cos\theta + i\sin\theta) + [x(\cos\theta + i\sin\theta)]^2 + \ldots +$$
$$+ [x(\cos\theta + i\sin\theta)]^n = z + z^2 + \ldots + z^n$$

where

$$z = x(\cos\theta + i\sin\theta) \qquad (5)$$

We have

$$C_n + i\,S_n = z + z^2 + \ldots + z^n \qquad (6)$$

$$z(C_n + i\,S_n) = z^2 + z^3 + \ldots + z^{n+1} \qquad (7)$$

Subtracting eq.(7) from eq.(6), we get

$$C_n + i\,S_n = \frac{z - z^{n+1}}{1 - z}$$

$$= \frac{x(\cos\theta + i\sin\theta) - x^{n+1}[\cos(n+1)\theta + i\sin(n+1)\theta]}{1 - x(\cos\theta + i\sin\theta)} \qquad (8)$$

The sum should be expressed in the form $\alpha + i\beta$.

$$C_n + i\,S_n =$$

$$= \frac{[x\cos\theta + ix\sin\theta - x^{n+1}\cos(n+1)\theta - ix^{n+1}\sin(n+1)\theta][(1-x\cos\theta) + ix\sin\theta]}{[(1-x\cos\theta) + ix\sin\theta][(1-x\cos\theta) - ix\sin\theta]} \qquad (9)$$

Thus

$$C_n = \frac{x\cos\theta(1-x\cos\theta)-x^2\sin^2\theta-x^{n+1}\cos(n+1)\theta\cdot(1-x\cos\theta)+x^{n+2}\sin\theta\cdot\sin(n+1)\theta}{(1-x\cos\theta)^2+x^2\sin^2\theta} \quad (10)$$

Finally

$$A = 1 + C_n$$

$$= \frac{1-x\cos\theta-x^{n+1}\cos(n+1)\theta\cdot(1-x\cos\theta)+x^{n+2}\sin\theta\cdot\sin(n+1)\theta}{1 - 2x\cos\theta + x^2} \quad (11)$$

Note that

$$\cos\theta\cdot\cos(\theta+n\theta) + \sin\theta\cdot\sin(\theta+n\theta) = \cos n\theta \quad (12)$$

and eq.(11) can be simplified to

$$A = \frac{1-x\cos\theta-x^{n+1}\cos(n+1)\theta+x^{n+2}\cos n\theta}{1 - 2x\cos\theta + x^2} \quad (13)$$

• **PROBLEM** 3-50

Prove that

1. $\displaystyle\sum_{k=0}^{n} \cos k\theta = \frac{1}{2} + \frac{\sin[(n+\frac{1}{2})\theta]}{2\sin\frac{\theta}{2}}$ \quad (1)

2. $\displaystyle\sum_{k=0}^{n} \sin k\theta = \frac{1}{2}\cot\frac{\theta}{2} - \frac{\cos[(n+\frac{1}{2})\theta]}{2\sin\frac{\theta}{2}}$ \quad (2)

where $0 < \theta < 2\pi$.

Hint: Use the formula

$$1 + z + z^2 + \ldots + z^n = \frac{1-z^{n+1}}{1-z} \quad (3)$$

Solution: We can use the results of Problem 3.48 to compute both sums. Observe that

$$\sum_{k=0}^{n} \cos k\theta + i \sum_{k=0}^{n} \sin k\theta$$

$$= \sum_{k=0}^{n} (\cos k\theta + i \sin k\theta) = \sum_{k=0}^{n} (\cos\theta + i \sin\theta)^k$$

$$= \frac{1-(\cos\theta + i\sin\theta)^{n+1}}{1-\cos\theta - i\sin\theta}$$

$$= \frac{[1-(\cos\theta + i\sin\theta)^{n+1}][1-\cos\theta + i\sin\theta]}{[1-\cos\theta - i\sin\theta][1-\cos\theta + i\sin\theta]} \quad (4)$$

$$= \frac{1}{2} + \frac{\sin[(n+1)\theta]\cos\frac{\theta}{2} - \cos[(n+1)\theta]\sin\frac{\theta}{2}}{2\sin\frac{\theta}{2}}$$

$$+ i\left[\frac{1}{2}\cot\frac{\theta}{2} - \frac{\sin[(n+1)\theta]\sin\frac{\theta}{2} + \cos[(n+1)\theta]\cos\frac{\theta}{2}}{2\sin\frac{\theta}{2}}\right]$$

Comparing real and imaginary parts and using some basic trigonometric identities we obtain

$$\sum_{k=0}^{n}\cos k\theta = \frac{1}{2} + \frac{\sin[(n+\frac{1}{2})\theta]}{2\sin\frac{\theta}{2}} \tag{5}$$

$$\sum_{k=0}^{n}\sin k\theta = \frac{1}{2}\cot\frac{\theta}{2} - \frac{\cos[(n+\frac{1}{2})\theta]}{2\sin\frac{\theta}{2}} \tag{6}$$

• **PROBLEM** 3-51

If n is a positive integer, show that

$$\cos n\theta = P_n(\cos\theta) \tag{1}$$

where $P_n(x)$ is an nth degree polynomial of x. Compute the polynomials for odd and even positive integers.

Solution: Let us start with eq.(5) of Problem 3-46.

$$\cos n\theta = \sum_{s=0}^{\ell}(-1)^s\binom{n}{2s}\cos^{n-2s}\theta\sin^{2s}\theta \tag{2}$$

$$\ell = \frac{(2n-1) + (-1)^n}{4} \tag{3}$$

We can substitute

$$\sin^2\theta = 1 - \cos^2\theta \tag{4}$$

into eq.(2)

$$\cos n\theta = \sum_{s=0}^{\ell}(-1)^s\binom{n}{2s}\cos^{n-2s}\theta(1-\cos^2\theta)^s \tag{5}$$

Thus, we expressed $\cos n\theta$ in the polynomial form of $\cos\theta$

$$\cos n\theta = P_n(\cos\theta).$$

Eq.(5) is an nth degree polynomial.

Let n be an even integer. Then, applying binomial expansion to $(1-\cos^2\theta)^s$ and substituting into eq.(5), we obtain

$$\cos n\theta = (-1)^\ell \cos^{n-2\ell}\theta + \sum_{s=2}^{\ell+1} \frac{(-1)^{\ell+s-1} \, 2^{2s-4} \, n}{s-1} \binom{\frac{n}{2}+s-2}{2s-3} \cos^{n+2s-2\ell-2}\theta \quad (6)$$

In the same manner, we can find an expression for $\cos n\theta$ for odd integers n.

$$\cos n\theta = \sum_{s=1}^{\ell+1} \frac{(-1)^{\ell+s-1} \, 2^{2s-2} \, n}{2s-1} \binom{\frac{n}{2}+s-\frac{3}{2}}{2s-2} \cos^{n+2s-2\ell-2}\theta \quad (7)$$

For example, for n = 6, we obtain from eqs.(6) and (3)

$$\ell = \frac{(2 \cdot 6) - 1 + (-1)^6}{4} = 3 \quad (8)$$

$$\cos 6\theta = (-1)^3 \cos^0\theta$$

$$+ \sum_{s=2}^{4} \frac{(-1)^{2+s} \, 2^{2s-4} \cdot 6}{s-1} \binom{s+1}{2s-3} \cos^{2s-2}\theta$$

$$= -1 + 6\binom{3}{1}\cos^2\theta - 48\cos^4\theta + 32\cos^6\theta \quad (9)$$

$$= 32\cos^6\theta - 48\cos^4\theta + 18\cos^2\theta - 1$$

• **PROBLEM** 3-52

The Chebyshev* polynomials are defined by

$$T_n(x) = 2^{1-n} \cos n\theta \qquad n = 1, 2, \ldots \quad (1)$$

where $x = \cos\theta$. Show that if

1. n is a positive even integer, then

$$T_n(x) = (-1)^{\frac{n}{2}} 2^{1-n} + \frac{n}{2^{n+3}} \sum_{s=2}^{\frac{n}{2}+1} \frac{(-1)^{\frac{n}{2}+s-1} 4^s \binom{\frac{n}{2}+s-2}{2s-3}}{s-1} x^{2s-2} \quad (2)$$

2. n is a positive odd integer, then

$$T_n(x) = \frac{n}{2^{n+1}} \sum_{s=1}^{\frac{n+1}{2}} \frac{(-1)^{\frac{n+1}{2}+s} 4^s \binom{\frac{n}{2}+s-\frac{3}{2}}{2s-2}}{2s-1} x^{2s-1} \quad (3)$$

Verify
$$T_1(x) = x$$
$$T_2(x) = \frac{1}{2}(2x^2 - 1)$$

$$T_3(x) = \frac{1}{4}(4x^3-3x) \tag{4}$$

$$T_4(x) = \frac{1}{8}(8x^4-8x^2+1)$$

$$T_5(x) = \frac{1}{16}(16x^5-20x^3+5x)$$

<u>Solution</u>: 1. From Problem 3-51 we take the expression for $\cos n\theta$, when n is an even integer

$$T_n(x) = 2^{1-n}\left[(-1)^k x^{n-2k} + \sum_{s=2}^{k+1} \frac{(-1)^{k+s-1} 2^{2s-4}}{s-1} n \binom{\frac{n}{2}+s-2}{2s-3} x^{n+2s-2k-2}\right] \tag{5}$$

Since n is even

$$k = \frac{2n - 1 + (-1)^n}{4} = \frac{2n}{4} = \frac{n}{2} \tag{6}$$

Hence

$$T_n(x) = 2^{1-n}(-1)^{\frac{n}{2}} + \sum_{s=2}^{\frac{n}{2}+1} \frac{2^{2s-4} \, 2^{1-n} n (-1)^{\frac{n}{2}+s-1}}{s-1} \binom{\frac{n}{2}+s-2}{2s-3} x^{2s-2}$$

$$= 2^{1-n}(-1)^{\frac{n}{2}} + \frac{n}{2^{n+3}} \sum_{s=2}^{\frac{n}{2}+1} \frac{4^s (-1)^{\frac{n}{2}+s-1}}{s-1} \binom{\frac{n}{2}+s-2}{2s-3} x^{2s-2} \tag{7}$$

From eq.(7), for n = 2

$$T_2(x) = \left[2^{-1} \cdot (-1)\right] + \frac{2}{2^5} \cdot \frac{4^2 \cdot (-1)^2}{2-1} \binom{1}{1} x^2 \tag{8}$$

$$= -\frac{1}{2} + x^2 = \frac{1}{2}(2x^2-1)$$

For n = 4, we obtain

$$T_4(x) = 2^{-3}(-1)^2 + \frac{4}{2^7} \sum_{s=2}^{3} \frac{4^s(-1)^{s+1}}{s-1} \binom{s}{2s-3} x$$

$$= \frac{1}{8} + \frac{1}{2^5}\left[\frac{4^2 \cdot (-1)^3}{2-1}\binom{2}{1}x^2 + \frac{4^3(-1)^4}{2}\binom{3}{3}x^4\right] \tag{9}$$

$$= \frac{1}{8} + \frac{1}{2^5}\left[-32x^2 + 32x^4\right] = \frac{1}{8}\left[8x^4 - 8x^2 + 1\right]$$

2. Again, from Problem 3-51, we have for an odd integer n

$$T_n(x) = 2^{1-n} \cos n\theta$$

$$= 2^{1-n} \sum_{s=1}^{k+1} \frac{(-1)^{k+s-1} n \, 2^{2s-2}}{2s-1} \binom{\frac{n}{2}+s-\frac{3}{2}}{2s-2} x^{n+2s-2k-2} \tag{10}$$

Since n is odd

$$k + 1 = \frac{2n-1+(-1)^n}{4} + 1 = \frac{n+1}{2} \tag{11}$$

and we have

$$T_n(x) = \frac{n}{2^{n+1}} \sum_{s=1}^{\frac{n+1}{2}} \frac{4^s(-1)^{\frac{n+1}{2}+s}}{2s-1} \binom{\frac{n}{2}+s-\frac{3}{2}}{2s-2} x^{2s-1} \tag{12}$$

From eq.(12) we compute

$$T_1(x) = \frac{1}{2^2} \frac{4 \cdot (-1)^2}{2-1} \binom{0}{0} x = x \tag{13}$$

Remember that $0! \equiv 1$.

$$T_3(x) = \frac{3}{2^4} \sum_{s=1}^{2} \frac{4^s(-1)^{s+2}}{2s-1} \binom{s}{2s-2} x^{2s-1}$$

$$= \frac{3}{2^4} \left[-4x + \frac{16}{3} x^3 \right] = \frac{1}{4}(4x^3 - 3x) \tag{14}$$

$$T_5(x) = \frac{5}{2^6} \sum_{s=1}^{3} \frac{(-1)^{s+3} 4^s}{2s-1} \binom{s+1}{2s-2} x^{2s-1}$$

$$= \frac{1}{16}(16x^5 - 20x^3 + 5x) \tag{15}$$

*P.L. Chebyshev (1821-1894), Russian mathematician, contributed to the theory of approximation of functions. This was later developed into the constructive theory of functions, mostly developed by S. Bernstein.

Chebyshev gave a complete answer to the question of expressing the integral

$$\int x^n (a+bx^m)^k \, dx$$

where n,m,k are rational numbers in terms of elementary functions.

● **PROBLEM 3-53**

Consider a sequence of numbers

$$a_n = \alpha_n^2 + \beta_n^2, \quad n = 1, 2, 3, \ldots \tag{1}$$

where α_n and β_n are positive integers. Prove that for every positive integer k there exist positive integers α and β such that

$$a_1 \cdot a_2 \cdot \ldots \cdot a_k = \alpha^2 + \beta^2 \qquad (2)$$

For example, let k = 3 and let

$$a_1 = 1^2 + 2^2 = 5$$
$$a_2 = 1^2 + 3^2 = 10$$
$$a_3 = 2^2 + 2^2 = 8$$

Then $\qquad a_1 \cdot a_2 \cdot a_3 = 400 \qquad (3)$

Indeed two numbers α and β exist such that

$$400 = \alpha^2 + \beta^2 = 16^2 + 12^2 \qquad (4)$$

Solution: Let k be a positive integer and let

$$a_1 \cdot a_2 \cdot \ldots \cdot a_k \qquad (5)$$

be the product of

$$\begin{aligned} a_1 &= \alpha_1^2 + \beta_1^2 \\ a_2 &= \alpha_2^2 + \beta_2^2 \\ &\vdots \\ a_k &+ \alpha_k^2 + \beta_k^2 \end{aligned} \qquad (6)$$

where $\alpha_1, \ldots, \alpha_k, \beta_1, \ldots, \beta_k$ are positive integers.

Form the sequence of complex numbers such that

$$\begin{aligned} z_1 &= \alpha_1 + i\beta_1 \\ z_2 &= \alpha_2 + i\beta_2 \\ &\vdots \\ z_k &= \alpha_k + \beta_k \end{aligned} \qquad (7)$$

Note that

$$z_1 \cdot \bar{z}_1 = a_1, \ldots, z_k \cdot \bar{z}_k = a_k \qquad (8)$$

Let z' be a complex number such that

$$\begin{aligned} z' &= z_1 \cdot z_2 \cdot \ldots z_k \\ &= (\alpha_1 + i\beta_1) \cdot (\alpha_2 + i\beta_2) \cdot \ldots \cdot (\alpha_k + i\beta_k) = r + is \end{aligned} \qquad (9)$$

Because all $\alpha_1, \ldots, \alpha_k, \beta_1, \ldots, \beta_k$ are positive integers, r and s must be integers. Hence

$$\begin{aligned} a_1 \cdot a_2 \cdot \ldots \cdot a_k &= z_1 \cdot \bar{z}_1 \cdot z_2 \cdot \bar{z}_2 \cdot \ldots \cdot z_k \bar{z}_k \\ &= (z_1 \cdot z_2 \cdot \ldots z_k)(\overline{z_1 \cdot z_2 \cdot \ldots z_k}) \qquad (10) \\ &= (r+is)(\overline{r+is}) = r^2 + s^2 \end{aligned}$$

q.e.d.

PROBLEM 3-54

Let Z be the set of complex numbers

$$Z \ni z = x + iy \tag{1}$$

and A be the set of matrices of the form

$$A \ni a = \begin{pmatrix} x & y \\ -y & x \end{pmatrix} \tag{2}$$

A one-to-one correspondence between Z and A can be established as follows

$$x + iy = z \longleftrightarrow a = \begin{pmatrix} x & y \\ -y & x \end{pmatrix} \tag{3}$$

We denote this correspondence by

$$a = z^* \tag{4}$$

Now, let

$$z_k = x_k + iy_k \quad , \quad k = 1,2 \tag{5}$$

and

$$a_k = \begin{pmatrix} x_k & y_k \\ -y_k & x_k \end{pmatrix}, \quad k = 1,2 \tag{6}$$

We define

$$a_1 + a_2 = \begin{pmatrix} x_1+x_2 & y_1+y_2 \\ -y_1-y_2 & x_1+x_2 \end{pmatrix} \tag{7}$$

$$a_1 a_2 = \begin{pmatrix} x_1 x_2 - y_1 y_2 & x_1 y_2 + x_2 y_1 \\ -x_1 y_2 - x_2 y_1 & x_1 x_2 - y_1 y_2 \end{pmatrix} \tag{8}$$

for α real,

$$\alpha a = \begin{pmatrix} \alpha x & \alpha y \\ -\alpha y & \alpha x \end{pmatrix} \tag{9}$$

Show that

$$(z_1 + z_2)^* = z_1^* + z_2^* \tag{10}$$

$$(z_1 z_2)^* = z_1^* z_2^* \tag{11}$$

$$(\alpha z)^* = \alpha z^* \tag{12}$$

The complex numbers form a linear algebra of order two over the field of real numbers. Also, the matrices of the type $\begin{pmatrix} x & y \\ -y & x \end{pmatrix}$ form a linear algebra of order two.

Eqs.(10), (11) and (12) show that the linear algebra Z is isomorphic to the linear algebra A.

Solution: We have

$$(z_1+z_2)^* = \left[(x_1+iy_1) + (x_2+iy_2)\right]^*$$

$$= \left[(x_1+x_2) + i(y_1+y_2)\right]^* = \begin{pmatrix} x_1+x_2 & y_1+y_2 \\ -y_1-y_2 & x_1+x_2 \end{pmatrix} \quad (13)$$

$$= \begin{pmatrix} x_1 & y_1 \\ -y_1 & x_1 \end{pmatrix} + \begin{pmatrix} x_2 & y_2 \\ -y_2 & x_2 \end{pmatrix} = z_1^* + z_2^*$$

$$(z_1z_2)^* = \left[(x_1+iy_1) \cdot (x_2+iy_2)\right]^*$$

$$= \left[(x_1x_2-y_1y_2) + i(x_1y_2+x_2y_1)\right]^*$$

$$= \begin{pmatrix} x_1x_2-y_1y_2 & x_1y_2+x_2y_1 \\ -x_1y_2-x_2y_1 & x_1x_2-y_1y_2 \end{pmatrix} \quad (14)$$

$$= \begin{pmatrix} x_1 & y_1 \\ -y_1 & x_1 \end{pmatrix}\begin{pmatrix} x_2 & y_2 \\ -y_2 & x_2 \end{pmatrix} = z_1^* z_2^*$$

$$(\alpha z)^* = \left[\alpha(x+iy)\right]^* = \left[\alpha x+i\alpha y\right]^*$$

$$= \begin{pmatrix} \alpha x & \alpha y \\ -\alpha y & \alpha x \end{pmatrix} = \alpha\begin{pmatrix} x & y \\ -y & x \end{pmatrix} = \alpha z^* \quad (15)$$

CHAPTER 4

COMPLEX NUMBERS AS METRIC SPACE

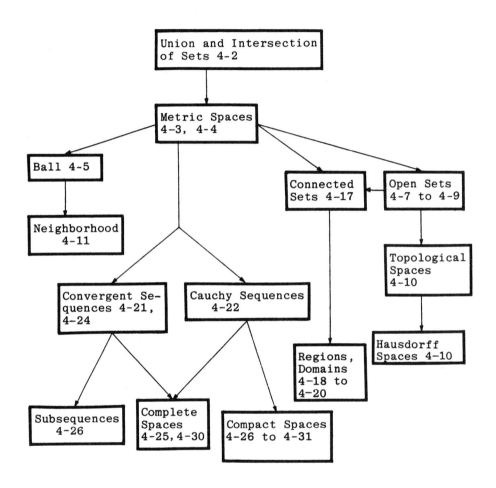

Logical Notation supports the whole structure but it is not crucial for its existence.

This chart is provided to facilitate rapid understanding of the interrelationships of the topics and subject matter in this chapter. Also shown are the problem numbers associated with the subject matter.

ELEMENTS OF SET THEORY

• **PROBLEM** 4-1

In mathematical logic, each sentence is either true or false. Certain operations can be performed on sentences. As a result, new sentences are obtained. In this problem, we shall introduce the basic elements of logical notation and the fundamental laws of logic.

Throughout this book we shall use the following symbols:

$\vee$ read as : or

$\wedge$ read as : and

$\bigvee\limits_{x}$ read as : there exists an x such that

$\bigwedge\limits_{x}$ read as : for all x there follows

$\neg$ read as : not

$\Rightarrow$ read as : if,...,then

$\Leftrightarrow$ read as : if and only if

$\equiv$ read as : equal by definition

Some of these symbols were explained earlier in the book.

The proof by contradiction (called reductio ad absurdum) is based on the law of contraposition:

$$(\alpha \Rightarrow \beta) \Leftrightarrow (\neg \beta \Rightarrow \neg \alpha) \qquad (1)$$

DeMorgan's* laws (called the rule of negation) can be written as

$$\neg(\alpha \wedge \beta) \Leftrightarrow (\neg \alpha \vee \neg \beta) \qquad (2)$$

$$\neg(\alpha \vee \beta) \Leftrightarrow (\neg \alpha \wedge \neg \beta) \qquad (3)$$

where α and β are sentences.

Choose two sentences of your choice and explain eq.(2).

In a similar way, we obtain

$$\neg \bigwedge\limits_{x} F \Leftrightarrow \bigvee\limits_{x} \neg F \qquad (4)$$

$$\neg \bigvee\limits_{x} F \Leftrightarrow \bigwedge\limits_{x} \neg F \qquad (5)$$

$$\neg \bigvee\limits_{x} \neg F \Leftrightarrow \bigwedge\limits_{x} F \qquad (6)$$

$$\neg \bigwedge\limits_{x} \neg F \Leftrightarrow \bigvee\limits_{x} F \qquad (7)$$

Solution: Let n be a natural number, $n \in N$. Sentence α is: n is greater than or equal to 1, $n \geq 1$.

Sentence β is: n is less than or equal to 2, $n \leq 2$.

$\alpha \wedge \beta$ is: n is greater than or equal to one and n is less than or equal to two, which is equivalent to: n is equal to one or two.

The negation of it is

$\neg(\alpha \wedge \beta)$, which is: n is different from one and from two. Since n is a natural number, we can say: n is greater than two.

Now we shall consider $\neg\alpha \vee \neg\beta$.

$\neg\alpha$ is: n is less than one.

$\neg\beta$ is: n is greater than two.

Thus

$\neg\alpha \vee \neg\beta$ is: n is less than one or n is greater than two, which is equivalent to n is greater than two.

*Augustus DeMorgan (born at Madura, India, 1806, died in London, 1871) was a brilliant and eccentric English mathematician known for his wide range of interests and lack of persistence.

Educated at Trinity College, Cambridge, DeMorgan became professor of mathematics at the then newly-established University of London. In 1849, he published Trigonometry and Double Algebra, a book which contained some elements of quaternions. Like with many other theories, DeMorgan didn't follow this one to any major conclusions. His books may be characterized as a labyrinth of mathematical knowledge compiled by a profound and disorganized mathematician. DeMorgan's main contributions are in the field of logic and probability. He is also the author of Budget or Paradoxes, a mathematical satire, or rather satirical mathematics edited by his wife after his death.

• **PROBLEM 4-2**

An open interval (a,b) is the set of all real numbers t such that $a < t < b$. The numbers a and b are called the boundary points of the interval.

A closed interval, denoted [a,b] is the set of all real numbers t such that $a \leq t \leq b$. If $(A_\omega)_{\omega \in \Omega}$ is a family of sets, then the union of sets is defined by

$$x \in \bigcup_{\omega \in \Omega} A_\omega \iff \bigvee_{\omega \in \Omega} x \in A_\omega \quad (1)$$

where Ω is an arbitrary set of indices

If $\Omega = \{1,2\}$, then
$$x \in A_1 \cup A_2 \iff \bigvee_{i \in \{1,2\}} x \in A_i \tag{2}$$

The intersection is defined by
$$x \in \bigcap_{\omega \in \Omega} A_\omega \iff \bigwedge_{\omega \in \Omega} x \in A_\omega \tag{3}$$

An open interval in the complex plane $I = (a_1, a_2; b_1, b_2)$ is the set of all numbers $x = a + ib$ such that
$$\begin{aligned} a_1 < a < a_2 \\ b_1 < b < b_2 \end{aligned} \tag{4}$$

as shown in Fig. 1.

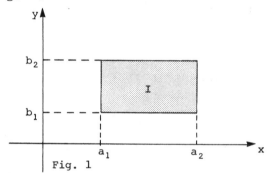

Fig. 1

The boundary of an open interval is the set of all points on the sides of the rectangle.

An open interval with its boundary is called a closed interval.

Find the union and intersection of the sets $A_n = \left[\frac{1}{n}, 3 - \frac{1}{n}\right]$, $n \in N$. $\tag{5}$

Find the intersection and union of the two open intervals in the complex plane
$$\begin{aligned} J_1 = (2,3;1,2) \\ J_2 = (3,4;1,2) \end{aligned} \tag{6}$$

Solution: The union is
$$\bigcup_{n \in N} \left[\frac{1}{n}, 3 - \frac{1}{n}\right] = (0,3) \tag{7}$$

The intersection is
$$\bigcap_{n \in N} \left[\frac{1}{n}, 3 - \frac{1}{n}\right] = [1,2] \tag{8}$$

Fig. 2 shows both intervals I_1 and I_2.

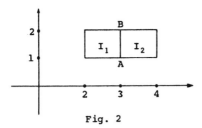

Fig. 2

Since both intervals are open, segment AB does not belong to I_1, nor does it belong to I_2. Thus, the intersection is the empty (void) set

$$I_1 \cap I_2 = \phi \tag{9}$$

The union is an open interval consisting of I_1 and I_2 with the segment AB removed

$$I_1 \cup I_2 = (2,4;1,2) - AB \tag{10}$$

METRIC SPACES

• **PROBLEM 4-3**

1. Show that the complex space with the metric defined as

$$d(z_1,z_2) \equiv |z_1 - z_2| \tag{1}$$

forms a metric space.

We shall denote the complex space by $(C, 1 \cdot 1)$.

2. Consider the transformation of complex space onto itself

$$z = x + iy \longrightarrow z' = 3x - iy \tag{2}$$

Is this transformation isometric?

<u>Solution</u>: 1. Let X be a set and d a non-negative function defined on the Cartesian* product

$X \times X$ (R^t is the set of all real numbers $t \geq 0$)

$$d : X \times X \longrightarrow R^t \tag{3}$$

The Cartesian product of sets A and B, denoted by $A \times B$ is

$$A \times B \equiv \{(a,b) : a \in A \wedge b \in B\} \tag{4}$$

where (a,b) is an ordered pair.

A function d satisfying the conditions

1. $d(x,y) = d(y,x)$ \hfill (5)

2. $d(x,y) \leq d(x,z) + d(z,y)$ \hfill (6)

3. $(d(x,y) = 0) \Longleftrightarrow (x = y)$ (7)

is called a distance, or a metric. A pair (X,d) is called a metric space. The following are examples of metric spaces

$X = R^1$, and $d(x,y) \equiv |x - y|$ (8)

$X = R^2$, and $d(x,y) \equiv \sqrt{(x_1-y_1)^2+(x_2-y_2)^2}$ (9)

or, for general $X = R^n$,

$$d(x,y) \equiv \sqrt{\sum_{j=1}^{n} (x_j-y_j)^2}$$ (10)

where

$x = (x_1, x_2, \ldots, x_n), \; y = (y_1, \ldots, y_n)$ (11)

Now, to show that the complex space is a metric space, let

$$\begin{aligned} z_1 &= x_1 + iy_1 \\ z_2 &= x_2 + iy_2 \end{aligned}$$ (12)

Then

$d(z_1, z_2) \equiv |z_1-z_2| = \sqrt{(x_1-x_2)^2+(y_1-y_2)^2}$ (13)

We have

$|z_1-z_2| = |z_2-z_1|$ (14)

$|z_1-z_2| \leq |z_1-z_3| + |z_3-z_2|$ (15)

$|z_1-z_2| = 0$ if and only if $z_1 = z_2$ (16)

Thus $(C, 1 \cdot 1)$ is a metric space.

2. Let (X_1, d_1) and (X_2, d_2) be two metric spaces and let

$f : X_1 \longrightarrow X_2$ (17)

f is an isometry if

$d_1(x_1, x_2) = d_2(f(x_1), f(x_2))$ (18)

(see Fig. 1).

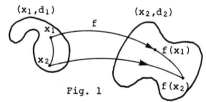

Fig. 1

The distance between x_1 and x_2, $d_1(x_1, x_2)$ is equal to the distance $d_2(f(x_1), f(x_2))$ between $f(x_1)$ and $f(x_2)$. Consider two numbers

$$z_1 = x_1+iy_1 \quad \text{and} \quad z_2 = x_2+iy_2 \tag{19}$$

Under the transformation given by eq.(2), z_1 becomes z_1',

$$z_1' = 3x_1 - iy_1 \tag{20}$$

and z_2 becomes z_2',

$$z_2' = 3x_2 - iy_2 \tag{21}$$

By comparing the distances

$$d(z_1,z_2) = |z_1-z_2| = |(x_1-x_2)+i(y_1-y_2)| \\ = \sqrt{(x_1-x_2)^2+(y_1-y_2)^2} \tag{22}$$

$$d(z_1',z_2') = |z_1'-z_2'| = |(3x_1-3x_2)+i(-y_1+y_2)| \\ = \sqrt{(3x_1-3x_2)^2+(y_1-y_2)^2} \tag{23}$$

we conclude that the transformation given by eq.(2) is not isometric.

Show that

$$f : z \longrightarrow \bar{z} \tag{24}$$

is isometric. Can you find some examples of isometric transformations of complex space?

● **PROBLEM 4-4**

Any set X to be promoted to the metric space has to have metric. There are many ways of defining the metric of a given set. Hence, from the same set we can obtain many different metric spaces.

Let X = C be the set of complex numbers. Show that (C,d) is a metric space when

1. $d(z_1,z_2) = |z_1-z_2|$ (1)

2. $d(z_1,z_2) = |x_1-x_2| + |y_1-y_2|$ (2)

3. $d(z_1,z_2) = \max\{|x_1-x_2|,|y_1-y_2|\}$ (3)

Here, $z_1 = x_1+iy_1$, $z_2 = x_2+iy_2$, and $z_1, z_2 \in C$.

4.
$$d(z_1,z_2) = \begin{cases} 0 & \text{when } z_1=z_2 \\ 1 & \text{when } z_1 \neq z_2 \end{cases} \tag{4}$$

Distance so defined is called discrete metric.

Solution: 1. It was shown in the previous problem that $|z_1-z_2|$ is a metric.

2. $$d(z_1,z_2) = |x_1-x_2| + |y_1-y_2| \\ = |x_2-x_1| + |y_2-y_1| = d(z_2,z_1) \tag{5}$$

$$d(z_1,z_2) = |x_1-x_2|+|y_1-y_2| \leq |x_1-x_3|+|x_3-x_2| \quad (6)$$
$$+ |y_1-y_3| + |y_3-y_2| = d(z_1,z_3) + d(z_3,z_2)$$

If $d(z_1,z_2) = 0$ then
$$|x_1-x_2| + |y_1-y_2| = 0 \text{ and } x_1 = x_2, \quad y_1 = y_2, \text{ and}$$
$$z_1 = z_2.$$

On the other hand if $z_1 = z_2$ then $d(z_1,z_2) = 0$.

3.
$$d(z_1,z_2) = \max\{|x_1-x_2|,|y_1-y_2|\} \quad (7)$$
$$= \max\{|x_2-x_1|,|y_2-y_1|\} = d(z_2,z_1)$$

$$d(z_1,z_2) = \max\{|x_1-x_2|,|y_1-y_2|\}$$
$$= \max\{|x_1-x_3+x_3-x_2|,|y_1-y_3+y_3-y_2|\} \quad (8)$$
$$\leq \max\{|x_1-x_3|+|x_3-x_2|,|y_1-y_3|+|y_3-y_2|\}$$

For any four real numbers $a, b, c, d \in R'$
$$\max\{a+b,c+d\} \leq \max\{a,c\} + \max\{b,d\} \quad (9)$$

Thus
$$\max\{|x_1-x_3|+|x_3-x_2|,|y_1-y_3|+|y_3-y_2|\}$$
$$\leq \max\{|x_1-x_3|,|y_1-y_3|\} + \max\{|x_3-x_2|,|y_3-y_2|\} \quad (10)$$
$$= d(z_1,z_3) + d(z_3,z_2)$$

where $z_3 = x_3 + iy_3$.

4.
$$d(z_1,z_2) = d(z_2,z_1) \quad (11)$$
$$d(z_1,z_2) \leq d(z_1,z_3) + d(z_3,z_2) \quad (12)$$

and
$$(d(z_1,z_2) = 0) \iff (z_1 = z_2) \quad (13)$$

Hence, $d(z_1,z_2)$ defined by eq. (4) is a metric.

• **PROBLEM 4-5**

Definition I

The set
$$K(x_0,r) \equiv \{x \in X: d(x,x_0) < r\} \quad (1)$$
is called a ball of radius r and center x_0.

Fig. 1 illustrates this definition.

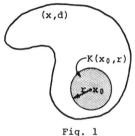

Fig. 1

Definition II

A set A of a metric space (X,d) is bounded when it is a subset of a ball.

1. Let X = N be the set of natural numbers and let

$$d(n,m) \equiv |n-m| \qquad (2)$$

(N,1·1) is a metric space.

Describe the ball K(7,11).

2. For X = R' and $d(x,y) \equiv |x-y|$ describe the ball $K(1,\frac{1}{2})$.

3. In the space (N,1·1), is the set of all even numbers bounded? Is the set of all prime numbers bounded?

4. Show that a subset of a bounded set is bounded.

5. Show that the sum of a finite number of bounded sets is bounded.

Solution: 1. By definition

$$K(7,11) \equiv \{n \in N: |7-n| < 11\} \qquad (3)$$

It can be easily seen that

$$K(7,11) = \{1,2,3,4,5,\ldots,17\} \qquad (4)$$

2.

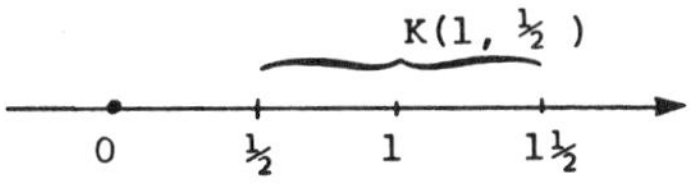

Fig. 2

The points $\frac{1}{2}$ and $1\frac{1}{2}$ do not belong to the ball $K(1,\frac{1}{2})$.

3. The set of all even numbers is not bounded. No ball exists that would contain this set. Also, the set of all prime numbers is not bounded.

4. Let A be a bounded set, then

$$A \subset K(x_0, r) \qquad (5)$$

If B is a subset of A, then

$$B \subset A \subset K(x_0, r). \tag{6}$$

Thus B is bounded.

5. Let $A_1, A_2, \ldots, A_n$ be bounded sets. Since A_1 is bounded,

$$A_1 \subset K_1(x_1, r_1) \tag{7}$$

as shown in Fig. 3.

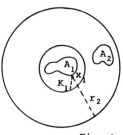

Fig. 3

Since A_2 is bounded, a ball $K_2(x_1, r_2)$ exists such that

$$A_2 \subset K_2(x_1, r_2) \tag{8}$$

Note that the center x_1 remains the same for K_1 and K_2. For each set A_k, $k=1,\ldots n$, we have

$$A_k \subset K_k(x_1, r_k) \tag{9}$$

Let

$$r = \max(r_1, r_2, \ldots, r_n) \tag{10}$$

then the ball $K(x_1, r)$ contains all $A_1, A_2, \ldots, A_n$. Hence $\bigcup_{k=1}^{n} A_k$ is bounded.

• **PROBLEM 4-6**

Definition

Let A be a subset of the real axis $A \subset R'$.

M is the least upper bound (supremum) of A (we write M = sup A) when

1. $\bigwedge_{x \in A} x \leq M$ (1)

2. $\bigwedge_{\varepsilon > 0} \bigvee_{x_1 \in A} x_1 + \varepsilon > M$ (2)

m is the greatest lower bound (infinum) of A (we write m = inf A) when

1. $\bigwedge_{x \in A} x \geq m$ (3)

2. $\bigwedge_{\varepsilon > 0} \bigvee_{x_1 \in A} m + \varepsilon > x_1$ (4)

1. Find the supremum and the infimum of the set $\{1, 2, 3\}$.

2. Show that the supremum (or infimum) does not have to be an element of the set.

3. Show that if the set has both a supremum and infimum, then it is bounded.

Solution: 1. The supremum of the set $\{1,2,3\}$ is 3. Therefore, both eq.(1) and eq.(2) are true.

$$\bigwedge_{x \in \{1,2,3\}} x \leq 3 \tag{5}$$

$$\bigwedge_{\varepsilon > 0} \bigvee_{x_1 \in \{1,2,3\}} x_1 + \varepsilon > M \tag{6}$$

For example, let $\varepsilon = 0.00001$. Then $x_1 = 3$ and

$$3 + 0.00001 > M = 3 \tag{7}$$

In the same manner, we can show that $1 = \inf\{1,2,3\}$.

2. Consider the set $\{\frac{1}{n} : n \in N\}$ which is the sequence $\{1, \frac{1}{2}, \frac{1}{3}, \frac{1}{4}, \ldots \frac{1}{1000000}, \ldots\}$. The infimum of this set is 0, but zero is not an element of the set,

$$0 \notin \{\frac{1}{n}\}. \tag{8}$$

The supremum of the set $\{1 - \frac{1}{n} : n \in N\}$ is $1 = \sup\{1 - \frac{1}{n}\}$, but one is not an element of this set. Thus, the supremum or the infimum does not necessarily have to be an element of the set.

3. Let m and M be the infimum and supremum of A, respectively. Then

$$A \subset K(m, |M - m| + 1) \tag{9}$$

Indeed, if $x \in A$ then

$x \in K(m, |M-m|+1)$, because

$$|x-m| < |M-m| + 1 \tag{10}$$

Since the set of real numbers R' is bounded and complete, the following can be proven:

A bounded set $A \subset R'$ has both an infimum and a supremum.

OPEN SETS

• PROBLEM 4-7

Let (X,d) be a metric space.

Definition

A subset A of (X,d) is an open set in X when it is empty or when

$$\bigwedge_{x \in A} \bigvee_{r > 0} K(x,r) \subset A \tag{1}$$

1. Show that $K(x_0, r_0)$ is an open set.
2. Show that the interval
$$(a,b) \equiv \{x \in R' : a < x < b\} \qquad (2)$$
is an open set.

Show that the interval
$$[a,b] \equiv \{x \in R' : a \leq x \leq b\} \qquad (3)$$
is not an open set.

3. Show that an interval given by eq.(2) in R^2 is not an open set.

Solution: 1. Let $x_1 \in K(x_0, r_0)$ that is
$$d(x_1, x_0) < r_0 \qquad (4)$$
We define the ball K'
$$K' := K'(x_1, r_0 - d(x_1, x_0)) \qquad (5)$$
as shown in Fig. 1.

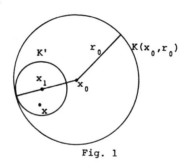

Fig. 1

We shall show that $K' \subset K$.

If $x \in K'$ then
$$d(x, x_1) < r_0 - d(x_1, x_0) \qquad (6)$$
or
$$d(x, x_1) + d(x_1, x_0) < r_0 \qquad (7)$$

Applying the triangle inequality
$$d(x, x_0) \leq d(x, x_1) + d(x_1, x_0) < r_0 \qquad (8)$$
Thus
$$x \in K(x_0, r_0). \qquad (9)$$

2. Let $x \in (a,b)$

Then
$$K(x, \min\{|x-a|, |x-b|\}) \subset (a,b) \qquad (10)$$
Thus interval (a,b) in R' is an open set.

Interval $[a,b]$ in R' is not an open set.

Taking $x = a$, no r exists such that $K(a,r) \subset [a,b]$.

3. If (a,b) is an interval as defined in eq.(2), then it is not an open set in R^2.

For $x \in (a,b)$, no ball K exists in R^2 such that $K \subset (a\ b)$.

• **PROBLEM 4-8**

Show that the set

$$\bigcup_{k=1}^{\infty} \left(-\frac{1}{k}, \frac{1}{k}\right) \tag{1}$$

is an open set in R'.

Solution: We shall prove the following theorem.

Theorem

The union of any number (cardinality) of open sets is an open set.

Let $O_\omega (\omega \in \Omega)$ be the family of open sets. Let

$$x \in \bigcup_{\omega \in \Omega} O_\omega \tag{2}$$

that is, there exists an open set O such that $x \in O_{\omega_0}$. Since O_{ω_0} is open, it contains a ball $K(x,r)$. Thus

$$K(x,r) \subset \bigcup_{\omega \in \Omega} O_\omega \tag{3}$$

q.e.d.

Each $\left(-\frac{1}{k}, \frac{1}{k}\right)$ is an open set in R'. Hence the union in eq.(1) is an open set.

• **PROBLEM 4-9**

Prove the theorem.

Theorem

The intersection (common part) of a finite number of open sets is an open set.

Solution: The theorem can be written as

$$\left(\begin{array}{c} O_k \text{ are open sets} \\ k=1,2,\ldots,s \end{array} \right) \Rightarrow \left(\bigcap_{k=1}^{s} O_k \text{ is an open set} \right) \tag{1}$$

It is enough to prove the theorem for $k = 2$.

Let O_1 and O_2 be two open sets, and let

$$x \in O_1 \cap O_2 \tag{2}$$

Since both sets are open, there exist the balls $K(x,r_1) \subset O_1$ and $K(x,r_2) \subset O_2$. Setting $r = \min(r_1, r_2)$ we obtain

$$K(x,r) \subset O_1 \cap O_2 \tag{3}$$

TOPOLOGICAL SPACES

● **PROBLEM 4-10**

Definition

Let X be a set and T a family of subsets of X, such that

1. The empty set and X belong to T, $\emptyset \in T$, $X \in T$.

2. If $O_\omega \in T$ ($\omega \in \Omega$ an indexing set) then

$$\bigcup_{\omega \in \Omega} O_\omega \in T$$

3. If $O_k \in T$ for $k = 1,\ldots,p$ then

$$\bigcap_{k=1}^{p} O_k \in T.$$

The family T is called the topology of the space X; its elements are called open sets.

The pair (X,T) is called a topological space.

Show that every metric space is a topological space.

A Hausdorff space is a topological space which has the following property

If $x_1 \neq x_2$, then there exist open sets O_1, O_2 such that

$$x_1 \in O_1, \quad x_2 \in O_2$$
$$O_1 \cap O_2 = \emptyset \tag{1}$$

Show that the metric space (X,d) is a Hausdorff space.

Solution: First we will show that a metric space (X,d) is a topological space. We take the set of all open sets in X and call it topology T of X. From the definition of an open set (see problem 4.7) we have $\emptyset \in T$, i.e., empty set is an open set. Also the whole space X is an open set.

From Problem 4-8: the union of any number of open sets is an open set. Thus, condition (2) is fulfilled.

From Problem 4-9: the intersection of a finite number of open sets is an open set. We conclude that every metric space is a topological space. Now, we will show that every metric space is a Hausdorff space. Indeed, if $x_1 \neq x_2$, then we take two open sets

$$O_1 = K(x_1, \tfrac{1}{2}d(x_1,x_2))$$
$$O_2 = K(x_2, \tfrac{1}{2}d(x_1,x_2))$$
(2)

Obviously $x_1 \in O_1$ and $x_2 \in O_2$ and

$$O_1 \cap O_2 = \phi \qquad (3)$$

We conclude that

$$\begin{pmatrix}\text{every metric}\\\text{space}\end{pmatrix} \text{ is } \begin{pmatrix}\text{Hausdorff}\\\text{space}\end{pmatrix} \text{ is } \begin{pmatrix}\text{topological}\\\text{space}\end{pmatrix}$$

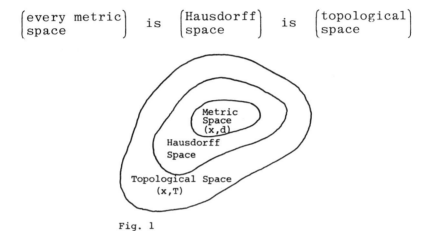

Fig. 1

Fig. 1 illustrates this relationship.

● **PROBLEM 4-11**

1. Find the accumulation point of the set $\{1, \tfrac{1}{2}, \tfrac{1}{3}, \tfrac{1}{4}, \ldots\}$ in R'.

2. What is the interior of an open interval $T(1,2,1,2)$ in the complex plane? What is the interior of a closed interval $T(1,2,1,2)$ in the complex plane?

Solution: 1. We shall apply the following definitions

Definition

A point x_0 is an accumulation point of A if every neighborhood of x_0 contains at least one point of A distinct from x_0. An accumulation point is sometimes called a cluster point.

Definition

A neighborhood $N(x)$ of a point x is a set which contains an open set $O \ni x$.

From the above definitions, we conclude that the accumulation point of $\{1,\frac{1}{2},\frac{1}{3},\ldots\}$ is $x_0 = 0$. Indeed any ball $K(0,\varepsilon)$ contains at least one (as a matter of fact it contains infinitely many) elements of the set

$$\frac{1}{k} < \varepsilon \tag{1}$$

Note that in this case the accumulation point of the set A does not belong to A.

2. Definition

The interior int A (or $\overset{0}{A}$) of a set A in (X,d) is the largest open set contained in A, i.e.

$$\text{int } A = U\, O_\omega \tag{2}$$

where O_ω runs over all open sets of (X,d) contained in A.

The interior of A is sometimes denoted by A^0.

The open interval T in the complex plane is shown in Fig.1

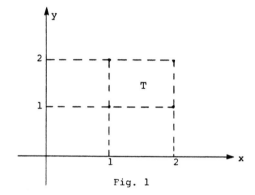

Fig. 1

Because interval T is open and $T \subset T$, we have

$$\text{int } T = T \tag{3}$$

If interval T is closed (see Fig. 2) then its interior is T without the boundary ABCD.

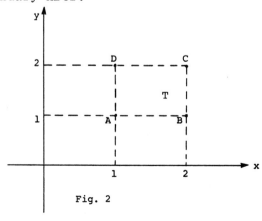

Fig. 2

Thus

$$\text{int } T = T - \{AB, BC, CD, DB\} \tag{4}$$

• **PROBLEM 4-12**

A point $a \in \text{int } A$ is called an interior point of the set A.

Show that a is an interior point if and only if

$$\bigvee_{r>0} K(a,r) \subset A \tag{1}$$

$$(a \in \text{int } A) \iff (\bigvee_{r>0} k(a,r) \subset A)$$

Solution:

$\Rightarrow$

Let us assume that

$$a \in \text{int } A \tag{2}$$

i.e.

$$a \in \bigcup_{\omega \in \Omega} O_\omega \tag{3}$$

such that

$$\bigwedge_\omega O_\omega \subset A \tag{4}$$

Thus, an open set $O_{\omega'} \subset A$ exists such that

$$a \in O_{\omega'} \tag{5}$$

Since $O_{\omega'}$ is an open set,

$$\bigvee_{r>0} K(a,r) \subset O_{\omega'} \tag{6}$$

Therefore

$$K(a,r) \subset A \tag{7}$$

$\Leftarrow$

Now, let us assume that

$$\bigvee_{r>0} K(a,r) \subset A \tag{8}$$

Since $K(a,r)$ is an open set which is contained in A, $K(a,r)$ is one of the sets in the union

$$\text{int } A = \bigcup_\omega O_\omega \qquad (9)$$

Thus

$$a \in \text{int } A \qquad (10)$$

q.e.d.

● **PROBLEM 4-13**

---

1. Show that:

   (a) A closed ball $\bar{K}$ with center x_0 and radius r

   $$\bar{K}(x_0, r) \equiv \{x : d(x, x_0) \leq r\} \qquad (1)$$

   is a closed set.

   (b) A one-point set is a closed set.

2. What is the closure of an open interval (a,b) in R'?

Solution: 1. Definition

A set $A \subset X$ is closed in (X,d) when its complement $X - A$ is an open set in (X,d).

An equivalent definition can be given as:

A set A is closed in (X,d) if every accumulation point of A belongs to A.

(a) To prove that $\bar{K}(x_0, r)$ is a closed set, we have to show that $X - \bar{K}$ is an open set.

$$C\bar{K} \equiv X - \bar{K} = \{x \in X : d(x_0, x) > r\} \qquad (2)$$

We shall use CA to denote the complement of A, $CA \equiv X - A$.

If $x \in C\bar{K}$ then $d(x_0, x) \equiv r_1 > r$, and

$$K(x, r_1 - r) \subset C\bar{K} \qquad (3)$$

The last relation follows from the triangle inequality.

2. The set $\{x\}$ is a closed set because $X - \{x\}$ is an open set.

3. Definition

The union of a set A with the set of accumulation points of A is called the closure of A. The closure of A is denoted by A^c. The accumulation points of an open interval (a,b) in R' are a and b (and of course all points belonging to this interval). Hence the closure of $(a,b) = (a,b) \cup \{a,b\} = [a,b]$. $\qquad (4)$

● **PROBLEM 4-14**

> 1. The empty set and the entire space X are closed.
> 2. The intersection $\cap_{\omega \in \Omega} A_\omega$ of the closed sets A_ω is closed.
> 3. The union of two closed sets, hence of a finite number of such sets, is itself a closed set.

Solution: 1. By definition, the empty set and the entire space are open. Thus

$$C\phi = X \tag{1}$$

$$CX = \phi \tag{2}$$

From eq.(1), we conclude that since ϕ is open, $C\phi$ is closed and X is a closed set.

From eq.(2), since X is open, ϕ is closed.

2. We shall utilize the following property of intersection:

$$X - \bigcap_\omega A_\omega = \bigcup_\omega (X - A_\omega) \tag{3}$$

Since all the sets A_ω are closed, all sets $X - A_\omega$ are open and their union is an open set. Hence $\cap_\omega A_\omega$ is a closed set.

3. Let A_1, A_2 be two closed sets. We have

$$X - (A_1 \cup A_2) = (X - A_1) \cap (X - A_2) \tag{4}$$

Since A_1, A_2 are closed, $X - A_1$ and $X - A_2$ are open. The intersection of two open sets is an open set. Hence $A_1 \cup A_2$ is a closed set.

● **PROBLEM 4-15**

> Theorem
>
> The union of a finite number of closed sets is a closed set (see Problem 4-14). You already know this theorem and a corresponding theorem for the intersection of a finite number of open sets.
>
> Find an example to illustrate that the word "finite" is important, i.e., that the union of an infinite number of closed sets can be an open set.

Solution: Consider closed intervals

$$S_n = \left[1 + \frac{1}{n},\ 3 - \frac{1}{n}\right] \tag{1}$$

for n = 1,2,...

The union of them is

$$\bigcup_{n=1}^{\infty} \left[1 + \frac{1}{n},\ 3 - \frac{1}{n}\right] = (1,3) \tag{2}$$

which is an open interval.

This example shows that the union of an infinite number of closed sets can be an open set.

In a similar way, it can be shown that the intersection of an infinite number of open sets can be a closed set.

$$\bigcap_{n=1}^{\infty} \left(-\frac{1}{n},\ 1 + \frac{1}{n}\right) = [0,1] \tag{3}$$

CLOSURE OF A SET

● **PROBLEM 4-16**

The closure of a set $A \subset X$ in (X,d) can be defined as follows:

Definition

The closure of a set $A \subset X$ in (X,d) is the set $\bar{A}$ which is the smallest closed set containing A. That is

$$\bar{A} = \bigcap_{\omega \in \Omega} Z_{\omega} \tag{1}$$

where Z_{ω} ranges over all closed sets containing A

$$A \subset Z_{\omega},\ \omega \in \Omega \tag{2}$$

When A is a closed set

$$A = \bar{A} \tag{3}$$

This definition is equivalent to the definition given in Problem 4-13. Show that the closure operation satisfies the following conditions:

1. $\bar{\bar{A}} = \bar{A}$ (4)
2. $A \subset \bar{A}$ (5)
3. $\bar{\phi} = \phi$ (6)
4. $\overline{A \cup B} = \bar{A} \cup \bar{B}$ (7)

Solution: 1. For a closed set A, the least closed set containing A is A itself. Hence

$$A = \overline{A} \tag{8}$$

The set $\overline{A}$ is closed by definition. Thus

$$(\overline{A}) = \overline{\overline{A}} = \overline{A} \tag{9}$$

2. Since each of the sets Z_ω in eq.(1) contains A,

$$A \subset \bigcap_{\omega \in \Omega} Z_\omega = \overline{A} \tag{10}$$

3. The smallest closed set containing ϕ is ϕ, therefore

$$\overline{\phi} = \phi \tag{11}$$

4. $\overline{A \cup B}$ is the smallest closed set containing $A \cup B$, thus it contains A and B. Since $\overline{A \cup B}$ is closed, it contains $\overline{A}$ and $\overline{B}$. Hence

$$\overline{A} \cup \overline{B} \subset \overline{A \cup B} \tag{12}$$

$\overline{A}$ is the smallest closed set containing A, and $\overline{B}$ is the smallest closed set containing B. Thus the smallest closed set containing $A \cup B$ is the subset of $\overline{A} \cup \overline{B}$,

$$\overline{A \cup B} \subset \overline{A} \cup \overline{B} \tag{13}$$

From eqs.(12) and (13), we obtain

$$\overline{A \cup B} = \overline{A} \cup \overline{B} \tag{14}$$

● **PROBLEM 4-17**

Let $\{A_\omega\}$ be a family of connected sets having a common point a_0. Show that the union

$$A = \cup A_\omega \tag{1}$$

is a connected set.

Solution: We shall start with the definition of a non-connected set (space).

Definition

A space (X,d) is not connected (non-connected) if there exist $X_1 \subset X$ and $X_2 \subset X$ such that

1. $X_1 \cup X_2 = X$ \hfill (2)

2. $X_1, X_2 \neq \phi$ \hfill (3)

3. $X_1 \cap X_2 = \phi$ \hfill (4)

4. X_1 and X_2 are open in X \hfill (5)

Otherwise, we say the space is connected.

A subset $A \subset X$ is connected (non-connected) if, when treated as a metric space with the metric induced from X, it is connected (non-connected).

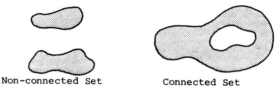

Non-connected Set   Connected Set

It can be shown that the only connected sets in R' are:

1. intervals $[a,b]$, (a,b), $[a,b)$, $(a,b]$
2. semi-lines
3. the entire line

To show that the set given by eq.(1) is connected, let us assume that it is not connected.

Thus, two sets B_1 and B_2 exist such that

$$B_1 \cup B_2 = A$$

$$B_1 \cap B_2 = \phi$$

B_1 and B_2 are open

$$B_1, B_2 \neq \phi$$

We assume that $a_0 \in B_1$. Let us consider any set $A_{\omega'}$ from the union in eq.(1).

$A_{\omega'}$ is connected and $a_0 \in A_{\omega'}$. The set $A_{\omega'} \cap B_2$ must be empty. If it is not, then $A_{\omega'} = (A_{\omega'} \cap B_1) \cup (A_{\omega'} \cap B_2)$ (6) cannot be connected.

Thus
$$A_{\omega'} \subset B_1 \tag{7}$$

This is true for all sets A_ω. Therefore, B_2 must be an empty set, which contradicts the assumption $B_2 \neq \phi$.

● **PROBLEM 4-18**

Here are some definitions.

Boundary Point

A boundary point of a set A is a point such that every neighborhood of that point contains at least one point of A and at least one point not of A. The boundary of A is denoted by ∂A.

Region

An open connected set together with all, some, or none of its boundary points is called a region.

Domain

A set is called a domain if and only if it is open and connected.

Sketch the annulus

$$M = \{z \in \mathbb{C} : 1 < |z| < 2\} \tag{1}$$

Show that M is a domain.

Is the set $N = \{z \in \mathbb{C} : 1 < |z| \leq 2\}$ a domain?

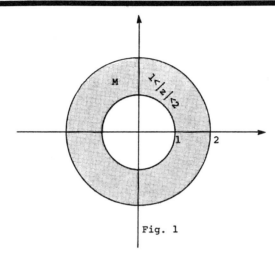

Fig. 1

Solution: The annulus M is shown in Fig. 1.

From eq.(1), we see that M is an open set. M is also connected, thus M is a domain.

The set $N = \{z \in \mathbb{C} : 1 < |z| \leq 2\}$ is not an open set. Hence it cannot be a domain.

The following theorem establishes a relationship between the closure and boundary of a set

$$\partial A = \overline{A^c} \cap \overline{(X - A)^c}$$

REGIONS, DOMAINS

• PROBLEM 4-19

Describe the region in the z plane defined by

$$|z + i| > |z - 1| \tag{1}$$

where $z = x + iy$.

Solution: Eq.(1) can be transformed as follows

$$|z + i|^2 = (z+i)(\overline{z+i}) = (z+i)(\bar{z}-i)$$

$$= z\bar{z} - i(z-\bar{z}) + 1 = z\bar{z} + 2y + 1 \qquad (2)$$

$$|z-1|^2 = (z-1)(\overline{z-1}) = z\bar{z} - 2x + 1 \qquad (3)$$

Hence eq.(1) is equivalent to

$$z\bar{z} + 2y + 1 > z\bar{z} - 2x + 1 \qquad (4)$$

or

$$y > -x \qquad (5)$$

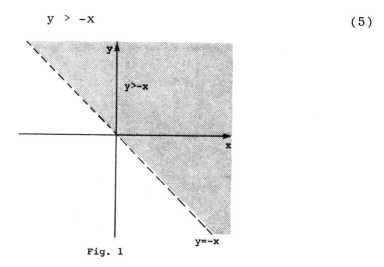

Fig. 1

The boundary line $y = -x$ does not belong to the region. Thus $y > -x$ is an open region, or a domain.

● **PROBLEM 4-20**

Sketch the following sets. Determine their closures and topological properties.

1. $|3z + 1| > 3$
2. $|z - 2 + i| \leq 3$
3. $|\text{Re } z| > 1$
4. $|z - 1| \geq |z|$
5. $|\text{Im } \frac{1}{2}| = 1$

Solution: 1. Let $z = x + iy$, then

$$\left|z + \frac{1}{3}\right| > 1 \qquad (1)$$

This is the whole space without a closed disc of radius 1 and center at $-\frac{1}{3}$, as shown in Fig. 1.

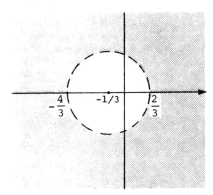

Fig. 1

The set is open and connected. The closure includes the circle $|3z+1| = 3$, hence it is a domain.

2. $\quad |z - 2 + i| = |z - (2 - i)| \leq 3$ (2)

This is a closed disc with center at $2 - i$ and radius 3.

The set is closed, bounded and connected. Since it is closed and bounded, it is also compact (see Problem 4-26).

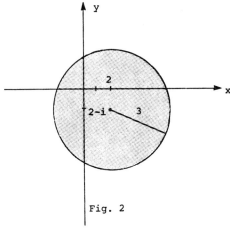

Fig. 2

3. $\quad |\text{Re } z| = |x| > 1$ (3)

Hence

$$x > 1 \quad \text{or} \quad x < -1$$

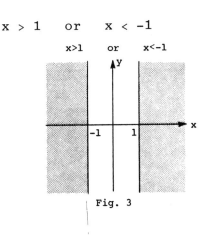

Fig. 3

This is an open set. It is not connected. The closure would include the lines x = 1 and x = -1.

4. $|z-1| \geq |z|$ (4)

Hence

$$(x-1)^2 + y^2 \geq x^2 + y^2$$ (5)

or

$$x \leq \frac{1}{2}$$

This region is closed and connected. It is not bounded.

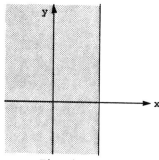

Fig. 4

5. $|Jm \frac{1}{z}| = 1$ (6)

Thus

$$\frac{|y|}{x^2+y^2} = 1$$ (7)

which is the union of two circles given by

$$x^2 + y^2 = \pm y$$ (8)

with the origin removed.

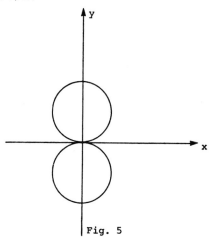

Fig. 5

This set is closed and bounded.

SEQUENCES

• **PROBLEM 4-21**

Show that the sequence z_n

$$z_n = 1 + \frac{1}{n^2} i \tag{1}$$

converges to $z = 1$. We say z is the limit of z_n.

Solution: A sequence (z_n) of complex numbers is defined by assigning to each natural number $n = 1, 2, 3, \ldots$ a complex number z_n. In the same way, we define a sequence (x_n) in any metric space (X, d). The limit of a sequence is defined as follows:

Definition

x is the limit of a sequence (x_n) or the sequence (x_n) converges (is convergent) to x when

$$\bigwedge_{\varepsilon > 0} \bigvee_{N \in N'} \bigwedge_{n > N'} d(x, x_n) < \varepsilon \tag{2}$$

This is written as

$$x \equiv \lim_{n \to \infty} x_n \tag{3}$$

or

$$x_n \longrightarrow x$$

This is a very important definition and should be understood clearly. What it means is that: it doesn't matter how small ε is, we can always find an integer N such that all the elements of the sequence with indices higher than N are located inside the ball $K(x, \varepsilon)$, see Fig. 1.

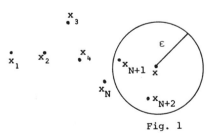

Fig. 1

Now, let C be the complex space with the usual metric $(C, 1 \cdot 1)$. We shall show that

$$1 + \frac{1}{n^2} i \longrightarrow 1 \tag{4}$$

Let ε be any real positive number. We have to find N such that for n > N

$$\left|1 + \frac{1}{n^2} i - 1\right| = \left|\frac{1}{n^2} i\right| = \frac{1}{n^2} < \varepsilon \tag{5}$$

Thus

$$N = \left[\frac{1}{\sqrt{\varepsilon}}\right] \tag{6}$$

That proves that

$$\lim_{n \to \infty}\left(1 + \frac{1}{n^2} i\right) = 1 \tag{7}$$

● **PROBLEM 4-22**

Show that every convergent sequence is a Cauchy* sequence.

Solution: Definition

A sequence $N \ni n \to x_n \in X$ in a metric space (X,d) is a Cauchy sequence when

$$\bigwedge_{\varepsilon > 0} \bigvee_{N} \bigwedge_{n,m > N} d(x_n, x_m) < \varepsilon \tag{1}$$

Let (z_n) be a convergent sequence in (X,d)

$$z_n \to z \tag{2}$$

and let ε be an arbitrary positive real number.

Since (z_n) is convergent, there exists an N_1 such that

$$d(z_n, z) < \frac{\varepsilon}{2} \quad \text{for } n > N_1 \tag{3}$$

By the triangle inequality

$$d(z_n, z_m) \leq d(z, z_n) + d(z, z_m) < \frac{\varepsilon}{2} + \frac{\varepsilon}{2} = \varepsilon \tag{4}$$

for $n > N_1$ and $m > N_1$.

Hence, (z_n) is a Cauchy sequence.

*Augustin Louis Cauchy born in Paris, 1789, died at Sceaux 1857). The great military and technical schools founded by Napoleon produced many brilliant minds of Napoleonic and post-napoleonic era, and had lasting effect on scientific and cultural life in France. Cauchy was educated at École Polytechnique in Paris and at École des Ponts. Due to the uncertain political situation in France he went in 1830 to Turin and later to Prag to undertake the education of the Comte de Chambord.

Upon his return to France he occupied a chair of mathematics and mathematical astronomy. Special Emperor's decree allowed him to become professor without taking the oath of allegiance.

The changing political situation and his own eccentricities frequently disturbed his life. Despite that he was able to publish over 700 papers.

His contributions to mathematics include differential equations, the theory of functions, theory of residues, convergence, imaginary numbers, theory of probability, theory of determinants. Cauchy was first to use the word determinant in its present sense.

● **PROBLEM 4-23**

1. Give an example of a sequence which is bounded but does not converge.

2. Give an example of a sequence which has an accumulation point but does not have a limit.

3. Prove that if (x_n) is a convergent sequence in (X,d), then (x_n) is bounded.

Solution: 1. A sequence (x_n) is bounded when the set $\{x_n\}$ is bounded, i.e., it is a subset of a ball.

The sequence
$$z_n = 1 + i^n \tag{1}$$
is bounded, see Fig. 1.

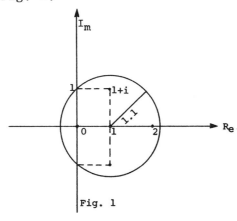

Fig. 1

All its elements are contained in a ball of center $(1,0)$ and radius 1.1. This sequence does not converge.

2. Definition

A sequence (x_n) is said to have an accumulation point x_0 if any neighborhood $N(x_0)$ of x_0 contains an infinite number of elements of (x_n).

For example, sequence $z_n = 1 + i^n$ has four accumulation points. One of them is $1 + i$, because

$$1 + i^{4k+1} = 1 + i \quad \text{for } k = 1, 2, \ldots \tag{2}$$

This sequence does not have a limit.

3. Assume that (x_n) is convergent. For any positive ε, all but a finite number of elements of (x_n) are located in a ball, i.e.

$$\text{for all } n > N \quad x_n \in K(x_0, \varepsilon) \tag{3}$$

We can find a ball $K(x_0, r)$ which contains all elements $x_1, x_2, \ldots x_{N-1}, x_N$. Thus

$$x_n \in K(x_0, R) \tag{4}$$

for all n, where

$$R = \max(\varepsilon, r) \tag{5}$$

● **PROBLEM 4-24**

Prove the theorem.

Theorem

$$\begin{pmatrix} A \text{ is closed} \\ \text{in } (X,d) \end{pmatrix} \iff \begin{bmatrix} \bigwedge_{A \ni a_n \to a} a \in A \end{bmatrix} \tag{1}$$

Solution:

=>

We shall use an indirect proof (see Problem 4-1)

Let $a_n \in A$ and $a_n \to a$, but $a \notin A$. Since A is closed, $X - A$ is open.

$a \in X - A$, which is an open set. Therefore, a ball exists $K(a, r) \subset X - A$, which contradicts the assumption $a_n \to a$.

<=

Suppose A is open. Then, there exists an $a_0 \notin A$ and a sequence $a_n \in K(a_0, \frac{1}{n})$ such that

$$a_n \notin X - A$$

that is, $a_n \in A$ and $a_n \to a_0$.

This contradicts the assumption. Thus, the theorem has been proven.

COMPLETE AND COMPACT SPACES

• **PROBLEM 4-25**

Why is the space of rational numbers (Q,1•1) not complete?

Solution: Definition

If in (X,d) every Cauchy sequence has a limit belonging to X, then (X,d) is complete.

Cauchy's theorem states that the space (R',1•1) of real numbers is complete under the natural distance

$$d(x,y) \equiv |x-y| \tag{1}$$

On the other hand, the space of rational numbers Q with the natural distance $|x-y|$ is not complete.

Consider the sequence of rational numbers

$$x_1 = 1$$
$$x_2 = 1.4$$
$$x_3 = 1.41$$
$$x_4 = 1.414$$
$$x_5 = 1.4142$$

etc.

which is the sequence of decimal expansions of $\sqrt{2}$. This sequence is (as can be easily shown) a Cauchy sequence, but it does not converge in Q since $\sqrt{2}$ is not a rational number. The space Q can be completed, i.e., embedded isometrically in a complete space - in this case the space R'. Hence, Q is dense in R'. Hausdorff observed that any space (X,d) can be completed by an analogous procedure.

As a simple exercise, you can prove that:

A closed subset of a complete space is a complete space.

• **PROBLEM 4-26**

Before formulating this problem, let us start with some definitions.

Subsequence

A subsequence of the sequence (x_n) is the sequence

$$N \ni k \to x_{n_k} \tag{1}$$

where $k \to n_k$ is an increasing sequence of natural numbers. For example, if the sequence is

$$1,2,3,4,5,\ldots$$

one of its subsequences is

$$2,4,6,\ldots$$

or

$$5,10,15,\ldots$$

It can be easily shown that: If the sequence (x_n) converges to x, each of its subsequences also converges to x.

Compact Set

A subset A of the metric space (X,d) is said to be compact if out of each sequence (x_n) of elements of A we can extract a subsequence (x_{n_k}) which converges to a limit contained within the set A.

Prove the Bolzano*-Weierstrass* theorem.

Theorem

An interval $[a,b] \subset R'$ is compact.

Solution: Proof by induction:

Let (x_k) be a sequence such that $x_k \in [a,b]$. Divide [a,b] into two equal parts. At least one part contains infinitely many elements from (x_k). Take this part and choose any x_k contained in this part and denote it by

$$x_{k_1}$$

Again divide this part of the interval into halves, choose the half with infinitely many elements from (x_k) and select x_{k_2}. Proceeding further, we obtain the subsequence (x_{k_n}) of the sequence (x_k). The subsequence is a Cauchy sequence.

$$\left| x_{k_m} - x_{k_n} \right| \leq \frac{a-b}{2^p}, \quad p = \min(m,n) \tag{1}$$

Since the interval $[a,b] \subset R'$ is complete, there exists an x such that

$$x = \lim_{k_n \to \infty} x_{k_n} \tag{2}$$

But $a \leq x_{k_n} \leq b$, therefore

$$a \leq x \leq b, \quad \text{i.e.} \quad x \in [a,b].$$

*Bernhard Bolzano (1781-1848) from Bohemia worked mostly in the field of infinite series.

Karl Weierstrass, born in Ostenfelde in 1815, died in Berlin in 1897. He studied law and finance in Bonn (1834). He taught at various schools and in 1864 became professor of mathematics at the University of Berlin. One of the leading mathematicians of the 19th century, he worked in the theory of elliptic and Abelian functions and in the theory of irrational numbers. He also constructed the theory of uniform analytic functions.

● **PROBLEM 4-27**

Prove the theorem

$$\begin{pmatrix} (X,d) \text{ is a metric space} \\ A \subset X, \ A \text{ is a compact set} \end{pmatrix} \Rightarrow \begin{pmatrix} A \text{ is bounded} \\ \text{and closed in } X \end{pmatrix}$$

Solution: First we shall prove that A is closed. Assume the opposite - A is not closed. Then there exists a point of accumulation x_0 of A which does not belong to A. Take the sequence

$$A \ni x_n \to x_0 \notin A \tag{1}$$

Each of the subsequences of x_n also converges to x_0; thus A is not compact. This is a contradiction. Now, suppose that A is not bounded, then

$$\begin{pmatrix} \bigwedge_{B \subset A} (B\text{-finite}) \end{pmatrix} \Rightarrow \begin{pmatrix} \bigvee_{x \in A} \bigwedge_{b \in B} d(x,b) > 1 \end{pmatrix} \tag{2}$$

Thus, by induction, we can construct a sequence (x_n) in A such that

$$d(x_n, x_m) > 1 \quad \text{for } n \neq m \tag{3}$$

No subsequence of this sequence is a Cauchy sequence; thus A is not compact. This is a contradiction.

We have shown that if A is compact, then it is bounded and closed.

The converse theorem is not, in general, true. However, it is true in R^n.

Theorem

A set A which is bounded and closed in R^n is compact.

• **PROBLEM 4-28**

Let A be a set consisting of a finite number of elements. Show that A is bounded, closed and compact.

Solution: Let

$$A = \{a_1, a_2, \ldots, a_k\} \tag{1}$$

and let (x_n) be a sequence of elements of A. Since x_n has infinitely many elements, at least one of $a_1, \ldots, a_k$ has to appear in (x_n) infinitely many times. Let us denote this element by a_ℓ.

We can extract a subsequence (x_{n_k}) of a sequence (x_n) in such a way that (x_{n_k}) consists only of elements a_ℓ. Obviously

$$x_{n_k} \to a_\ell \tag{2}$$

i.e., (x_{n_k}) converges to a limit contained in A.

Since A is compact, by the theorem outlined in Problem 4-27, we conclude that A is bounded and closed.

• **PROBLEM 4-29**

1. Show that the union of a finite number of compact sets is compact.

2. Let (B_n) be a sequence of compact sets such that

$$B_{n+1} \subset B_n, \quad n = 1, 2, \ldots. \tag{1}$$

Show that there exists at least one point b_0 common to all B_n.

Solution: 1. We shall show that if

$$A = A_1 \cup A_2 \tag{2}$$

where A_1, A_2 are compact, then A is a compact set. Let (x_n) be a sequence of elements of A. Either A_1 or A_2 (or both) contains infinitely many elements of (x_n). Hence, we can extract a subsequence (x_{n_k}) of (x_n) such that all elements of (x_{n_k}) belong to A_1 (or A_2).

A_1 is a compact set. Out of sequence (x_{n_k}) of elements of

A_1 we can extract a subsequence $(x_{n_{k_\ell}})$, which converges to a limit contained within the set A_1. Thus, the union of two compact sets is a compact set; hence, the union of a finite number of compact sets is a compact set.

2. Since all sets B_n are compact, they are also bounded. We form a bounded sequence (x_n) such that

$$x_n \in B_n \quad \text{for } n = 1, 2, 3, \ldots \quad (3)$$

Since (x_n) is bounded, it has an accumulation point x_0. Thus, for each n, x_0 is an accumulation point of the subsequence $x_{n+1}, x_{n+2}, x_{n+3}, \ldots$ contained in B_n. Remember that $B_{n+1} \subset B_n$. Therefore, x_0 is an accumulation point of B_n. B_n is a compact set; therefore, it is closed. A closed set contains its accumulation points. Hence x_0 belongs to B_n.

• **PROBLEM 4-30**

Show that every compact space is complete.

Show that the converse is not true, and give an example.

Solution: Let X be a compact space, thus out of each sequence (x_n) of elements of X we can extract a subsequence (x_{n_k}) which converges to a limit contained in X. That implies that every Cauchy sequence in X has a limit belonging to X. That is, X is complete.

The converse theorem is not, in general, true. For example, R' is complete but is not compact because it is not bounded. Note that a compact set in R' contains its bounds since only a point of accumulation or an element of a set can be a bound.

THEOREMS AND PROPERTIES

• **PROBLEM 4-31**

Let $I = [x_1, x_2] \subset R'$ be a closed interval Assume that

$$\bigwedge_{x \in I} \bigvee_{N(x)} x \in N(x) \quad \text{and} \quad N(x) \in \{N(\cdot)\} \quad (1)$$

i.e., for each x belonging to interval I there exists a neighborhood $N(x)$ of x, which belongs to a given family $\{N(\cdot)\}$ of neighborhoods with respect to I.

Show that there exists a finite subfamily $N(x_1)$, $N(x_2)$..., $N(x_k)$ of the family $\{N(\cdot)\}$ which covers I.

<u>Solution</u>: We shall begin with a definition.

<u>Definition</u>

If $\{A_\ell\}_{\ell \in L}$ is a family of subsets of a set X such that

$$X = \bigcup_{\ell \in L} A_\ell \tag{2}$$

then $\{A_\ell\}$ is called a covering of the set X. If all A_ℓ are open (closed) we speak about an open (closed) covering of the set X.

The following theorem will be helpful:

<u>Theorem</u> (Borel*, Lebesgue sometimes called Heine-Borel)

Let (X,d) be a metric space, then X is compact if and only if for every open covering $\{A_\ell\}_{\ell \in L}$ of X there exists a finite set $(i_1, i_2, \ldots, i_k)$ such that $\{A_{i_n}\}_{n=1,2,\ldots k}$ is a covering of X.

Now, we can proceed with our problem.

I is a compact set and $\{N(\cdot)\}$ is a covering of I. Applying the quoted theorem, we deduce that a subset of the family $\{N(\cdot)\}$ can be chosen which has a finite number of elements and which is a covering of I.

*E. Borel, a French mathematician, extended the general rules for investigating the divergency and convergency of infinite series.

• **PROBLEM 4-32**

Let $S_1, S_2, S_3, \ldots$ be a sequence of sets such that diameter $(S_n) \to 0$ as $n \to \infty$. Show that

$$\bigcap_{n=1}^{\infty} S_n = \begin{cases} \{z\} \\ \phi \end{cases} \tag{1}$$

i.e., the intersection of all the sets S_n is either an empty set or a set consisting of one element.

<u>Solution</u>: Let us assume that statement (1) is not true and that

$$z \in \bigcap_{n=1}^{\infty} S_n \tag{2}$$

and
$$z_1 \in \bigcap_{n=1}^{\infty} S_n \tag{3}$$

where $z \neq z_1$

Then
$$|z-z_1| = r > 0 \tag{4}$$

But:

Definition

The diameter of a set S of complex numbers is the least upper bound of the set
$$r_{\omega\lambda} = |z_\omega - z_\lambda| \tag{5}$$

for all z_ω, z_λ in S.

From eq.(4) we conclude that
$$\text{diameter } S_n \geq |z-z_1| = r > 0 \tag{6}$$

for all n, which contradicts the assumption
$$\text{diameter } (S_n) \xrightarrow[n \to \infty]{} 0 \tag{7}$$

Hence, (1) must hold true.

● **PROBLEM 4-33**

Let $\{I_n\}$ be a sequence of closed intervals in the z plane such that
$$I_{n+1} \subset I_n \quad \text{for } n = 1,2,3\ldots \tag{1}$$

Show that there exists at least one point common to all intervals.

Solution: For definition of an interval see Problem 4.2.

Interval I_n consists of all $z=x+iy$ such that
$$\begin{aligned} a_{n,1} \leq x \leq a_{n,2} \\ b_{n,1} \leq y \leq b_{n,2} \end{aligned} \tag{2}$$

Since
$$I_{n+1} \subset I_n$$

we have

$$a_{n,1} \leq a_{n+1,1} \leq a_{n,2} \qquad b_{n,1} \leq b_{n+1,1} \leq b_{n,2}$$
$$a_{n,1} \leq a_{n+1,2} \leq b_{n,2} \qquad b_{n,1} \leq b_{n+1,2} \leq b_{n,2} \qquad (3)$$

At this point we can utilize the following property of real numbers, sometimes called the nested intervals property:

Let I_n be a sequence of closed nested (i.e., such that $I_{n+1} \subset I_n$ for all n) intervals, then there exists at least one point common to all intervals. Let us define the sequence of intervals I_n such that

$$I_n := \{x \in R' : a_{n,1} \leq x \leq a_{n,2}\} \qquad (4)$$

I_n is a sequence of nested closed intervals. Hence, there exists a real number x_0 that belongs to all intervals. Similarly for the sequence of closed, nested intervals

$$K_n := \{y \in R' : b_{n,1} \leq y \leq b_{n,2}\} \qquad (5)$$

a real number y_0 exists that belongs to all intervals K_n. Hence, the point

$$z_0 = x_0 + iy_0 \qquad (6)$$

is contained in all intervals I_n.

• **PROBLEM 4-34**

Let (I_n) be a sequence of closed nested intervals in the z plane such that

$$\text{diameter}(I_n) \xrightarrow[n \to \infty]{} 0 \qquad (1)$$

Show that there exists a unique point z_0 common to all I_n.

Solution: Since (I_n) is a sequence of closed-nested intervals, then there exists at least one point common to all intervals.

We shall show that this is a unique point. Let us assume the opposite. (As your mathematical knowledge increases you feel more comfortable with "assuming the opposite". This method, called reductio ad absurdum, is a powerful mathematical tool.) We assume that there exist two distinct points z_0 and z_1 which are common to all I_n. Since $z_1 \neq z_0$

$$|z_1 - z_0| = r > 0 \qquad (2)$$

That contradicts the assumption

$$\text{diameter } (I_n) \to 0 \tag{3}$$

because for sufficiently high N,

$$\text{diameter } (I_n) < r \quad \text{for all } n > N \tag{4}$$

Such sets cannot contain z_0 and z_1. Therefore, z_0 is a unique point.

● **PROBLEM 4-35**

Show that if a sequence (z_n) converges to z_0, then (z'_n), where

$$z'_n = \frac{1}{n} \sum_{k=1}^{n} z_k \tag{1}$$

also converges to z_0.

<u>Solution</u>: (z_n) converges to z_0. Thus for any $\varepsilon > 0$, there exists $N_1(\varepsilon)$ such that

$$|z_k - z_0| < \frac{\varepsilon}{2} \quad \text{for } k > N_1 \tag{2}$$

With fixed N_1 we choose $N_2 > N_1$ such that

$$\frac{1}{n}\left| \sum_{k=1}^{N_1} (z_k - z_0) \right| < \frac{\varepsilon}{2} \quad \text{for } n > N_2 \tag{3}$$

It is possible to do so because

$$\left| \sum_{k=1}^{N_1} (z_k - z_0) \right| \text{ is a finite number.}$$

We have

$$\left(\frac{1}{n} \sum_{k=1}^{n} z_k \right) - z_0 = \frac{1}{n} \sum_{k=1}^{n} (z_k - z_0) \tag{4}$$

Then, for $n > N_2$, we get

$$|z'_n - z_0| = \left| \left(\frac{1}{n} \sum_{k=1}^{n} z_k \right) - z_0 \right|$$

$$= \left| \frac{1}{n} \sum_{k=1}^{n} (z_k - z_0) \right|$$

$$= \left| \frac{1}{n} \left[\sum_{k=1}^{N_1} (z_k - z_0) + \sum_{k=N_1+1}^{n} (z_k - z_0) \right] \right|$$

$$\leq \left| \frac{1}{n} \sum_{k=1}^{N_1} (z_k - z_0) \right| + \left| \frac{1}{n} \sum_{k=N_1+1}^{n} (z_k - z_0) \right| \quad (5)$$

$$< \frac{\varepsilon}{2} + \frac{1}{n} \sum_{k=N_1+1}^{n} |z_k - z_0|$$

$$\leq \frac{\varepsilon}{2} + \frac{1}{n} \cdot n \cdot \frac{\varepsilon}{2} = \varepsilon$$

Thus $\frac{1}{n} \sum_{k=1}^{n} z_k$ converges to z_0.

• **PROBLEM 4-36**

Let (z_n) be a sequence of complex numbers where

$$z_n = x_n + iy_n, \quad n = 1, 2, \ldots \quad (1)$$

A necessary and sufficient condition that (z_n) converge to $z_0 = x_0 + iy_0$ is that

$$x_0 = \lim_{n \to \infty} x_n \quad \text{and} \quad y_0 = \lim_{n \to \infty} y_n \quad (2)$$

Prove the above theorem.

Solution: First, we shall show that (2) is a necessary condition.

If z_0 is the limit of (z_n) then for any given $\varepsilon > 0$

$$|z_n - z_0| < \varepsilon \quad \text{for } n > N \quad (3)$$

But

$$|x_n - x_0| \leq |z_n - z_0| < \varepsilon \quad \text{for } n > N \quad (4)$$

$$|y_n - y_0| \leq |z_n - z_0| < \varepsilon \quad \text{for } n > N \quad (5)$$

Hence

$$\lim x_n = x_0$$
$$\lim y_n = y_0 \quad (6)$$

To show sufficiency, let us assume that both limits in eq.(6) exist.

For any given $\frac{\varepsilon}{2}$ there exists N such that

$$|x_n - x_0| < \frac{\varepsilon}{2} \quad \text{for } n > N \quad (7)$$

$$|y_n - y_0| < \frac{\varepsilon}{2}$$

Thus

$$|z_n - z_0| \leq |x_n - x_0| + |y_n - y_0| < \varepsilon \quad \text{for } n > N \tag{8}$$

Hence, z_0 is the limit of the sequence (z_n).

● **PROBLEM 4-37**

Summarize briefly all essential properties of the set of complex numbers.

Solution: The set of complex numbers consists of all numbers

$$z = x + iy \tag{1}$$

where x and y are real numbers, and some basic arithmetic operations such as addition, subtraction, multiplication and division are defined on these numbers.

Many properties of real numbers can be extended to the system of complex numbers.

I. The space of real numbers and the space of complex numbers are complete. This fact can be formulated as follows:

Every Cauchy sequence of real (complex) numbers has a limit that is a real (complex) number.

II. Now, the so-called Bolzano-Weierstrass property:

Every bounded set of real (complex) numbers containing an infinite number of elements has an accumulation point.

III. The nested intervals property deals with a sequence of intervals.

Let (I_n) be a sequence of closed-nested (i.e., such that $I_{n+1} \subset I_n$ for all n) intervals. Then there exists at least one element common to all intervals.

This is true when by intervals we understand subsets of the space of real numbers or subsets of the complex plane.

These three properties are equivalent, thus one of them can be derived from any one of the others.

CHAPTER 5

SEQUENCES AND SERIES OF COMPLEX NUMBERS

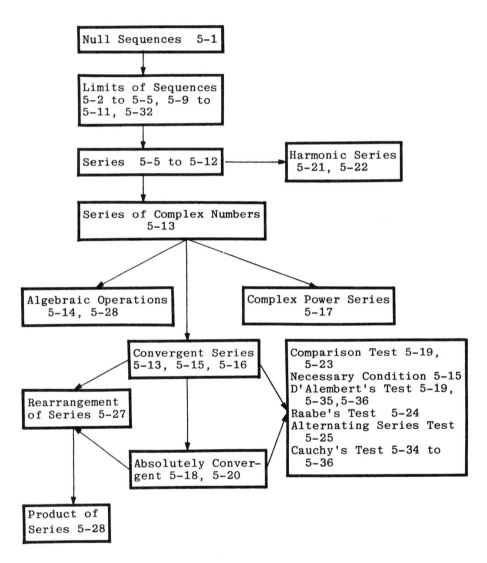

This chart is provided to facilitate rapid understanding of the interrelationships of the topics and subject matter in this chapter. Also shown are the problem numbers associated with the subject matter.

SEQUENCES

• **PROBLEM** 5-1

Definition

A sequence which has a limit equal to zero is called a null sequence.

From this definition, we conclude that if (z_n) converges to z_0, then $(z_n - z_0)$ converges to zero, that is $(z_n - z_0)$ is a null sequence.

Prove the following:

1. The product of a bounded sequence and a null sequence is a null sequence.

2. The sum (difference) of two null sequences is a null sequence.

Solution: 1. Let (a_n) be a null sequence and (b_n) a bounded sequence. Thus, there exists a constant K such that

$$|b_n| < K \quad \text{for all } n. \tag{1}$$

Since (a_n) is a null sequence,

$$\bigwedge_{\varepsilon > 0} \bigvee_{N} \quad |a_n| < \frac{\varepsilon}{K} \quad \text{for } n > N \tag{2}$$

Thus, $\quad |a_n b_n| < K|a_n| < K\frac{\varepsilon}{K} = \varepsilon \quad \text{for } n > N \tag{3}$

Hence, $(a_n b_n)$ is a null sequence.

2. Now, let (a_n) and (b_n) be two null sequences. We have

$$|a_n| < \frac{\varepsilon}{2} \quad \text{for } n > N_1$$
and $\tag{4}$
$$|b_n| < \frac{\varepsilon}{2} \quad \text{for } n > N_2$$

Therefore, for

$$n > \max(N_1, N_2) \tag{5}$$

we obtain

$$|a_n + b_n| \leq |a_n| + |b_n| < \frac{\varepsilon}{2} + \frac{\varepsilon}{2} = \varepsilon \tag{6}$$

and $(a_n + b_n)$ is a null sequence.

In a similar manner we can show that the product of two null sequences is a null sequence.

• **PROBLEM 5-2**

1. Prove the theorem

$$\left. \begin{array}{l} \lim_{n\to\infty} z_n = z \\ \lim_{n\to\infty} \omega_n = \omega \end{array} \right\} \Rightarrow \left[\lim_{n\to\infty} (z_n \pm \omega_n) = z \pm \omega \right] \quad (1)$$

2. Let

$$z_n = \sqrt[n]{p} + i n^2, \quad p > 0, \quad p \in R'$$

$$\omega_n = 4 + i(1-n^2) \quad (2)$$

Compute $\lim(z_n + \omega_n)$ and $\lim(z_n - \omega_n)$.

Solution: 1. The theorem can be stated as follows:

If (z_n) is a convergent sequence with limit z and (ω_n) is a convergent sequence with limit ω, then $(z_n + \omega_n)$ is convergent and its limit is $z + \omega$. Because $z_n \to z$, the sequence $(z_n - z)$ is a null sequence, also, because $\omega_n \to \omega$, $(\omega_n \to \omega)$ is a null sequence.

The sum of two null sequences is a null sequence (see Problem 5-1). Therefore,

$(z_n - z + \omega_n - \omega) = ([z_n + \omega_n] - [z + \omega])$ is a null sequence. Therefore, $(z_n + \omega_n)$ converges to $z + \omega$.

2. Note that we cannot apply theorem (1). The sequences z_n and ω_n are not convergent. Thus, the limits do not exist. But, the sum

$$z_n + \omega_n = \sqrt[n]{p} + in^2 + 4 + i(1-n^2)$$

$$= \sqrt[n]{p} + i \quad (3)$$

is a convergent sequence. Its limit is

$$\lim_{n\to\infty}(z_n + \omega_n) = 1 + i \quad (4)$$

The difference

$$z_n - \omega_n = \sqrt[n]{p} + in^2 - 4 - i(1-n^2)$$

is not a convergent sequence.

● **PROBLEM 5-3**

1. Prove the theorem

$$\left.\begin{array}{l}\lim z_n = z \\ \lim \omega_n = \omega\end{array}\right\} \Rightarrow (\lim z_n \cdot \omega_n = z \cdot \omega) \quad (1)$$

2. Compute $\lim_{n \to \infty} z_n \omega_n$ where

$$z_n = \frac{n^3 + 2n^2 - 1}{(n+1)(n^2 - n + 6)} + a\left(1 + \frac{a}{n^2}\right)^n i, \quad a \in R' \quad (2)$$

$$\omega_n = \frac{n+1}{n-1} + 3i$$

Solution: 1. Because $z_n \to z$, $(z_n - z)$ is a null sequence and it can be denoted as

$$z_n = z + (z_n - z) = z + a_n \quad (3)$$

where (a_n) is a null sequence.

By the same method,

$$\omega_n = \omega + b_n \quad (4)$$

where (b_n) is a null sequence.

Multiplying, we find

$$z_n \omega_n = z\omega + (z\, b_n + \omega\, a_n + a_n b_n) \quad (5)$$

By theorem (1) in Problem 5-1, we conclude that $(zb_n + \omega a_n + a_n b_n)$ is a null sequence. Therefore,

$$z_n \omega_n \to z\omega \quad (6)$$

and the proof is completed.

2. We have

$$\lim_{n \to \infty} \frac{n^3 + 2n^2 - 1}{(n+1)(n^2 - n + 6)} = \lim_{n \to \infty} \frac{1 + \frac{2}{n} - \frac{1}{n^3}}{(1 + \frac{1}{n})(1 - \frac{1}{n} + \frac{6}{n^2})} \quad (7)$$

Since

$$\lim_{n \to \infty} \frac{1}{n^\alpha} = 0 \quad \text{for } \alpha > 0 \quad (8)$$

eq.(7) yields

$$\frac{1 + 0 - 0}{(1+0)(1-0+0)} = 1 \quad (9)$$

179

We shall show that
$$\lim a\left(1 + \frac{a}{n^2}\right)^n = a \qquad (10)$$
or
$$\lim \left(1 + \frac{a}{n^2}\right)^n = 1 \qquad (11)$$

Therefore,
$$\left|\left(1 + \frac{a}{n^2}\right)^n - 1\right| = \left|\sum_{k=0}^{n} \binom{n}{k}\left(\frac{a}{n^2}\right)^k - 1\right|$$

$$= \left|\sum_{k=1}^{n} \binom{n}{k}\left(\frac{a}{n^2}\right)^k\right| \leq \sum_{k=1}^{n} \frac{n^k}{k!}\left(\frac{|a|}{n^2}\right)^k \qquad (12)$$

$$\leq \sum_{k=1}^{n} \left(\frac{|a|}{n}\right)^k$$

We applied the inequalities
$$\binom{n}{k} \leq \frac{n^k}{k!} \qquad (13)$$
and
$$\frac{n^k}{k!\,n^k} < 1 \qquad (14)$$

On the other hand,
$$\sum_{k=1}^{n} \left(\frac{|a|}{n}\right)^k = \frac{1 - \left(\frac{|a|}{n}\right)^{n+1}}{1 - \frac{a}{n}} - 1 = \frac{\frac{|a|}{n} - \left(\frac{|a|}{n}\right)^{n+1}}{1 - \frac{a}{n}}$$

$$= \frac{\frac{|a|}{n}\left[1 - \left(\frac{|a|}{n}\right)^{n-1}\right]}{1 - \frac{a}{n}} < \frac{\frac{|a|}{n}}{1 - \frac{a}{n}} \qquad (15)$$

because for $n > |a|$,
$$1 - \left(\frac{|a|}{n}\right)^n < 1 \qquad (16)$$

Therefore,
$$\left|\left(1 + \frac{a}{n^2}\right)^n - 1\right| < \frac{\frac{|a|}{n}}{1 - \frac{a}{n}} = \frac{|a|}{n - |a|} \qquad (17)$$

where
$$\frac{|a|}{n - |a|} \xrightarrow[n \to \infty]{} 0 \qquad (18)$$

Hence,
$$\lim_{n \to \infty} a\left(1 + \frac{a}{n^2}\right)^n = a \qquad (19)$$

and
$$z_n \to z = 1 + ai \tag{20}$$

For ω_n, we compute
$$\lim \frac{n+1}{n-1} = \lim_{n \to \infty} \frac{1 + \frac{1}{n}}{1 - \frac{1}{n}} = 1 \tag{21}$$

and
$$\omega_n \to \omega = 1 + 3i \tag{22}$$

Applying eq.(1), we find
$$\lim_{n \to \infty} z_n \omega_n = z\omega = (1+ai)(1+3i)$$
$$= (1-3a) + i(3+a). \tag{23}$$

● **PROBLEM 5-4**

1. Prove the theorem
$$\left\{ \begin{matrix} \lim z_n = z \\ \lim \omega_n = \omega \\ \omega_n, \omega \neq 0 \end{matrix} \right\} \Rightarrow \left(\lim \frac{z_n}{\omega_n} = \frac{z}{\omega} \right) \tag{1}$$

2. Calculate $\lim_{n \to \infty} \frac{z_n}{\omega_n}$ where
$$z_n = (n\sqrt{n^2+1} - n^2) + ni \text{ (or replace } ni \text{ by } 2i) \tag{2}$$
$$\omega_n = \left(1 + \frac{1}{n}\right)^n - 3i \tag{3}$$

Solution: 1. Let
$$\varepsilon = \left|\frac{\omega}{2}\right| > 0 \tag{4}$$

Then, since $\omega_n \to \omega$, we obtain
$$|\omega_n - \omega| < \varepsilon \quad \text{for } n > N$$

or
$$|\omega_n| = |\omega + (\omega_n - \omega)| \geq |\omega| - |\omega_n - \omega|$$
$$> |\omega| - \varepsilon = |\omega| - \left|\frac{\omega}{2}\right| = \left|\frac{\omega}{2}\right| \tag{5}$$

Thus, the sequence $\left\{\frac{1}{\omega_n}\right\}$ is bounded.

Since
$$z_n = z + a_n$$
$$\omega_n = \omega + b_n \tag{6}$$

where a_n and b_n are null sequences, we obtain

$$\frac{z_n}{\omega_n} - \frac{z}{\omega} = \frac{z+a_n}{\omega+b_n} - \frac{z}{\omega} = \frac{\omega a_n - z b_n}{\omega(\omega+b_n)} \tag{7}$$

$\omega \neq 0$ and $\left(\frac{1}{\omega_n}\right)$ is bounded. Therefore,

$$\left(\frac{1}{\omega(\omega+b_n)}\right) = \left(\frac{1}{\omega \omega_n}\right) \tag{8}$$

is a bounded sequence.

The sequence
$$(\omega a_n - z b_n) \tag{9}$$

is a null sequence. Hence,

$$\left(\frac{z_n}{\omega_n}\right) = \frac{z}{\omega} + \text{null sequence.} \tag{10}$$

We proved that
$$\frac{z_n}{\omega_n} \to \frac{z}{\omega} \tag{11}$$

2. $\lim\left[n\sqrt{n^2+1} - n^2\right] = \lim\left(\frac{n\sqrt{n^2+1} + n^2}{n\sqrt{n^2+1} + n^2}\right)\left[n\sqrt{n^2+1} - n^2\right]$

$$= \lim \frac{n^2}{n\sqrt{n^2+1} + n^2} = \lim \frac{1}{\sqrt{1 + \frac{1}{n^2}} + 1} = \frac{1}{2} \tag{12}$$

Since
$$\lim_{n \to \infty} \frac{1}{n^2} = 0 \tag{13}$$

The problem is with the term ni.

<u>Definition</u>

A sequence of complex numbers (z_n) is said to diverge to infinity. We can denote this by $\lim_{n \to \infty} z_n = \infty$. (14)

if for any real number M, there exists a number N, such that

$$|z_n| > M \quad \text{for } n > N \tag{15}$$

The sequence given by eq.(2) diverges to ∞. If we replace ni by 2i, we obtain

$$z_n \to \tfrac{1}{2} + 2i \tag{16}$$

Remembering that

$$\lim_{n \to \infty} \left(1 + \frac{1}{n}\right)^n = e \tag{17}$$

we find

$$\omega_n \to e - 3i \tag{18}$$

Hence,

$$\lim_{n \to \infty} \frac{z_n}{\omega_n} = \frac{\tfrac{1}{2} + 2i}{e - 3i} \tag{19}$$

● **PROBLEM 5-5**

Show that if

$$\lim_{n \to \infty} z_n = 0 \tag{1}$$

then

$$\lim_{n \to \infty} \frac{z_1 + z_2 + \ldots + z_n}{n} = 0 \tag{2}$$

Solution: Choose an arbitrary positive real number ε. Then, there exists N_1 such that

$$\bigwedge_{k > N_1} |z_k| < \frac{\varepsilon}{2} \tag{3}$$

There exists N_2, such that

$$N_2 > N_1 \quad \text{and}$$

$$\frac{z_1 + z_2 + \ldots + z_{N_1}}{n} < \frac{\varepsilon}{2} \tag{4}$$

for every $n > N_2$.

Consider the sum

$$\frac{z_1 + z_2 + \ldots + z_n}{n} = \frac{z_1 + \ldots + z_{N_1}}{n} + \frac{z_{N_1+1} + \ldots + z_n}{n} \tag{5}$$

Applying eq.(3) and eq.(4) to eq.(5) we find

$$\frac{z_1 + \ldots + z_n}{n} < \frac{\varepsilon}{2} + \frac{1}{n} \cdot (n - N_1) \cdot \frac{\varepsilon}{2} \qquad (6)$$

$$= \frac{\varepsilon}{2} + \frac{\varepsilon}{2} - \frac{N_1 \varepsilon}{2n} < \varepsilon$$

Note that all terms $z_{N_1}, z_{N_1+1}, z_{N_1+2}, \ldots, z_n$ are smaller than $\frac{\varepsilon}{2}$. That completes the proof that

$$\underset{\varepsilon > 0}{\wedge} \underset{N}{\vee} \underset{n > N}{\wedge} \frac{z_1 + z_2 + \ldots + z_n}{n} < \varepsilon \qquad (7)$$

i.e.

$$\lim_{n \to \infty} \frac{z_1 + z_2 + \ldots + z_n}{n} = 0 \qquad (8)$$

• **PROBLEM 5-6**

Can you make any generalizations of the results obtained in Problem 5-5?

Solution: The most natural way to obtain a more general theorem than in Problem 5-5 would be to consider a sequence

$$z_n \to a \qquad (1)$$

In such case

$$\lim_{n \to \infty} (z_n - a) = 0 \qquad (2)$$

Applying the results of Problem 5-5 to the sequence $(z_n - a) \to 0$ we find

$$\lim_{n \to \infty} \frac{(z_1 - a) + (z_2 - a) + \ldots + (z_n - a)}{n} = \lim_{n \to \infty} \left(\frac{z_1 + \ldots + z_n}{n} - a \right) = 0 \qquad (3)$$

Hence, if $\lim_{n \to \infty} z_n = a$, then

$$\lim_{n \to \infty} \frac{z_1 + \ldots + z_n}{n} = a \qquad (4)$$

• **PROBLEM 5-7**

Let
$$\lim_{n \to \infty} z_n = a \qquad (1)$$

and (α_n) be any sequence of positive numbers such that

$$\alpha_1 + \ldots + \alpha_n \xrightarrow[n \to \infty]{} + \infty \qquad (2)$$

Show that

$$\frac{\alpha_1 z_1 + \alpha_2 z_2 + \ldots + \alpha_n z_n}{\alpha_1 + \alpha_2 + \ldots + \alpha_n} \to a \qquad (3)$$

Solution: First, we assume that $a = 0$, i.e.

$$\lim_{n \to \infty} z_n = 0 \qquad (4)$$

Thus, there exists N_1 such that

$$\bigwedge_{k > N_1} |z_k| < \frac{\varepsilon}{2} \qquad (5)$$

A natural number $N_2 > N_1$ can be chosen such that

$$\frac{|\alpha_1 z_1 + \ldots + \alpha_{N_1} z_{N_1}|}{\alpha_1 + \alpha_2 + \ldots \alpha_n} < \frac{\varepsilon}{2} \qquad (6)$$

for $n > N_2$.

Therefore,

$$\frac{|\alpha_1 z_1 + \ldots + \alpha_n z_n|}{\alpha_1 + \ldots + \alpha_n} \leq \frac{|\alpha_1 z_1 + \ldots + \alpha_{N_1} z_{N_1}|}{\alpha_1 + \ldots + \alpha_n} +$$

$$+ \frac{|\alpha_{N_1+1} z_{N_1+1} + \ldots + \alpha_n z_n|}{\alpha_1 + \ldots + \alpha_n} < \frac{\varepsilon}{2} + \frac{\alpha_{N_1+1} |z_{N_1+1}| + \ldots + \alpha_n |z_n|}{\alpha_1 + \ldots + \alpha_n}$$

$$< \frac{\varepsilon}{2} + \frac{\alpha_{N_1+1} \frac{\varepsilon}{2} + \ldots + \alpha_n \frac{\varepsilon}{2}}{\alpha_1 + \ldots + \alpha_n} < \varepsilon \qquad (7)$$

Because

$$\frac{\alpha_{N_1+1} + \ldots + \alpha_n}{\alpha_1 + \ldots + \alpha_n} < 1 \qquad (8)$$

all α_j are positive.

We have proved that the theorem is true for $a = 0$.

Using the same method as the one described in Problem 5-6 we can show that the theorem is true for any a.

Note that setting

$$\alpha_1 = \alpha_2 = \ldots = \alpha_n = \ldots = 1 \qquad (9)$$

we obtain from eq.(3)

$$\frac{z_1 + \ldots + z_n}{n} \to a \qquad (10)$$

which is eq.(4) of Problem 5-6.

● **PROBLEM 5-8**

In the theorem proven in Problem 5-7, replace the positive numbers (α_j) with complex numbers (β_j) such that

$$\bigwedge_{n \in N} \frac{|\beta_1 + \ldots + \beta_n|}{|\beta_1| + \ldots + |\beta_n|} > b \tag{1}$$

where b is some fixed positive number. Moreover, let

$$|\beta_1| + \ldots + |\beta_n| \xrightarrow[n \to \infty]{} +\infty \tag{2}$$

Prove the theorem.

Solution: First, assume that

$$z_n \to 0 \tag{3}$$

and choose $\varepsilon > 0$.

We can find N_1 such that

$$\bigwedge_{k > N_1} |z_k| < \frac{\varepsilon b}{2} \tag{4}$$

A number $N_2 > N_1$ can be chosen such that for $n > N_2$

$$\frac{|\beta_1 z_1 + \ldots + \beta_{N_1} z_{N_1}|}{|\beta_1 + \ldots + \beta_n|} < \frac{\varepsilon}{2} \tag{5}$$

The numerator of the fraction is constant and the denominator converges to infinity.

$$|\beta_1 + \ldots + \beta_n| > b\left[|\beta_1| + \ldots + |\beta_n|\right] \tag{6}$$

For any $n > N_2$ (and $n > N_1$)

$$\left|\frac{\beta_1 z_1 + \ldots + \beta_n z_n}{\beta_1 + \ldots + \beta_n}\right| \leq \frac{|\beta_1 z_1 + \ldots + \beta_{N_1} z_{N_1}|}{|\beta_1 + \ldots + \beta_n|} + \frac{|\beta_{N_1+1} z_{N_1+1} + \ldots + \beta_n z_n|}{|\beta_1 + \ldots + \beta_n|} \tag{7}$$

The first term on the right-hand side of eq.(7) is smaller than $\frac{\varepsilon}{2}$, see eq.(5).

$$\frac{|\beta_{N_1+1} z_{N_1+1} + \ldots + \beta_n z_n|}{|\beta_1 + \ldots + \beta_n|} \leq \frac{|\beta_{N_1+1} z_{N_1+1}| + \ldots + |\beta_n z_n|}{|\beta_1 + \ldots + \beta_n|} \tag{8}$$

$$< \frac{\varepsilon b}{2} \frac{|\beta_{N_1+1}|+\ldots+|\beta_n|}{|\beta_1+\ldots+\beta_n|}$$

From eq.(1),

$$\frac{1}{b} > \frac{|\beta_1|+\ldots+|\beta_n|}{|\beta_1+\ldots+\beta_n|} \geq \frac{|\beta_{N_1+1}|+\ldots+|\beta_n|}{|\beta_1+\ldots+\beta_n|} \qquad (9)$$

Combining eq.(8) with eq.(9) we obtain

$$\frac{|\beta_{N_1+1}z_{N_1+1}+\ldots+\beta_n z_n|}{|\beta_1+\ldots+\beta_n|} \leq \frac{\varepsilon b}{2} \cdot \frac{1}{b} = \frac{\varepsilon}{2} \qquad (10)$$

Hence, we have proved the theorem.

If

$$z_n \to 0, \qquad (11)$$

then

$$\frac{\beta_1 z_1+\ldots+\beta_n z_n}{\beta_1+\ldots+\beta_n} \to 0. \qquad (12)$$

where (β_k) is a sequence of complex numbers satisfying eq.(1) and (2). It is easy to show that if

$$z_n \to a, \qquad (13)$$

then

$$\frac{\beta_1 z_1+\ldots+\beta_n z_n}{\beta_1+\ldots+\beta_n} \to a \qquad (14)$$

with the same conditions on (β_k).

● **PROBLEM 5-9**

Two sequences are given (z_n) and (ω_n) such that

$$\lim_{n\to\infty} z_n = a \qquad (1)$$

and

$$\lim_{n\to\infty} \omega_n = b \qquad (2)$$

Show that

$$\lim_{n\to\infty} \frac{z_1\omega_1+z_2\omega_2+\ldots+z_n\omega_n}{n} = ab \qquad (3)$$

Solution: We shall apply the following theorem about sequences:
Let (z_n) and (ω_n) be two sequences of complex numbers such that $z_n \to a$, $\omega_n \to b$ and

$$z_n \omega_n \xrightarrow[n\to\infty]{} ab \qquad (4)$$

Consider a sequence $(z_n \omega_n)$ which converges to ab. We shall apply the results of Problem 5-6. Hence,

$$\lim_{n\to\infty} \frac{z_1\omega_1 + z_2\omega_2 + \ldots + z_n\omega_n}{n} = ab \qquad (5)$$

q.e.d.

• **PROBLEM 5-10**

The arrangement of numbers a_{nm} is given as shown:

a_{11}
$a_{21}\ a_{22}$
$a_{31}\ a_{32}\ a_{33}$
$a_{41}\ a_{42}\ a_{43}\ a_{44}$
$a_{51}\ a_{52}\ \ldots$
$\vdots\ \vdots$

Each vertical sequence tends to zero. That is, for every fixed ℓ

$$a_{n\ell} \xrightarrow[n\to\infty]{} 0 \qquad (1)$$

there exists a positive constant K. Such that,

$$|a_{n1}| + |a_{n2}| + \ldots + |a_{nn}| < K \qquad (2)$$

for all n.
Show that if

$$z_n \to 0 \qquad (3)$$

then

$$a_{n1}z_1 + a_{n2}z_2 + \ldots + a_{nn}z_n \to 0 \qquad (4)$$

Solution: Let $\varepsilon > 0$ be an arbitrary number. Then there exists N such that for every $k > N$

$$|z_k| < \frac{\varepsilon}{2K} \qquad (5)$$

For n > N

$$|a_{n_1}z_1 + a_{n_2}z_2 + \ldots + a_{nn}z_n|$$

$$\leq |a_{n_1}||z_1| + |a_{n_2}||z_2| + \ldots + |a_{nN}||z_N| + \ldots + |a_{nn}||z_n|$$

$$\leq (|a_{n_1}||z_1| + \ldots + |a_{nN}||z_N|) + |a_{n(N+1)}|\frac{\varepsilon}{2K} + \ldots + |a_{nn}|\frac{\varepsilon}{2K}$$

by eq.(2)
$$\leq (|a_{n_1}||z_1| + \ldots + |a_{nN}||z_N|) + \frac{\varepsilon}{2K} \cdot K \quad (6)$$

Hence, for n > N

$$|a_{n_1}z_1 + \ldots + a_{nn}z_n| < |a_{n_1}||z_1| + \ldots + |a_{nN}||z_N| + \frac{\varepsilon}{2} \quad (7)$$

Keeping N fixed and letting $n \to \infty$ we have

$$a_{n_1} \xrightarrow[n \to \infty]{} 0$$
$$\vdots \qquad\qquad\qquad (8)$$
$$a_{nN} \xrightarrow[n \to \infty]{} 0$$

Thus, each term of the sum of N elements converges to zero. We can choose N_2 such that, for every $n > N_2$

$$|a_{n_1}||z_1| + \ldots + |a_{nN}||z_N| < \frac{\varepsilon}{2} \quad (9)$$

Hence, for

$$n > \max(N_1 N_2) \quad (10)$$

$$|a_{n_1}z_1 + \ldots + a_{nn}z_n| < \varepsilon \quad (11)$$

and

$$a_{n_1}z_1 + \ldots + a_{nn}z_n \to 0 \quad (12)$$

● **PROBLEM 5-11**

Consider the system of numbers $a_{\ell k}$ as described in Problem 5-10 with one additional condition

$$A_n = a_{n_1} + a_{n_2} + \ldots + a_{nn} \to 1. \quad (1)$$

Show that if

$$z_n \to a$$

then

$$\omega_n = a_{n_1}z_1 + a_{n_2}z_2 + \ldots + a_{nn}z_n \to a. \quad (2)$$

Solution:
$$\omega_n - A_n a = a_{n_1}(z_1-a)+a_{n_2}(z_2-a)+\ldots+a_{nn}(z_n-a) \quad (3)$$

The sequence $z_n - a$ converges to zero. We can apply the results of Problem 5-10 to the sequence in eq.(3). Hence,

$$\omega_n - A_n a \to 0 \quad (4)$$

But

$$A_n a \to a \quad (5)$$

therefore,

$$\omega_n \to a \quad (6)$$

Note that the theorem obtained in Problem 5-5 can be easily derived from the results of this problem by setting

$$a_{nk} = \frac{1}{n} \quad (7)$$

Then, if

$$z_n \to a \quad (8)$$

$$\omega_n = \frac{z_1+\ldots+z_n}{n} \to a \quad (9)$$

• **PROBLEM 5-12**

Show that if

$$z_n \to a \quad (1)$$

then

$$\frac{\binom{n}{1}z_1 + \binom{n}{2}z_2 + \ldots + \binom{n}{n}z_n}{2^n} \to a \quad (2)$$

Solution: We shall use the results of Problem 5-10. Let

$$a_{nk} = \frac{1}{2^n}\binom{n}{k} \quad (3)$$

We have to show that all conditions imposed on a_{nk} are fulfilled.

$$a_{nk} = \frac{1}{2^n}\frac{n!}{k!(n-k)!} \xrightarrow[n\to\infty]{} 0 \quad (4)$$

Let $K = 1$.

Then, for every n

$$a_{n_1} + a_{n_2} + \ldots + a_{nn}$$

$$= \frac{1}{2^n}\binom{n}{1} + \frac{1}{2^n}\binom{n}{2} + \ldots + \frac{1}{2^n}\binom{n}{n} = \frac{1}{2^n}(2^n-1) < 1 \qquad (5)$$

We used the identity

$$\binom{n}{0} + \binom{n}{1} + \binom{n}{2} + \ldots + \binom{n}{n} = 2^n \qquad (6)$$

Furthermore, for every n

$$a_{n_1} + a_{n_2} + \ldots + a_{nn} = \frac{2^n-1}{2^n} \xrightarrow[n\to\infty]{} 1 \qquad (7)$$

Thus, if

$$z_n \to a \qquad (8)$$

then

$$\frac{1}{2^n}\binom{n}{1}z_1 + \frac{1}{2^n}\binom{n}{2}z_2 + \ldots + \frac{1}{2^n}\binom{n}{n}z_n \to a \qquad (9)$$

SERIES

• **PROBLEM 5-13**

Let $z_1, z_2, \ldots, z_n, \ldots$ be a sequence of complex numbers. An infinite series is defined by

$$z_1 + z_2 + \ldots + z_n + \ldots \equiv \sum_{n=1}^{\infty} z_n \qquad (1)$$

We can look at it from a different point of view.

Let (S_n) be a sequence of partial sums

$$S_1 := z_1$$
$$S_2 := z_1+z_2$$
$$S_3 := z_1+z_2+z_3$$
$$\vdots$$
$$S_n := z_1+z_2+\ldots+z_n$$
$$\vdots$$

The infinite series is defined by two sequences (z_n) and (S_n). Prove the theorem:

<u>Theorem</u>

A necessary and sufficient condition for the series $\sum_{n=1}^{\infty} z_n$ to converge to the sum S such that,

$$\sum_{n=1}^{\infty} \text{Re}(z_n) = \text{Re}(S) \qquad (2)$$

$$\sum_{n=1}^{\infty} \text{Im}(z_n) = \text{Im}(S) \qquad (3)$$

Solution: We shall apply the following theorem:

If (z_n) is a sequence where $z_n = x_n + iy_n$, then a necessary and sufficient condition such that (z_n) converge to $z_0 = x_0 + iy_0$ is

$$\begin{aligned} \lim x_n &= x_0 \\ \lim y_n &= y_0 \end{aligned} \qquad (4)$$

Substituting z_n by S_n we find that a necessary and sufficient condition for S_n to converge to S is

$$\begin{aligned} \lim_{n\to\infty} \text{Re}(S_n) &= \text{Re}(S) \\ \lim_{n\to\infty} \text{Im}(S_n) &= \text{Im}(S) \end{aligned} \qquad (5)$$

But,

$$\begin{aligned} \lim_{n\to\infty} \text{Re}(S_n) &= \lim_{n\to\infty} \text{Re}(z_1 + \ldots + z_n) \\ &= \lim_{n\to\infty} \sum_{1}^{n} \text{Re}(z_n) \end{aligned} \qquad (6)$$

Hence,

$$\begin{aligned} \sum_{n=1}^{\infty} \text{Re}(z_n) &= \text{Re}(S) \\ \sum_{n=1}^{\infty} \text{Im}(z_n) &= \text{Im}(S) \end{aligned} \qquad (7)$$

Note that the series $\sum_{n=1}^{\infty} z_n$ is said to be convergent, or to converge if there exists a finite number S such that, $\lim_{n\to\infty} S_n = S$.

S is called the sum of the series. If $\lim S_n$ does not exist or is infinite, the series is said to be divergent.

• **PROBLEM 5-14**

Prove the theorem:

Theorem

If
$$\sum_{n=1}^{\infty} z_n = S \quad \text{and} \quad \sum_{n=1}^{\infty} \omega_n = W \qquad (1)$$

then

$$\sum_{n=1}^{\infty} (z_n \pm \omega_n) = S \pm W \qquad (2)$$

and

$$\sum_{n=1}^{\infty} a z_n = aS, \quad a \text{ is constant} \qquad (3)$$

Solution: From the definition of a convergent series, we have

$$\sum_{n=1}^{\infty} z_n = \lim_{n\to\infty} S_n = S$$

and

$$\sum_{n=1}^{\infty} \omega_n = \lim_{n\to\infty} W_n = W \qquad (4)$$

where

$$S_n = z_1 + \ldots + z_n$$
$$W_n = \omega_1 + \ldots + \omega_n \qquad (5)$$

Applying the theorem about the limit of sum of sequences, we obtain

$$\sum_{n=1}^{\infty} (z_n \pm \omega_n) = \lim_{n\to\infty} (S_n \pm W_n) = S \pm W \qquad (6)$$

and

$$\sum_{n=1}^{\infty} a z_n = \lim_{n\to\infty} (a S_n) = a \lim_{n\to\infty} S_n = aS \qquad (7)$$

● **PROBLEM 5-15**

Show that a necessary condition for a series $\sum_{n=1}^{\infty} z_n$ of complex numbers to converge is

$$\lim_{n\to\infty} z_n = 0 \qquad (1)$$

Is it a sufficient condition?

Solution: Assume that the series converges to S, that is

$$\lim_{n\to\infty} S_n = S \qquad (2)$$

Then

$$\lim_{n\to\infty} z_n = \lim_{n\to\infty}(S_n - S_{n-1}) = S - S = 0 \qquad (3)$$

That completes the proof.

The condition (1) is not sufficient; the series $\sum_{n=1}^{\infty} S_n$ converge. For example, the harmonic series $\sum_{n=1}^{\infty} \frac{1}{n}$ is not convergent, but

$$\frac{1}{n} \to 0 \tag{4}$$

This theorem enables us to determine quickly that some series are not convergent. For example, $\left(\frac{n+1}{n}\right)$ is not convergent because

$$\frac{n+1}{n} \to 1$$

● **PROBLEM 5-16**

Prove the theorem

$$\left(\sum_{n=1}^{\infty} z_n \text{ is convergent}\right) \Rightarrow \left(\bigwedge_{\substack{(k_n) \\ k_n \in N}} \lim_{n \to \infty}(z_{n+1} + \ldots + z_{n+k_n}) = 0\right) \tag{1}$$

<u>Solution</u>: We have to show that the series $\sum_{n=1}^{\infty} z_n$ is convergent if and only if, for every sequence of natural numbers (k_n) the following condition is satisfied

$$z_{n+1} + z_{n+2} + \ldots + z_{n+k_n} \xrightarrow[n \to \infty]{} 0 \tag{2}$$

$\Rightarrow$

If the series converges, then

$$\bigwedge_{\varepsilon > 0} \bigvee_{N \in N} \bigwedge_{n > N} \bigwedge_{k \in N}$$

$$|z_{n+1} + z_{n+2} + \ldots + z_{n+k}| < \varepsilon \tag{3}$$

Hence, for $n > N$

$$|z_{n+1} + z_{n+2} + \ldots + z_{n+k_n}| < \varepsilon \tag{4}$$

Therefore,

$$\lim_{n \to \infty}(z_{n+1} + z_{n+2} + \ldots + z_{n+k_n}) = 0 \tag{5}$$

$\Leftarrow$

Assume that the series diverges. Then, there exists $\varepsilon > 0$ such that for any N there exists at least one pair of numbers $n > N$ and k such that

$$|z_{n+1} + z_{n+2} + \ldots + z_{n+k}| \geq \varepsilon \tag{6}$$

Using logical notation, we can denote the above statement

$$\bigvee_{\varepsilon>0} \bigwedge_{N\in N} \bigvee_{n>N} \bigvee_{k} |z_{n+1}+\ldots+a_{n+k}| \geq \varepsilon \qquad (7)$$

Thus, there exists a sequence of natural numbers (k_n) such that $z_{n+1}+\ldots+z_{n+k_n}$ does not converge to zero. This is in contradiction with the assumption.

q.e.d.

● **PROBLEM 5-17**

Definition

A series of the form

$$c_0 + c_1(z-z_0) + c_2(z-z_0)^2 + c_3(z-z_0)^3 + \ldots \qquad (1)$$

where $c_0, c_1, c_2 \ldots$ are complex constants and z is a complex variable, is called a complex power series.

A separate chapter will be devoted to the complex power series. Prove the following theorem dealing with the simplest power series - the geometric series:

The series $\sum_{n=1}^{\infty} z^n$ converges if and only if $|z| < 1$ and its sum is $\frac{1}{1-z}$.

Solution: Let

$$S_n = 1 + z + \ldots + z^n \qquad (2)$$

Then

$$zS_n = z + z^2 + \ldots + z^{n+1} \qquad (3)$$

Subtracting we obtain

$$S_n = \frac{1-z^{n+1}}{1-z} \qquad (4)$$

Hence, if $|z| < 1$, then $z^n \to 0$ and

$$S_n = \frac{1-z^{n+1}}{1-z} \to \frac{1}{1-z} \qquad (5)$$

If $|z| = 1$, but $z \neq 1$, then

$$z = \cos \alpha + i \sin \alpha, \qquad 0 < \alpha < 2\pi \qquad (6)$$

and eq.(4) can be written

$$S_n = \frac{1-\cos(n+1)\alpha - i \sin(n+1)\alpha}{1 - \cos \alpha - i \sin \alpha} \qquad (7)$$

S_n does not converge to any limit. Therefore, the series does not converge for $|z| = 1$, $z \neq 1$.

If $|z| > 0$, then

$$|1-z^{n+1}| \geq |z|^{n+1} - 1 \tag{8}$$

and

$$|1-z| \leq 1 + |z| \tag{9}$$

S_n given by eq.(4) becomes

$$|S_n| = \frac{|1-z^{n+1}|}{|1-z|} \geq \frac{|z|^n - 1}{1+|z|} \tag{10}$$

Hence, for any number M there exists $N \in \mathbb{N}$ such that for every $n > N$

$$|S_n| > M \tag{11}$$

S_n does not converge.

If $z=1$, then $S_n \to \infty$

● **PROBLEM** 5-18

Definition

A series of complex numbers $\sum_{n=1}^{\infty} z_n$ is said to converge absolutely or to be absolutely convergent if the series $\sum_{n=1}^{\infty} |z_n|$ is convergent.

Prove the theorem:
If the series $\sum_{n=1}^{\infty} z_n$ of complex numbers converges absolutely, then $\sum_{n=1}^{\infty} z_n$ converges.

Solution: Observe that a necessary and sufficient condition for a series $\sum_{n=1}^{\infty} z_n$ of complex numbers to converge is that (1)

$$\bigwedge_{\varepsilon>0} \bigvee_{N \in \mathbb{N}} \bigwedge_{m>n>N} |S_m - S_n| = |z_{n+1} + z_{n+2} + \ldots + z_m| < \varepsilon$$

If $\sum_{n=1}^{\infty} z_n$ converges absolutely, then for each $\varepsilon > 0$, an N exists such that for $m > n > N$

$$|z_{n+1}| + |z_{n+2}| + \ldots + |z_m| < \varepsilon \tag{2}$$

Hence,

$$|z_{n+1} + z_{n+2} + \ldots + z_m| \leq |z_{n+1}| + \ldots + |z_m| < \varepsilon \tag{3}$$

and the series converges.

The converse is not true.

Take for example, the series $\sum_{n=1}^{\infty} \frac{(-1)^{n+1}}{n}$ which is convergent but not absolutely convergent.

TESTS FOR CONVERGENCE

● **PROBLEM** 5-19

There are many methods of determining whether a series is convergent, absolutely convergent or divergent. Prove these two.

Comparison Test for Convergence

If $|z_n| \leq |\omega_n|$ for all $n > N$ and the series $\sum_{n-1}^{\infty} |\omega_n|$ converges, then the series $\sum_{n=1}^{\infty} z_n$ is absolutely convergent.

Ratio Test (called D'Alembert's* Test)

If
$$\lim_{n \to \infty} \left| \frac{z_{n+1}}{z_n} \right| = K \qquad (1)$$

then the series $\sum_{n=1}^{\infty} z_n$ is absolutely convergent if $K < 1$.

The series is divergent, if $K > 1$.

Solution: Since $\sum_{n=1}^{\infty} \omega_n$ converges absolutely, for every $\varepsilon > 0$, there exists $N_1 > N$ such that for every $n > N_1$

$$|z_{n+1}| + \ldots + |z_k| \leq |\omega_{n+1}| + \ldots + |\omega_k| < \varepsilon \qquad (2)$$

for all k.

Hence, the series $\sum_{n=1}^{\infty} z_n$ also converges absolutely.

D'Alembert's Test

$$\left| \frac{z_{n+1}}{z_n} \right| \to k \qquad (3)$$

Hence,

$$\bigwedge_{\varepsilon > 0} \bigvee_{N} \bigwedge_{n > N} \left| \left| \frac{z_{n+1}}{z_n} \right| - K \right| < \varepsilon \qquad (4)$$

or

$$-\varepsilon < \left|\frac{z_{n+1}}{z_n}\right| - K < \varepsilon \tag{5}$$

For K < 1 we can choose ε such that

$$L := K + \varepsilon < 1 \tag{6}$$

For n > N

$$\left|\frac{z_{n+1}}{z_n}\right| < L < 1 \tag{7}$$

or

$$|z_{n+1}| < L|z_n| \tag{8}$$

Now we will show that the series $\sum_{n=1}^{\infty} |z_n|$ is convergent.

$$\sum_{n=1}^{\infty} |z_n| = \underbrace{|z_1| + \ldots + |z_{N+1}|}_{M} + |z_{N+2}| + \ldots + |z_n| + \ldots \tag{9}$$

Let M be the sum of the first N+1 elements.

$$\begin{aligned}
|z_{N+2}| &< L|z_{N+1}| \\
|z_{N+3}| &< L|z_{N+2}| < L^2|z_{N+1}| \\
&\vdots \qquad\qquad \vdots \qquad\qquad \vdots \\
|z_{N+n+1}| &< L|z_{N+n}| < \ldots < L^n|z_{N+1}|
\end{aligned} \tag{10}$$

Combining eq.(9) and eq.(10) we find

$$\begin{aligned}
\sum_{n=1}^{\infty} |z_n| &= M + |z_{N+1}| + |z_{N+2}| + \ldots \\
&< M + |z_{N+1}| + L|z_{N+1}| + L^2|z_{N+1}| + L^3|z_{N+1}| + \ldots \\
&= M + |z_{N+1}| \sum_{j=1}^{\infty} L^j
\end{aligned} \tag{11}$$

From Problem 5-17, $\sum_{j=1}^{\infty} L^j$ converges to $\frac{1}{1-L}$ for $|L| < 1$. Therefore, $\sum_{n=1}^{\infty} |z_n|$ converges and $\sum_{n=1}^{\infty} z_n$ converges absolutely.

For

$$\left|\frac{z_{n+1}}{z_n}\right| \to K > 1 \tag{12}$$

we have

$$\bigwedge_{\varepsilon>0} \bigvee_{N} \bigwedge_{n>N} \quad K-\varepsilon < \left|\frac{z_{n+1}}{z_n}\right| < K+\varepsilon \tag{13}$$

Since $K > 1$, we can choose ε such that

$$1 < L := K - \varepsilon \tag{14}$$

Thus,

$$|z_{N+1}| \leq |z_n| < |z_{n+1}| \tag{15}$$

and choosing ε such that $|z_{N+1}| > 0$ we obtain

$$\lim_{n\to\infty} |z_n| \neq 0 \tag{16}$$

Based on results of Problem 5-15, the series $\sum_{n=1}^{\infty} |z_n|$ is not convergent.

D'Alembert's test cannot be applied to test for convergence when $K = 1$.

*Jean-le-Rond D'Alembert who was born and who died in Paris (1717-1783), was the illegitimate child of the chevalier Destouches (who later became general) and Madame de Tencin (sister of a cardinal). D'Alembert was abandoned by his mother on the steps of the little church-St. Jean-le-Rond. As it was customary in those days the child was given the name of the place where he had been found. In 1738 he wrote his first mathematical paper on integral calculus and in 1740 was elected a member of the French Academy. In 1743 D'Alembert published "Traite de Dynamique" where he describes the principle known by his name. He also wrote mechanics of fluids and the motion of air. His research brought him to an equation of the form:

$$\frac{\partial^2 f}{\partial x^2} = \frac{\partial^2 f}{\partial t^2}$$

He succeeded in showing that the solution was of the form

$$f = \Psi(x+t) + \phi(x-t)$$

where Ψ and ϕ are arbitrary functions. D'Alembert was one of the major contributors to the great French encyclopedia, for which he wrote an introduction and numerous articles on philosophy and mathematics.

Known for his capricious character and bad manners D'Alembert had more enemies than friends.

S.G. Tallentrye in his Friends of Voltaire, London, 1907, writes:

"In himself D'Alembert was always rather a great intelligence than a great character. To the magnificence of the one he owed all that has made him immortal, and to the weakness of the other the sorrows and the failures of his life."

• **PROBLEM 5-20**

Test the series $\sum_{n=1}^{\infty} \frac{i^n}{n}$ and $\sum_{n=1}^{\infty} \frac{i^n}{n!}$ for absolute convergence.

Solution: Let us try the ratio test.

$$\left|\frac{z_{n+1}}{z_n}\right| = \left|\frac{i^{n+1} n}{(n+1) i^n}\right| = \frac{n}{n+1} \quad (1)$$

Hence,

$$\lim_{n \to \infty} \left|\frac{z_{n+1}}{z_n}\right| = 1 \quad (2)$$

Thus, the ratio test fails.

We can try to solve this problem by separating z_n into its real and imaginary parts.

Let

$$z_n = x_n + iy_n \quad (3)$$

where

$$z_n = \frac{i^n}{n}$$

Then

$$\sum_{n=1}^{\infty} x_n = \sum_{n=1}^{\infty} \frac{(-1)^n}{2n} \quad (4)$$

and

$$\sum_{n=1}^{\infty} y_n = \sum_{n=1}^{\infty} \frac{(-1)^{n-1}}{2n-1} \quad (5)$$

Each of these series is convergent, hence $\sum_{n=1}^{\infty} z_n$ is convergent. But, neither of the series converges absolutely. Therefore, $\sum_{n=1}^{\infty} z_n$ does not converge absolutely.

$$\left|\frac{z_{n+1}}{z_n}\right| = \left|\frac{i^{n+1} n!}{(n+1)! \, i^n}\right| = \frac{1}{n+1} \quad (6)$$

and

$$\lim_{n \to \infty} \left|\frac{z_{n+1}}{z_n}\right| = 0 \quad (7)$$

and the series $\sum_{n=1}^{\infty} \frac{i^n}{n}$ converges absolutely.

• **PROBLEM** 5-21

Determine the convergence of the series

$$1 + \frac{1}{2} + \frac{1}{3} + \frac{1}{4} + \frac{1}{5} + \ldots \quad (1)$$

called harmonic series.

Solution: Observe that

$$\frac{1}{n+1} + \frac{1}{n+2} + \ldots + \frac{1}{2n} < n\,\frac{1}{2n} = \frac{1}{2} \quad (2)$$

Neglecting the first two terms, we shall assemble the terms of the series in groups consisting of $2, 4, 8, \ldots, 2^{n-1} \ldots$ terms

$$1 + \frac{1}{2} + \underbrace{\frac{1}{3} + \frac{1}{4}}_{2} + \underbrace{\frac{1}{5} + \frac{1}{6} + \frac{1}{7} + \frac{1}{8}}_{2^2} + \underbrace{\frac{1}{9} + \ldots + \frac{1}{16}}_{2^3} \quad (3)$$

$$+ \ldots + \underbrace{\frac{1}{2^{n-1}+1} + \ldots + \frac{1}{2^n}}_{2^{n-1}} + \ldots$$

Each of the partial sums is larger than $\frac{1}{2}$. Since there are infinitely many groups in (3), the series $\sum_{n=1}^{\infty} \frac{1}{n}$ is divergent.

• **PROBLEM** 5-22

Determine the convergence of the series

$$\sum_{n=1}^{\infty} \frac{1}{n^s} \quad (1)$$

where s is a real number.

Solution: Observe that for $s < 1$

$$\frac{1}{n} < \frac{1}{n^s} \quad (2)$$

Thus,

$$\sum_{n=1}^{\infty} \frac{1}{n} \leq \sum_{n=1}^{\infty} \frac{1}{n^s} \quad (3)$$

Since the harmonic series is divergent (see Problem 5-21) then, the series on the right hand side in (3) is also divergent.

Thus, for $s \leq 1$ the series $\sum_{n=1}^{\infty} \frac{1}{n^s}$ is divergent.

Consider $s > 1$. Let $s = 1 + \sigma$, $\sigma > 0$. We have

$$\frac{1}{(n+1)^s} + \frac{1}{(n+2)^s} + \ldots + \frac{1}{(2n)^s} < n \cdot \frac{1}{n^s} = \frac{1}{n^\sigma} \tag{4}$$

Grouping the terms we obtain

$$1 + \frac{1}{2^s} + \underbrace{\frac{1}{3^s} + \frac{1}{4^s}}_{2} + \underbrace{\frac{1}{5^s} + \ldots + \frac{1}{8^s}}_{2^2} + \underbrace{\frac{1}{9^s} + \ldots + \frac{1}{16^s}}_{2^3} + \ldots$$

$$+ \underbrace{\frac{1}{(2^{k-1}+1)^s} + \ldots + \frac{1}{(2^k)^s}}_{2^{k-1}} + \ldots \tag{5}$$

Each sum is smaller than the respective element of the geometrical sequence

$$\frac{1}{2^\sigma}, \; \frac{1}{4^\sigma} = \frac{1}{(2^\sigma)^2}, \; \frac{1}{8^\sigma} = \frac{1}{(2^\sigma)^3}, \ldots, \frac{1}{(2^{k-1})^\sigma} = \frac{1}{(2^\sigma)^{k-1}} \tag{6}$$

Thus,

$$1 + \frac{1}{2^s} + \frac{1}{3^s} + \frac{1}{4^s} + \ldots < 1 + \frac{1}{2^{1+\sigma}} + \frac{1}{2^\sigma} + \frac{1}{(2^\sigma)^2} + \frac{1}{(2^\sigma)^3} + \ldots$$

$$= 1 + \frac{1}{2^{1+\sigma}} + \frac{1}{1 - \frac{1}{2^\sigma}} \tag{7}$$

Hence, the series (1) is convergent for $s > 1$.

● **PROBLEM 5-23**

Determine the convergence of the series

$$\sum_{n=1}^{\infty} \left(\frac{1}{5n!} + \frac{2i}{5n!} \right) \tag{1}$$

Solution: Let

$$z_n = \frac{1}{5n!} + \frac{2i}{5n!} \tag{2}$$

Then

$$|z_n| = \sqrt{\frac{1}{25(n!)^2} + \frac{4}{25(n!)^2}} = \frac{1}{\sqrt{5}\, n!} \tag{3}$$

Observe that

$$\frac{1}{\sqrt{5}\, n!} < \frac{1}{n!} \tag{4}$$

Let
$$\omega_n = \frac{1}{n!} \qquad (5)$$

The series $\sum_{n=1}^{\infty} \omega_n$ is convergent because

$$e = 1 + \frac{1}{1!} + \frac{1}{2!} + \frac{1}{3!} + \ldots + \frac{1}{n!} + \ldots \qquad (6)$$

where e is Euler's constant

$$e = 2.71828182845\ldots \qquad (7)$$

Because

$$|z_n| \leq |\omega_n|$$

($\sum_{n=1}^{\infty} |\omega_n|$ is convergent), the series $\sum_{n=1}^{\infty} \left(\frac{1}{5n!} + \frac{2i}{5n!}\right)$ is absolutely convergent (see Problem 5-19), it is therefore convergent (see Problem 5-18).

● **PROBLEM 5-24**

Determine the convergence of the series $\sum_{n=1}^{\infty} \frac{1}{n!}\left(\frac{n}{e}\right)^n$. (1)

Solution: We shall apply Raabe's* test to determine convergence of series (1). If

$$\lim_{n \to \infty} n\left(\frac{|z_n|}{|z_{n+1}|} - 1\right) = K, \qquad (2)$$

then $\sum_{n=1}^{\infty} z_n$ converges absolutely if $K > 1$ and diverges or converges conditionally if $K < 1$.

If $K = 1$, the test fails.

If $\sum_{n=1}^{\infty} z_n$ converges but $\sum_{n=1}^{\infty} |z_n|$ does not converge, then $\sum_{n=1}^{\infty} z_n$ converges conditionally.

Let
$$R_n = n\left(\frac{|z_n|}{|z_{n+1}|} - 1\right) \qquad (3)$$

where
$$z_n = \frac{1}{n!}\left(\frac{n}{e}\right)^n \qquad (4)$$

then
$$R_n = n\left[\frac{e}{\left(1 + \frac{1}{n}\right)^n} - 1\right] \qquad (5)$$

To evaluate the limit $\lim R_n$ consider a more general expression

$$\frac{1}{x}\left[\frac{e}{(1+x)^{\frac{1}{x}}} - 1\right], \quad x \to 0 \tag{6}$$

Applying de l'Hospital's theorem we find

$$-\frac{e}{\left[(1+x)^{\frac{1}{x}}\right]^2} \cdot \left\{(1+x)^{\frac{1}{x}} + \ln(1+x) \cdot \left(-\frac{1}{x^2}\right)\right.$$

$$\left. + \frac{1}{x}(1+x)^{\frac{1}{x}-1}\right\} = \frac{e}{(1+x)^{\frac{1}{x}}} \cdot \frac{\ln(1+x) - \frac{x}{1+x}}{x^2} \tag{7}$$

Substituting

$$\ln(1+x) = x - \tfrac{1}{2}x^2 + O(x^2) \tag{8}$$

and

$$\frac{x}{1+x} = x - x^2 + O(x^2) \tag{9}$$

we find

$$\lim_{n\to\infty} R_n = \frac{1}{2} \tag{10}$$

Hence, the series diverges. It does not converge conditionally because all elements are real and positive.

*J.L. Raabe (1801-1859) of Zürich. His work was almost entirely devoted to the question of convergency of series.

• **PROBLEM 5-25**

Show that the series

$$\sum_{n=1}^{\infty} \frac{(-1)^n}{n} \tag{1}$$

converges.

<u>Solution</u>: We shall use the following test called Alternating series test.

If $a_n \geq 0$ and $a_{n+1} \leq a_n$ for $n = 1,2,3\ldots$ and $\lim_{n\to\infty} a_n = 0$, then

$$a_1 - a_2 + a_3 - a_4 \ldots = \Sigma(-1)^{n+1} a_n \tag{2}$$

converges.

In our case
$$a_n = \frac{1}{n} \tag{3}$$

Hence, $a_n \geq 0$, $a_{n+1} \leq a_n$ and $\lim a_n = \lim \frac{1}{n} = 0$.

We have
$$\sum_{n=1}^{\infty} \frac{(-1)^n}{n} = (-1)\Sigma(-1)^{n+1} a_n \tag{4}$$

Since $\sum_{n=1}^{\infty} (-1)^{n+1} a_n$ converges also $(-1) \sum_{n=1}^{\infty} (-1)^{n+1} a_n$ converges. Thus, $\sum_{n=1}^{\infty} \frac{(-1)^n}{n}$ is a convergent series.

• PROBLEM 5-26

Extend the triangle inequality to infinite series.

<u>Solution</u>: The triangle inequality states that
$$|z_1 + \ldots + z_n| \leq |z_1| + \ldots + |z_n| \tag{1}$$

where $z_1, \ldots, z_n$ are complex numbers. We shall attempt to prove the following:

<u>Theorem</u>

If the series $\sum_{n=1}^{\infty} z_n$ of complex numbers converges absolutely, then
$$\left| \sum_{n=1}^{\infty} z_n \right| \leq \sum_{n=1}^{\infty} |z_n| \tag{2}$$

Let
$$S = \sum_{n=1}^{\infty} z_n \quad \text{and} \tag{3}$$

$$S_N = \sum_{n=1}^{N-1} z_n, \quad T_N = \sum_{n=N}^{\infty} z_n \tag{4}$$

Thus,
$$S = S_N + T_N \quad \text{for all N.} \tag{5}$$

Let $\varepsilon > 0$. The series $\sum_{n=1}^{\infty} z_n$ converges thus, for sufficiently large N
$$|S - S_N| = |T_N| < \varepsilon \tag{6}$$

Hence,

$$\left| \sum_{n=1}^{\infty} z_n \right| = |S| = |S_N + T_N| \leq |S_N| + |T_N|$$

$$= \left| \sum_{n=1}^{N-1} z_n \right| + |T_N| \leq \sum_{n=1}^{N-1} |z_n| + |T_N| \qquad (7)$$

$$< \sum_{n=1}^{\infty} |z_n| + \varepsilon$$

● **PROBLEM 5-27**

A series $\sum_{m=1}^{\infty} b_m$ is said to be a rearrangement of a series $\sum_{n=1}^{\infty} a_n$ if there exists a one-to-one correspondence between the indices n and m such that $a_n = b_m$ for corresponding indices. Prove the following:

Theorem

If $\sum_{n=1}^{\infty} a_n$ is absolutely convergent and $\sum_{m=1}^{\infty} b_m$ is a rearrangement of $\sum_{n=1}^{\infty} a_n$, then $\sum_{m=1}^{\infty} b_m$ is absolutely convergent and has the same sum as $\sum_{n=1}^{\infty} a_n$.

Solution: We have

$$\sum_{m=1}^{M} |b_m| \leq \sum_{n=1}^{\infty} |a_n| \qquad (1)$$

Let us denote the partial sum and sum

$$S_n = a_1 + \ldots + a_n$$

$$S = \sum_{n=1}^{\infty} a_n \qquad (2)$$

$$S'_m = b_1 + \ldots + b_m$$

$$S' = \sum_{m=1}^{\infty} b_m$$

We choose an arbitrary $\varepsilon > 0$ and N such that for $n > N$ and $k \geq 1$

$$|S_n - S| < \frac{\varepsilon}{2} \quad \text{and} \qquad (3)$$

$$|a_{n+1}| + |a_{n+2}| + \ldots + |a_{n+k}| < \frac{\varepsilon}{2} \qquad (4)$$

N can be found since Σa_n converges to S.

For m sufficiently large, S'_m will include all of $a_1 \ldots a_n$ and perhaps more

$$S'_m = S_n + a_{k_1} + \ldots + a_{k_r}, \qquad (5)$$

where $k_1, \ldots, k_r$ are all larger than n. Let $K = n+k$ be the largest of these indices. Then

$$\begin{aligned}|S'_m - S_n| &\leq |a_{k_1}| + \ldots + |a_{k_r}| \\ &\leq |a_{n+1}| + \ldots + |a_{n+k}| < \frac{\varepsilon}{2}\end{aligned} \qquad (6)$$

We have

$$|S'_m - S| = |S'_m - S_n + S_n - S| \leq |S'_m - S_n| \\ + |S_n - S| \leq \frac{\varepsilon}{2} + \frac{\varepsilon}{2} = \varepsilon \qquad (7)$$

Hence, S'_m converges to S and

$$S' = S \qquad (8)$$

● **PROBLEM 5-28**

Prove the following.

Theorem

If the series $\sum_{n=1}^{\infty} z_n = S$, $\sum_{m=1}^{\infty} \omega_m = T$ and are absolutely convergent, then the series

$$\sum_{n=1}^{\infty} (z_1 \omega_n + z_2 \omega_{n-1} + \ldots + z_{n-1} \omega_2 + z_n \omega_1) \qquad (1)$$

also converges absolutely and its sum is $S \cdot T$.

Solution: The terms of the product of two series

$$\sum_{n=1}^{\infty} z_n \cdot \sum_{m=1}^{\infty} \omega_m \qquad (2)$$

can be arranged in many ways. Of the many possible arrangements the Cauchy product is commonly used. Fig. 1 illustrates its principle

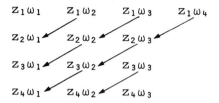

Fig. 1

Another possible arrangement is shown in Fig. 2.

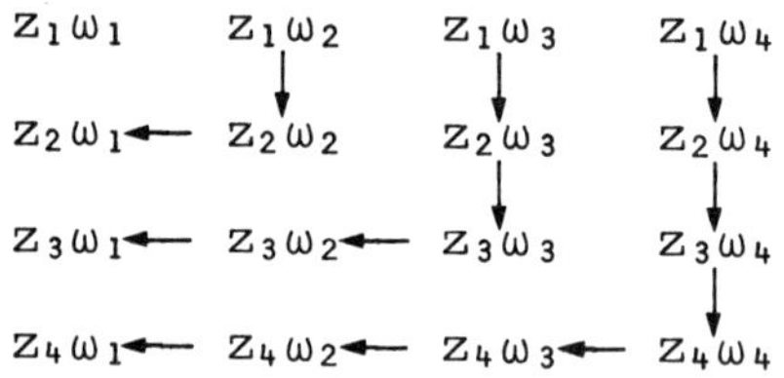

Fig. 2

Let
$$\sum_{n=1}^{\infty} z_n \sum_{m=1}^{\infty} \omega_m = \sum_{k=1}^{\infty} \nu_k \qquad (3)$$

We choose the arrangement shown in Fig. 2, that is

$$z_1\omega_1+z_1\omega_2+z_2\omega_2+z_2\omega_1+z_1\omega_3+z_2\omega_3+z_3\omega_3+z_3\omega_2+z_3\omega_1+\ldots \qquad (4)$$

Hence,
$$\nu_1 = z_1\omega_1$$
$$\nu_1 + \nu_2 + \nu_3 + \nu_4 = (z_1+z_2)(\omega_1+\omega_2) \qquad (5)$$

and in general

$$\nu_1 + (\nu_2+\nu_3+\nu_4) + \ldots + (\ldots + \nu_{n^2})$$
$$= (z_1 + \ldots + z_n)(\omega_1 + \ldots + \omega_n) \qquad (6)$$

Hence,

$$|\nu_1| + (|\nu_2|+|\nu_3|+|\nu_4|) + \ldots + (\ldots + |\nu_{n^2}|)$$
$$= (|z_1| + \ldots + |z_n|)(|\omega_1| + \ldots + |\omega_n|) \qquad (7)$$

Since the series $\Sigma|z_n|$ and $\Sigma|\omega_n|$ converge, from eq.(7) we conclude that the left-hand side approaches a limit as $n \to \infty$. On the basis of the theorem about rearrangement of a series, we can drop the parentheses.

Hence, $\Sigma|\nu_k|$ converges so that $\Sigma\nu_k$ is absolutely convergent.

Furthermore,

$$\sum_{k=1}^{\infty} \nu_k = \lim_{n\to\infty}(\nu_1+\ldots+\nu_{n^2}) = \sum_{n=1}^{\infty} z_n \cdot \sum_{m=1}^{\infty} \omega_m = S \cdot T \qquad (8)$$

The theorem is valid for any arrangement of (ν_k).

If the series $\sum_{n=1}^{\infty} z_n$ and $\sum_{n=1}^{\infty} \omega_n$ converge, but are not both absolutely convergent, we still can form the Cauchy product and show that if $\Sigma\nu_k$ converges, then

$$\sum_{n=1}^{\infty} z_n \sum_{m=1}^{\infty} \omega_m = \sum_{k=1}^{\infty} \nu_k \qquad (9)$$

Furthermore, if one of the series Σz_n, $\Sigma\omega_n$ converges absolutely, then $\Sigma\nu_k$ must converge.

● **PROBLEM 5-29**

Let (x_n) be a sequence of real numbers bounded from above

$$\left(\bigwedge_{n\in N} \quad x_n < M \right) \qquad (1)$$

that has one or more accumulation points. The least upper bound of all accumulation points is called the upper limit or limit superior of the sequence (x_n). It is denoted by

$$\overline{\lim_{n\to\infty}} \, x_n \qquad (2)$$

If (x_n) is bounded from below and has one or more accumulation points, then the greatest lower bound of all accumulation points is called the lower limit or limit inferior of (x_n) and is denoted by

$$\underline{\lim_{n\to\infty}} \, x_n \qquad (3)$$

If the sequence is unbounded from above and

$$\bigwedge_{K\in R'} \quad \bigvee \text{ infinitely many } x_n \quad x_n > K,$$

then we write

$$\overline{\lim_{n\to\infty}} \, x_n = +\infty \qquad (4)$$

If for every real number K, $x_n < K$ for infinitely many n, we denote

$$\lim_{n \to \infty} x_n = -\infty \qquad (5)$$

Compute the upper and lower limit of the sequence

$$x_n = 1 + \sin\left(\frac{n\pi}{2}\right) \qquad (6)$$

Solution: Compute the first few elements of sequence (6)

$$\begin{aligned} x_1 &= 2 \\ x_2 &= 1 \\ x_3 &= 0 \\ x_4 &= 1 \end{aligned} \qquad (7)$$

Observe that

$$x_n = 1 + \sin\left(\frac{n\pi}{2}\right) = \begin{cases} 1 & \text{for } n = 2k \\ 2 & \text{for } n = 4k+1 \\ 0 & \text{for } n = 4k+3 \end{cases} \qquad (8)$$

where $k = 0, 1, 2, \ldots$

Hence,

$$\overline{\lim_{n \to \infty}} \left[1 + \sin\left(\frac{n\pi}{2}\right)\right] = 2$$

$$\underline{\lim_{n \to \infty}} \left[1 + \sin\left(\frac{n\pi}{2}\right)\right] = 0 \qquad (9)$$

● **PROBLEM 5-30**

Find the upper and lower limits of the sequences
1. $n - (-1)^{n+1} n$ \qquad (1)
2. $\frac{1}{n} + \sin\left(\frac{n\pi}{2}\right)$ \qquad (2)

Solution: 1. Let $x_n = n - (-1)^{n+1} n$

Then

$$x_n = \begin{cases} 0 & \text{for } n = 2k+1 \\ 2n & \text{for } n = 2k \end{cases} \qquad (3)$$

Hence,

$$\overline{\lim_{n \to \infty}} \left[n - (-1)^{n+1} n\right] = +\infty \qquad (4)$$

$$\lim_{n\to\infty}\left[n-(-1)^{n+1}n\right] = 0 \qquad (5)$$

2. Let
$$x_n = \frac{1}{n} + \sin\left(\frac{n\pi}{2}\right) \qquad (6)$$

Observe that

$$\sin\left(\frac{n\pi}{2}\right) = \begin{cases} 0 & \text{for } n = 2k \\ 1 & \text{for } n = 4k+1 \\ -1 & \text{for } n = 4k+3 \end{cases} \qquad (7)$$

The sequence (6) has three accumulation points $0, 1, -1$.

The sequence $\frac{1}{n}$ converges to zero. Thus, term $\frac{1}{n}$ does not change the situation.

$$\overline{\lim_{n\to\infty}}\left[\frac{1}{n} + \sin\left(\frac{n\pi}{2}\right)\right] = 1$$
$$\underline{\lim_{n\to\infty}}\left[\frac{1}{n} + \sin\left(\frac{n\pi}{2}\right)\right] = -1 \qquad (8)$$

● **PROBLEM 5-31**

Let A be the finite upper limit of the sequence (x_n). Show that for any given $\varepsilon > 0$

1. $A - \varepsilon < x_n$ for an infinite number of elements of (x_n).

2. $x_n < A + \varepsilon$ for all except, at most, a finite number of elements of (x_n).

Solution:
$$\overline{\lim_{n\to\infty}} x_n = A < \infty \qquad (1)$$

A is the least upper bound of all the accumulation points.

Hence, for any $\varepsilon > 0$ the ball $K(A, \varepsilon)$ contains infinitely many elements of (x_n).

$$|x_n - A| < \varepsilon \qquad (2)$$

or
$$x_n - A < \varepsilon, \quad x_n < A + \varepsilon \qquad (3)$$

Since A is the least upper bound of all the accumulation points eq.(3) is true for all, except at most, finite number of elements of (x_n).

The ball $K(A, \varepsilon)$ contains infinitely many elements (x_n). Hence,

$$|x_n - A| < \varepsilon \tag{4}$$

or

$$-\varepsilon < x_n - A < \varepsilon \tag{5}$$

is true for infinitely many elements of (x_n)

$$A - \varepsilon < x_n \tag{6}$$

• **PROBLEM 5-32**

Show that a sequence of real numbers (x_n) has a limit $\lim_{n \to \infty} x_n = a$, if and only if,

$$\overline{\lim_{n \to \infty}} \, x_n = \underline{\lim_{n \to \infty}} \, x_n = a \tag{1}$$

We can denote it as

$$\left(\lim_{n \to \infty} x_n = a \right) \Rightarrow \left(\overline{\lim_{n \to \infty}} \, x_n = \underline{\lim_{n \to \infty}} \, x_n = a \right) \tag{2}$$

Solution: $\Rightarrow$ The sequence (x_n) is convergent to a. Thus, it has one accumulation point and

$$\overline{\lim} \, x_n = \underline{\lim} \, x_n \tag{3}$$

Because

$$\bigwedge_{\varepsilon > 0} \bigvee_{N} \bigwedge_{n > N} |x_n - a| < \varepsilon \tag{4}$$

we have

$$\overline{\lim} \, x_n = \underline{\lim} \, x_n = \lim x_n = a \tag{5}$$

$\Leftarrow$

Let us assume that

$$\overline{\lim_{n \to \infty}} \, x_n = \underline{\lim_{n \to \infty}} \, x_n = a \tag{6}$$

Since the upper and lower limits of (x_n) are equal the sequence has only one accumulation point and

$$\bigwedge_{\varepsilon} \bigvee_{N \in \mathbb{N}} \bigwedge_{n > N} |x_n - a| < \varepsilon \tag{7}$$

Hence,

$$\lim_{n \to \infty} x_n = a \tag{8}$$

• **PROBLEM 5-33**

Let (z_n) be a sequence of complex numbers such that

$$\overline{\lim_{n \to \infty}} \, \sqrt[n]{|z_n|} = K \tag{1}$$

Show that if $K < 1$, $\sum_{n=1}^{\infty} z_n$ is absolutely convergent, if $K > 1$, $\sum_{n=1}^{\infty} z_n$ is divergent.

Solution: Assume that $K < 1$. A positive number α exists such that

$$K < \alpha < 1 \tag{2}$$

A number N exists such that

$$\bigwedge_{n>N} \sqrt[n]{|z_n|} < \alpha \tag{3}$$

or

$$|z_n| < \alpha^n \tag{4}$$

The geometric series $\alpha^N + \alpha^{N+1} + \alpha^{N+2} + \ldots$ converges for $\alpha < 1$. Therefore, the series $|z_N| + |z_{N+1}| + \ldots$ also converges (see Problem 5-19 - comparison test).

Consequently $\sum_{n=1}^{\infty} |z_n|$ converges and $\sum_{n=1}^{\infty} z_n$ converges absolutely.

For $K > 1$, there exists an infinite number of z_n such that

$$\sqrt[n]{|z_n|} > 1 \tag{5}$$

or

$$|z_n| > 1 \tag{6}$$

Hence, the series is divergent.

● **PROBLEM 5-34**

Let (z_n) be a sequence of complex numbers such that

$$\lim_{n \to \infty} \sqrt[n]{|z_n|} = K \tag{1}$$

Show that

if $K < 1$, $\sum_{n=1}^{\infty} z_n$ is absolutely convergent,

if $K > 1$, $\sum_{n=1}^{\infty} z_n$ is divergent,

if $K = 1$, the test fails.

Solution: This theorem known as root test, is very similar to the theorem proved in Problem 5-33.

Suppose $K < 1$. A positive number α can be found such that $K < \alpha < 1$.

There exists N such that

$$\bigwedge_{n > N} \sqrt[n]{|z_n|} < \alpha \quad \text{or} \tag{2}$$

$$|z_n| < \alpha^n$$

The geometrical series converges for $\alpha < 1$, therefore, $\sum_{n=1}^{\infty} |z_n|$ also converges and $\sum_{n=1}^{\infty} z_n$ converges absolutely.

For $\alpha > 1$ the proof is the same as in Problem 5-33.

For

$$\sqrt[n]{|z_n|} \xrightarrow[n \to \infty]{} 1 \tag{3}$$

we cannot establish a geometrical series $\alpha + \alpha^2 + \alpha^3 + \ldots$ for the purpose of comparison.

In this case, other methods have to be applied to determine if the series is convergent or divergent. This test is sometimes called the Cauchy's test.

• **PROBLEM 5-35**

Consider the series

$$1 + a + ab + a^2 b + a^2 b^2 + \ldots + a^n b^{n-1} + a^n b^n + \ldots \tag{1}$$

where a and b are two different positive numbers. What are the conditions on a and b so that the series will be convergent?

Solution: Let us apply d'Alembert's test. If

$$\alpha_1 + \alpha_2 + \ldots + \alpha_n + \ldots \tag{2}$$

is a series, then we denote

$$D_n := \frac{\alpha_{n+1}}{\alpha_n} \tag{3}$$

Thus, for the series (1)

$$D_{2n-1} = a \quad D_{2n} = b \tag{4}$$

and

$$\lim D_{2n-1} = a$$
$$\lim D_{2n} = b \tag{5}$$

Hence, the series (1) is convergent when both $a < 1$ and $b < 1$ and divergent when both $a > 1$ and $b > 1$.

Now, we apply Cauchy's test. For series (2) we define the sequence
$$C_n := \sqrt[n]{\alpha_n} \tag{6}$$

For series (1)
$$C_{2n-1} = \sqrt[2n-1]{a^{n-1}b^{n-1}}, \quad C_{2n} = \sqrt[2n]{a^n b^{n-1}} \tag{7}$$

the limit is
$$\lim C_{2n-1} = \sqrt{ab}$$
$$\lim C_{2n} = \sqrt{ab} \tag{8}$$

Therefore, the series (1) is convergent for $ab < 1$ and divergent for $ab > 1$. Note that Cauchy's test yields more general results in this case than D'Alembert's test.

• **PROBLEM 5-36**

Test for absolute convergence, convergence or divergence.

1. $\sum_{n=1}^{\infty} \dfrac{n}{2} i^{n+1}$ (1)

2. $\sum_{n=1}^{\infty} \dfrac{n! \, i^n}{n^n}$ (2)

3. $\sum_{n=1}^{\infty} \left(\dfrac{1+i}{2}\right)^n$ (3)

Solution: 1. Let us denote
$$a_n = \frac{n}{2} i^{n+1} \tag{4}$$

Observe that $\lim a_n \neq 0$. In Problem 5-15 we proved that a necessary condition for a series $\sum_{n=1}^{\infty} z_n$ to converge is that $\lim z_n = 0$.

Hence, series (1) does not converge.

2. It appears that d'Alembert's test would be appropriate in this case (see Problem 5-19).

$$\left|\frac{z_{n+1}}{z_n}\right| = \left|\frac{(n+1)!i^{n+1}}{(n+1)^{n+1}} \cdot \frac{n^n}{n!i^n}\right|$$

$$= \left|\frac{1\cdot 2 \cdot \ldots \cdot n(n+1)\cdot n^n i}{1\cdot 2 \cdot \ldots \cdot n(n+1)^n(n+1)}\right|$$

$$= \frac{n^n}{(n+1)^n} = \frac{1}{\left(\frac{n+1}{n}\right)^n} = \frac{1}{\left(1+\frac{1}{n}\right)^n} \tag{5}$$

$$\lim_{n\to\infty}\left|\frac{z_{n+1}}{z_n}\right| = \lim_{n\to\infty} \frac{1}{\left(1+\frac{1}{n}\right)^n} = \frac{1}{e} < 1 \tag{6}$$

Hence, the series is absolutely convergent.

3. Applying the root test we find:

$$z_n := \left(\frac{1+i}{2}\right)^n \tag{7}$$

$$\lim_{n\to\infty}\sqrt[n]{\left|\frac{1+i}{2}\right|^n} = \lim_{n\to\infty}\left|\frac{1+i}{2}\right| = \sqrt{\tfrac{1}{4}+\tfrac{1}{4}} < 1 \tag{8}$$

Hence, the series is absolutely convergent.

CHAPTER 6

CONTINUOUS MAPPINGS, CONTINUOUS CURVES, STEREOGRAPHIC PROJECTION

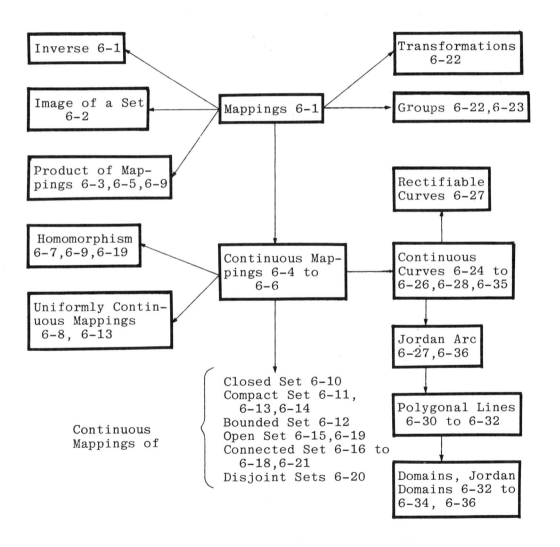

This chart is provided to facilitate rapid understanding of the interrelationships of the topics and subject matter in this chapter. Also shown are the problem numbers associated with the subject matter.

MAPPINGS

● **PROBLEM** 6-1

Fingerprints lead to the arrest and conviction of a criminal. The judge reaches the verdict assuming that there exists a relationship between every human being and the set of fingerprints. Can you describe this relationship in mathematical terms?

Solution: First let us note the following definitions.

Definition

If to every point x of a set A, there corresponds a unique point y in a set B, then we call this correspondence a mapping T of A onto B, and denote it:

$$T : A \to B \tag{1}$$

or $\qquad T(x) = y$

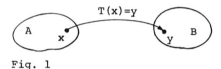

Fig. 1

The point y is called the image of x, and the point x is called an antecedent of y.

Note that any point of A has only one image by definition.

It is possible for a point y to have more than one antecedent.

Definition

T is said to be a one-to-one mapping of A onto B if any two distinct points of A have distinct images in B. We can write

$$\bigwedge_{x_1, x_2 \in A} (T(x_1) = T(x_2)) \Rightarrow (x_1 = x_2) \tag{2}$$

When T maps A onto B, there may be points in B which have no antecedents in A.

Definition

If every point y of B is the image under T of some point x of A, then T is said to map A onto B.

Of special importance are the mappings T of A which are one-to-one and onto. We can then define a mapping which assigns to every point y in B its unique antecedent x, under the mapping T. We denote

$$T^{-1}(y) = x. \tag{3}$$

T^{-1} is called the inverse of T. Note that

$$T^{-1} : B \to A \tag{4}$$

maps B one-to-one onto A.

Let A be the set of all human beings, and B be the set of fingerprints. T is a mapping which assigns every human being to its fingerprint. This mapping is onto and one-to-one. Therefore, for every fingerprint, there exists its owner. If two fingerprints are the same, then they belong to the same person.

Hence, the mapping is onto and one-to-one, and there exists an inverse mapping which assigns every fingerprint to its owner.

● **PROBLEM 6-2**

T maps a set X onto a set Y. Let Y_1 and Y_2 be two disjoint subsets of Y, and let X_1 and X_2 be their respective inverse images under T. Prove X_1 and X_2 are disjoint.

Solution: Definition

Let T be the mapping of X onto Y, $T : X \to Y$. The set

$$T(A) := \{T(x) : x \in A \subset X\} \tag{1}$$

is called the image of a set $A \subset X$.

For any $B \subset Y$, the set

$$T^{-1}(B) := \{x : T(x) \in B\} \tag{2}$$

is called the inverse image of the set B.

To solve the problem, assume that X_1 and X_2 are not disjoint, and that $a \in X_1 \cap X_2$. Then

$$T(a) = b \tag{3}$$

and $b \in Y_1$ and $b \in Y_2$.

This contradicts the hypothesis that Y_1 and Y_2 are disjoint.

● **PROBLEM 6-3**

Let
$$T_1 : X \to Y \tag{1}$$
$$T_2 : Y \to W. \tag{2}$$

The composition, or product,

$$T_2 \circ T_1 : X \to W \tag{3}$$

is defined by

$$\bigwedge_{x \in X} (T_2 \circ T_1)(x) = T_2(T_1(x)). \tag{4}$$

Sometimes composition is denoted by $T_2 T_1$.

Show that

$$(T_2 \circ T_1)^{-1} = T_1^{-1} \circ T_2^{-1}. \tag{5}$$

Solution: If both T_1 and T_2 are one-to-one and onto, then the inverse mappings exist.

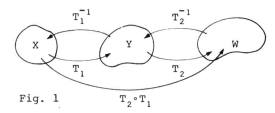

Fig. 1

Also, $T_2 \circ T_1$ is onto and one-to-one, hence $(T_2 \circ T_1)^{-1}$ exists.

Two mappings

$$\begin{aligned} S_1 &: X_1 \to Y_1 \\ S_2 &: X_2 \to Y_2 \end{aligned} \tag{6}$$

are identical when

$$X_1 = X_2, \quad Y_1 = Y_2 \tag{7}$$

$$\bigwedge_{x \in X_1} S_1(x) = S_2(x).$$

If $w \in W$, then

$$(T_2 \circ T_1)^{-1}(w) = x \tag{8}$$

such that

$$(T_2 \circ T_1)(x) = w. \tag{9}$$

Denoting $T_1(x) = y$, we can write

$$(T_2 \circ T_1)(x) = T_2(T_1(x)) = T_2(y) = w. \tag{10}$$

Thus,

$$\begin{aligned} (T_1^{-1} \circ T_2^{-1})(w) &= T_1^{-1}(T_2^{-1}(w)) \\ &= T_1^{-1}(y) = x = (T_2 \circ T_1)^{-1}(w) \end{aligned} \tag{11}$$

we have

$$\bigwedge_{w \in W} (T_2 \circ T_1)^{-1}(w) = (T_1^{-1} \circ T_2^{-1})(w). \tag{12}$$

CONTINUOUS MAPPINGS, HOMOMORPHISMS

• **PROBLEM 6-4**

Whenever a mapping $T : X_1 \to X_2$ of topological spaces is defined, we can determine if it is continuous or not. We shall start with the Neighborhood Definition, which can be applied to general topological spaces.

Definition

Let (X_1, d_1) and (X_2, d_2) be metric spaces with topologies τ_1 and τ_2, correspondingly. The mapping

$$T : X_1 \to X_2 \qquad (1)$$

is continuous when

$$\bigwedge_{\sigma_2 \in \tau_2} T^{-1}(\sigma_2) \in \tau_1 \qquad (2)$$

i.e., the inverse image of any open set is an open set.

Note that every open set in a metric space is the union of balls. Thus, in order to demonstrate the continuity of $T : X_1 \to X_2$, it is sufficient to show that for every ball $K(x_2, r_2)$ in X_2, $T^{-1}(K(x_2, r_2))$ contains a ball.

Definition

T is continuous at a point x_0 if for every neighborhood σ_2 of the point $T(x_0)$, the set $T^{-1}(\sigma_2)$ is a neighborhood of the point x_0. We have

$$\begin{pmatrix}\text{The mapping T} \\ \text{is continuous}\end{pmatrix} <=> \begin{pmatrix}\text{T is continuous} \\ \text{at every point}\end{pmatrix}.$$

Please note the Sequential Definition of continuous mappings:

$T : X_1 \to X_2$ is continuous at $x_0 \in X_1$ if for every sequence $x_n \to x_0$, the sequence $T(x_n) \to T(x_0)$. T is continuous if it is continuous at every point.

Show that the two definitions of continuity are equivalent.

Solution: We have to prove

$$\begin{pmatrix}\text{Neighborhood} \\ \text{definition}\end{pmatrix} <=> \begin{pmatrix}\text{Sequential} \\ \text{definition}\end{pmatrix}.$$

$\Rightarrow$

Let $x_n \to x_0$, that is

$$\bigwedge_{\delta > 0} \bigvee_{N \in \mathbb{N}} \bigwedge_{n > N} d(x_n, x_0) < \delta \qquad (3)$$

or
$$x_n \in K(x_0, \delta) \tag{4}$$

Applying the neighborhood definition, we have

$$\bigwedge_{\varepsilon>0} \bigvee_{\delta>0} T(K(x_0,\delta)) \subset K(T(x_0),\varepsilon) \tag{5}$$

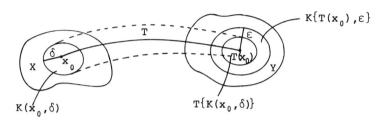

Hence,
$$T(x_n) \in K(T(x_0),\varepsilon) \tag{6}$$

That is
$$T(x_n) \to T(x_0) \tag{7}$$

Since x_0 is an arbitrary element of X, the mapping is continuous in the sequential meaning.

$\Longleftarrow$

Let $\quad x_n \to x_0 \quad$ and
$$T(x_n) \to T(x_0) \tag{8}$$

Assume that T is neighborhood-wise non-continuous at x_0. That is

$$\bigvee_{\varepsilon>0} \bigwedge_{\delta>0} \bigvee_{x_\delta \in X_1} d_1(x_\delta, x_0) < \delta \quad \text{and}$$
$$d_2(T(x_\delta), T(x_0)) \geq \varepsilon. \tag{9}$$

Taking $\delta = \frac{1}{n}$ and the corresponding points $x_\delta = x_n$, such that (9) holds, we find that
$$x_n \to x_0$$
while
$$T(x_n) \not\to T(x_0) \tag{10}$$

This is a contradiction.

• PROBLEM 6-5

> Prove that a composition of continuous mappings is a continuous mapping.

Solution: Let $T_1 : X \to Y$ and $T_2 : Y \to W$, where T_1 and T_2 are continuous. We have

$$T_2 \circ T_1 : X \to W. \qquad (1)$$

However,

$$(T_2 \circ T_1)^{-1}(A) = T_1^{-1}(T_2^{-1}(A)) \qquad (2)$$

where $A \subset W$.

When A is an open set, then $T_2^{-1}(A)$ is an open set in Y. Hence,

$$T_1^{-1}(T_2^{-1}(A)) \text{ is open in } X.$$

• PROBLEM 6-6

> Let $f, g : X \to C$ be continuous functions on X, where C is the set of complex numbers. Show that
>
> 1. $f + g$ (1)
> 2. $f - g$ (2)
> 3. fg (3)
> 4. $\dfrac{f}{g}$ (4)
>
> are continuous functions whenever g does not vanish.

Solution: 1. Let $x_n \to x_0$ and $y_n \to y_0$. Applying the theorem of the sum of the limits, we find

$$\lim(x_n + y_n) = x_0 + y_0. \qquad (5)$$

Since f and g are continuous, we have

$$f(x_n) \to f(x_0)$$
$$g(x_n) \to g(x_0). \qquad (6)$$

Then

$$(f+g)(x_n) = f(x_n) + g(x_n) \to f(x_0) + g(x_0)$$
$$= (f+g)(x_0). \qquad (7)$$

Hence, $f + g$ is continuous.

2. In the same way we can show that f - g is continuous.

3. Using the theorem of the product of the limits of sequences,

if
$$\lim x_n = x_0$$
$$\lim y_n = y_0 \qquad (8)$$

then
$$\lim x_n y_n = x_0 y_0 \qquad (9)$$

Therefore,
$$\lim(fg)(x_n) = (fg)(x_0). \qquad (10)$$

Hence, fg is a continuous function.

4. In the same way, we can show that $\frac{f}{g}$ is continuous whenever $g \neq 0$.

● **PROBLEM 6-7**

Consider the function
$$f : C \to C \qquad (1)$$
where $\quad f(z) = z^2 + 1. \qquad (2)$

Is this function a homomorphism?

Solution: We shall start with the following definition:

If T is a continuous one-to-one and onto mapping of A to B, such that the inverse T^{-1} mapping of B onto A is also continuous, then T is said to be a homomorphism or a topological mapping of A onto B. The sets A and B are said to be homomorphic.

The function f given by (2) is continuous, but it is not one-to-one. For example,

$$f(1+i) = f(-1-i). \qquad (3)$$

Therefore, the inverse function does not exist, and $f(z) = z^2+1$ is not a homomorphism.

● **PROBLEM 6-8**

Let
$$f : R' \to R' \qquad (1)$$
where $\quad f(x) = x^2. \qquad (2)$

Is the function f uniformly continuous on R'?

Solution: **Definition**

A mapping $T : X \to Y$ is said to be uniformly continuous on X if

$$\bigwedge_{\varepsilon > 0} \bigvee_{\delta > 0} \bigwedge_{x_1, x_2 \in X} (d(x_1, x_2) < \delta) \Rightarrow (d(T(x_1), T(x_2)) < \varepsilon). \tag{3}$$

A uniformly continuous mapping of the set X is continuous on X. The function $f(x) = x^2$ is continuous.

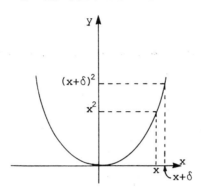

We have

$$(x+\delta)^2 - x^2 = 2\delta x + \delta^2. \tag{4}$$

This expression may assume any value for a particular δ. Hence, this function is not uniformly continuous on R'.

● **PROBLEM 6-9**

Show that if
$$T_1 : X \to Y$$
$$T_2 : Y \to W \tag{1}$$
where T_1 and T_2 are homomorphisms, then $T_2 \circ T_1 : X \to W$ is also a homomorphism.

Solution: Since both T_1 and T_2 are one-to-one and onto, the product $T_2 \circ T_1$ is also one-to-one and onto. Because both mappings T_1 and T_2 are continuous, $T_2 \circ T_1$ is also continuous (see Problem 6-5). The inverse mapping $(T_2 \circ T_1)^{-1}$ exists, and

$$(T_2 \circ T_1)^{-1} = T_1^{-1} \circ T_2^{-1} \tag{2}$$

Since both T_1^{-1} and T_2^{-1} are continuous, their product $T_1^{-1} \circ T_2^{-1}$ is a continuous mapping.

Hence, both $T_2 \circ T_1$ and $(T_2 \circ T_1)^{-1}$ are continuous.

Thus, $T_2 \circ T_1$ is a homomorphism.

PROPERTIES OF TOPOLOGICAL SPACES

● **PROBLEM 6-10**

Prove the following theorem:

$$\begin{pmatrix} T \text{ is continuous} \\ \text{on } X \end{pmatrix} \iff \begin{pmatrix} \text{The inverse image of every} \\ \text{closed set in Y is a closed} \\ \text{set in X} \end{pmatrix}$$

where $T : X \to Y$.

Solution: $\Rightarrow$

If T is continuous, and A is a closed set in Y, the Y-A is an open set in Y. The inverse image $T^{-1}(Y-A)$ of an open set is an open set, because T is continuous. Hence, $X - T^{-1}(A)$ is open in X. Therefore, $T^{-1}(A)$ is closed in X.

$\Leftarrow$

Let $A \subset Y$. We have

$$T^{-1}(Y-A) = X - T^{-1}(A).$$

If the set A is closed in Y, and $T^{-1}(A)$ is closed in X, then Y - A is open in Y, and $T^{-1}(Y-A)$ is open in X. Hence, T is continuous.

● **PROBLEM 6-11**

Show that if X is a compact set, and $T : X \to Y$ is a continuous mapping on X, then $T(X) \subset Y$ is compact.

Solution: We shall repeat the definition of a compact set:

A set $A \subset (X,d)$ is said to be compact if out of each sequence (x_n) of elements of the set A we can extract a subsequence (x_{n_k}) which converges to a limit contained in A.

Let $y_n \in T(X) \subset Y$. Then, for any y_n there exists $x_n \in X$, such that

$$T(x_n) = y_n \tag{1}$$

X is compact. Hence, from the sequence (x_n), we can extract a subsequence (x_{n_k}) which converges to $x \in X$.

$$x_{n_k} \to x \tag{2}$$

Since T is continuous on X,

$$y_{n_k} = T(x_{n_k}) \to T(x) \in T(X) \qquad (3)$$

and $T(X)$ is compact in Y.

• PROBLEM 6-12

Prove that a continuous mapping of a compact set is bounded.

Solution: Definition

A mapping $T : X \to Y$ is said to be bounded if the set $T(X) \subset Y$ is a bounded set.

We shall apply the following theorem:

A compact set in a metric space X is bounded and closed in X.

If $T : X \to Y$ is a continuous mapping of a compact set, then $T(X)$ is a compact set (see Problem 6-11). Since $T(X)$ is a compact set, it is bounded.

We can prove the following theorem in a similar way:

A continuous function f on a compact set X attains its bounds, i.e.,

$$\underset{x_0 \in X}{V} f(x_0) = \sup f(X)$$

$$\underset{x_1 \in X}{V} f(x_1) = \inf f(X).$$

• PROBLEM 6-13

Prove that a continuous mapping on a compact set is uniformly continuous on this set.

Solution: To prove this theorem, we shall apply the reductio ad absurdum method.

$$\begin{pmatrix} T : X \to Y \\ T \text{ is continuous} \\ X \text{ is compact} \end{pmatrix} \Rightarrow \begin{pmatrix} T \text{ is uniformly continuous} \\ \text{on } X \end{pmatrix}. (1)$$

In the future, to indicate that this method was applied, we shall write r.a.a.

$\Rightarrow$ r.a.a.

Assume that T is not uniformly continuous on X, i.e.,

$$\underset{\varepsilon > 0}{V} \underset{0 < \delta = \frac{1}{n}}{\Lambda} \underset{x_n \in X}{V} \underset{x'_n \in X}{V} d_1(x_n, x'_n) < \frac{1}{n} \text{ and }$$

$$d_2(T(x_n), T(x'_n)) \geq \varepsilon. \qquad (2)$$

Here, d_1 is the metric of X, (X,d_1), and d_2 the metric of Y, (Y,d_2).

We have two sequences: (x_n) and (x'_n). Since the space X is compact, subsequences can be selected such that

$$x_{n_k} \to x_0 \in X \qquad (3)$$

$$x'_n \to x'_0 \in X. \qquad (4)$$

However, $d_1(x_n, x'_n) < \frac{1}{n}$, so the subsequences can be chosen so that $x_0 = x'_0$. Since T is continuous,

$$\bigwedge_{\varepsilon>0} \bigvee_{N \in N'} \bigwedge_{k>N} d_2(T(x_{n_k}), T(x'_{n_k})) < \varepsilon \qquad (5)$$

which contradicts the assumption (2).

• **PROBLEM 6-14**

Prove the following theorem:

$$\begin{cases} T : X \to Y \\ T \text{ is one-to-one and onto} \\ T \text{ is continuous on } X \\ X \text{ is compact} \end{cases} \Rightarrow \begin{cases} T^{-1} : Y \to X \\ T^{-1} \text{ is continuous on } Y \end{cases}$$

Solution: Since $T : X \to Y$ is one-to-one and onto, the inverse mapping exists. Let (y_n) be a sequence in Y, such that

$$y_n \to y_0 = T(x_0) \qquad (1)$$

Since T is onto and one-to-one,

$$y_n = T(x_n), \quad x_n \in X \qquad (2)$$

$$x_n = T^{-1}(T(x_n)). \qquad (3)$$

Assume

$$x_n \not\to x_0 \qquad (4)$$

Then, either

$$x_n \to x' \neq x_0 \qquad (5)$$

or (x_n) is not convergent.

The first possibility is incompatible with the continuity of T because

$$T(x_n) \to T(x_0) \neq T(x') \tag{6}$$

If (x_n) is not convergent, we still can select a subsequence convergent to $x \in X$, where $x \neq x_0$, which is incompatible with the continuity of T.

● **PROBLEM 6-15**

Let T be a one-to-one mapping of X onto Y:

$$T : X \to Y \tag{1}$$

If A is an open subset of X, $A \subset X$, then the image $T(A)$ is an open subset of Y.

Show that the inverse mapping

$$T^{-1} : Y \to X \tag{2}$$

is continuous.

Solution: T is a one-to-one mapping of X onto Y. Therefore, an inverse mapping exists of Y onto X:

$$T^{-1} : Y \to X \tag{3}$$

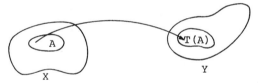

If A is open, then $T(A)$ is open. We shall apply the following theorem: (4)

$$\begin{pmatrix} \text{T is continuous} \\ \text{on X} \end{pmatrix} \Longleftrightarrow \begin{pmatrix} \text{The inverse image of every} \\ \text{open set in Y is an open} \\ \text{set in X} \end{pmatrix}$$

(See Problem 6-4)

Let B be any open set in X. Under the transformation T^{-1}, its inverse image is $T(B)$, which is an open set in Y. Hence, T^{-1} is a continuous mapping.

● **PROBLEM 6-16**

Let f be a continuous function

$$f : X \to R' \tag{1}$$

where X is a connected space, and $f(x_1) < f(x_2)$. Then, for every $a \in R'$, such that

$$f(x_1) < a < f(x_2), \quad (2)$$

there exists $x_0 \in X$ such that

$$f(x_0) = c. \quad (3)$$

Therefore, a continuous function on a connected set assumes all intermediate values.

Prove this statement.

Solution: We shall repeat the definition of the connected space:

A metric space (X,d) is not connected if there exist $X_1 \subset X$ and $X_2 \subset X$ such that

1. $X_1 \cup X_2 = X$

2. $X_1 \cap X_2 = \phi$

3. X_1 and X_2 are open in X.

Otherwise, the space is connected.

Using r.a.a.:

Let
$$\bigwedge_{x \in X} f(x) \neq a \quad (4)$$

Then for every x, either $f(x) > a$, or $f(x) < a$. The sets

$$\begin{aligned} X_1 &= \{x \in X : f(x) > a\} \\ X_2 &= \{x \in X : f(x) < a\} \end{aligned} \quad (5)$$

are not empty for $x_1 \in X_2$, $x_2 \in X_1$. X_1 and X_2 are disjoint and open sets. Their union is the entire space X. Thus, X is not connected, which is a contradiction.

• **PROBLEM 6-17**

Prove the following theorem:

$$\begin{pmatrix} X \text{ is a connected space} \\ T: X \to Y \text{ is a continu-} \\ \text{ous mapping} \end{pmatrix} \Rightarrow \begin{bmatrix} T(X) \text{ is connected} \end{bmatrix}. \quad (1)$$

Solution: Using r.a.a.:

Assume that $T(X)$ is not connected. Then there exist sets Y_1 and Y_2, such that

$$Y_1 \cap Y_2 = \phi, \quad Y_1, Y_2 \neq \phi \quad (2)$$

$$Y_1 \cup Y_2 = T(X) \quad (3)$$

Y_1 and Y_2 are open sets. (4)

Since T is continuous, we have

$$T^{-1}(Y_1) \text{ is an open set in } X$$

$$T^{-1}(Y_2) \text{ is an open set in } X$$

$$T^{-1}(Y_1), T^{-1}(Y_2) \neq \phi$$

$$T^{-1}(Y_1) \cup T^{-1}(Y_2) = X$$

$$T^{-1}(Y_1) \cap T^{-1}(Y_2) = \phi.$$

Hence, X is not connected.

This is a contradiction.

● **PROBLEM 6-18**

Prove that if T is a homomorphism,

$$T : X \to Y \qquad (1)$$

then

$$\begin{pmatrix} A \subset X \text{ is} \\ \text{connected} \end{pmatrix} \Longleftrightarrow \left[T(A) \text{ is connected} \right] \qquad (2)$$

Solution: $\Longrightarrow$

Since T is a homomorphism, then $T : X \to Y$ is a continuous mapping. If A is a connecting set, then T(A) is also connected (see Problem 6-17).

$\Longleftarrow$

Since T is a homomorphism, T^{-1} exists and is continuous. Again, if T(A) is connected, then

$$T^{-1}(T(A)) = A \qquad (3)$$

is also connected.

● **PROBLEM 6-19**

Let
$$T : X \to Y \qquad (1)$$
where T is continuous, one-to-one and onto.

Assume that if $A \subset X$ is open, then its image is open in Y. Show that T is a homomorphism.

Solution: Since T is one-to-one and onto, T^{-1} exists. T is continuous, hence to show that T is a homomorphism, it is enough to show that T^{-1} is continuous. We have

$$\begin{pmatrix} A \subset X \\ A \text{ open set} \end{pmatrix} \Rightarrow \begin{bmatrix} T(A) \text{ is open} \end{bmatrix} \qquad (2)$$

For T^{-1}, the inverse image of an open set A in X is $T(A)$, which is open. Hence, T^{-1} is continuous and T is a homeomorphism.

• **PROBLEM 6-20**

Let
$$X = \bigcup_{\ell=1}^{m} A_\ell \qquad (1)$$

where A_ℓ are disjoint sets, and

$$Y = \bigcup_{\ell=1}^{m} B_\ell \qquad (2)$$

where B_ℓ are disjoint sets. T maps X one-to-one and onto Y.

$$T : X \to Y \qquad (3)$$

Assume further that

$$\bigwedge_\ell T : A_\ell \to B_\ell \qquad (4)$$

Show that T maps each A_ℓ one-to-one onto B_ℓ.

Solution: From condition (4), we see that T maps each A_ℓ into B_ℓ. T maps X onto Y

$$T : X \to Y \qquad (5)$$

Thus,

$$\bigwedge_{y \in Y} \bigvee_{x \in X} T(x) = y \qquad (6)$$

A_ℓ are disjoint sets. We have

$$\bigwedge_{y \in B_\ell} \bigvee_{x \in A_\ell} T(x) = y \qquad (7)$$

Thus, T maps A_ℓ onto B_ℓ.

We will show that $T : A_\ell \to B_\ell$ is one-to-one.

Assume that $T : A_\ell \to B_\ell$ is not one-to-one. Thus,

$$\bigwedge_{\substack{x_1, x_2 \in A_\ell \\ x_1 \neq x_2}} T(x_1) = T(x_2) \in B_\ell \qquad (8)$$

Hence, $T : X \to Y$ is not one-to-one. This is a contradiction.

• PROBLEM 6-21

Let
$$X = A_1 \cup A_2 \qquad (1)$$
X is separated into two connected sets, and
$$Y = B_1 \cup B_2 \qquad (2)$$
Y is separated into two connected sets. Let
$$T : X \to Y \qquad (3)$$
be a continuous onto mapping. Show that

$$T : A_1 \xrightarrow{onto} B_1$$
$$T : A_2 \xrightarrow{onto} B_2 \qquad (4)$$

or

$$T : A_1 \xrightarrow{onto} B_2$$
$$T : A_2 \xrightarrow{onto} B_1 \qquad (5)$$

Solution: The images of A_1 and A_2 are $T(A_1)$ and $T(A_2)$, respectively. Since both A_1 and A_2 are connected, and T is continuous, $T(A_1)$ and $T(A_2)$ are connected (see Problem 6-17). Hence, $T(A_1)$ must be contained in B_1 (or B_2), and $T(A_2)$ must be contained in B_2 (or B_1). $T(A_1)$ and $T(A_2)$ cannot both be contained in B_1 (or B_2).

Assume that
$$T(A_1) \subset B_1$$
$$T(A_2) \subset B_2. \qquad (6)$$

Then, according to Problem 6-20,
$$T : A_1 \xrightarrow{onto} B_1$$
$$T : A_2 \xrightarrow{onto} B_2 \qquad (7)$$

• PROBLEM 6-22

Let A be a non-empty set. Any one-to-one mapping of the set A onto itself is called a transformation of the set A. Show that the collection M of all transformations of A forms a group.

Solution: If $T_1 \in M$ and $T_2 \in M$, then their product $T_1 \circ T_2$ is also a transformation.

We shall now define a group.

A nonempty set G is called a group if:

1. An operation is defined on G such that each ordered pair of elements a,b of G is associated with a unique element c of G. The operation is called multiplication, and the element c is called the product of a and b. We denote ab = c. Note that in general,

$$ab \neq ba \qquad (1)$$

If for any two elements a and b of G

$$ab = ba, \qquad (2)$$

then the group is called Abelian* or commutative.

2. Multiplication is associative, i.e.,

$$\bigwedge_{a,b,c \in G} (ab)c = a(bc) \qquad (3)$$

3. An identity element $e \in G$ exists such that

$$\bigwedge_{a \in G} ae = ea = a \qquad (4)$$

4. $$\bigwedge_{a \in G} \bigvee_{a^{-1} \in G} aa^{-1} = a^{-1}a = e \qquad (5)$$

For every $a \in G$, there exists an inverse element $a^{-1} \in G$.

For the set M of all transformations with multiplication defined as a product $T_1 \circ T_2$ of transformations, condition (1) is fulfilled. Multiplication is associative. Hence,

$$(T_3 \circ T_2) \circ T_1 = T_3 \circ (T_2 \circ T_1). \qquad (6)$$

Therefore,

$$(T_3 \circ T_2) \circ T_1(z) = T_3(T_2(T_1(z)))$$
$$= T_3(T_2 T_1)(z) = T_3 \circ (T_2 \circ T_1)(z) \qquad (7)$$

We define the identity transformation I by

$$\bigwedge_{a \in A} I(a) = a \qquad (8)$$

Hence,

$$\bigwedge_{T \in M} TI = IT = T \qquad (9)$$

The inverse transformation is T^{-1}, such that

$$\bigwedge_{T \in M} T^{-1}T = TT^{-1} = I \qquad (10)$$

Therefore, the set of all transformations of a nonempty set

forms a group.

*Niels Henrick Abel, born in Findoe, Norway, 1802, died in Arendal, 1829. He worked on elliptic functions. The results were published in Crelle's Journal. His major discovery, known as Abel's theorem, was printed in 1841 after his death. Abel proved the binomial theorem for $(1+x)^n$, when x and n are complex. Abel proved that it is impossible to express a root of the general algebraic equation

$$ax^5 + bx^4 + cx^3 + dx^2 + ex + f = 0$$

in terms of its coefficients by means of a finite number of rational functions and radicals. He also investigated the higher transcendents of multiple periodicity, known today as Abelian functions.

● **PROBLEM 6-23**

Let A be a non-empty set, and M a collection of transformations of A, such that if $T_1, T_2 \in M$, then T_1^{-1} and $T_2 \circ T_1 \in M$.

Show that M is a group.

Solution: The multiplication is defined as in Problem 6-22.

Multiplication is associative. The identity transformation I belongs to M.

If $T \in M$, then $T^{-1} \in M$. Since $T, T^{-1} \in M$,

$$T \circ T^{-1} = I \in M \tag{1}$$

For each $T \in M$, the inverse element $T^{-1} \in M$.

Hence, M is a group.

CURVES

● **PROBLEM 6-24**

Let $I = [t_1, t_2] \subset R'$ be a closed interval. Let C be a continuous mapping of I onto the complex plane. C is said to represent a continuous curve, or path in the plane. We denote

$$z = C(t) \tag{1}$$

or

$$z = z(t). \tag{2}$$

Give the "mechanical" interpretation of a curve.

Find the curve for which

$$I = [0,1] \tag{3}$$

$$z(t) = \cos 2\pi t + i \sin 2\pi t \tag{4}$$

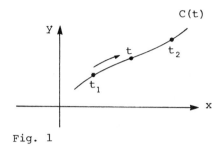

Fig. 1

Solution: Assume that a point z moves continuously in the plane during the time from t_1 to t_2. At each instant $t_1 \leq t \leq t_2$, the point occupies a position in the plane determined by $z = z(t)$. Hence, the mapping $z = z(t)$ gives the motion of the point during the time interval $I = [t_1, t_2]$.

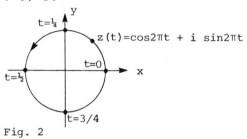

Fig. 2

For $t = 0$, we have

$$z(0) = \cos 0 + i \sin 0 = 1 \tag{5}$$

For $t = \frac{1}{4}$,

$$z(\tfrac{1}{4}) = \cos \frac{\pi}{2} + i \sin \frac{\pi}{2} = i \tag{6}$$

$t = \tfrac{1}{2}$, $z(\tfrac{1}{2}) = \cos \pi + i \sin \pi = -1$

$t = \tfrac{3}{4}$, $z = -i$

$t = 1$, $z = 1$

The point moves once around a unit circle, counterclockwise.

In most cases, the curve $z = z(t)$ has a piecewise smooth parametrization. Italian mathematician Peano found a continuous mapping T

$$T : [0,1] \to [0,1] \times [0,1]$$

i.e., a mapping (continuous) of an interval [0,1] onto a unit square. In Peano's example, the coordinates $x(t)$ and $y(t)$ were continuous but not differentiable functions. $(x(t), y(t))$ is a "space-filling" curve.

● **PROBLEM 6-25**

Consider two continuous curves

$$z(t) = \cos 2\pi t + i \sin 2\pi t, \quad 0 \leq t \leq 1 \qquad (1)$$

and

$$z(t) = \cos 4\pi t + i \sin 4\pi t, \quad 0 \leq t \leq 1. \qquad (2)$$

Why are they different?

Solution: For eq.1, as t varies from 0 to 1, the point moves counterclockwise once around the unit circle. For eq.2, as t varies from 0 to 1, the point travels around the unit circle twice in a counterclockwise direction.

Although the locus of the moving point, i.e., the unit circle, is the same in both cases, the path is different. In (1), the point traverses the unit circle once, while in (2) the point traverses the unit circle twice.

Consider another case:

$$z(t) = \cos 2\pi t + i \sin 2\pi t, \quad 0 \leq t \leq 1 \qquad (3)$$

$$z(t) = \cos \pi t + i \sin \pi t, \quad 0 \leq t \leq 2. \qquad (4)$$

The path traversed by the moving point is the same for (3) and (4). The unit circle traversed twice in the counterclockwise direction.

In general, if

$$C : I \to \text{Complex Plane}$$
$$C' : I' \to \text{Complex Plane} \qquad (5)$$

are continuous mappings of closed intervals, $I = [a,b]$ and $I' = [a',b']$, respectively.

If there exists a homomorphism

$$T : I \to I' \qquad (6)$$

with $T(a) = a'$ and $T(b) = b'$, $\qquad (7)$

such that

$$C = C' \circ T, \qquad (8)$$

then C and C' are said to represent the same curve.

• **PROBLEM 6-26**

Let $z = C(t)$ be a continuous curve, and $z_0 = C(t_0)$ a point on this curve. If $M(z_0)$ is a neighborhood of z_0 in the plane, then there exists a neighborhood $N(z_0)$ of z_0 along the curve which is contained in $M(z_0)$.

Solution: We define a neighborhood of z_0 along the curve C:

Let $z = C(t)$ be a continuous mapping of the closed interval $I =$

$[t_1, t_2]$ onto the plane. The image of $t_0 \in I$ is $z_0 = C(t_0)$. A neighborhood of z_0 along the curve C is the set of all $z = C(t)$, with t in a neighborhood $N(t_0)$ of t_0 in the interval I. (See Fig. 1.)

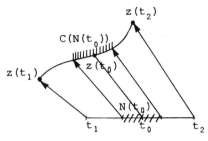

Fig. 1

$z_0 = C(t_0)$ is a point on the curve, and $M(z_0)$ is its neighborhood in the plane, i.e.,

$$(z \in M(z_0)) \implies (|z - z_0| < \varepsilon', \quad \varepsilon' > 0) \qquad (1)$$

The mapping

$$C : [t_1, t_2] \to C(t)$$

is continuous; hence;

$$\bigwedge_{\delta > 0} \bigvee_{\varepsilon > 0} |t - t_0| < \varepsilon \implies |C(t) - C(t_0)| < \delta . \qquad (2)$$

Setting $\delta = \frac{\varepsilon'}{2}$, we obtain

$$N(z_0) \subset M(z_0) \qquad (3)$$

● PROBLEM 6-27

Consider an arc of the continuous curve containing the origin

$$y = \begin{cases} x \sin \frac{\pi}{x} & \text{for } x \neq 0 \\ 0 & \text{for } x = 0 \end{cases} \qquad (1)$$

Show that this curve is not rectifiable.

Solution: A continuous curve which has no multiple points is called a Jordan* arc.

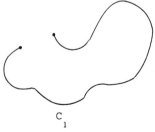
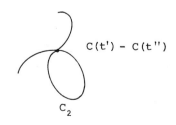

Fig. 1

C_1 is a Jordan arc. C_2 is not because two different $t' \neq t''$ are mapped onto the same point of the curve

$$C(t') = C(t'') \tag{2}$$

This is a multiple point.

If a Jordan arc has a definite length, the arc is said to be rectifiable, and is then called a path segment.

Let $z = z(t)$, $t \in [a,b]$ be a continuous curve. We divide the interval $[a,b]$ into n parts

$$a = t_1 < t_2 < t_3 \ldots < t_n = b \tag{3}$$

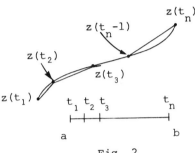

Fig. 2

The points $z(t_1), z(t_2), \ldots, z(t_n)$ are marked on the curve and connected in order by the straight line segments. If the set of the lengths of all such segments is bounded, then the curve is rectifiable. The length of the curve is defined as the least upper bound of this set.

We shall show that a segmental arc of arbitrarily large length can be inscribed in an arc of the curve containing the origin.

Let P_k be the points on the curve, such that

$$y = 0, \quad x = \frac{1}{k} \tag{4}$$

and Q_k be the points on the curve such that

$$x = \frac{2}{2k-1}, \quad |y| = x \tag{5}$$

Hence, for Q_k,

$$x = \frac{2}{2k-1}$$

and

$$y = \frac{2}{2k-1} \sin\left[\frac{\pi}{2} \cdot (2k-1)\right] \tag{6}$$

such that $|y| = x$.

Let the inscribed path be $P_{k-1} Q_k P_k$, for $k = 2, 3, 4, \ldots$.

The length of $P_{k-1} Q_k P_k$ for any given k exceeds the double ordinate of Q_k.

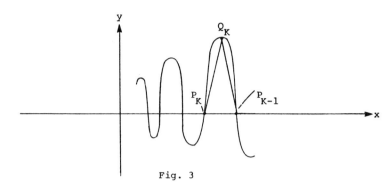

Fig. 3

Hence,

$$\text{Length of } P_{k-1}Q_kP_k > \frac{2}{k} > \frac{1}{k} \tag{7}$$

The series $\sum_{1}^{\infty} \frac{1}{k}$ diverges, thus

$$\underset{M \in R'}{\wedge} \underset{k' \in N}{\vee} \underset{k > k'}{\wedge} \text{ length of the inscribed path} > M \tag{8}$$

*M.E.C. Jordan (1838-1922) was a French mathematician. His work was mainly concerned with finite discontinuous substitution groups.

• **PROBLEM 6-28**

Let $C(t)$ be a continuous curve, where $t \in [t_1, t_2]$. Show that C is a connected set.

Solution: The continuous curve is the image of an interval under a continuous mapping. We shall apply the following theorem:

The image of a connected set under a continuous mapping is a connected set.

An interval is a connected set. Since $C(t)$ is continuous, the set of all points of a continuous curve is connected.

CONVEX SETS, DOMAINS

• **PROBLEM 6-29**

Definition

A set A in the plane is said to be convex if the line segment joining any two points $z_1, z_2 \in A$ is also contained in A.

Show that the interior of a circle is convex.

Solution: The interior of a circle is given by

$$|z - z_0| < r \tag{1}$$

Let z_1 and z_2 belong to the interior of the circle.

The points z on the segment joining z_1 and z_2 are

$$z = z_1 + t(z_2-z_1), \quad 0 \le t \le 1. \qquad (2)$$

Substituting eq.(2) into eq.(1), we obtain

$$|z_1+t(z_2-z_1)-z_0| = |(z_1-z_0)(1-t)+t(z_2-z_0)| \qquad (3)$$
$$< R(1-t)+Rt = R$$

● **PROBLEM 6-30**

Let A be an open set, and S the interior of a circle in A. Points z_1 and z_2 belong to S. Show that a point $z_0 \in A$ can be joined to z_1 by a polygonal line in A if z_0 can also be joined to z_2 by a polygonal line in A.

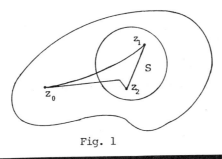

Fig. 1

Solution: Assume z_0 can be joined to z_1 by a polygonal line in A. Attaching one straight line segment z_1z_2, to this line, we obtain a polygonal line joining z_0 to z_2. Since $S \subset A$, the new polygonal line is in A.

The proof is identical in the opposite direction.

● **PROBLEM 6-31**

Let A be an open set, and z_0 a fixed point in A.

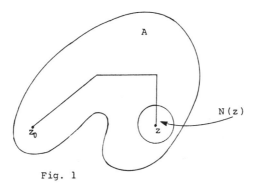

Fig. 1

Let z be a point in A, which can be joined to z_0 by a polygonal line in A. Show that there exists a neighborhood N(z) of z in A, such that every point z_1 in N(z) can be joined to z_0 by a polygonal line in A.

Solution: A is an open set and $z \in A$. Hence, a ball $K(z,\varepsilon)$ exists such that

$$K(z,\varepsilon) \subset A \tag{1}$$

Point z_0 can be joined to $z \in K(z,\varepsilon)$ by a polygonal line. Hence (see Problem 6-30), z_0 can be joined to any point of $K(z,\varepsilon)$ by a polygonal line.

Similarly, we can show that if A is open, $z_0 \in A$,

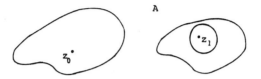

Fig. 2

and $z \in A$ cannot be joined to z_0 by a polygonal line in A, then a neighborhood $N(z)$ of z exists in A, such that none of the points of $N(z)$ can be joined to z_0 by a polygonal line in A.

● **PROBLEM 6-32**

Prove the following theorem:

$\begin{pmatrix} \text{An open set D} \\ \text{is a domain} \end{pmatrix}$ <=> $\begin{pmatrix} \text{Any two points in D can be} \\ \text{joined by a polygonal line} \\ \text{lying in D} \end{pmatrix}$

Solution: An open connected set is called a domain.

<=

Assume that any two points in D can be joined by a polygonal line lying in D. Taking any point z_0 of D, we can consider D to be the union of all polygonal lines in D joining z_0 with the points of D. Each polygonal line is a connected set (see Problem 6-28). Since a union of a family of connected sets is a connected set, D is connected.

=>

Assume D is a domain. Let z_1 be any point in D. We will show that any other point $z \in D$ can be joined with z_1 by a polygonal line in D. Assume the opposite.

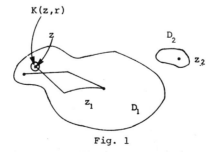

Fig. 1

Point $z_2 \in D$ cannot be joined with z_1 by a polygonal line in D. Now, let D_1 and D_2 denote, respectively, the sets of all points in D, which can and which cannot be joined to z_1.

$$z_1 \in D_1, \quad z_2 \in D_2$$

If a point $z \in D$ can be joined to z_1, then $K(z,r)$ exists, such that $K(z,r) \subset D$ and every point of $K(z,r)$ can be joined to z_1 by a polygonal line in D.

If $z \in D_1$ then there exists $K(z,r)$ such that all points of $K(z,r)$ are in D_1. The set D is a union of D_1 and D_2 such that $D_1 \cap D_2 = \phi$. Both are open sets. That contradicts the assumption that D is connected.

● **PROBLEM 6-33**

$\{D_\omega\}$ is a family of domains, such that

$$\bigcap_\omega D_\omega = A \tag{1}$$

where A is a non-empty set

Show that

$$\bigcup_\omega D_\omega \tag{2}$$

is a domain.

Solution: Since D_ω are domains, they are open and connected.

The union of any family of open sets is an open set. Hence,

$$\bigcup_\omega D_\omega$$

is an open set. We will show that $\bigcup_\omega D_\omega$ is path-connected. The intersection of $\{D_\omega\}$ is a non-empty set. Assume that

$$\bigcap_\omega D_\omega = z_0 \tag{3}$$

Taking any two elements of the family $\{D_\omega\}$, for example, D_ω and $D_{\omega'}$, we see that

$$D_\omega \cup D_{\omega'} \tag{4}$$

is path-connected. Therefore, D_ω and $D_{\omega'}$ are path-connected and are not disjoined. Hence, $D_\omega \cup D_{\omega'}$ is path-connected. We conclude that any two points in $\bigcup_\omega D_\omega$ can be joined by a polygonal line lying in $\bigcup_\omega D_\omega$. Therefore, $\bigcup_\omega D_\omega$ is a domain.

CONTINUOUS CURVES

• PROBLEM 6-34

Let $z = z(t)$, $t \in [a,b]$ be a continuous curve. Each point of the curve $z(t)$ belongs to some domain D_t. Show that a finite subfamily $D_{t_1}, \ldots, D_{t_k}$ of $\{D_t\}$ exists such that $D_{t_m} \cap D_{t_{m+1}} \neq \phi$, $m = 1, 2, \ldots, k-1$.

Furthermore, a subdivision of $[a,b]$ is given by $a = t'_0 < t'_1 \ldots < t'_k = b$, such that if C_m is part of the curve given by $z = z(t)$; $t'_{m-1} \leq t \leq t'_m$, then $C_m \subset D_{t_m}$.

<u>Solution</u>: We assign a neighborhood $N(t)$ to each $a < t < b$, such that the image of $N(t)$ on the curve is contained in $\bar{D}_t$ (see Fig. 1).

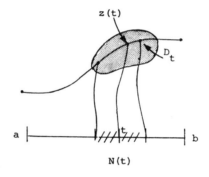

Fig. 1

We can choose a finite number of neighborhoods.

$$N(t_1), N(t_2), \ldots, N(t_k), \qquad (1)$$

that cover $[a,b]$. The corresponding domains are

$$D_{t'_1}, D_{t'_2}, \ldots, D_{t'_k} \qquad (2)$$

For $k = 1, 2$, the theorem holds.

Assume that $k \geq 3$.

The neighborhoods $N(t_m)$ are open sets that cover $[a,b]$.

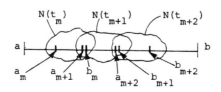

Fig. 2

We choose two sets of points $\{a_m\}$ and $\{b_m\}$, such that for

$$N(t_m) : a_m < t < b_m, \tag{3}$$

$$a = a_1 \leq a_2 < \ldots < a_k \tag{4}$$

Then, we obtain

$$b_1 < \ldots < b_{k-1} < b_k = b \tag{5}$$

and

$$a_{m+1} < b_m \tag{6}$$

$$b_m \leq a_{m+2} \tag{7}$$

(see Fig. 2).

Setting

$$t'_m = \frac{a_{m+1} + b_m}{2} \tag{8}$$

$$m = 1, 2, \ldots, k-1$$

we obtain the required subdivision of $[a,b]$.

Note that we utilized the theorem of Bolzano-Weierstrass:

An interval $[a,b] \subset R'$ is compact.

• **PROBLEM 6-35**

A continuous curve

$$z = z(t), \quad t \in [a,b] \tag{1}$$

is given. Show that the set of all points of the curve is compact.

Solution: We shall apply the following theorem:

If $T: A \to B$ is a continuous mapping of a compact set A onto B, then B is also compact.

An interval $I = [a,b]$ is both bounded and closed. Therefore, it is compact.

Hence, the set of all points of a continuous curve given by eq.(1) is compact.

• **PROBLEM 6-36**

1. Which of the following are Jordan curves?

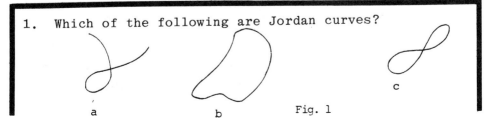

Fig. 1

2. Which are Jordan domains?

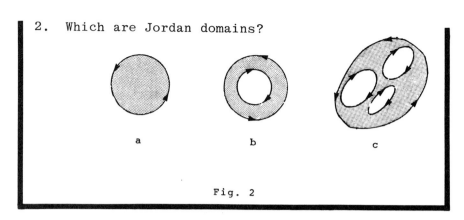

Fig. 2

Solution: 1. A Jordan curve is a simple closed curve (piecewise-smooth). Thus, it is a loop which does not cross itself.

a. is not closed

b. is a Jordan curve

c. crosses itself

The following theorem establishes an important property of Jordan curves:

A Jordan curve decomposes the plane into two separated regions, one lying inside the curve and the other outside the curve.

Intuitively, it is an obvious theorem.

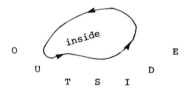

Fig. 3

A rigorous complete proof is difficult (Jordan's proof was not correct), and requires advanced topology.

Note that if the interior lies to the left of the curve, then the curve has positive orientation. Otherwise the orientation is negative.

2. A Jordan domain is a bounded domain D, whose boundary ∂D is the union of a finite number of disjoint Jordan curves, with each curve parametrized so that the boundary curves are positively oriented. a, b and c are Jordan domains.

STEREOGRAPHIC PROJECTION, RIEMANN SPHERE

• **PROBLEM 6-37**

Let S be a sphere with radius $\frac{1}{2}$ and center at $(0,0,\frac{1}{2})$, and let (ξ,η,ζ) be a point on this sphere. Show that

$$\xi^2 + \eta^2 = \zeta(1-\zeta) \tag{1}$$

Solution: The most general equation of a sphere is

$$(x_1-a)^2 + (x_2-b)^2 + (x_3-c)^2 = r^2 \tag{2}$$

where r is the radius, and (a,b,c) the coordinates of the center of sphere. In our case

$$(a,b,c) = (0,0,\tfrac{1}{2}) \tag{3}$$

and

$$r = \tfrac{1}{2}, \quad (\xi,\eta,\zeta) = (x_1,x_2,x_3).$$

Thus,

$$\xi^2 + \eta^2 + (\zeta-\tfrac{1}{2})^2 = (\tfrac{1}{2})^2 \tag{4}$$

or

$$\xi^2 + \eta^2 = \zeta(1-\zeta) \tag{5}$$

● **PROBLEM 6-38**

Describe the Riemann sphere and stereographic projection.

Solution: Consider the sphere Σ with center at $(\xi,\eta,\zeta) = (0,0,\tfrac{1}{2})$ and radius $\tfrac{1}{2}$.

At the point $S:(0,0,0)$, the sphere is tangent to the plane π whose equation is $\zeta = 0$.

Point $N:(0,0,1)$ of Σ is called the north pole (see Fig. 1).

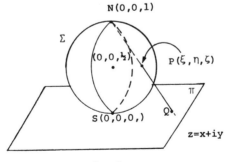

Fig. 1

This sphere is often called a Riemann sphere.

Let $P(\xi,\eta,\zeta)$ be any point on the sphere other than N. The line joining N to P crosses the plane π at the point Q.

If π is the complex plane, then Q is a complex number $z = x+iy$. However, if Q is a complex number, then the line QN crosses the sphere at exactly one point: $P(\xi,\eta,\zeta)$.

We conclude that stereographic projection T establishes one-to-one mapping of a sphere Σ (without N) onto the complex plane

$$T : \Sigma - N \to C$$

● **PROBLEM 6-39**

Let Σ be a Riemann sphere (as described in Problem 6-38), and let $N : (0,0,1)$ and $P : (\xi,\eta,\zeta)$ be points on Σ. Line NP crosses the plane π, $\zeta = 0$ at $Q : z = x+iy$.

Compute:

1. $z = z(\xi,\eta,\zeta)$ \hfill (1)

2. $\xi = \xi(z,\bar{z})$, $\eta = \eta(z,\bar{z})$, $\zeta = \zeta(z,\bar{z})$. \hfill (2)

Solution: 1. Line QN crosses Σ at (ξ,η,ζ).

Since $N : (0,0,1)$, the equations of QN are given by

$$x' = 0 + \xi t$$
$$y' = 0 + \eta t \qquad (3)$$
$$z' = 1 + (\zeta-1)t$$

Setting $z' = 0$, we find

$$t = \frac{1}{1-\zeta} \qquad (4)$$

Hence,

$$z = x+iy = \xi t + i\eta t = \frac{\xi+i\eta}{1-\zeta} \qquad (5)$$

2. Eq.(5) can be written

$$\xi + i\eta = z(1-\zeta) \qquad (6)$$

Thus, since ξ, η and ζ are real,

$$\xi - i\eta = \bar{z}(1-\zeta) \qquad (7)$$

(ξ,η,ζ) is a point on the sphere. Thus,

$$\xi^2 + \eta^2 = \zeta(1-\zeta) \qquad (8)$$

(See Problem 6-37).

Multiplying eq.(6) by eq.(7), we find

$$\xi^2 + \eta^2 = z\bar{z}(1-\zeta)^2 \qquad (9)$$

Applying eq.(8),

$$\zeta = z\bar{z}(1-\zeta)$$

Solving for ζ,

$$\zeta = \frac{z\bar{z}}{z\bar{z}+1} \qquad (10)$$

Adding (6) and (7),

$$\xi = \frac{z(1-\zeta)+\bar{z}(1-\zeta)}{2} = \frac{z+\bar{z}}{2(z\bar{z}+1)} \qquad (11)$$

Subtracting (6) from (7),

$$\eta = \frac{z-\bar{z}}{2i(z\bar{z}+1)} \qquad (12)$$

For $z = x+iy$, we denote

$$(\xi,\eta,\zeta) = \left(\frac{x}{x^2+y^2+1},\ \frac{y}{x^2+y^2+1},\ \frac{x^2+y^2}{x^2+y^2+1}\right) \qquad (13)$$

● **PROBLEM 6-40**

Show that if the sequence $z_n = x_n + iy_n$ in the plane converges to $z_0 = x_0 + iy_0$, then corresponding (ξ_n,η_n,ζ_n) on the sphere converges to (ξ_0,η_0,ζ_0).

<u>Solution</u>: Since

$$z_n \to z_0, \qquad (1)$$

both sequences (x_n) and (y_n) converge.

$$x_n \to x_0 \qquad (2)$$
$$y_n \to y_0$$

From Problem 6-39 we find

$$(\xi_n,\eta_n,\zeta_n) = \left(\frac{x_n}{x_n^2+y_n^2+1},\ \frac{y_n}{x_n^2+y_n^2+1},\ \frac{x_n^2+y_n^2}{x_n^2+y_n^2+1}\right) \qquad (3)$$

Applying the thoerem about the limits of the sequences,

$$\frac{x_n}{x_n^2+y_n^2+1} \to \frac{x_0}{x_0^2+y_0^2+1} \qquad (4)$$

$$\frac{y_n}{x_n^2+y_n^2+1} \to \frac{y_0}{x_0^2+y_0^2+1} \qquad (5)$$

$$\frac{x_n^2+y_n^2}{x_n^2+y_n^2+1} \to \frac{x_0^2+y_0^2}{x_0^2+y_0^2+1} \qquad (6)$$

If E^k is k-dimensional Euclidean space

$$(x_1,x_2,\ldots,x_k) \in E^k,$$

then the sequence $(P^{(n)})$ of points in E^k,
$$P^{(n)} = (x_1^{(n)}, \ldots, x_k^{(n)}), \tag{7}$$
converges to $P^{(o)} \in E^k$
$$P^{(o)} = (x_1^{(o)}, \ldots, x_k^{(o)}) \tag{8}$$
if
$$\lim_{n \to \infty} x_\ell^{(n)} = x_\ell^{(o)} \quad \ell = 1, 2, \ldots, k \tag{9}$$
Thus,
$$(\xi_n, \eta_n, \zeta_n) \xrightarrow[n \to \infty]{} (\xi_0, \eta_0, \zeta_0) \tag{10}$$

• **PROBLEM 6-41**

Let (ξ_n, η_n, ζ_n) be a sequence of points of the sphere Σ (without the north pole N) which converges to the point
$$(\xi_0, \eta_0, \zeta_0), \quad \zeta_0 \neq 1 \tag{1}$$
Show that the sequence of points (x_n, y_n) on the complex plane corresponding to (ξ_n, η_n, ζ_n) converges to (x_0, y_0).

Solution: We have
$$(\xi_n, \eta_n, \zeta_n) \xrightarrow[n \to \infty]{} (\xi_0, \eta_0, \zeta_0), \quad \zeta_0 \neq 1 \tag{2}$$
From Problem 6-39, the points on the complex plane are
$$\left(\frac{\xi_n}{1 - \zeta_n}, \frac{\eta_n}{1 - \zeta_n}, 0 \right) \tag{3}$$
Hence, since
$$\xi_n \to \xi_0$$
$$\eta_n \to \eta_0 \tag{4}$$

$$\frac{\xi_n}{1 - \zeta_n} \to \frac{\xi_0}{1 - \zeta_0} \tag{5}$$

$$\frac{\eta_n}{1 - \zeta_n} \to \frac{\eta_0}{1 - \zeta_0} \tag{6}$$

Therefore,
$$(x_n, y_n) \to (x_0, y_0) \tag{7}$$

● **PROBLEM 6-42**

Show that the mapping

$$T : \Sigma-N \to \pi \tag{1}$$

and the inverse mapping

$$T^{-1} : \pi \to \Sigma-N \tag{2}$$

are continuous.

Solution: We shall repeat the sequential definition of continuous mapping:

A mapping $T : A \to B$ is said to be continuous if

$$\lim_{n \to \infty} a_n = a, \quad a_n \in A \tag{3}$$

implies that

$$\lim_{n \to \infty} T(a_n) = T(a) \tag{4}$$

We will show that the transformation T of the sphere (without the north pole N) onto the complex plane π is continuous.

If

$$(\xi_n, \eta_n, \zeta_n) \in \Sigma-N \tag{5}$$

and

$$(\xi_n, \eta_n, \zeta_n) \to (\xi_0, \eta_0, \zeta_0) \in \Sigma-N, \tag{6}$$

then

$$T : (\xi_n, \eta_n, \zeta_n) \to (x_n, y_n)$$
$$T : (\xi_0, \eta_0, \zeta_0) \to (x_0, y_0) \tag{7}$$

According to Problem 6-41,

$$T(\xi_n, \eta_n, \zeta_n) = (x_n, y_n) \to (x_0, y_0) = T(\xi_0, \eta_0, \zeta_0) \tag{8}$$

Therefore, T is continuous.

If

$$(x_n, y_n) \to (x_0, y_0), \tag{9}$$

then, according to Problem 6-40,

$$T^{-1}(x_n, y_n) = (\xi_n, \eta_n, \zeta_n) \to (\xi_0, \eta_0, \zeta_0)$$
$$= T^{-1}(x_0, y_0) \tag{10}$$

and transformation T^{-1} is continuous.

• **PROBLEM** 6-43

Describe the relationship between point N, called the north pole of the Riemann sphere, and the point at infinity of the extended complex plane.

Solution: The complex sequence (z_n) approaches infinity

$$\lim_{n \to \infty} z_n = \infty \qquad (1)$$

if for any given $M > 0$, there exists $N \in N'$ such that for $n > N$,

$$|z_n| > M \qquad (2)$$

We can express the above statement as:

$$\lim z_n = \infty \text{ if } \bigwedge_{M>0} \bigvee_{N \in N'}, \bigwedge_{n>N} |z_n| > M \qquad (3)$$

Let (z_n) be a complex sequence approaching infinity. The corresponding sequence of points on the sphere is (P_n).

As $z_n \to \infty$, P_n approaches the pole N.

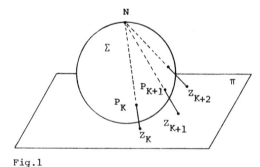

Fig.1

We say that N corresponds to the point at infinity of the complex plane. A complex plane equipped with the point at infinity is called the extended plane. The sphere Σ (with N) is called the Riemann sphere. The mapping of the Riemann sphere onto the extended complex plane is called stereographic projection.

The stereographic projection establishes a one-to-one correspondence between the extended complex plane and the Riemann sphere.

• **PROBLEM** 6-44

Show that (z_n) converges to z_0 when z_n and z_0 are regarded as points in the plane π, if (z_n) converges to z_0 when z_n and z_0 are regarded as the points of the sphere Σ-N.

Solution: (z_n) and z_0 are complex numbers which we denote by Q_n and Q_0, the points in π representing z_n and z_0.

The points representing z_n and z_0 on the sphere are denoted by P_n and P_0.

The stereographic projection

$$T : \Sigma \to \pi \tag{1}$$

maps

$$T(P_n) = Q_n \tag{2}$$

$$T(P_0) = Q_0 \tag{3}$$

We now have to show that $P_n \to P_0$ if $Q_n \to Q_0$.

Since T is continuous and $P_n \to P_0$, we obtain

$$Q_n = T(P_n) \to T(P_0) = Q_0 \tag{4}$$

The inverse transformation

$$T^{-1}(Q_n) = P_n \tag{5}$$

$$T^{-1}(Q_0) = P_0 \tag{6}$$

is also continuous; therefore, if $Q_n \to Q_0$, then

$$P_n = T^{-1}(Q_n) \to T^{-1}(Q_0) = P_0 \tag{7}$$

● **PROBLEM 6-45**

A point $P(\xi, \eta, \zeta)$ on the sphere represents a complex number z. Show that

$$|z| > \alpha \quad \text{if} \tag{1}$$

$$1 - \zeta < \frac{1}{1+\alpha^2} \tag{2}$$

where $\alpha > 0$.

Solution: Since $P(\xi, \eta, \zeta)$ is located on the sphere,

$$\xi^2 + \eta^2 = \zeta(1-\zeta) \tag{3}$$

(see Problem 6-37).

The coordinates of z are given by

$$z = \frac{\xi + i\eta}{1 - \zeta} \tag{4}$$

(see Problem 6-39).

Hence,

$$|z|^2 = \frac{\xi^2 + \eta^2}{(1-\zeta)^2} = \frac{\zeta(1-\zeta)}{(1-\zeta)^2} = \frac{\zeta}{1-\zeta} \tag{5}$$

$$= \frac{1}{1-\zeta} - 1$$

From

or
$$|z| > \alpha$$
$$|z|^2 > \alpha^2 \tag{6}$$

we obtain
$$\frac{1}{1-\zeta} - 1 > \alpha^2 \tag{7}$$

or
$$\frac{1}{1+\alpha^2} > 1-\zeta \tag{8}$$

• **PROBLEM 6-46**

Show that the distance of a point $P(\xi,\eta,\zeta)$ on the sphere from $N(0,0,1)$ is $\sqrt{1-\zeta}$.

Solution: All points are shown in Fig. 1.

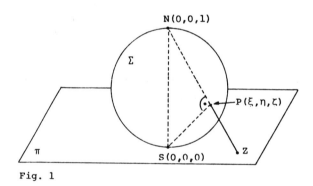

Fig. 1

Since P is on the sphere,
$$(NS)^2 = (NP)^2 + (PS)^2 \tag{1}$$

For further information, check additional text books.
$$NS = \tfrac{1}{2} + \tfrac{1}{2} = 1 \tag{2}$$

The radius of the sphere is $\tfrac{1}{2}$.
$$SP = \sqrt{\xi^2+\eta^2+\zeta^2} \tag{3}$$

Hence,
$$1 = \xi^2 + \eta^2 + \zeta^2 + (NP)^2 \tag{4}$$

Since P is located on the sphere,
$$\xi^2 + \eta^2 = \zeta(1-\zeta) \tag{5}$$

(see Problem 6-37).

Therefore,
$$1 = \zeta(1-\zeta) + \zeta^2 + (NP)^2 \tag{6}$$
and
$$NP = \sqrt{1-\zeta} \tag{7}$$

● **PROBLEM 6-47**

Consider a part of π plane defined by
$$|z| > R \tag{1}$$
Show what part of the sphere Σ corresponds to that part of π^2.

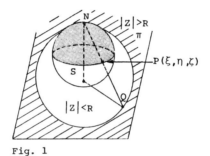

Fig. 1

Solution: Let P be a point on the sphere, such that its projection on π is point Q, which represents a complex number z, such that
$$|z| = R \tag{2}$$
Hence, all points of a spherical cap (shaded part of the sphere-see Fig. 1) are projected onto $|z| > R$. We shall evaluate distance NP,
$$NP = \sqrt{1-\zeta} \tag{3}$$
(see Problem 6-46).

$|z| > R$ if
$$1-\zeta < \frac{1}{1+R^2} \tag{4}$$
(see Problem 6-45).

Therefore,
$$NP = \sqrt{1-\zeta} = \frac{1}{\sqrt{1+R^2}} \tag{5}$$

• **PROBLEM 6-48**

Let $T: \Sigma-N \to \pi$ be a stereographic projection. Prove that a set A on the sphere $\Sigma-N$ is connected and open with respect to the sphere Σ if its image in the plane π is a domain.

Solution: In Problem 6-42 we proved that both mappings

$$T: \Sigma-N \to \pi \tag{1}$$

and

$$T^{-1}: \pi \to \Sigma-N \tag{2}$$

are continuous.

A connected set on the sphere Σ, which is open with respect to Σ, is called a domain on the sphere. Let A be an open and connected set on the sphere.

T is a homeomorphism; therefore, A is connected if T(A) is connected. Furthermore, A is open if T(A) is open. We conclude that A is a domain on the sphere if its image T(A) in the plane π is a domain.

• **PROBLEM 6-49**

Sometimes stereographic projections are presented in a slightly different way.

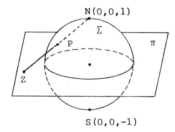

Fig. 1

Σ is a sphere of radius one, described by

$$x_1^2 + x_2^2 + x_3^2 = 1 \tag{1}$$

The complex plane is identified with the $x_1 x_2$ plane. There is a one-to-one correspondence between the complex numbers and the points of the sphere.

$$\begin{aligned}(\xi, \eta, \zeta) &= \left(\frac{2x_1}{x_1^2 + x_2^2 + 1}, \frac{2x_2}{x_1^2 + x_2^2 + 1}, \frac{x_1^2 + x_2^2 - 1}{x_1^2 + x_2^2 + 1}\right) \\ &= \left(\frac{z + \bar{z}}{|z|^2 + 1}, \frac{z - \bar{z}}{i(|z|^2 + 1)}, \frac{|z|^2 - 1}{|z|^2 + 1}\right)\end{aligned} \tag{2}$$

Express this correspondence and its inverse in spherical polar coordinates.

Solution: Let
$$T : \Sigma \to \pi \tag{3}$$
and
$$T^{-1} : \pi \to \Sigma$$

where Σ is the Riemann sphere (without the north pole), and π is the plane of complex numbers.

For $z \in \pi$, we have
$$z = re^{i\theta}, \quad 0 \le \theta < 2\pi \tag{4}$$

A point (ξ, η, ζ) on the sphere can be expressed by
$$\begin{aligned} \xi &= \rho \sin\phi \cos\theta \\ \eta &= \rho \sin\phi \sin\theta \\ \zeta &= \rho \cos\phi \end{aligned} \tag{5}$$

Since $\rho = 1$, the points on the sphere are described by $(1, \phi, \theta)$ with $0 \le \phi \le \pi$, $0 \le \theta < 2\pi$. We have
$$T^{-1}(z) = \left(\frac{z + \bar{z}}{|z|^2 + 1}, \frac{z - \bar{z}}{i(|z|^2 + 1)}, \frac{|z|^2 - 1}{|z|^2 + 1} \right) \tag{6}$$

and
$$T(\xi, \eta, \zeta) = \frac{\xi + i\eta}{1 - \zeta} \tag{7}$$

Note that
$$\begin{aligned} r &= |z| = \sqrt{x_1^2 + x_2^2} \\ x_1 &= r\cos\theta \\ x_2 &= r\sin\theta \end{aligned} \tag{8}$$

Hence,
$$\begin{aligned} T^{-1}(re^{i\theta}) &= \left(\frac{2r\cos\theta}{r^2 + 1}, \frac{2r\sin\theta}{r^2 + 1}, \frac{r^2 - 1}{r^2 + 1} \right) \\ &= (\xi, \eta, \zeta) \end{aligned} \tag{9}$$

From eq.(5),
$$\phi = \cos^{-1}\zeta = \cos^{-1}\left(\frac{r^2 - 1}{r^2 + 1} \right) \tag{10}$$

and
$$\tan\theta = \frac{\sin\theta}{\cos\theta} = \frac{\eta}{\xi} = \frac{\frac{2r\sin\theta}{r^2+1}}{\frac{2r\cos\theta}{r^2+1}} = \tan\theta \tag{11}$$

where $\rho = 1$.

Hence,
$$T^{-1}(re^{i\theta}) = \left(1, \cos^{-1}\left(\frac{r^2 - 1}{r^2 + 1} \right), \theta \right) \tag{12}$$

Eq.(11) shows that θ in the spherical representation is the same as θ in the polar representation.

so
$$\phi = \cos^{-1}\left(\frac{r^2-1}{r^2+1}\right)$$
$$\cos\phi = \frac{r^2-1}{r^2+1} \tag{13}$$

$$r^2\cos\phi + \cos\phi = r^2 - 1$$

$$r^2 = \frac{1+\cos\phi}{1-\cos\phi}$$

Since $r \geq 0$

$$r = \sqrt{\frac{1+\cos\phi}{1-\cos\phi}} = \cot\frac{\phi}{2} \tag{14}$$

Hence,
$$T(1,\phi,\theta) = \cos\frac{\phi}{2} e^{i\theta} \tag{15}$$

• **PROBLEM 6-50**

Compute the metric on the extended complex plane induced by the geometry of the Riemann sphere.

Solution: Let $C \cup \{\infty\}$ be the extended complex plane, and Σ the Riemann sphere. We denote

$$T^{-1}: C \cup \{\infty\} \to \Sigma \tag{1}$$

Then,

$$T^{-1}(z) = T^{-1}(x_1 + ix_2) = \left(\frac{z+\bar{z}}{|z|^2+1}, \frac{z-\bar{z}}{i(|z|^2+1)}, \frac{|z|^2-1}{|z|^2+1}\right) \tag{2}$$

and
$$T^{-1}(\infty) = (0,0,1) \tag{3}$$

We shall compute the distance between z and z' by computing the distance between their projections (ξ,η,ζ) and (ξ',η',ζ').

$$(\xi-\xi')^2 + (\eta-\eta')^2 + (\zeta-\zeta')^2$$
$$= \xi^2 + \eta^2 + \zeta^2 + \xi'^2 + \eta'^2 + \zeta'^2 - 2(\xi\xi'+\eta\eta'+\zeta\zeta') \tag{4}$$
$$= 2 - 2(\xi\xi'+\eta\eta'+\zeta\zeta')$$

From eq.(2),

$$\xi\xi'+\eta\eta'+\zeta\zeta' = \frac{(z+\bar{z})(z'+\bar{z'})-(z-\bar{z})(z'-\bar{z'})+(|z|^2-1)(|z'|^2-1)}{(1+|z|^2)(1+|z'|^2)}$$
$$= \frac{(1+|z|^2)(1+|z'|^2)-2|z-z'|^2}{(1+|z|^2)(1+|z'|^2)} \tag{5}$$

Hence, we may define

$$d(z,z') := \sqrt{(\xi-\xi')^2+(\eta-\eta')^2+(\zeta-\zeta')^2}$$

$$= \sqrt{2-2(\xi\xi'+\eta\eta'+\zeta\zeta')} \qquad (6)$$

$$= \sqrt{\frac{4|z-z'|^2}{(1+|z|^2)(1+|z'|^2)}} = \frac{2|z-z'|}{\sqrt{(1+|z|^2)(1+|z'|^2)}}$$

Note that for $z' = \infty$, eq.(6) yields

$$d(z,\infty) = \sqrt{(\xi-0)^2+(\eta-0)^2+(\zeta-1)^2}$$

$$= \sqrt{\xi^2+\eta^2+\zeta^2-2\zeta+1} = \sqrt{2-2\zeta} \qquad (7)$$

$$= \sqrt{2\left(1 - \frac{|z|^2-1}{|z|^2+1}\right)} = \sqrt{\frac{4}{|z|^2+1}} = \frac{2}{\sqrt{1+|z|^2}}$$

● **PROBLEM 6-51**

Let

$$T_+ : \Sigma \to \pi \qquad (1)$$

be the usual stereographic projection from the north pole $N(0,0,1)$.

Let

$$T_- : \Sigma' \to \pi' \qquad (2)$$

be the stereographic projection from the south pole $S(0,0,0)$.

Show that

$$T_- \circ T_+^{-1} : \pi-\{0\} \to \pi-\{0\} \qquad (3)$$

is defined by

$$T_- \circ T_+^{-1}(z) = \frac{1}{z} \qquad (4)$$

Solution: T_+ indicates "ordinary" stereographic projection of Σ onto π. Hence,

$$T_+(\xi,\eta,\zeta) = \frac{\xi+i\eta}{1-\zeta} \qquad (5)$$

and

$$T_+^{-1}(x_1+ix_2) = \left(\frac{x_1}{x_1^2+x_2^2+1}, \frac{x_2}{x_1^2+x_2^2+1}, \frac{x_1^2+x_2^2}{x_1^2+x_2^2+1}\right) \qquad (6)$$

Let

$$T_- : \Sigma' \to \pi' \qquad (7)$$

be the stereographic projection of a point on the sphere through the south pole $S(0,0,0)$ onto the surface $\zeta = 1$. It is easy to show that

$$T_-(\xi,\eta,\zeta) = \frac{\xi+i\eta}{\zeta} \tag{8}$$

$$T_-^{-1}(x_1+ix_2) = \left(\frac{x_1}{x_1^2 + x_2^2 + 1}, \frac{x_2}{x_1^2 + x_2^2 + 1}, 1 - \frac{x_1^2 + x_2^2}{x_1^2 + x_2^2 + 1}\right) \tag{9}$$

From eq.(6) and (8), we find

$$\begin{aligned}
T_- \circ T_+^{-1}(z) &= T_-\left[T_+^{-1}(x_1+ix_2)\right] \\
&= T_-\left(\frac{x_1}{x_1^2 + x_2^2 + 1}, \frac{x_2}{x_1^2 + x_2^2 + 1}, \frac{x_1^2 + x_2^2}{x_1^2 + x_2^2 + 1}\right) \\
&= \frac{x_1 + ix_2}{x_1^2 + x_2^2 + 1} : \frac{x_1^2 + x_2^2}{x_1^2 + x_2^2 + 1} \\
&= \frac{x_1 + ix_2}{(x_1 + ix_2)(x_1 - ix_2)} = \frac{1}{x_1 - ix_2} = \frac{1}{\bar{z}}
\end{aligned} \tag{10}$$

CHAPTER 7

FUNCTIONS, LIMITS AND CONTINUITY

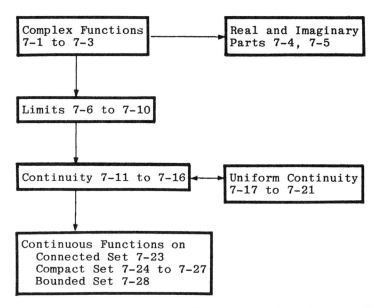

This chart is provided to facilitate rapid understanding of the interrelationships of the topics and subject matter in this chapter. Also shown are the problem numbers associated with the subject matter.

COMPLEX FUNCTIONS

● **PROBLEM 7-1**

Let C be the complex plane and D ⊂ C be a domain.

A complex function

$$f : D \to \mathbb{C} \tag{1}$$

of a complex variable assigns to each $z \in D$ a complex number $f(z)$.

In most cases, the functions are single-valued (i.e. to each z there is only one value $f(z)$ assigned).

In some cases, we will deal with the multiple-valued functions. In multiple-valued functions, some (or all) values of z are assigned many (or even infinite) values $f(z)$.
Let

$$f(z) = z^2 \tag{2}$$

Compute f(z) for

$$z = 2 + 3i$$
$$z = -3 - 2i \quad (3)$$

How would you represent eq.(2) and (3) graphically?

Solution: We have

$$f(z) = f(2+3i) = (2+3i)^2 = -5 + 12i \quad (4)$$

$$f(z) = f(-3-2i) = (-3-2i)^2 = 5 + 12i \quad (5)$$

f(z) is a single-valued function, but it is not one-to-one, since

$$f(-1) = f(1) \quad (6)$$

Since C is two-dimensional, the graph f(z) would require a four-dimensional (z,f(z)) space. Thus, we cannot graph complex functions. However, we can draw domain D and its image f(D).

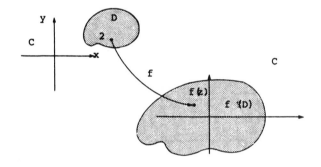

Fig. 1

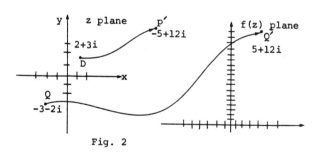

Fig. 2

Transformation $f(z) = z^2$ maps P into P' and Q into Q'.

● **PROBLEM 7-2**

The line joining points P = 2+3i and Q = -3-2i is mapped by the function $f(z) = z^2$ into a curve joining P' = -5+12i and Q' = 5+12i (see Problem 7-1).
Find the equation of this curve.

Solution: The coordinates of P and Q are (2,3) and (-3,-2) correspondingly. Hence, the parametric equation of the line joining P and Q is

$$\frac{x-2}{-3-2} = \frac{y-3}{-2-3} = t \tag{1}$$

or
$$x = -5t + 2 \tag{2}$$
$$y = -5t + 3$$

The equation of the line joining P and Q is

$$z = -5t + 2 + i(-5t+3) \tag{3}$$

Therefore, the equation of the curve $f(z) = z^2$ is

$$f(z) = z^2 = \left[-5t+2+i(-5t+3)\right]^2$$
$$= (-5t+2)^2 - (-5t+3)^2 + 2i(-5t+2)(-5t+3) \tag{4}$$
$$= 10t - 5 + i(50t^2-50t+12)$$

● **PROBLEM 7-3**

Consider a point P moving in the z plane in a counterclockwise direction around a circle of radius one and center at the origin.

Let $\qquad f(z) = z^5 \tag{1}$

be a function which maps point P in the z plane into point P' in the f(z) plane.

What happens with P' when P makes one full revolution?

Solution: Point P is shown in Fig. 1.

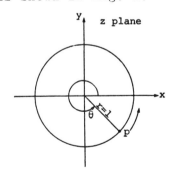

Fig. 1

A complex number z can be represented by

$$z = re^{i\theta} \tag{2}$$

Since r = 1, point P can be represented by a complex number

$$z = e^{i\theta} \qquad (3)$$

Denote by ω the position of P'

$$\omega = f(z) = z^5 \qquad (4)$$

ω can be represented by

$$\omega = r' e^{i\phi} = z^5 = e^{i5\theta} \qquad (5)$$

Then

$$r' = 1 \quad \text{and} \quad \phi = 5\theta \qquad (6)$$

P' moves around the unit circle with center at the origin in the ω plane.

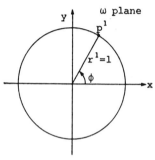

Fig. 2

When P moves counterclockwise through an angle θ, P' also moves counterclockwise through an angle $\phi = 5\theta$.

Hence, when P makes one revolution, P' makes five revolutions.

● **PROBLEM 7-4**

A complex function

$$\omega = f(z) \qquad (1)$$

can be written in the form

$$\omega = f(z) = u(z) + i\, v(z) \qquad (2)$$

where $u(z)$ and $v(z)$ are real valued functions of the complex variable z. They are equal, respectively, to the real and imaginary parts of $f(z)$. Since $z = x+iy$, both functions u and v may be regarded as functions of two real variables (x,y).

There is one-to-one and onto correspondence between z and (x,y).

Hence, we denote

$$\omega = f(z) = u(x,y) + i\, v(x,y) \qquad (3)$$

Determine $u(x,y)$ and $v(x,y)$ when

$$f(z) = z^2 + z + 3 \qquad (4)$$

Solution: Since $z = x+iy$ we obtain

$$\begin{aligned} f(z) &= (x+iy)^2 + (x+iy) + 3 \\ &= x^2 - y^2 + 2ixy + x + iy + 3 \\ &= x^2 - y^2 + x + 3 + i(2xy+y) \end{aligned} \qquad (5)$$

Hence,

$$\begin{aligned} u(x,y) &= x^2 - y^2 + x \\ v(x,y) &= 2xy + y \end{aligned} \qquad (6)$$

● **PROBLEM 7-5**

Express the following functions in the form:

$$f(z) = u(x,y) + i\, v(x,y) \qquad (1)$$

1. $f(z) = z + \bar{z}$
2. $f(z) = z + \dfrac{1}{z}$
3. $f(z) = z^{\frac{1}{2}}$
4. $f(z) = \dfrac{1-z}{1+z}$

Solution: 1. $f(z) = x + iy + x - iy = 2x \qquad (2)$

Hence,
$$f(z) = u(x,y) + i\, v(x,y) = 2x$$
$$u(x,y) = 2x$$
$$v(x,y) = 0$$

2.
$$f(z) = z + \frac{1}{z} = z + \frac{\bar{z}}{z\bar{z}} \qquad (3)$$

$$= x + iy + \frac{x-iy}{x^2+y^2}$$

$$= x + \frac{x}{x^2+y^2} + i\left(y - \frac{y}{x^2+y^2}\right)$$

$$u(x,y) = x + \frac{x}{x^2+y^2}$$

$$v(x,y) = y - \frac{y}{x^2+y^2}$$

3.
$$f(z) = z^{\frac{1}{2}} = (re^{i\theta})^{\frac{1}{2}} = \sqrt{r}\, e^{\frac{i\theta}{2}}$$
$$= r^{\frac{1}{2}}\left[\cos\frac{\theta}{2} + i\sin\frac{\theta}{2}\right] \tag{4}$$

where
$$x = r\cos\theta$$
$$y = r\sin\theta \tag{5}$$

From eq.(5) we can evaluate
$$r = r(x,y) = \sqrt{x^2+y^2}$$
$$\theta = \theta(x,y) = \arccos\frac{x}{\sqrt{x^2+y^2}} \tag{6}$$

Substituting eq.(6) into eq.(4) we find u and v.

4.
$$f(z) = \frac{1-z}{1+z} = \frac{1-x-iy}{1+x+iy}$$
$$= \frac{(1-x-iy)(1+x-iy)}{(1+x+iy)(1+x-iy)}$$
$$= \frac{1+x-iy-x-x^2+ixy-iy-ixy-y^2}{(1+x)^2+y^2} \tag{7}$$
$$= \frac{1-x^2-y^2}{(1+x)^2+y^2} + i\frac{-2y}{(1+x)^2+y^2}$$

$$u(x,y) = \frac{1-x^2-y^2}{(1+x)^2+y^2}$$
$$v(x,y) = \frac{-2y}{(1+x)^2+y^2} \tag{8}$$

LIMITS

• **PROBLEM 7-6**

Definition
Let $f(z) = \omega$ be a single-valued function defined in a domain D. The limit of the function $f(z)$ as z approaches z_0 equals ω_0, we denote

$$\lim_{z \to z_0} f(z) = \omega_0 \tag{1}$$

if

$$\bigwedge_{\varepsilon>0} \bigvee_{\delta>0} (|z-z_0| < \delta) \Rightarrow (|f(z)-\omega_0| < \varepsilon) \tag{2}$$

Show that

$$\lim_{z \to 3i} \frac{z^2+6-iz}{z-3i} = 5i \tag{3}$$

Solution: Note that the function

$$f(z) = \frac{z^2+6-iz}{z-3i} \tag{4}$$

is defined everywhere except at $z = 3i$. The definition of the limit of $f(z)$ does not require that $f(z)$ be defined at z_0. All that matters is the behavior of $f(z)$ as $z \to z_0$.

$$f(z) = \frac{z^2+6-iz}{z-3i} = \frac{(z-3i)(z+2i)}{z-3i} = z + 2i \qquad (5)$$

for $z \neq 3i$

Thus, for $z \neq 3i$

$$|f(z)-5i| = |z+2i-5i| = |z-3i| \qquad (6)$$

Taking $0 < \varepsilon = \delta$ we have

$$|z-3i| < \delta \Rightarrow \left|\frac{z^2+6-iz}{z-3i} - 5i\right| < \varepsilon \qquad (7)$$

Hence,

$$\lim_{z \to 3i} \frac{z^2+6-iz}{z-3i} = 5i \qquad (8)$$

In general, let $P(z)$ be

$$P(z) = a_n z^n + a_{n-1} z^{n-1} + \ldots + a_1 z + a_0 \qquad (9)$$

then

$$\lim_{z \to z_0} P(z) = P(z_0) \qquad (10)$$

• **PROBLEM 7-7**

Show that if a limit of a function $f(z)$ exists, then it is unique.

Solution: Assume the opposite

$$\lim_{z \to z_0} f(z) = \omega_1 \qquad (1)$$

and

$$\lim_{z \to z_0} f(z) = \omega_2 \qquad (2)$$

Then for any $\varepsilon > 0$ there exist $\delta_1 > 0$ and $\delta_2 > 0$ such that

$$0 < |z-z_0| < \delta_1 \Rightarrow |f(z)-\omega_1| < \frac{\varepsilon}{2} \qquad (3)$$

and

$$0 < |z-z_0| < \delta_2 \Rightarrow |f(z)-\omega_2| < \frac{\varepsilon}{2} \qquad (4)$$

Taking

$$\delta = \min(\delta_1, \delta_2) \qquad (5)$$

we obtain

$$|\omega_2-\omega_1| = \left|\left[f(z)-\omega_1\right]-\left[f(z)-\omega_2\right]\right| \tag{6}$$

$$\leq |f(z)-\omega_1| + |f(z)-\omega_2| < \frac{\varepsilon}{2} + \frac{\varepsilon}{2} = \varepsilon$$

for
$$0 < |z-z_0| < \delta \tag{7}$$

Since ε is an arbitrarily small number, it follows that

$$|\omega_1-\omega_2| = 0 \tag{8}$$

or
$$\omega_1 = \omega_2$$

That contradicts our assumption.

● **PROBLEM 7-8**

Evaluate
$$\lim_{z \to 2i} \frac{(3z+1)(z-1)}{z^2+z-1} \tag{1}$$

using theorems on limits. Show each step.

Solution: We have

$$\lim_{z \to 2i} \frac{(3z+1)(z-1)}{z^2+z-1} = \frac{\lim_{z \to 2i} (3z+1)(z-1)}{\lim_{z \to 2i} (z^2+z-1)}$$

$$= \frac{\lim_{z \to 2i}(3z+1)\lim_{z \to 2i}(z-1)}{\lim_{z \to 2i} z^2+\lim_{z \to 2i} z-\lim_{z \to 2i} 1}$$

$$= \frac{\left[(3 \lim_{z \to 2i} z)+\lim_{z \to 2i} 1\right]\left[\lim_{z \to 2i} z-\lim_{z \to 2i} 1\right]}{\lim_{z \to 2i} z \cdot \lim_{z \to 2i} z+\lim_{z \to 2i} z-\lim_{z \to 2i} 1} \tag{2}$$

$$= \frac{[3 \cdot 2i + 1][2i - 1]}{2i \cdot 2i + 2i - 1}$$

$$= \frac{-12 - 6i + 2i - 1}{-4 - 1 + 2i} = \frac{-13 - 4i}{-5 + 2i}$$

● **PROBLEM 7-9**

Show that the limit
$$\lim_{z \to 0} \frac{\bar{z}}{z} \tag{1}$$

does not exist.

Solution: The limit, if it exists, must be independent of the manner in which z approaches the point 0. Let $z \to 0$ along the y-axis. Then,

$$x = 0 \text{ and } z = iy, \quad \bar{z} = -iy \qquad (2)$$

so that the limit is

$$\lim_{z \to 0} \frac{\bar{z}}{z} = \lim_{y \to 0} \frac{-iy}{iy} = -1 \qquad (3)$$

Now, let $z \to 0$ along the x-axis. Then,

$$y = 0, \quad z = x, \quad \bar{z} = x \qquad (4)$$

and the limit is

$$\lim_{z \to 0} \frac{\bar{z}}{z} = \lim_{x \to 0} \frac{x}{x} = 1 \qquad (5)$$

● **PROBLEM 7-10**

Let $f(z)$ be defined on a set A and let z_0 be the accumulation point of A.

Prove that if

$$\lim_{\substack{\alpha \to z_0 \\ \beta \to z_0}} \left[f(\alpha) - f(\beta) \right] = 0 \qquad (1)$$

then, $\lim_{z \to z_0} f(z)$ exists.

Solution: Let $z_n \to z_0$ be a sequence of points in A.

$(f(z_n))$ is a Cauchy sequence, thus it has a limit. We have

$$f(z_n) \to \omega \qquad (2)$$

and

$$\lim_{\substack{\alpha \to z_0 \\ \beta \to z_0}} \left[f(\alpha) - f(\beta) \right] = 0 \qquad (3)$$

Therefore,

$$\underset{\epsilon > 0}{\wedge} \; \underset{N_1 \in \mathbb{N}}{\vee} \; \underset{K(z_0, \epsilon)}{\vee} \; \underset{n > N_1}{\wedge} \text{ we have}$$

$$|f(z) - \omega| = |f(z) - f(z_n) + f(z_n) - \omega|$$
$$\leq |f(z) - f(z_n)| + |f(z_n) - \omega| < \frac{\epsilon}{2} + \frac{\epsilon}{2} = \epsilon \qquad (4)$$

Thus,

$$\lim_{z \to z_0} f(z) = \omega \qquad (5)$$

CONTINUITY

• PROBLEM 7-11

Show that the function

$$\omega = z^n \qquad n = 1, 2, 3, \ldots \qquad (1)$$

is continuous in C.

Solution: Earlier in this book, we discussed in detail continuous mappings on metric spaces. All the definitions and theorems can be applied to the space of complex numbers. Here, we shall only repeat two definitions of continuity. They are equivalent.

Definition I

$f(z)$ is continuous at z_0 when for every $z \to z_0$

$$\lim_{z \to z_0} f(z) = f(z_0) \qquad (2)$$

Definition II

$f(z)$ is continuous at z_0 if

$$\bigwedge_{\varepsilon > 0} \bigvee_{\delta > 0} \quad |z - z_0| < \delta \Rightarrow |f(z) - f(z_0)| < \varepsilon \qquad (3)$$

Remember that $\delta = \delta(\varepsilon)$.

If $f(z)$ is continuous at every point z_0 of the domain D, we say that $f(z)$ is continuous in D.

Let

$$\omega_0 = z_0^n \quad \text{where } z_0 \in C \qquad (4)$$

then

$$\omega - \omega_0 = z^n - z_0^n = (z - z_0)(z^{n-1} + z^{n-2} z_0 + z^{n-3} z_0^2 + \ldots + z_0^{n-1}) \qquad (5)$$

Thus

$$|\omega - \omega_0| = |z - z_0||z^{n-1} + z^{n-2} z_0 + \ldots + z_0^{n-1}|$$

$$\leq |z - z_0|(r^{n-1} + r^{n-2} r_0 + \ldots + r_0^{n-1}) \qquad (6)$$

where

$$|z| = r, \quad |z_0| = r_0 \qquad (7)$$

If $|z - z_0| < \delta$ then,

$$r = |z| = |z_0 + (z - z_0)| \leq |z_0| + |z - z_0| < r_0 + \delta \qquad (8)$$

Therefore,

$$|\omega - \omega_0| < |z - z_0|\left[(r_0 + \delta)^{n-1} + (r_0 + \delta)^{n-2} r_0 + \ldots + r_0^{n-1}\right]$$

$$< n\delta(r_0 + \delta)^{n-1} \qquad (9)$$

Setting

$$\varepsilon = n\delta(r_0+\delta)^{n-1} \qquad (10)$$

we complete the proof that $\omega = z^n$ is continuous in C.

● **PROBLEM 7-12**

Prove that if $f(z)$ is continuous in a region G, then $|f(z)|$ is also continuous.

Solution: Note, that the function $f(z)$ is continuous in a region G.

A set D in the complex plane is said to be a domain if it is open and connected.

G is called a region if it consists of a domain D and some (possibly all or none) of its boundary points. Hence any open domain D or closed $\overline{D}$ is a region.

Suppose $f(z)$ is defined in a region G containing boundary points of G. The continuity of $f(z)$ "inside" G is defined as usual.

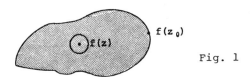

Fig. 1

Let z_0 be the boundary point of G. Then continuity at z_0 is defined as follows

$$\underset{\varepsilon>0}{\wedge} \underset{\delta>0}{\vee} \quad \underset{z\in G}{|z-z_0|} < \delta \Rightarrow |f(z)-f(z_0)| < \varepsilon \qquad (1)$$

We denote

$$\lim_{\substack{z \to z_0 \\ z \in G}} f(z) = f(z_0) \qquad (2)$$

Let $f(z)$ be continuous at $z_0 \in G$ then, $|f(z_0)|$ is also continuous.

$$\Big| |f(z)|-|f(z_0)| \Big| \leq |f(z)-f(z_0)| \qquad (3)$$

and for $|z-z_0| < \delta$

$$\Big| |f(z)|-|f(z_0)| \Big| \leq |f(z)-f(z_0)| < \varepsilon \qquad (4)$$

● **PROBLEM 7-13**

Function $f(z)$ is defined by

$$f(z) = \begin{cases} \dfrac{[\operatorname{Re}(z^2)]^2}{|z^2|} & \text{if } z \neq 0 \\ 0 & \text{if } z = 0 \end{cases} \qquad (1)$$

Show that $f(z)$ is continuous at $z = 0$.

Solution: We have to show that for every $\varepsilon > 0$ there exists $\delta > 0$ such that whenever

$$|z| < \delta \qquad (2)$$

$$\frac{[\operatorname{Re}(z^2)]^2}{|z^2|} < \varepsilon \qquad (3)$$

Therefore since

$$|\operatorname{Re}(z)| \leq |z| \qquad (4)$$

we can replace z by z^2

$$|\operatorname{Re}(z^2)| \leq |z^2| \qquad (5)$$

to obtain

$$\frac{[\operatorname{Re}(z^2)]^2}{|z^2|} \leq \frac{|z^2|^2}{|z^2|} = |z^2| \qquad (6)$$

Setting $\delta = \sqrt{\varepsilon}$ we obtain

$$\frac{[\operatorname{Re}(z^2)]^2}{|z^2|} \leq |z|^2 < \delta^2 = \varepsilon \qquad (7)$$

• **PROBLEM 7-14**

If
$$\lim_{z \to z_0} f(z) = \alpha \neq 0 \qquad (1)$$

show that there exists $\delta > 0$ such that

$$|f(z)| > \tfrac{1}{2}|\alpha| \qquad (2)$$

for

$$0 < |z - z_0| < \delta \qquad (3)$$

Solution: Setting $\varepsilon = \tfrac{1}{2}|\alpha|$ and applying the definition of continuity we find

$$\lim_{z \to z_0} f(z) = \alpha \qquad (4)$$

Hence $\delta > 0$ exists such that

$$|f(z) - \alpha| < \varepsilon = \tfrac{1}{2}|\alpha| \qquad (5)$$

whenever

$$0 < |z-z_0| < \delta$$

Obviously

$$\alpha = \alpha - f(z) + f(z) \tag{6}$$

then

$$|\alpha| \leq |\alpha-f(z)|+|f(z)| < \tfrac{1}{2}|\alpha|+|f(z)| \tag{7}$$

or

$$\tfrac{1}{2}|\alpha| < |f(z)|$$

In the same manner, we can show that if a function $f(z)$ is continuous at z_0 in some domain D and $f(z) \neq 0$, then there exists some neighborhood of z_0 throughout which $f(z) \neq 0$.

● **PROBLEM 7-15**

Consider the following functions

1. $\dfrac{\text{Re}z}{|z|}$

2. $\dfrac{\text{Re}z^2}{|z|^2}$

3. $\dfrac{z\text{Re}z}{|z|}$

which are defined and continuous everywhere except $z = 0$.

Defining the values of the functions at $z = 0$, can you make them continuous?

Solution: If the function is to be continuous at $z = 0$ then

$$f(z) \to \alpha \tag{1}$$

for all $z \to 0$. That is, the value of α cannot depend on the way which z approaches 0.

For $\dfrac{\text{Re}z}{|z|}$, if $z \to 0$ along x^+-axis

$$\lim_{z \to 0} \frac{\text{Re}z}{|z|} = \frac{x}{|x|} = 1 \tag{2}$$

along x-axis

$$\lim_{z \to 0} \frac{\text{Re}z}{|z|} = \lim_{x \to 0} \frac{-|x|}{|x|} = -1 \tag{3}$$

Nothing can be done to make this function continuous.

For $\dfrac{\text{Re}z^2}{|z|^2}$, if $z \to 0$ along x^--axis

$$\lim_{z \to 0} \frac{\text{Re}z^2}{|z|^2} = \lim_{\substack{x \to 0 \\ x > 0}} \frac{x^2}{|x|^2} = 1 \tag{4}$$

if $z \to 0$ along y (imaginary) axis

$$\lim_{z \to 0} \frac{\text{Re}z^2}{|z|^2} = \lim_{y \to 0} \frac{0}{y^2} = 0 \tag{5}$$

Again, we cannot make this function continuous.

$$\text{For } \frac{z \text{ Re}z}{|z|} = \frac{x(x+iy)}{\sqrt{x^2+y^2}} = \frac{x^2+ixy}{\sqrt{x^2+y^2}} \tag{6}$$

if we assign $f(0) = 0$, then

$$\frac{z \text{ Re}z}{|z|} \to 0 \tag{7}$$

for all $z \to 0$.

Hence $\frac{z \text{ Re}z}{|z|}$ with $f(0) = 0$ is a continuous function everywhere in C.

• **PROBLEM 7-16**

> Investigate continuity of the following function:
>
> $$f(z) = \begin{cases} 0 & \text{for } z = 0 \\ 0 & \text{when } |z| \text{ is an irrational number} \\ \frac{1}{n} & \text{when } |z| = \frac{m}{n} \text{ is a rational number} \end{cases}$$
>
> m,n are positive relatively prime integers.

Solution: We will illustrate that $f(z)$ is continuous at $z = 0$ and at all points z for which $|z| \neq \frac{m}{n}$ is irrational.

Let $|z_0|$ be irrational. For any $\varepsilon > 0$ we can obtain a sufficiently small circle $K(z_0, \delta)$ with center at z_0, such that for all points z within this circle and with rational $|z| = \frac{m}{n}$,

$$n > \frac{1}{\varepsilon} \tag{1}$$

It follows that for all points z within this circle

$$|f(z_0) - f(z)| < \varepsilon \tag{2}$$

If z is such that $|z|$ is irrational, then

$$|f(z_0) - f(z)| = |0 - 0| < \varepsilon \tag{3}$$

If, $z \in K(z_0, \delta)$ and is such that $|z| = \frac{m}{n}$ is rational,

$$|z| = \frac{m}{n}, \quad \frac{1}{n} < \varepsilon \tag{4}$$

Then

$$|f(z_0) - f(z)| = \left|0 - \frac{1}{n}\right| = \left|\frac{1}{n}\right| < \varepsilon \tag{5}$$

Hence $f(z)$ is continuous at all points z for which $|z|$ is irrational or $z = 0$.

Let z_0 be such that

$$|z_0| = \frac{m_0}{n_0} \tag{6}$$

is a rational number.

$\varepsilon > 0$ can be chosen such that

$$0 < \varepsilon < \frac{1}{n_0} \tag{7}$$

The appropriate δ cannot be found.

There are points z arbitrarily close to z_0 (all those for which $|z|$ is irrational) for which

$$|f(z_0) - f(z)| = \left|\frac{1}{n_0} - 0\right| = \frac{1}{n_0} > \varepsilon \tag{8}$$

Hence $f(z)$ is not continuous for all z such that z is rational.

UNIFORM CONTINUITY

• **PROBLEM 7-17**

Prove that

$$f(z) = z^2 \tag{1}$$

is uniformly continuous in an open subset of $\mathbb{C}$ defined by

$$|z| < 1 \tag{2}$$

Solution: Uniformly continuous mappings were defined in Chapter 6 Problems 6-8 and 6-13.

Nevertheless, let us repeat the definition.

<u>Definition</u>

A function $f(z)$ defined in a domain D is said to be uniformly continuous in D if for every $\varepsilon > 0$, there exists $\delta = \delta(\varepsilon) > 0$ such that for all z_1 and z_2 in D

$$|f(z_1) - f(z_2)| < \varepsilon \tag{3}$$

when

$$|z_1 - z_2| < \delta \tag{4}$$

Using logical notation we denote : $f(z)$ is uniformly continuous in D if

$$\bigwedge_{\varepsilon>0} \bigvee_{\delta>0} \bigwedge_{z_1,z_2 \in D} (|z_1-z_2|<\delta) \Rightarrow (|f(z_1)-f(z_2)|<\varepsilon) \quad (5)$$

Let $\varepsilon > 0$ be any arbitrarily small positive number. Since for both z_1 and z_2

$$|z_1| < 1, \quad |z_2| < 1 \quad (6)$$

we can denote

$$|z_1^2-z_2^2| = |z_1+z_2||z_1-z_2| \leq (|z_1|+|z_2|)|z_1-z_2| \\ < 2|z_1-z_2| \quad (7)$$

Choosing $\delta = \tfrac{1}{2}\varepsilon$ we obtain

$$(|z_1-z_2|<\delta) \Rightarrow (|z_1^2-z_2^2|<2|z_1-z_2|<2\delta = \varepsilon) \quad (8)$$

Note that δ depends only on ε and not on z_1 or z_2. Hence $f(z) = z^2$ is uniformly continuous.

• **PROBLEM 7-18**

Prove that

$$f(z) = \frac{1}{1-z} \quad (1)$$

is not uniformly continuous for $|z| < 1$.

Solution: Suppose $f(z)$ is uniformly continuous. We select $\varepsilon = \frac{1}{10}$. Then $0 < \delta < 1$ can be obtained.

Now, we select z_1 and z_2 such that

$$z_1 = 1 - \delta \\ z_2 = 1 - \frac{9}{10}\delta \quad (2)$$

$$|z_1| < 1, \quad |z_2| < 1 \quad (3)$$

We have

$$|z_1-z_2| = |(1-\delta)-(1-\tfrac{9}{10}\delta)| = \tfrac{\delta}{10} < \delta \quad (4)$$

However,

$$|f(z_1)-f(z_2)| = \left|\frac{1}{1-z_1} - \frac{1}{1-z_2}\right|$$

$$= \left|\frac{1}{1-(1-\delta)} - \frac{1}{1-(1-\tfrac{9}{10}\delta)}\right| \quad (5)$$

$$= \left|\frac{1}{\delta} - \frac{1}{\tfrac{9}{10}\delta}\right| = \frac{1}{9\delta} > \frac{1}{10} = \varepsilon$$

Thus, $f(z) = \frac{1}{1-z}$ is not uniformly continuous in the domain D defined by $|z| < 1$.

● **PROBLEM** 7-19

Show that
$$f(z) = \frac{1}{z}$$
is not uniformly continuous in the region $0 < |z| < 1$.

Solution: Assume that $f(z)$ is uniformly continuous, then for any $\varepsilon > 0$ we can find $\delta > 0$ such that
$$|z-z_0| < \delta \implies |f(z)-f(z_0)| < \varepsilon \tag{1}$$
for all z and z_0 such that
$$|z| < 1, \quad |z_0| < 1$$
We can choose ε in such a way that corresponding δ will be $0 < \delta < 1$.

Let $\quad z = \delta \quad$ and $\quad z_0 = \dfrac{\delta}{1+\varepsilon} \tag{2}$

then $\quad |z| < 1 \quad$ and $\quad |z_0| < 1$

We have
$$|z-z_0| = \left|\delta - \frac{\delta}{1+\varepsilon}\right| = \frac{\varepsilon}{1+\varepsilon}\delta < \delta \tag{3}$$

On the other hand
$$\left|\frac{1}{z} - \frac{1}{z_0}\right| = \left|\frac{1}{\delta} - \frac{1+\varepsilon}{\delta}\right| = \frac{\varepsilon}{\delta} > \varepsilon \tag{4}$$
where $0 < \delta < 1$

We obtained a contradiction, hence $f(z) = \dfrac{1}{z}$ cannot be uniformly continuous in the region $0 < |z| < 1$.

We can use a different method to prove this fact.

Let z and $z+\alpha$ be any two points of the region such that
$$|z+\alpha-z| = |\alpha| = \delta \tag{5}$$

Then
$$\left|\frac{1}{z} - \frac{1}{z+\alpha}\right| = \left|\frac{z+\alpha-z}{z(z+\alpha)}\right| = \frac{|\alpha|}{|z||z+\alpha|}$$
$$= \frac{\delta}{|z||z+\alpha|} \tag{6}$$

Choosing z sufficiently close to zero we can make $\dfrac{\delta}{|z||z+\alpha|}$ arbitrarily large.

Hence $f(z) = \frac{1}{z}$ cannot be uniformly continuous in $0 < |z| < 1$.

• **PROBLEM 7-20**

Prove that if $f(z)$ is uniformly continuous in a domain D, then $f(z)$ is continuous in D.

Solution: If $f(z)$ is uniformly continuous in D, then

$$\bigwedge_{\varepsilon>0} \bigvee_{\delta>0} \bigwedge_{z_1,z_2 \in D} |z_1-z_2| < \delta \text{ implies that}$$

$$|f(z_1)-f(z_2)| < \varepsilon$$

Hence,

$$\lim_{z \to z_1} f(z) = f(z_1)$$

for any $z_1 \in D$. Function $f(z)$ is continuous.

We see that the condition of uniform continuity is stronger than condition of continuity. There are functions, which are continuous but not uniformly continuous.

• **PROBLEM 7-21**

Let $f(z)$ be a uniformly continuous function in the domain D

$$D := \{z : |z| < 1\} \qquad (1)$$

Prove that for every sequence (z_n) of points in D, such that

$$z_n \xrightarrow[n \to \infty]{} \omega, \quad |\omega| = 1 \qquad (2)$$

the limit exists

$$\lim_{n \to \infty} f(z_n) \qquad (3)$$

and depends only on ω.

Solution:

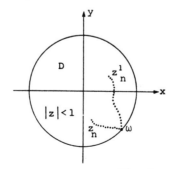

Fig. 1

$f(z)$ is uniformly continuous in D. Hence for every $\varepsilon > 0$ there exists $\delta > 0$ such that

$$|z'-z| < \delta \implies |f(z')-f(z)| < \varepsilon \qquad (4)$$

where

$$|z| < 1 \quad \text{and} \quad |z'| < 1 \qquad (5)$$

Let (z_n) be a sequence in D such that

$$z_n \xrightarrow[n \to \infty]{} \omega, \quad |\omega| = 1 \qquad (6)$$

That is (z_n) converges to a point ω of the unit circle.

There exists $N_1 \in N$ such that

$$\bigwedge_{n > N_1} |z_n - \omega| < \frac{\delta}{2} \qquad (7)$$

Thus, for $n > N_1$ and $m > N_1$ we have

$$|z_n - z_m| = |(z_n - \omega) - (z_m - \omega)| \leq |z_n - \omega| + |z_m - \omega| < \frac{\delta}{2} + \frac{\delta}{2} = \delta \qquad (8)$$

Therefore
$$|f(z_n) - f(z_m)| < \varepsilon \qquad (9)$$

It follows that

$$\lim_{n \to \infty} f(z_n) = \alpha \qquad (10)$$

exists.

We will show that α depends only upon ω.

Consider another sequence (z_n') in D

$$z_n' \to \omega \qquad (11)$$

and let
$$\lim_{n \to \infty} f(z_n') = \alpha' \qquad (12)$$

The sequence $z_1, z_1', z_2, z_2', z_3, z_3', \ldots$ converges to ω, $|\omega| = 1$, hence
$$f(z_1), f(z_1'), f(z_2), f(z_2'), f(z_3), \ldots \qquad (13)$$
converges to, let say α''.

But all subsequences of a convergent sequence, converge to the same limit. Thus
$$\alpha = \alpha' = \alpha'' \qquad (14)$$

The boundary values of $f(z)$ are uniquely determined.

• **PROBLEM 7-22**

Show that the boundary values of the function f(z) defined in Problem 7.21 are continuous.

Solution: Let ω and ω' be two boundary points, as shown in Fig. 1.

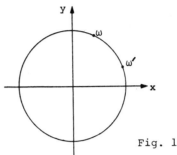

Fig. 1

Their corresponding limit values for $f(z)$ are $f(\omega)$ and $f(\omega')$. Since $f(z)$ is uniformly continuous in D, for every $\varepsilon > 0$ there exists $\delta > 0$ such that

$$|z'-z| < \delta \implies |f(z')-f(z)| < \varepsilon \qquad (1)$$

We will show that

$$|\omega-\omega'| < \frac{\delta}{2} \implies |f(\omega)-f(\omega')| \leq \varepsilon \qquad (2)$$

Let

$$z_n \xrightarrow[n \to \infty]{} \omega,$$
$$z'_n \longrightarrow \omega' \qquad (3)$$

then for all sufficiently large n

$$|\omega-\omega'| < \frac{\delta}{2} \implies |z_n-z'_n| < \delta \qquad (4)$$

Then

$$|f(z_n)-f(z'_n)| < \varepsilon \qquad (5)$$

and since

$$f(z_n) \to f(\omega)$$
$$f(z'_n) \to f(\omega') \qquad (6)$$

also

$$|f(\omega')-f(\omega)| \leq \varepsilon \qquad (7)$$

Hence

$$f(z)\Big|_{|z|=1} = f(\omega) \text{ is continuous.}$$

FUNCTIONS CONTINUOUS OR CONNECTED, COMPACT, AND BOUNDED SETS

● **PROBLEM 7-23**

Let f(z) be a continuous function on a connected set A. Show that, if f(z) takes on integral values only, then f(z) is constant on A.

Solution: Remember the theorem:

If $T : B \to G$ is a continuous mapping of a connected set B onto set G, then G is also a connected set.

Since f(z) is continuous on a connected set A, then the image f(A) of A is also a connected set. f(A) is the set of some integers, since it is connected, it must consist of one element only. Hence f(z) is constant on A.

● **PROBLEM 7-24**

Is the function

$$f(z) = z^4 - 3z^3 + \frac{1}{z} \tag{1}$$

defined on the set $A \subset C$ by

$$1 \leq |z| \leq 2 \tag{2}$$

uniformly continuous?

Solution: It is a cumbersome task to show that this function is uniformly continuous, using ε, δ method.

To avoid it we shall apply the following theorem.

Theorem
If f(z) is continuous on a compact set A, then f(z) is uniformly continuous on A.

Since A defined by $1 \leq |z| \leq 2$ is both bounded and closed, A is a compact set.

Function f(z) is a sum of three functions z^4, $3z^3$, $\frac{1}{z}$. They are all continuous on A, hence f(z) is continuous on A. Since A is a compact set f(z) is uniformly continuous.

● **PROBLEM 7-25**

Let A be a compact set and $\rho(z)$ be the minimum distance from z to A.

If $z \in A$ then $\rho(z) = 0$.

Show that $\rho(z)$ is a continuous function.

Solution: Let z_1 and z_2 be any two points in the plane and z' and z'' be any two points in A such that

$$\rho(z_1) = |z_1-z'|$$
$$\rho(z_2) = |z_2-z''| \tag{1}$$

as shown in Fig. 1.

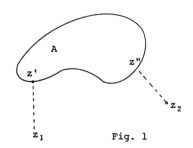

Fig. 1

Then

$$\rho(z_2) \leq |z_2-z'| \tag{2}$$

Hence

$$\rho(z_2)-\rho(z_1) \leq |z_2-z'|-|z_1-z'| \leq |z_2-z_1| \tag{3}$$

In the same way, we show

$$\rho(z_1)-\rho(z_2) \leq |z_1-z_2| \tag{4}$$

Hence

$$|\rho(z_1)-\rho(z_2)| \leq |z_1-z_2| \tag{5}$$

Thus

$$\bigwedge_{\varepsilon>0} \bigvee_{\delta>0} \quad |z_1-z_2| < \delta \Rightarrow |\rho(z_1)-\rho(z_2)| < \varepsilon \tag{6}$$

Setting $\varepsilon = \delta$ we show that $\rho(z)$ is continuous.

• **PROBLEM 7-26**

Let A and B be two disjoint compact sets, see Fig. 1.

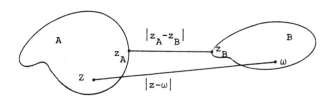

Show that two points exist $z_A \in A$ and $z_B \in B$ such that

$$\bigwedge_{z \in A} \bigwedge_{\omega \in B} \quad |z-\omega| \geq |z_A-z_B| \tag{1}$$

Solution: Let $\rho(z)$ denote the minimum distance from z to A. $\rho(z)$ is a continuous function. Hence a point $z_B \in B$ exists such that

$$\bigwedge_{\omega \in B} \rho(z_B) \le \rho(\omega) \qquad (2)$$

Now, let z_A be a point in A such that

$$\rho(z_B) = |z_B - z_A| \qquad (3)$$

Therefore, if $z \in A$ and $\omega \in B$, we have

$$|z_A - z_B| \le \rho(\omega) \le |z-\omega| \qquad (4)$$

● **PROBLEM 7-27**

Let A be a closed and bounded set, $A \subset C$ and let $z_0 \notin A$. Show that there exists a point $z_1 \in A$ such that

$$\bigwedge_{z \in A} |z_1 - z_0| \le |z - z_0| \qquad (1)$$

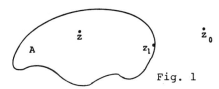

Fig. 1

Consider the case when A is a continuous curve

$$A = \{z(t) : a \le t \le b\} \qquad (2)$$

Solution: We define a function

$$f(z) := |z - z_0| \qquad (3)$$

$f(z)$ is a continuous real-valued function defined on A.

A is a closed and bounded set, hence it is a compact set. We shall apply the following.

Theorem

A continuous function $f : K \to R$ on a compact set K attains its bounds, i.e.,

$$\begin{array}{c} \bigvee_{x_0 \in K} f(x_0) = \sup f(K) \\ \bigvee_{x_1 \in K} f(x_1) = \inf f(K) \end{array} \qquad (4)$$

$f(z)$ is continuous on a compact set A, therefore z_1 exists, $z_1 \in A$ such that

$$f(z_1) = |z_1 - z_0| = \inf f(A)$$

Thus

$$\bigwedge_{z \in A} |z_1 - z_0| \leq |z - z_0| \qquad (5)$$

Let A be a continuous curve

$$A = \{z(t) : a \leq t \leq b\}$$

then the set of all points of A is compact.

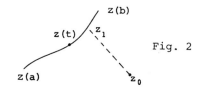

Fig. 2

If $z_0 \notin A$, then there exists a point $z \in A$ such that

$$\bigwedge_{z \in A} |z_1 - z_0| \leq |z - z_0|$$

● **PROBLEM 7-28**

Consider a function f(z) defined in a domain D, which is continuous at $z_0 \in D$.

Show that there exists a neighborhood of z_0 where f(z) is bounded.

Solution: A ball $K(z_0, \varepsilon)$ exists such that f(z) is continuous in $K(z_0, \varepsilon)$, i.e., for every z such that

$$|z - z_0| < \varepsilon \qquad (1)$$

Hence f(z) is continuous on a closed bounded set

$$\overline{K\left(z_0, \frac{\varepsilon}{2}\right)} = \{z : |z - z_0| \leq \frac{\varepsilon}{2}\} \qquad (2)$$

We proved that if f(z) is continuous on a compact set A, then f(z) is bounded on A, that is, there exists a constant M such that $|f(z)| \leq M$ for all $z \in A$.

Therefore a constant M exists such that

$$|f(z)| \leq M \qquad (3)$$

for all

$$z \in \overline{K\left(z_0, \frac{\varepsilon}{2}\right)} \qquad (4)$$

Hence if $|z - z_0| < \frac{\varepsilon}{2}$ then

$$|f(z)| \leq M \qquad (5)$$

CHAPTER 8

DIFFERENTIATION, ANALYTIC FUNCTIONS, CAUCHY-RIEMANN CONDITIONS

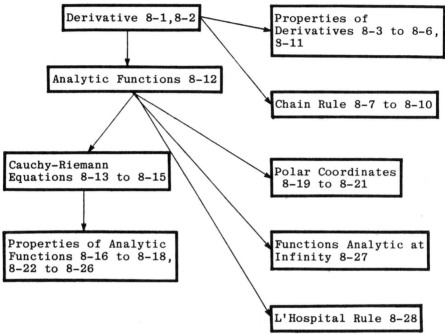

This chart is provided to facilitate rapid understanding of the interrelationships of the topics and subject matter in this chapter. Also shown are the problem numbers associated with the subject matter.

DERIVATIVES

● **PROBLEM 8-1**

Applying the definition of the derivative of a function, compute

$$\frac{df}{dz}, \quad \text{where } f(z) = z^2. \tag{1}$$

Solution: Let $f : D \to C, z_0 \in D$ where D is a domain. The first derivative $f'(z_0)$ of f at z_0 is defined as

$$f'(z_0) := \frac{df(z_0)}{dz} := \lim_{z \to z_0} \frac{f(z)-f(z_0)}{z-z_0} \tag{2}$$

$z \in D$

provided the limit exists and is a finite complex number. The limit in eq.(2), must equal the same number, regardless of the manner in which the variable z approaches z_0. Remember, $f'(z_0)$ is not to be thought of as the slope of some curve, because the complex function cannot be sketched in the usual way.

Directly from the definition we find

$$\frac{dz^2}{dz} = \lim_{z \to z_0} \frac{z^2 - z_0^2}{z - z_0} = \lim_{z \to z_0} \frac{(z-z_0)(z+z_0)}{z-z_0}$$
$$= \lim_{z \to z_0} (z+z_0) = 2z_0 \tag{3}$$

We can express z as

$$z = z_0 + \Delta z \tag{4}$$

Then eq.(2) becomes

$$f'(z_0) := \lim_{\Delta z \to 0} \frac{f(z_0 + \Delta z) - f(z_0)}{\Delta z} \tag{5}$$

In our case

$$\frac{dz^2(z_0)}{dz} = \lim_{\Delta z \to 0} \frac{(z_0 + \Delta z)^2 - z_0^2}{\Delta z}$$
$$= \lim_{\Delta z \to 0} \frac{z_0^2 + 2z_0 \Delta z + (\Delta z)^2 - z_0^2}{\Delta z} \tag{6}$$
$$= 2z_0$$

● **PROBLEM 8-2**

Prove the following theorem:

If $f(z)$ has a derivative at a point z_0, then $f(z)$ is continuous at z_0.

Show that a continuous function at z_0 does not have to have a derivative at z_0.

Solution: We have

$$\lim_{z \to z_0} \left[f(z) - f(z_0) \right] = \lim_{z \to z_0} (z - z_0) \lim_{z \to z_0} \frac{f(z) - f(z_0)}{z - z_0}$$
$$= 0 \cdot f'(z_0) \tag{1}$$
$$= 0$$

Hence,

$$\lim_{z \to z_0} f(z) = \lim_{z \to z_0} \left[f(z_0) + f(z) - f(z_0) \right] \tag{2}$$

$$= f(z_0) + 0 = f(z_0)$$

and the function $f(z)$ is continuous at z_0.

Consider the function

$$f(z) = |z|^2 \qquad (3)$$

which is continuous for all values of z. We will show that $f'(z_0)$ does not exist for any $z_0 \neq 0$.

Observe

$$\frac{f(z)-f(z_0)}{z-z_0} = \frac{|z|^2-|z_0|^2}{z-z_0} = \frac{z\bar{z} - z_0\bar{z_0}}{z-z_0}$$

$$= \bar{z} + z_0 \left(\frac{\bar{z}-\bar{z_0}}{z-z_0}\right) \qquad (4)$$

Now let

$$z - z_0 = re^{i\theta}, \qquad r > 0 \qquad (5)$$

Then eq.(4) becomes

$$\frac{f(z)-f(z_0)}{z-z_0} = \bar{z} + z_0 \frac{re^{-i\theta}}{re^{i\theta}} = \bar{z} + z_0 e^{-2i\theta}$$

$$= \bar{z} + z_0(\cos 2\theta - i \sin 2\theta) \qquad (6)$$

From eq.(6), we conclude that the limit $\lim_{z \to z_0} \frac{f(z)-f(z_0)}{z-z_0}$ does not exist.

For example, for $\theta = 0$

$$\lim_{z \to z_0} \frac{f(z)-f(z_0)}{z-z_0} = \bar{z_0} + z_0 = 2 \text{ Re}(z_0) \qquad (7)$$

while for $\theta = \frac{\pi}{2}$

$$\lim_{z \to z_0} \frac{f(z)-f(z_0)}{z-z_0} = \bar{z_0} - z_0 = -2 \text{ Im}(z_0) \qquad (8)$$

• **PROBLEM** 8-3

Let $f(z)$ and $g(z)$ be two complex functions defined on a domain D. Prove that

1. $(f(z) \pm g(z))' = f'(z) \pm g'(z)$ \hfill (1)

2. $(f(z)g(z))' = f'(z)g(z) + f(z)g'(z)$ \hfill (2)

3. $\left(\frac{f(z)}{g(z)}\right)' = \frac{f'(z)g(z)-g'(z)f(z)}{(g(z))^2}$ \hfill (3)

Solution: 1. Let z_0 be an arbitrary point in D. Then,

$$(f(z) + g(z))' = \lim_{z \to z_0} \frac{f(z)+g(z)-f(z_0)-g(z_0)}{z-z_0}$$

$$= \lim_{z \to z_0} \frac{f(z)-f(z_0)}{z-z_0} + \lim_{z \to z_0} \frac{g(z)-g(z_0)}{z-z_0} \quad (4)$$

$$= f'(z_0) + g'(z_0)$$

2. $(f(z)g(z))' = \lim_{z \to z_0} \frac{f(z)g(z)-f(z_0)g(z_0)}{z-z_0}$

$$= \lim_{z \to z_0} \frac{f(z)g(z)-f(z)g(z_0)+f(z)g(z_0)-f(z_0)g(z_0)}{z-z_0} \quad (5)$$

$$= \lim_{z \to z_0} f(z)\frac{g(z)-g(z_0)}{z-z_0} + \lim_{z \to z_0} g(z_0)\frac{f(z)-f(z_0)}{z-z_0}$$

$$= f(z_0)g'(z_0) + g(z_0)f'(z_0)$$

Note, since f and g are continuous

$$\lim_{z \to z_0} f(z) = f(z_0)$$
$$\lim_{z \to z_0} g(z) = g(z_0) \quad (6)$$

3. First, assume $g(z_0) \neq 0$, $z_0 \in D$. Since $g'(z_0)$ exists, $g(z)$ is continuous at z_0. Hence, a ball $K(z_0, \varepsilon)$ exists such that

$$\bigwedge_{z \in K(z_0, \varepsilon)} g(z) \neq 0 \quad (7)$$

$$\left(\frac{f(z)}{g(z)}\right)' = \lim_{z \to z_0} \frac{\frac{f(z)}{g(z)} - \frac{f(z_0)}{g(z_0)}}{z-z_0}$$

$$= \lim_{z \to z_0} \frac{1}{z-z_0} \frac{f(z)g(z_0)-f(z_0)g(z)}{g(z)g(z_0)} \quad (8)$$

$$= \lim_{z \to z_0} \frac{1}{z-z_0} \frac{f(z)g(z_0)-f(z_0)g(z_0)+f(z_0)g(z_0)-f(z_0)g(z)}{g(z)g(z_0)}$$

$$= \lim_{z \to z_0} \frac{f(z)-f(z_0)}{z-z_0} \frac{g(z_0)}{g(z)g(z_0)} - \lim_{z \to z_0} \frac{g(z)-g(z_0)}{z-z_0} \cdot \frac{f(z_0)}{g(z)g(z_0)}$$

$$= \frac{f'(z_0)g(z_0)-g'(z_0)f(z_0)}{(g(z_0))^2}$$

• **PROBLEM 8-4**

Show that

1. $\frac{da}{dz} = 0$ \quad (1)

2. $\dfrac{dz}{dz} = 1$ \hfill (2)

3. $\dfrac{dz^n}{dz} = nz^{n-1}$ \hfill (3)

4. $\dfrac{d\bar{z}}{dz}$ does not exist.

Solution: 1. From the definition of the derivative we have

$$\frac{da}{dz} = \lim_{z \to z_0} \frac{a-a}{z-z_0} = 0 \tag{4}$$

2. Similarly,

$$\frac{dz}{dz} = \lim_{z \to z_0} \frac{z-z_0}{z-z_0} = 1 \tag{5}$$

3. Observe

$$z^n = z \cdot z^{n-1} \tag{6}$$

Applying eq.(2) of Problem 8-3, we obtain

$$\frac{dz^n}{dz} = \frac{d}{dz}(z \cdot z^{n-1}) = \frac{dz}{dz} z^{n-1} + z \frac{dz^{n-1}}{dz}$$

$$= z^{n-1} + z \frac{dz^{n-1}}{dz} \tag{7}$$

$$= z^{n-1} + z \frac{d}{dz}(z \cdot z^{n-2}) = z^{n-1} + z^{n-1} + z^2 \frac{dz^{n-2}}{dz}$$

$$= 2z^{n-1} + z^2 \frac{dz^{n-2}}{dz} = \ldots$$

$$= (n-1)z^{n-1} + z^{n-1}\frac{dz}{dz} = nz^{n-1}$$

4. $\dfrac{d\bar{z}}{dz} = \lim\limits_{z \to z_0} \dfrac{\bar{z}-\bar{z}_0}{z-z_0} = \lim\limits_{\Delta z \to 0} \dfrac{\overline{\Delta z}}{\Delta z}$ \hfill (8)

This limit does not exist.

If $\Delta z = \Delta x$, then

$$\frac{d\bar{z}}{dz} = 1 \tag{9}$$

On the other hand, if

$$\Delta z = i\Delta y \tag{10}$$

then

$$\frac{d\bar{z}}{dz} = -1 \tag{11}$$

Hence, $\frac{d\bar{z}}{dz}$ does not exist.

• **PROBLEM 8-5**

Often, especially in physical applications, we deal with the function f(z), such that

$$f'(z) = 0 \quad \text{for every } z \in D \qquad (1)$$

where D is some domain.

Show that if f'(z) = 0 in D, then

$$f(z) = a = \text{const} \qquad (2)$$

in domain D.

<u>Solution</u>: This can be proved in different ways. One is separating f(z) into real and imaginary parts.

Here we shall utilize the fact that D is a domain.

Hence, any two points of D can be joined by a polygonal path.

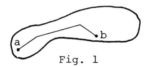

Fig. 1

It is enough to show that f(z) has the same value at the endpoints of any straight segment (α', α'') located in D.

Let s be the length of the segment. Then, for any $\varepsilon > 0$ and any z' which belongs to the segment, there exists a neighborhood of z',

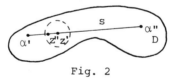

Fig. 2

such that for any z" in this neighborhood

$$\left|\frac{f(z'')-f(z')}{z''-z'}\right| < \frac{\varepsilon}{2s} \qquad (3)$$

Applying the Heine-Borel theorem, we conclude that a finite number of such neighborhoods covers the segment. Therefore, a sequence exists.

$$\alpha' = z_0, z_1, z_2, \ldots, z_n = \alpha'' \qquad (4)$$

such that two consecutive elements belong to one neighborhood.

We can express two consecutive elements as

$$|f(z_k) - f(z_{k-1})| < \frac{\varepsilon}{s}|z_k - z_{k-1}| \qquad (5)$$

where k = 1, 2, ..., n.

Adding up ℓ inequalities we find (6)

$$|f(\alpha'')-f(\alpha')| < \frac{\varepsilon}{s}\left[|z_1-z_0|+|z_2-z_1|+\ldots+|z_n-z_{n-1}|\right] = \frac{\varepsilon}{s} \cdot s = \varepsilon$$

Since ε was an arbitrary number, we conclude

$$f(\alpha') = f(\alpha'') \tag{7}$$

● **PROBLEM 8-6**

Compute $\frac{df}{dz}$ for the following functions:

1. $f(z) = z^2 + 4$ (1)
2. $f(z) = 3z^3 + (z+1)^2 + z$ (2)
3. $f(z) = \frac{z-1}{z+1}$ (3)
4. $f(z) = \left(z + \frac{1}{z}\right)^2$ (4)

Solution: 1.
$$\frac{d}{dz}(z^2+4) = \frac{dz^2}{dz} + \frac{d}{dz}(4)$$
$$= 2z \tag{5}$$

2.
$$f(z) = 3z^3 + (z+1)^2 + z$$
$$= 3z^3 + z^2 + 3z + 1 \tag{6}$$

Hence,
$$\frac{df}{dz} = \frac{d}{dz}(3z^3+z^2+3z+1)$$
$$= 9z^2 + 2z + 3 \tag{7}$$

3. From the definition of the derivative

$$f'(z) = \lim_{\Delta z \to 0} \frac{f(z+\Delta z)-f(z)}{\Delta z}$$

$$= \lim_{\Delta z \to 0} \frac{\frac{z+\Delta z-1}{z+\Delta z+1} - \frac{z-1}{z+1}}{\Delta z} \tag{8}$$

$$= \lim_{\Delta z \to 0} \frac{(z+\Delta z-1)(z+1)-(z-1)(z+\Delta z+1)}{(z+\Delta z+1)(z+1)\Delta z}$$

$$= \lim_{\Delta z \to 0} \frac{2\Delta z}{\Delta z(z+1)(z+\Delta z+1)} = \frac{2}{(z+1)^2}$$

We can also apply eq.(3) of Problem 8-3

$$\frac{df}{dz} = \frac{d}{dz}\left(\frac{z-1}{z+1}\right) = \frac{\frac{d}{dz}(z-1)\cdot(z+1)-(z-1)\frac{d}{dz}(z+1)}{(z+1)^2}$$

$$= \frac{z+1-z+1}{(z+1)^2} = \frac{2}{(z+1)^2} \tag{9}$$

4.
$$f(z) = (z + \frac{1}{z})^2 = z^2 + 2z \cdot \frac{1}{z} + \frac{1}{z^2}$$
$$= z^2 + \frac{1}{z^2} + 2 \tag{10}$$

$$\frac{df}{dz} = \frac{d}{dz}(z^2 + \frac{1}{z^2} + 2) = \frac{d}{dz}(z^2) + \frac{d}{dz}(z^{-2}) + \frac{d}{dz}(2)$$
$$= 2z - 2z^{-3} \tag{11}$$

We applied the formula

$$\frac{d}{dz}(z^n) = nz^{n-1} \tag{12}$$

● **PROBLEM 8-7**

Let $g(z)$ be a function defined and differentiable in a domain D, with values in a domain G and let $f(z)$ be a differentiable function defined in a domain G.

Show that the mapping

$$\omega = f(g(z)) \tag{1}$$

is differentiable and

$$\frac{d\omega}{dz} = f'(g(z))g'(z) \tag{2}$$

This equation is sometimes called the chain rule.

Solution: From the definition of the derivative, it follows that $f(z)$ has a derivative at $z = z_0$ if

$$f(z) - f(z_0) = (z-z_0)\left[f'(z_0) + R(z)\right] \tag{3}$$

where

$$R(z) \xrightarrow[z \to z_0]{} 0 \tag{4}$$

Let $z_0 \in D$, then $g(z_0) \in G$

We shall denote

$$\Delta z = z - z_0 \tag{5}$$

$$\Delta g = g(z) - g(z_0) \tag{6}$$

and

$$\Delta \omega = f\left[g(z)\right] - f\left[g(z_0)\right] \tag{7}$$

Applying eq.(3) to the functions g and ω, we can denote

$$\Delta g = \Delta z \left[g'(z_0) + R_1(z) \right] \qquad (8)$$

$$\Delta \omega = \Delta g \left[f'(g(z_0)) + R_2(g(z)) \right] \qquad (9)$$

from eq.(8) and (9) we find

$$\frac{\Delta \omega}{\Delta z} = \frac{\Delta g}{\Delta z} \left[f'(g(z_0)) + R_2 \right]$$
$$= \left[f'(g(z_0)) + R_2 \right] \left[g'(z_0) + R_1 \right] \qquad (10)$$

Hence,

$$\left. \frac{d\omega}{dz} \right|_{z=z_0} = \lim_{\Delta z \to 0} \frac{\Delta \omega}{\Delta z} = f'(g(z_0)) g'(z_0) \qquad (11)$$

for all $z_0 \in D$.

● **PROBLEM 8-8**

Find $\frac{df}{dz}$ using the chain rule derived in Problem 8-7.

1. $f(z) = ((z+2)^3 + z^3)^4$ \qquad (1)

2. $f(z) = 2z \left(\frac{1}{(z+1)^2} + (z+1)^2 \right)^2$ \qquad (2)

Solution: 1. $\frac{d}{dz} \left[(z+2)^3 + z^3 \right]^4$

$$= 4 \left[(z+2)^3 + z^3 \right]^3 \frac{d}{dz} \left[(z+2)^3 + z^3 \right]$$
$$= 4 \left[(z+2)^3 + z^3 \right]^3 \left[3(z+2)^2 + 3z^2 \right] \qquad (3)$$

2. $\frac{d}{dz} \left[2z \left(\frac{1}{(z+1)^2} + (z+1)^2 \right)^2 \right]$

$$= 2 \left(\frac{1}{(z+1)^2} + (z+1)^2 \right)^2 + 2z \frac{d}{dz} \left(\frac{1}{(z+1)^2} + (z+1)^2 \right)^2$$

$$= 2 \left(\frac{1}{(z+1)^2} + (z+1)^2 \right)^2 \qquad (4)$$

$$+ 2z \cdot 2 \left(\frac{1}{(z+1)^2} + (z+1)^2 \right) \cdot \left[\frac{-2}{(z+1)^3} + 2(z+1) \right]$$

● **PROBLEM 8-9**

Applying mathematical induction, derive the chain rule for a composite function of n functions

$$\omega = f_1\{f_2[\ldots f_n(z)]\ldots\} \qquad (1)$$

Solution: We have to make all necessary assumptions concerning the domains and differentiability of the functions $f_1, f_2, \ldots, f_n$.

Applying n-1 times the chain rule derived in Problem 8-7, we find

$$\frac{d\omega}{dz} = \left[f_1(f_2(\ldots f_n(z)\ldots))\right]'$$

$$= f_1'(f_2(\ldots f_n(z)\ldots))\left[f_2(\ldots f_n(z))\right]'$$

$$= f_1'(f_2(\ldots f_n(z)\ldots))f_2'(f_3(\ldots f_n(z)))\left[f_3(\ldots f_n(z))\right]' \qquad (2)$$

$$= f_1'(f_2(\ldots f_n(z)\ldots))\ldots f_\ell'(f_{\ell+1}(\ldots f_n(z)))\left[f_{\ell+1}(\ldots f_n(z))\right]'$$

$$= f_1'(f_2(\ldots f_n(z)\ldots))f_2'(f_3(\ldots f_n(z)))\ldots f_{n-1}'(f_n(z))f_n'(z)$$

● **PROBLEM 8-10**

The logarithmic differential is defined by

$$D[f(z)] := \frac{f'(z)}{f(z)} \qquad (1)$$

where $f(z) \neq 0$.

The name derives from the formula

$$\frac{d}{dz} \ln z = \frac{1}{z} \qquad (2)$$

which you will learn later. Hence,

$$\frac{d}{dz} \ln f(z) = \frac{f'(z)}{f(z)} \qquad (3)$$

which resembles eq.(1).

Let $f_1, f_2, \ldots, f_n$ be the set of differentiable functions. Prove that

$$D\left[f_1 \cdot f_2 \cdot \ldots \cdot f_n\right] = \sum_{k=1}^{n} D(f_k) \qquad (4)$$

Solution: Consider the case of two functions f_1 and f_2. Then

$$D\left[f_1 \cdot f_2\right] = \frac{(f_1 \cdot f_2)'}{f_1 \cdot f_2} = \frac{f_1' f_2 + f_1 f_2'}{f_1 \cdot f_2}$$

$$= \frac{f_1'}{f_1} + \frac{f_2'}{f_2} = D(f_1) + D(f_2) \qquad (5)$$

Assume that eq.(4) is true for k functions, i.e.,

$$D\left[f_1 \cdot \ldots \cdot f_k\right] = D(f_1) + \ldots + D(f_k) \tag{6}$$

We will show that it is true for k+1 as well.

$$D\left[f_1 \cdot \ldots \cdot f_k \cdot f_{k+1}\right] = D(f_1) + \ldots + D(f_{k-1}) + D(f_k \cdot f_{k+1}) \tag{7}$$

$$= D(f_1) + D(f_2) + \ldots + D(f_k) + D(f_{k+1})$$

Hence, mathematical induction leads to the expected results – eq.(4).

• **PROBLEM 8-11**

A smooth curve $C(t)$ is given by the equations

$$x = x(t), \quad y = y(t) \quad a \le t \le b \tag{1}$$

or $\quad z(t) = x(t) + iy(t)$.

Show that the complex number

$$\frac{dz}{dt} = \lim_{\Delta t \to 0} \frac{\Delta x + i \Delta y}{\Delta t} = \frac{dx}{dt} + i \frac{dy}{dt} \tag{2}$$

represents a vector tangent to the curve.

$\frac{dz}{dt}$ is a derivative of a complex function of a real variable t.

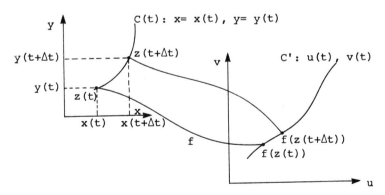

Let f be a differentiable function of z in a domain containing $C(t)$

$$f = u + iv \tag{3}$$

Thus, f maps $C(t)$ into its image C' in uv plane. Show that

$$\frac{df}{dt} = f'(z) \frac{dz}{dt} \tag{4}$$

Solution: Let $\bar{r}(t) = [x(t), y(t)]$ be a vector in R^2. As t changes, $\bar{r}(t)$ defines a curve in R^2. A vector tangent to the curve $\bar{r}(t)$ is

$$\frac{d\bar{r}}{dt} = \left[\frac{dx}{dt}, \frac{dy}{dt}\right] \tag{5}$$

Hence, in the space of complex numbers if

$$z(t) = x(t) + iy(t) \tag{6}$$

describes a curve, then

$$\frac{dz}{dt} = \frac{dx}{dt} + i\frac{dy}{dt} \tag{7}$$

is a vector tangent to $z(t)$. Remember this parallel is possible only because t is a real parameter. If

$$f = f(z(t)) \tag{8}$$

then applying the chain rule

$$\frac{df}{dt} = f'(z(t))\frac{dz}{dt} \tag{9}$$

Eq.(9) defines a tangent vector to the image of curve C(t).

ANALYTIC FUNCTIONS, CAUCHY - RIEMANN EQUATIONS

• **PROBLEM 8-12**

Definition

A single-valued function f(z) defined and differentiable throughout a domain D, is called an analytic function in D. D is called a domain of regularity of the function.

Sometimes the term regular, regular analytic, or holomorphic is used instead of analytic.

Almost all functions used in the applications of mathematics to physical problems are analytic functions. The theory of functions of complex variables is almost entirely the study of analytic functions. Prove the theorem:

Theorem

If the function $f(z) = u(x,y) + iv(x,y)$ is differentiable at the point $z_0 = x_0 + iy_0$, then the four partial derivatives

$$\frac{\partial u}{\partial x} = u_x(x_0, y_0) \qquad \frac{\partial v}{\partial x} = v_x(x_0, y_0) \tag{1}$$

$$\frac{\partial u}{\partial y} = u_y(x_0, y_0) \qquad \frac{\partial v}{\partial y} = v_y(x_0, y_0)$$

of u and v with respect to x_0 and y_0 exist.

Solution: Since $f'(z_0)$ exists, we have

$$f'(z_0) = \lim_{\Delta z \to 0} \frac{f(z_0 + \Delta z) - f(z_0)}{\Delta z} \tag{2}$$

Separating $f(z)$ into real and imaginary parts

$$f(z) = u(x,y) + iv(x,y) \tag{3}$$

$$z = x + iy$$
$$\Delta z = \Delta x + i\Delta y \tag{4}$$

we find

$$f'(z_0) = \lim_{z \to z_0} \frac{[u(x,y) + iv(x,y)] - [u(x_0,y_0) + iv(x_0,y_0)]}{(x+iy) - (x_0+iy_0)} \tag{5}$$

$f'(z_0)$ does not depend on the way z approaches z_0.

Let z approach z_0 along a line parallel to the real axis $y=y_0$, then

$$z - z_0 = x - x_0 = \Delta x \tag{6}$$

Eq.(5) yields

$$f'(z_0) = \lim_{\Delta x \to 0} \frac{u(x_0 + \Delta x, y_0) - u(x_0, y_0)}{\Delta x}$$
$$+ i \lim_{\Delta x \to 0} \frac{v(x_0 + \Delta x, y_0) - v(x_0, y_0)}{\Delta x} \tag{7}$$

Hence, $u_x(x_0,y_0)$ and $v_x(x_0,y_0)$ exist and

$$f'(z_0) = u_x(x_0,y_0) + iv_x(x_0,y_0) \tag{8}$$

On the other hand, for z approaching z_0 along a line $x = x_0$ we obtain

$$z - z_0 = i(y-y_0) = i\Delta y \tag{9}$$

$$f'(z_0) = \lim_{\Delta y \to 0} \frac{u(x_0, y_0 + \Delta y) - u(x_0, y_0)}{i\Delta y}$$
$$+ \lim_{\Delta y \to 0} \frac{v(x_0, y_0 + \Delta y) - v(x_0, y_0)}{\Delta y} \tag{10}$$

Hence, the partial derivatives exist and

$$f'(z_0) = -iu_y(x_0,y_0) + v_y(x_0,y_0) \tag{11}$$

• PROBLEM 8-13

Show that a necessary condition for a function $f(z)$ to be analytic in a domain D, is that all four partial derivatives exist and satisfy the Cauchy-Riemann condition

$$u_x = v_y \tag{1}$$

$$u_y = -v_x \tag{2}$$

for all points of D.

These equations play a very important part in the theory of functions. They are called the Cauchy-Riemann partial differential equations.

Is function $f(z) = \text{Re}\,z$ analytic? Is $f(z) = z^2 + 1$?

<u>Solution</u>: $f(z)$ is analytic in a domain D. Hence, for any $z_0 \in D$, a number exists $f'(z_0)$ which is defined as a limit

$$f'(z_0) = \lim_{z \to z_0} \frac{f(z) - f(z_0)}{z - z_0} \tag{3}$$

This limit is independent of the manner z approaches z_0.

Utilizing the results of Problem 8-12, eq.(8) and eq.(11)

$$f'(z_0) = u_x(x_0, y_0) + iv_x(x_0, y_0) \tag{4}$$

$$f'(z_0) = -iu_y(x_0, y_0) + v_y(x_0, y_0) \tag{5}$$

we find, comparing both equations

$$\bigwedge_{z_0 \in D} \begin{array}{l} u_x(x_0, y_0) = v_y(x_0, y_0) \\ \\ u_y(x_0, y_0) = -v_x(x_0, y_0) \end{array} \tag{6}$$

or

$$\begin{array}{l} \dfrac{\partial u}{\partial x} = \dfrac{\partial v}{\partial y} \\ \\ \dfrac{\partial u}{\partial y} = -\dfrac{\partial v}{\partial x} \end{array} \tag{7}$$

Note, that the Cauchy-Riemann equations impose restrictions on the real and imaginary parts of $f(z)$. u and v are not arbitrary functions.

For
$$f(z) = \text{Re}\,z = x \tag{8}$$

and
$$u = x, \quad v = 0$$

$$\frac{\partial u}{\partial x} = 1 \neq \frac{\partial v}{\partial y} = 0 \tag{9}$$

Thus, $f(z) = \text{Re}\,z$ is not analytic. For

$$f(z) = z^2 + 1 = (x+iy)^2 + 1 = x^2 - y^2 + 1 + i2xy \tag{10}$$

$$u(x,y) = x^2 - y^2 + 1$$
$$v(x,y) = 2xy \tag{11}$$

and

$$\frac{\partial u}{\partial x} = 2x = \frac{\partial v}{\partial y} = 2x \tag{12}$$

$$\frac{\partial u}{\partial y} = -2y = -\frac{\partial v}{\partial x} = -2y \tag{13}$$

The Cauchy-Riemann conditions are fulfilled. At this point we don't know if these conditions are sufficient, (as a matter of fact they are not, as you will see in the next problem). We can suspect that $f(z) = z^2+1$ is an analytic function.

• **PROBLEM 8-14**

Let $f(z)$ be defined by

$$f(z) = \begin{cases} \dfrac{x^3+y^3}{x^2+y^2} + i\,\dfrac{y^3-x^3}{x^2+y^2} & \text{for } x^2+y^2 \neq 0 \\ 0 & \text{for } x^2+y^2 = 0 \end{cases} \tag{1}$$

Show that Cauchy-Riemann conditions are satisfied at the origin, but $f'(0)$ does not exist.

Solution: We have

$$u(x,y) = \frac{x^3+y^3}{x^2+y^2} \tag{2}$$

$$v(x,y) = \frac{y^3-x^3}{x^2+y^2} \tag{3}$$

and

$$\left.\frac{\partial u}{\partial x}\right|_{(x,y)=(0,0)} = \lim_{x \to 0} \frac{u(x,0)-u(0,0)}{x} = \lim_{x \to 0} \frac{x^3}{x^3} = 1 \tag{4}$$

$$\left.\frac{\partial u}{\partial y}\right|_{(x,y)=(0,0)} = \lim_{y \to 0} \frac{u(0,y)-u(0,0)}{y} = \lim_{y \to 0} \frac{y^3}{y^3} = 1 \tag{5}$$

$$\left.\frac{\partial v}{\partial x}\right|_{(x,y)=(0,0)} = \lim_{x \to 0} \frac{v(x,0)-v(0,0)}{x} = \lim_{x \to 0} \frac{-x^3}{x^3} = -1 \tag{6}$$

$$\left.\frac{\partial v}{\partial y}\right|_{(x,y)=(0,0)} = \lim_{y \to 0} \frac{v(0,y)-v(0,0)}{y} = \lim_{y \to 0} \frac{y^3}{y^3} = 1 \tag{7}$$

All partial derivatives exist and the Cauchy-Riemann equations are satisfied.

$$\frac{\partial u}{\partial x} = 1 = \frac{\partial v}{\partial y} = 1 \tag{8}$$

$$\frac{\partial u}{\partial y} = 1 = -\frac{\partial v}{\partial x} = 1 \tag{9}$$

We can suspect function $f(z)$ of being analytic at $z = 0$.

Compute $f'(0)$, when z approaches 0 along the x axis, then $y = 0$ and

$$f'(0) = u_x(0,0) + iv_x(0,0) = 1 - i \tag{10}$$

When $z \to 0$ along $y = x$ line

$$f(z) = \frac{x^3 + x^3}{x^2 + x^2} = x \tag{11}$$

and

$$f'(0) = \lim_{z \to 0} \frac{f(z) - f(0)}{z - 0} = \lim_{x \to 0} \frac{x}{x} = 1 \tag{12}$$

In both cases we obtain different values for $f'(0)$. Hence, $f'(0)$ does not exist and the function $f(z)$ is not analytic at $z = 0$.

• **PROBLEM 8-15**

Prove the following theorem:

If the four partial derivatives of $u(x,y)$ and $v(x,y)$, that is

$$\frac{\partial u}{\partial x}, \frac{\partial u}{\partial y}, \frac{\partial v}{\partial x}, \frac{\partial v}{\partial y}$$

exist in a domain D and if they are continuous and satisfy the Cauchy-Riemann equations, then

$$f(z) = u(x,y) + iv(x,y) \tag{1}$$

is an analytic function in D.

Solution: Let $z, z' \in D$, then

$$f(z) - f(z') = \left[u(x,y) + iv(x,y)\right] - \left[u(x',y') + iv(x',y')\right] \tag{2}$$

$$z = x + iy, \quad z' = x' + iy'$$

Applying the theorem on the total differential of real functions of two real variables to $u(x,y)$ and $v(x,y)$ we find

$$u(x,y) - u(x',y')$$
$$= \left[u_x(x',y') + \alpha(x,y)\right](x - x') + \left[u_y(x',y') + \beta(x,y)\right](y - y') \tag{3}$$

and

$$v(x,y) - v(x',y')$$
$$= \left[v_x(x',y') + \gamma(x,y)\right](x - x') + \left[v_y(x',y') + \delta(x,y)\right](y - y') \tag{4}$$

$\alpha, \beta, \gamma, \delta$ denote the functions which tend toward zero as $(x,y) \to (x',y')$.

Note that

$$\left|\frac{x-x'}{z-z'}\right| \leq 1, \quad \left|\frac{y-y'}{z-z'}\right| \leq 1 \tag{5}$$

and

$$u_x = v_y$$
$$u_y = -v_x \tag{6}$$

Then from eq.(2), (3) and (4) we obtain

$$\frac{f(z)-f(z')}{z-z'} = \frac{u(x,y)-u(x',y')}{z-z'} + i\frac{v(x,y)-v(x',y')}{z-z'}$$

$$= \frac{[u_x(x',y')+\alpha(x,y)](x-x')}{z-z'} + \frac{[-v_x(x',y')+\beta(x,y)](y-y')}{z-z'}$$

$$+ i\frac{[v_x(x',y')+\gamma(x,y)](x-x')}{z-z'} + i\frac{[u_x(x',y')+\delta(x,y)](y-y')}{z-z'}$$

$$\xrightarrow[(x,y)\to(x',y')]{} \frac{[u_x(x',y')+iv_x(x',y')][(x-x')+i(y-y')]}{(x-x')+i(y-y')}$$

$$\xrightarrow[z\to z']{} u_x(x',y') + iv_x(x',y') \tag{7}$$

Therefore, f(z) is differentiable at the point z', and hence z' was arbitrary, f(z) is differentiable everywhere in D.

We conclude that the Cauchy-Riemann equations determine which pairs of functions u(x,y) and v(x,y) can be used as the components of an analytic function

$$f(z) = u(x,y) + iv(x,y) \tag{8}$$

PROPERTIES OF ANALYTIC FUNCTIONS
● **PROBLEM 8-16**

1. Find function f(z) which is differentiable in a domain D, its derivative is zero everywhere in D and

$$f(1+i) = 2 \tag{1}$$

2. Let $f(z) = z^2+1$ be a function defined in a domain D and g(z) be analytic in D, such that

$$f'(z) = g'(z) \text{ everywhere in D} \tag{2}$$

If g(1+i) = 5 + 2i, find g(z).

Solution: 1. Let us prove the following theorem:

If a function f(z) is differentiable in a domain D, and if its derivative is zero everywhere in D, then

$$f(z) \equiv a \quad \text{in D.}$$

From Problem 8-12, we have

$$f'(z) = u_x + iv_x$$

$$f'(z) = v_y - iu_y \qquad (3)$$

Hence,

$$u_x = u_y = 0 \qquad (4)$$

and

$$v_x = v_y = 0 \qquad (5)$$

Thus,

$$u \equiv a_1$$
$$v \equiv a_2$$

everywhere in D.

That completes the proof.

Since $f'(z) = 0$ and $f(1+i) = 2$ we conclude that

$$f(z) \equiv 2 \qquad (6)$$

everywhere in D.

2. From the theorem proved in part 1 we obtain the following theorem:

Two functions which are analytic in the same domain D and whose derivatives are equal in D, differ only by an additive constant. Since $f(z)$ and $g(z)$ are analytic in D, and $f'(z) = g'(z)$ in D, we obtain

$$g(z) = f(z) + \alpha \qquad (7)$$

where α is an additive constant.

and

$$g(z) = z^2 + 1 + \alpha \qquad (8)$$

$$g(1+i) = (1+i)^2 + 1 + \alpha = 5 + 2i \qquad (9)$$

$$\alpha = 4$$

$$g(z) = z^2 + 5 \qquad (10)$$

• **PROBLEM 8-17**

We can summarize the results of Problem 8-15 and 8-13 in the following theorem:

$$\begin{pmatrix} f(z) \text{ is analytic in} \\ \text{a domain D} \end{pmatrix} \Longleftrightarrow \begin{pmatrix} u_x, u_y, v_x, v_y \text{ exist and are} \\ \text{continuous in D} \\ u_x = v_y \text{ and } u_y = -v_x \text{ every-} \\ \text{where in D} \end{pmatrix}$$

Show that $f(z) = z^3$ is analytic and $f(z) = |z|^2 + 1$ is not.

Solution:
$$f(z) = u + iv = (x+iy)^3 \qquad (1)$$

Hence,
$$u = x^3 - 3xy^2$$
$$v = 3x^2y - y^3 \qquad (2)$$

The partial derivatives
$$\frac{\partial u}{\partial x} = 3x^2 - 3y^2$$
$$\frac{\partial u}{\partial y} = -6xy \qquad (3)$$
$$\frac{\partial v}{\partial x} = 6xy$$
$$\frac{\partial v}{\partial y} = 3x^2 - 3y^2$$

exist and are continuous in C.

Moreover, the Cauchy-Riemann equations are satisfied
$$\frac{\partial u}{\partial x} = 3x^2 - 3y^2 = \frac{\partial v}{\partial y}$$
$$\frac{\partial u}{\partial y} = -6xy = -\frac{\partial v}{\partial x} \qquad (4)$$

Hence, $f(z) = z^3$ is analytic in C.

For $f(z) = |z|^2 + 1$
$$f(z) = u + iv = x^2 + y^2 + 1$$
and
$$\frac{\partial u}{\partial x} = 2x \neq \frac{\partial v}{\partial y} = 0 \qquad (5)$$

Hence, $|z|^2 + 1$ is not analytic.

• **PROBLEM 8-18**

Which of the functions are analytic?

1. $f(z) = z^3 + 3i$
2. $f(z) = (\overline{z}+2i)^2 - 1$
3. $f(z) = \dfrac{z+1}{z+4}$
4. $f(z) = \dfrac{z+i}{z-i}$

Solution: In the definition of an analytic function, the requirement of continuity of the derivative is sometimes included. Actually, that is not necessary.

By theorem of Goursat (sometimes called the Cauchy-Goursat Theorem), if a complex derivative exists, then it is continuous. Therefore, both definitions of analytic functions are equivalent.

1. $f(z) = u+iv = (x+iy)^3 + 3i$

$$= (x^3-3xy^2) + i(3x^2y-y^3+3) \tag{1}$$

and

$$\frac{\partial u}{\partial x} = 3x^2-3y^2 = \frac{\partial v}{\partial y} = 3x^2-3y^2 \tag{2}$$

$$\frac{\partial u}{\partial y} = -6xy = -\frac{\partial v}{\partial x} = -6xy \tag{3}$$

$f(z) = z^3+3i$ is analytic everywhere.

2. $f(z) = u+iv = (x-iy+2i)^2 - 1$

$$= (x^2-y^2+4y-5) + i(4x-2xy) \tag{4}$$

$$\frac{\partial u}{\partial x} = 2x$$

$$\frac{\partial v}{\partial y} = -2x \tag{5}$$

This function is not analytic, because

$$u_x \neq v_y \tag{6}$$

3. $f(z) = u+iv = \frac{x+iy+1}{x+iy+4}$

$$= \frac{(x+1+iy)(x+4-iy)}{(x+4+iy)(x+4-iy)}$$

$$= \frac{x^2+y^2+5x+4}{x^2+8x+16+y^2} + i\frac{3y}{x^2+8x+16+y^2} \tag{7}$$

$$\frac{\partial u}{\partial x} = \frac{(2x+5)(x^2+8x+16+y^2)-(2x+8)(x^2+y^2+5x+4)}{(x^2+8x+16+y^2)^2} \tag{8}$$

$$\frac{\partial v}{\partial y} = \frac{3(x^2+8x+16+y^2)-3y \cdot 2y}{(x^2+8x+16+y^2)^2} \tag{9}$$

and

$$\frac{\partial u}{\partial x} = \frac{\partial v}{\partial y} \tag{10}$$

$$\frac{\partial u}{\partial y} = \frac{2y(x^2+8x+16+y^2)-2y(x^2+y^2+5x+4)}{(x^2+8x+16+y^2)^2} \tag{11}$$

$$\frac{\partial v}{\partial x} = \frac{-3y \cdot (2x+8)}{(x^2+8x+16+y^2)^2} \tag{12}$$

$$\frac{\partial u}{\partial x} = -\frac{\partial v}{\partial x} \tag{13}$$

$f(z) = \frac{z+1}{z+4}$ is analytic everywhere except $z = -4$.

4. $f(z) = u+iv = \frac{z+i}{z-i}$

$$= \frac{x+iy+i}{x-iy-i} = \frac{(x+iy+i)(x+i(y+1))}{(x-i(y+1))(x+i(y+1))} \tag{14}$$

$$= \frac{x^2-y^2-2y-1}{x^2+y^2+2y+1} + i \frac{2xy+2x}{x^2+y^2+2y+1}$$

$$\frac{\partial u}{\partial x} = \frac{2x(x^2+y^2+2y+1)-2x(x^2-y^2-2y-1)}{(x^2+y^2+2y+1)^2}$$

$$\frac{\partial v}{\partial y} = \frac{2x(x^2+y^2+2y+1)-(2y+2)(2xy+2x)}{(x^2+y^2+2y+1)^2} \tag{15}$$

Since $\frac{\partial u}{\partial x} \ne \frac{\partial v}{\partial y}$, $f(z) = \frac{z+i}{z-i}$ is not analytic.

E. Goursat, a French mathematician, gave a proof of the integral theorem which requires only that $f'(z)$ exists.

In Paris (1895), E. Goursat, together with P.E. Appell, published a fundamental textbook on the theory of functions entitled, <u>Théorie des Functions Algébriques</u>.

● **PROBLEM** 8-19

Let $f(z)$ be defined in a domain D and let $u = u(r,\theta)$, $v = v(r,\theta)$ be expressed in polar coordinates and have continuous partial derivatives in D (D does not include $z = 0$).

Prove that

$$\begin{Bmatrix} f(z) = u+iv \text{ is} \\ \text{analytic} \end{Bmatrix} \iff \begin{vmatrix} \dfrac{\partial u}{\partial r} = \dfrac{1}{r}\dfrac{\partial v}{\partial \theta} \\ \\ \dfrac{\partial v}{\partial r} = -\dfrac{1}{r}\dfrac{\partial u}{\partial \theta} \end{vmatrix} \tag{1}$$

<u>Solution</u>: We have

$$x = r\cos\theta \tag{2}$$
$$y = r\sin\theta$$

or

$$r = \sqrt{x^2+y^2}$$
$$\theta = \arctan \frac{y}{x}$$

$$\frac{\partial u}{\partial x} = \frac{\partial u}{\partial r}\frac{\partial r}{\partial x} + \frac{\partial u}{\partial \theta}\frac{\partial \theta}{\partial r}$$

$$= \frac{\partial u}{\partial r}\frac{x}{\sqrt{x^2+y^2}} + \frac{\partial u}{\partial \theta}\frac{-y}{x^2+y^2} = \frac{\partial u}{\partial r}\cos\theta - \frac{1}{r}\frac{\partial u}{\partial \theta}\sin\theta \tag{3}$$

$$\frac{\partial u}{\partial y} = \frac{\partial u}{\partial r}\frac{\partial r}{\partial y} + \frac{\partial u}{\partial \theta}\frac{\partial \theta}{\partial y} = \frac{\partial u}{\partial r}\frac{y}{\sqrt{x^2+y^2}} + \frac{\partial u}{\partial \theta}\frac{x}{x^2+y^2}$$

$$= \frac{\partial u}{\partial r}\sin\theta + \frac{1}{r}\frac{\partial u}{\partial \theta}\cos\theta \tag{4}$$

Similarly,

$$\frac{\partial v}{\partial x} = \frac{\partial v}{\partial r}\cos\theta - \frac{1}{r}\frac{\partial v}{\partial \theta}\sin\theta \tag{5}$$

$$\frac{\partial v}{\partial y} = \frac{\partial v}{\partial r}\sin\theta + \frac{1}{r}\frac{\partial v}{\partial \theta}\cos\theta \qquad (6)$$

From the Cauchy-Riemann equation

$$\frac{\partial u}{\partial x} = \frac{\partial v}{\partial y} \qquad (7)$$

substituting eq.(3) and (6) we find

$$\left(\frac{\partial u}{\partial r} - \frac{1}{r}\frac{\partial v}{\partial \theta}\right)\cos\theta - \left(\frac{\partial v}{\partial r} + \frac{1}{r}\frac{\partial u}{\partial \theta}\right)\sin\theta = 0 \qquad (8)$$

From

$$\frac{\partial u}{\partial y} = -\frac{\partial v}{\partial x} \qquad (9)$$

substituting eq.(4) and (5) we obtain

$$\left(\frac{\partial u}{\partial r} - \frac{1}{r}\frac{\partial v}{\partial \theta}\right)\sin\theta + \left(\frac{\partial v}{\partial r} + \frac{1}{r}\frac{\partial u}{\partial \theta}\right)\cos\theta = 0 \qquad (10)$$

From eq.(8) and eq.(10) we obtain

$$\frac{\partial u}{\partial r} = \frac{1}{r}\frac{\partial v}{\partial \theta}$$
$$\frac{\partial v}{\partial r} = -\frac{1}{r}\frac{\partial u}{\partial \theta} \qquad (11)$$

• **PROBLEM 8-20**

Show that if f(z) is an analytic function, then

$$\frac{df}{dz} = (\cos\theta - i\sin\theta)\frac{\partial f}{\partial r} \qquad (1)$$

Solution: Since f(z) is an analytic function, we can denote

$$\frac{df}{dz} = \left(\frac{df}{dz}\right)_\theta = \frac{\partial f}{\partial r}\left(\frac{\partial r}{\partial z}\right)_\theta \qquad (2)$$

where θ is held fixed, or

$$\frac{df}{dz} = \left(\frac{df}{dz}\right)_r = \frac{\partial f}{\partial \theta}\left(\frac{\partial \theta}{\partial z}\right)_r, \qquad (3)$$

where r is held fixed.

Since
$$z = x+iy = r\cos\theta + ir\sin\theta \qquad (4)$$

we have
$$\left(\frac{dr}{dz}\right)_\theta = \cos\theta - i\sin\theta \qquad (5)$$

and
$$\left(\frac{d\theta}{dz}\right)_r = -\frac{i}{r}(\cos\theta - i\sin\theta) \qquad (6)$$

From eq.(2) and eq.(5), we obtain

$$\frac{df}{dz} = \frac{\partial f}{\partial r}(\cos\theta - i\sin\theta) \qquad (7)$$

which is the required eq.(1).

From eq.(3) and eq.(6), we find

$$\frac{df}{dz} = \frac{\partial f}{\partial \theta} \cdot \frac{-i}{r} (\cos\theta - i\sin\theta)$$

$$= -\frac{\partial f}{\partial \theta}\left(\frac{\sin\theta}{r} + i\frac{\cos\theta}{r}\right) \tag{8}$$

● **PROBLEM 8-21**

Which of the functions are analytic?

1. $f(z) = \sqrt[n]{z}$ (1)
2. $f(z) = \frac{y-ix}{x^2+y^2}$ (2)
3. $f(z) = \frac{x-iy}{x^2+y^2}$ (3)

Solution: 1. Expressing $f(z) = \sqrt[n]{z}$ in polar coordinates,

$$f(z) = \sqrt[n]{z} = \sqrt[n]{r}\left(\cos\frac{\theta}{n} + i\sin\frac{\theta}{n}\right) \tag{4}$$

$$0 < r, \quad 0 \le \theta < 2\pi$$

and applying the Cauchy-Riemann conditions,

$$\frac{\partial u}{\partial r} = \frac{1}{r}\frac{\partial v}{\partial \theta}, \quad \frac{\partial v}{\partial r} = -\frac{1}{r}\frac{\partial u}{\partial \theta} \tag{5}$$

we find

$$\frac{\partial u}{\partial r} = \frac{1}{n} r^{\frac{1}{n}-1} \cdot \cos\frac{\theta}{n} \tag{6}$$

$$\frac{\partial v}{\partial r} = \frac{1}{n} r^{\frac{1}{n}-1} \cdot \sin\frac{\theta}{n} \tag{7}$$

$$\frac{\partial u}{\partial \theta} = \frac{-1}{n} \sqrt[n]{r} \sin\frac{\theta}{n} \tag{8}$$

$$\frac{\partial v}{\partial \theta} = \frac{1}{n} \sqrt[n]{r} \cos\frac{\theta}{n} \tag{9}$$

we have

$$\frac{\partial u}{\partial r} = \frac{1}{n} r^{\frac{1}{n}-1} \cos\frac{\theta}{n} = \frac{1}{r}\frac{1}{n}\sqrt[n]{r}\cos\frac{\theta}{n} = \frac{1}{r}\frac{\partial v}{\partial \theta} \tag{10}$$

and

$$\frac{\partial v}{\partial r} = \frac{1}{n} r^{\frac{1}{n}-1} \sin\frac{\theta}{n} = -\frac{1}{r} \cdot \frac{-1}{n}\sqrt[n]{r}\sin\frac{\theta}{n} = -\frac{1}{r}\frac{\partial u}{\partial \theta} \tag{11}$$

Hence, $f(z) = \sqrt[n]{z}$ is analytic.

2. $f(z) = \frac{y-ix}{x^2+y^2} = \frac{r\sin\theta}{r^2} - i\frac{r\cos\theta}{r^2} = \frac{\sin\theta}{r} - i\frac{\cos\theta}{r}$ (12)

Then,

$$\frac{\partial u}{\partial r} = -\frac{\sin\theta}{r^2}, \quad \frac{\partial v}{\partial r} = \frac{\cos\theta}{r^2} \tag{13}$$

$$\frac{\partial u}{\partial \theta} = \frac{\cos\theta}{r}, \quad \frac{\partial v}{\partial \theta} = \frac{\sin\theta}{r} \tag{14}$$

$$\frac{\partial u}{\partial r} = -\frac{\sin\theta}{r^2} \neq \frac{1}{r} \cdot \frac{\sin\theta}{r} = \frac{1}{r}\frac{\partial v}{\partial \theta} \tag{15}$$

Thus, $f(z)$ is not analytic.

3. $\quad f(z) = \frac{x-iy}{x^2+y^2} = \frac{\cos\theta}{r} - i\frac{\sin\theta}{r}, \quad r > 0 \tag{16}$

Here,
$$\frac{\partial u}{\partial r} = \frac{1}{r}\frac{\partial v}{\partial \theta}$$
$$\frac{\partial v}{\partial r} = -\frac{1}{r}\frac{\partial u}{\partial \theta} \tag{17}$$

and $f(z)$ is an analytic function.

● **PROBLEM 8-22**

Suppose $f(z)$ and $\overline{f(z)}$ are analytic functions in a domain D. Show that $f(z) = $ const in D.

Solution: $f(z) = u + iv$ is analytic in D. The Cauchy-Riemann conditions are satisfied
$$\frac{\partial u}{\partial x} = \frac{\partial v}{\partial y} \tag{1}$$

$$\frac{\partial u}{\partial y} = -\frac{\partial v}{\partial x} \tag{2}$$

Also, $\overline{f(z)} = \overline{u+iv} = u - iv = u + i(-v) \tag{3}$

is analytic in D, and

$$\frac{\partial u}{\partial x} = \frac{\partial(-v)}{\partial y} = -\frac{\partial v}{\partial y} \tag{4}$$

$$\frac{\partial u}{\partial y} = -\frac{\partial(-v)}{\partial x} = \frac{\partial v}{\partial x} \tag{5}$$

Adding eq.(1) and (4), we find

$$\frac{\partial u}{\partial x} = 0 \tag{6}$$

Subtracting (1) and (4),

$$\frac{\partial v}{\partial y} = 0 \tag{7}$$

Adding eq.(2) and (5)

$$\frac{\partial u}{\partial y} = 0 \tag{8}$$

Subtracting eq.(2) and (5),

$$\frac{\partial v}{\partial x} = 0 \tag{9}$$

Therefore, $\quad u(x,y) = a \tag{10}$

$$v(x,y) = b \tag{11}$$

where a and b are real constants, and

$$f(z) = a+ib = \text{const in D} \tag{12}$$

• **PROBLEM 8-23**

Show that if $f(z)$ is analytic in a domain D and one of the functions $u(x,y)$, $v(x,y)$ is constant, then the other one is also constant.

Solution: Let us assume that $f(z)$ is analytic in D and $u(x,y)$ is constant in D.

$$f(z) = a + iv(x,y) \tag{1}$$

Then,
$$\frac{\partial u}{\partial x} = 0 = \frac{\partial v}{\partial y} \tag{2}$$

$$\frac{\partial u}{\partial y} = 0 = -\frac{\partial v}{\partial x} \tag{3}$$

Since
$$\frac{\partial v}{\partial x} = \frac{\partial v}{\partial y} = 0 \tag{4}$$

function $v(x,y)$ is constant in D

$$v(x,y) = b \tag{5}$$

and
$$f(z) \equiv a + ib \tag{6}$$

On the other hand, if $f(z)$ is analytic and $v(x,y)$ is constant

$$f(z) = u(x,y) + ib \tag{7}$$

then,
$$\frac{\partial u}{\partial x} = \frac{\partial v}{\partial y} = 0 \tag{8}$$

$$\frac{\partial u}{\partial y} = -\frac{\partial v}{\partial x} = 0 \tag{9}$$

and $u(x,y)$ is constant.

Thus,
$$f(z) \equiv a + ib \tag{10}$$

for all $z \in D$.

• **PROBLEM 8-24**

Show that if $f(z)$ is analytic in a domain D, and

$$f(z) = u + iv \tag{1}$$

and
$$u^2 + v^2 = \text{const in D}, \tag{2}$$

then $f(z) = \text{const in D}$.

Solution: Since $u^2 + v^2$ is constant in D, we have

$$\frac{\partial}{\partial x}(u^2+v^2) = 2u\frac{\partial u}{\partial x} + 2v\frac{\partial v}{\partial x} = 0 \tag{3}$$

and
$$\frac{\partial}{\partial y}(u^2+v^2) = 2u\frac{\partial u}{\partial y} + 2v\frac{\partial v}{\partial y} = 0 \qquad (4)$$
or
$$u\frac{\partial u}{\partial x} + v\frac{\partial v}{\partial x} = 0 \qquad (5)$$
$$u\frac{\partial u}{\partial y} + v\frac{\partial v}{\partial y} = 0 \qquad (6)$$

Since f(z) is analytic

$$\frac{\partial u}{\partial x} = \frac{\partial v}{\partial y} \quad \text{and} \quad \frac{\partial u}{\partial y} = -\frac{\partial v}{\partial x} \qquad (7)$$

Substituting eq.(7) into eq.(5) and (6) we obtain

$$\left. \begin{array}{l} u\dfrac{\partial u}{\partial x} - v\dfrac{\partial u}{\partial y} = 0 \\[6pt] u\dfrac{\partial u}{\partial y} + v\dfrac{\partial u}{\partial x} = 0 \end{array} \right\} \qquad (8)$$

Since $u^2 + v^2 = \text{const} > 0$, we obtain from eq.(8)

$$\frac{\partial u}{\partial x} = \frac{\partial u}{\partial y} = 0 \qquad (9)$$

and
$$u(x,y) = a = \text{const}$$

From eq.(5) and (6), substituting eq.(7) we obtain

$$u\frac{\partial v}{\partial y} + v\frac{\partial v}{\partial x} = 0$$
$$-u\frac{\partial v}{\partial x} + v\frac{\partial v}{\partial y} = 0 \qquad (10)$$

Hence,
$$\frac{\partial v}{\partial x} = \frac{\partial v}{\partial y} = 0 \qquad (11)$$

and $v(x,y) = b = \text{const}$. Thus,

$$f(z) = u + iv = a + ib = \text{const} \qquad (12)$$

• **PROBLEM 8-25**

Show that if $f(z) = u + iv$ is an analytic function in a domain D and u and v have continuous second partial derivatives in D, then

$$\left(\frac{\partial^2}{\partial x^2} + \frac{\partial^2}{\partial y^2}\right)|f(z)|^2 = 4\left|\frac{df}{dz}\right|^2 \qquad (1)$$

Solution: Note that

$$|f(z)|^2 = |u+iv|^2 = u^2 + v^2 \qquad (2)$$

The left-hand side of eq.(2) is equal to

$$\left(\frac{\partial^2}{\partial x^2} + \frac{\partial^2}{\partial y^2}\right)(u^2+v^2)$$

$$= 2\left(\frac{\partial u}{\partial x}\right)^2 + 2\left(\frac{\partial u}{\partial y}\right)^2 + 2\left(\frac{\partial v}{\partial x}\right)^2 + 2\left(\frac{\partial v}{\partial y}\right)^2 \quad (3)$$

$$+ 2u\frac{\partial^2 u}{\partial x^2} + 2u\frac{\partial^2 u}{\partial y^2} + 2v\frac{\partial^2 v}{\partial x^2} + 2v\frac{\partial^2 v}{\partial y^2} = \ldots$$

Since f(z) is analytic, we can substitute

$$\frac{\partial v}{\partial y} \quad \text{by} \quad \frac{\partial u}{\partial x} \quad (4)$$

$$\frac{\partial u}{\partial y} \quad \text{by} \quad -\frac{\partial v}{\partial x} \quad (5)$$

$$\ldots = 4\left(\frac{\partial u}{\partial x}\right)^2 + 4\left(\frac{\partial v}{\partial x}\right)^2 + 2u\left(\frac{\partial^2 u}{\partial x^2} + \frac{\partial^2 u}{\partial y^2}\right)$$

$$+ 2v\left(\frac{\partial^2 v}{\partial x^2} + \frac{\partial^2 v}{\partial y^2}\right) = \ldots \quad (6)$$

Differentiating the Cauchy-Riemann conditions, we find

$$\frac{\partial u}{\partial x} = \frac{\partial v}{\partial y} \rightarrow \frac{\partial^2 u}{\partial x^2} = \frac{\partial^2 v}{\partial x \partial y}$$

$$\frac{\partial u}{\partial y} = -\frac{\partial v}{\partial x} \rightarrow \frac{\partial^2 u}{\partial y^2} = -\frac{\partial^2 v}{\partial y \partial x} \quad (7)$$

Since the second partial derivatives of u and v exist and are continuous

$$\frac{\partial^2 v}{\partial x \partial y} = \frac{\partial^2 v}{\partial y \partial x} \quad (8)$$

and

$$\frac{\partial^2 u}{\partial x^2} + \frac{\partial^2 u}{\partial y^2} = \frac{\partial^2 v}{\partial x \partial y} - \frac{\partial^2 v}{\partial y \partial x} = 0 \quad (9)$$

In the same manner, we can show that

$$\frac{\partial^2 v}{\partial x^2} + \frac{\partial^2 v}{\partial y^2} = 0 \quad (10)$$

Referring back to eq.(6),

$$\ldots = 4\left(\frac{\partial u}{\partial x}\right)^2 + 4\left(\frac{\partial v}{\partial x}\right)^2 = \ldots \quad (11)$$

Since f(z) is an analytic function in D,

$$\frac{df}{dz} = \frac{\partial u}{\partial x} + i\frac{\partial v}{\partial x} \quad (12)$$

Hence,

$$\ldots = 4\left[\left(\frac{\partial u}{\partial x}\right)^2 + \left(\frac{\partial v}{\partial x}\right)^2\right] = 4\left|\frac{df}{dz}\right|^2 \quad (13)$$

• **PROBLEM 8-26**

Let f(z) be analytic in a domain D, and let $\bar{D}$ be the mirror image of D in the x axis

$$\bar{D} := \{z : \bar{z} \in D\} \tag{1}$$

Show that

$$F(z) = \overline{f(\bar{z})} \tag{2}$$

is analytic in $\bar{D}$.

__Solution__: Let $z_0 \in \bar{D}$ be an arbitrary point in $\bar{D}$ (see Fig. 1).

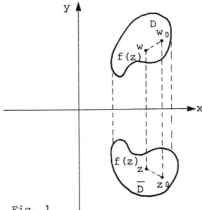

Fig. 1

We have to show that $F'(z_0)$ exists.

If

$$\begin{aligned} \bar{z}_0 &= \omega_0 \\ \bar{z} &= \omega, \end{aligned} \tag{3}$$

then

$$\frac{F(z) - F(z_0)}{z - z_0} = \frac{\overline{f(\bar{z})} - \overline{f(\bar{z}_0)}}{z - z_0}$$

$$= \overline{\left(\frac{f(\bar{z}) - f(\bar{z}_0)}{\bar{z} - \bar{z}_0}\right)} = \overline{\left(\frac{f(\omega) - f(\omega_0)}{\omega - \omega_0}\right)} \tag{4}$$

Note that if $z, z_0 \in \bar{D}$, then $\omega, \omega_0 \in D$.

f(z) is analytic in D, thus

$$F'(z_0) = \lim_{z \to z_0} \frac{F(z) - F(z_0)}{z - z_0} = \lim_{\omega \to \omega_0} \overline{\left(\frac{f(\omega) - f(\omega_0)}{\omega - \omega_0}\right)}$$

$$= \overline{f'(\omega_0)} = \overline{f'(\bar{z}_0)} \tag{5}$$

SOME APPLICATIONS

• **PROBLEM 8-27**

A function $f(z)$ is said to be analytic at infinity if the function

$$F(\omega) := f\left(\frac{1}{\omega}\right) \qquad (1)$$

is analytic at $\omega = 0$.

If $f(z)$ is analytic at infinity, then the value of $f(z)$ at infinity, $f(\infty)$, is defined as

$$f(\infty) := F(0) \qquad (2)$$

Show that

$$\lim_{z \to \infty} f(z) = f(\infty) \qquad (3)$$

$f'(\infty)$ — the derivative of $f(z)$ at infinity — is defined as

$$f'(\infty) := F'(0) \qquad (4)$$

Show that in general

$$f'(\infty) \neq \lim_{z \to \infty} f'(z) \qquad (5)$$

Solution: Since $f(z)$ is analytic at infinity, $F(\omega)$ is analytic at $\omega = 0$, hence, it is continuous at $\omega = 0$. We have

$$\lim_{z \to \infty} f(z) = \lim_{z \to \infty} F\left(\frac{1}{z}\right) = F(0) = f(\infty) \qquad (6)$$

Consider the function

$$f(z) = \frac{1}{z} + 1 \qquad (7)$$

Then, by definition,

$$F(\omega) = \omega + 1 \qquad (8)$$

where $\omega = \frac{1}{z}$.

$F(\omega) = \omega + 1$ is analytic at $\omega = 0$, hence, $f(z) = \frac{1}{z} + 1$ is analytic at infinity. We have

$$f'(\infty) = F'(0) = 1 \qquad (9)$$

On the other hand,

$$f'(z) = -\frac{1}{z^2} \qquad (10)$$

and

$$\lim_{z \to \infty} f'(z) = \lim_{z \to \infty} -\frac{1}{z^2} = 0 \tag{11}$$

Thus,

$$f'(\infty) = 1 \neq \lim_{z \to \infty} f'(z) = 0 \tag{12}$$

• **PROBLEM 8-28**

Prove the following rule, known as l'Hospital rule:

If $f(z)$ and $g(z)$ are analytic at z_0, and $f(z_0) = g(z_0) = 0$ but $g'(z_0) \neq 0$, then

$$\lim_{z \to z_0} \frac{f(z)}{g(z)} = \frac{f'(z_0)}{g'(z_0)} \tag{1}$$

Remember that $f'(z) := \frac{df}{dz}$.

Evaluate

$$\lim_{z \to i} \frac{z^6+1}{z^2+1} \tag{2}$$

Solution: Both $f'(z_0)$ and $g'(z_0)$ exist

$$f'(z_0) = \lim_{z \to z_0} \frac{f(z)-f(z_0)}{z-z_0} \tag{3}$$

$$g'(z_0) = \lim_{z \to z_0} \frac{g(z)-g(z_0)}{z-z_0} \tag{4}$$

Hence,

$$\lim_{z \to z_0} \frac{f(z)}{g(z)} = \lim_{z \to z_0} \frac{f(z)-f(z_0)}{z-z_0} \bigg/ \frac{g(z)-g(z_0)}{z-z_0}$$

$$= \left[\lim_{z \to z_0} \frac{f(z)-f(z_0)}{z-z_0}\right] \bigg/ \left[\lim_{z \to z_0} \frac{g(z)-g(z_0)}{z-z_0}\right]$$

$$= \frac{f'(z_0)}{g'(z_0)} \tag{5}$$

For
$$f(z) = z^6 + 1, \quad f(i) = 0$$
$$g(z) = z^2 + 1, \quad g(i) = 0 \tag{6}$$

Both $f(z)$ and $g(z)$ are analytic at $z = i$.

Thus by l'Hospital rule,

$$\lim_{z \to i} \frac{z^6+1}{z^2+1} = \lim_{z \to i} \frac{6z^5}{2z} = \lim_{z \to i} \frac{6}{2} z^4 = 3 \tag{7}$$

*Guillaume Francois Antoine L'Hospital, Marquis de St.-Mesme was born and died in Paris (1661-1704). As a child prodigy in mathematics, he solved one of Pascal's problems when he was only fifteen. The first book on infinitesimal calculus was published by L'Hospital in 1696. This book, under the title <u>Analyse des Infiniment Petits</u>, brought the differential notation to general use in Europe.

L'Hospital wrote on analytical conics, algebra, and mechanics, and gave a solution of the brachistochrone, a curve joining two points P_1 and P_2 and such that a bead subject only to the gravitational force will traverse the distance in the shortest amount of time.

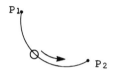

● **PROBLEM 8-29**

Let P be a particle moving along a curve in the complex plane

$$z = \alpha \cos\omega t + \beta i \sin\omega t$$

where α, β, ω are positive constants, and t is a real parameter.

1. Find a unit tangent vector to the curve.

2. Compute the velocity and speed of the particle.

3. Compute the acceleration of the particle.

Solution: 1. Particle P moves along the ellipse in the counterclockwise direction as t increases.

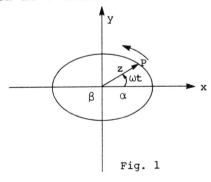

Fig. 1

A tangent vector to the curve is

$$\frac{dz}{dt} = -\alpha\omega \sin\omega t + \beta\omega i \cos\omega t \qquad (1)$$

Hence, the unit tangent vector is

$$\frac{\frac{dz}{dt}}{\left|\frac{dz}{dt}\right|} = \frac{-\alpha\omega\sin\omega t + \beta\omega i\cos\omega t}{|-\alpha\omega\sin\omega t + \beta\omega i\cos\omega t|} \qquad (2)$$

$$= \frac{-\alpha\sin\omega t + \beta i\cos\omega t}{\sqrt{\alpha^2\sin^2\omega t + \beta^2\cos^2\omega t}}$$

(See Problem 8-11).

2. Velocity $= \frac{dz}{dt}$ is given by eq.(1).

$$\text{Speed} = |\text{velocity}| = \left|\frac{dz}{dt}\right| = \omega\sqrt{\alpha^2\sin^2\omega t + \beta^2\cos^2\omega t} \qquad (3)$$

3. Acceleration of the particle is

$$\frac{d^2z}{dt^2} = -\alpha\omega^2\cos\omega t - \beta\omega^2 i\sin\omega t$$
$$= -\omega^2(\alpha\cos\omega t + \beta i\sin\omega t) = -\omega^2 z \qquad (4)$$

From eq.(4), we see that acceleration is always directed toward the origin 0, and is proportional to the distance from 0.

CHAPTER 9

DIFFERENTIAL OPERATORS, HARMONIC FUNCTIONS

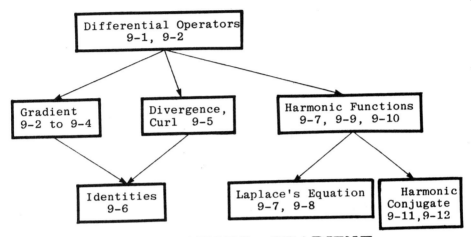

DIFFERENTIAL OPERATORS, GRADIENT, DIVERGENCE, AND CURL

● **PROBLEM 9-1**

Prove that

$$\frac{\partial}{\partial x} = \frac{\partial}{\partial z} + \frac{\partial}{\partial \bar{z}} \qquad (1)$$

$$\frac{\partial}{\partial y} = i\left(\frac{\partial}{\partial z} - \frac{\partial}{\partial \bar{z}}\right) \qquad (2)$$

where

$$z = x + iy, \quad \bar{z} = x - iy.$$

Solution: Note that both equations (1) and (2) represent equivalence of the operators. Hence, we have to show that if $f(z)$ is any continuously differentiable function, then

$$\frac{\partial f}{\partial x} = \frac{\partial f}{\partial z} + \frac{\partial f}{\partial \bar{z}} \qquad (3)$$

Therefore,

$$\frac{\partial f}{\partial x} = \frac{\partial f}{\partial z}\frac{\partial z}{\partial x} + \frac{\partial f}{\partial \bar{z}}\frac{\partial \bar{z}}{\partial x} = \frac{\partial f}{\partial z} + \frac{\partial f}{\partial \bar{z}} \qquad (4)$$

In the same manner, we show equivalence of operators in eq.(2)

$$\frac{\partial f}{\partial y} = \frac{\partial f}{\partial z}\frac{\partial z}{\partial y} + \frac{\partial f}{\partial \bar{z}}\frac{\partial \bar{z}}{\partial y}$$

$$= \frac{\partial f}{\partial z}\frac{\partial}{\partial y}(x+iy) + \frac{\partial f}{\partial \bar{z}}\frac{\partial}{\partial y}(x-iy) \qquad (5)$$

$$= i\frac{\partial f}{\partial z} - i\frac{\partial f}{\partial \bar{z}} = i\left(\frac{\partial f}{\partial z} - \frac{\partial f}{\partial \bar{z}}\right)$$

• **PROBLEM 9-2**

The operators ∇(del) and $\bar{\nabla}$ (del bar) are defined by

$$\nabla := \frac{\partial}{\partial x} + i\frac{\partial}{\partial y} \qquad (1)$$

$$\bar{\nabla} := \frac{\partial}{\partial x} - i\frac{\partial}{\partial y} \qquad (2)$$

1. Prove that

$$\nabla = 2\frac{\partial}{\partial \bar{z}} \qquad (3)$$

$$\bar{\nabla} = 2\frac{\partial}{\partial z} \qquad (4)$$

2. The gradient of a real function $f(x,y)$ is defined by

$$\text{grad } f := \nabla f := \frac{\partial f}{\partial x} + i\frac{\partial f}{\partial y} \qquad (5)$$

Express grad f in terms of z and $\bar{z}$.

Solution: 1. From Problem 9-1 we have

$$\nabla = \frac{\partial}{\partial x} + i\frac{\partial}{\partial y} = \left(\frac{\partial}{\partial z} + \frac{\partial}{\partial \bar{z}}\right) + i^2\left(\frac{\partial}{\partial z} - \frac{\partial}{\partial \bar{z}}\right)$$
$$= 2\frac{\partial}{\partial \bar{z}} \qquad (6)$$

Similarly,

$$\bar{\nabla} = \frac{\partial}{\partial x} - i\frac{\partial}{\partial y} = \left(\frac{\partial}{\partial z} + \frac{\partial}{\partial \bar{z}}\right) - i^2\left(\frac{\partial}{\partial z} - \frac{\partial}{\partial \bar{z}}\right)$$
$$= 2\frac{\partial}{\partial z} \qquad (7)$$

2. From
$$z = x + iy$$
$$\bar{z} = x - iy \qquad (8)$$

we have

$$x = \frac{z+\bar{z}}{2}$$
$$y = \frac{z-\bar{z}}{2i} \qquad (9)$$

Hence,
$$f(x,y) = f\left(\frac{z+\bar{z}}{2}, \frac{z-\bar{z}}{2i}\right) = g(z,\bar{z}) \tag{10}$$
and
$$\text{grad } f = \frac{\partial f}{\partial x} + i\frac{\partial f}{\partial y} = 2\frac{\partial g}{\partial \bar{z}} \tag{11}$$

• **PROBLEM 9-3**

Let $f(x,y) = a$ be a curve in the xy plane, where a is a real constant, and f is continuously differentiable.

Show that ∇f is a vector normal to the curve.

Solution: grad f is defined as
$$\nabla f = \frac{\partial f}{\partial x} + i\frac{\partial f}{\partial y} \tag{1}$$

We have
$$df = \frac{\partial f}{\partial x} dx + \frac{\partial f}{\partial y} dy = 0 \tag{2}$$

Eq.(2) can be written as the dot product of z_1 and z_2 where
$$z_1 = \frac{\partial f}{\partial x} + i\frac{\partial f}{\partial y} \tag{3}$$

$$z_2 = dx + i\,dy$$

Remember that
$$z_1 \circ z_2 = |z_1||z_2|\cos\theta = x_1 x_2 + y_1 y_2 \tag{4}$$

Hence,
$$\left(\frac{\partial f}{\partial x} + i\frac{\partial f}{\partial y}\right) \circ (dx + i\,dy) = 0 \tag{5}$$

Since $(dx + i\,dy)$ is a vector tangent to the curve,
$$\nabla f = \frac{\partial f}{\partial x} + i\frac{\partial f}{\partial y}$$
is a vector normal to the curve.

• **PROBLEM 9-4**

In Problem 9-2 the gradient of a real function was defined
$$\nabla f = \frac{\partial f}{\partial x} + i\frac{\partial f}{\partial y} \tag{1}$$
in such a way that the result was a complex function.

Similarly, the gradient of a complex function $F(z)$ where

$$F(z) = u(x,y) + i\,v(x,y) \qquad (2)$$

can be defined

$$\text{grad } F := \nabla F := \left(\frac{\partial}{\partial x} + i\frac{\partial}{\partial y}\right)(u + iv)$$

$$= \frac{\partial u}{\partial x} - \frac{\partial v}{\partial y} + i\left(\frac{\partial u}{\partial y} + \frac{\partial v}{\partial x}\right) \qquad (3)$$

The result is a complex function. Express ∇F in terms of independent variables z and $\bar{z}$.

Show that if $F(z,\bar{z})$ is analytic, then $\nabla F = 0$.

Solution: From Problem 9-2

$$\nabla = 2\frac{\partial}{\partial \bar{z}} \qquad (4)$$

Hence,

$$\nabla F = \left(\frac{\partial}{\partial x} + i\frac{\partial}{\partial y}\right)(u + iv) = 2\frac{\partial}{\partial \bar{z}} F(z,\bar{z}) \qquad (5)$$

where

$$F(x,y) = F\left(\frac{z+\bar{z}}{2}, \frac{z-\bar{z}}{2}\right) \qquad (6)$$

We obtain

$$\nabla F = \frac{\partial u}{\partial x} - \frac{\partial v}{\partial y} + i\left(\frac{\partial u}{\partial y} + \frac{\partial v}{\partial x}\right) = 2\frac{\partial}{\partial \bar{z}} F(z,\bar{z}) \qquad (7)$$

If $F = u + iv$ is an analytic function, then

$$\frac{\partial u}{\partial x} = \frac{\partial v}{\partial y}$$

$$\frac{\partial u}{\partial y} = -\frac{\partial v}{\partial x} \qquad (8)$$

and $\frac{\partial F}{\partial \bar{z}} = 0$, and the gradient of F is zero.

• **PROBLEM 9-5**

1. The dot product of two complex numbers is defined as

$$z_1 \circ z_2 = x_1 x_2 + y_1 y_2 \qquad (1)$$

Define the divergence of a complex function F as a dot product of ∇ and F, and derive a formula for div F.

2. The cross product of z_1 and z_2 is defined as

$$z_1 \times z_2 = x_1 y_2 - y_1 x_2 \qquad (2)$$

Define the curl of a complex function F as a cross product of ∇ and F.

Solution: 1. Let F be a complex function

$$F = u + iv \tag{3}$$

The divergence of F is defined by

$$\text{div } F := \nabla \circ F := \left(\frac{\partial}{\partial x} + i\frac{\partial}{\partial y}\right) \circ (u + iv)$$

$$= \frac{\partial u}{\partial x} + \frac{\partial v}{\partial y} \tag{4}$$

Since

$$\overline{\nabla} = \frac{\partial}{\partial x} - i\frac{\partial}{\partial y} \tag{5}$$

we can express

$$\text{div } F = \text{Re}(\overline{\nabla}F) = \text{Re}\left[\left(\frac{\partial}{\partial x} - i\frac{\partial}{\partial y}\right)(u + iv)\right]$$

$$= \frac{\partial u}{\partial x} + \frac{\partial v}{\partial y} = 2\,\text{Re}\left(\frac{\partial F(z,\bar{z})}{\partial z}\right) \tag{6}$$

2. The curl of a complex function f is defined by

$$\text{curl } F := \nabla \times F \tag{7}$$

Combining eq.(2) and (7), we find

$$\text{curl } F := \nabla \times F = \left(\frac{\partial}{\partial x} + i\frac{\partial}{\partial y}\right) \times (u + iv)$$

$$= \frac{\partial v}{\partial x} - \frac{\partial u}{\partial y} \tag{8}$$

We can also express

$$\text{curl } F = \text{Im}(\overline{\nabla}F) = \text{Im}\left[\left(\frac{\partial}{\partial x} - i\frac{\partial}{\partial y}\right)(u + iv)\right]$$

$$= \frac{\partial v}{\partial x} - \frac{\partial u}{\partial y} = 2\,\text{Im}\left(\frac{\partial F(z,\bar{z})}{\partial z}\right) \tag{9}$$

Note that the divergence (or curl) of a complex or real function is always a real function.

• **PROBLEM 9-6**

1. Show that for any continuously differentiable complex function f, such that Im(f) is harmonic,

$$\text{curl grad } f = 0 \tag{1}$$

2. Show that for any continuously differentiable complex function f, such that Re(f) is harmonic,

$$\text{div grad } f = 0 \tag{2}$$

A function g(x,y) which satisfies the equation

$$\frac{\partial^2 g}{\partial x^2} + \frac{\partial^2 g}{\partial y^2} = 0 \qquad (3)$$

is called a harmonic function.

Solution: 1. Let $f = u + iv$ \hfill (4)

Then
$$\text{grad } f = \left(\frac{\partial}{\partial x} + i\frac{\partial}{\partial y}\right)(u + iv)$$
$$= \frac{\partial u}{\partial x} - \frac{\partial v}{\partial y} + i\left(\frac{\partial u}{\partial y} + \frac{\partial v}{\partial x}\right) \qquad (5)$$

Hence,
$$\text{curl grad } f = \frac{\partial}{\partial x}\left(\frac{\partial u}{\partial y} + \frac{\partial v}{\partial x}\right) - \frac{\partial}{\partial y}\left(\frac{\partial u}{\partial x} - \frac{\partial v}{\partial y}\right)$$
$$= \frac{\partial^2 u}{\partial x \partial y} - \frac{\partial^2 u}{\partial y \partial x} + \frac{\partial^2 v}{\partial x^2} + \frac{\partial^2 v}{\partial y^2} \qquad (6)$$

Since Im f is harmonic,
$$\frac{\partial^2 v}{\partial x^2} + \frac{\partial^2 v}{\partial y^2} = 0 \qquad (7)$$

and we obtain
$$\text{curl grad } f = 0 \qquad (8)$$

2.
$$\text{grad } f = \frac{\partial u}{\partial x} - \frac{\partial v}{\partial y} + i\left(\frac{\partial u}{\partial y} + \frac{\partial v}{\partial x}\right) \qquad (9)$$
$$\text{div grad } f = \frac{\partial}{\partial x}\left(\frac{\partial u}{\partial x} - \frac{\partial v}{\partial y}\right) + \frac{\partial}{\partial y}\left(\frac{\partial u}{\partial y} + \frac{\partial v}{\partial x}\right)$$
$$= \frac{\partial^2 u}{\partial x^2} + \frac{\partial^2 u}{\partial y^2} - \frac{\partial^2 v}{\partial x \partial y} + \frac{\partial^2 v}{\partial y \partial x} \qquad (10)$$

Since $u = \text{Re}(f)$ is a harmonic function, we find that
$$\text{div grad } f = 0 \qquad (11)$$

HARMONIC FUNCTIONS

● **PROBLEM 9-7**

Prove that if $f(z) = u + iv$ is analytic in a domain D, then u and v are harmonic in D.

Solution: Since $f(z)$ is analytic, $f'(z)$ exists in D, and the Cauchy-Riemann equations are satisfied.

It will be shown later that if $f(z)$ is analytic in D, then all its derivatives exist and are continuous in D.

$$\frac{\partial u}{\partial x} = \frac{\partial v}{\partial y} \qquad (1)$$

$$\frac{\partial u}{\partial y} = -\frac{\partial v}{\partial x} \tag{2}$$

Since u and v have continuous second partial derivatives, we obtain differentiating eq.(1) and (2)

$$\frac{\partial^2 u}{\partial x^2} = \frac{\partial^2 v}{\partial x \partial y} \tag{3}$$

$$\frac{\partial^2 u}{\partial y^2} = -\frac{\partial^2 v}{\partial y \partial x} \tag{4}$$

$$\frac{\partial^2 u}{\partial x^2} + \frac{\partial^2 u}{\partial y^2} = \frac{\partial^2 v}{\partial x \partial y} - \frac{\partial^2 v}{\partial y \partial x} = 0 \tag{5}$$

Equation

$$\frac{\partial^2 f}{\partial x^2} + \frac{\partial^2 f}{\partial y^2} := \nabla^2 f = 0 \tag{6}$$

where

$$\nabla^2 f := \frac{\partial^2 f}{\partial x^2} + \frac{\partial^2 f}{\partial y^2} \tag{7}$$

is called Laplace's equation.

The operator $\nabla^2 := \frac{\partial^2}{\partial x^2} + \frac{\partial^2}{\partial y^2}$ is called the Laplacian.

Functions which satisfy $\nabla^2 f = 0$ in a domain D, are called harmonic functions in D.

Differentiating eq.(1) and (2) with respect to y and x, and subtracting results we find

$$\frac{\partial^2 v}{\partial x^2} + \frac{\partial^2 v}{\partial y^2} = 0 \tag{8}$$

Hence, if f(z) is analytic, both u and v are harmonic.

Pierre-Simon Marquis de Laplace was born in 1749 in Normandy, and died in Paris in 1827. He studied at Beaumont, and later took part in the organization of the École Polytechnique and the École Normale. In 1799 Napoleon appointed him minister of the interior. Laplace's major contributions are in the fields of astronomy, celestial mechanics, calculus, probability and differential equations. As for his mathematical style, American astronomer Nathaniel Bowditch, once remarked, "I never come across one of Laplace's 'Thus it plainly appears' without feeling sure that I have hours of hard work before me to fill up the chasm and find out and show how it plainly appears."

• **PROBLEM 9-8**

Let $f(z) = u + iv$ be an analytic function. Show that u and v, when expressed in polar form, satisfy the equation

$$\frac{\partial^2 u}{\partial r^2} + \frac{1}{r}\frac{\partial u}{\partial r} + \frac{1}{r^2}\frac{\partial^2 u}{\partial \theta^2} = 0 \tag{1}$$

Solution: Eq.(1) is Laplace's equation in polar form. Since $f(z)$ is analytic, u and v satisfy the Cauchy-Riemann equations in polar form

$$\frac{\partial u}{\partial r} = \frac{1}{r}\frac{\partial v}{\partial \theta} \qquad (2)$$

$$\frac{\partial v}{\partial r} = -\frac{1}{r}\frac{\partial u}{\partial \theta} \qquad (3)$$

Differentiating eq.(2) with respect to r, and eq.(3) with respect to θ, we find

$$\frac{\partial^2 v}{\partial r \partial \theta} = \frac{\partial}{\partial r}\left(r\frac{\partial u}{\partial r}\right) = r\frac{\partial^2 u}{\partial r^2} + \frac{\partial u}{\partial r} \qquad (4)$$

$$\frac{\partial^2 v}{\partial \theta \partial r} = \frac{\partial}{\partial \theta}\left(-\frac{1}{r}\frac{\partial u}{\partial \theta}\right) = -\frac{1}{r}\frac{\partial^2 u}{\partial \theta^2} \qquad (5)$$

Since the second partial derivatives are continuous

$$\frac{\partial^2 v}{\partial r \partial \theta} = \frac{\partial^2 v}{\partial \theta \partial r} \qquad (6)$$

Hence, from eq.(4) and (5)

$$\frac{\partial^2 u}{\partial r^2} + \frac{1}{r}\frac{\partial u}{\partial r} + \frac{1}{r^2}\frac{\partial^2 u}{\partial \theta^2} = 0 \qquad (7)$$

Similarly, we can show that $v(r,\theta)$ satisfies the same equation.

● **PROBLEM 9-9**

Let $u(x,y)$ be a function harmonic in a domain D, such that all partial derivatives exist and are continuous in D.

Show that the functions

$$\frac{\partial u}{\partial x}, \frac{\partial u}{\partial y}, \frac{\partial^2 u}{\partial x^2}, \frac{\partial^2 u}{\partial x \partial y}, \text{ etc.}$$

are harmonic in D.

Solution: $u(x,y)$ is harmonic in D

$$\frac{\partial^2 u}{\partial x^2} + \frac{\partial^2 u}{\partial y^2} = 0 \qquad (1)$$

We will show that $\frac{\partial u}{\partial x}$ is also harmonic.

$$\frac{\partial^2}{\partial x^2}\left(\frac{\partial u}{\partial x}\right) + \frac{\partial^2}{\partial y^2}\left(\frac{\partial u}{\partial x}\right) = \frac{\partial}{\partial x}\left(\frac{\partial^2 u}{\partial x^2} + \frac{\partial^2 u}{\partial y^2}\right)$$

$$= \frac{\partial}{\partial x}(0) = 0 \qquad (2)$$

Note that since all partial derivatives exist and are continuous, we can always change the order of differentiation.

$$\frac{\partial}{\partial x}\frac{\partial}{\partial y} = \frac{\partial}{\partial y}\frac{\partial}{\partial x} \tag{3}$$

Similarly for $\frac{\partial^2 u}{\partial x^2}$

$$\frac{\partial^2}{\partial x^2}\left(\frac{\partial^2 u}{\partial x^2}\right) + \frac{\partial^2}{\partial y^2}\left(\frac{\partial^2 u}{\partial x^2}\right) = \frac{\partial^2}{\partial x^2}\left(\frac{\partial^2 u}{\partial x^2} + \frac{\partial^2 u}{\partial y^2}\right) = 0 \tag{4}$$

In the same way, we can show that any other derivative of u is a harmonic function in D.

• **PROBLEM 9-10**

Prove the following statement.

$$\left\{\begin{array}{l}u, v \text{ are harmonic} \\ \text{in a domain } D\end{array}\right\} \Rightarrow \left(\left(\frac{\partial u}{\partial y} - \frac{\partial v}{\partial x}\right) + i\left(\frac{\partial u}{\partial x} + \frac{\partial v}{\partial y}\right)\right) \text{ is analytic in } D \tag{1}$$

Solution: We shall show that the function on the right-hand side of eq.(1) satisfies the Cauchy-Riemann conditions.

Therefore,

$$\frac{\partial}{\partial x}\left(\frac{\partial u}{\partial y} - \frac{\partial v}{\partial x}\right) = \frac{\partial^2 u}{\partial x \partial y} - \frac{\partial^2 v}{\partial x^2} \overset{?}{=} \frac{\partial}{\partial y}\left(\frac{\partial u}{\partial x} + \frac{\partial v}{\partial y}\right)$$
$$= \frac{\partial^2 u}{\partial y \partial x} + \frac{\partial^2 v}{\partial y^2} \tag{2}$$

The second derivatives of u and v exist and are continuous, hence,

$$\frac{\partial^2 u}{\partial x \partial y} = \frac{\partial^2 u}{\partial y \partial x} \tag{3}$$

v is a harmonic function, thus

$$\frac{\partial^2 v}{\partial x^2} + \frac{\partial^2 v}{\partial y^2} = 0 \tag{4}$$

We can remove the question mark from eq.(2).

The next step is

$$\frac{\partial}{\partial y}\left(\frac{\partial u}{\partial y} - \frac{\partial v}{\partial x}\right) = \frac{\partial^2 u}{\partial y^2} - \frac{\partial^2 v}{\partial y \partial x} \overset{?}{=}$$
$$= -\frac{\partial}{\partial x}\left(\frac{\partial u}{\partial x} + \frac{\partial v}{\partial y}\right) = \frac{\partial^2 u}{\partial x^2} - \frac{\partial^2 v}{\partial x \partial y} \tag{5}$$

As before,

$$\frac{\partial^2 v}{\partial x \partial y} = \frac{\partial^2 v}{\partial y \partial x} \tag{6}$$

and

$$\frac{\partial^2 u}{\partial x^2} + \frac{\partial^2 u}{\partial y^2} = 0 \tag{7}$$

That proves (1).

• **PROBLEM 9-11**

Show that
$$u(x,y) = e^y \cos x \tag{1}$$
is harmonic, and find its harmonic conjugate $v(x,y)$.

Solution: **Definition**: If $u(x,y)$ and $v(x,y)$ are harmonic functions in a domain D, and $f(z) = u(x,y) + iv(x,y)$ is analytic in D, then $v(x,y)$ is said to be the harmonic conjugate of $u(x,y)$.

First, we will show that $e^y \cos x$ is a harmonic function.

$$\frac{\partial^2}{\partial x^2}(e^y \cos x) + \frac{\partial^2}{\partial y^2}(e^y \cos x) = -e^y \cos x + e^y \cos x$$
$$= 0 \tag{2}$$

Since $f(z)$ is to be an analytic function, the Cauchy-Riemann equations must be satisfied.

$$\frac{\partial u}{\partial x} = -e^y \sin x = \frac{\partial v}{\partial y} \tag{3}$$

$$\frac{\partial u}{\partial y} = e^y \cos x = -\frac{\partial v}{\partial x} \tag{4}$$

Integrating $-e^y \sin x = \frac{\partial v}{\partial y}$ with respect to y, we find

$$-e^y \sin x + F(x) = v(x,y) \tag{5}$$

where $F(x)$ is an arbitrary function of x.

Differentiating eq.(5) with respect to x,

$$-e^y \cos x + \frac{\partial F}{\partial x} = \frac{\partial v}{\partial x} \tag{6}$$

From eq.(4) and (6)

$$F(x) = a \tag{7}$$

where a is a real constant. Hence, the harmonic conjugate of $u(x,y)$ is

$$v(x,y) = -e^y \sin x + a \tag{8}$$

and the analytic function $f(z)$ is given by

$$f(z) = e^y \cos x - i\, e^y \sin x + ai \tag{9}$$

Therefore, $v(x,y)$ is harmonic.

• **PROBLEM 9-12**

u(x,y) is harmonic in (x,y), and x(α,β) is harmonic in (α,β). y(α,β) is a harmonic conjugate of x(α,β). Show that

$$u = u[x(\alpha,\beta), y(\alpha,\beta)] \qquad (1)$$

is harmonic in (α,β).

Solution: We have

$$\frac{\partial^2 u}{\partial x^2} + \frac{\partial^2 u}{\partial y^2} = 0 \qquad (2)$$

and

$$\frac{\partial^2 x}{\partial \alpha^2} + \frac{\partial^2 x}{\partial \beta^2} = 0 \qquad (3)$$

Since y(α,β) is a harmonic conjugate of x(α,β)

$$\frac{\partial x}{\partial \alpha} = \frac{\partial y}{\partial \beta} \qquad (4)$$

$$\frac{\partial x}{\partial \beta} = - \frac{\partial y}{\partial \alpha} \qquad (5)$$

We obtain

$$\frac{\partial^2 u}{\partial \alpha^2} + \frac{\partial^2 u}{\partial \beta^2} = \frac{\partial}{\partial \alpha}\left[\frac{\partial u}{\partial x}\frac{\partial x}{\partial \alpha} + \frac{\partial u}{\partial y}\frac{\partial y}{\partial \alpha}\right] + \frac{\partial}{\partial \beta}\left[\frac{\partial u}{\partial x}\frac{\partial x}{\partial \beta} + \frac{\partial u}{\partial y}\frac{\partial y}{\partial \beta}\right]$$

$$= \frac{\partial^2 x}{\partial \alpha^2}\frac{\partial u}{\partial x} + \frac{\partial x}{\partial \alpha}\frac{\partial}{\partial \alpha}\left(\frac{\partial u}{\partial x}\right) + \frac{\partial^2 y}{\partial \alpha^2}\frac{\partial u}{\partial y} + \frac{\partial y}{\partial \alpha}\frac{\partial}{\partial \alpha}\left(\frac{\partial u}{\partial y}\right)$$

$$+ \frac{\partial^2 x}{\partial \beta^2}\frac{\partial u}{\partial x} + \frac{\partial x}{\partial \beta}\frac{\partial}{\partial \beta}\left(\frac{\partial u}{\partial x}\right) + \frac{\partial^2 y}{\partial \beta^2}\frac{\partial u}{\partial y} + \frac{\partial y}{\partial \beta}\frac{\partial}{\partial \beta}\left(\frac{\partial u}{\partial y}\right)$$

$$= \left(\frac{\partial^2 x}{\partial \alpha^2} + \frac{\partial^2 x}{\partial \beta^2}\right)\frac{\partial u}{\partial x} + \left(\frac{\partial^2 y}{\partial \alpha^2} + \frac{\partial^2 y}{\partial \beta^2}\right)\frac{\partial u}{\partial y}$$

$$+ 2\frac{\partial^2 u}{\partial x \partial y}\left(\frac{\partial x}{\partial \alpha}\frac{\partial y}{\partial \alpha} + \frac{\partial x}{\partial \beta}\frac{\partial y}{\partial \beta}\right)$$

$$+ \frac{\partial^2 u}{\partial x^2}\left(\frac{\partial x}{\partial \alpha}\frac{\partial x}{\partial \alpha} + \frac{\partial x}{\partial \beta}\frac{\partial x}{\partial \beta}\right) + \frac{\partial^2 u}{\partial y^2}\left(\frac{\partial y}{\partial \alpha}\frac{\partial y}{\partial \alpha} + \frac{\partial y}{\partial \beta}\frac{\partial y}{\partial \beta}\right)$$

$$= 0\frac{\partial u}{\partial x} + 0\frac{\partial u}{\partial y} + 2\frac{\partial^2 u}{\partial x \partial y}\cdot 0 \qquad (6)$$

$$+ \left(\frac{\partial x}{\partial \alpha}\frac{\partial x}{\partial \alpha} + \frac{\partial x}{\partial \beta}\frac{\partial x}{\partial \beta}\right)\left(\frac{\partial^2 u}{\partial x^2} + \frac{\partial^2 u}{\partial y^2}\right) = 0$$

Thus u[x(α,β), y(α,β)] is harmonic in (α,β).

DIFFERENTIAL EQUATIONS

• **PROBLEM 9-13**

Let $f(z) = u(x,y) + i\,v(x,y)$ be a complex function, such that u and v have continuous first order partial derivatives with respect to x and y in a domain D. Since

$$x = \frac{z+\bar{z}}{2} \qquad (1)$$

$$y = \frac{z-\bar{z}}{2i} \qquad (2)$$

$f(z)$ can be expressed in the form

$$f = u\left(\frac{z+\bar{z}}{2}, \frac{z-\bar{z}}{2i}\right) + iv\left(\frac{z+\bar{z}}{2}, \frac{z-\bar{z}}{2i}\right) \qquad (3)$$

f is independent of $\bar{z}$ if

$$\frac{\partial f}{\partial \bar{z}} = 0 \qquad (4)$$

Show that eq.(4) is equivalent to

$$\frac{\partial u}{\partial x} = \frac{\partial v}{\partial y}, \quad \frac{\partial u}{\partial y} = -\frac{\partial v}{\partial x} \qquad (5)$$

Show that if $f(z)$ is analytic, and $f(\bar{z})$ is also analytic, then $f(z)$ is constant.

Solution: We have

$$\frac{\partial f}{\partial \bar{z}} = \frac{\partial u}{\partial \bar{z}} + i\frac{\partial v}{\partial \bar{z}} = \frac{\partial u}{\partial x}\frac{\partial x}{\partial \bar{z}} + \frac{\partial u}{\partial y}\frac{\partial y}{\partial \bar{z}} + i\frac{\partial v}{\partial x}\frac{\partial x}{\partial \bar{z}} + i\frac{\partial v}{\partial y}\frac{\partial y}{\partial \bar{z}}$$

$$= \frac{1}{2}\left(\frac{\partial u}{\partial x} - \frac{\partial v}{\partial y}\right) + \frac{i}{2}\left(\frac{\partial v}{\partial x} + \frac{\partial u}{\partial y}\right) \qquad (6)$$

Hence (4) and (5) are equivalent. If $f(z)$ is analytic, then

$$\frac{\partial u}{\partial x} = \frac{\partial v}{\partial y} \quad \text{and} \quad \frac{\partial u}{\partial y} = -\frac{\partial v}{\partial x} \qquad (7)$$

From (6) we conclude that $\frac{\partial f}{\partial \bar{z}} = 0$. In the same way, we conclude that since $f(\bar{z})$ is analytic, $\frac{\partial f}{\partial z} = 0$.

Thus $f(z)$ is constant.

• **PROBLEM 9-14**

Solve the partial differential equation

$$\frac{\partial^2 f}{\partial x^2} + \frac{\partial^2 f}{\partial y^2} = 2x^2 + y^2 \qquad (1)$$

Solution: We shall replace the independent variables (x,y) by $(z,\bar{z})$ where

$$x = \frac{z+\bar{z}}{2} \qquad (2)$$

$$y = \frac{z-\bar{z}}{2i} \qquad (3)$$

Then,

$$2x^2 + y^2 = 2\left(\frac{z+\bar{z}}{2}\right)^2 + \left(\frac{z-\bar{z}}{2i}\right)^2$$

$$= \frac{z^2}{4} + \frac{3}{2} z\bar{z} + \frac{\bar{z}^2}{4} \qquad (4)$$

$$\frac{\partial^2 f}{\partial x^2} + \frac{\partial^2 f}{\partial x^2} = \nabla^2 f = 4 \frac{\partial^2 f}{\partial z \partial \bar{z}} \qquad (5)$$

Hence, eq.(1) becomes

$$\frac{\partial^2 f}{\partial z \partial \bar{z}} = \frac{1}{16} z^2 + \frac{3}{8} z\bar{z} + \frac{1}{16} \bar{z}^2 \qquad (6)$$

Integrating (6) with respect to z, we obtain

$$\frac{\partial f}{\partial \bar{z}} = \frac{1}{48} z^3 + \frac{3}{16} z^2 \bar{z} + \frac{1}{16} \bar{z}^2 z + F_1(\bar{z}) \qquad (7)$$

where $F_1(\bar{z})$ is an arbitrary function of $\bar{z}$.

Now, integrating (7) with respect to $\bar{z}$,

$$f(z,\bar{z}) = \frac{1}{48} z^3\bar{z} + \frac{3}{32} z^2\bar{z}^2 + \frac{1}{48} \bar{z}^3 z + F_2(z) + F_3(\bar{z}) \qquad (8)$$

where

$$\int F_1(\bar{z}) d\bar{z} = F_3(\bar{z}) \qquad (9)$$

and $F_2(z)$ is an arbitrary function of z.

Returning to (x,y) variables we find

$$f(x,y) = \frac{1}{48}(x+iy)^3(x-iy) + \frac{3}{32}(x+iy)^3(x-iy)^2$$

$$+ \frac{1}{48}(x-iy)^3(x+iy) + F_2(x+iy) + F_3(x-iy) \qquad (10)$$

● **PROBLEM 9-15**

Solve the equation

$$L \frac{d^2q}{dt^2} + R \frac{dq}{dt} + \frac{q}{C} = E_0 \cos \omega t \tag{1}$$

which appears in the theory of alternating electric currents. L, R, C, E_0 and ω are constants. Replace $E_0 \cos \omega t$ by $E_0 e^{i\omega t}$, and assume that $q(t) = \alpha e^{i\omega t}$ is the solution of the "new" equation.

Solution: Substituting

$$q(t) = \alpha e^{i\omega t} \tag{2}$$

$$\frac{dq}{dt} = i\omega \alpha e^{i\omega t} \tag{3}$$

$$\frac{d^2q}{dt^2} = -\omega^2 \alpha e^{i\omega t} \tag{4}$$

into

$$L \frac{d^2q}{dt^2} + R \frac{dq}{dt} + \frac{q}{C} = E_0 e^{i\omega t} \tag{5}$$

we find

$$-L\alpha\omega^2 e^{i\omega t} + iR\alpha\omega e^{i\omega t} + \frac{\alpha}{C} e^{i\omega t} \tag{6}$$

$$= E_0 e^{i\omega t}$$

Hence,

$$\alpha = \frac{E_0}{-L\omega^2 + iR\omega + \frac{1}{C}} \tag{7}$$

and the solution of (5) is

$$q = \frac{E_0}{-L\omega^2 + iR\omega + \frac{1}{C}} e^{i\omega t} \tag{8}$$

Taking the real part of (8),

$$\text{Re}(q) = \text{Re} \left[\frac{E_0 e^{i\omega t}}{-L\omega^2 + \frac{1}{C} + iR\omega} \right] \tag{9}$$

we find the solution of (1). Note that

$$e^{i\omega t} = \cos \omega t + i \sin \omega t \tag{10}$$

● **PROBLEM 9-16**

Explain the relationship between analytic and harmonic functions. List fundamental definitions and theorems.

Solution: First we shall define analytic functions.

Definition

A function f(z), is said to be analytic in a domain D, if f(z) is differentiable at every point of D. Each function f(z), analytic or not, can be represented uniquely in the form

$$f(z) = u(x,y) + i\, v(x,y) \tag{1}$$

This leads to an important theorem.

Theorem

The function f(z) = u + iv is analytic in the domain D if u and v are continuously differentiable in D, and satisfy the Cauchy-Riemann equations

$$\begin{aligned}\frac{\partial u}{\partial x} &= \frac{\partial v}{\partial y}\\ \frac{\partial u}{\partial y} &= -\frac{\partial v}{\partial x}\end{aligned} \tag{2}$$

The component functions u and v of an analytic function f(z) are not arbitrary; they must satisfy the Cauchy-Riemann equations.

Definition

A real function u = u(x,y), is said to be harmonic in a domain D if it has continuous second partial derivatives and

$$\frac{\partial^2 u}{\partial x^2} + \frac{\partial^2 u}{\partial y^2} = 0 \tag{3}$$

This theorem describes the connection between analytic and harmonic functions.

Theorem

If f(z) = u(x,y) + i v(x,y) is a complex function defined in a domain D, then f(z) is analytic in D, if u(x,y) and v(x,y) are conjugate harmonic functions in D.

CHAPTER 10

ELEMENTARY FUNCTIONS, MULTIPLE-VALUED FUNCTIONS

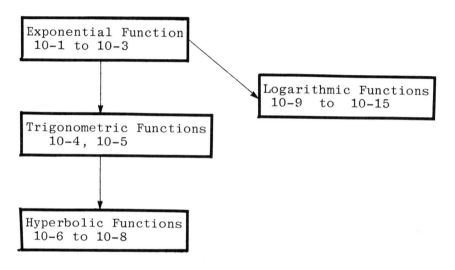

This chart is provided to facilitate rapid understanding of the interrelationships of the topics and subject matter in this chapter. Also shown are the problem numbers associated with the subject matter.

EXPONENTIAL FUNCTIONS

● **PROBLEM** 10-1

Definition

For $z = x + iy$, the function e^z is defined by

$$e^z := e^{x+iy} := e^x(\cos y + i \sin y) \qquad (1)$$

For z real, (1) becomes e^x; the exponential function known from basic calculus. Show that

$$e^{z_1} e^{z_2} = e^{z_1 + z_2} \qquad (2)$$

$$|e^z| = e^x$$

e^z is called the exponential function.

Solution: From the definition (1), we obtain

$$e^{z_1} e^{z_2} = e^{x_1}(\cos y_1 + i \sin y_1) e^{x_2}(\cos y_2 + i \sin y_2)$$

$$= e^{x_1+x_2}[(\cos y_1 \cos y_2 - \sin y_1 \sin y_2)$$

$$+ i(\sin y_1 \cos y_2 + \sin y_2 \cos y_1)] \qquad (3)$$

$$= e^{x_1+x_2}[\cos(y_1+y_2) + i \sin(y_1+y_2)] = e^{z_1+z_2}$$

$$|e^z| = |e^x||e^{iy}| = e^x|\cos y + i \sin y| \qquad (4)$$

$$= e^x\sqrt{\cos^2 y + \sin^2 y} = e^x$$

Note that for all z,

$$e^z \neq 0 \qquad (5)$$

• **PROBLEM 10-2**

Definition

Let $f(z)$ be a single-valued function defined in a domain D. Let $\alpha \neq 0$ be a constant, such that for every $z \in D$, $z+\alpha \in D$.

The function $f(z)$ is said to be periodic of period α, if for all $z \in D$

$$f(z) = f(z+\alpha) \qquad (1)$$

1. Find the period of the function e^z.

2. Show that e^z is analytic for all values of z and

$$\frac{d}{dz}(e^z) = e^z \qquad (2)$$

Solution: 1. From the definition of the exponential function, we have

$$e^{z+2\pi i} = e^z \cdot e^{2\pi i} = e^z(\cos 2\pi + i \sin 2\pi) = e^z \qquad (3)$$

Hence, e^z is of period 2π.

2. $\quad e^z = e^{x+iy} = e^x \cos y + i e^x \sin y = u + iv \qquad (4)$

$$\frac{\partial u}{\partial x} = e^x \cos y = \frac{\partial v}{\partial y} \qquad (5)$$

$$\frac{\partial u}{\partial y} = -e^x \sin y = -\frac{\partial v}{\partial x} \qquad (6)$$

Hence, e^z is analytic for all values of z.

$$\frac{d}{dz}(e^z) = \frac{\partial u}{\partial x} + i\frac{\partial v}{\partial x} = e^x \cos y + i e^x \sin y = e^z \qquad (7)$$

• **PROBLEM 10-3**

1. Show that if $\omega = f(z)$ is an analytic function of z, then e^ω is an analytic function, and

$$\frac{d}{dz}(e^\omega) = e^\omega \frac{d\omega}{dz} \qquad (1)$$

2. Show that

$$\overline{e^z} = e^{\overline{z}} \qquad (2)$$

and that $e^{\overline{z}}$ is not analytic.

Solution: 1. We shall utilize the following theorem. If $g(z)$ is an analytic function of z, in a domain D, with values in a domain G, and if $f(\omega)$ is an analytic function of ω in G, then $h = f[g(z)]$ is an analytic function of z in D, and

$$\frac{dh}{dz} = \frac{dh}{dg} \cdot \frac{dg}{dz} \qquad (3)$$

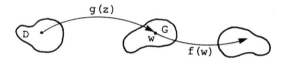

Hence,

$$\frac{d}{dz}(e^\omega) = \frac{de^\omega}{d\omega}\frac{d\omega}{dz} = e^\omega \frac{d\omega}{dz} \qquad (4)$$

2. We have

$$\overline{e^z} = \overline{e^x(\cos y + i \sin y)} = e^x(\cos y - i \sin y)$$
$$= e^{\overline{z}} \qquad (5)$$

$$e^{\overline{z}} = u + iv = e^x \cos y + i(-e^x \sin y)$$

Hence,

$$\frac{\partial u}{\partial x} \neq \frac{\partial v}{\partial y} \qquad (6)$$

$$\frac{\partial u}{\partial y} \neq -\frac{\partial v}{\partial x} \qquad (7)$$

and $e^{\overline{z}}$ is not analytic.

TRIGONOMETRIC FUNCTIONS

• **PROBLEM** 10-4

From
$$e^{iy} = \cos y + i \sin y \quad (1)$$
$$e^{-iy} = \cos y - i \sin y$$

we evaluate
$$\sin y = \frac{e^{iy} - e^{-iy}}{2i}$$
$$\cos y = \frac{e^{iy} + e^{-iy}}{2} \quad (2)$$

Eq.(2) can be extended to the domain of complex numbers:

$$\sin z := \frac{e^{iz} - e^{-iz}}{2i}$$
$$\cos z := \frac{e^{iz} + e^{-iz}}{2} \quad (3)$$

1. Show that
$$\frac{d}{dz}(\sin z) = \cos z$$
$$\frac{d}{dz}(\cos z) = -\sin z \quad (4)$$

2. **Definition**

A point z_0 is called a zero of the function $f(z)$ if $f(z_0) = 0$. Show that the zeros of $\sin z$ are $z = k\pi$, and the zeros of $\cos z$ are $z = \frac{\pi}{2} + k\pi$, where $k = 0, \pm 1, \pm 2, \ldots$.

Solution: 1.
$$\frac{d}{dz}(\sin z) = \frac{d}{dz}\left(\frac{e^{iz} - e^{-iz}}{2i}\right)$$
$$= \frac{1}{2i}\frac{d}{dz}(e^{iz} - e^{-iz}) = \frac{e^{iz} + e^{-iz}}{2} = \cos z \quad (5)$$

$$\frac{d}{dz}(\cos z) = \frac{d}{dz}\left(\frac{e^{iz} + e^{-iz}}{2}\right)$$
$$= -\frac{e^{iz} - e^{-iz}}{2i} = -\sin z \quad (6)$$

2. For $\sin z = 0$ we obtain
$$e^{2iz} = 1 \quad (7)$$

Hence,
$$2iz = 2k\pi i$$

and
$$z = k\pi \tag{8}$$
For $\cos z = 0$, we have
$$e^{2iz} = -1 \tag{9}$$
and
$$z = \frac{\pi}{2} + k\pi$$
where $k = 0, \pm 1, \pm 2, \ldots$.

● **PROBLEM 10-5**

1. Definition

Let $f(z)$ be defined in a domain D, symmetric with respect to the origin. If $f(-z) = f(z)$ for all $z \in D$, then $f(z)$ is called an even function. If $f(-z) = -f(z)$ for all $z \in D$, then $f(z)$ is called an odd function.

Show that $\sin z$ is an odd function, and $\cos z$ an even function.

2. We define

$$\tan z = \frac{\sin z}{\cos z} \quad \text{for } z \ne \frac{\pi}{2} + k\pi \tag{1}$$

$$\cot z = \frac{\cos z}{\sin z}, \quad z \ne k\pi \tag{2}$$

$$\sec z = \frac{1}{\cos z}, \quad z \ne \frac{\pi}{2} + k\pi \tag{3}$$

$$\csc z = \frac{1}{\sin z}, \quad z \ne k\pi \tag{4}$$

where $k = 0, \pm 1, \pm 2, \ldots$.

Evaluate the derivatives of the functions (1) (2) (3) and (4).

Solution: 1. From the definition

$$\sin(-z) = \frac{e^{-iz} - e^{iz}}{2i} = -\frac{e^{iz} - e^{-iz}}{2i} \tag{5}$$
$$= -\sin z$$

$$\cos(-z) = \frac{e^{-iz} + e^{iz}}{2} = \cos z \tag{6}$$

2. We have

$$\frac{d}{dz}(\tan z) = \sec^2 z, \quad z \ne \frac{\pi}{2} + k\pi \tag{7}$$

$$\frac{d}{dz}(\cot z) = -\csc^2 z, \quad z \neq k\pi \tag{8}$$

$$\frac{d}{dz}(\sec z) = \sec z \tan z, \quad z \neq \frac{\pi}{2} + k\pi \tag{9}$$

$$\frac{d}{dz}(\csc z) = -\csc z \cot z, \quad z \neq k\pi \tag{10}$$

For example:

$$\frac{d}{dz}\left(\frac{\sin z}{\cos z}\right) = \frac{\cos^2 z + \sin^2 z}{\cos^2 z} = \frac{1}{\cos^2 z}$$
$$= \sec^2 z \tag{11}$$

HYPERBOLIC FUNCTIONS

• **PROBLEM 10-6**

For real x, sin hx and cos hx are defined by

$$\sinh x := \frac{e^x - e^{-x}}{2} \tag{1}$$

$$\cosh x := \frac{e^x + e^{-x}}{2} \tag{2}$$

Show that

1. $\sin z = \sin x \cosh y + i \cos x \sinh y$
2. $\cos z = \cos x \cosh y - i \sin x \sinh y$
3. For $z = x + iy$

   $\sin iy = i \sinh y$

   $\cos iy = \cosh y$

Solution: 1. We have
$$\sin z = \frac{e^{iz} - e^{-iz}}{2i} \tag{3}$$

Hence,

$$2i \sin z = e^{iz} - e^{-iz} = e^{i(x+iy)} - e^{-i(x+iy)}$$

$$= e^{ix-y} - e^{-ix+y} = e^{-y}(\cos x + i \sin x)$$

$$- e^y(\cos x - i \sin x) \tag{4}$$

$$= i \sin x(e^y + e^{-y}) - \cos x(e^y - e^{-y})$$

$$= 2i \sin x \cosh y - 2 \cos x \sinh y$$

Thus,
$$\sin z = \sin x \cosh y + i \cos x \sinh y \tag{5}$$

2. Similarly,
$$\cos z = \frac{e^{iz} + e^{-iz}}{2}$$

$$= \frac{e^{ix-y} + e^{-ix+y}}{2}$$

$$= \frac{1}{2}\left[e^{-y}(\cos x + i \sin x) + e^{y}(\cos x - i \sin x)\right]$$

$$= \cos x \left(\frac{e^{-y} + e^{y}}{2}\right) - i \sin x \left(\frac{e^{y} - e^{-y}}{2}\right) \tag{6}$$

$$= \cos x \cosh y - i \sin x \sinh y$$

3. Substituting $z = 0 + iy$ into (5), we find
$$\sin iy = i \sinh y \tag{7}$$

Substituting $z = 0 + iy$ into (6),
$$\cos iy = \cosh y \tag{8}$$

• **PROBLEM 10-7**

For any complex number z, the hyperbolic functions are defined by

$$\sinh z := \frac{e^{z} - e^{-z}}{2} \tag{1}$$

$$\cosh z := \frac{e^{z} + e^{-z}}{2} \tag{2}$$

For real z, we obtain eq.(1) and (2) of Problem 10-6.

1. Show that the functions sinh z and cosh z are analytic for all values of z, and

$$\frac{d}{dz}(\sinh z) = \cos hz \tag{3}$$

$$\frac{d}{dz}(\cosh z) = \sin hz \tag{4}$$

2. Show that the zeros of sinh z are
$$z = n\pi i \tag{5}$$

and that the zeros of cosh z are
$$z = (n + \tfrac{1}{2})\pi i \tag{6}$$

Solution: 1. Since the exponential function e^z is analytic for all values of z, and

$$\frac{d}{dz}(e^z) = e^z \tag{7}$$

we find

$$\frac{d}{dz}(\sinh z) = \frac{d}{dz}\left[\frac{e^z - e^{-z}}{2}\right]$$
$$= \tfrac{1}{2}(e^z + e^{-z}) = \cosh z \tag{8}$$

$$\frac{d}{dz}(\cosh z) = \frac{d}{dz}\left[\frac{e^z + e^{-z}}{2}\right]$$
$$= \frac{e^z - e^{-z}}{2} = \sinh z \tag{9}$$

2. From eq.(1),

$$\sinh z = \frac{e^z - e^{-z}}{2} = 0 \tag{10}$$

Hence,

$$e^z = e^{-z} \quad \text{or} \quad e^{2z} = 1 \tag{11}$$

and

$$2z = 2n\pi i \quad \text{or} \quad z = n\pi i \tag{12}$$

From (2),

$$\cosh z = \frac{e^z + e^{-z}}{2} = 0 \tag{13}$$

or

$$e^{2z} = -1 \tag{14}$$

$$2z = 2n\pi i + \pi i$$
$$z = (n+\tfrac{1}{2})\pi i \tag{15}$$

● **PROBLEM 10-8**

If $z = x + iy$, then

1. $\sinh z = \cos y \sinh x + i \sin y \cosh x$
2. $\cosh z = \cos y \cosh x + i \sin y \sinh x$
3. $|\sinh z|^2 = \sin^2 y + \sinh^2 x$

Prove 1, 2, and 3.

Solution: 1. We have

$$\cos y \sinh x + i \sin y \cosh x$$

$$= \cos y \left(\frac{e^x - e^{-x}}{2}\right) + i \sin y \left(\frac{e^x + e^{-x}}{2}\right)$$

$$= e^x \left(\frac{\cos y + i \sin y}{2} \right) - e^{-x} \left(\frac{\cos y - i \sin y}{2} \right) \qquad (1)$$

$$= \frac{e^x e^{iy}}{2} - \frac{e^{-x} e^{-iy}}{2} = \frac{e^z - e^{-z}}{2} = \sinh z$$

2. $\qquad \cos y \cosh x + i \sin y \sinh x$

$$= \left(\frac{e^x + e^{-x}}{2} \right) \cos y + \left(\frac{e^x - e^{-x}}{2} \right) i \sin y$$

$$= \frac{e^z + e^{-z}}{2} = \cosh z \qquad (2)$$

3. $\qquad |\sinh z|^2 = |\cos y \sinh x + i \sin y \cosh x|^2$

$\qquad = \cos^2 y \sinh\ x + \sin^2 y \cosh\ x$

$\qquad = (1-\sin^2 y)\sinh\ x + \sin^2 y(1+\sinh\ x) \qquad (3)$

$\qquad = \sinh\ x + \sin^2 y$

LOGARITHMIC FUNCTIONS

• **PROBLEM 10-9**

Note that for all $z \in C$,

$$e^z \neq 0 \qquad (1)$$

Prove the following theorem:

For any complex number $z \in C$, $z \neq 0$, there exist complex numbers ω, such that

$$e^\omega = z \qquad (2)$$

where ω are given by

$$\omega = \ln|z| + i \arg z + 2n\pi i, \qquad (3)$$

n is an integer.

This theorem leads to the following definition:

If $z \neq 0$, $z \in C$ and $\omega \in C$, such that

$$e^\omega = z, \qquad (4)$$

then ω is called a logarithm of z, and is denoted

$$\omega := \ln z \qquad (5)$$

Solution: In polar form,

$$z = x + iy = |z|e^{i\theta}$$

$$|z| = \sqrt{x^2+y^2}, \quad \theta = \arg z, \quad -\pi < \theta \leq \pi \qquad (6)$$

Hence,
$$e^{\ln|z|+i \arg z} = e^{\ln|z|} e^{i \arg z} = |z|e^{i\theta} = z \qquad (7)$$

Thus,
$$\omega = \ln|z| + i \arg z$$

is one of the solutions of the equation $e^\omega = z$. If ω' is another solution, then
$$e^{\omega'-\omega} = 1$$

and
$$\omega' - \omega = 2n\pi i \qquad (8)$$

That proves the theorem.

● **PROBLEM 10-10**

If $e^\omega = z$, then
$$\omega = \ln z = \ln|z| + i(\arg z + 2n\pi) \qquad (1)$$

For $n = 0$, ω is called the principal value of the logarithm of z
$$\omega = \ln|z| + i \arg z \qquad (2)$$

1. Show that $\ln i = i \frac{\pi}{2}$ $\qquad (3)$

2. Evaluate $\ln(-1-i)$, $\ln(-1+i)$

3. Show that
$$\ln(z_1 \cdot z_2) = \ln z_1 + \ln z_2 \qquad (4)$$

in general is not true.

Hint: Use $\ln(-1+i)$ and $\ln i$ for example $\ln(-1-i)$.

Solution: 1. By definition,
$$\ln i = \ln|i| + i \arg i$$
$$= \ln 1 + i \frac{\pi}{2} = i \frac{\pi}{2} \qquad (5)$$

2. We have
$$\ln(-1-i) = \ln|-1-i| + i \arg(-1-i)$$
$$= \ln\sqrt{2} - i \frac{3}{4}\pi = \frac{1}{2} \ln 2 - i \frac{3}{4}\pi \qquad (6)$$

$$\ln(-1+i) = \ln|-1+i| + i \arg(-1+i)$$
$$= \ln\sqrt{2} + i \frac{3}{4}\pi = \frac{1}{2} \ln 2 + i \frac{3}{4}\pi \qquad (7)$$

3. Note that

$$i(-1+i) = -1 - i \qquad (8)$$

and

$$\ln(-1-i) = \tfrac{1}{2} \ln 2 - i \tfrac{3}{4} \pi$$

$$\neq \ln i + \ln(-1+i) = i \tfrac{\pi}{2} + \tfrac{1}{2} \ln 2 + i \tfrac{3}{4} \pi \qquad (9)$$

• **PROBLEM 10-11**

The principal value of the logarithm of z is

$$\omega = \ln z = \ln|z| + i \arg z \qquad (1)$$

For $\omega = u + iv, \quad z = re^{i\theta}$

$$u + iv = \ln z = \ln r + i\theta \qquad (2)$$

Hence,

$$u = \ln r \quad \text{and} \quad v = \theta \qquad (3)$$

The function ln r is not continuous at the origin, and arg z is not continuous on the negative part of the real axis. For $x, y < 0$,

$$\lim_{y \to 0} \arg(x+iy) = \pi \qquad (4)$$

and

$$\lim_{y \to 0} \arg(x-iy) = -\pi \qquad (5)$$

Show that:

The function ln z (principal value), is single-valued and analytic in the domain C, except for points on the non-positive real axis; i.e., is analytic in the domain D.

$$D = C - \{z : \operatorname{Im} z = 0, \operatorname{Re} z \leq 0\} \qquad (6)$$

and

$$\tfrac{d}{dz} (\ln z) = \tfrac{1}{z} \qquad (7)$$

Solution: In the domain D given by (6), u and v and their partial derivatives are continuous functions of r and θ. Moreover, the Cauchy-Riemann equations are satisfied.

$$\frac{\partial u}{\partial r} = \frac{1}{r}, \quad \frac{\partial u}{\partial \theta} = 0 \qquad (8)$$

$$\frac{\partial v}{\partial r} = 0, \quad \frac{\partial v}{\partial \theta} = 1$$

$$\frac{\partial u}{\partial r} = \frac{1}{r} \frac{\partial v}{\partial \theta}, \quad \frac{\partial v}{\partial r} = -\frac{1}{r} \frac{\partial u}{\partial \theta} \qquad (9)$$

Thus, ln z is analytic in D.

If $\omega = f(z)$ is analytic, then

$$\frac{d\omega}{dz} = (\cos\theta - i\sin\theta)\frac{\partial\omega}{\partial r} \qquad (10)$$

Utilizing (10), we find

$$\frac{d}{dz}(\ln z) = (\cos\theta - i\sin\theta)\frac{\partial}{\partial r}(\ln r + i\theta) \qquad (11)$$

$$= \frac{\cos\theta - i\sin\theta}{r} = \frac{1}{z}$$

• **PROBLEM 10-12**

Definition

For any two complex numbers z and ω, $z \neq 0$, we define

$$z^\omega := e^{\omega \ln z} \qquad (1)$$

We can also denote

$$e^{\omega \ln z} = e^{\omega[\ln|z| + i\arg z + 2n\pi i]}, \qquad (2)$$

n is an integer.

The value of (2) for $n = 0$ is called the principal value of z^ω. Evaluate

1. $(-i)^i$
2. $(-1)^{2i}$
3. $(1+i)^i$

Solution: 1. Applying definition (1), we find

$$(-i)^i = e^{i\ln(-i)} = e^{i(-i\frac{\pi}{2})} = e^{\frac{\pi}{2}} \qquad (3)$$

2. $(-1)^{2i} = e^{2i\ln(-1)} = e^{2i(\ln|-1| + i\arg -1)}$

$$= e^{2i(0+i\pi)} = e^{-2\pi} \qquad (4)$$

3. $(1+i)^i = e^{i\ln(1+i)} = e^{i[\ln|1+i| + i\arg(1+i)]}$

$$= e^{i[\ln\sqrt{2} + i\frac{\pi}{4}]} = e^{-\frac{\pi}{4} + i\frac{1}{2}\ln 2} \qquad (5)$$

• **PROBLEM 10-13**

---

1. Prove that if $z, \omega_1, \omega_2 \in C$ and $z \neq 0$, then

$$z^{\omega_1} z^{\omega_2} = z^{\omega_1 + \omega_2} \qquad (1)$$

2. Prove that if $z_1 \neq 0$, $z_2 \neq 0$, $z_1, z_2, \omega \in C$, then

$$(z_1 z_2)^\omega = z_1^\omega z_2^\omega e^{2\pi i \omega \alpha(z_1, z_2)} \qquad (2)$$

where

$$\alpha(z_1, z_2) = \begin{cases} -1, & \text{if } \pi < \arg z_1 + \arg z_2 \leq 2\pi \\ 0, & \text{if } -\pi < \arg z_1 + \arg z_2 \leq \pi \\ 1, & \text{if } -2\pi < \arg z_1 + \arg z_2 \leq -\pi \end{cases} \qquad (3)$$

Solution: 1. We have

$$z^{\omega_1} z^{\omega_2} = e^{\omega_1 \ln z} e^{\omega_2 \ln z} = e^{(\omega_1 + \omega_2) \ln z} = z^{\omega_1 + \omega_2} \qquad (4)$$

2.
$$\begin{aligned} (z_1 z_2)^\omega &= e^{\omega \ln(z_1 z_2)} \\ &= e^{\omega [\ln z_1 + \ln z_2 + 2\pi i \alpha(z_1, z_2)]} \\ &= e^{\omega \ln z_1} e^{\omega \ln z_2} e^{2\pi i \omega \alpha(z_1, z_2)} \\ &= z_1^\omega z_2^\omega e^{2\pi i \omega \alpha(z_1, z_2)} \end{aligned} \qquad (5)$$

In the same way, we can show that

$$\left(\frac{z_1}{z_2}\right)^\omega = \frac{z_1^\omega}{z_2^\omega} e^{2\pi i \omega \beta(z_1, z_2)} \qquad (6)$$

where

$$\beta(z_1, z_2) = \begin{cases} -1, & \text{if } \pi < \arg z_1 - \arg z_2 \leq 2\pi \\ 0, & \text{if } -\pi < \arg z_1 - \arg z_2 \leq \pi \\ 1, & \text{if } -2\pi < \arg z_1 - \arg z_2 \leq -\pi \end{cases} \qquad (7)$$

We utilized the following property:

$$\ln(z_1 \cdot z_2) = \ln z_1 + \ln z_2 + 2\pi i\, \alpha(z_1, z_2) \qquad (8)$$

and

$$\ln\left(\frac{z_1}{z_2}\right) = \ln z_1 - \ln z_2 + 2\pi i\, \beta(z_1, z_2) \qquad (9)$$

• **PROBLEM 10-14**

In what domain is the function z^ω analytic?

Solution: Note that if
$$\omega = f[g(z)], \quad (1)$$
then
$$\frac{d\omega}{dz} = \frac{d\omega}{dg}\frac{dg}{dz} \quad (2)$$

We denote
$$z^\omega = e^{\omega \ln z} = e^{\omega\alpha} \quad (3)$$
where
$$\alpha = \ln z \quad (4)$$

Then,
$$\frac{d}{dz}(z^\omega) = \frac{d}{d\alpha}(e^{\omega\alpha})\frac{d\alpha}{dz}$$
$$= \omega e^{\omega\alpha} \cdot \frac{1}{z} = \omega z^\omega \frac{1}{z} = \omega z^{\omega-1} \quad (5)$$

Theorem

The function z^ω, $\omega \in C$ is fixed, is analytic in the domain of complex numbers C, except the nonpositive real axis, and
$$\frac{d}{dz}(z^\omega) = \omega z^{\omega-1} \quad (6)$$

We have to "remove" the nonpositive real axis because of the function $\ln z$ (see Problem 10-11).

• **PROBLEM 10-15**

Show that
$$\ln(z^\omega) = \omega \ln z + 2\pi i k, \quad z \neq 0, \quad \omega \in C \quad (1)$$
$$(z^\omega)^\alpha = z^{\omega\alpha} e^{2\pi i \alpha k}, \quad \alpha \in C \quad (2)$$
where
$$k = \left[\frac{1}{2} - \frac{\text{Im}(\omega)\ln|z| + \text{Re}(\omega)\arg z}{2\pi}\right] \quad (3)$$
is the integer given by the bracket function.

Solution: Let
$$\omega = u + iv \quad (4)$$
Then,
$$z^\omega = e^{\omega \ln z} = e^{(u+iv)(\ln|z|+i \arg z)}$$
$$= e^{u \ln|z| - v \arg z} e^{i(v \ln|z| + u \arg z)} \quad (5)$$
and
$$|z^\omega| = e^{u \ln|z| - v \arg z} \quad (6)$$

$$\arg z^\omega = v \ln|z| + u \arg z + 2\pi k$$

k is the integer, such that

$$-\pi < v \ln|z| + u \arg z + 2\pi k \leq \pi \tag{7}$$

Solving eq.(7) for k, we find

$$p < k \leq p + 1,$$

where

$$p = -\frac{1}{2} - \frac{v \ln|z| + u \arg z}{2\pi} \tag{8}$$

Therefore,

$$k = [p+1] = \left[\frac{1}{2} - \frac{v \ln|z| + u \arg z}{2\pi}\right]$$
$$= \left[\frac{1}{2} - \frac{\text{Im}(\omega)\ln|z| + \text{Re}(\omega)\arg z}{2\pi}\right] \tag{9}$$

Hence,

$$\ln(z^\omega) = \ln|z^\omega| + i \arg z^\omega$$
$$= u \ln|z| - v \arg z + i(v \ln|z| + u \arg z + 2\pi k)$$
$$= (u+iv)(\ln|z| + i \arg z) + 2\pi i k \tag{10}$$
$$= \omega \ln z + 2\pi i k$$

Verification of (2) is similar.

MULTIPLE - VALUED FUNCTIONS

● **PROBLEM** 10-16

Definition

A function f(z) is said to be univalent in a domain D if it is one-to-one and analytic in D.

Prove the following theorem:

Let
$$f(z) = u(x,y) + i\, v(x,y) \tag{1}$$

be univalent in a domain D, and let E be the image of D, E = f(D): then E is also a domain.

Solution: To prove that E is a domain, we have to show that it is connected and open.

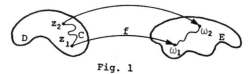

Fig. 1

Let $\omega_1, \omega_2 \in E$ be any two points in E. Their corresponding points in D are z_1, z_2, $\omega_1 = f(z_1)$, and $\omega_2 = f(z_2)$.

Let C be the curve joining z_1 and z_2, and lying entirely in D. The corresponding curve f(C) in E joins ω_1 and ω_2, and is lying entirely in E.

Hence, E is connected.

Now, we will show that E is open. Let $\omega_0 = u_0 + iv_0$ be any point in E, and $z_0 = x_0 + iy_0$ be a point in D, such that $f(z_0) = \omega_0$. We apply the implicit function theorem to the system of equations

$$\left. \begin{array}{l} u - u(x,y) = 0 \\ v - v(x,y) = 0 \end{array} \right\} \tag{2}$$

Thus, $u = u_0$, and $v = v_0$ are continuous, and have continuous partial derivatives with the nonvanishing Jacobian.

$$\begin{vmatrix} \frac{\partial u}{\partial x} & \frac{\partial u}{\partial y} \\ \frac{\partial v}{\partial x} & \frac{\partial v}{\partial y} \end{vmatrix} = \frac{\partial u}{\partial x}\frac{\partial v}{\partial y} - \frac{\partial u}{\partial y}\frac{\partial v}{\partial x}$$

$$= \left(\frac{\partial u}{\partial x}\right)^2 + \left(\frac{\partial u}{\partial y}\right)^2 = \left|\frac{df}{dz}\right|^2 \tag{3}$$

Since f(z) is analytic, we used the Cauchy-Reimann equations. Thus for any (u,v) sufficiently close to (u_0, v_0), there exists a unique pair of functions, $x = x(u,v)$ and $y = y(u,v)$, which are continuous, and satisfy the system (2), such that

$$x_0 = x(u_0, v_0), \quad y_0 = y(u_0, v_0) \tag{4}$$

Hence, (x(u,v), y(u,v)) will lie in any given neighborhood of (x_0, y_0), provided (u_0, v_0) and (u,v) are close enough. Therefore, E is open.

● **PROBLEM 10-17**

Find the derivative of the inverse of the function

$$\omega = z^n \tag{1}$$

__Solution__: We shall use the following theorem:

Let $\omega = f(z)$ be univalent in a domain D, and E be the image of D under f, let

$$z = f^{-1}(\omega) \tag{2}$$

be the inverse of the function f(z), then $f^{-1}(\omega)$ is univalent in E, and its derivative is

$$\frac{d\, f^{-1}(\omega)}{d\omega} = \frac{1}{f'(z)} \tag{3}$$

For any point in the ω-plane
$$\omega = z^n = r\, e^{i\theta} \tag{4}$$
there are n distinct points $z_1, z_2, \ldots z_n$ of the z-plane, such that
$$z_k^n = \omega \qquad k = 1,\ldots n \tag{5}$$
These points are given by
$$z = \sqrt[n]{r}\, e^{\frac{i(\theta + 2k\pi)}{n}} \qquad k = 0, 1, \ldots n-1, \tag{6}$$
and are located on the vertices of a regular n-gon.

Thus z^n is univalent in any wedge (infinite triangle) of the form
$$\theta < \arg z < \theta + \frac{2\pi}{n} \tag{7}$$
Note that each of the n wedges is a domain of univalence for z^n,

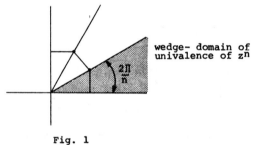

Fig. 1

which is also the largest (maximal) domain. Making any of the n domains larger would destroy the univalence of z^n.

The function z^n maps the wedge shown in Fig. 1 into the domain

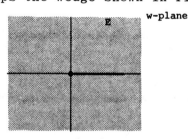

Fig. 2

consisting of all the points of ω-plane with the nonnegative real numbers removed (see Fig. 2).

The inverse of $\omega = z^n$ is
$$z = \sqrt[n]{\omega} \tag{8}$$
and its derivative is

$$\frac{d}{d\omega}(\sqrt[n]{\omega}) = \frac{1}{(z^n)'} = \frac{1}{nz^{n-1}} = \frac{\sqrt[n]{\omega}}{n\omega} \tag{9}$$

● **PROBLEM 10-18**

Find the maximal domains of univalence of
$$\omega = e^z \tag{1}$$
and the derivative of the inverse function.

Solution: Let
$$z_1 = x_1 + iy_1 \tag{2}$$
$$z_2 = x_2 + iy_2$$

be such that
$$e^{z_1} = e^{z_2} \tag{3}$$

Then,
$$|e^{z_1}| = e^{x_1}, \quad |e^{z_2}| = e^{x_2} \tag{4}$$

For (3) to be true, x_1 must equal x_2.
$$x_1 = x_2 = x \tag{5}$$

Then,
$$e^{z_1} - e^{z_2} = e^x(e^{iy_1} - e^{iy_2})$$
$$= e^x\left[(\cos y_1 + i \sin y_1) - (\cos y_2 + i \sin y_2)\right] \tag{6}$$
$$= e^x\left[-2 \sin \frac{y_1+y_2}{2} \sin \frac{y_1-y_2}{2} + 2i \cos \frac{y_1+y_2}{2} \sin \frac{y_1-y_2}{2}\right]$$
$$= 2i\, e^x\, e^{\frac{i(y_1+y_2)}{2}} \sin \frac{y_1-y_2}{2}$$

Since $e^\alpha \neq 0$, we have
$$\sin \frac{y_1-y_2}{2} = 0 \tag{7}$$

and
$$y_1 - y_2 = 2k\pi \tag{8}$$
$$k = 0, \pm 1, \pm 2, \ldots$$

Therefore, e^z is univalent in any strip of the form
$$a < \text{Im } z < a + 2\pi, \quad a \text{ real} \tag{9}$$

In particular, we can set $a = 0$ and
$$0 < \text{Im } z < 2\pi \tag{10}$$

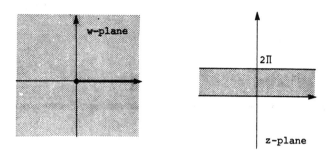

e^z maps the strip $0 < \text{Im } z < 2\pi$ into ω-plane without the non-negative part of the real axis.

The inverse function is

$$z = \ln \omega \tag{11}$$

and

$$\frac{d}{d\omega}(\ln\omega) = \frac{1}{\frac{d}{dz}(e^z)} = \frac{1}{e^z} = \frac{1}{\omega} \tag{12}$$

• **PROBLEM 10-19**

Find the branches of the multiple-valued function

$$z = \sqrt[n]{\omega} \tag{1}$$

Solution: We discussed this function in Problem 10-17. $\sqrt[n]{\omega}$ has n domains of univalence described by

$$\frac{2k\pi}{n} < \arg z < \frac{2(k+1)\pi}{n}, \tag{2}$$

which we denote by $D_0, D_1, \ldots, D_{n-1}$, as shown in Fig. 1.

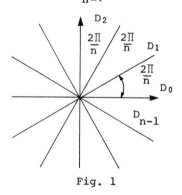

Fig. 1

The domains and their boundaries fill up the whole z-plane.

Each of the domains is mapped by $\omega = z^n$ into the same domain E, given by

$$0 < \arg \omega < 2\pi \tag{3}$$

Therefore, (2) leads to

$$2k\pi < \arg \omega = n \arg z < 2(k+1)\pi, \qquad (4)$$

which yields (3) for $k = 0, 1, \ldots n-1$. If we consider $\omega = z^n$ in a domain D_k and range E, then this function has a single-valued inverse with domain E and range D_k. We define this inverse function by

$$z = (\sqrt[n]{\omega})_k \qquad (5)$$

For n domains we obtain n single-valued functions

$$(\sqrt[n]{\omega})_0, (\sqrt[n]{\omega})_1, \ldots, (\sqrt[n]{\omega})_{n-1}, \qquad (6)$$

which are called branches of the multiple-valued function $z = \sqrt[n]{\omega}$.

● **PROBLEM 10-20**

Let
$$\omega = \sqrt{z^2+1} \qquad (1)$$

Show that $z = \pm i$ are branch points of ω.

Solution: We have

$$\omega = (z^2+1)^{\frac{1}{2}} = (z+i)^{\frac{1}{2}}(z-i)^{\frac{1}{2}} \qquad (2)$$

Then,

$$\arg \omega = \tfrac{1}{2} \arg(z+i) + \tfrac{1}{2} \arg(z-i) \qquad (3)$$

We will investigate changes in $\arg \omega$. Note that if Δ indicates the change, then

$$\Delta \arg \omega = \tfrac{1}{2}\Delta\arg(z+i) + \tfrac{1}{2}\Delta\arg(z-i) \qquad (4)$$

Assume that point z moves once around a closed curve C, enclosing the point i in the counterclockwise direction (Fig. 1).

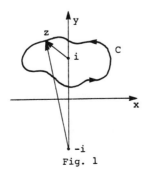

Fig. 1

Then,
$$\Delta \arg(z+i) = 0 \tag{5}$$
$$\Delta \arg(z-i) = 2\pi$$
so that
$$\Delta \arg \omega = \pi \tag{6}$$

Thus, ω does not return to its original value, and changes its branches. Since a complete circuit about $z = i$ changes the branches of the function, $z = i$ is a branch point.

For $z = -i$,

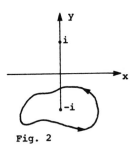

Fig. 2

$$\Delta \arg(z+i) = 2\pi \tag{7}$$
$$\Delta \arg(z-i) = 0$$
and
$$\Delta \arg \omega = \pi \tag{8}$$

Hence, $z = -i$ is a branch point. Enclosing both $z = \pm i$, we find

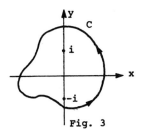

Fig. 3

$$\Delta \arg(z+i) = 2\pi \tag{9}$$
$$\Delta \arg(z-i) = 2\pi$$
and
$$\Delta \arg \omega = 2\pi \tag{10}$$

Thus, a circuit around both branch points does not change the branches of the function.

● **PROBLEM 10-21**

Construct the Riemann surface of the function $z = \sqrt[n]{\omega}$.

Solution: It has been shown in Problem 10-17 that $\omega = z^n$ is a multiple-valued function, and that each of the n wedges is a domain of univalence for z^n (see Fig. 1).

352

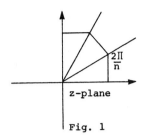

Fig. 1

Each wedge is mapped by z^n into the domain E,

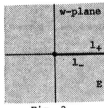

Fig. 2

which is the ω-plane cut along the positive real axis. Let ℓ_+ and ℓ_- denote the upper and lower edges of the cut. Now, we return to the multiple-valued function $z = \sqrt[n]{\omega}$.

Take n planes E and arrange them in such a way that $\omega = 0$ is the same for all of them. We shall denote the planes $E_0, E_1, \ldots, E_{n-1}$. Splice together the edges

$$\ell_0^- \text{ with } \ell_1^+$$
$$\ell_1^- \text{ with } \ell_2^+$$
$$\vdots$$
$$\ell_{n-2}^- \text{ with } \ell_{n-1}^+$$
$$\ell_{n-1}^- \text{ with } \ell_0^+$$

This procedure joins E_0 to E_1, E_1 to $E_2, \ldots, E_{n-2}$ to E_{n-1} and E_{n-1} back to E_0.

Fig. 3 shows the case for n = 4.

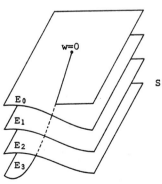

Fig. 3

This structure of n sheets (E planes) is called the Riemann surface of the function $z = \sqrt[n]{\omega}$.

The function $z = \sqrt[n]{\omega}$ defined on the Riemann surface is one-to-one (single-valued).

We choose the value of $\sqrt[n]{\omega}$ on the kth sheet E_k of S to be the value corresponding to the kth branch $(\sqrt[n]{\omega})_k$. The function $\sqrt[n]{\omega}$ defined on the Riemann surface S is single-valued (one-to-one), and maps S onto the whole z-plane.

• **PROBLEM 10-22**

Let the principal branch of arc sin z be the one for which

$$\text{arc sin } 0 = 0 \tag{1}$$

Show that

$$\text{arc sin } z = -i \ln(iz + \sqrt{1-z^2}) \tag{2}$$

<u>Solution</u>: If $\quad$ arc sin $z = \omega$, $\tag{3}$

then

$$z = \sin \omega = \frac{e^{i\omega} - e^{-i\omega}}{2i} \tag{4}$$

or

$$e^{i\omega} - e^{-i\omega} - 2iz = 0 \tag{5}$$

Since $e^{-i\omega}$ is always different from zero, we denote

$$e^{2i\omega} - 2iz\, e^{i\omega} - 1 = 0 \tag{6}$$

Solving (6) for $e^{i\omega}$,

$$e^{i\omega} = \frac{2iz \pm 2\sqrt{1-z^2}}{2} = iz \pm \sqrt{1-z^2} \tag{7}$$

$\sqrt{1-z^2}$ is a two-valued function, hence

$$e^{i\omega} = iz + \sqrt{1-z^2} \tag{8}$$

Note that

$$e^{i\omega} = e^{i(\omega - 2k\pi)}, \quad k = 0, \pm 1, \pm 2, \ldots \tag{9}$$

and

$$e^{i(\omega - 2k\pi)} = iz + \sqrt{1-z^2} \tag{10}$$

Or,

$$\omega = 2k\pi - i \ln(iz + \sqrt{1-z^2}) \tag{11}$$

To find the branch for which $\omega = 0$ when $z = 0$, we set $k = 0$ in (11), and obtain

$$\omega = \arcsin z = -i \ln(iz + \sqrt{1-z^2}) \tag{12}$$

● **PROBLEM 10-23**

Show that if we choose the principal branch of arc tanh z to be the one for which arc tanh $0 = 0$, then

$$\text{arc tanh } z = \frac{1}{2} \ln \frac{1+z}{1-z} \tag{1}$$

Solution: Let

$$\text{arc tanh } z = \omega \tag{2}$$

Then,

$$z = \tanh \omega = \frac{\sinh \omega}{\cosh \omega} = \frac{e^\omega - e^{-\omega}}{e^\omega + e^{-\omega}} \tag{3}$$

Eq.(3) can be written as

$$(1-z)e^\omega = (1+z)e^{-\omega} \tag{4}$$

or

$$e^{2\omega} = \frac{1+z}{1-z}, \tag{5}$$

where $z \neq 1$.

Since

$$e^{2\omega} = e^{2(\omega - k\pi i)}, \tag{6}$$

(5) can be written as

$$e^{2(\omega - k\pi i)} = \frac{1+z}{1-z} \tag{7}$$

or

$$\omega = k\pi i + \frac{1}{2} \ln \frac{1+z}{1-z} \tag{8}$$

Choosing the branch for which arc tanh $0 = 0$, we obtain

$$\text{arc tanh } z = \frac{1}{2} \ln \frac{1+z}{1-z} \tag{9}$$

Sometimes the notation $\sin^{-1} z$ or $\tanh^{-1} z$ is used to denote the inverse trigonometric functions. These symbols can be easily confused with $\frac{1}{\sin z}$ or $\frac{1}{\tanh z}$. Therefore, we will use arc sin z, arc cos z, etc.

CHAPTER 11

COMPLEX INTEGRALS, CAUCHY'S THEOREM

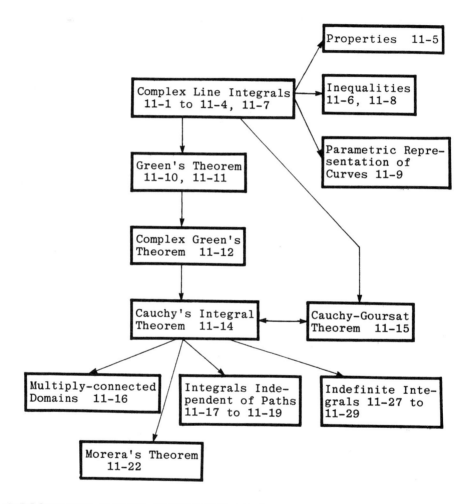

COMPLEX LINE INTEGRALS

● **PROBLEM 11-1**

Let C be a smooth curve with parametric equation

$$z = z(t), \quad a \leq t \leq b. \tag{1}$$

C is called smooth when it has a continuous nonvanishing derivative $z'(t) \neq 0$. Further, let D be a domain and $f(z)$ a function defined in D.

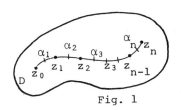

Fig. 1

We choose $z_0, z_1, \ldots, z_n$ which is the set of points on C, arranged consecutively in the direction of increasing t (positive direction). Let α_k be an arbitrary point on the arc $z_{k-1} z_k$, see Fig. 1 and ℓ_k denote the length of the arc. The integral of $f(z)$ along C is defined by

$$\int_C f(z)dz \equiv \lim_{\max\{\ell_1, \ldots, \ell_n\} \to 0} \sum_{k=1}^{n} f(\alpha_k) \Delta z_k \qquad (2)$$

where

$$\Delta z_k = z_k - z_{k-1} \qquad (3)$$

Evaluate the integral

$$\int_C (2+i)dz$$

where C is the straight line segment shown in Fig. 2.

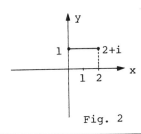

Fig. 2

Solution: From (2) we obtain

$$\int_C f(z)dz = \int_C (2+i)dz$$
$$= \lim \sum_{k=1}^{n} (2+i)\Delta z_k = (2+i)\lim \sum_{k=1}^{n} \Delta z_k \qquad (4)$$

$$\sum_{k=1}^{n} \Delta z_k = (2+i) - i = 2. \qquad (5)$$

Hence,

$$\int_C (2+i)dz = (2+i) \cdot 2 = 4 + 2i. \qquad (6)$$

● **PROBLEM 11-2**

Theorem

If $f(z)$ is continuous in a domain D containing a smooth curve C, then $f(z)$ is integrable along C, i.e.,

$$\int_C f(z)dz$$

exists, and (1)

$$\int_C F(z)dz = \int_C udx - vdy + i \int_C vdx + udy$$

Prove this theorem.

Solution: Let

$$z_k = x_k + iy_k \tag{2}$$

$$\alpha_k = a_k + ib_k \tag{3}$$

$$\Delta z_k = \Delta x_k + i\Delta y_k \tag{4}$$

and

$$f(z) = u(x,y) + i\,v(x,y). \tag{5}$$

Then the sum in eq.(2) of Problem 11-1 can be written in the form

$$\sum_{k=1}^{n} f(\alpha_k)\Delta z_k = \sum_{k=1}^{n} (u_k + iv_k)(\Delta x_k + i\Delta y_k)$$

$$= \sum_{k=1}^{n} (u_k\Delta x_k - v_k\Delta y_k) + i\sum_{k=1}^{n} (v_k\Delta x_k + u_k\Delta y_k) \tag{6}$$

where

$$u_k = u(a_k, b_k)$$
$$v_k = v(a_k, b_k). \tag{7}$$

Let

$$L = \max\{\ell_1, \ldots, \ell_n\} \tag{8}$$

where ℓ_k indicates the length of the arc $z_k z_{k-1}$.

As $L \to 0$, the first sum in (6) approaches

$$\sum_{k=1}^{n} (u_k\Delta x_k - v_k\Delta y_k) \to \int_C udx - vdy \tag{9}$$

358

and the second sum

$$\sum_{k=1}^{n} (v_k \Delta x_k + u_k \Delta y_k) \to \int_C v\,dx + u\,dy \tag{10}$$

Hence,

$$\int_C f(z)\,dz = \int_C u\,dx - v\,dy + i \int_C v\,dx + u\,dy$$

$$= \int_C (u+iv)(dx+idy). \tag{11}$$

● **PROBLEM 11-3**

Derive the formula for $\int_C f(z)\,dz$, when the curve C is given by the parametric equation

$$z = z(t) = x(t) + iy(t), \quad a \le t \le b. \tag{1}$$

Solution: We have

$$\int_C f(z)\,dt = \int_a^b f(z(t)) \frac{dz(t)}{dt}\,dt$$

$$= \int_a^b \left[u(z(t))x'(t) - v(z(t))y'(t) \right] dt \tag{2}$$

$$+ i \int_a^b \left[v(z(t))x'(t) + u(z(t))y'(t) \right] dt.$$

Let us denote

$$R(t) = \mathrm{Re}\{f(z(t)) \cdot \tfrac{dz}{dt}\} \tag{3}$$

$$I(t) = \mathrm{Im}\{f(z(t)) \cdot \tfrac{dz}{dt}\} \tag{4}$$

Then eq.(2) can be written in the form

$$\int_C f(z)\,dz = \int_a^b R(t)\,dt + i \int_a^b I(t)\,dt. \tag{5}$$

Thus, the calculation of a complex integral is reduced to the calculation of two real integrals. Note that in general, the function f(z) does not have to be continuous everywhere in the domain D. It is sufficient when it is continuous along the curve C. In that case all calculations of Problem 11-2 and 11-3 remain true.

• **PROBLEM 11-4**

A curve C is piecewise smooth if it consists of smooth arcs $C_1, C_2, \ldots, C_n$ joined together in such a way that the final point of C_k coincides with the initial point of C_{k+1} for $k = 1, 2, \ldots n-1$. Evaluate the integral

$$\int_C z^n \, dz \tag{1}$$

where n is an integer different from -1 and C is a piecewise smooth curve joining two points z_a and z_b. For negative n, the curve C does not pass through the point $z = 0$.

Solution: Assume C has the parametric equation

$$z = z(t) = x(t) + iy(t), \quad a \leq t \leq b. \tag{2}$$

Then,

$$\int_C z^n dz = \int_a^b z^n(t) z'(t) dt = \int_a^b \frac{1}{n+1} \frac{d}{dz} (z^{n+1}(t)) dt$$

$$= \frac{1}{n+1} \left[z^{n+1}(t) \right] \bigg|_{t=a}^{t=b} = \frac{1}{n+1} (z_b^{n+1} - z_a^{n+1}) \tag{3}$$

where $n \neq -1$.

Note that if the curve C is closed, then $z_a = z_b$ and (3) reduces to

$$\int_C z^n dz = 0 \tag{4}$$

Integral (3) does not depend on the curve C joining the points z_a and z_b.

• **PROBLEM 11-5**

The following theorems establish the basic properties of the line integrals.

Theorem

If f(z) is continuous on a piecewise smooth curve C, then

$$\int_C f(z) dz = - \int_{C^-} f(z) dz. \tag{1}$$

C^- denotes the curve C traversed in the negative (i.e., reverse of the positive) direction.

Theorem

If $f(z)$ and $g(z)$ are continuous on a piecewise smooth curve C, then

$$\int_C \left[\alpha f(z) + \beta g(z)\right] dz = \alpha \int_C f(z) dz + \beta \int_C g(z) dz \qquad (2)$$

where α and β are arbitrary complex numbers.

Prove both theorems.

Solution: From the definition of the integral we obtain (see Problem 11-1)

$$-\int_C f(z) dz = -\lim_{L \to 0} \sum_{k=1}^{n} f(\alpha_{n-k+1})(z_{n-k} - z_{n-k+1})$$

$$= \lim_{L \to 0} \sum_{k=1}^{n} f(\alpha_k)(z_k - z_{k-1}) = \int_C f(z) dz. \qquad (3)$$

Where α_k is any point inside the arc $z_{k-1} z_k$ and L is defined in eq.(8) Problem 11-2. This completes the proof of the first theorem.

$$\int_C \left[\alpha f(z) + \beta g(z)\right] dz = \lim_{L \to 0} \sum_{k=1}^{n} \left[\alpha f(\alpha_k) + \beta g(\alpha_k)\right] \Delta z_k$$

$$= \alpha \lim_{L \to 0} \sum_{k=1}^{n} f(\alpha_k) \Delta z_k + \beta \lim_{L \to 0} \sum_{k=1}^{n} g(\alpha_k) \Delta z_k \qquad (4)$$

$$= \alpha \int_C f(z) dz + \beta \int_C g(z) dz$$

● **PROBLEM 11-6**

Let C be a piecewise smooth curve with parametric equation $z = z(t)$, $a \leq t \leq b$. Any polygonal curve with consecutive vertices at the points $z_k = z(t_k)$ where $a = t_0 < t_1 < t_2 < \ldots < t_{n-1} < t_n = b$ is said to be inscribed in C.

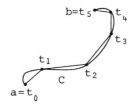

Fig. 1

Theorem

If $f(z)$ is continuous on a piecewise smooth curve C and such that

$$|f(z)| \leq M \quad \text{for all } z \in C \tag{1}$$

then,

$$\left| \int_C f(z)dz \right| \leq M\ell \tag{2}$$

where ℓ is the length of C. Prove this theorem.

Solution: We denote

$$\left| \sum_{k=1}^{n} f(\alpha_k) \Delta z_k \right| \leq \sum_{k=1}^{n} \left| f(\alpha_k) \Delta z_k \right| \leq M \sum_{k=1}^{n} \left| \Delta z_k \right| \tag{3}$$

Since the length of any polygonal curve inscribed in C is less than the length of C, we denote

$$\sum_{k=1}^{n} |\Delta z_k| \leq \ell \tag{4}$$

(3) becomes

$$\left| \sum_{k=1}^{n} f(\alpha_k) \Delta z_k \right| \leq M\ell \tag{5}$$

Hence, taking the limit, as $L \to 0$ we obtain

$$\left| \int_C f(z)dz \right| \leq M\ell. \tag{6}$$

● **PROBLEM 11-7**

Compute the value of the integral

$$\int_C (z-z_0)^n dz \tag{1}$$

where n is any integer and C is the circle with center at z_0 and radius r described in the counterclockwise direction.

Solution: The equation of the circle C is

$$z = z(t) = z_0 + re^{it}, \quad 0 \leq t \leq 2\pi. \tag{2}$$

Note that

$$z' = \frac{dz}{dt} = ir\, e^{it} \tag{3}$$

and

$$\int_C (z-z_0)^n dz = ir^{n+1} \int_0^{2\pi} e^{i(n+1)t}\, dt$$

$$= ir^{n+1} \int_0^{2\pi} \left[\cos(n+1)t + i\sin(n+1)t\right] dt. \tag{4}$$

For $n = -1$, (4) leads to
$$\int_C (z-z_0)^{-1} dz = i \int_0^{2\pi} dt = 2\pi i \tag{5}$$

For $n \neq -1$, we find
$$\int_C (z-z_0)^n dz = \frac{ir^{n+1}}{n+1}\left[\sin(n+1)t - i\cos(n+1)t\right]\bigg|_{t=0}^{2\pi} = 0. \tag{6}$$

Hence,
$$\int_C (z-z_0)^n dz = \begin{cases} 2\pi i & \text{for } n = -1 \\ 0 & \text{for } n \neq -1. \end{cases} \tag{7}$$

• **PROBLEM 11-8**

Show that
$$\left|\int_C (x^2 + iy^2) dz\right| \leq \pi \tag{1}$$
where C is a semicircle of radius one shown in Fig. 1.

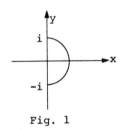

Fig. 1

Solution: We shall apply the theorem proved in Problem 11-6. Curve C can be parametrized by
$$z = z(t) = e^{it}, \quad -\frac{\pi}{2} \leq t \leq \frac{\pi}{2}. \tag{2}$$

The length of the curve is π.
$$\begin{aligned} x &= \cos t \\ y &= \sin t \end{aligned} \tag{3}$$

Then,
$$\begin{aligned} |x^2 + iy^2| &= |\cos^2 t + i\sin^2 t| \\ &= \sqrt{\cos^4 t + \sin^4 t} \leq 1. \end{aligned} \tag{4}$$

Hence, we have proved that

$$\left| \int_C (x^2+iy^2)dz \right| \leq \pi. \tag{5}$$

GREEN'S THEOREM

• **PROBLEM 11-9**

Let C be a curve represented by the functions

$$z = f(t), \quad a \leq t \leq b \tag{1}$$

$$z = g(t), \quad c \leq t \leq d. \tag{2}$$

Show that if F(z) is a continuous function on C, then

$$\int_a^b F\bigl[f(t)\bigr]f'(t)dt$$

$$= \int_c^d F\bigl[g(t)\bigr]g'(t)dt = \int_C F(z)dz. \tag{3}$$

<u>Solution</u>: Note that since f(t) and g(t) represent the same curve, there exists a function h(t) defined on the interval $a \leq t \leq b$ such that

$$f(t) = g\bigl[h(t)\bigr], \quad a \leq t \leq b \tag{4}$$

and

$$h(a) = c, \quad h(b) = d$$

h(t) is continuously differentiable.

Differentiating eq.(4), we find

$$f'(t) = g'\bigl[h(t)\bigr]h'(t) \tag{5}$$

which is true everywhere on the interval [a,b], except for at most a finite number of points.

Hence,

$$\int_a^b F\bigl[f(t)\bigr]f'(t)dt$$

$$= \int_a^b F\bigl[g(h(t))\bigr]g'\bigl[h(t)\bigr]h'(t)dt. \tag{6}$$

Let

$$h(t) = s \tag{7}$$

then,

$$ds = h'(t)dt$$

and (6) yields

$$\int_a^b F[f(t)]f'(t)dt = \int_c^d F[g(s)]g'(s)ds \qquad (8)$$

$$= \int_C F(z)dz.$$

Hence, the value of the integral $\int_C F(z)dz$ does not depend on the parametrization of the curve.

● **PROBLEM 11-10**

Green's theorem in the xy-plane can be formulated as follows.

Theorem

If $f(x,y)$ and $g(x,y)$ are continuous and have continuous partial derivatives in a domain D and on its boundary C, then

$$\iint_D \left(\frac{\partial g}{\partial x} - \frac{\partial f}{\partial y} \right) dxdy = \oint_C fdx + gdy. \qquad (1)$$

Verify Green's theorem in the plane for

$$\oint_C (2xy+y^2)dx + (x^2+y)dy \qquad (2)$$

where C is the closed curve consisting of $y = x^2$ and $y^2 = x$ as shown in Fig. 1.

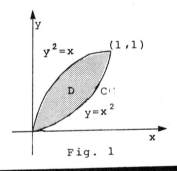

Fig. 1

Solution: The curves $y = x^2$ and $y^2 = x$ intersect at (0,0) and (1,1). The positive direction is indicated by the arrows.

We have along $y = x^2$

$$\int_{x=0}^1 \left[2x \cdot x^2 + x^4 \right] dx + 2x^2 \cdot 2xdx$$

$$= \int_0^1 (6x^3+x^4)dx = \left(\frac{3}{2}x^4 + \frac{x^5}{5} \right) \Big|_0^1 = \frac{3}{2} + \frac{1}{5} \, . \qquad (3)$$

Along $y^2 = x$

$$\int_{y=1}^{0} \left[(2y^2 \cdot y + y^2) \cdot 2y \, dy + (y^4 + y) \, dy \right] \tag{4}$$

$$= \int_{1}^{0} (5y^4 + 2y^3 + y) \, dy = \left[y^5 + \frac{y^4}{2} + \frac{y^2}{2} \right] \Big|_{y=1}^{0}$$

$$= -1 - \tfrac{1}{2} - \tfrac{1}{2} = -2.$$

Then,

$$\tfrac{3}{2} + \tfrac{1}{5} - 2 = -\tfrac{3}{10} \tag{5}$$

$$\iint_D \left(\frac{\partial g}{\partial x} - \frac{\partial f}{\partial y} \right) dx \, dy = \iint_D (2x - 2x - 2y) \, dx \, dy$$

$$= -2 \iint_D y \, dx \, dy = -2 \int_{x=0}^{1} \int_{y=x^2}^{\sqrt{x}} y \, dy \, dx$$

$$= -2 \int_{x=0}^{1} \left[\frac{y^2}{2} \Big|_{y=x^2}^{\sqrt{x}} \right] dx = -2 \int_{x=0}^{1} \left(\frac{x}{2} - \frac{x^4}{2} \right) dx \tag{6}$$

$$= -2 \left[\frac{x^2}{4} - \frac{x^5}{10} \right] \Big|_0^1 = -2 \left(\frac{1}{4} - \frac{1}{10} \right) = -\frac{3}{10}$$

This verifies Green's theorem. Note that Green's theorem is valid for multiply-connected domains. The symbol $\oint$, is reserved for integrals evaluated in the positive direction along a closed curve.

G. Green (1793-1841) an English mathematician from Cambridge, worked in the field of potential and its properties. A collected edition of his papers published in 1871 included topics on attraction in space of n dimensions, equilibrium of fluids, motion of waves in a canal and the reflexion and refraction of sound and light.

• **PROBLEM 11-11**

Let $f(x,y)$ and $g(x,y)$ be continuous and have continuous first partial derivatives in a simply-connected domain D. Prove that

$$\left\{ \begin{array}{l} \oint_C f \, dx + g \, dy = 0 \\ \text{where C is any closed} \\ \text{curve in D} \end{array} \right\} \quad \Longleftrightarrow \quad \left\{ \begin{array}{l} \dfrac{\partial f}{\partial y} = \dfrac{\partial g}{\partial x} \text{ everywhere} \\ \text{in D} \end{array} \right\}$$

Solution: Assume

⇐

$$\frac{\partial f}{\partial y} = \frac{\partial g}{\partial x}, \qquad (1)$$

then by Green's theorem

$$\oint_C f\,dx + g\,dy = \iint_D \left(\frac{\partial g}{\partial x} - \frac{\partial f}{\partial y}\right) dx\,dy = 0. \qquad (2)$$

⇒ Assume

$$\oint_C f\,dx + g\,dy = 0 \qquad (3)$$

for every closed curve in D. Assume that at some point $(x_0, y_0) \in D$

$$\frac{\partial f}{\partial y} - \frac{\partial g}{\partial x} > 0. \qquad (4)$$

Both partial derivatives are continuous in D, so there exists some domain $D_0 \subset D$, such that $(x_0, y_0) \in D_0$ and (4) is true in D_0. If C_0 is the boundary of D_0 then,

$$\oint_{C_0} f\,dx + g\,dy = \iint_{D_0} \left(\frac{\partial g}{\partial x} - \frac{\partial f}{\partial y}\right) dx\,dy > 0. \qquad (5)$$

This contradicts the hypothesis that $\oint_C f\,dx + g\,dy = 0$, for all closed curves in D.

The proof is completed. The following similar theorem is sometimes useful. Let $f(x,y)$ and $g(x,y)$ be continuous and have continuous partial derivatives in D. Then

$$\left(\int_A^B f\,dx + g\,dy \text{ is independent of the curve in D joining A and B} \right) \iff \left(\frac{\partial f}{\partial y} = \frac{\partial g}{\partial x} \text{ everywhere in D} \right)$$

• PROBLEM 11-12

Let $f(z,\bar{z})$ be continuous and have continuous partial derivatives in a domain D and on its boundary C. Prove that

$$\oint_C f(z,\bar{z})\,dz = 2i \iint_D \frac{\partial f}{\partial \bar{z}}\,dx\,dy \qquad (1)$$

which is Green's theorem for the complex plane.

Solution: Let

$$f(z,\bar{z}) = u(x,y) + iv(x,y). \qquad (2)$$

Applying Green's theorem we obtain

$$\oint_C f(z,\bar{z})dz = \oint_C (u+iv)(dx+idy)$$

$$= \oint_C udx - vdy + i \oint_C vdx + udy$$

$$= - \iint_D \left(\frac{\partial v}{\partial x} + \frac{\partial u}{\partial y}\right)dxdy + i \iint_D \left(\frac{\partial u}{\partial x} - \frac{\partial v}{\partial y}\right)dxdy$$

$$= i \iint_D \left[\left(\frac{\partial u}{\partial x} - \frac{\partial v}{\partial y}\right) + i\left(\frac{\partial v}{\partial x} + \frac{\partial u}{\partial y}\right)\right]dxdy \qquad (3)$$

$$= 2i \iint_D \frac{\partial f}{\partial \bar{z}} dxdy.$$

Remember that

$$2\frac{\partial f}{\partial \bar{z}} = \frac{\partial u}{\partial x} - \frac{\partial v}{\partial y} + i\left(\frac{\partial v}{\partial x} + \frac{\partial u}{\partial y}\right) \qquad (4)$$

where $f(z,\bar{z}) = u(x,y) + iv(x,y)$.

● **PROBLEM 11-13**

Let C be a simple closed curve enclosing a domain of area S, then

$$S = \frac{1}{2i} \oint_C \bar{z}\, dz. \qquad (1)$$

Show that the centroid of this domain is given in conjugate coordinates $(z,\bar{z})$ by

$$z_c = -\frac{1}{4Si} \oint_C z^2 d\bar{z} \qquad (2)$$

$$\bar{z}_c = \frac{1}{4Si} \oint_C \bar{z}^2 dz. \qquad (3)$$

Solution: Let D be a domain in the xy-plane of area S. Then, the coordinates of the centroid are

$$x_c = \frac{1}{S} \iint_D x\, dxdy \qquad (4)$$

$$y_c = \frac{1}{S} \iint_D y \, dx dy. \tag{5}$$

We shall use the following theorem: if $f(z,\bar{z})$ and $g(z,\bar{z})$ are continuous and have continuous partial derivatives in a domain D and on its boundary C then

$$\oint_C f(z,\bar{z})dz + g(z,\bar{z})d\bar{z} = 2i \iint_D \left(\frac{\partial f}{\partial \bar{z}} - \frac{\partial g}{\partial z}\right) dS. \tag{6}$$

Let us first verify that

$$\frac{z_c + \bar{z}_c}{2} = x_c \tag{7}$$

$$\frac{1}{2}\left[-\frac{1}{4Si} \oint_C z^2 d\bar{z} + \frac{1}{4Si} \oint_C \bar{z}^2 dz\right]$$

$$= \frac{1}{8Si} \oint \bar{z}^2 dz - z^2 d\bar{z} = \frac{2i}{8Si} \iint_D (2\bar{z} + 2z) dS \tag{8}$$

$$= \frac{1}{S} \iint_D x \, dx dy = x_c.$$

Now, we will denote that

$$\frac{z_c - \bar{z}_c}{2i} = y_c \tag{9}$$

$$\frac{1}{2i}\left[-\frac{1}{4Si} \oint_C z^2 d\bar{z} - \frac{1}{4Si} \oint_C \bar{z}^2 dz\right]$$

$$= -\frac{1}{2i} \cdot \frac{1}{4Si} \oint_C z^2 d\bar{z} + \bar{z}^2 dz \tag{10}$$

$$= -\frac{1}{2i} \cdot \frac{1}{4Si} \cdot 2i \iint_D 2(\bar{z}-z) dS = \frac{1}{S} \iint_D y \, dx dy.$$

CAUCHY'S THEOREM, THE CAUCHY - GORSAT THEOREM

● **PROBLEM 11-14**

One of the key theorems of complex analysis is Cauchy's integral theorem.

Theorem (Cauchy's integral theorem)

Let $f(z)$ be analytic in a simply-connected domain D. Then,

$$\oint_C f(z)\,dz = 0 \tag{1}$$

for every piecewise smooth closed curve C contained in D.

Adding the restriction that f'(z) be continuous in D, this theorem can be proved by use of Green's theorem. Goursat gave a proof with the above restriction removed. This version is often called the Cauchy-Goursat theorem.

Prove Cauchy's theorem, assuming that f(z) is analytic and f'(z) is continuous everywhere in D.

Solution: Since f(z) = u + iv is analytic, its derivative is continuous and

$$f'(z) = \frac{\partial u}{\partial x} + i\frac{\partial v}{\partial x} = \frac{\partial v}{\partial y} - i\frac{\partial u}{\partial y} \tag{2}$$

$$\frac{\partial u}{\partial x} = \frac{\partial v}{\partial y}, \quad \frac{\partial u}{\partial y} = -\frac{\partial v}{\partial x}. \tag{3}$$

Let C be any simple closed curve inside D. By Green's theorem, we have

$$\oint_C f(z)\,dz = \oint_C (u+iv)(dx+i\,dy)$$

$$= \oint_C u\,dx - v\,dy + i\oint_C v\,dx + u\,dy \tag{4}$$

$$= \iint_D \left(-\frac{\partial v}{\partial x} - \frac{\partial u}{\partial y}\right) dx\,dy + i\iint_D \left(\frac{\partial u}{\partial x} - \frac{\partial v}{\partial y}\right) dx\,dy$$

$$= 0.$$

This theorem can be extended to multiply-connected domains, for which Green's theorem is valid.

The shorter method to prove Cauchy's theorem is to apply the complex form of Green's theorem. Thus

$$f(z,\bar{z}) = f(z)$$

$$\frac{\partial f}{\partial \bar{z}} = 0 \quad \text{and}$$

$$\oint_C f(z)\,dz = 2i\iint_D \frac{\partial f}{\partial \bar{z}} dx\,dy = 0. \tag{5}$$

● **PROBLEM 11-15**

Prove Cauchy's theorem for the case of a triangle, see Problem 11-14. Assume that f(z) is analytic, but do not make any assumptions about f'(z).

Solution: Let ABC be a triangle in the z plane (See Fig. 1). We shall denote it by Δ. Joining the midpoints DEF, we obtain four triangles Δ_1, Δ_2, Δ_3 and Δ_4. Since f(z) is analytic inside Δ and on its boundary, we have

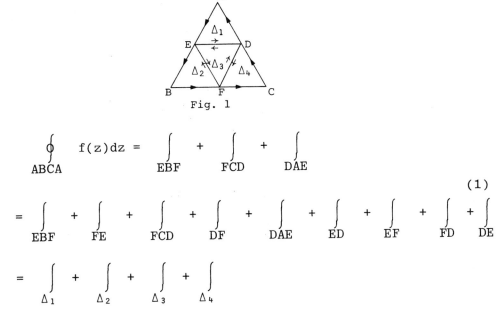

Fig. 1

$$\oint_{ABCA} f(z)dz = \int_{EBF} + \int_{FCD} + \int_{DAE} \quad (1)$$

$$= \int_{EBF} + \int_{FE} + \int_{FCD} + \int_{DF} + \int_{DAE} + \int_{ED} + \int_{EF} + \int_{FD} + \int_{DE}$$

$$= \int_{\Delta_1} + \int_{\Delta_2} + \int_{\Delta_3} + \int_{\Delta_4}$$

All of the integrals have the same integrand f(z), which for clarity we omit.

We used the property of the line integrals

$$\int_{PQ} f(z)dz = \int_{-QP} f(z)dz \quad (2)$$

in (1).

From (1) we obtain

$$\left| \oint_{\Delta} f(z)dz \right| \leq \left| \oint_{\Delta_1} \right| + \left| \oint_{\Delta_2} \right| + \left| \oint_{\Delta_3} \right| + \left| \oint_{\Delta_4} \right|. \quad (3)$$

Assume that triangle Δ, has the largest absolute value

$$\left| \oint_{\Delta_1} f(z)dz \right|$$

$$\left| \oint_{\Delta} f(z)dz \right| \leq 4 \left| \oint_{\Delta_1} f(z)dz \right|. \quad (4)$$

Let us repeat the procedure. We join the midpoints of the sides of triangle Δ_1, choose the traingle $\Delta_{1,2}$ with the highest absolute value and obtain

$$\left| \oint_\Delta f(z)dz \right| \leq 4^2 \left| \oint_{\Delta_{1,2}} f(z)dz \right|. \qquad (5)$$

This procedure can be repeated ad infinitum. For n steps

$$\left| \oint_\Delta f(z)dz \right| \leq 4^n \left| \oint_{\Delta_{1,2\ldots n}} f(z)dz \right|. \qquad (6)$$

$\Delta_1, \Delta_{1,2}, \Delta_{1,2,3} \ldots \Delta_{1,2,\ldots n}$ are a sequence of triangles which are nested, i.e., each is contained in the preceding one. Let z_0 be a point common for all the triangles, $z_0 \in \Delta$. $f(z)$ is analytic at z_0. Then,

$$f(z) = f(z_0) + f'(z_0)(z-z_0) + p(z-z_0) \qquad (7)$$

where for any $\varepsilon > 0$ there exists δ such that

$$|p| < \varepsilon \quad \text{whenever} \quad |z-z_0| < \delta. \qquad (8)$$

Integrating (7), we obtain

$$\oint_{\Delta_{1,2\ldots n}} f(z)dz = \oint_{\Delta_{1,2,\ldots n}} p(z-z_0)dz. \qquad (9)$$

Let A be the perimeter of Δ, then the perimeter of $\Delta_{1,2,\ldots n}$ is $A_n = \dfrac{A}{2^n}$.

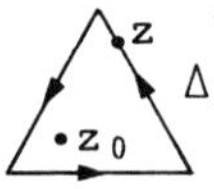

Fig. 2

If z is any point on $\Delta_{1,2,\ldots n}$ then,

$$|z-z_0| < \frac{A}{2^n} < \delta. \qquad (10)$$

From (9) we then obtain

$$\left| \oint_{\Delta_{1,2,\ldots n}} f(z)dz \right| = \left| \oint_{\Delta_{1,2,\ldots n}} p(z-z_0)dz \right| \leq \varepsilon \cdot \frac{A}{2^n} \cdot \frac{A}{2^n}$$

$$= \varepsilon \cdot \frac{A^2}{4^n} \qquad (11)$$

Combining eq.(6) and (11) we find

$$\left| \oint_\Delta f(z)dz \right| \leq 4^n \cdot \varepsilon \frac{A^2}{4^n} = \varepsilon A^2. \qquad (12)$$

Since $\varepsilon > 0$ is an arbitrary number, we conclude that

$$\oint_\Delta f(z)dz = 0. \tag{13}$$

Note that nowhere in this proof we used the fact that $f'(z)$ is continuous. This variation of the Cauchy's theorem is sometimes called the Cauchy-Goursat theorem.

● **PROBLEM 11-16**

Prove the Cauchy-Goursat theorem for multiply-connected domains.

Solution: In Problem 11-15 we have proved the Cauchy-Goursat theorem for a triangle. Utilizing that theorem, we can easily extend the proof to the case of any closed polygon and then to any simple closed curve. Let D be a multiply-connected domain bounded by the simple closed curves C_1 and C_2, shown in Fig. 1.

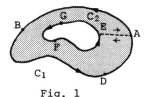

Fig. 1

AE is a cross-cut of the domain. The domain bounded by the curve ABDAEFGEA is simply-connected. Thus,

$$0 = \oint_{ABDAEFGEA} f(z)dz = \oint_{ABDA} f(z)dz + \int_{AE} f(z)dz$$

$$+ \oint_{EFGE} f(z)dz + \int_{EA} f(z)dz$$

$$= \oint_{ABDA} f(z)dz + \oint_{EFGE} f(z)dz = 0 \tag{1}$$

Note that we used the fact that

$$\int_{AE} f(z)dz + \int_{EA} f(z)dz = 0 \tag{2}$$

Thus, from (1) we conclude that

$$\oint_C f(z)dz = 0 \tag{3}$$

where C is the boundary of the domain, consisting of the curves C_1 and C_2 traversed in the positive direction, i.e., such direction that an observer walking on the boundary always has the domain D on his left.

INTEGRALS INDEPENDENT OF PATH

● **PROBLEM 11-17**

Let $f(z)$ be an analytic function in a simply-connected domain D. One of the most interesting consequences of Cauchy's theorem is that the integral

$$\int_{z_1}^{z_2} f(z)dz$$

is independent of the path in D joining any two points z_1 and z_2 in D. Prove this.

Solution: Let C_1 and C_2 be any two paths in D joining z_1 and z_2.

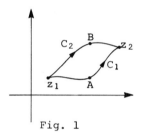

Fig. 1

By Cauchy's theorem, we have

$$\oint_{z_1 A z_2 B z_1} f(z)dz = 0 \tag{1}$$

or

$$\int_{z_1 A z_2} f(z)dz + \int_{z_2 B z_1} f(z)dz = 0. \tag{2}$$

Hence,

$$\int_{z_1 A z_2} f(z)dz = - \int_{z_2 B z_1} f(z)dz = \int_{z_1 B z_2} f(z)dz \tag{3}$$

Thus,

$$\int_{C_1} f(z)dz = \int_{C_2} f(z)dz = \int_{z_1}^{z_2} f(z)dz \tag{4}$$

q.e.d.

• **PROBLEM 11-18**

Let $f(z)$ be analytic in a simply-connected domain D and let z_0 and z be points in D. Show that the function defined by

$$F(z) = \int_{z_0}^{z} f(u)du \tag{1}$$

is analytic in D and

$$F'(z) = f(z). \tag{2}$$

Solution: We have

$$\frac{F(z+\Delta z) - F(z)}{\Delta z} - f(z)$$

$$= \frac{1}{\Delta z}\left\{\int_{z_0}^{z+\Delta z} f(u)du - \int_{z_0}^{z} f(u)du\right\} - f(z)$$

$$= \frac{1}{\Delta z}\int_{z}^{z+\Delta z}\left[f(u)-f(z)\right]du \tag{3}$$

Since $f(z)$ is analytic in D, the last integral is independent of the path joining z and $z+\Delta z$, assuming the path is contained in D.

Choosing Δz sufficiently small, we obtain the straight line segment $[z, z+\Delta z]$ which is contained in D.

Since $f(z)$ is continuous, then

whenever
$$|f(u) - f(z)| < \varepsilon \tag{4}$$
$$|u-z| < \delta.$$

For
$$|\Delta z| < \delta \tag{5}$$

we obtain
$$\left|\int_{z}^{z+\Delta z}\left[f(u)-f(z)\right]du\right| < \varepsilon|\Delta z|. \tag{6}$$

Thus, (3) leads to

$$\left|\frac{F(z+\Delta z)-F(z)}{\Delta z} - f(z)\right| = \frac{1}{|\Delta z|}\left|\int_{z}^{z+\Delta z}\left[f(u)-f(z)\right]dz\right| < \varepsilon \tag{7}$$

whenever $|\Delta z| < \delta$.

From (7) we conclude that

$$\lim_{\Delta z \to 0} \frac{F(z+\Delta z)-F(z)}{\Delta z} = f(z) \tag{8}$$

or

$$F'(z) = f(z)$$

$F(z)$ is analytic.

● **PROBLEM 11-19**

Let $f(z)$ be analytic in a domain D bounded by two simple closed curves C_1 and C_2. $f(z)$ is also analytic on C_1 and C_2. Prove that

$$\oint_{C_1} f(z)dz = \oint_{C_2} f(z)dz \tag{1}$$

C_1 and C_2 are traversed in the positive direction relative to their interiors.

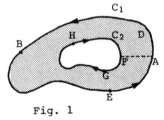

Fig. 1

Solution: Let AF be a cross-cut as shown in Fig. 1. Since $f(z)$ is analytic in D, we obtain by Cauchy's theorem

$$\oint_{ABEAFGHF} f(z)dz = 0 \tag{2}$$

or

$$\oint_{ABEA} f(z)dz + \int_{AF} f(z)dz + \int_{FGHF} f(z)dz + \int_{FA} f(z)dz \tag{3}$$

$$= 0.$$

Since

$$\int_{AF} f(z)dz = -\int_{FA} f(z)dz \tag{4}$$

we obtain

$$\oint_{ABEA} f(z)dz = -\oint_{FGHF} f(z)dz = \oint_{FHGF} f(z)dz \tag{5}$$

or
$$\oint_{C_1} f(z)dz = \oint_{C_2} f(z)dz. \qquad (6)$$

APPLICATIONS

• **PROBLEM 11-20**

Evaluate
$$\oint_C \frac{dz}{z-a} \qquad (1)$$

where C is any simple closed curve. Consider two cases when

1. $z = a$ is outside C
2. $z = a$ is inside C.

Solution: 1. If a is outside C, then the function $f(z) = \frac{1}{z-a}$ is analytic everywhere inside and on C. Then, by Cauchy's theorem,

$$\oint_C \frac{dz}{z-a} = 0. \qquad (2)$$

2. Now, assume a is inside C. Let C_1 be a circle of radius ε with center at $z = a$, such that C_1 is inside C.

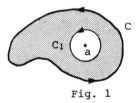

Fig. 1

$f(z) = \frac{1}{z-a}$ is analytic in the domain bounded by C and C_1.

Hence, by Problem 11-19

$$\oint_C \frac{dz}{z-a} = \oint_{C_1} \frac{dz}{z-a} \qquad (3)$$

On C_1 $|z-a| = \varepsilon$ or

$$z - a = \varepsilon\, e^{i\theta}$$
$$z = a + \varepsilon e^{i\theta}, \qquad 0 \leq \theta \leq 2\pi \qquad (4)$$

and
$$dz = i\varepsilon e^{i\theta} d\theta \qquad (5)$$

$$\oint_{C_1} \frac{dz}{z-a} = \int_{\theta=0}^{2\pi} \frac{i\varepsilon e^{i\theta}}{\varepsilon e^{i\theta}} d\theta = i \int_0^{2\pi} d\theta = 2\pi i. \qquad (6)$$

Hence,

$$\oint_C \frac{dz}{z-a} = 2\pi i. \qquad (7)$$

It is easy to show that for $\oint_C \frac{dz}{(z-a)^n}$, $n = 2,3,4,\ldots$ we always obtain

$$\oint_C \frac{dz}{(z-a)^n} = 0. \qquad (8)$$

• **PROBLEM 11-21**

Let C be the curve

$$y = x^3 + 2x^2 - 2x \qquad (1)$$

joining points (0,0) and (1,1). Evaluate

$$\int_C (4z^2 - 2iz) dz. \qquad (2)$$

Solution: We shall solve this problem using two methods.

I. Since integral (2) is independent of the path joining (0,0) and (1,1), we can choose the path shown in Fig. 1.

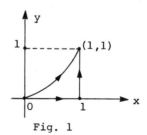

Fig. 1

Thus,

$$\int_C (4z^2-2iz)dz = \int_{(0,0)}^{(1,0)} (4z^2-2iz)dz + \int_{(1,0)}^{(1,1)} (4z^2-2iz)dz \qquad (3)$$

Along the path from (0,0) to (1,0) $y = 0$, $dy = 0$, $dz = dx$, $z = x$.

$$\int_0^1 (4x^2-2ix)dx = \left[\frac{4}{3} x^3 - ix^2\right]\Big|_0^1$$

$$\qquad (4)$$

$$= \frac{4}{3} - i.$$

Along the path from (1,0) to (1,1) $x = 1$, $dx = 0$, $dz = i\,dy$, $z = 1 + iy$.

$$\int_0^1 \left[4(1+iy)^2 - 2i(1+iy)\right] i\,dy$$

$$= \left[4iy - 4y^2 - \frac{4}{3} y^3 i + 2y + y^2 i\right]\Big|_0^1 \qquad (5)$$

$$= 5i - \frac{4}{3} i - 2.$$

Hence,

$$\int_C (4z^2 - 2iz)\,dz = \frac{4}{3} - i + 5i - \frac{4}{3} i - 2$$

$$= -\frac{2}{3} + \frac{8}{3} i. \qquad (6)$$

II. The shorter method is shown here.

$$\int_0^{1+i} (4z^2 - 2iz)\,dz = \left(\frac{4}{3} z^3 - iz^2\right)\Big|_0^{1+i}$$

$$= \frac{4}{3}(1+i)^3 - i(1+i)^2 = -\frac{2}{3} + \frac{8}{3} i \qquad (7)$$

● **PROBLEM 11-22**

The converse of Cauchy's theorem is sometimes called Morera's theorem.

Theorem (Morera's theorem)

If $f(z)$ is continuous in a simply-connected domain D and

$$\oint_C f(z)\,dz = 0 \qquad (1)$$

around every simply closed curve C in D, then $f(z)$ is analytic in D.

Assume that $f'(z)$ is continuous in D. Prove the theorem.

Solution: If $f'(z)$ is continuous in D, then we can apply Green's theorem.

$$\oint_C f(z)\,dz = \oint_C u\,dx - v\,dy + i \oint_C v\,dx + u\,dy$$

$$= \iint_D \left(-\frac{\partial v}{\partial x} - \frac{\partial u}{\partial y}\right) dxdy \qquad (2)$$

$$+ i \iint_D \left(\frac{\partial u}{\partial x} - \frac{\partial v}{\partial y}\right) dxdy$$

where
$$f = u + iv. \qquad (3)$$

If
$$\oint_C f(z)dz = 0 \qquad (4)$$

then both line integrals in (2) are zero.

$$\oint_C udx - vdy = \oint_C vdx + udy = 0 \qquad (5)$$

Hence, from Problem 11-11 we obtain

$$\frac{\partial u}{\partial x} = \frac{\partial v}{\partial y}$$
$$\frac{\partial u}{\partial y} = -\frac{\partial v}{\partial x} \qquad (6)$$

which are the Cauchy-Riemann equations. Thus $f(z) = u + iv$ is analytic in D.

This theorem, which will be shown in the next chapter, can be proved without assuming that $f'(z)$ is continuous.

• **PROBLEM 11-23**

Evaluate
$$\oint_C \frac{z^2 + \sinh 2z}{z^3 - 3iz^2 + 4z - 12i} dz \qquad (1)$$

where C is the circle $|z| = 1$.

Solution: We shall find the singular points of the function

$$f(z) = \frac{z^2 + \sinh 2z}{z^3 - 3iz^2 + 4z - 12i} \qquad (2)$$

The singular points are the solutions of the equation

$$z^3 - 3iz^2 + 4z - 12i = 0. \qquad (3)$$

One of the solutions is $z_1 = 3i$. Hence,

$$z^3 - 3iz^2 + 4z - 12i = (z-3i)(z^2+4). \qquad (4)$$

Thus, the solutions of eq.(3) are

$$z_1 = 3i, \quad z_2 = 2i, \quad z_3 = -2i \tag{5}$$

These points are not contained within or on C.

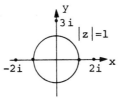

Thus, f(z) is analytic within and on C. Therefore, from the Cauchy theorem we conclude that

$$\oint_C \frac{z^2 + \sinh 2z}{z^3 - 3iz^2 + 4z - 12i} \, dz = 0. \tag{6}$$

● **PROBLEM 11-24**

Evaluate

1. $\displaystyle\int_{z_0}^{z_1} a^z \, dz \tag{1}$

2. $\displaystyle\int_{-ki}^{ki} \frac{dz}{z}$, where k is a positive number $\tag{2}$

3. $\displaystyle\int_{z_0}^{z_1} \frac{dz}{z\sqrt{a^2 \pm z^2}}. \tag{3}$

Solution: 1. An indefinite integral of the analytic function

$$f(z) = a^z \tag{4}$$

is given by the analytic function

$$F(z) = \frac{a^z}{\ln a}. \tag{5}$$

Thus, for any contour joining z_0 and z_1, we have

$$\int_{z_0}^{z_1} a^z \, dz = \frac{a^z}{\ln a} \bigg|_{z=z_0}^{z_1} = \frac{a^{z_1} - a^{z_0}}{\ln a}. \tag{6}$$

2. Integral (2) is evaluated for any curve joining ki and -ki lying in the simply-connected domain D consisting of the complex plane with the nonpositive half of the real axis removed.

An indefinite integral of $f(z) = \frac{1}{z}$, which is analytic in D, is given by the analytic function

$$F(z) = \ln z. \tag{7}$$

Thus,

$$\int_{-ki}^{ki} \frac{dz}{z} = \ln ki - \ln(-ki) = (\ln k + \frac{\pi}{2} i) - (\ln k - \frac{\pi}{2} i)$$
$$= \pi i. \tag{8}$$

3. We have

$$\int \frac{dz}{z\sqrt{a^2 \pm z^2}} = \frac{1}{a} \ln \left(\frac{z}{a + \sqrt{a^2 \pm z^2}} \right) \tag{9}$$

and

$$\int_{z_0}^{z_1} \frac{dz}{z\sqrt{a^2 \pm z^2}} = \frac{1}{a} \ln \frac{z_1}{a + \sqrt{a^2 \pm z_1^2}} - \frac{1}{a} \ln \frac{z_0}{a + \sqrt{a^2 \pm z_0^2}} \tag{10}$$

● **PROBLEM 11-25**

Evaluate

$$\oint_C \frac{dz}{1+z^2} \tag{1}$$

where C is the circle

1. $|z+i| = 1$ (2)
2. $|z-i| = 1$ (3)
3. $|z| = 3$. (4)

Solution: Function

$$f(z) = \frac{1}{1+z^2} \tag{5}$$

can be written in the form

$$\frac{1}{z^2+1} = \frac{1}{2i}\left[\frac{1}{z-i} - \frac{1}{z+i}\right]. \tag{6}$$

1. The circle $|z+i| = 1$ is shown in Fig. 1.

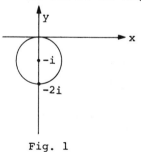

Fig. 1

The singularities of $\frac{1}{z^2+1}$ are located at $z = i$, $z = -i$. Only one singularity $z = -i$ is located inside the circle.

Hence, by Problem 11-20, we obtain

$$\oint_C \frac{1}{z^2+1} dz = \frac{1}{2i} \oint_C \frac{1}{z-i} dz - \frac{1}{2i} \oint_C \frac{1}{z+i} dz$$

$$= -\frac{1}{2i} \oint_C \frac{1}{z+i} dz = -\frac{2\pi i}{2i} = -\pi. \qquad (7)$$

2. For $|z-i| = 1$, the singularity $z = i$ is inside the circle. Hence,

$$\oint_C \frac{dz}{z^2+1} = \frac{1}{2i} \oint_C \frac{1}{z-i} dz = \frac{2\pi i}{2i} = \pi \qquad (8)$$

3. The circle $|z| = 3$ contains both singularities

$$\oint_C \frac{dz}{z^2+1} = \frac{1}{2i} \oint_C \frac{1}{z-i} dz - \frac{1}{2i} \oint_C \frac{1}{z+i} dz$$

$$= \frac{2\pi i}{2i} - \frac{2\pi i}{2i} = 0. \qquad (9)$$

● **PROBLEM 11-26**

Let C be a path joining $z = 0$ and $z = 1$. Find all possible values of

$$\int_C \frac{dz}{z^2+1}. \qquad (1)$$

Consider only such paths along which $\frac{1}{z^2+1}$ is continuous.

Solution: Let C_0 be a path consisting of the circle $|z+i| = 1$, and the circle $|z-i| = 1$ and the straight segment from 0 to 1. Both circles can be traversed any number of times in any direction (i.e., positive or negative).

The Cauchy's theorem states that any path from 0 to 1, not containing the points $z = \pm i$ may be replaced by C_0.

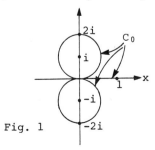

Fig. 1

By Problem 11-25 we conclude that the integrals taken along the circles will be multiples of π. They will be equal to $k\pi$, $k = 0, \pm 1, \pm 2, \ldots$. Along the segment from 0 to 1 we have

$$\int_0^1 \frac{dz}{z^2+1} = \int_0^1 \frac{dx}{x^2+1} = (\arctan x)\Big|_{x=0}^{1} = \frac{\pi}{4}. \tag{2}$$

Hence,

$$\int_0^1 \frac{dz}{z^2+1} = k\pi + \frac{\pi}{4}, \quad k = 0, \pm 1, \pm 2, \ldots. \tag{3}$$

INDEFINITE INTEGRALS

• **PROBLEM 11-27**

In evaluating real integrals, the fundamental theorem of calculus is useful.

<u>Theorem</u>

Let $f(x)$ be a continuous real-valued function, $a \leq x \leq b$, and let $F(x)$ be defined by

$$F(x) = \int_a^x f(t)\,dt, \quad a \leq x \leq b. \tag{1}$$

Then,

1. $F(x)$ is continuously differentiable on $a < x < b$.
2. $F'(x) = f(x)$. \hfill (2)
3. if a function $G(x)$ exists, such that $G'(x) = f(x)$. Then $F(x)$ and $G(x)$ differ by a constant and

$$\int_a^b f(t)\,dt = G(b) - G(a). \tag{3}$$

A similar theorem exists in complex analysis.

<u>Theorem</u>

Let $f(z)$ be continuous in a domain D and let $F(z)$ be a function defined and analytic in D and such that

$$F'(z) = f(z) \quad \text{for all } z \in D. \tag{4}$$

if C is a curve in D from z_0 to z_1. Then

$$\int_C f(z)\,dz = \int_C F'(z)\,dz = F(z_1) - F(z_0). \tag{5}$$

Notice that the value of the integral depends only on the end points of the curve, but not on C itself.

Prove this theorem.

Solution: Let
$$f(z) = u + iv \text{ and} \quad (6)$$
$$F(z) = u_1 + iv_1.$$

Since u_1 and v_1 satisfy the Cauchy-Riemann equations and $u = \frac{\partial u_1}{\partial x}$ and $v = \frac{\partial v_1}{\partial x}$, we obtain

$$\int_C f(z)dz = \int_C (udx - vdy) + i\int_C (vdx + udy)$$

$$= \int_C \left(-\frac{\partial u_1}{\partial x}dx + \frac{\partial u_1}{\partial y}dy\right) + i\int_C \left(\frac{\partial v_1}{\partial x}dx + \frac{\partial v_1}{\partial y}dy\right) \quad (7)$$

$$= \int_C du_1 + i\int_C dv_1.$$

The exact differential is

$$dU_1 = \frac{\partial u_1}{\partial x}dx + \frac{\partial u_1}{\partial y}dy$$
$$dV_1 = \frac{\partial v_1}{\partial x}dx + \frac{\partial v_1}{\partial y}dy. \quad (8)$$

Finally, we conclude that

$$\int_C f(z)dz = \int_C du_1 + i\int_C dV_1 = u_1(z_1) - u_1(z_0)$$
$$+ i\left[v_1(z_1) - v_1(z_0)\right] = F(z_1) - F(z_0). \quad (9)$$

● PROBLEM 11-28

Compute the integral

$$\int_C z^2 dz \quad (1)$$

where C is the straight-line segment from $z_0 = 0$ to $z_1 = 2 + 2i$.

Solution: We shall use two methods to solve this problem.

I. If C can be parametrized by the variable x, then the points on C have the form

$$z = x + ix = (1+i)x. \quad (2)$$

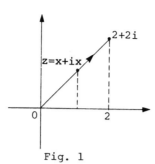

Fig. 1

We find

$$\int_C z^2 dz = \int_{x=0}^{x=2} (1+i)^2 x^2 \, d\left[(1+i)x\right]$$

$$= (1+i)^3 \int_0^2 x^2 dx = (1+i)^3 \cdot \frac{8}{3}. \quad (3)$$

II. In this case, it is easy to find the function F(z) such that

$$F'(z) = z^2 \quad (4)$$

$$F(z) = \frac{z^3}{3}. \quad (5)$$

Then,

$$\int_C z^2 dz = \left(\frac{z^3}{3}\right)\bigg|_{z_0=0}^{z_1=2+2i} = \frac{(2+2i)^3}{3}$$

$$= \frac{8}{3}(1+i)^3. \quad (6)$$

● **PROBLEM 11-29**

1. Evaluate

$$\int \sin 3z \cos 3z \, dz \quad (1)$$

2. Prove that

$$\int f(z)g'(z)dz = f(z)g(z) - \int f'(z)g(z)dz \quad (2)$$

This formula is known as integration by parts.

3. Find
$$\int z\, e^{2z}\, dz. \qquad (3)$$

Solution: 1. We substitute
$$\sin 3z = u. \qquad (4)$$

Then,
$$du = 3 \cos 3z\, dz \qquad (5)$$

and
$$\int \sin 3z \cos 3z\, dz = \frac{1}{3} \int u\, du$$
$$= \frac{1}{3} \frac{u^2}{2} + C = \frac{1}{6} \sin^2 3z + C. \qquad (6)$$

2. Note that
$$\left[f(z)g(z) \right]' = f(z)g'(z) + f'(z)g(z). \qquad (7)$$

Then,
$$\int \left[f(z)g(z) \right]' = f(z)g(z)$$
$$= \int f(z)g'(z)\,dz + \int f'(z)g(z)\,dz \qquad (8)$$

(2) follows from (8).

3. We shall use eq.(2). Let
$$f(z) = z$$
$$g'(z) = e^{2z}. \qquad (9)$$

Then
$$f'(z) = 1$$
$$g(z) = \tfrac{1}{2} e^{2z}. \qquad (10)$$

Hence,
$$\int z\, e^{2z}\, dz = z \cdot \tfrac{1}{2} e^{2z} - \int \tfrac{1}{2} e^{2z}\, dz$$
$$= \tfrac{1}{2} z\, e^{2z} - \tfrac{1}{4} e^{2z} + c. \qquad (11)$$

CHAPTER 12

CAUCHY'S INTEGRAL FORMULA AND RELATED THEOREMS

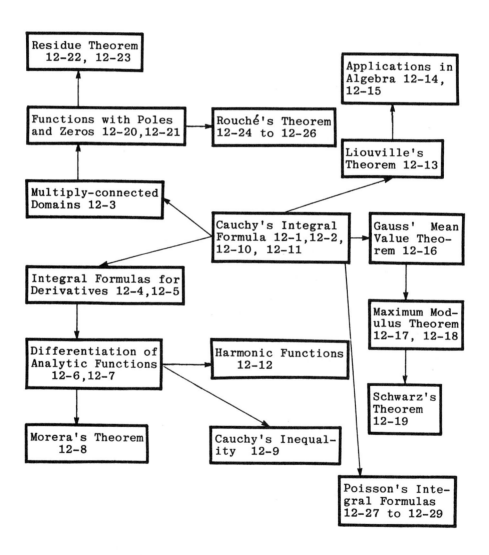

This chart is provided to facilitate rapid understanding of the interrelationships of the topics and subject matter in this chapter. Also shown are the problem numbers associated with the subject matter.

CAUCHY'S INTEGRAL FORMULA

• **PROBLEM** 12-1

Cauchy's integral formula can be stated as follows:

<u>Theorem</u>

Let f(z) be analytic in a domain D, and let C be a simple closed curve inside D whose interior is also contained in D. If z_0 is any point interior to C, then the value of $f(z_0)$ is given by

$$f(z_0) = \frac{1}{2\pi i} \int_C \frac{f(z)}{z-z_0} dz \qquad (1)$$

1. Give the interpretation of this theorem. Why is Cauchy's integral formula so remarkable?

2. Note, that (1) can be written in the form

$$\int_C \frac{f(z)}{z-z_0} dz = 2\pi i\, f(z_0) \qquad (2)$$

Applying (2), evaluate

$$\int_C \frac{z}{(11-z^2)(z+2i)}\, dz \qquad (3)$$

where C is the circle $|z| = 3$, positively oriented.

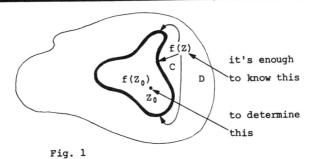

Fig. 1

<u>Solution</u>: 1. The theorem states that if a function is analytic in a domain D and its values are known along a simple closed curve C contained in D, then the values of the function in the interior of C can be uniquely determined and that is remarkable.

2.
$$\int_{\substack{C \\ |z|=3}} \frac{z}{(11-z^2)(z+2i)}\, dz = \int_C \frac{\frac{z}{11-z^2}}{z-(-2i)}\, dz$$

$$= 2\pi i\, \frac{-2i}{11-(-2i)^2} = \frac{4\pi}{15} \qquad (4)$$

Point $z_0 = -2i$ is located inside the circle $|z| = 3$ and $f(z) = \frac{z}{11-z^2}$ is analytic within and on C. Note that, whenever it does not lead to confusion, instead of $\oint_C$, we shall use $\int_C$ to indicate an integral over a closed curve.

● **PROBLEM 12-2**

Prove Cauchy's integral formula, see Problem 12-1.

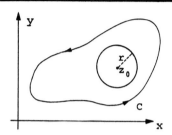

Fig. 1

Solution: We have to prove that

$$f(z_0) = \frac{1}{2\pi i} \int_C \frac{f(z)}{z-z_0} dz \qquad (1)$$

where $f(z)$ is analytic in some domain D, C is a simple closed curve in D and z_0 is an interior point in the region bounded by C.

(1) can be written as

$$\frac{1}{2\pi i} \int_C \frac{f(z)}{z-z_0} dz = \frac{1}{2\pi i} \int_C \frac{f(z_0)}{z-z_0} dz + \frac{1}{2\pi i} \int_C \frac{f(z)-f(z_0)}{z-z_0} dz \qquad (2)$$

We have

$$\int_C \frac{f(z_0)}{z-z_0} dz = f(z_0) \int_C \frac{1}{z-z_0} dz \qquad (3)$$

Since z_0 is inside C

$$\int_C \frac{1}{z-z_0} dz = 2\pi i \qquad (4)$$

and (3) becomes

$$\int_C \frac{f(z_0)}{z-z_0} dz = 2\pi i \, f(z_0) \qquad (5)$$

To evaluate $\int_C \frac{f(z)-f(z_0)}{z-z_0} dz$, the curve C can be replaced by any other curve in the interior of C and enclosing z_0.

Let us take a circle $C(r,z_0)$ radius r and center at z_0

$$\int_C \frac{f(z)-f(z_0)}{z-z_0} dz = \int_{C(r,z_0)} \frac{f(z)-f(z_0)}{z-z_0} dz \qquad (6)$$

Since $f(z)$ is analytic in D including z_0, we have

$$f(z) = f(z_0) + f'(z_0)(z-z_0) + R(z-z_0) \qquad (7)$$

where

$$R \to 0 \quad \text{as} \quad z \to z_0 \qquad (8)$$

Hence,

$$\int_{C(r,z_0)} \frac{f(z)-f(z_0)}{z-z_0} dz = \int_{C(r,z_0)} f'(z_0) dz + \int_{C(r,z_0)} R\, dz \qquad (9)$$

$$= \int_{C(r,z_0)} R\, dz$$

$C(r,z_0)$ can be chosen sufficiently small, so that for all points on $C(r,z_0)$ we have

$$|R| < \frac{r}{2\pi} \qquad (10)$$

Hence,

$$\left| \int_{C(r,z_0)} R\, dz \right| < \frac{r}{2\pi} \cdot 2\pi r = r \qquad (11)$$

and

$$\int_{C(r,z_0)} R\, dz = 0 \qquad (12)$$

that completes the proof.

● **PROBLEM 12-3**

Prove Cauchy's integral formula for the domain D shown in Fig. 1.

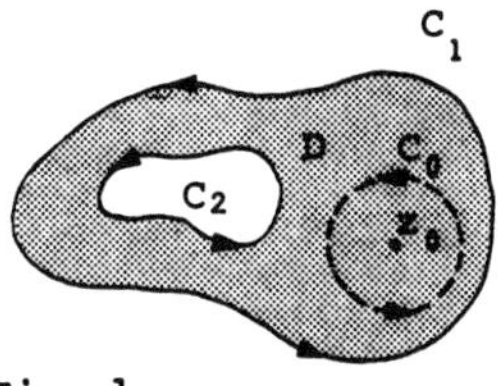

Fig. 1

Solution: Domain D is multiply-connected and bounded by two simple closed curves C_1 and C_2. Let z_0 be a point inside D such that

$$z_0 \in D, \quad z_0 \not\in C_1, \quad z_0 \not\in C_2 \qquad (1)$$

We can construct a circle C_0, with center at z_0 such that

$$C_0 \subset D \tag{2}$$

The function $\frac{f(z)}{z-z_0}$ is analytic inside and on the boundary of D - int C_0. Thus, by Cauchy's theorem for multiply-connected domains,

$$\int_{C_1} \frac{f(z)}{z-z_0} dz - \int_{C_2} \frac{f(z)}{z-z_0} dz - \int_{C_0} \frac{f(z)}{z-z_0} dz = 0 \tag{3}$$

By Cauchy's integral formula for a simply-connected domain

$$f(z_0) = \frac{1}{2\pi i} \int_{C_0} \frac{f(z)}{z-z_0} dz \tag{4}$$

Substituting (4) into (1)

$$f(z_0) = \frac{1}{2\pi i} \int_{C_1} \frac{f(z)}{z-z_0} dz - \frac{1}{2\pi i} \int_{C_2} \frac{f(z)}{z-z_0} dz \tag{5}$$

Note that, in (5), both integrals are traversed in the counterclockwise direction. Denoting by C the boundary of D (consisting of C_1 and C_2) with positive orientation, we denote

$$f(z_0) = \frac{1}{2\pi i} \int_C \frac{f(z)}{z-z_0} dz \tag{6}$$

This proof can be extended to any other multiply-connected domain.

● **PROBLEM 12-4**

Let C be an arbitrary curve in a domain D, and $\phi(\omega)$ a continuous function defined along C. The function

$$F(z) = \frac{1}{2\pi i} \int_C \frac{\phi(\omega)}{\omega - z} d\omega \tag{1}$$

is defined for every $z \notin C$.

Theorem

The function $F(z)$, defined by (1), is analytic everywhere on D except the points on C, and its derivative is given by

$$F'(z) = \frac{1}{2\pi i} \int_C \frac{\phi(\omega)}{(\omega-z)^2} d\omega \tag{2}$$

Prove this theorem.

Solution: For any fixed $z \in D - C$, we must show that for any $z_n \to z$ where $z_n \in D - C$

$$\lim_{n \to \infty} \left\{ \frac{F(z_n)-F(z)}{z_n-z} - \frac{1}{2\pi i} \int_C \frac{\phi(\omega)}{(\omega-z)^2} d\omega \right\} = 0 \tag{3}$$

Since

$$F(z) = \frac{1}{2\pi i} \int_C \frac{\phi(\omega)}{\omega - z} d\omega \tag{4}$$

and

$$F(z_n) = \frac{1}{2\pi i} \int_C \frac{\phi(\omega)}{\omega-z_n} d\omega \tag{5}$$

then

$$\frac{F(z_n)-F(z)}{z_n-z} = \frac{1}{2\pi i} \int_C \frac{\phi(\omega)}{z_n-z}\left[\frac{1}{\omega-z_n} - \frac{1}{\omega-z}\right] d\omega$$

$$= \frac{1}{2\pi i} \int_C \frac{\phi(\omega)}{(\omega-z)(\omega-z_n)} d\omega \tag{6}$$

Denoting by A_n the expression in the brackets in (3), we find

$$A_n = \frac{1}{2\pi i} \int_C \phi(\omega) \left[\frac{1}{(\omega-z)(\omega-z_n)} - \frac{1}{(\omega-z)^2}\right] d\omega$$

$$= \frac{z_n-z}{2\pi i} \int_C \frac{\phi(\omega)}{(\omega-z)^2(\omega-z_n)} d\omega \tag{7}$$

Now, we have to show that

$$A_n \xrightarrow[n \to \infty]{} 0 \tag{8}$$

Let M be an upper bound of the values $|\phi(\omega)|$ along C

$$M = \sup_{\omega \in C} |\phi(\omega)| \tag{9}$$

Let d denote the distance from curve C to the point z. Choosing n sufficiently large

$$|z-z_n| < \frac{d}{2} \tag{10}$$

we obtain

$$|A_n| < \left|\frac{z_n-z}{2\pi i}\right| \frac{2M \cdot \ell}{d^2 \cdot d} = |z_n-z| \frac{M \cdot \ell}{\pi d^3} \tag{11}$$

393

where ℓ is the length of C.

Hence,

$$\lim_{n\to\infty} A_n \to 0 \qquad \text{q.e.d.} \tag{12}$$

Note that, this theorem leads directly to the following: If $f(z)$ is analytic inside and on the boundary C, then by Cauchy's formula for any z_0 inside C

$$f(z_0) = \frac{1}{2\pi i} \oint_C \frac{f(z)}{z-z_0} dz \tag{13}$$

and by (2)

$$f'(z_0) = \frac{1}{2\pi i} \oint_C \frac{f(z)}{(z-z_0)^2} dz \tag{14}$$

Observe that (14) resembles Leibnitz's rule for differentiating under the integral sign

$$\begin{aligned}\frac{df(z)}{dz} &= \frac{d}{dz}\left[\frac{1}{2\pi i} \int_C \frac{f(\omega)}{\omega - z} d\omega\right] \\ &= \frac{1}{2\pi i} \int_C \frac{\partial}{\partial z}\left[\frac{f(\omega)}{\omega - z}\right] d\omega\end{aligned} \tag{15}$$

In the next problem we will show that a similar formula exists for higher derivatives

$$\begin{aligned}\frac{d^n}{dz^n} f(z) &= \frac{d^n}{dz^n}\left[\frac{1}{2\pi i} \int_C \frac{f(\omega)}{\omega - z} d\omega\right] \\ &= \frac{1}{2\pi i} \int_C \frac{\partial^n}{\partial z^n}\left[\frac{f(\omega)}{\omega - z}\right] d\omega\end{aligned} \tag{16}$$

• **PROBLEM 12-5**

Consider the function $F(z)$ defined in Problem 12-4, eq.(1). Prove the following:

Theorem

The function $F(z)$ defined by (1) of Problem 12-4 has derivatives in D of every order which are given by

$$F''(z) = \frac{2!}{2\pi i} \int_C \frac{\phi(\omega)}{(\omega-z)^3} d\omega \tag{1}$$

or by the general formula

$$F^{(n)}(z) = \frac{n!}{2\pi i} \int_C \frac{\phi(\omega)}{(\omega-z)^{n+1}} d\omega \qquad (2)$$

for $n = 0, 1, 2, 3, \ldots$, remember that $0! = 1$.

Solution: We shall prove (1). Let

$$B_n = \frac{F'(z_n) - F'(z)}{z_n - z} - \frac{2!}{2\pi i} \int_C \frac{\phi(\omega)}{(\omega-z)^3} d\omega$$

$$= \frac{1}{2\pi i} \int_C \phi(\omega) \left[\frac{1}{z_n - z} \left(\frac{1}{(\omega-z_n)^2} - \frac{1}{(\omega-z)^2} \right) - \frac{2}{(\omega-z)^3} \right] d\omega \qquad (3)$$

$$= \frac{1}{2\pi i} \int_C \phi(\omega) \left\{ \frac{1}{z_n - z} \left[\frac{2\omega(z_n - z) - (z_n - z)(z_n + z)}{(\omega-z)^2 (\omega-z_n)^2} \right] - \frac{2}{(\omega-z)^3} \right\} d\omega$$

$$= \frac{1}{2\pi i} \int_C \phi(\omega) \left[\frac{(2\omega - z_n - z)(\omega-z) - 2(\omega-z_n)^2}{(\omega-z)^3 (\omega-z_n)^2} \right] d\omega$$

$$= \frac{1}{2\pi i} \int_C \phi(\omega) \frac{(3\omega - z - 2z_n)(z_n - z)}{(\omega-z)^3 (\omega-z_n)^2} d\omega$$

We used the expression for the first derivative

$$F'(z) = \frac{1}{2\pi i} \int_C \frac{\phi(\omega)}{(\omega-z)^2} d\omega \qquad (4)$$

in (3).

Hence, (1) is equivalent to

$$B_n \xrightarrow[n \to \infty]{} 0. \qquad (5)$$

Let ℓ denote the length of C and d the distance of the point z from C. Then

$$|B_n| < \frac{4M'\ell}{d^5} \cdot \frac{|z_n - z|}{2\pi} \qquad (6)$$

Therefore, $B_n \to 0$.

q.e.d.

In the similar manner we prove (2) for $n = 3, 4, \ldots$.

APPLICATIONS

● **PROBLEM 12-6**

Using the results of Problem 12-4 and 12-5, we shall derive an important property of regular functions. As previously mentioned, a single-valued function is analytic if it has a derivative. In the case of functions of a real variable that does not imply any information about the nature of the derivative, it need not be continuous. The situation becomes quite different for the analytic functions of a complex variable.

Theorem

If a single-valued function f(z) is defined in a domain D and has a first derivative there (i.e., f(z) is analytic in D), then all higher derivatives exist (and are thus continuous) in D.

Prove this theorem.

Solution: We have to prove that if $f(z)$ is analytic in D, then $f'(z)$, $f''(z)$, $f'''(z)$... are analytic in D.

Let z be an arbitrary point in D and let C be any simple closed curve which contains z and only points of interior of D. Then, by Problem 12-4, since f(z) is continuous along C, the function

$$\frac{1}{2\pi i} \int_C \frac{f(\omega)}{\omega - z} d\omega \qquad (1)$$

is therefore analytic and differentiable any number of times everywhere in C. By Cauchy's integral formula

$$f(z) = \frac{1}{2\pi i} \int_C \frac{f(\omega)}{\omega - z} d\omega \qquad (2)$$

Thus, f(z) is analytic and differentiable any number of times everywhere in C. Since z was chosen arbitrarily, the conclusion holds true for any z in D. q.e.d.

Note that, from Problem 12-5 we conclude that

$$f^{(n)}(z) = \frac{n!}{2\pi i} \int_C \frac{f(\omega)}{(\omega-z)^{n+1}} d\omega \qquad (3)$$

for n = 0,1,2,....

• **PROBLEM 12-7**

Evaluate

1. $$\int_C \frac{\sin\pi(z+1) + \cos\pi z}{(z-1)(z-2)} \, dz \qquad (1)$$

   where C is the circle $|z| = 4$.

2. $$\int_C \frac{e^{3z}}{(z+2)^{10}} \, dz \qquad (2)$$

   where C is the circle $|z| = 3$.

Solution: 1. Note that

$$\frac{1}{(z-1)(z-2)} = \frac{1}{z-2} - \frac{1}{z-1} \qquad (3)$$

and we obtain

$$\int_C \frac{\sin\pi(z+1) + \cos\pi z}{(z-1)(z-2)} \, dz$$

$$= \int_C \frac{\sin\pi(z+1) + \cos\pi z}{z-2} \, dz - \int_C \frac{\sin\pi(z+1) + \cos\pi z}{z-1} \, dz. \qquad (4)$$

By Cauchy's integral formula we find

$$\int_C \frac{\sin\pi(z+1) + \cos\pi z}{z-2} \, dz = 2\pi i \left[\sin 3\pi + \cos 2\pi\right]$$

$$= 2\pi i \qquad (5)$$

$$\int_C \frac{\sin\pi(z+1) + \cos\pi z}{z-1} \, dz = 2\pi i(\sin 2\pi + \cos\pi)$$

$$= -2\pi i \qquad (6)$$

Both z=1 and z=2 are inside C and $f(z) = \sin\pi(z+1) + \cos\pi z$ is analytic inside C.

Hence,

$$\int_C \frac{\sin\pi(z+1) + \cos\pi z}{(z-1)(z-2)} \, dz = 2\pi i - (-2\pi i) = 4\pi i \qquad (7)$$

2. To evaluate $\int_C \frac{e^{3z}}{(z+2)^{10}} \, dz$ where C is the circle $|z| = 3$, we

shall use the expression for the nth derivative of an analytic function

$$f^{(n)}(a) = \frac{n!}{2\pi i} \int_C \frac{f(z)}{(z-a)^{n+1}} dz \qquad (8)$$

For $f(z) = e^{3z}$, $n = 9$ and $a = -2$ we obtain

$$f^{(9)}(-2) = \frac{9!}{2\pi i} \int_C \frac{e^{3z}}{(z+2)^{10}} dz \qquad (9)$$

on the other hand

$$f^{(9)}(z) = 3^9 e^{3z} \qquad (10)$$

Substituting (10) into (9), we obtain

$$3^9 e^{-6} = \frac{9!}{2\pi i} \int_C \frac{e^{3z}}{(z+2)^{10}} dz \qquad (11)$$

or

$$\int_C \frac{e^{3z}}{(z+2)^{10}} dz = \frac{2\pi i}{9!} 3^9 e^{-6} \qquad (12)$$

● **PROBLEM 12-8**

In Chapter 11, we have proved Morera's theorem (the converse of Cauchy's theorem):

<u>Theorem</u>

If $f(z)$ is continuous in a simply-connected domain D, and if

$$\int_C f(z)dz = 0 \qquad (1)$$

for every closed path C lying within D, then $f(z)$ is analytic in D. Prove this using results of this chapter.

<u>Solution</u>: Observe that the integral

$$\int_{z_0}^{z} f(z)dz \qquad (2)$$

is independent of the path because for every closed path C

$$\int_C f(z)dz = 0 \qquad (3)$$

Hence, (2) represents a function

$$F(z) = \int_{z_0}^{z} f(z)dz \tag{4}$$

analytic in D and such that

$$F'(z) = f(z) \tag{5}$$

Thus, F(z) is an analytic function and has a second derivative in D

$$F''(z) = f'(z) \tag{6}$$

and f(z) has a first derivative in D. We conclude that f(z) is analytic in D.

● **PROBLEM 12-9**

Let f(z) be analytic inside and on a circle C of radius r and center at z = a. Prove Cauchy's inequality

$$|f^{(n)}(a)| \leq \frac{M \cdot n!}{r^n} \quad n=0,1,2,\ldots \tag{1}$$

where M is a constant such that

$$|f(z)| < M \tag{2}$$

Solution: We shall utilize the following inequality:

$$\left| \int_C f(z)dz \right| \leq ML \tag{3}$$

where L is the length of a path C and M is a positive number such that

$$|f(z)| \leq M \quad \text{on } C \tag{4}$$

By Cauchy's integral formula

$$f^{(n)}(a) = \frac{n!}{2\pi i} \int_C \frac{f(z)}{(z-a)^{n+1}} dz \tag{5}$$

Hence, applying (3) to (5) we find

$$|f^{(n)}(a)| = \frac{n!}{2\pi} \left| \int_C \frac{f(z)}{(z-a)^{n+1}} dz \right| \tag{6}$$

$$\leq \frac{n!}{2\pi} \cdot \frac{M \cdot L}{r^{n+1}} = \frac{n!}{2\pi} \cdot \frac{M \cdot 2\pi r}{r^{n+1}} = \frac{n!M}{r^n}$$

• **PROBLEM 12-10**

Prove the formula

$$\int_0^{2\pi} \cos^{2n}\theta\, d\theta = \frac{1\cdot 3\cdot 5\cdot\ldots\cdot(2n-1)}{2\cdot 4\cdot 6\cdot\ldots\cdot(2n)}\, 2\pi \quad (1)$$

where $n = 1, 2, 3, 4, \ldots$.

Solution: Let $z = e^{i\theta}$, then

$$dz = ie^{i\theta}d\theta = iz\,d\theta \quad (2)$$

or

$$d\theta = \frac{dz}{iz} \quad (3)$$

We have

$$\cos\theta = \frac{e^{i\theta}+e^{-i\theta}}{2} = \frac{z+\frac{1}{z}}{2} \quad (4)$$

Thus, if C is the unit circle $|z| = 1$,
Then,

$$\int_0^{2\pi}\cos^{2n}\theta\, d\theta = \int_C \left[\frac{1}{2}\left(z+\frac{1}{z}\right)\right]^{2n}\frac{dz}{iz}$$

$$= \frac{1}{2^{2n}i}\int_C \frac{1}{z}\left[z^{2n} + \binom{2n}{1}z^{2n-1}\frac{1}{z} + \ldots + \binom{2n}{\ell}z^{2n-\ell}\left(\frac{1}{z}\right)^{\ell}\right.$$

$$\left. + \ldots + \left(\frac{1}{z}\right)^{2n}\right]dz \quad (5)$$

$$= \frac{1}{2^{2n}i}\int_C \left[z^{2n-1} + \binom{2n}{1}z^{2n-3} + \ldots + \binom{2n}{\ell}z^{2n-2\ell-1}\right.$$

$$\left. + \ldots + z^{-2n}\right]dz$$

$$= \frac{1}{2^{2n}i}\cdot 2\pi i\binom{2n}{n} = \frac{1}{2^{2n}}\cdot 2\pi\,\frac{(2n)!}{n!n!}$$

$$= \frac{1\cdot\ldots\cdot n(n-1)\ldots(2n-2)(2n-1)\cdot 2n}{2^{2n}\,n!n!}\,2\pi$$

$$= \frac{1\cdot 3\cdot 5\cdot\ldots\cdot(2n-1)}{2\cdot 4\cdot 6\cdot\ldots\cdot 2n}\,2\pi$$

• **PROBLEM 12-11**

Let $f(z)$ be analytic in a region bounded by two concentric circles C_1 and C_2 and on its boundary. Show that if z_0 is any point in the region, then

$$f(z_0) = \frac{1}{2\pi i} \int_{C_1} \frac{f(z)}{z-z_0} dz - \frac{1}{2\pi i} \int_{C_2} \frac{f(z)}{z-z_0} dz \qquad (1)$$

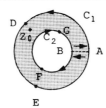

Fig. 1

Solution: Construct cross-cut AB connecting circles C_1 and C_2, Fig. 1.

The function $f(z)$ is analytic in the region bounded by ADEABFGBA. Thus, by Cauchy's integral formula,

$$f(z_0) = \frac{1}{2\pi i} \int_{ADEABFGBA} \frac{f(z)}{z-z_0} dz$$

$$= \frac{1}{2\pi i} \int_{ADEA} \frac{f(z)}{z-z_0} dz + \frac{1}{2\pi i} \int_{AB} \frac{f(z)}{z-z_0} dz + \frac{1}{2\pi i} \int_{BFGB} \frac{f(z)}{z-z_0} dz$$

$$+ \frac{1}{2\pi i} \int_{BA} \frac{f(z)}{z-z_0} dz \qquad (2)$$

$$= \frac{1}{2\pi i} \int_{C_1} \frac{f(z)}{z-z_0} dz - \frac{1}{2\pi i} \int_{C_2} \frac{f(z)}{z-z_0} dz.$$

The integrals along AB and BA cancel and

$$\int_{BFGB} = -\int_{BGFB} = -\int_{C_2} \qquad (3)$$

Compare this with Problem 12-3.

THEOREMS

• **PROBLEM 12-12**

Earlier we had proved the following:

<u>Theorem I</u>

If $f(z) = u + iv$ is analytic in a domain D and if u and v

have continuous second partial derivatives in D, then u and v are harmonic in D.

Function u(x,y) which satisfies Laplace's equation in a domain D

$$\frac{\partial^2 u}{\partial x^2} + \frac{\partial^2 u}{\partial y^2} = 0 \quad \text{or} \quad \nabla^2 u = 0 \tag{1}$$

is harmonic in D.

Applying the results of Problem 12-6 prove stronger version of the theorem.

Theorem II

If $f(z) = u + iv$ is analytic in a domain D, then u and v are harmonic in D.

Solution: Since f(z) is analytic in D, then all the derivatives of f(z) exist and are continuous in D. Thus, the second partial derivatives of u(x,y) and v(x,y) exist and are continuous in D. The assumptions of Theorem I are fulfilled, therefore u and v are harmonic in D.

q.e.d.

● **PROBLEM 12-13**

Theorem (Liouville's Theorem)

If f(z) is analytic in the entire complex plane and bounded, then f(z) is a constant function.

Prove this theorem.

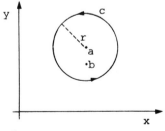

Fig. 1

Solution: Let a and b be any two points in the z plane, and let C be a circle with the center a and radius r such that

$$2|a-b| < r \tag{1}$$

From Cauchy's integral formula we have

$$f(b) - f(a) = \frac{1}{2\pi i} \int_C \frac{f(z)}{z-b} dz - \frac{1}{2\pi i} \int_C \frac{f(z)}{z-a} dz$$

$$= \frac{b-a}{2\pi i} \int_C \frac{f(z)}{(z-b)(z-a)} dz \tag{2}$$

Since $|z-a| = r$, applying (1) we find

$$|z-b| = |z-a+a-b| \geq |z-a|-|a-b| \qquad (3)$$
$$= r-|a-b| \geq \frac{r}{2}$$

Since $f(z)$ is bounded

$$\bigvee_{M>0} \bigwedge_{z} \; |f(z)| < M \qquad (4)$$

The length of C is $2\pi r$, thus

$$|f(b)-f(a)| = \frac{|b-a|}{2\pi} \left| \int_C \frac{f(z)}{(z-b)(z-a)} dz \right|$$

$$\leq \frac{|b-a|}{2\pi} \cdot \frac{M}{\frac{r}{2} \cdot r} \cdot 2\pi r = \frac{2M|b-a|}{r} \qquad (5)$$

Inequality (5) holds for any r, hence letting $r \to \infty$ we conclude that

$$|f(b)-f(a)| = 0$$

or

$$f(b) = f(a) \qquad (6)$$

Function $f(z)$ must be constant.

• **PROBLEM 12-14**

Using the methods of complex analysis prove

<u>Fundamental Theorem of Algebra</u>

Every polynomial equation

$$P_n(z) = a_n z^n + a_{n-1} z^{n-1} + \ldots + a_1 z + a_0 \qquad (1)$$

where $a_n \neq 0$ and $n \geq 1$ has at least one root.

<u>Solution</u>: Assume that $P_n(z) = 0$ has no roots, then

$$f(z) = \frac{1}{P_n(z)} \qquad (2)$$

is analytic in the entire complex plane. The function

$$|f(z)| = \frac{1}{|P_n(z)|} \qquad (3)$$

is bounded. By Liouville's theorem we conclude that $f(z)$ must be constant and hence, $P_n(z)$ must be constant. This is a contradiction. Therefore, $P_n(z) = 0$ must have at least one root.

• **PROBLEM 12-15**

Consider the polynomial equation

$$P_n(z) = a_n z^n + a_{n-1} z^{n-1} + \ldots + a_1 z + a_0 = 0 \qquad (1)$$

where $n \geq 1$ and $a_n \neq 0$.

Show that eq.(1) has exactly n roots.

Solution: By the fundamental theorem of algebra (see Problem 12-14), $P_n(z) = 0$ has at least one root. Denote it by z_1. Then

$$P_n(z_1) = 0 \qquad (2)$$

and

$$\begin{aligned}P_n(z) - P_n(z_1) &= a_n z^n + a_{n-1} z^{n-1} + \ldots + a_1 z + a_0 \\ &\quad - (a_n z_1^n + a_{n-1} z_1^{n-1} + \ldots + a_1 z_1 + a_0) \qquad (3) \\ &= a_n(z^n - z_1^n) + a_{n-1}(z^{n-1} - z_1^{n-1}) + \ldots + a_2(z^2 - z_1^2) \\ &\quad + a_1(z - z_1) = (z - z_1) \cdot Q_{n-1}(z)\end{aligned}$$

$Q_{n-1}(z)$ is a polynomial of degree n-1. By the same argument, we conclude that $Q_{n-1}(z)$ has at least one root, which we shall denote by z_2, i.e.,

$$Q_{n-1}(z_2) = 0 \qquad (4)$$

Hence, $P_n(z)$ can be written in the form

$$P_n(z) = (z-z_1)(z-z_2) R_{n-2}(z) \qquad (5)$$

Repeating this procedure n times we find

$$P_n(z) = (z-z_1)(z-z_2) \ldots (z-z_n) = 0 \qquad (6)$$

Hence, $P_n(z) = 0$ has exactly n roots.

• **PROBLEM 12-16**

Prove

Gauss' mean value theorem

If f(z) is analytic inside and on a circle C with center at a, then the mean value of f(z) on C is f(a).

Solution: By Cauchy's integral formula

$$f(a) = \frac{1}{2\pi i} \int_C \frac{f(z)}{z - a} dz \qquad (1)$$

If r is the radius of C then, the equation of C is

$$r = |z-a| \text{ or}$$
$$z = a + re^{i\theta} \tag{2}$$

Hence, the mean value of $f(z)$ on C is given by

$$\frac{1}{2\pi} \int_0^{2\pi} f(a+re^{i\theta})d\theta \tag{3}$$

From (1) we obtain

$$f(a) = \frac{1}{2\pi i} \int_0^{2\pi} \frac{f(a+re^{i\theta})}{re^{i\theta}} \cdot ri\, e^{i\theta} d\theta$$

$$= \frac{1}{2\pi} \int_0^{2\pi} f(a+re^{i\theta})d\theta \tag{4}$$

q.e.d.

● **PROBLEM 12-17**

Prove the following theorem.

The Maximum Modulus Theorem

If $f(z)$ is not constant and is analytic inside and on a simple closed curve C, then the maximum value of $|f(z)|$ occurs on C.

Solution: The function $f(z)$ is analytic inside and on C. Hence, $|f(z)|$ is continuous inside and on C and has a maximum value M for at least one point inside or on C.

Contradictory to the theorem, let us assume that the maximum value is attained at an interior point a of C, i.e.,

$$|f(a)| = M \tag{1}$$

Draw the circle C_1 inside C with the center at $z = a$. It is possible, since a is an interior point of C.

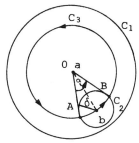

Fig. 1

At a, $|f(z)|$ attains its maximum. Thus, there must be a point b inside C_1 such that

$$|f(b)| < M \qquad (2)$$

and a number $\varepsilon > 0$ exists such that

$$|f(b)| = M - \varepsilon, \qquad \varepsilon > 0 \qquad (3)$$

The function $|f(z)|$ is continuous at b

$$\underset{\varepsilon > 0}{\wedge} \underset{\delta > 0}{\vee} \quad |z-b| < \delta \Rightarrow \left||f(z)| - |f(b)|\right| < \frac{\varepsilon}{2} \qquad (4)$$

or

$$|f(z)| < |f(b)| + \frac{\varepsilon}{2} = M - \varepsilon + \frac{\varepsilon}{2} = M - \frac{\varepsilon}{2} \qquad (5)$$

Inequality (2) holds true for all points inside the circle C_2 of radius δ and center at $z = b$.

Now, let C_3 be the circle with center a and passing through b. Since, arc AB is inside C_2, for all its points inequality (5) is true

$$|f(z)| < M - \frac{\varepsilon}{2} \quad \text{along AB} \qquad (6)$$

Along the remaining part

$$|f(z)| \leq M \qquad (7)$$

Indicating by α the angle $\sphericalangle AOB = \alpha$, and measuring θ counter-clockwise, and finally applying the results of Problem 12-16, we obtain

$$f(a) = \frac{1}{2\pi} \int_0^{2\pi} f(a+re^{i\theta}) d\theta$$

$$= \frac{1}{2\pi} \int_0^{\alpha} f(a+re^{i\theta}) d\theta + \frac{1}{2\pi} \int_{\alpha}^{2\pi} f(a+re^{i\theta}) d\theta \qquad (8)$$

Hence,

$$|f(a)| \leq \frac{1}{2\pi} \int_0^{\alpha} |f(a+re^{i\theta})| d\theta + \frac{1}{2\pi} \int_{\alpha}^{2\pi} |f(a+re^{i\theta})| d\theta$$

$$\leq \frac{1}{2\pi} \int_0^{\alpha} (M - \frac{\varepsilon}{2}) d\theta + \frac{1}{2\pi} \int_{\alpha}^{2\pi} M d\theta \qquad (9)$$

$$= \frac{\alpha}{2\pi}(M - \frac{\varepsilon}{2}) + \frac{M}{2\pi}(2\pi - \alpha) = M - \frac{\alpha \varepsilon}{4\pi}$$

Combining (1) and (9) we obtain

$$|f(a)| = M \leq M - \frac{\alpha \varepsilon}{4\pi} \qquad (10)$$

which is a contradiction.

Therefore, $|f(z)|$ attains its maximum value on C.

q.e.d.

● **PROBLEM 12-18**

> Utilizing the maximum modulus theorem, prove the following.
>
> <u>Minimum modulus theorem</u>
>
> If $f(z)$ is analytic inside and on a simple closed curve C and if
>
> $$f(z) \neq 0 \quad \text{inside C,} \tag{1}$$
>
> then $|f(z)|$ attains its minimum value on C.

<u>Solution</u>: Since the function $f(z)$ is analytic inside and on C and $f(z) \neq 0$ inside C, the function $\frac{1}{f(z)}$ is analytic inside C. By the maximum modulus theorem, the function $\frac{1}{|f(z)|}$ cannot attain its maximum value inside C, hence $|f(z)|$ cannot attain its minimum value inside C.

Since $f(z)$ is analytic, $|f(z)|$ is continuous inside and on C and attains its minimum inside or on C. Therefore, since $|f(z)|$ has a minimum, this minimum is attained on C.

● **PROBLEM 12-19**

> Prove Schwarz's theorem:
>
> If $f(z)$ is analytic for $|z| \leq R$, $f(0) = 0$ and $|f(z)| \leq M$, then
>
> $$|f(z)| \leq \frac{M}{R} |z| \tag{1}$$

<u>Solution</u>: Consider the function $\frac{f(z)}{z}$ in the region $|z| \leq R$.

Since $f(z)$ is analytic at $z = 0$, it is also continuous at $z = 0$. Hence, by l'Hospital's rule

$$\lim_{z \to 0} \frac{f(z)}{z} = \frac{f'(0)}{1} = f'(0) \tag{2}$$

The function $\frac{f(z)}{z}$ is analytic in $|z| \leq R$. By the maximum modulus theorem for $|z| = R$, we have

$$\left| \frac{f(z)}{z} \right| \leq \frac{M}{R} \tag{3}$$

This inequality must hold for all points inside $|z| = R$. Therefore, for $|z| \leq R$, we have

$$|f(z)| \leq \frac{M}{R} |z| \tag{4}$$

q.e.d.

FUNCTIONS WITH POLES AND ZEROS

● **PROBLEM 12-20**

Let f(z) be analytic inside and on a simple closed curve C except for a pole z = a of order n inside C.

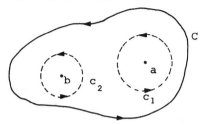

Fig. 1

Function f(z) has inside C only one zero at z = b of order p and no zeros on C. Show that

$$\frac{1}{2\pi i} \int_C \frac{f'(z)}{f(z)} dz = p - n \tag{1}$$

Solution: Let C_1 and C_2 be two disjoint circles lying inside C with the center a and b respectively. By Cauchy's integral formula for multiply-connected domains, we have

$$\frac{1}{2\pi i} \int_C \frac{f'(z)}{f(z)} dz = \frac{1}{2\pi i} \int_{C_1} \frac{f'(z)}{f(z)} dz + \frac{1}{2\pi i} \int_{C_2} \frac{f'(z)}{f(z)} dz \tag{2}$$

Function f(z) has a pole of order n at z = a if

$$\lim_{z \to a} (z-a)^n f(z) = \alpha \neq 0 \tag{3}$$

Since f(z) has a pole of order n at z = a

$$f(z) = \frac{F(z)}{(z-a)^n} \tag{4}$$

then F(z) is an analytic function different from zero inside and on C_1. From (4) we get

$$\ln f(z) = \ln F(z) - n \ln(z-a) \tag{5}$$

Differentiating (5)

$$\frac{f'(z)}{f(z)} = \frac{F'(z)}{F(z)} - \frac{n}{z-a} \tag{6}$$

Then,

$$\frac{1}{2\pi i} \int_{C_1} \frac{f'(z)}{f(z)} dz = \frac{1}{2\pi i} \int_{C_1} \frac{F'(z)}{F(z)} - \frac{n}{2\pi i} \int_{C_1} \frac{dz}{z-a}$$
$$= 0 - n = -n \tag{7}$$

Since $f(z)$ has a zero at $z = b$ of order p, hence

$$f(z) = (z-b)^p G(z) \tag{8}$$

where $G(z)$ is analytic and different from zero inside and on C_2. Taking logarithm of (8) and differentiating we find

$$\frac{f'(z)}{f(z)} = \frac{G'(z)}{G(z)} + p\frac{1}{z-b} \tag{9}$$

Thus,

$$\frac{1}{2\pi i} \int_{C_2} \frac{f'(z)}{f(z)} dz = \frac{1}{2\pi i} \int_{C_2} \frac{G'(z)}{G(z)} dz + \frac{p}{2\pi i} \int_{C_2} \frac{1}{z-b} dz$$
$$= 0 + p = p \tag{10}$$

Substituting (7) and (10) into (2), we obtain

$$\frac{1}{2\pi i} \int_C \frac{f'(z)}{f(z)} dz = p - n \tag{11}$$

● **PROBLEM 12-21**

Let $f(z)$ be analytic inside and on a simple closed curve C except for poles $a_1, a_2, \ldots, a_k$ inside C. Let $n_1, n_2, \ldots, n_k$ denote the order (multiplicity) of poles. Assume that $f(z)$ has zeros at $b_1, b_2, \ldots, b_\ell$ of order $p_1, p_2, \ldots, p_\ell$ inside C and $f(z) \neq 0$ on C.

Show that

$$\frac{1}{2\pi i} \int_C \frac{f'(z)}{f(z)} dz = P - N \tag{1}$$

where

$$P = p_1 + p_2 + \ldots + p_\ell$$
$$N = n_1 + n_2 + \ldots + n_k \tag{2}$$

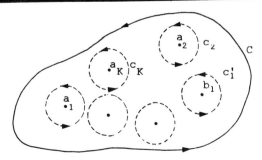

Fig.1

Solution: Let us enclose each pole and zero by disjoint circles $C_1, C_2, \ldots, C_k$ and $C'_1, C'_2, \ldots, C'_\ell$.

Applying the results of Problem 12-20, we obtain

$$\frac{1}{2\pi i} \oint_C \frac{f'(z)}{f(z)} dz = \sum_{s=1}^{\ell} \frac{1}{2\pi i} \oint_{C'_s} \frac{f'(z)}{f(z)} dz$$

$$+ \sum_{r=1}^{k} \frac{1}{2\pi i} \oint_{C_r} \frac{f'(z)}{f(z)} dz \qquad (3)$$

$$= \sum_{s=1}^{\ell} p_s - \sum_{r=1}^{k} n_r = P - N$$

• **PROBLEM 12-22**

> Let $f(z)$ be analytic inside and on a simple closed curve C except for a pole of order k at $z = a$ inside C. Show that
>
> $$\frac{1}{2\pi i} \oint_C f(z) dz = \lim_{z \to a} \frac{1}{(k-1)!} \frac{d^{k-1}}{dz^{k-1}} \left[(z-a)^k f(z) \right] \qquad (1)$$
>
> Generalize the result for the case of two or more poles inside C.

Solution: Since $f(z)$ has a pole of order k at $z = a$ inside C, then

$$f(z) = \frac{F(z)}{(z-a)^k} \qquad (2)$$

where $F(z)$ is analytic inside and on C and $F(a) \neq 0$.

By Cauchy's integral formula

$$\frac{1}{2\pi i} \oint_C f(z) dz = \frac{1}{2\pi i} \oint_C \frac{F(z)}{(z-a)^k} dz$$

$$= \frac{1}{(k-1)!} \frac{d^{k-1}}{dz^{k-1}} F(a) = \lim_{z \to a} \frac{1}{(k-1)!} \frac{d^{k-1}}{dz^{k-1}} \left[(z-a)^k f(z) \right] \qquad (3)$$

Now, assume there are two poles inside C, $z = a_1$ and $z = a_2$ of orders k_1 and k_2 respectively. Let C_1 and C_2 be the circles inside C with centers at a_1 and a_2 of radii r_1 and r_2. C_1 and C_2 are disjoint.

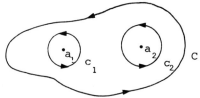

Fig. 1

Then,

$$\frac{1}{2\pi i}\int_C f(z)dz = \frac{1}{2\pi i}\int_{C_1} f(z)dz + \frac{1}{2\pi i}\int_{C_2} f(z)dz \tag{4}$$

Since $f(z)$ has a pole of order k_1 at $z = a_1$

$$f(z) = \frac{F_1(z)}{(z-a_1)^{k_1}} \tag{5}$$

where $F_1(z)$ is analytic and $F_1(a_1) \neq 0$.

Furthermore $f(z)$ has a pole of order k_2 at $z = a_2$

$$f(z) = \frac{F_2(z)}{(z-a_2)^{k_2}} \tag{6}$$

where $F_2(z)$ is analytic and $F_2(a_2) \neq 0$.

From (4), (5) and (6) we obtain

$$\frac{1}{2\pi i}\int_C f(z)dz = \frac{1}{2\pi i}\int_{C_1}\frac{F_1(z)}{(z-a_1)^{k_1}}dz + \frac{1}{2\pi i}\int_{C_2}\frac{F_2(z)}{(z-a_2)^{k_2}}dz$$

$$= \lim_{z \to a_1} \frac{1}{(k_1-1)!}\frac{d^{k_1-1}}{dz^{k_1-1}}\left[(z-a_1)^{k_1}f(z)\right] \tag{7}$$

$$+ \lim_{z \to a_2} \frac{1}{(k_2-1)!}\frac{d^{k_2-1}}{dz^{k_2-1}}\left[(z-a_2)^{k_2}f(z)\right]$$

Denoting the limits on the right-hand side of (7) by R_1 and R_2, we can denote

$$\int_C f(z)dz = 2\pi i(R_1+R_2) \tag{8}$$

R_1 and R_2 are called the residues of $f(z)$ at the poles a_1 and a_2. If $f(z)$ has ℓ poles, (8) becomes

$$\int_C f(z)dz = 2\pi i(R_1+R_2+\ldots+R_\ell) \tag{9}$$

Eq.(9) is called the residue theorem.

● **PROBLEM 12-23**

Evaluate

$$\int_C \frac{z^2 - 2z}{(z+1)^2(z^2+4)}dz \tag{1}$$

where C is the circle $|z| = 10$.

Solution: The function

$$f(z) = \frac{z^2 - 2z}{(z+1)^2(z^2+4)} \quad (2)$$

has a pole of order two at $z = -1$ and two poles $z = \pm 2i$ of order one.

We shall evaluate the residues. Residue at $z = -1$ is

$$\lim_{z \to -1} \frac{1}{1!} \frac{d}{dz}\left[(z+1)^2 \frac{z^2-2z}{(z+1)^2(z^2+4)}\right]$$

$$= \lim_{z \to -1} \frac{(z^2+4)(2z-2)-(z^2-2z)2z}{(z^2+4)^2} = -\frac{14}{25} \quad (3)$$

Residue at $z = 2i$ is

$$\lim_{z \to 2i}\left[(z-2i)\frac{z^2-2z}{(z+1)^2(z-2i)(z+2i)}\right]$$

$$= \frac{-4i-4}{4i(2i+1)^2} = \frac{7+i}{25} \quad (4)$$

Remember that $0! = 1$.

Residue at $z = -2i$ is

$$\lim_{z \to -2i}\left[(z+2i)\frac{z^2-2z}{(z+1)^2(z-2i)(z+2i)}\right]$$

$$= \frac{4i-4}{-4i(-2i+1)^2} = \frac{7-i}{25} \quad (5)$$

Since all poles are inside C

$$\int_C \frac{z^2-2z}{(z+1)^2(z^2+4)} dz = 2\pi i\left(-\frac{14}{25} + \frac{7+i}{25} + \frac{7-i}{25}\right) = 0 \quad (6)$$

● **PROBLEM 12-24**

Prove Rouche's theorem:

If $f(z)$ and $g(z)$ are analytic inside and on a simple closed curve C and if

$$|g(z)| < |f(z)| \quad \text{on C}, \quad (1)$$

then both functions

$$f(z) + g(z) \quad \text{and} \quad f(z)$$

have the same number of zeros inside C.

Solution: Let us define the new function

$$F(z) = \frac{g(z)}{f(z)} \qquad (2)$$

Then,

$$g(z) = F(z) f(z) \qquad (3)$$

Let P_1 and P_2 denote the number of zeros inside C of $f(z)+g(z)$ and $f(z)$ correspondingly.

Since $f(z)+g(z)$ and $f(z)$ have no poles inside C, by Problem 12-21 we can denote

$$P_1 = \frac{1}{2\pi i} \int_C \frac{f'+g'}{f+g} dz \qquad (4)$$

$$P_2 = \frac{1}{2\pi i} \int_C \frac{f'}{f} dz \qquad (5)$$

Hence,

$$P_1 - P_2 = \frac{1}{2\pi i} \int_C \frac{f'+g'}{f+g} dz - \frac{1}{2\pi i} \int_C \frac{f'}{f} dz$$

$$= \frac{1}{2\pi i} \int_C \left(\frac{f'+F'f+Ff'}{f+Ff} - \frac{f'}{f} \right) dz \qquad (6)$$

$$= \frac{1}{2\pi i} \int_C \frac{F'}{F+1} dz$$

Since $|g(z)| < |f(z)|$ on C, $|F(z)| < 1$, on C and $\frac{1}{1+F}$ can be expressed as a uniformly convergent series on C

$$\frac{1}{1+F} = 1 - F + F^2 - F^3 + \ldots \qquad (7)$$

Substituting (7) into (6) and integrating term by term

$$P_1 - P_2 = \frac{1}{2\pi i} \int_C \frac{F'}{1+F} dz$$

$$= \frac{1}{2\pi i} \int_C F'(1-F+F^2-F^3+\ldots) dz = 0 \qquad (8)$$

Hence,

$$P_1 = P_2 \qquad (9)$$

• **PROBLEM 12-25**

Applying Rouche's theorem, show that every polynomial of degree n has exactly n zeros.

<u>Solution</u>: Let P_n be the polynomial of degree n

$$P_n(z) = a_n z^n + a_{n-1} z^{n-1} + \ldots + a_1 z + a_0 \tag{1}$$

where $a_n \neq 0$.

We denote

$$f(z) = a_n z^n \tag{2}$$

and

$$g(z) = a_{n-1} z^{n-1} + \ldots + a_1 z + a_0 \tag{3}$$

For z on a circle C of radius $r > 1$ and center at the origin we have

$$\left|\frac{g(z)}{f(z)}\right| = \frac{|a_{n-1} z^{n-1} + \ldots + a_2 z^2 + a_1 z + a_0|}{|a_n z^n|}$$

$$\leq \frac{|a_{n-1}| r^{n-1} + \ldots + |a_2| r^2 + |a_1| r + |a_0|}{|a_n| r^n} \tag{4}$$

$$\leq \frac{r^{n-1}(|a_{n-1}| + \ldots + |a_2| + |a_1| + |a_0|)}{|a_n| r^n}$$

$$= \frac{|a_{n-1}| + \ldots + |a_2| + |a_1| + |a_0|}{|a_n| r}$$

Choosing r sufficiently large we obtain

$$\left|\frac{g(z)}{f(z)}\right| < 1 \quad \text{or}$$

$$|g(z)| < |f(z)| \quad \text{on C} \tag{5}$$

The assumptions of Rouché's theorem are met, therefore

$$f(z) + g(z) = P_n(z) \quad \text{and}$$

$$f(z) = a_n z^n \tag{6}$$

have the same number of zeros inside C. We can choose C large enough to include all zeros of $f(z)+g(z)$ and $f(z)$. Since $f(z) = a_n z^n$ has n zeros located at $z = 0$, we conclude that $f(z)+g(z) = P_n(z)$ also has n zeros.

• **PROBLEM 12-26**

Show that all the roots of

$$2z^5 - z^3 + z + 7 = 0 \tag{1}$$

are located between the circles

$$|z| = 1 \quad \text{and} \quad |z| = 2.$$

Solution: First, we shall show that (1) has no solutions inside circle $|z| = 1$. Indeed, let

$$f(z) = 7 \quad \text{and} \quad g(z) = 2z^5 - z^3 + z \tag{2}$$

On $|z| = 1$ we have

$$|g(z)| = |2z^5 - z^3 + z| \leq 2|z^5| + |z^3| + |z|$$
$$= 4 < 7 = |f(z)| \tag{3}$$

Thus, by Rouché's theorem $f(z) + g(z) = 2z^5 - z^3 + z + 7$ has the same number of zeros inside $|z| = 1$ as $f(z) = 7$, that is there are no zeros inside $|z| = 1$.

For the circle $|z| = 2$ let

$$f(z) = 2z^5 \quad \text{and} \quad g(z) = -z^3 + z + 7 \tag{4}$$

Thus,

$$|g(z)| = |-z^3 + z + 7| \leq |z^3| + |z| + 7$$
$$= 8 + 2 + 7 < 2 \cdot 2^5 = |f(z)| \tag{5}$$

Again, by Rouché's theorem $f(z) + g(z) = 2z^5 - z^3 + z + 7$ has the same number of zeros inside $|z| = 2$ as $f(z) = 2z^5$, i.e., five zeros. We conclude that all roots of (1) are located between the circles $|z| = 1$ and $|z| = 2$.

• **PROBLEM 12-27**

Prove Poisson's integral formulas for the circle.

I. If $f(z)$ is analytic inside and on the circle C given by $|z| = R$, then for any point $z = re^{i\theta}$ inside C

$$f(z) = f(re^{i\theta}) = \frac{1}{2\pi} \int_0^{2\pi} \frac{(R^2 - r^2)f(Re^{i\alpha})}{R^2 - 2Rr\cos(\theta - \alpha) + r^2} d\alpha \tag{1}$$

II. If $u(r,\theta)$ and $v(r,\theta)$ are real and imaginary parts of $f(r,\theta)$, then

$$u(r,\theta) = \frac{1}{2\pi} \int_0^{2\pi} \frac{(R^2 - r^2)u(R,\alpha)}{R^2 - 2Rr\cos(\theta - \alpha) + r^2} d\alpha \tag{2}$$

$$v(r,\theta) = \frac{1}{2\pi} \int_0^{2\pi} \frac{(R^2-r^2)v(R,\alpha)}{R^2-2Rr\cos(\theta-\alpha)+r^2} d\alpha \qquad (3)$$

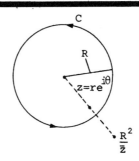

Fig. 1

<u>Solution</u>: I. Let $z = re^{i\theta}$ be any point inside C. By Cauchy's integral formula

$$f(z) = f(re^{i\theta}) = \frac{1}{2\pi i} \int_C \frac{f(\omega)}{\omega - z} d\omega \qquad (4)$$

For the inverse of the point $z = re^{i\theta}$ with respect to C, we have

$$0 = \frac{1}{2\pi i} \int_C \frac{f(\omega)}{\omega - \frac{R^2}{\bar{z}}} d\omega \qquad (5)$$

Note that the inverse $\frac{R^2}{\bar{z}}$ lies outside C. Subtracting (5) from (4) we find

$$f(z) = \frac{1}{2\pi i} \int_C \left[\frac{f(\omega)}{\omega - z} - \frac{f(\omega)}{\omega - \frac{R^2}{\bar{z}}} \right] d\omega$$

$$= \frac{1}{2\pi i} \int_C \frac{z - \frac{R^2}{\bar{z}}}{(\omega - z)(\omega - \frac{R^2}{\bar{z}})} f(\omega) d\omega \qquad (6)$$

Substituting

$$z = re^{i\theta}, \quad \omega = Re^{i\alpha}$$
$$\bar{z} = re^{-i\theta} \qquad (7)$$

into (6) we obtain

$$f(re^{i\theta}) = \frac{1}{2\pi i} \int_0^{2\pi} \frac{(re^{i\theta} - \frac{R^2}{r}e^{i\theta})f(Re^{i\alpha})iRe^{i\alpha}}{(Re^{i\alpha}-re^{i\theta})(Re^{i\alpha}-\frac{R^2}{r}e^{i\theta})} d\alpha$$

$$= \frac{1}{2\pi} \int_0^{2\pi} \frac{(r^2-R^2)e^{i(\theta+\alpha)}f(Re^{i\alpha})}{(Re^{i\alpha}-re^{i\theta})(re^{i\alpha}-Re^{i\theta})} d\alpha \qquad (8)$$

$$= \frac{1}{2\pi} \int_0^{2\pi} \frac{(R^2-r^2)f(Re^{i\alpha})}{R^2-2Rr\cos(\theta-\alpha)+r^2} \, d\alpha$$

q.e.d.

II. The function $f(re^{i\theta})$ can be written as

$$f(re^{i\theta}) = u(r,\theta) + iv(r,\theta) \qquad (9)$$

For $r = R$, (9) becomes

$$f(Re^{i\alpha}) = u(R,\alpha) + iv(R,\alpha) \qquad (10)$$

Substituting (9) and (10) into (8) we get

$$u(r,\theta) + iv(r,\theta)$$

$$= \frac{1}{2\pi} \int_0^{2\pi} \frac{(R^2-r^2)[u(R,\alpha)+iv(R,\alpha)]}{R^2-2Rr\cos(\theta-\alpha)+r^2} \, d\alpha \qquad (11)$$

$$= \frac{1}{2\pi} \int_0^{2\pi} \frac{(R^2-r^2)u(R,\alpha)}{R^2-2Rr\cos(\theta-\alpha)+r^2} \, d\alpha + \frac{i}{2\pi} \int_0^{2\pi} \frac{(R^2-r^2)v(R,\alpha)}{R^2-2Rr\cos(\theta-\alpha)+r^2} \, d\alpha$$

Equating real and imaginary parts in (11), we obtain (2) and (3).

q.e.d.

● **PROBLEM 12-28**

Let $f(z)$ be analytic in the upper half $y \geq 0$ of the z plane. Then, for any point $\rho = \zeta + i\eta$ in the upper half

$$f(\rho) = \frac{1}{\pi} \int_{-\infty}^{\infty} \frac{\eta f(x)}{(x-\zeta)^2+\eta^2} \, dx \qquad (1)$$

This equation can be written in terms of real and imaginary parts of $f(\rho)$

$$u(\zeta,\eta) = \frac{1}{\pi} \int_{-\infty}^{\infty} \frac{\eta u(x,0)}{(x-\zeta)^2+\eta^2} \, dx \qquad (2)$$

$$v(\zeta,\eta) = \frac{1}{\pi} \int_{-\infty}^{\infty} \frac{\eta v(x,0)}{(x-\zeta)^2+\eta^2} \, dx \qquad (3)$$

Note that, these formulas express the values of a harmonic function in the upper half of the z plane in terms of the values on the x axis - the boundary of the half plane.

Solution: Let $\rho = \zeta + i\eta$, be a point in the upper half plane and C the boundary of a semicircle of radius R containing ρ as an interior point. The point $\bar{\rho} = \zeta - i\eta$ is located in the lower half plane.

By Cauchy's integral formula

$$f(\rho) = \frac{1}{2\pi i} \int_C \frac{f(z)}{z - \rho} dz \quad (4)$$

$$0 = \frac{1}{2\pi i} \int_C \frac{f(z)}{z - \bar{\rho}} dz \quad (5)$$

Subtracting (5) from (4) we get

$$f(\rho) = \frac{1}{2\pi i} \int_C \left[\frac{1}{z-\rho} - \frac{1}{z-\bar{\rho}} \right] f(z) dz$$

$$= \frac{1}{2\pi i} \int_C \frac{(\rho-\bar{\rho})f(z)}{(z-\rho)(z-\bar{\rho})} dz \quad (6)$$

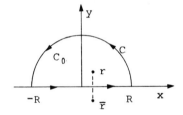

Fig. 1

Semicircle C consists of the arc C_0 and the segment $[-R, +R]$ on the x axis. Hence

$$f(\rho) = \frac{1}{\pi} \int_{-R}^{R} \frac{\eta f(x)}{(x-\zeta)^2 + \eta^2} dx + \frac{1}{\pi} \int_{C_0} \frac{\eta f(z)}{(z-\rho)(z-\bar{\rho})} dz \quad (7)$$

As $R \to \infty$, it can be shown, the last integral approaches zero.

Thus, (7) becomes

$$f(\rho) = \frac{1}{\pi} \int_{-\infty}^{\infty} \frac{\eta f(x)}{(x-\zeta)^2 + \eta^2} dx \quad (8)$$

For the real and imaginary parts

$$f(\rho) = f(\zeta + i\eta) = u(\zeta, \eta) + iv(\zeta, \eta)$$
$$f(x) = u(x, 0) + iv(x, 0) \quad (9)$$

eq.(8) yields

$$u(\zeta,\eta) = \frac{1}{\pi} \int_{-\infty}^{\infty} \frac{\eta u(x,0)}{(x-\zeta)^2+\eta^2} dx \tag{10}$$

$$v(\zeta,\eta) = \frac{1}{\pi} \int_{-\infty}^{\infty} \frac{\eta v(x,0)}{(x-\zeta)^2+\eta^2} dx \tag{11}$$

• **PROBLEM 12-29**

Let C, C_j, $j=1,\ldots,n$ be simple closed curves, each described in the positive direction and such that each C_j is interior to C and exterior to C_ℓ for all $j \neq \ell$, $j,k = 1,\ldots,n$.

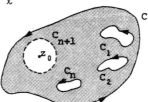

Fig. 1

Let $f(z)$ be analytic on each of the curves C, C_j, $j=1,\ldots,n$ and at each point interior to C and exterior to all the C_j. Show that for any point z_0 interior to C and exterior to each C_j we have (1)

$$f^{(k)}(z_0) = \frac{k!}{2\pi i} \int_C \frac{f(z)}{(z-z_0)^{k+1}} dz - \sum_{j=1}^{n} \frac{k!}{2\pi i} \int_{C_j} \frac{f(z)}{(z-z_0)^{k+1}} dz$$

for $k = 0,1,2,\ldots$.

Solution: We draw a small circle C_{n+1} with center z_0, lying inside C and exterior to $C_1, C_2, \ldots, C_n$, Fig. 1.

Consider the function $\frac{f(z)}{(z-z_0)^{k+1}}$, which is analytic on each of the curves $C, C_1, \ldots, C_n, C_{n+1}$ and at each point interior to C and exterior to $C_1, C_2, \ldots, C_{n+1}$. By Problem 12-3 we have (2)

$$\int_C \frac{f(z)}{(z-z_0)^{k+1}} dz = \sum_{j=1}^{n} \int_{C_j} \frac{f(z)}{(z-z_0)^{k+1}} dz + \int_{C_{n+1}} \frac{f(z)}{(z-z_0)^{k+1}} dz$$

The function $f(z)$ is analytic inside and on C_{n+1}, hence since $z_0 \in C_{n+1}$

$$f^{(k)}(z_0) = \frac{k!}{2\pi i} \int_{C_{n+1}} \frac{f(z)}{(z-z_0)^{k+1}} dz \tag{3}$$

Combining eq.(2) and (3), we find (4)

$$f^{(k)}(z_0) = \frac{k!}{2\pi i} \int_C \frac{f(z)}{(z-z_0)^{k+1}} dz - \sum_{j=1}^{n} \frac{k!}{2\pi i} \int_{C_j} \frac{f(z)}{(z-z_0)^{k+1}} dz$$

CHAPTER 13

POWER SERIES

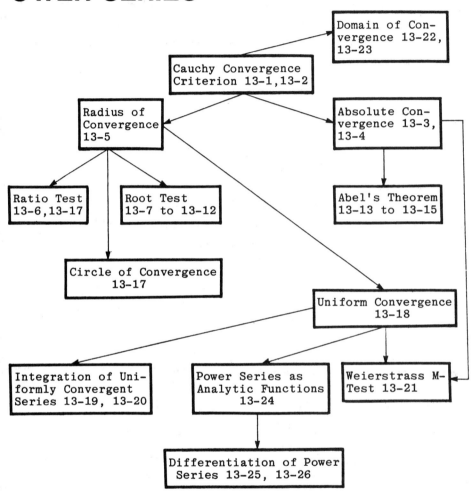

This chart is provided to facilitate rapid understanding of the interrelationships of the topics and subject matter in this chapter. Also shown are the problem numbers associated with the subject matter.

CAUCHY CONVERGENCE CRITERION

● **PROBLEM 13-1**

One of the most important properties of convergent series is the Cauchy convergence criterion. For the real series it can be formulated as follows:

The real series $\sum_{n=0}^{\infty} x_n$ is convergent if for every $\varepsilon > 0$

there exists a positive number N, such that for every $n > N$ and for every positive m

$$|x_{n+1} + x_{n+2} + \ldots + x_{n+m}| < \varepsilon \qquad (1)$$

Note that this criterion does not make use of the notion of sum of the series. Find its analogue for the complex series and prove it.

Solution: <u>Cauchy Convergence Criterion for Complex Series</u>

$$\left(\begin{array}{l}\text{The complex series } \sum_{n=0}^{\infty} z_n \\ \text{is convergent}\end{array}\right) \iff \left(\begin{array}{l} \wedge \ \vee \ \wedge \ \wedge \\ \varepsilon>0 \ N \ n>N \ m \\ |z_{n+1} + z_{n+2} + \ldots + z_{n+m}| < \varepsilon \end{array}\right) \qquad (2)$$

$\Rightarrow$

Since $z_n = u_n + iv_n$ and $\sum_{n=0}^{\infty} z_n$ converge, both series Σu_n and Σv_n converge. By Cauchy convergence criterion for real series, taking $\frac{\varepsilon}{2} > 0$ we obtain

$$|u_{n+1} + \ldots + u_{n+m}| < \frac{\varepsilon}{2} \quad \text{for } n > N_1 \qquad (3)$$

$$|v_{n+1} + \ldots + v_{n+m}| < \frac{\varepsilon}{2} \quad \text{for } n > N_2 \qquad (4)$$

Taking

$$N > \max\{N_1, N_2\}$$

it follows from (3) and (4)

$$|z_{n+1} + \ldots + z_{n+m}| \leq |u_{n+1} + \ldots + u_{n+m}| + |v_{n+1} + \ldots + v_{n+m}| \qquad (5)$$

$$< \frac{\varepsilon}{2} + \frac{\varepsilon}{2} = \varepsilon$$

$\Leftarrow$

Let $\varepsilon > 0$, then N exists such that for all $n > N$ and for all m

$$|z_{n+1} + \ldots + z_{n+m}| < \varepsilon \qquad (6)$$

Hence,

$$|u_{n+1} + \ldots + u_{n+m}| \leq |z_{n+1} + \ldots + z_{n+m}| < \varepsilon, \qquad (7)$$

where u_n is the real part of z_n. By Cauchy criterion for real series, we conclude that $\sum_{n=0}^{\infty} u_n$ is convergent. By the same token, $\sum_{n=0}^{\infty} v_n$ is convergent.

Therefore, $\sum_{n=0}^{\infty} z_n$ is convergent.

• **PROBLEM 13-2**

Prove the following theorem:

$$\left(\sum_{n=0}^{\infty} z_n \text{ is convergent}\right) \Rightarrow \begin{cases} 1. & \lim_{n \to \infty} z_n = 0 \\ 2. & \bigvee_{M>0} \bigwedge_n |z_n| < M \end{cases}$$

<u>Solution</u>: 1. We shall apply Cauchy criterion (see Problem 13-1). Taking m=1, we obtain

$$|z_{n+1}| < \varepsilon \tag{1}$$

for all $n > N$.

Hence, all but a finite number of terms $\{z_1, z_2, \ldots, z_N\}$ satisfy $|z_n| < \varepsilon$. Therefore,

$$\lim_{n \to \infty} z_n = 0 \tag{2}$$

2. Since all but a finite number of terms satisfy

$$|z_n| < \varepsilon, \tag{3}$$

we can take M

$$M > \max\{|z_1|, |z_2|, \ldots, |z_N|, \varepsilon\} \tag{4}$$

Then

$$\bigwedge_n |z_n| < M \tag{5}$$

i.e., the terms z_n are bounded.

• **PROBLEM 13-3**

Show that:

If the series $\sum_{n=0}^{\infty} c_n z^n$ converges at the point z_1, then it converges absolutely for all points z, such that

$$|z| < |z_1| \tag{1}$$

<u>Solution</u>: Applying the theorem proved in Problem 13-2, we arrive at the conclusion that since $\sum c_n z_1^n$ converges, its terms are bounded; i.e., M exists such that

$$|c_n z_1^n| < M \tag{2}$$

We have
$$\left|c_n z^n\right| = \left|c_n z_1^n \frac{z^n}{z_1^n}\right| < M\left|\frac{z}{z_1}\right|^n = Mr^n \tag{3}$$

We denoted
$$r = \left|\frac{z}{z_1}\right| \tag{4}$$

Since $r < 1$, the geometric series $\sum_{n=0}^{\infty} r^n$ converges.

Therefore, the series $\sum_{n=0}^{\infty} |c_n z^n|$ converges.

$\sum_{n=0}^{\infty} c_n z^n$ converges absolutely for $|z| < |z_1|$.

Observe that we can replace z by $z - z_0$, and z_1 by $z_1 - z_0$ everywhere in the proof. Nothing will change. Let us illustrate the theorem (see Fig. 1). The series $\sum_{n=0}^{\infty} c_n (z-z_0)^n$ converges at z_1.

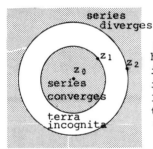

Series converges at z_1, hence it converges everywhere inside the circle $C(z_0, |z_1-z_0|)$. i.e. for all z, $|z_0-z| < |z_0-z_1|$
If the series diverges at z_2, then it diverges everywhere outside the circle $C(z_0, |z_2-z_0|)$.

Fig. 1

We do not know what happens to the series in the region between the circles. We can take z_3 such that
$$|z_1| < |z_3| < |z_2| \tag{5}$$
and determine whether the series converges or diverges at z_3. Depending on the result, the region of convergence or divergence will increase.

● **PROBLEM 13-4**

Show that if $\sum_{n=0}^{\infty} a_n$ is convergent, then $\sum_{n=0}^{\infty} a_n z^n$ is convergent for $|z| < 1$.

<u>Solution</u>: Since $\sum_{n=0}^{\infty} a_n$ is convergent, the series $\sum_{n=0}^{\infty} a_n z^n$ is also convergent for $z_1 = 1$.

By the theorem proved in Problem 13-3, we concluded that the series $\sum_{n=0}^{\infty} a_n z^n$ is absolutely convergent for all z, such that

$$|z| < |z_1| = 1 \tag{1}$$

Hence, the series

$$\sum_{n=0}^{\infty} a_n z^n \tag{2}$$

is convergent for $|z| < 1$.

● **PROBLEM 13-5**

> The following theorem establishes the existence of the radius of convergence of the power series:
>
> If $\sum_{n=0}^{\infty} c_n(z-z_0)^n$ is a power series, then R exists, $0 \leq R \leq \infty$, such that:
>
> 1. the series converges absolutely for all z in the open disk $C(z_0, R)$ if $R > 0$, or at z_0 if $R = 0$.
>
> 2. the series diverges at all points z outside the closed disk $\overline{C(z_0, R)}$, such that
>
> $$|z-z_0| > R \tag{1}$$
>
> Prove this theorem.

<u>Solution</u>: If the series converges only at $z = z_0$, then we set $R = 0$ and the proof is complete. Assume now that $\sum_{n=0}^{\infty} c_n(z-z_0)^n$ converges at some $z_1 \neq z_0$. By Problem 13-3, it converges in the open disk $C(z_0, |z_1-z_0|)$. We can form the union of all open disks $C(z_0, |z-z_0|)$, throughout which the series converges. This union is an open disk $C(z_0, R)$ with the radius R (which can be infinite). The series converges (absolutely) at each point of the disk $C(z_0, R)$, and diverges outside the disk; i.e., for all $|z_0-z| > R$. The series may or may not diverge or converge at z such that $|z_0-z| = R$. Each case must be studied individually.

The circle of convergence can be illustrated as in Fig. 1.

$C(z_0, R)$

At each point of the circle the series can converge or diverge.

series diverges

Fig. 1

Compare it with the drawing in Problem 13-3.

SOME USEFUL THEOREMS • PROBLEM 13-6

Prove the following theorem:

If the power series

$$\sum_{n=0}^{\infty} c_n z^n \quad (1)$$

is such that

$$\lim_{n \to \infty} \left| \frac{c_{n+1}}{c_n} \right| = L \quad (2)$$

exists, then (1) has radius of convergence

$$R = \frac{1}{L} \quad \text{if } L \neq 0 \quad (3)$$

and $R = \infty$ if $L = 0$.

Apply this theorem to show that

$$1 + z + \frac{z^2}{2!} + \frac{z^3}{3!} + \cdots \quad (4)$$

converges in the entire plane.

Solution: We form the ratio

$$\left| \frac{c_{n+1} z^{n+1}}{c_n z^n} \right| = \left| \frac{c_{n+1}}{c_n} \right| |z| \quad (5)$$

Let

$$\alpha = \lim_{n \to \infty} \left| \frac{c_{n+1} z^{n+1}}{c_n z^n} \right| = L|z| \quad (6)$$

If L exists, then α exists. By the ratio test (see Problem 5-19), $\alpha < 1$ if $|z| < \frac{1}{L}$. Hence, the radius of convergence is $R = \frac{1}{L}$. q.e.d.

The coefficients for series (4) are

$$c_n = \frac{1}{n!} \quad (7)$$

Hence,

$$\left| \frac{c_{n+1}}{c_n} \right| = \left| \frac{n!}{(n+1)!} \right| = \frac{1}{n+1} \quad (8)$$

and

$$L = \lim_{n \to \infty} \frac{1}{n+1} = 0 \quad (9)$$

Therefore, the radius of convergence is $R = \infty$.

● **PROBLEM 13-7**

An infinite series of the form

$$c_0 + c_1(z-z_0) + c_2(z-z_0)^2 + \ldots + c_n(z-z_0)^n + \ldots \quad (1)$$

$$= \sum_{n=0}^{\infty} c_n(z-z_0)^n,$$

where z is a complex variable and $z_0, c_0, c_1, c_2, \ldots$ are complex constants, is called a power series in powers of $(z-z_0)$. The numbers $c_0, c_1, c_2, \ldots$ are called the coefficients.

Prove the following Cauchy-Hadamard theorem:

Every power series $\sum_{n=0}^{\infty} c_n(z-z_0)^n$ has the radius of convergence R. When $R = 0$, the series converges only for $z = z_0$. When $0 < R < \infty$, the series converges absolutely for $|z-z_0| < R$, and diverges for $|z-z_0| > R$. When $R = \infty$ the series converges for all z. The radius of convergence is given by

$$R = \frac{1}{\overline{\lim_{n \to \infty}} \sqrt[n]{|c_n|}} \quad (2)$$

Solution: For the definition of the upper limit, see Problem 5-29.

We denote
$$A_n = c_n(z-z_0)^n, \quad (3)$$
then
$$\overline{\lim_{n \to \infty}} \sqrt[n]{|A_n|} = \overline{\lim_{n \to \infty}} \sqrt[n]{|c_n|} \, |z-z_0| = \frac{|z-z_0|}{R} \quad (4)$$

Applying the root test (see Problem 5-33), we conclude that the power series converges absolutely when $\frac{|z-z_0|}{R} < 1$ (that is, when $|z-z_0| < R$), and diverges when

$$\frac{|z-z_0|}{R} > 1, \quad \text{or} \quad |z-z_0| > R \quad (5)$$

When $R = 0$, the series converges only for $z = z_0$. When $R = \infty$, it converges for all values of z. The circle $|z-z_0| = R$ is called the circle of convergence. Note that the Cauchy-Hadamard theorem states nothing about the convergence or divergence of the series for the boundary points of the circle of convergence. For these points, the convergence has to be investigated separately for each case.

● **PROBLEM 13-8**

Find the radius of convergence for:

1. $z + \frac{2^2}{2!} z^2 + \ldots + \frac{n^n}{n!} z^n + \ldots$ (1)

2. $$z + \frac{z^2}{2!}^2 + \ldots + \frac{z^{n^2}}{n!} + \ldots \qquad (2)$$

Hint: Prove and apply this formula:

$$\lim_{n \to \infty} \sqrt[n]{\frac{n!}{n^n}} = \frac{1}{e} \qquad (3)$$

Solution: We have

$$e^n = 1 + \frac{n}{1!} + \frac{n^2}{2!} + \ldots + \frac{n^n}{n!} + \ldots$$

$$= 1 + \frac{n}{1!} + \ldots + \frac{n^{n-1}}{(n-1)!} + \frac{n^n}{n!}\left[1 + \frac{n}{n+1} + \frac{n^2}{(n+1)(n+2)} + \ldots\right] \qquad (4)$$

From (4) we obtain

$$e^n < n\frac{n^n}{n!} + \frac{n^n}{n!}\left[1 + \frac{n}{n+1} + \left(\frac{n}{n+1}\right)^2 + \ldots\right]$$

$$= \frac{n^n}{n!}\left[n + \frac{1}{1 - \frac{n}{n+1}}\right] = (2n+1)\frac{n^n}{n!} \qquad (5)$$

From (4) it follows that

$$e^n > \frac{n^n}{n!} \qquad (6)$$

Thus,

$$\frac{1}{e^n} < \frac{n!}{n^n} < \frac{2n+1}{e^n} \qquad (7)$$

and

$$\frac{1}{e} < \sqrt[n]{\frac{n!}{n^n}} < \frac{\sqrt[n]{2n+1}}{e} \qquad (8)$$

It follows from (8) that

$$\lim_{n \to \infty} \sqrt[n]{\frac{n!}{n^n}} = \frac{1}{e} \qquad (9)$$

For series (1), the radius of convergence is

$$R = \frac{1}{\overline{\lim_{n \to \infty}} \sqrt[n]{|c_n|}} = \frac{1}{\overline{\lim_{n \to \infty}} \sqrt[n]{\frac{n^n}{n!}}} = \frac{1}{e} \qquad (10)$$

Now consider

$$z + \frac{z^2}{2!}^2 + \ldots + \frac{z^{n^2}}{n!} + \ldots \qquad (11)$$

Note that

$$c_{n^2} = \frac{1}{n!}, \qquad (12)$$

and the radius of convergence is

$$\frac{1}{R} = \overline{\lim_{n\to\infty}} \sqrt[n^2]{|C_{n^2}|} = \lim_{n\to\infty} \sqrt[n^2]{\frac{1}{n!}}$$

$$= \lim_{n\to\infty} \sqrt[n]{\sqrt[n]{\frac{n^n}{n!} \cdot \frac{1}{n^n}}} = \lim_{n\to\infty} \sqrt[n]{\sqrt[n]{\frac{n^n}{n!} \cdot \frac{1}{n}}} = 1 \qquad (13)$$

● **PROBLEM 13-9**

Find the radius of convergence of the series.

1. $\sum_{n=0}^{\infty} z^{n^2}$ (1)

2. $\sum_{n=0}^{\infty} \frac{1}{n^s} z^n, \quad s \geq 0$ (2)

3. $\sum_{n=0}^{\infty} \frac{1}{n!} z^n$ (3)

Solution: 1. The coefficients of series (1) $1+z+z^4+z^9+\ldots$ are

$$c_n = \begin{cases} 1 & \text{if } n = m^2 \\ 0 & \text{otherwise} \end{cases} \qquad (4)$$

Thus

$$\sqrt[n]{|c_n|} = \begin{cases} 1 \\ 0 \end{cases} \qquad (5)$$

Sequence

$$|c_1|, \sqrt{|c_2|}, \sqrt[3]{|c_3|}, \ldots, \sqrt[n]{|c_n|} \qquad (6)$$

has two limit points 0 and 1. Thus,

$$R = \frac{1}{\overline{\lim_{n\to\infty}} \sqrt[n]{|c_n|}} = 1 \qquad (7)$$

2. For series (2),

$$c_n = \frac{1}{n^s} \qquad (8)$$

and

$$\sqrt[n]{|c_n|} = \frac{1}{n^{\frac{s}{n}}} = \frac{1}{e^{\frac{s \ln n}{n}}} \qquad (9)$$

Since
$$\frac{\ln n}{n} \to 0 \qquad (10)$$

as $n \to \infty$, from (9) we get

$$\sqrt[n]{|c_n|} \to 1, \qquad (11)$$

and
$$R = 1$$

3. In the case of series (3), we will show that
$$\lim_{n\to\infty} \sqrt[n]{|c_n|} = 0 \qquad (12)$$

Note that
$$(n!)^2 = (1 \cdot 2 \cdot 3 \cdot \ldots \cdot (n-1)n)(n \cdot (n-1) \cdot \ldots \cdot 2 \cdot 1) \qquad (13)$$
$$= (1 \cdot n)(2(n-1))(3 \cdot (n-2)) \cdot \ldots \cdot (n \cdot 1)$$

For k-th term in (13), we have
$$k \cdot (n-k+1) \geq n \qquad (14)$$

because
$$k(n-k+1) - n = (k-1)(n-k) \geq 0 \qquad (15)$$

Hence, from (13) it follows
$$(n!)^2 \geq n^n$$

or
$$n! \geq (\sqrt{n})^n$$

or
$$\sqrt[n]{n!} \geq \sqrt{n}$$

and finally
$$\sqrt[n]{\frac{1}{n!}} \leq \frac{1}{\sqrt{n}} \qquad (16)$$

Thus,
$$\lim_{n\to\infty} \sqrt[n]{|c_n|} = 0 \qquad (17)$$

and the radius of convergence is $R = +\infty$.

● PROBLEM 13-10

Find the radius of convergence of the series.

1. $1 - \frac{z^2}{2!} + \frac{z^4}{4!} - \frac{z^6}{6!} + \ldots$ \hfill (1)

2. $z - \frac{z^3}{3!} + \frac{z^5}{5!} - \frac{z^7}{7!} + \ldots$ \hfill (2)

3. $1 + z + 2!z^2 + 3!z^3 + \ldots$ \hfill (3)

Solution: 1. The coefficients of series (1) are

$$1, 0, -\frac{1}{2!}, 0, \frac{1}{4!}, 0, -\frac{1}{6!}, 0 \ldots \qquad (4)$$

Hence,

$$R = \frac{1}{\varlimsup_{n\to\infty} \sqrt[n]{|c_n|}} = \frac{1}{\lim_{n\to\infty} \sqrt[n]{\frac{1}{n!}}} = \infty \qquad (5)$$

Series (1) converges in the whole complex plane.

2. For series (2), the radius of convergence is

$$R = \frac{1}{\varlimsup_{n\to\infty} \sqrt[n]{|c_n|}} = \frac{1}{\lim_{n\to\infty} \sqrt[n]{\frac{1}{n!}}} = \infty \qquad (6)$$

In both cases, we utilized the inequality

$$\sqrt[n]{n!} \geq \sqrt{n} \qquad (7)$$

3. $$R = \frac{1}{\varlimsup_{n\to\infty} \sqrt[n]{|c_n|}} = \frac{1}{\lim_{n\to\infty} \sqrt[n]{n!}} = 0 \qquad (8)$$

Series (3) converges only at the point $z = 0$.

● **PROBLEM 13-11**

The radius of convergence of the series

$$\sum_{n=0}^{\infty} c_n z^n \qquad (1)$$

is $0 < R < \infty$.

Find the radii of convergence of the following series:

1. $\sum_{n=0}^{\infty} c_n z^{kn}$ (2)

2. $\sum_{n=0}^{\infty} \frac{c_n}{n!} z^n$ (3)

3. $\sum_{n=0}^{\infty} c_n^k z^n$ (4)

k is a natural number.

Solution: 1. By definition,

$$R = \frac{1}{\overline{\lim} \sqrt[n]{|c_n|}} \tag{5}$$

Hence, for series (2) we obtain

$$R_{(2)} = \frac{1}{\overline{\lim_{n\to\infty}} \sqrt[nk]{|c_n|}} = \frac{1}{\overline{\lim_{n\to\infty}}\left[\sqrt[n]{|c_n|}\right]^{\frac{1}{k}}}$$

$$= \left(\frac{1}{\overline{\lim_{n\to\infty}} \sqrt[n]{|c_n|}}\right)^{\frac{1}{k}} = R^{\frac{1}{k}} \tag{6}$$

2. $$R_{(3)} = \frac{1}{\overline{\lim_{n\to\infty}} \sqrt[n]{\frac{|c_n|}{n!}}} = \frac{1}{\overline{\lim_{n\to\infty}} \frac{\sqrt[n]{|c_n|}}{\sqrt[n]{n!}}} = \infty \tag{7}$$

Remember that $\lim_{n\to\infty} \sqrt[n]{n!} = \infty$.

3. $$R_{(4)} = \frac{1}{\overline{\lim_{n\to\infty}} \sqrt[n]{|c_n^k|}} = \frac{1}{\overline{\lim_{n\to\infty}}\left[\sqrt[n]{|c_n|}\right]^k} = R^k \tag{8}$$

• **PROBLEM 13-12**

The two power series

$$\sum_{n=0}^{\infty} c_n z^n \quad \text{and} \quad \sum_{n=0}^{\infty} c_n' z^n \tag{1}$$

have radii of convergence R and R' respectively. What is the relationship between R, R' and the radius of convergence r of each of the following series:

1. $$\sum_{n=0}^{\infty} (c_n + c_n') z^n \tag{2}$$

2. $$\sum_{n=0}^{\infty} c_n c_n' z^n \tag{3}$$

3. $$\sum_{n=0}^{\infty} \frac{c_n}{c_n'} z^n \quad (c_n' \neq 0) \tag{4}$$

Solution: 1. The circles of convergence are shown in Fig. 1.

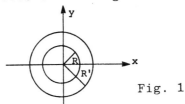

Fig. 1

The series $\sum_{n=0}^{\infty} (c_n+c_n')z^n$ converges for all $|z| < \min\{R,R'\}$.
Hence,

$$r \geq \min\{R,R'\} \tag{5}$$

We can reach the same conclusion from

$$r = \frac{1}{\overline{\lim_{n\to\infty}} \sqrt[n]{|c_n+c_n'|}} \geq \frac{1}{\overline{\lim_{n\to\infty}} \left(\sqrt[n]{|c_n|} + \sqrt[n]{|c_n'|}\right)} \tag{6}$$

2. From the Cauchy-Hadamard theorem,

$$r = \frac{1}{\overline{\lim_{n\to\infty}} \sqrt[n]{|c_n c_n'|}} \geq \frac{1}{\overline{\lim_{n\to\infty}} \sqrt[n]{|c_n|}} \cdot \frac{1}{\overline{\lim_{n\to\infty}} \sqrt[n]{|c_n'|}} \tag{7}$$

Thus,

$$r \geq R \cdot R' \tag{8}$$

3. We have

$$r = \frac{1}{\overline{\lim_{n\to\infty}} \sqrt[n]{\left|\frac{c_n}{c_n'}\right|}} = \frac{1}{\overline{\lim_{n\to\infty}} \frac{\sqrt[n]{|c_n|}}{\sqrt[n]{|c_n'|}}} \leq \frac{R}{R'} \tag{9}$$

• **PROBLEM 13-13**

> Prove the following theorem:
>
> If the power series
>
> $$s(z) = \sum_{n=0}^{\infty} c_n z^n \tag{1}$$
>
> whose circle of convergence is R converges at z_0, $|z_0| = R$, then $s(z)$ approaches $s(z_0)$ as $z \to z_0$ along the radius $0z_0$
>
> $$\lim_{\substack{z \to z_0 \\ \text{along } 0z_0}} s(z) = s(z_0) \tag{2}$$
>
> This theorem is known as Abel's theorem.

Solution: Without losing generality of this theorem we can assume that $s(z_0) = 0$ and that $z_0 = 1 = R$. Then $z \to 1$, that is z approaches 1 along the segment [0,1] on the real axis. Indeed if $s(z_0) = A \neq 0$ we can write

$$S'(z) = C_0 - S(z_0) + \sum_{n=1}^{\infty} c_n z^n \tag{3}$$

and then
$$s'(z_0) = 0 \qquad (4)$$

We shall apply

Theorem

If two series

$$z_1+z_2+\ldots+z_n+\ldots$$

and
$$z_1'+z_2'+\ldots+z_n'+\ldots$$

are absolutely convergent with sums s and s', respectively, then the series

$$z_1z_1' + (z_1z_2'+z_2z_1') + \ldots + (z_1z_n'+z_2z_{n-1}'+\ldots+z_nz_1')+\ldots$$

is absolutely convergent with the sum ss'.

Thus, multiplying (1) by the geometric series

$$\frac{1}{1-z} = 1 + z + z^2 + \ldots + z^n + \ldots \qquad (5)$$

we obtain
$$\frac{s(z)}{1-z} = a_0 + a_1z + a_2z^2 + \ldots + a_nz^n + \ldots \qquad (6)$$

where
$$a_n = c_0 + c_1 + \ldots + c_n \qquad (7)$$

Since $s(z)$ has radius of convergence one, the series (6) has radius of convergence one.

Let $\varepsilon > 0$ be an arbitrary positive number, then $N(\varepsilon)$ exists such that

$$|a_n| < \frac{\varepsilon}{2} \quad \text{for every } n > N(\varepsilon)$$

we have
$$|s(z)| = \left|(1-z) \sum_{n=0}^{\infty} a_n z^n\right|$$

$$= \left|(1-z) \sum_{n=0}^{N(\varepsilon)} a_n z^n + (1-z) \sum_{n=N(\varepsilon)+1}^{\infty} a_n z^n\right|$$

$$\leq (1-z) \sum_{n=0}^{N(\varepsilon)} |a_n z^n| + (1-z) \sum_{n=N(\varepsilon)+1}^{\infty} |a_n z^n| \qquad (8)$$

Denoting
$$K = |a_0| + |a_1| + \ldots + |a_{N(\varepsilon)}| \qquad (9)$$

and keeping in mind that $0 < z < 1$ and that $\sum_{n=N(\varepsilon)+1}^{\infty} z^n = \frac{z^{N(\varepsilon)+1}}{1-z} \qquad (10)$

we obtain from (8)

$$s(z) < (1-z)K + (1-z)\frac{\varepsilon}{2}\sum_{n=N(\varepsilon)+1}^{\infty} z^n \quad (11)$$

$$= (1-z)K + \frac{\varepsilon}{2} z^{N(\varepsilon)+1} < (1-z)K + \frac{\varepsilon}{2}$$

Taking z sufficiently close to 1, so that

$$1 - z < \frac{\varepsilon}{2k} \quad (12)$$

we get from (11)

$$s(z) < (1-z)K + \frac{\varepsilon}{2} < \frac{\varepsilon}{2} + \frac{\varepsilon}{2} = \varepsilon \quad (13)$$

Hence

$$\lim_{z \to 1_-} s(z) = 0 = s(z_0) \quad (14)$$
$$\text{q.e.d.}$$

Note that $z \to 1_-$ indicates that $z \to 1$ in such way that $z < 1$.

● **PROBLEM 13-14**

Show that

$$\ln 2 = 1 - \frac{1}{2} + \frac{1}{3} - \frac{1}{4} + \ldots \quad (1)$$

Hint: Use Abel's theorem.

Solution: We shall use the formula for expansion of $\ln$ around 1:

$$\ln(1+x) = x - \frac{x^2}{2} + \frac{x^3}{3} - \ldots, \quad (2)$$

where $|x| < 1$.

The circle of convergence of series (2) is $R = 1$.

Since series (2) is convergent for $x_0 = 1$, then $\ln(1+x)$ approaches $\ln(1+1)$ as $x \to x_0 = 1$ along the segment $[0,1]$.

$$\lim_{x \to 1_-} \ln(1+x) = \ln 2 \quad (3)$$

Hence, $\ln 2 = 1 - \frac{1}{2} + \frac{1}{3} - \frac{1}{4} + \ldots \quad (4)$

● **PROBLEM 13-15**

Use the series

$$\frac{1}{1+z} = 1 - z + z^2 - \ldots \quad (1)$$

to prove that the converse of Abel's theorem is false.

Solution: The radius of convergence is

$$R = \frac{1}{\overline{\lim} \sqrt[n]{|c_n|}} = 1 \tag{2}$$

Taking the limit of $\frac{1}{1+z}$ as $z \to 1$, we obtain

$$\lim_{z \to 1} \frac{1}{1+z} = \frac{1}{2} \tag{3}$$

However, the power series

$$1 - z + z^2 - z^3 + \ldots \tag{4}$$

does not converge to $\frac{1}{2}$ as z approaches 1. Hence, the converse of Abel's theorem is not true.

● **PROBLEM 13-16**

Prove Tauber's theorem:

Let $s(z) = \sum_{n=0}^{\infty} c_n z^n$ be the power series, such that

$$\lim_{n \to \infty} n c_n = 0, \tag{1}$$

and

$$\lim_{z \to 1} s(z) = M \quad (0 < z < 1) \tag{2}$$

Then $\sum_{n=0}^{\infty} c_n$ converges to M. $\tag{3}$

Solution: We have

$$\sum_{n=0}^{m} c_n - s(z) = \sum_{n=0}^{m} c_n(1-z^n) - \sum_{n=m+1}^{\infty} c_n z^n \tag{4}$$

Since

$$1 + z + \ldots + z^n = \frac{1-z^{n+1}}{1-z}, \tag{5}$$

eq.(4) becomes

$$= \sum_{n=0}^{m} c_n(1-z)(1+z+\ldots+z^{n-1}) - \sum_{n=m+1}^{\infty} c_n z^n$$

$$\leq (1-z) \sum_{n=0}^{m} |c_n|(1+z+\ldots+z^{n-1}) + \sum_{n=m+1}^{\infty} |c_n| z^n \tag{6}$$

$$< (1-z)m \frac{1}{m} \sum_{n=0}^{m} n|c_n| + \sum_{n=m+1}^{\infty} n|c_n|\frac{z^n}{n}$$

It was proved in Problem 5-5 that if $\lim_{n\to\infty} z_n = 0$, then

$$\lim_{n\to\infty} \frac{z_1+\ldots+z_n}{n} = 0 \tag{7}$$

Since

$$\lim_{n\to\infty} n|c_n| = 0, \tag{8}$$

then

$$\lim_{m\to\infty} \left[\frac{1}{m} \sum_{n=0}^{m} n|c_n| \right] = 0 \tag{9}$$

Therefore,

$$\lim_{m\to\infty} \sum_{n=0}^{m} c_n = \lim_{z\to 1} s(z) = M \tag{10}$$

q.e.d.

● **PROBLEM 13-17**

Investigate the series

$$s(z) = \sum_{n=1}^{\infty} \frac{z^n}{n} \tag{1}$$

on its circle of convergence.

Solution: The circle of convergence can be evaluated from the ratio test

$$L = \lim_{n\to\infty} \left|\frac{c_{n+1}}{c_n}\right| = \lim_{n\to\infty} \frac{n}{n+1} = 1 \tag{2}$$

Hence, the radius of convergence is one.

On the circle $|z| = 1$, the series is not absolutely convergent, because the series $\Sigma \frac{1}{n}$ diverges.

Let z be a complex number on the circle $|z| = 1$, then

$$z = \cos\theta + i\sin\theta \tag{3}$$

and

$$\sum_{n=1}^{\infty} \frac{z^n}{n} = \sum_{n=1}^{\infty} \frac{(\cos\theta + i\sin\theta)^n}{n}$$

$$= \sum_{n=1}^{\infty} \frac{\cos n\theta}{n} + i \sum_{n=1}^{\infty} \frac{\sin n\theta}{n} \tag{4}$$

Both series in (4) are Fourier series.

The series $\sum_{n=1}^{\infty} \frac{\cos n\theta}{n}$ diverges for $\theta = 0$. For $\theta \neq 0$, the series is the Fourier series of the function

$$\frac{1}{2} \log \frac{1}{2-2\cos\theta} \qquad (5)$$

It converges to (5) for all values of θ, except for $\theta = 0$.

The series $\sum_{n=1}^{\infty} \frac{\sin n\theta}{n}$ converges everywhere.

$$\sum_{n=1}^{\infty} \frac{\sin n\theta}{n} = \begin{cases} -\frac{\pi}{2} - \frac{\theta}{2}, & \text{for } -\pi \leq \theta < 0 \\ \frac{\pi}{2} - \frac{\theta}{2}, & 0 < \theta < \pi \\ 0, & \text{for } \theta = 0 \end{cases} \qquad (6)$$

Thus, the series $\sum_{n=1}^{\infty} \frac{z^n}{n}$ converges for $|z| \leq 1$, except for $z = 1$.

It can be shown that this is the Taylor series of

$$\ln \frac{1}{1-z} = \sum_{n=1}^{\infty} \frac{z^n}{n}, \quad |z| \leq 1, \quad z \neq 1. \qquad (7)$$

UNIFORM CONVERGENCE

• **PROBLEM 13-18**

Definition

A power series

$$s(z) = c_0 + c_1 z + c_2 z^2 + \ldots \qquad (1)$$

is said to be uniformly convergent in a set S, if

$$\bigwedge_{\epsilon > 0} \bigvee_{N(\epsilon)} \bigwedge_{z \in A} \bigwedge_{n \geq N(\epsilon)}$$

$$\left| \sum_{k=n}^{\infty} c_k z^k \right| = \left| c_n z^n + c_{n+1} z^{n+1} + \ldots \right| < \epsilon \qquad (2)$$

Prove the following theorem:

Let the power series $\sum_{n=0}^{\infty} c_n (z-z_0)^n$ have a radius of convergence $R > 0$. Then, for any disk $\overline{C(z_0, r)}$, $r < R$, the power series $\sum_{n=0}^{\infty} c_n (z-z_0)^n$ converges uniformly within and on $\overline{C}$.

Solution: Let us draw the circle $C_1(z_0, r_1)$ such that $r < r_1 < R$.

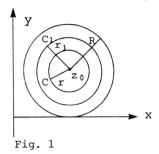

Fig. 1

Since $r < r_1$, $r = \alpha r_1$ where $0 < \alpha < 1$. For z such that $|z - z_0| \le r$,

$$|c_n(z-z_0)^n| \le |c_n| r^n = |c_n| \alpha^n r_1^n \tag{3}$$

Since the power series $\sum_{n=0}^{\infty} c_n(z-z_0)^n$ converges for $|z - z_0| = r_1$, it follows that for sufficiently large n

$$|c_n| r_1^n < 1 \tag{4}$$

Furthermore, since $0 < \lambda < 1$, for any $\varepsilon > 0$

$$\lambda^n < \varepsilon(1-\lambda) \tag{5}$$

for sufficiently large n.

Taking n so large that both (4) and (5) hold, we obtain

$$\left| \sum_{k=n}^{\infty} c_n(z-z_0)^n \right| \le \sum_{k=n}^{\infty} |c_n(z-z_0)^n| \tag{6}$$

$$\le \sum_{k=n}^{\infty} |c_n| \alpha^n r_1^n < \sum_{k=n}^{\infty} \alpha^n$$

$$= \lambda^k \left[1 + \lambda + \lambda^2 + \ldots \right] = \frac{\lambda^k}{1-\lambda} < \frac{\varepsilon(1-\lambda)}{1-\lambda} = \varepsilon$$

for $|z - z_0| \le r$

● **PROBLEM 13-19**

Consider the sequence of functions $\{f_n\}$, which converges uniformly to f on some set S. In most cases, if the functions f_n have some interesting property, one can expect the function of $f = \lim_{n \to \infty} f_n$ to have this property.

For example:

Theorem

If the sequence of continuous functions defined on S converges uniformly to a function f, then f is also a continuous function.

Prove the following theorem:

If $\{f_n\}$ is a sequence of functions continuous on a curve C, which converges uniformly on C to a function $f = \lim_{n \to \infty} f_n$, then

$$\lim_{n \to \infty} \int_C f_n(z)dz = \int_C \lim_{n \to \infty} f_n(z)dz = \int_C f(z)dz \qquad (1)$$

Solution: To prove (1) we have to show that

$$\lim_{n \to \infty} \left| \int_C (f_n(z) - f(z))dz \right| = 0 \qquad (2)$$

Let M be an arbitrarily small positive number. Since f_n converges uniformly to f, N exists such that

$$\bigwedge_{n > N} \bigwedge_{z \in C} |f_n(z) - f(z)| < M \qquad (3)$$

Applying the basic inequality for the integral, we obtain

$$\left| \int_C (f_n(z) - f(z))dz \right| < ML, \qquad (4)$$

where L is the length of the path. Since M can be made arbitrarily small, we have

$$\lim_{n \to \infty} \left| \int_C (f_n(z) - f(z))dz \right| = 0 \qquad (5)$$

This proves (1).

q.e.d.

• **PROBLEM 13-20**

Compute:

$$\lim_{n \to \infty} \int_{z_0}^{z_1} \left[1 + z + \frac{z^2}{2!} + \ldots + \frac{z^n}{n!} \right] dz \qquad (1)$$

Solution: The series $\sum_{n=0}^{\infty} \frac{z^n}{n!}$ converges for all z to e^z:

$$\sum_{n=0}^{\infty} \frac{z^n}{n!} = e^z \qquad (2)$$

Hence, the radius of convergence is $R = \infty$. Let us choose r, such that the disc $\overline{C(z_0, r)}$ contains z_1.

Each integral in (1) is independent of the path from z_0 to z_1. We can choose path C from z_0 to z_1, such that $C \subseteq \overline{C(z_0,r)}$.

Since $\sum_{n=0}^{\infty} \frac{z^n}{n!}$ (see Problem 13-18) converges everywhere, $R = \infty$. It converges uniformly on $\overline{C(z_0,r)}$. Hence, as shown in Problem 13-19,

$$\lim_{n \to \infty} \int_{z_0}^{z_1} \left[1 + z + \frac{z^2}{2!} + \ldots + \frac{z^n}{n!} \right] dz$$

$$= \int_{z_0}^{z_1} \lim_{n \to \infty} \left(1 + z + \frac{z^2}{2!} + \ldots + \frac{z^n}{n!} \right) dz$$

$$= \int_{z_0}^{z_1} \sum_{n=0}^{\infty} \frac{z^n}{n!} \, dz = \int_{z_0}^{z_1} e^z \, dz \qquad (3)$$

$$= e^{z_1} - e^{z_0}$$

● **PROBLEM 13-21**

Prove the following important test for uniform convergence known as the Weierstrass M-Test:

If the positive numbers $M_0, M_1, \ldots, M_n \ldots$ are such that

$$|f_n(z)| \leq M_n, \quad n = 0, 1, 2, \ldots \qquad (1)$$

for all $z \in D$, and the series $\sum_{n=0}^{\infty} M_n$ converges, then the series

$$f_0(z) + f_1(z) + \ldots + f_n(z) + \ldots \qquad (2)$$

converges uniformly and absolutely throughout D.

Solution: Since the series $\sum_{n=0}^{\infty} M_n$ is independent of z, we talk about uniform convergence of the series $\sum_{n=0}^{\infty} f_n(z)$.

Let $z \in D$, then

$$|f_{n+1}(z) + \ldots + f_{n+m}(z)| \leq |f_{n+1}(z)| + \ldots + |f_{n+m}(z)|$$
$$\leq M_{n+1} + \ldots + M_{n+m} \qquad (3)$$

Since the series ΣM_n is convergent, for any positive $\varepsilon < 0$, there exists $N(\varepsilon)$ such that for any $n > N(\varepsilon)$ and $m > 0$

$$|f_{n+1}(z) + \ldots + f_{n+m}(z)| < \varepsilon \qquad (4)$$

Since inequality (4) is independent of z, the series $\sum_{n=0}^{\infty} f_n(z)$ converges uniformly and absolutely throught D.

● **PROBLEM 13-22**

Find the domain (or domains) of convergence of the series

$$\frac{1}{2} + \sum_{n=1}^{\infty} \left(\frac{1}{1+z^n} - \frac{1}{1+z^{n-1}} \right) \qquad (1)$$

What function (or functions) does this series represent?

Solution: Note that (1) can be written as

$$\lim_{n \to \infty} \frac{1}{1+z^n} = \frac{1}{2} + \sum_{n=1}^{\infty} \left(\frac{1}{1+z^n} - \frac{1}{1+z^{n-1}} \right) \qquad (2)$$

For the disc $|z| < 1$,

$$\lim_{n \to \infty} \frac{1}{1+z^n} = 1 \qquad (3)$$

Hence, for $|z| < 1$, the series (1) is convergent and represents the function 1.

For $|z| > 1$, we obtain

$$\lim_{n \to \infty} \frac{1}{1+z^n} = 0 \qquad (4)$$

The series (1) is convergent for $|z| > 1$, and represents the function 0.

Thus, the series $\frac{1}{2} + \sum_{n=1}^{\infty} \left(\frac{1}{1+z^n} - \frac{1}{1+z^{n-1}} \right)$ has two disjoint domains of convergence D_1 and D_2, $D_1 \cap D_2 = \phi$. The same series in these domains represents two different functions.

● **PROBLEM 13-23**

Find the domain of convergence of each of the following series:

1. $\sum_{n=0}^{\infty} \left(\frac{z+2i}{z+1} \right)^n$ \qquad (1)

2. $\sum_{n=0}^{\infty} \frac{n^i}{(2i)^n}$ \qquad (2)

3. $\sum_{n=0}^{\infty} \frac{n^3}{z^n}$ \qquad (3)

Solution: 1. Note that (1) represents a geometric series which is convergent for

$$\left|\frac{z+2i}{z+1}\right| < 1, \qquad (4)$$

or

$$2y + \frac{3}{2} < x \qquad (5)$$

to

$$\left(1 - \frac{z+2i}{z+1}\right)^{-1} = \frac{z+1}{1-2i} \qquad (6)$$

The region of convergence (shaded area) is shown in Fig. 1.

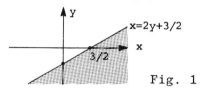

Fig. 1

2. Applying the ratio test, we find

$$L = \lim_{n\to\infty} \left[\frac{\frac{(n+1)^i}{(2i)^{n+1}}}{\frac{n^i}{(2i)^n}}\right] = \lim_{n\to\infty} \frac{(1+\frac{1}{n})^i}{2i}$$

$$= \frac{1^i}{2i} = \frac{1}{2i} \qquad (7)$$

Since $|L| = \frac{1}{2} < 1$, the series converges absolutely.

3. Applying the root test, we find

$$L = \lim_{n\to\infty} \left(\frac{n^3}{z^n}\right)^{\frac{1}{n}} = \frac{1}{z} \lim_{n\to\infty} n^{\frac{3}{n}} = \frac{1}{z} \qquad (8)$$

Hence, the series is absolutely convergent outside the unit circle $\frac{1}{|z|} < 1$, and divergent inside the unit circle.

POWER SERIES OF ANALYTIC FUNCTIONS

● **PROBLEM 13-24**

Prove the following theorem:

Let $\{f_n\}$ be a sequence of analytic functions, defined on a domain D, which converges uniformly on D to $f(z)$.

$$f(z) = \lim_{n\to\infty} f_n(z) \qquad (1)$$

Then f is also analytic on D.

Based on this theorem can you draw any conclusions about a power series?

Solution: Since $\{f_n\}$ are analytic functions, they are also continuous. The functions $\{f_n\}$ converge uniformly on D to f. Referring to Problem 13-19, we have

$$\lim_{n \to \infty} \int_C f_n(z)dz = \int_C f(z)dz \qquad (2)$$

where C is any closed curve contained in any disc inside D. Since all f_n are analytic, the integral $\int_C f_n(z)dz$ vanishes.

Thus, the integral $\int_C f(z)dz$ vanishes. Since f(z) is a continuous function on D, and C is a simple closed curve in D, we conclude from Morera's theorem that f is analytic on D.

From this theorem we can also deduct the following:

A power series defines an analytic function at each point inside its disc of convergence.

We shall prove it.

A power series is the uniform limit of a sequence of polynomials which are analytic functions. Let $z \in D$, where D is the disc of convergence, then $\bar{D}_1$ (a closed subdisc), exists such that

$$z \in \bar{D}_1, \quad \bar{D}_1 \subset D \qquad (3)$$

Thus a function defined by a power series is analytic.

● **PROBLEM 13-25**

Theorem

Let $\{f_n\}$ be a sequence of functions analytic in the domain D, which converge uniformly on each closed subdisc of D, to the analytic function f.

$$f(z) = \lim_{n \to \infty} f_n(z), \quad z \in \bar{D}_0 \subset D \qquad (1)$$

Then the sequence $\{f'_n\}$ of derivatives converges uniformly on each closed subdisc of D to the function f'

$$f'(z) = \lim_{n \to \infty} f'_n(z) \qquad (2)$$

This theorem will lead us to some important conclusions about differentiation of power series. Prove this theorem by applying the integral formula.

Solution: Let $\bar{D}_0$ be any closed subdisk in the domain D, $\bar{D}_0 \subset D$. To prove (2), we have to show that

$$\bigwedge_{\varepsilon>0} \bigvee_{N(\varepsilon)} \bigwedge_{n>N(\varepsilon)} \bigwedge_{\omega\in\bar{D}_0} |f'(\omega)-f'_n(\omega)| < \varepsilon \qquad (3)$$

Applying the integral formula, we obtain

$$f'(\omega)-f'_n(\omega) = \frac{1}{2\pi i} \int_C \frac{f(z)-f_n(z)}{(z-\omega)^2} dz, \qquad (4)$$

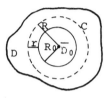

Fig. 1

where C is a circle in D with the same center as $\bar{D}_0$, and radius larger than $\bar{D}_0$.

We define

$$r = R - R_0 > 0 \qquad (5)$$

See Fig. 1, where R is the radius of C, and R_0 the radius of $\bar{D}_0$. Denoting the length of C by L, we obtain the following from (4):

$$|f'(\omega)-f'_n(\omega)| \leq \frac{1}{2\pi} \max_{z\in C} \left|\frac{f(z)-f_n(z)}{(z-\omega)^2}\right| \cdot L \qquad (6)$$

Let

$$\varepsilon' = \frac{2\pi r^2 \varepsilon}{L} \qquad (7)$$

Since $\{f_n\}$ converges uniformly on a closed disc bounded by C to the function f,

$$\bigvee_{N'} \bigwedge_{n>N'} \bigwedge_{z\in C} |f(z)-f_n(z)| < \varepsilon' \qquad (8)$$

for ε'.
Therefore, for $n > N'$, we obtain the following from (6) and (8) for $n > N'$:

$$|f'(\omega)-f'_n(\omega)| \leq \frac{1}{2\pi} \max_{z\in C} \frac{|f(z)-f_n(z)|}{|z-\omega|^2} \cdot L \qquad (9)$$

$$< \frac{1}{2\pi} \cdot \frac{\varepsilon'}{r^2} \cdot L = \varepsilon$$

Eq.(9) is true for every ω in $\bar{D}_0$. Hence, $\{f'_n\}$ converges uniformly to f' on each closed subdisc $\bar{D}_0$, $\bar{D}_0 \subset D$.

q.e.d.

● **PROBLEM 13-26**

We may differentiate the power series

$$c_0 + c_1(z-z_0) + c_2(z-z_0)^2 + \ldots + c_n(z-z_0)^n + \ldots \quad (1)$$

term by term to obtain another power series,

$$c_1 + 2c_2(z-z_0) + \ldots + nc_n(z-z_0)^{n-1} + \ldots \quad (2)$$

Series (1) represents an analytic function $f(z)$ in its circle of convergence. What function does series (2) represent?

Is it $f'(z)$?

Is it possible to evaluate the radius of convergence of (2) knowing the radius of convergence of (1)?

The following theorem answers these questions:

Let R be the radius of convergence of the series

$$\sum_{n=0}^{\infty} c_n(z-z_0)^n = f(z) \quad (3)$$

Then, the series obtained by differentiation of (3) term by term converges to $f'(z)$, i.e.,

$$f'(z) = \sum_{n=1}^{\infty} nc_n(z-z_0)^{n-1} \quad (4)$$

and has the same radius of convergence.

Prove this theorem.

Solution: Since $f(z)$ is an analytic function, $f'(z)$ is defined inside the circle of convergence.

By Problem 13-25, the partial sums

$$c_1 + 2c_2(z-z_0) + \ldots + kc_K(z-z_0)^{k-1} \quad (5)$$

converges to $f'(z)$ inside the circle of convergence. Hence,

$$f'(z) = \sum_{n=1}^{\infty} nc_n(z-z_0)^{n-1} \quad (6)$$

for all $|z| < R$, where R is the radius of convergence of series (3). We have shown that $R < R'$, where R' is the radius of convergence of (6). We will show that $R = R'$.

Let z_1 be such that

$$|z_1 - z_0| > R \tag{7}$$

For $m > |z_1-z_0|$, we have

$$|c_m(z_1-z_0)^m| < |mc_m(z_1-z_0)^{m-1}| \tag{8}$$

The series obtained from the terms of the left side of (8) diverges, therefore the series of the terms on the right side diverges. Hence, both series (3) and (4) have the same radius of convergence.

CHAPTER 14

TAYLOR SERIES

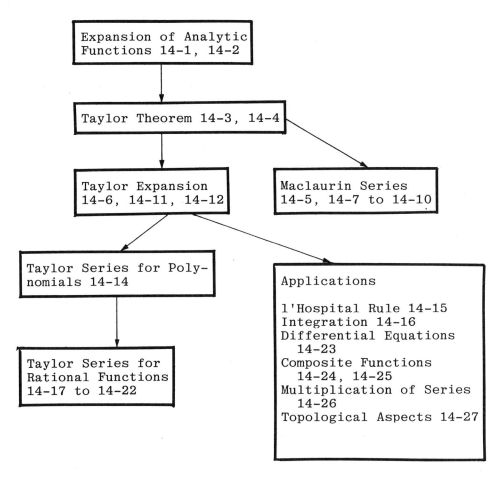

This chart is provided to facilitate rapid understanding of the interrelationships of the topics and subject matter in this chapter. Also shown are the problem numbers associated with the subject matter.

TAYLOR THEOREM • PROBLEM 14-1

Later in this chapter we will show that every function $f(z)$ is analytic in a neighborhood of z_0 and can be represented by a power series

$$f(z) = c_0 + c_1(z-z_0) + c_2(z-z_0)^2 + \ldots \tag{1}$$

The following theorem gives the interpretation of the coefficients $c_0, c_1, c_2, \ldots$.

Theorem

If an analytic function $f(z)$ has a power series representation

$$f(z) = c_0 + c_1(z-z_0) + c_2(z-z_0)^2 + \ldots \qquad (2)$$

in some open disk with center at z_0, then the coefficients c_n are given by

$$c_n = \frac{f^{(n)}(z_0)}{n!} \qquad (3)$$

for $n = 0, 1, 2, \ldots$.

Solution: Let $z = z_0$, then from (2) we obtain

$$f(z_0) = c_0 = \frac{f^{(0)}(z_0)}{0!} \qquad (4)$$

Differentiating (2), we get

$$f'(z) = c_1 + 2c_2(z-z_0) + \ldots \qquad (5)$$

Setting $z = z_0$ in (5), we get

$$c_1 = \frac{f^{(1)}(z_0)}{1!} = f'(z_0) \qquad (6)$$

Repeating this procedure, we obtain

$$f^{(n)}(z) = (1 \cdot 2 \cdot \ldots \cdot n)c_n + \left(2 \cdot 3 \cdot \ldots \cdot (n+1)\right)c_{n+1}(z-z_0) + \ldots \qquad (7)$$

From (7) we find

$$c_n = \frac{f^n(z_0)}{n!} \qquad (8)$$

● **PROBLEM 14-2**

Expand in the power series

$$f(z) = \sum_{n=0}^{\infty} c_n(z-z_0)^n \qquad (1)$$

the function

$$f(z) = (1+z)e^z \qquad (2)$$

at $z_0 = 0$.

Solution: We shall apply (3) of Problem 14-1 to evaluate the coefficients c_n.

We have

$$f(z)\Big|_{z=0} = (1+0)e^0 = 1$$

$$f'(z)\Big|_{z=0} = (2e^z + ze^z)\Big|_{z=0} = 2$$

$$f^{(2)}(z)\Big|_{z=0} = (3e^z + ze^z)\Big|_{z=0} = 3 \qquad (3)$$

$$f^{(n)}(z)\Big|_{z=0} = ((n+1)e^z + ze^z)\Big|_{z=0} = n+1$$

and

$$c_n = \frac{f^{(n)}(z_0)}{n!} = \frac{f^{(n)}(0)}{n!} = \frac{n+1}{n!} \qquad (4)$$

Thus,

$$(1+z)e^z = 1 + \frac{2}{1!}z + \frac{3}{2!}z^2 + \ldots \qquad (5)$$

$$= \sum_{n=0}^{\infty} \frac{n+1}{n!} z^n$$

● **PROBLEM 14-3**

Prove Taylor's theorem.

Theorem

If $f(z)$ is analytic inside a circle C with a center at z_0, then there is always one and only one power series, for all z inside C

$$f(z) = f(z_0) + f'(z_0)(z-z_0) + \frac{f''(z_0)}{2!}(z-z_0)^2$$
$$+ \frac{f'''(z_0)}{3!}(z-z_0)^3 + \ldots + \frac{f^{(k)}(z_0)}{k!}(z-z_0)^k + \ldots \qquad (1)$$

Series (1) converges for all z such that

$$|z-z_0| < R \qquad (2)$$

where the radius of convergence R is the distance from z_0 to the nearest singularity of the function $f(z)$. On $|z-z_0| = R$ the series (1) may or may not converge.

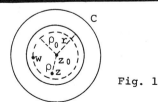

Fig. 1

Solution: Let z be an arbitrary interior point of $C(z_0, r)$, see Fig. 1.

Since

$$|z-z_0| = \rho < r \qquad (3)$$

we can find ρ_0, such that

$$\rho < \rho_0 < r \tag{4}$$

Now, let ω be an arbitrary point of the circumference of the circle $C(z_0, \rho_0)$. Then

$$\frac{1}{\omega - z} = \frac{1}{(\omega - z_0) - (z - z_0)} = \frac{1}{\omega - z_0} \cdot \frac{1}{1 - \frac{z - z_0}{\omega - z_0}}$$

$$= \sum_{n=0}^{\infty} \frac{(z - z_0)^n}{(\omega - z_0)^{n+1}} \tag{5}$$

By the Weierstrass test, series (5) is uniformly convergent with respect to ω along $C(z_0, \rho_0)$ because

$$\left|\frac{z - z_0}{\omega - z_0}\right| = \frac{\rho}{\rho_0} < 1 \tag{6}$$

By the same token, the series

$$\frac{f(\omega)}{\omega - z} = \sum_{n=0}^{\infty} \frac{f(\omega)}{(\omega - z_0)^{n+1}} (z - z_0)^n \tag{7}$$

is uniformly convergent. We shall integrate both sides of (7) along $C(z_0, \rho_0)$. Integration of the right-hand side of (7) is carried out term by term and the resulting series is convergent. Multiplying by $\frac{1}{2\pi i}$ we obtain

$$\frac{1}{2\pi i} \int_{C(z_0, \rho_0)} \frac{f(\omega)}{\omega - z} d\omega = \sum_{n=0}^{\infty} \frac{1}{2\pi i} \int_{C(z_0, \rho_0)} \frac{f(\omega)}{(\omega - z_0)^{n+1}} (z - z_0)^n d\omega \tag{8}$$

By Cauchy's integral formula, we obtain from (8)

$$f(z) = \sum_{n=0}^{\infty} \frac{1}{n!} f^{(n)}(z_0)(z - z_0)^n$$

$$= f(z_0) + f'(z_0)(z - z_0) + \frac{f''(z_0)}{2!} (z - z_0)^2 + \ldots \tag{9}$$

q.e.d.

In the next problem we shall prove that the expansion obtained is unique.

● PROBLEM 14-4

Show that the expansion of an analytic function in a power series is unique.

Solution: Assume that $f(z)$ is an analytic function, which has two expansions in its power series

$$f(z) = \sum_{n=0}^{\infty} a_n (z - z_0)^n = \sum_{n=0}^{\infty} b_n (z - z_0)^n \tag{1}$$

Both series have a positive radius of convergence and their sums coincide for an infinite number of points. For $z = z_0$ we obtain

$$a_0 = b_0 \tag{2}$$

Assume that the first k coefficients are equal

$$a_1 = b_1, \quad a_2 = b_2, \ldots, a_k = b_k \tag{3}$$

Then, from (1) and (3) we get

$$a_{k+1} + a_{k+2}(z-z_0) + \ldots = b_{k+1} + b_{k+2}(z-z_0) + \ldots \tag{4}$$

which is true for infinite points. Let z approach z_0 in (4). Since the power series represent continuous functions, it follows that

$$a_{k+1} = b_{k+1} \tag{5}$$

Thus, both expansions are identical. We can put this result in the form of a theorem.

Theorem

If two power series

$$\sum_{n=0}^{\infty} a_n (z-z_0)^n \quad \text{and} \quad \sum_{n=0}^{\infty} b_n (z-z_0)^n \tag{6}$$

have a positive radius of convergence, and if their sums are the same for an infinite number of points (or all points of a neighborhood of z_0) with z_0 as the limit point, then they are identical, i.e., $a_n = b_n$.

MACLAURIN SERIES

• PROBLEM 14-5

Find a Maclaurin series of $\sin z$, $\cos z$ and e^z.

Solution: The Taylor expansion of an analytic function for $z_0 = 0$ is called a Maclaurin series.

For $f(z) = \sin z$ we have

$$f'(z) = \cos z, \quad f''(z) = -\sin z, \quad f'''(z) = -\cos z,$$
$$f^{(4)}(z) = \sin z, \ldots \tag{1}$$

Maclaurin series is given by (2)

$$f(z) = f(0) + f'(0)z + \frac{f''(0)}{2!} z^2 + \ldots + \frac{f^{(k)}(0)}{k!} z^k + \ldots$$

Setting $z = 0$ in (1) we find

$$f(0) = 0, \quad f'(0) = 1, \quad f''(0) = 0, \quad f'''(0) = -1 \tag{3}$$

$$f^{(4)}(0) = 0, \quad f^{(5)}(0) = 1, \ldots$$

Substituting (3) into (2) we obtain

$$\sin z = z - \frac{z^3}{3!} + \frac{z^5}{5!} + \ldots + (-1)^{k-1} \frac{z^{2k-1}}{(2k-1)!} + \ldots \tag{4}$$

In the similar way we find for

$$f(z) = \cos z \tag{5}$$

$$f(z) = \cos z, \quad f'(z) = -\sin z, \quad f''(z) = -\cos z, \quad f'''(z) = \sin z$$
$$f^{(4)}(z) = \cos z, \ldots \tag{6}$$

Hence, for $z = 0$ we find

$$\cos z = 1 - \frac{z^2}{2!} + \frac{z^4}{4!} + \ldots + (-1)^{k-1} \frac{z^{2k-2}}{(2k-2)!} + \ldots \tag{7}$$

Since both functions are analytic in the z plane, the radius of convergence is infinite, i.e., the series converge for all z, $|z| < \infty$.

For
$$f(z) = e^z \tag{8}$$

we have

$$f(z) = e^z, \quad f'(z) = e^z, \quad \ldots, \quad f^{(k)}(z) = e^z, \ldots \tag{9}$$

Thus,

$$e^z = 1 + z + \frac{z^2}{2!} + \frac{z^3}{3!} + \ldots + \frac{z^k}{k!} + \ldots \tag{10}$$

• **PROBLEM 14-6**

Expand

$$f(z) = \frac{1}{z^2} \tag{1}$$

around $z_0 = 2$. Find the radius of convergence.

Solution: We have

$$f(z) = \frac{1}{z^2}, \quad f'(z) = \frac{-2}{z^3}, \quad f''(z) = \frac{2 \cdot 3}{z^4},$$
$$\ldots, \quad f^{(k)}(z) = \frac{(-1)^k (k+1)!}{z^{k+2}} \tag{2}$$

For $z = 2$

$$f^{(k)}(2) = \frac{(-1)^k (k+1)!}{2^{k+2}} \tag{3}$$

Hence,

$$\frac{1}{z^2} = \frac{1}{4} - \frac{1}{4}(z-2) + \ldots + \frac{(-1)^k(k+1)!}{2^{k+2} k!}(z-2)^k + \ldots \quad (4)$$

$$= \sum_{k=0}^{\infty} \frac{(-1)^k(k+1)!}{2^{k+2} k!}(z-2)^k$$

$$= \frac{1}{4} \sum_{k=0}^{\infty} (-1)^k(k+1)\left(\frac{z-2}{2}\right)^k$$

The function $f(z) = \frac{1}{z^2}$ is analytic for all values of z such that $|z-2| < 2$. Hence, the radius of convergence of the series (4) is R = 2.

● **PROBLEM 14-7**

Find the Maclaurin expansion for

$$u(x,y) = e^x \cos y \quad (1)$$

Express the result in polar coordinates.

Solution: Since an analytic function f(z) can be expanded in a Taylor series, we conclude that

$$u(x,y) = \text{Re}[f(z)] \quad \text{and}$$

$$v(x,y) = \text{Im}[f(z)] \quad \text{can be expanded in a Taylor}$$

series in x and y. This series will converge in the same domain as the series for f(z). From

$$e^z = e^{x+iy} = e^x(\cos y + i \sin y) \quad (2)$$

we find

$$\text{Re } e^z = e^x \cos y = u(x,y) \quad (3)$$

Since

$$e^z = 1 + z + \frac{z^2}{2!} + \frac{z^3}{3!} + \ldots + \frac{z^n}{n!} \quad (4)$$

we obtain for $e^x \cos y$

$$u = e^x \cos y = \text{Re}(e^z) = 1 + x + \frac{x^2-y^2}{2!} + \ldots + \frac{\text{Re}(x+iy)^n}{n!} + \ldots \quad (5)$$

Eq.(2) in polar coordinates becomes

$$e^z = 1 + r(\cos\theta + i\sin\theta) + \ldots + \frac{r^n(\cos n\theta + i \sin n\theta)}{n!} + \ldots \quad (6)$$

Hence, for $e^x \cos y$ we obtain

$$u = e^{r\cos\theta}\cos(r\sin\theta) = 1 + r\cos\theta + \frac{r^2 \cos 2\theta}{2!} + \ldots$$
$$+ \frac{r^n \cos n\theta}{n!} + \ldots \quad (7)$$

• **PROBLEM 14-8**

Find the Maclaurin series expansion of the following functions and determine the radius of convergence:

1. sinh z
2. cosh z

__Solution__: 1. By definition

$$\sinh z = \frac{e^z - e^{-z}}{2} \tag{1}$$

We shall apply

$$e^z = 1 + z + \frac{z^2}{2!} + \ldots \tag{2}$$

Eq.(2) for $-z$ becomes

$$e^{-z} = 1 - z + \frac{z^2}{2!} - \frac{z^3}{3!} + \ldots \tag{3}$$

Thus,

$$\sinh z = \frac{1}{2}\left[2z + 2\frac{z^3}{3!} + 2\frac{z^5}{5!} + \ldots\right]$$

$$= \sum_{n=1}^{\infty} \frac{z^{2n-1}}{(2n-1)!} \tag{4}$$

The radius of convergence is $R = \infty$.

2. Since

$$\cosh z = \frac{e^z + e^{-z}}{2} \tag{5}$$

substituting (2) and (3) into (5), we obtain

$$\cosh z = \frac{1}{2}\left[2 \cdot 1 + 2\frac{z^2}{2!} + 2\frac{z^4}{4!} + \ldots\right]$$

$$= \sum_{n=0}^{\infty} \frac{z^{2n}}{(2n)!} \tag{6}$$

The radius of convergence is $R = \infty$.

• **PROBLEM 14-9**

Derive the Maclaurin series for

$$f(z) = \frac{1}{1+z^2} \tag{1}$$

Use the formula

$$\frac{1}{1-q} = \sum_{n=0}^{\infty} q^n \tag{2}$$

where $|q| < 1$.

Solution: Let us forget for a moment about (2) and try to derive the Maclaurin series for (1) using the old fashioned method, i.e., computing derivatives.

We get

$$f'(z) = \frac{-2z}{(1+z^2)^2} \tag{3}$$

$$f''(z) = \frac{6z^2-2}{(1+z^2)^3} \tag{4}$$

etc.

As the order of derivatives increases, the calculations become more and more complicated. There is a much faster method. Set $-z^2 = q$ in (2) then

$$\frac{1}{1+z^2} = \sum_{n=0}^{\infty} (-z^2)^n = \sum_{n=0}^{\infty} (-1)^n z^{2n} \tag{5}$$

The merits of this method are obvious. In the next problems, we shall develop a general method for deriving the Taylor series for rational functions.

• **PROBLEM 14-10**

Prove that an odd function has an odd Maclaurin series representation. Also, show that an even function has an even series.

Solution: Assume that $f(z)$ is analytic in a neighborhood of the origin and that f is an odd function, i.e.,

$$f(-z) = -f(z) \tag{1}$$

Since $f(z)$ is analytic, the Maclaurin expansion is

$$f(z) = f(0) + \frac{f'(0)}{1!} z + \frac{f''(0)}{2!} z^2 + \frac{f'''(0)}{3!} z^3 + \ldots \tag{2}$$

For $-z$ the expansion is

$$f(-z) = f(0) - \frac{f'(0)}{1!} z + \frac{f''(0)}{2!} z^2 - \frac{f'''(0)}{3!} z^3 + \ldots \tag{3}$$

Adding (2) and (3) we find

$$f(z)+f(-z) = 2\left[f(0) + \frac{f''(0)}{2!} z^2 + \frac{f^{(4)}(0)}{4!} z^4 + \ldots \right] \tag{4}$$

Since $f(z)$ is an odd function $f(z)+f(-z) = 0$ and hence,

$$f(0) + \frac{f''(0)}{2!} z^2 + \frac{f^{(4)}(0)}{4!} z^4 + \ldots = 0 \tag{5}$$

Substituting (5) into (2) we obtain

$$f(z) = \frac{f'(0)}{1!} z + \frac{f^{(3)}(0)}{3!} z^3 + \ldots + \frac{f^{(2n+1)}(0)}{(2n+1)!} z^{2n+1} + \ldots \tag{6}$$

Thus, the Maclaurin series of an odd function contains only odd powers of z.

For an even function, i.e., $f(z) = f(-z)$ subtracting (3) from (2) we find

$$\frac{f'(0)}{1!} z + \frac{f^{(3)}(0)}{3!} z^3 + \frac{f^{(5)}(0)}{5!} z^5 + \ldots = 0 \tag{7}$$

Hence, from (2) and (7) we get

$$f(z) = f(0) + \frac{f''(0)}{2!} z^2 + \frac{f^{(4)}(0)}{4!} z^4 + \ldots \tag{8}$$

TAYLOR EXPANSION

• **PROBLEM 14-11**

Many interesting conclusions can be drawn from the Taylor series of expansion of an analytic function. Here are some of them.

1. If any nonzero term of Taylor series is omitted, its loss cannot be compensated for by adjusting the coefficients of the series.

2. If $f(z_0) \neq 0$ and $f(z)$ is analytic in the neighborhood of $z = z_0$, then

$$f(z) \neq \sum_{n=1}^{\infty} \frac{f^{(n)}(z_0)}{n!} (z-z_0)^n \tag{1}$$

3. If $f(z)$ and $g(z)$ are analytic in a domain which includes a point z_0, and $f(z) = g(z)$ for all z which lie on a line segment through z_0, then

$$f(z) = g(z) \tag{2}$$

for all z belonging to the domain of an analytical function.

Solution: 1. Let

$$f(z) = \sum_{n=0}^{\infty} c_n (z-z_0)^n \tag{3}$$

If we omit any particular term and compensate for it, then

$$f(z) = \sum_{n=0}^{\infty} b_n (z-z_0)^n \qquad (4)$$

Since the expansion is unique, we get

$$c_n = b_n \quad \text{for } u = 0,1,2,\ldots \qquad (5)$$

2. This follows directly from part 1 of this problem.

3. At the point $z = z_0$ we can compute all derivatives of f and g from their values, which are identical, on the line segment passing through z_0. Hence, all derivatives of the function

$$F = f - g$$

at $z = z_0$ are zero.

Therefore,

$$F(z) = \sum_{n=0}^{\infty} \frac{F^{(n)}(z_0)}{n!} (z-z_0)^n = 0 \qquad (6)$$

and $f = g$ in their domain.

• **PROBLEM 14-12**

For

$$f(z) = \ln(z+1) \qquad (1)$$

consider the branch which has the value zero for $z = 0$. Find the Taylor expansion of $f(z)$ in the neighborhood of $z = 0$. What is the radius of convergence of this series?

Solution: First we compute the derivatives

$$f(z) = \ln(z+1)$$

$$f'(z) = \frac{1}{z+1}$$

$$f''(z) = -(z+1)^{-2}$$

$$\vdots$$

$$f^{(k)}(z) = (-1)^{k-1}(k-1)!(z+1)^{-k} \qquad (2)$$

For $z = 0$ \hfill (3)

$$f(0) = 0, \; f'(0) = 1, \; f''(0) = -1, \ldots f^{(k)}(0) = (-1)^{k-1}(k-1)!$$

Thus,

$$\ln(z+1) = z - \frac{z^2}{2} + \frac{z^3}{3} - \frac{z^4}{4} + \ldots = \sum_{k=1}^{\infty} \frac{(-1)^{k+1}}{k} z^k \qquad (4)$$

The same result can be obtained by observing that

$$\text{for } |z| < 1 \quad \frac{1}{z+1} = 1 - z + z^2 - z^3 + \ldots \qquad (5)$$

and integrating (5) from 0 to z

$$\ln(z+1) = z - \frac{z^2}{2} + \frac{z^3}{3} - \frac{z^4}{4} + \ldots \tag{6}$$

To find the radius of convergence, note that, the k-th term of (4) is

$$u_k = \frac{(-1)^{k+1} z^k}{k} \tag{7}$$

Applying the ratio test we find

$$\lim_{k \to \infty} \left| \frac{u_{k+1}}{u_k} \right| = \lim_{k \to \infty} \left| \frac{kz}{k+1} \right| = |z| \tag{8}$$

Hence, the series converges for $|z| < 1$. The radius of convergence of (4) is one.

APPLICATIONS

• **PROBLEM 14-13**

Consider a truncated Taylor series consisting of the first n terms

$$c_0 + c_1(z-z_0) + c_2(z-z_0)^2 + \ldots + c_{n-1}(z-z_0)^{n-1} \tag{1}$$

Thus, the remainder R_n is

$$R_n \equiv f(z) - \sum_{k=0}^{n-1} c_k(z-z_0)^k = f(z) - \sum_{k=0}^{n-1} \frac{f^{(k)}(z_0)}{k!}(z-z_0)^k \tag{2}$$

Show that

$$R_n = \frac{(z-z_0)^n}{2\pi i} \int_C \frac{f(\omega) d\omega}{(\omega - z_0)^n (\omega - z)} \tag{3}$$

The curve C is located inside the domain of convergence.

Solution: For z inside the domain of convergence we have

$$f(z) = \frac{1}{2\pi i} \int_C \frac{f(\omega)}{(\omega - z)} d\omega \tag{4}$$

$$f^{(k)}(z) = \frac{k!}{2\pi i} \int_C \frac{f(\omega)}{(\omega-z)^{k+1}} d\omega \tag{5}$$

Substituting (4) and (5) into (2), we obtain

$$R_n = \frac{1}{2\pi i}\left[\int_C \frac{f(\omega)}{\omega-z}d\omega - \sum_{k=0}^{n-1}\int_C \frac{(z-z_0)^k}{(\omega-z)^{k+1}}f(\omega)d\omega\right] \quad (6)$$

$$= \frac{1}{2\pi i}\int_C \left[\frac{1}{\omega-z} - \sum_{k=0}^{n-1}\frac{(z-z_0)^k}{(\omega-z_0)^{k+1}}\right]f(\omega)d\omega$$

To simplify (6) we shall use the following identity

$$\frac{1}{1-q} = \frac{q^n}{1-q} + \sum_{k=0}^{n-1} q^k \quad (7)$$

Substituting

$$q = \frac{z-z_0}{\omega-z_0} \quad (8)$$

we obtain

$$\frac{\omega-z_0}{\omega-z} = \frac{(z-z_0)^n}{(\omega-z_0)^n} \cdot \frac{\omega-z_0}{\omega-z} + \sum_{k=0}^{n-1}\frac{(z-z_0)^k}{(\omega-z_0)^k} \quad (9)$$

or

$$\frac{1}{\omega-z} - \sum_{k=0}^{n-1}\frac{(z-z_0)^k}{(\omega-z_0)^{k+1}} = \frac{(z-z_0)^n}{(\omega-z_0)^n(\omega-z)} \quad (10)$$

Finally, substituting (10) into (6), we find

$$R_n = \frac{1}{2\pi i}\int_C \frac{(z-z_0)^n f(\omega)}{(\omega-z_0)^n(\omega-z)} d\omega \quad (11)$$

$$= \frac{(z-z_0)^n}{2\pi i}\int_C \frac{f(\omega)d\omega}{(\omega-z_0)^n(\omega-z)}$$

● **PROBLEM 14-14**

Derive the Taylor series of $z_0 = 1 - i$ for the polynomial

$$f(z) = (1+i)z^2 - 2z + 4i \quad (1)$$

Solution: The Taylor series of z_0 for a polynomial of degree n is a polynomial in $(z-z_0)$ of degree n. Thus, in this case

$$f(z) = f(z_0) + f'(z_0)(z-z_0) + \frac{f''(z_0)}{2}(z-z_0)^2 \quad (2)$$

From (1) we compute

$$f'(z) = 2(1+i)z-2 \quad (3)$$

$$f''(z) = 2(1+i) \quad (4)$$

and
$$f(1-i) = 4i \tag{5}$$
$$f'(1-i) = 2 \tag{6}$$

Hence, the Taylor expansion is

$$f(z) = 4i + 2(z-1+i) + (1+i)(z-1+i)^2 \tag{7}$$

Note that the polynomial in (7) is the same polynomial as in (1) It is easy to verify it by performing necessary algebraic operations on (7) or on (1).

• **PROBLEM 14-15**

Let $f(z)$ and $g(z)$ functions be analytic in a domain including the point $z = z_0$, and such that

$$f(z_0) = g(z_0) = 0 \text{ and } g'(z_0) \neq 0 \tag{1}$$

Prove l'Hospital rule

$$\lim_{z \to z_0} \frac{f(z)}{g(z)} = \frac{f'(z_0)}{g'(z_0)} \tag{2}$$

applying the Taylor expansion.

Solution: Since $f(z)$ and $g(z)$ are analytic in a domain containing z_0 we can find Taylor series expansion about $z = z_0$.

$$f(z) = f(z_0) + f'(z_0)(z-z_0) + \frac{f''(z_0)}{2!}(z-z_0)^2 + \ldots + \frac{f^{(n)}(z_0)}{n!}(z-z_0)^n + \ldots \tag{3}$$

$$g(z) = g(z_0) + g'(z_0)(z-z_0) + \frac{g''(z_0)}{2!}(z-z_0)^2 + \ldots + \frac{g^{(n)}(z_0)}{n!}(z-z_0)^n + \ldots \tag{4}$$

Remembering that $f(z_0) = g(z_0) = 0$ we find, substituting (3) and (4)

$$\lim_{z \to z_0} \frac{f(z)}{g(z)} = \lim_{z \to z_0} \frac{f'(z_0)(z-z_0) + \frac{f''(z_0)}{2!}(z-z_0)^2 + \ldots}{g'(z_0)(z-z_0) + \frac{g''(z_0)}{2!}(z-z_0)^2 + \ldots}$$

$$= \lim_{z \to z_0} \frac{f'(z_0) + \frac{f''(z_0)}{2!}(z-z_0) + \ldots}{g'(z_0) + \frac{g''(z_0)}{2!}(z-z_0) + \ldots} = \frac{f'(z_0)}{g'(z_0)} \tag{5}$$

• **PROBLEM 14-16**

The function

$$f(z) = \int_{z_0}^{z} \frac{d\omega}{\omega} \tag{1}$$

is an analytic function of z, if the path of integration from z_0 to z is located in the interior of the right half-plane.

Expand f(z) in a power series for a neighborhood of the point $z_0 = 1$.

Solution: We shall evaluate the coefficients of the series

$$f(z) = f(z_0) + \frac{f'(z_0)}{1!}(z-z_0) + \frac{f''(z_0)}{2!}(z-z_0)^2 + \ldots$$
$$\ldots + \frac{f^{(k)}(z_0)}{k!}(z-z_0)^k + \ldots \qquad (2)$$

From (1) we obtain

$$f'(z) = \frac{1}{z}, \quad f''(z) = -\frac{1}{z^2}, \quad f'''(z) = \frac{2}{z^3},$$
$$f^{(k)}(z) = (-1)^{k-1} \frac{(k-1)!}{z^k} \qquad (3)$$

For z = 1

$$f'(1) = 1, \quad f''(1) = -1, \quad f'''(1) = 2, \quad f^{(k)}(1) = (-1)^{k-1}(k-1)! \qquad (4)$$

$$f(1) = 0$$

Hence,

$$f(z) = (z-1) - \frac{1}{2}(z-1)^2 + \frac{1}{3}(z-1)^3 + \ldots$$
$$\ldots + \frac{(-1)^{k-1}}{k}(z-1)^k + \ldots \qquad (5)$$

The radius of convergence is r = 1.

TAYLOR SERIES FOR RATIONAL FUNCTIONS

• **PROBLEM 14-17**

Compute the Taylor series of the function

$$f(z) = \frac{1}{1-z^2} \qquad (1)$$

around the point $z_0 = 2$.

Solution: The function f(z) can be written as a sum of partial fractions

$$\frac{1}{1-z^2} = \frac{1}{2}\left[\frac{1}{1-z} + \frac{1}{1+z}\right]$$

$$= -\frac{1}{2}\left[\frac{1}{1+(z-2)}\right] + \frac{1}{6} \cdot \frac{1}{\left[1+\frac{1}{3}(z-2)\right]} \qquad (2)$$

Note, each of the fractions are in the form of $\frac{1}{1+q}$. Hence, we can apply the geometric series formula

$$\frac{1}{1+(z-2)} = \sum_{n=0}^{\infty} (-1)^n (z-2)^n \qquad (3)$$

for $|z-2| < 1$

$$\frac{1}{1+\frac{1}{3}(z-2)} = \sum_{n=0}^{\infty} \frac{(-1)^n}{3^n} (z-2)^n \qquad (4)$$

for $|z-2| < 2$

From (2), (3) and (4) we obtain

$$f(z) = \frac{1}{1-z^2} = -\frac{1}{2} \sum_{n=0}^{\infty} (-1)^n (z-2)^n + \frac{1}{6} \sum_{n=0}^{\infty} \frac{(-1)^n}{3^n} (z-2)^n$$

$$= \frac{1}{2} \sum_{n=0}^{\infty} (-1)^n \left[\frac{1}{3^{n+1}} - 1\right] (z-2)^n \qquad (5)$$

Since (3) converges in the circle $|z-2| < 1$ and (4) converges in the circle $|z-2| < 2$, their sum will converge in the circle $|z-2| < 1$.

Try to obtain the first few terms of the Taylor series of $\frac{1}{1-z^2}$ by differentiating $f(z)$. The expressions quickly become very complicated.

In the next problems we shall discuss the method of partial fractions in detail.

● **PROBLEM 14-18**

Using

$$\sum_{n=0}^{\infty} q^n = \frac{1}{1-q}, \qquad |q| < 1 \qquad (1)$$

derive the Taylor series for

$$\frac{1}{\omega-z}, \frac{1}{(\omega-z)^2}, \ldots, \frac{1}{(\omega-z)^m}$$

around z_0.

Solution: Since

$$\frac{1}{1-\left(\frac{z-z_0}{\omega-z_0}\right)} \cdot \frac{1}{\omega-z_0} = \frac{1}{\omega-z} \qquad (2)$$

we find

$$\frac{1}{\omega-z} = \sum_{n=0}^{\infty} \left(\frac{z-z_0}{\omega-z_0}\right)^n \cdot \frac{1}{\omega-z_0} = \sum_{n=0}^{\infty} \frac{(z-z_0)^n}{(\omega-z_0)^{n+1}} \qquad (3)$$

for $\left|\frac{z-z_0}{\omega-z_0}\right| < 1$ or $|z-z_0| < |\omega-z_0|$ \qquad (4)

Differentiating (3) inside the circle of convergence we obtain

$$\frac{1}{(\omega-z)^2} = \sum_{n=1}^{\infty} \frac{n(z-z_0)^{n-1}}{(\omega-z_0)^{n+1}} = \sum_{n=0}^{\infty} \frac{(n+1)(z-z_0)^n}{(\omega-z_0)^{n+2}} \qquad (5)$$

and

$$\frac{1}{(\omega-z)^3} = \frac{1}{2} \sum_{n=2}^{\infty} \frac{n(n-1)(z-z_0)^{n-2}}{(\omega-z_0)^{n+1}}$$

$$= \frac{1}{2} \sum_{n=0}^{\infty} \frac{(n+1)(n+2)(z-z_0)^n}{(\omega-z_0)^{n+3}} \qquad (6)$$

$$= \sum_{n=0}^{\infty} \binom{n+2}{2} \frac{(z-z_0)^n}{(\omega-z_0)^{n+3}}$$

where $\binom{n}{m} \equiv \frac{n!}{m!(n-m)!}$ \qquad (7)

is the binomial coefficient.

From (5) and (6) we can deduce the formula for $\frac{1}{(\omega-z)^m}$

$$\frac{1}{(\omega-z)^m} = \sum_{n=0}^{\infty} \binom{n+m-1}{m-1} \frac{(z-z_0)^n}{(\omega-z_0)^{n+m}} \qquad (8)$$

for $|z-z_0| < |\omega-z_0|$

• **PROBLEM 14-19**

Applying the general formula

$$\frac{1}{(\omega-z)^m} = \sum_{n=0}^{\infty} \binom{n+m-1}{m-1} \frac{(z-z_0)^n}{(\omega-z_0)^{n+m}} \qquad (1)$$

expand $\frac{1}{z}$ in a Taylor series about $z_0 = -1$.

Solution: Since in the case of $\frac{1}{z}$, $m = 1$ and $\omega = 0$, (1) leads to

$$-\frac{1}{z} = \sum_{n=0}^{\infty} \binom{n}{0} \frac{(z-z_0)^n}{(-z_0)^{n+1}} \qquad (2)$$

Function $\frac{1}{z}$ is expanded around $z_0 = -1$. Therefore,

$$\frac{1}{z} = -\frac{1}{-z} = -\sum_{n=0}^{\infty} \frac{(z+1)^n}{(1)^{n+1}}$$

$$= -\sum_{n=0}^{\infty} (z+1)^n \qquad (3)$$

for $|z+1| < 1$.

Here, we obtained the Taylor series of an analytic function without computing derivatives.

● **PROBLEM 14-20**

Expand the rational function

$$f(z) = \frac{-3z - 30}{(z+4)^2(z-2)} \qquad (1)$$

in a Taylor series about $z = 1$.

<u>Solution</u>: Let $F(z)$ be a rational function of z whose numerator and denominator contain no common factors. Then, we can denote

$$F(z) = \frac{g(z)}{h(z)} + f(z) \qquad (2)$$

where f, g, h are polynomials and the degree of h exceeds that of g. Hence, $\frac{g(z)}{h(z)}$ is a proper rational function. We shall apply the following theorem :

The proper rational function

$$\frac{g(z)}{h(z)} = \frac{g(z)}{(z-a)^n k(z)} \qquad (3)$$

can be denoted in the form

$$\frac{g(z)}{h(z)} = \frac{A}{(z-a)^n} + \frac{\ell(z)}{(z-a)^{n-1} k(z)} \qquad (4)$$

where the constant A and the polynomial $\ell(z)$ are uniquely determined for $k(a) \neq 0$.

$$A = \frac{g(a)}{k(a)} = \lim_{z \to a}\left[(z-a)^n \frac{g(z)}{h(z)}\right] \qquad (5)$$

We shall determine the partial fraction. For the expression $F(z)$

$$f(z) = \frac{A}{(z+4)^2} + \frac{B}{z+4} + \frac{C}{z-2} \qquad (6)$$

Thus, applying (5) we obtain

$$A = \left[f(z)(z+4)^2\right]\bigg|_{z=-4} = \frac{-3z-30}{z-2}\bigg|_{z=-4} \quad (7)$$

$$= 3$$

In the same manner, we shall compute C

$$C = \left[f(z)(z-2)\right]\bigg|_{z=2} = \frac{-3z-30}{(z+4)^2}\bigg|_{z=2} \quad (8)$$

$$= -1$$

Substituting (1), (7) and (8) into (6), we obtain

$$\frac{-3z-30}{(z+4)^2(z-2)} = \frac{3}{(z+4)^2} + \frac{B}{z+4} - \frac{1}{z-2} \quad (9)$$

Solving (9) for B, we find B = 1. Thus,

$$f(z) = \frac{3}{(z+4)^2} + \frac{1}{z+4} - \frac{1}{z-2} \quad (10)$$

We have proved the following formula

$$\frac{1}{(\omega-z)^m} = \sum_{n=0}^{\infty} \binom{n+m-1}{m-1} \frac{(z-z_0)^n}{(\omega-z_0)^{n+m}} \quad (11)$$

for $|z-z_0| < |\omega-z_0|$.

Hence,

$$\frac{1}{(z+4)^2} = \sum_{n=0}^{\infty} \binom{n+1}{1} \frac{(z-1)^n}{(-5)^{n+2}} \quad (12)$$

$$= \sum_{n=0}^{\infty} (n+1) \frac{(z-1)^n}{(-5)^{n+2}} \quad \text{for } |z-1| < 5.$$

$$\frac{1}{z+4} = -\frac{1}{-4-z} = -\sum_{n=0}^{\infty} \binom{n}{0} \frac{(z-1)^n}{(-5)^{n+1}} \quad (13)$$

$$= -\sum_{n=0}^{\infty} \frac{(z-1)^n}{(-5)^{n+1}} \quad \text{for } |z-1| < 5$$

and

$$\frac{1}{z-2} = -\frac{1}{2-z} = -\sum_{n=0}^{\infty} (z-1)^n \quad (14)$$

for $|z-1| < 1$

Substituting (12), (13) and (14) into (10) we obtain

$$f(z) = \frac{-3z-30}{(z+4)^2(z-2)} = \sum_{n=0}^{\infty} \left[\frac{3(n+1)}{(-5)^{n+2}} - \frac{1}{(-5)^{n+1}} - 1\right](z-1)^n \quad (15)$$

for $|z-1| < 1$

● **PROBLEM 14-21**

Find the Taylor series of the function
$$f(z) = \frac{z^4+z^3(2-3i)-6iz^2+2}{z(z+2)} \quad (1)$$
for $z_0 = 1$.

Solution: First, we shall write (1) in a different form
$$f(z) = \frac{z^4+z^3(2-3i)-6iz^2}{z(z+2)} + \frac{2}{z(z+2)}$$

$$= \frac{z^3+z^2(2-3i)-6iz}{z+2} + \frac{1}{z} - \frac{1}{z+2} \quad (2)$$

$$= z^2 - 3iz + \frac{1}{z} - \frac{1}{z+2}$$

We denote
$$z^2 - 3iz = g(z), \text{ then} \quad (3)$$

$$g'(z) = 2z - 3i, \quad g'(1) = 2 - 3i \quad (4)$$

$$g''(z) = 2, \quad g''(1) = 2 \quad (5)$$

Then,
$$g(z) = (z-1)^2 + (2-3i)(z-1) + (1-3i) \quad (6)$$

$$\frac{1}{z} = -\sum_{n=0}^{\infty} \frac{(z-1)^n}{(-1)^{n+1}} \quad \text{for } |z-1| < 1 \quad (7)$$

$$-\frac{1}{z+2} = \frac{1}{-2-z} = \sum_{n=0}^{\infty} \frac{(z-1)^n}{(-3)^{n+1}} \quad (8)$$

$$\text{for } |z-1| < 3$$

Therefore,
$$f(z) = (1-3i) + (2-3i)(z-1) + (z-1)^2$$
$$+ \sum_{n=0}^{\infty} \left[\frac{-1}{(-1)^{n+1}} + \frac{1}{(-3)^{n+1}} \right] (z-1)^n \quad (9)$$
$$\text{for } |z-1| < 1$$

● **PROBLEM 14-22**

Expand the function
$$f(z) = \text{arc tan}(z+i) \quad (1)$$
in a Taylor series for $z_0 = 3$.

Solution: We shall expand the derivative of f(z) in a Taylor series and then integrate the result.

$$f'(z) = \frac{d}{dz}\left[\arctan(z+i)\right] \qquad (2)$$

$$= \frac{1}{1+(z+i)^2}$$

The function $f'(z)$ can be written in partial fractions

$$\frac{1}{1+(z+i)^2} = \frac{1}{z^2+2iz} = \frac{1}{z(z+2i)} \qquad (3)$$

$$= \frac{1}{2i}\left(\frac{1}{z} + \frac{1}{-2i-z}\right)$$

In the general formula

$$\frac{1}{(\omega-z)^m} = \sum_{n=0}^{\infty} \binom{n+m-1}{m-1} \frac{(z-z_0)^n}{(\omega-z_0)^{n+m}} \qquad (4)$$

$$\text{for } |z-z_0| < |\omega-z_0|$$

we set $m = 1$, $\omega = 0$ and $z_0 = 3$, then

$$\frac{1}{z} = -\sum_{n=0}^{\infty} \frac{(z-3)^n}{(-3)^{n+1}} \qquad \text{for } |z-3| < 3 \qquad (5)$$

Similarly, $\dfrac{1}{-2i-z} = \sum_{n=0}^{\infty} \dfrac{(z-3)^n}{(-2i-3)^{n+1}} \quad \text{for } |z-3| < |-2i-3| = \sqrt{13}$ (6)

Thus, from (3), (5) and (6) we obtain

$$f'(z) = \frac{1}{2i} \sum_{n=0}^{\infty} \left[\frac{-1}{(-3)^{n+1}} + \frac{1}{(-2i-3)^{n+1}}\right](z-3)^n \qquad (7)$$

$$\text{for } |z-3| < 3$$

Integrating both sides of (7) from 3 to z, we obtain

$$\int_3^z f'(z)dz = f(z) - f(3)$$

$$= \frac{1}{2i} \sum_{n=0}^{\infty} \left[\frac{-1}{(-3)^{n+1}} + \frac{1}{(-2i-3)^{n+1}}\right]\frac{(z-3)^{n+1}}{n+1} \qquad (8)$$

Hence,

$$\arctan(z+i) = \arctan(3+i)$$

$$+ \frac{1}{2i} \sum_{n=0}^{\infty} \left[\frac{-1}{(-3)^{n+1}} + \frac{1}{(-2i-3)^{n+1}}\right]\frac{(z-3)^{n+1}}{n+1} \qquad (9)$$

MORE APPLICATIONS

● **PROBLEM 14-23**

Consider the differential equation

$$\frac{d^2f}{dz^2} - f = 0 \tag{1}$$

with conditions

$$f(0) = 1, \quad f'(0) = 0 \tag{2}$$

Apply the Taylor series method to solve this equation.

Solution: Assume that $f(z)$ is an analytic function. Then,

$$f(z) = c_0 + c_1 z + \ldots + c_n z^n + \ldots \tag{3}$$

where

$$c_n = \frac{f^{(n)}(0)}{n!} \tag{4}$$

From (2) we have

$$c_0 = f(0) = 1 \tag{5}$$

$$c_1 = f'(0) = 0 \tag{6}$$

Since

$$f''(z) = f(z) \tag{7}$$

we get

$$f''(0) = f(0) = 1 \tag{8}$$

and

$$c_2 = \frac{1}{2!}$$

Differentiating (7)

$$f'''(z) = f'(z) \tag{9}$$

we find

$$c_3 = \frac{f'''(0)}{3!} = 0 \tag{10}$$

Thus, we obtain

$$c_{2n} = \frac{1}{(2n)!} \tag{11}$$

$$c_{2n+1} = 0$$

Function f can be written as

$$f(z) = 1 + \frac{z^2}{2!} + \frac{z^4}{4!} + \ldots + \frac{z^{2n}}{(2n)!} + \ldots \quad (12)$$

Note that series (12) represents hyperbolic cosine (see Problem 14-8)

$$f(z) = \cosh z \quad (13)$$

● **PROBLEM 14-24**

Find the Maclaurin series for

1. $\dfrac{\sin z}{1-z}$ \hfill (1)

2. $e^{\sin z}$ \hfill (2)

Solution: 1. Both functions $\sin z$ and $\frac{1}{1-z}$ are analytic for $|z|<1$. From Problem 14-5 we get

$$\sin z = z - \frac{z^3}{3!} + \frac{z^5}{5!} - \frac{z^7}{7!} + \ldots \quad (3)$$

The function $\frac{1}{1-z}$ can be expanded as a geometric series

$$\frac{1}{1-z} = 1 + z + z^2 + z^3 + \ldots \quad (4)$$

Combining (3) and (4) we obtain

$$\frac{\sin z}{1-z} = (z - \frac{z^3}{3!} + \frac{z^5}{5!} - \frac{z^7}{7!} + \ldots)(1+z+z^2+z^3+\ldots)$$

$$= z+z^2+(1-\frac{1}{3!})z^3+(1-\frac{1}{3!})z^4+(1-\frac{1}{3!}+\frac{1}{5!})z^5 + \ldots \quad (5)$$

for $|z| < 1$.

2. Here, we have to compute the Maclaurin series for a composite function. We know Maclaurin expansions for both functions e^z and $\sin z$.

$$\sin z = z - \frac{z^3}{3!} + \frac{z^5}{5!} - \ldots \quad (6)$$

$$e^\omega = 1 + \omega + \frac{\omega^2}{2!} + \frac{\omega^3}{3!} + \ldots + \frac{\omega^n}{n!} + \ldots \quad (7)$$

Setting $\omega = \sin z$ we obtain

$$\exp(\sin z) = 1+(\sin z) + \frac{1}{2!}(\sin z)^2 + \frac{1}{3!}(\sin z)^3 + \ldots$$

$$= 1+(z - \frac{z^3}{6} + \frac{z^5}{120} - \ldots) + \frac{1}{2}(z - \frac{z^3}{6} + \ldots)^2$$

$$+ \frac{1}{6}(z - \frac{z^3}{6} + \ldots)^3 + \ldots \quad (8)$$

$$= 1 + z + \frac{z^2}{2} + (\frac{1}{6} - \frac{1}{6})z^3 + \ldots$$

• **PROBLEM 14-25**

Find the Maclaurin expansion for the function sec z.

Solution: Since

$$\sec z = \frac{1}{\cos z} \tag{1}$$

the function sec z has singularities at the zeros of cos z, namely at

$$z = (2n+1)\frac{\pi}{2} i, \quad n = 0,1,2 \tag{2}$$

The function cos z is analytic in the entire plane.
From the general formula

$$f(z) = \sum_{n=0}^{\infty} \frac{f^{(n)}(z_0)}{n!} (z-z_0)^n \tag{3}$$

we obtain for $f(z) = \cos z$ and $z_0 = 0$.

$$\cos z = 1 - \frac{z^2}{2!} - \frac{z^4}{4!} + \frac{z^6}{6!} - \cdots \tag{4}$$

Hence,

$$\sec z = \frac{1}{\cos z} = \frac{1}{1 - \frac{z^2}{2!} + \frac{z^4}{4!} - \frac{z^6}{6!} + \cdots}$$

$$= \frac{1}{1 - \left(\frac{z^2}{2!} - \frac{z^4}{4!} + \cdots\right)} \tag{5}$$

The Maclaurin expansion for cos z converges for $|z| < \frac{\pi}{2}$, since it converges in the entire plane. Setting

$$\omega = \frac{z^2}{2!} - \frac{z^4}{4!} + \cdots \tag{6}$$

we have

$$\sec z = \frac{1}{\cos z} = \frac{1}{1-\omega} \tag{7}$$

$$\frac{1}{1-\omega} = 1 + \omega + \omega^2 + \omega^3 + \cdots \tag{8}$$

$$= 1 + \left(\frac{z^2}{2!} - \frac{z^4}{4!} + \cdots\right) + \left(\frac{z^2}{2!} - \frac{z^4}{4!} + \cdots\right)^2$$

$$+ \left(\frac{z^2}{2!} - \frac{z^4}{4!} + \cdots\right)^3 + \cdots = 1 + \frac{z^2}{2} + \frac{z^4}{8} + \cdots \tag{9}$$

$$\text{for } |z| < \frac{\pi}{2}$$

• **PROBLEM 14-26**

Let $f(z)$ and $g(z)$ be analytic for $|z| < R$, then

$$f(z) = \sum_{n=0}^{\infty} a_n z^n \qquad (1)$$

$$g(z) = \sum_{n=0}^{\infty} b_n z^n \qquad (2)$$

converge for $|z| < R$.

Compute the Maclaurin series expansion of the product $f(z) g(z)$, $|z| < R$.

Solution: Since both $f(z)$ and $g(z)$ are analytic for $|z| < R$, the Maclaurin series expansion of their product exists

$$f(z)g(z) = \sum_{n=0}^{\infty} c_n z^n, \quad |z| < R \qquad (3)$$

Consecutive differentiation of

$$\left(\sum_{n=0}^{\infty} a_n z^n\right)\left(\sum_{n=0}^{\infty} b_n z^n\right) = \sum_{n=0}^{\infty} c_n z^n \qquad (4)$$

yields

$$c_0 = f(0)g(0) = a_0 b_0$$

$$c_1 = f(0)g'(0) + f'(0)g(0) = a_0 b_1 + a_1 b_0$$

$$c_2 = \frac{1}{2}\left[f(0)g''(0) + 2f'(0)g'(0) + f''(0)g(0)\right]$$

$$= a_0 b_2 + a_1 b_1 + a_2 b_0$$

$$c_n = a_0 b_n + a_1 b_{n-1} + \ldots + a_{n-1} b_1 + a_n b_0 \qquad (5)$$

Thus,

$$f(z)g(z) = a_0 b_0 + (a_0 b_1 + a_1 b_0)z \\ + (a_0 b_2 + a_1 b_1 + a_2 b_0)z^2 + \ldots + \left(\sum_{k=0}^{\infty} a_k b_{n-k}\right)z^n + \ldots \qquad (6)$$

for $|z| < R$

Note that, series (6) can be obtained by multiplying series (1) and (2) term by term, and assembling the terms with the same power of z. The result is called the Cauchy product of two series.

● PROBLEM 14-27

1. Is it true that two analytic functions f and g defined in the same domain Ω and having identical power series expansions around $z_0 \in \Omega$, are equal throughout Ω, f=g?

2. If $f(z_0) = g(z_0)$ for some $z_0 \in \Omega$ and f and g are analytic in a domain Ω, does that imply that f=g on Ω?

3. If f and g are analytic in a domain Ω, and for a sequence $\{z_n\}$ of distinct points of Ω $f(z_n) = g(z_n)$, does that imply that f=g identically on Ω?

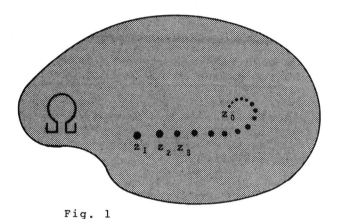

Fig. 1

Solution: 1. We shall quote here without a proof (which is not very difficult) the following theorem.

Theorem

If $f(z)$ is an analytic function in a domain Ω and $\{z_n\}$ is an infinite sequence of distinct points of Ω such that

1. $\lim_{n \to \infty} z_n = z_0$ exists and belongs to Ω, $z_0 \in \Omega$

2. $f(z_n) = 0$ for n = 1,2,3,... then f = 0 identically in the entire domain Ω.

Since f and g have identical power series around $z_0 \in \Omega$

$$f(z) = g(z) = C_0 + c_1(z-z_0) + \ldots + c_n(z-z_0)^n + \ldots \quad (1)$$

Let $\{z_n\}$ be an infinite sequence of distinct points of Ω such that

$$z_n \to z \in \Omega \qquad (2)$$

Then, for each z_n

$$(f-g)(z_n) = 0, \quad n = 1, 2, \ldots \qquad (3)$$

Therefore, by the above theorem

$$(f-g)(z) = 0 \quad \text{or}$$
$$f(z) = g(z) \qquad (4)$$

Since z was an arbitrary element of Ω, that proves that $f = g$ throughout Ω.

2. This condition does not imply that $f = g$ everywhere on Ω. Let $f(z) = z$ and $g(z) = z^2$ and let

$$\Omega = \{z : |z| < 1\} \qquad (5)$$

Then $f(0) = g(0) = 0$ but $f(z) \neq g(z)$ for all $z \neq 0$.

3. This condition does not imply that $f = g$ throughout Ω. From the above quoted theorem we conclude that an analytic function is not "elastic". If the function is changed at one point, then in order to remain analytic it has to be changed (adjusted) everywhere.

CHAPTER 15

LAURENT EXPANSION

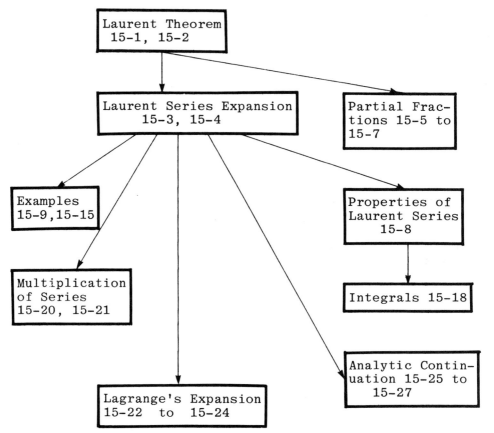

This chart is provided to facilitate rapid understanding of the interrelationships of the topics and subject matter in this chapter. Also shown are the problem numbers associated with the subject matter.

LAURENT SERIES EXPANSION

• **PROBLEM 15-1**

If the function $f(z)$ is analytic at $z = z_0$, then it can be expanded into a Taylor series about z_0. This series has a finite circle of convergence. Outside this circle, the series will not converge. Can we find a series representation for $f(z)$ outside the circle of convergence?

In most cases, the answer is yes. This new series includes

the negative powers of $(z-z_0)$.

Consider the function
$$f(z) = \frac{1}{1-z} \tag{1}$$

Find its series representation around $z_0 = 0$ for $|z| < 1$ and $|z| > 1$.

Solution: In Chapter XIV, we proved the following formula:
$$\frac{1}{\omega-z} = \sum_{n=0}^{\infty} \frac{(z-z_0)^n}{(\omega-z_0)^{n+1}} \quad \text{for} \quad |z-z_0| < |\omega-z_0| \tag{2}$$

Multiplying (2) by minus one,
$$\frac{1}{z-\omega} = -\sum_{n=0}^{\infty} \frac{(z-z_0)^n}{(\omega-z_0)^{n+1}} \quad \text{for} \quad |z-z_0| < |\omega-z_0| \tag{3}$$

Interchanging ω with z in (2), we obtain
$$\frac{1}{z-\omega} = \sum_{n=0}^{\infty} \frac{(\omega-z_0)^n}{(z-z_0)^{n+1}} \quad \text{for} \quad |\omega-z_0| < |z-z_0| \tag{4}$$

Note that (4) can be written in the form
$$\frac{1}{z-\omega} = \sum_{n=-1}^{-\infty} \frac{(z-z_0)^n}{(\omega-z_0)^{n+1}} \quad \text{for} \quad |\omega-z_0| < |z-z_0|, \tag{5}$$

by replacing n by $-n-1$.

Setting $z_0 = 0$ and $\omega = 0$, we have (from (2)),
$$\frac{1}{1-z} = \sum_{n=0}^{\infty} z^n \quad \text{for} \quad |z| < 1 \tag{6}$$

From (4) we obtain
$$\frac{1}{1-z} = -\sum_{n=1}^{\infty} z^{-n-1} \quad \text{for} \quad |z| > 1 \tag{7}$$
$$= -\sum_{n=-1}^{-\infty} z^n$$

Both series represent different functions whose domains of definition are disjoint. But two series, (6) and (7), converge to the same expression $\frac{1}{1-z}$. We say that each series is the analytic continuation of the other series.

• PROBLEM 15-2

<u>Theorem</u> (Laurent)

If $f(z)$ is analytic in an annulus D with center $z = z_0$, then

$$f(z) = \sum_{n=-\infty}^{\infty} c_n (z-z_0)^n \quad \text{for } z \in D, \tag{1}$$

where

$$c_n = \frac{1}{2\pi i} \int_C \frac{f(\omega) d\omega}{(\omega - z_0)^{n+1}} \tag{2}$$

An annulus D is an open connected domain bounded by two concentric circles, and C is any simple closed path lying in D and enclosing z_0 (see Fig. 1).

Prove this theorem.

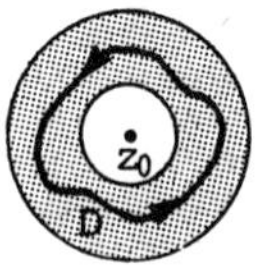

Fig. 1

<u>Solution</u>: Let z be any point in D, and C_1 and C_2 two circles between z and the boundaries of D (see Fig. 2).

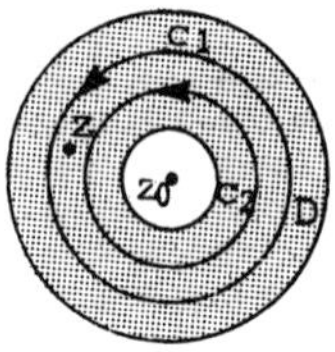

Fig. 2

The function $f(z)$ is analytic between and on C_1 and C_2. Hence,

$$f(z) = \frac{1}{2\pi i} \int_{C_1} \frac{f(\omega)}{\omega - z} d\omega - \frac{1}{2\pi i} \int_{C_2} \frac{f(\omega)}{\omega - z} d\omega \tag{3}$$

In $\int_{C_1} \frac{f(\omega)}{\omega - z} d\omega$ we replace $\frac{1}{\omega - z}$ by

$$\frac{1}{\omega - z} = \sum_{n=0}^{\infty} \frac{(z-z_0)^n}{(\omega - z_0)^{n+1}} \tag{4}$$

Note that for the points on C_1,

$$|z - z_0| < |\omega - z_0| \tag{5}$$

Similarly, in $\int_{C_2} \frac{f(\omega)}{\omega - z} d\omega$, because $|z - z_0| > |\omega - z_0|$, we can replace

$$\frac{1}{\omega-z} = -\sum_{n=-1}^{-\infty} \frac{(z-z_0)^n}{(\omega-z_0)^{n+1}} \tag{6}$$

Substituting (4) and (6) into (3), we obtain

$$f(z) = \sum_{n=0}^{\infty} \left[\frac{1}{2\pi i} \int_{C_1} \frac{f(\omega)d\omega}{(\omega-z_0)^{n+1}} \right] (z-z_0)^n$$

$$+ \sum_{n=-1}^{-\infty} \left[\frac{1}{2\pi i} \int_{C_2} \frac{f(\omega)d\omega}{(\omega-z_0)^{n+1}} \right] (z-z_0)^n \tag{7}$$

The function $\dfrac{f(\omega)}{(\omega-z_0)^{n+1}}$ is analytic everywhere in D, therefore we can replace (deform) both paths C_1 and C_2 with C to obtain

$$f(z) = \sum_{n=0}^{\infty} \left[\frac{1}{2\pi i} \int_{C} \frac{f(\omega)d\omega}{(\omega-z_0)^{n+1}} \right] (z-z_0)^n$$

$$+ \sum_{n=-1}^{-\infty} \left[\frac{1}{2\pi i} \int_{C} \frac{f(\omega)d\omega}{(\omega-z_0)^{n+1}} \right] (z-z_0)^n \tag{8}$$

$$= \sum_{n=0}^{\infty} c_n (z-z_0)^n + \sum_{n=-1}^{-\infty} c_n (z-z_0)^n$$

$$= \sum_{n=-\infty}^{+\infty} c_n (z-z_0)^n$$

Series (8) is called the Laurent series.

● **PROBLEM 15-3**

For all finite z the function e^z has the Taylor expansion

$$e^z = \sum_{n=0}^{\infty} \frac{z^n}{n!}, \quad z \neq 0 \tag{1}$$

Find the series expansion of $e^{\frac{1}{z}}$.

Solution: Replacing z by $\frac{1}{z}$ in (1), we obtain

$$e^{\frac{1}{z}} = \sum_{n=0}^{\infty} \frac{\left(\frac{1}{z}\right)^n}{n!} = \sum_{n=0}^{\infty} \frac{1}{n!} z^{-n}$$

$$= \sum_{n=0}^{-\infty} \frac{1}{(-n)!} z^n \quad \text{for } |z| > 0. \tag{2}$$

The series (2) is called the Laurent series for $e^{\frac{1}{z}}$. It converges for $|z| > 0$. One may say that it converges in the infinite annulus $|z| > 0$.

• **PROBLEM 15-4**

Find the Laurent series for

$$f(z) = \frac{1}{z-2i} \qquad (1)$$

about $z_0 = 1$. Establish domains of convergence.

Solution: Function $f(z)$ has only one singularity at $z = 2i$ (see Fig. 1).

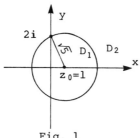

Fig. 1

The domains of convergence are

$$D_1 = \{z: |z-1| < |2i-1| = \sqrt{5}\} \qquad (2)$$

$$D_2 = \{z: |z-1| > \sqrt{5}\} \qquad (3)$$

From eq.(3) of Problem 15-1, we obtain

$$\frac{1}{z-2i} = -\sum_{n=0}^{\infty} \frac{(z-1)^n}{(2i-1)^{n+1}} \qquad (4)$$

for domain D_1.

For D_2 we have (applying (4) of Problem 15-1),

$$\frac{1}{z-2i} = \sum_{n=0}^{\infty} \frac{(2i-1)^n}{(z-1)^{n+1}} = \sum_{n=-1}^{-\infty} \frac{(z-1)^n}{(2i-1)^{n+1}} \qquad (5)$$

Note that the coefficients C_n in D_1 are

$$C_n = -\frac{1}{(2i-1)^{n+1}} = \frac{f^{(n)}(1)}{n!} \qquad (6)$$

where $f(z) = \frac{1}{z-2i}$.

Series (4) is a Taylor series representation of $\frac{1}{z-2i}$ in D_1.

Series (5) is a Laurent series representation of $\frac{1}{z-2i}$ in D_2.

• **PROBLEM 15-5**

Find the Laurent series for

$$f(z) = \frac{1}{1-z^2} \qquad (1)$$

about $z_0 = 1$.

Solution: We shall show two methods of solving this problem.

1. Function $f(z)$ has two singular points: $z = 1$, $z = -1$ (see Fig. 1).

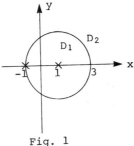

Fig. 1

$$f(z) = \frac{-1}{z-1} \cdot \frac{1}{z+1} \qquad (2)$$

For $D_1 = \{z: 0 < |z-1| < 2\}$, we have

$$\frac{1}{z+1} = -\sum_{n=0}^{\infty} \frac{(z-1)^n}{(-1-1)^{n+1}} \qquad \text{for } |z-1| < 2 \qquad (3)$$

Hence,

$$f(z) = \sum_{n=0}^{\infty} \frac{(z-1)^{n-1}}{(-2)^{n+1}} = \frac{1}{4} \sum_{n=-1}^{\infty} \frac{(z-1)^n}{(-2)^n}$$

$$= \frac{1}{4} \sum_{n=-1}^{\infty} \frac{(1-z)^n}{2^n} \qquad (4)$$

For $D_2 = \{z: |z-1| > 2\}$, we obtain

$$\frac{1}{z+1} = \sum_{n=-1}^{-\infty} \frac{(z-1)^n}{(-1-1)^{n+1}} \qquad \text{for } |z-1| > 2 \qquad (5)$$

Thus,

$$f(z) = -\sum_{n=-1}^{-\infty} \frac{(z-1)^{n-1}}{(-2)^{n+1}} = -\frac{1}{4} \sum_{n=-2}^{-\infty} \frac{(z-1)^n}{(-2)^n}$$

$$= -\frac{1}{4} \sum_{n=-2}^{-\infty} \frac{(1-z)^n}{2^n} \qquad (6)$$

2. The second method involves the partial fraction expansion.

$$f(z) = \frac{-1}{2(z-1)} + \frac{1}{2(z+1)} \qquad (7)$$

The first term $-\frac{1}{2} \frac{1}{z-1}$ is already a Laurent series.

Applying (3) for D_1, we obtain

$$f(z) = -\frac{1}{2}\frac{1}{z-1} + \frac{1}{2}\left[-\sum_{n=0}^{\infty}\frac{(z-1)^n}{(-2)^{n+1}}\right]$$

$$= \frac{1}{4}\sum_{n=-1}^{\infty}\frac{(1-z)^n}{2^n} \qquad (8)$$

Applying (5) for D_2, we obtain

$$f(z) = -\frac{1}{2}\frac{1}{z-1} + \frac{1}{2}\sum_{n=-1}^{-\infty}\frac{(z-1)^n}{(-2)^{n+1}}$$

$$= -\frac{1}{4}\sum_{n=-2}^{-\infty}\frac{(1-z)^n}{2^n} \qquad (9)$$

● PROBLEM 15-6

Find the Laurent series about $z = 2$ for the function

$$f(z) = \frac{1}{(z-2)(z-3)} \qquad (1)$$

<u>Solution</u>: The function $f(z)$ can be decomposed into partial fractions:

$$\frac{1}{(z-2)(z-3)} = -\frac{1}{z-2} + \frac{1}{z-3} \qquad (2)$$

The singularities of $f(z)$ are located at $z = 2$ and $z = 3$ (see Fig. 1).

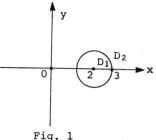

Fig. 1

First consider domain D_1, $0 < |z-2| < 1$. The first fraction in (2) is already the Laurent series (consisting of one element). The second fraction can be expanded as follows:

$$\frac{1}{z-3} = -\sum_{n=0}^{\infty}\frac{(z-2)^n}{(3-2)^{n+1}} = -\sum_{n=0}^{\infty}(z-2)^n \qquad (3)$$

From (2) and (3), we obtain

$$\frac{1}{(z-2)(z-3)} = -\frac{1}{z-2} - \sum_{n=0}^{\infty}(z-2)^n$$

$$= -\sum_{n=-1}^{\infty}(z-2)^n \qquad (4)$$

Series (4) converges for $0 < |z-2| < 1$. Domain D_2 consists of z such that $|z-2| > 1$. We obtain

$$\frac{1}{z-3} = \sum_{n=0}^{\infty} \frac{(3-2)^n}{(z-2)^{n+1}} = \sum_{n=0}^{\infty} \frac{1}{(z-2)^{n+1}} \tag{5}$$

Hence,

$$\frac{1}{(z-2)(z-3)} = -\frac{1}{z-2} + \sum_{n=0}^{\infty} \frac{1}{(z-2)^{n+1}} = \sum_{n=2}^{\infty} \frac{1}{(z-2)^n} \tag{6}$$

● **PROBLEM 15-7**

Expand the function

$$f(z) = \frac{z}{(z-2)(z-3)} \tag{1}$$

in the Laurent series for $2 < |z| < 3$.

Solution: The annulus $2 < |z| < 3$ is shown in Fig. 1.

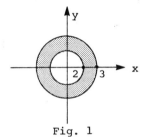

Fig. 1

Since the annulus is centered at $z_0 = 0$, the series must be in power of z. Function (1) can be written as

$$\frac{z}{(z-2)(z-3)} = \frac{-2}{z-2} + \frac{3}{z-3} \tag{2}$$

where

$$\frac{-2}{z-2} = -2 \sum_{n=-1}^{-\infty} \frac{z^n}{2^{n+1}} = - \sum_{n=-1}^{-\infty} \frac{z^n}{2^n} \tag{3}$$

$$= - \sum_{n=1}^{\infty} \frac{2^n}{z^n} \quad \text{for } 2 < |z|$$

and

$$\frac{3}{z-3} = -3 \sum_{n=0}^{\infty} \frac{z^n}{3^{n+1}} = - \sum_{n=0}^{\infty} \frac{z^n}{3^n} \tag{4}$$

for $|z| < 3$

Putting together (3) and (4), we obtain the Laurent series

$$\frac{z}{(z-2)(z-3)} = - \sum_{n=1}^{\infty} \frac{2^n}{z^n} - \sum_{n=0}^{\infty} \frac{z^n}{3^n} \tag{5}$$

for $2 < |z| < 3$.

● PROBLEM 15-8

> Show that the Laurent series can be regarded as a generalization of the Taylor series.

Solution: If $f(z)$ is analytic in an annulus D, then the Laurent series for $f(z)$ converges in D. Suppose there are some singular points outside D. Then D can be increased until singular points of f are on both boundaries of D.

Let $\{z_1, z_2, \ldots, z_k\}$ be the singular points of $f(z)$ ordered in increasing distance from z_0. We define

$$D_\ell \equiv \{z : |z_{\ell-1} - z_0| < |z - z_0| < |z_\ell - z_0|\} \qquad (1)$$

for $\ell = 1, 2, \ldots, k$

and

$$D_{k+1} \equiv \{z : |z_\ell - z_0| < |z - z_0|\} \qquad (2)$$

Note that the last D_{k+1} is the unbounded outer region.

In each domain D_ℓ there is exactly one Laurent series in powers of $(z-z_0)$ that converges to $f(z)$.

Suppose $f(z)$ is analytic at $z = z_0$. Then, for the negative coefficients,

$$C_n = C_{-|n|} \qquad (3)$$

and

$$C_n = \frac{1}{2\pi i} \int_C f(\omega)(\omega - z_0)^{|n|-1} d\omega = 0 \qquad (4)$$

For $n \leq -1$ and C is a simple closed path about z_0, including no singular points of $f(z)$.

Then the Laurent series reduces to the Taylor series, which contains only positive powers of $(z-z_0)$.

EXAMPLES

● PROBLEM 15-9

> Compute the Laurent series about $z = 0$ for
>
> 1. $\sin z$ (1)
>
> 2. $\dfrac{\sin z}{z}$ (2)
>
> 3. $\sin \dfrac{1}{z}$ (3)
>
> 4. $\dfrac{\sin z}{z^4}$ (4)

Solution: 1. The function sin z is analytic everywhere. Therefore, the Laurent series is the same as the Taylor expansion. Since $z_0 = 0$, it is a Maclaurin series.

$$\sin z = z - \frac{z^3}{3!} + \frac{z^5}{5!} - \ldots + (-1)^{n-1} \frac{z^{2n-1}}{(2n-1)!} + \ldots \quad (5)$$

$$= \sum_{n=0}^{\infty} (-1)^n \frac{z^{2n+1}}{(2n+1)!}$$

2. Dividing (5) by z, we find

$$\frac{\sin z}{z} = \sum_{n=0}^{\infty} (-1)^n \frac{z^{2n}}{(2n+1)!} \quad (6)$$

This is also a Maclaurin series.

3. Substituting $\frac{1}{z}$ for z in (5), we obtain

$$\sin \frac{1}{z} = \frac{1}{z} - \frac{1}{z^3} \cdot \frac{1}{3!} + \frac{1}{z^5} \frac{1}{5!} + \ldots + (-1)^{n-1} \frac{1}{z^{2n-1}} \cdot \frac{1}{(2n-1)!}$$

$$+ \ldots = \sum_{n=0}^{\infty} (-1)^n z^{-2n-1} \cdot \frac{1}{(2n+1)!} \quad (7)$$

for all $z \neq 0$.

4. Dividing (5) by z^4, we find

$$\frac{\sin z}{z^4} = \sum_{n=0}^{\infty} (-1)^n \frac{z^{2n-3}}{(2n+1)!} \quad (8)$$

for all $z \neq 0$.

● **PROBLEM 15-10**

Derive the Laurent series for the function

$$f(z) = \frac{1}{z^2(z+2)} \quad (1)$$

about $z_0 = -2$.

Solution Function (1) has to be expressed as a series of powers of z+2. Hence, it is enough to derive the Laurent series for $\frac{1}{z^2}$, and multiply the result by $\frac{1}{z+2}$.

Function (1) has singularities at $z = 0$ and $z = -2$.

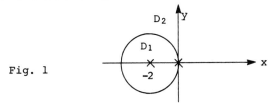

Fig. 1

We have

$$\frac{1}{z} = -\sum_{n=0}^{\infty} \frac{(z+2)^n}{2^{n+1}} \qquad \text{for } |z+2| < 2 \qquad (2)$$

and

$$\frac{1}{z} = \sum_{n=-1}^{-\infty} \frac{(z+2)^n}{2^{n+1}} \qquad \text{for } |z+2| > 2 \qquad (3)$$

Differentiating (2) term by term, we find

$$\frac{1}{z^2} = \sum_{n=0}^{\infty} \frac{(n+1)}{2^{n+2}} (z+2)^n \qquad \text{for } |z+2| < 2 \qquad (4)$$

Differentiating (3), we obtain

$$\frac{1}{z^2} = -\sum_{n=-2}^{-\infty} \frac{n+1}{2^{n+2}} (z+2)^n \qquad \text{for } |z+2| > 2 \qquad (5)$$

Multiplying by $\frac{1}{z+2}$, we obtain

$$f(z) = \sum_{n=0}^{\infty} \frac{n+1}{2^{n+2}} (z+2)^{n-1} = \sum_{n=-1}^{\infty} \frac{n+2}{2^{n+3}} (z+2)^n \qquad (6)$$

$$\text{for domain } D_1 : |z+2| < 2$$

$$f(z) = -\sum_{n=-2}^{-\infty} \frac{n+1}{2^{n+2}} (z+2)^{n-1} = \sum_{n=-3}^{-\infty} \frac{n+2}{2^{n+3}} (z+2)^n \qquad (7)$$

$$\text{for domain } D_2 : |z+2| > 2$$

● **PROBLEM 15-11**

Find the Laurent expansions about $z_0 = i$ for the function

$$f(z) = \frac{z^2 + z(2i+1) - 1}{z(z-i)^2(z+i)^2} \qquad (1)$$

Solution: Function (1) has three singular points at $z = 0$, $z = i$, and $z = -i$. The Laurent expansions should be found for the three domains shown in Fig. 1.

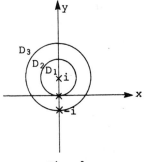

Fig. 1

Function (1) can be written in the form

$$f(z) = \frac{1}{(z-i)^2}\left[\frac{1}{z} - \frac{1}{(z+i)^2}\right] \qquad (2)$$

We shall find the series expansions of $\frac{1}{z}$ and $\frac{1}{(z+i)^2}$, and multiply them by $\frac{1}{(z-i)^2}$. We have

$$\frac{1}{z} = -\sum_{n=0}^{\infty} \frac{(z-i)^n}{(-i)^{n+1}} \qquad \text{for } |z-i| < 1 \qquad (3)$$

which is domain D_1.

$$\frac{1}{z} = \sum_{n=-1}^{\infty} \frac{(z-i)^n}{(-i)^{n+1}} \qquad \text{for } |z-i| > 1 \qquad (4)$$

(that is, for domains D_2, D_3 and their boundary).

$$\frac{1}{z+i} = -\sum_{n=0}^{\infty} \frac{(z-i)^n}{(-2i)^{n+1}} \qquad \text{for } |z-i| < 2 \qquad (5)$$

which is domain D_1 and D_2.

$$\frac{1}{z+i} = \sum_{n=-1}^{-\infty} \frac{(z-i)^n}{(-2i)^{n+1}} \qquad \text{for } |z-i| > 2 \qquad (6)$$

which is domain D_3.

Term by term differentiation of (5) yields

$$\frac{1}{(z+i)^2} = \sum_{n=0}^{\infty} \frac{(n+1)(z-i)^n}{(-2i)^{n+2}} \qquad \text{for } D_1 \text{ and } D_2 \qquad (7)$$

and of (6) yields

$$\frac{1}{(z+i)^2} = -\sum_{n=-2}^{-\infty} \frac{(n+1)(z-i)^n}{(-2i)^{n+2}} \qquad \text{for } D_3 \qquad (8)$$

Putting together all results we obtain the following:

For domain $D_1 : |z-i| < 1$

$$f(z) = \frac{-1}{(z-i)^2} \cdot \sum_{n=0}^{\infty} \frac{(z-i)^n}{(-i)^{n+1}} - \frac{1}{(z-i)^2} \sum_{n=0}^{\infty} \frac{(n+1)(z-i)^n}{(-2i)^{n+2}}$$

$$= \sum_{n=0}^{\infty} \left[\frac{-1}{(-i)^{n+1}} - \frac{n+1}{(-2i)^{n+2}}\right] (z-i)^{n-2} \qquad (9)$$

For domain $D_2 : 1 < |z-1| < 2$

$$f(z) = \frac{1}{(z-i)^2} \sum_{n=-1}^{-\infty} \frac{(z-i)^n}{(-i)^{n+1}} - \frac{1}{(z-i)^2} \sum_{n=0}^{\infty} \frac{(n+1)(z-i)^n}{(-2i)^{n+2}}$$

$$= \sum_{n=-1}^{-\infty} \frac{(z-i)^{n-2}}{(-i)^{n+1}} - \sum_{n=0}^{\infty} \frac{n+1}{(-2i)^{n+2}} \cdot (z-i)^{n-2} \qquad (10)$$

For D_3 : $|z-i| > 2$

$$f(z) = \frac{1}{(z-i)^2} \sum_{n=-1}^{-\infty} \frac{(z-i)^n}{(-i)^{n+1}} + \frac{1}{(z-i)^2} \sum_{n=-2}^{-\infty} \frac{(n+1)(z-i)^n}{(-2i)^{n+2}}$$

$$= \sum_{n=-1}^{-\infty} \frac{(z-i)^{n-2}}{(-i)^{n+1}} + \sum_{n=-2}^{-\infty} \frac{n+1}{(-2i)^{n+2}} (z-i)^{n-2} \quad (11)$$

PROBLEM 15-12

Derive the Laurent expansion for

$$\csc z = \frac{1}{\sin z} \quad (1)$$

about $z = 0$.

Solution: Since

$$\sin z = 0 \quad \text{for} \quad z = \pm n\pi, \quad (2)$$

the function $\csc z = \frac{1}{\sin z}$ is analytic in the domain $0 < |z| < \pi$.
The expansion of $\sin z$ is

$$\sin z = \sum_{n=1}^{\infty} (-1)^{n+1} \frac{z^{2n-1}}{(2n-1)!} \quad (3)$$

We shall apply the following theorem:

If the power series

$$f(z) = \sum_{n=0}^{\infty} a_n (z-z_0)^n \quad (4)$$

$$g(z) = \sum_{n=0}^{\infty} b_n (z-z_0)^n \quad (5)$$

have non-zero radii of convergence, and $g(z_0) \neq 0$, then there exists a power series $\sum_{n=0}^{\infty} c_n (z-z_0)^n$ and a number $\alpha > 0$, such that

$$\frac{f(z)}{g(z)} = \sum_{n=0}^{\infty} c_n (z-z_0)^n, \quad |z-z_0| < \alpha \quad (6)$$

The coefficients c_n are given by

$$c_n = \frac{a_n - b_n c_0 - b_{n-1} c_1 - \ldots - b_1 c_{n-1}}{b_0} \quad (7)$$

Substituting (3) into (1), we obtain

$$\csc z = \frac{1}{\sin z} = \frac{1}{z - \frac{z^3}{3!} + \frac{z^5}{5!} - \ldots} \quad (8)$$

We cannot apply the theorem in this form, because for $z = 0$ the denominator is equal to zero.

To obviate this difficulty, we can express (8) in the form

$$\csc z = \frac{1}{z} \cdot \frac{1}{1 - \frac{z^2}{3!} + \frac{z^4}{5!} - \frac{z^6}{7!} + \ldots} = \frac{1}{z} \cdot \frac{f(z)}{g(z)} \quad (9)$$

Since $f(z) = 1$, we have

$$a_0 = 1, \quad a_1 = a_2 = a_3 = \ldots = 0$$

For $g(z)$ we have

$$b_0 = 1, \quad b_1 = 0, \quad b_2 = -\frac{1}{3!}, \quad b_3 = 0, \quad b_4 = \frac{1}{5!}, \quad b_5 = 0, \ldots$$

From (7) we obtain

$$c_0 = \frac{a_0}{b_0} = 1$$

$$c_1 = \frac{a_1 - b_1 c_0}{b_0} = 0$$

$$c_2 = \frac{a_2 - b_2 c_0 - b_1 c_1}{b_0} = \frac{1}{3!} \quad (10)$$

$$c_3 = -b_3 c_0 - b_2 c_1 - b_1 c_2 = 0$$

$$c_4 = -b_4 c_0 - b_2 c_2 = -\frac{1}{5!} + \frac{1}{3!} \cdot \frac{1}{3!}$$

$$c_5 = -b_5 c_0 - b_4 c_1 - b_3 c_2 - b_2 c_3 - b_1 c_4 = 0$$

Thus,

$$\csc z = \frac{1}{z} \cdot \frac{1}{g(z)} = \frac{1}{z} \cdot \left[1 + \frac{1}{3!} z^2 + \left(\frac{1}{3!} \cdot \frac{1}{3!} - \frac{1}{5!}\right) z^4 + \ldots\right] \quad (11)$$

$$= \frac{1}{z} + \frac{1}{6} z + \frac{7}{360} z^3 + \ldots$$

Note that from

$$\csc z = \frac{1}{\sin z} = \frac{1}{z - \frac{z^3}{3!} + \frac{z^5}{5!} - \frac{z^7}{7!} + \ldots} \quad (12)$$

we can obtain series (10) by carrying out the division.

● **PROBLEM 15-13**

The function

$$e^{\alpha\left(z + \frac{1}{z}\right)} \quad (1)$$

is analytic in

$$0 < |z| < \infty$$

Find the Laurent expansion of this function.

Solution: We have

$$e^{\alpha(z+\frac{1}{z})} = e^{\alpha z} \cdot e^{\frac{\alpha}{z}} \qquad (2)$$

Thus,

$$e^{\alpha(z+\frac{1}{z})} = \left(1 + \alpha z + \frac{\alpha^2}{2!} z^2 + \ldots\right) \cdot \left(1 + \frac{\alpha}{z} + \frac{\alpha^2}{2!} \frac{1}{z^2} + \ldots\right) \qquad (3)$$

$$= \sum_{n=-\infty}^{+\infty} c_n z^n$$

Note that we can avoid writing negative n by

$$\sum_{n=-\infty}^{+\infty} c_n z^n = c_0 + \sum_{n=1}^{\infty} c_n z^n + \sum_{n=-1}^{-\infty} c_n z^n \qquad (4)$$

$$= c_0 + \sum_{n=1}^{\infty} c_n \left(z^n + \frac{1}{z^n}\right)$$

where

$$c_n = c_{-n} = \sum_{m=0}^{\infty} \frac{\alpha^{n+m}}{(n+m)!} \cdot \frac{\alpha^m}{m!} \qquad (5)$$

for $n \geq 0$

● **PROBLEM 15-14**

Show that the Laurent series expansion of the function

$$f(z) = \sinh\left(z + \frac{1}{z}\right) \qquad (1)$$

in the domain $|z| > 0$ is given by

$$f(z) = \sum_{n=-\infty}^{\infty} c_n z^n \qquad (2)$$

where

$$c_n = \frac{1}{2\pi} \int_0^{2\pi} \cos n\theta \, \sinh(2\cos\theta) d\theta \qquad (3)$$

Solution: Since $\sinh\left(z + \frac{1}{z}\right)$ is analytic for all $z \neq 0$, we have

$$\sinh\left(z + \frac{1}{z}\right) = \sum_{n=-\infty}^{\infty} c_n z^n \qquad (4)$$

$$c_n = \frac{1}{2\pi i} \int_C \frac{\sinh\left(\omega + \frac{1}{\omega}\right)}{\omega^{n+1}} d\omega \qquad (5)$$

for $n = 0, \pm 1, \pm 2, \ldots$

We can choose the curve C as the circle $|z| = 1$, then for ω on C

$$\omega = e^{i\theta}$$

and

$$c_n = \frac{1}{2\pi i} \int_0^{2\pi} \frac{\sinh(e^{i\theta} + e^{-i\theta})}{e^{i\theta n + i\theta}} \cdot ie^{i\theta} d\theta$$

$$= \frac{1}{2\pi} \int_0^{2\pi} \sinh(2\cos\theta)(\cos n\theta - i\sin n\theta)d\theta \qquad (6)$$

$$= \frac{1}{2\pi} \int_0^{2\pi} \cos n\theta \sinh(2\cos\theta)d\theta$$

$$- \frac{i}{2\pi} \int_0^{2\pi} \sin n\theta \sinh(2\cos\theta)d\theta$$

However,

$$\int_0^{2\pi} \sin n\theta \sinh(2\cos\theta)d\theta = 0 \qquad (7)$$

for $n = 0, \pm 1, \pm 2, \ldots$

because for $\theta = 2\pi - \alpha$ we have

$$\int_0^{2\pi} \sin n\theta \sinh(2\cos\theta)d\theta = -\int_0^{2\pi} \sin n\alpha \sinh(2\cos\alpha)d\alpha \qquad (8)$$

Replacing α by θ in (8), we obtain (7).

Thus,

$$c_n = \frac{1}{2\pi} \int_0^{2\pi} \cos n\theta \sinh(2\cos\theta)d\theta \qquad (9)$$

Note that since

$$c_n = c_{-n}$$

we can express

$$\sinh\left(z + \frac{1}{z}\right) = c_0 + \sum_{n=1}^{\infty} c_n \left(z^n + \frac{1}{z^n}\right) \qquad (10)$$

• **PROBLEM 15-15**

Derive the Laurent series for

$$f(z) = \frac{1}{z - \omega} \qquad (1)$$

about $z_0 = 0$, in the domain $|z| > |\omega|$. Substituting $z = e^{i\theta}$ in the expansion, derive the formulas

$$\sum_{n=0}^{\infty} \omega^n \cos\left[(n+1)\theta\right] = \frac{\cos\theta - \omega}{1 - 2\omega\cos\theta + \omega^2} \tag{2}$$

$$\sum_{n=0}^{\infty} \omega^n \sin\left[(n+1)\theta\right] = \frac{\sin\theta}{1 - 2\omega\cos\theta + \omega^2} \tag{3}$$

where ω is real, and $0 < |\omega| < 1$.

Solution: We have

$$\frac{1}{z-\omega} = \sum_{n=-1}^{-\infty} \frac{z^n}{\omega^{n+1}} = \sum_{n=0}^{\infty} \frac{\omega^n}{z^{n+1}} \tag{4}$$

for $|z| > |\omega|$.

Substituting $z = e^{i\theta} = \cos\theta + i\sin\theta$, we obtain

$$\frac{1}{z-\omega} = \frac{1}{\cos\theta - \omega + i\sin\theta} = \frac{\cos\theta - \omega - i\sin\theta}{1 - 2\omega\cos\theta + \omega^2} \tag{5}$$

Substituting (5) into (4), we find

$$\sum_{n=0}^{\infty} \frac{\omega^n}{z^{n+1}} = \sum_{n=0}^{\infty} \frac{\omega^n}{\cos[(n+1)\theta] + i\sin[(n+1)\theta]}$$

$$= \sum_{n=0}^{\infty} \omega^n \left[\cos\left[(n+1)\theta\right] - i\sin\left[(n+1)\theta\right]\right] \tag{6}$$

$$= \frac{\cos\theta - \omega - i\sin\theta}{1 - 2\omega\cos\theta + \omega^2}$$

Comparing the real and imaginary parts in (6), we obtain equations (2) and (3).

APPLICATIONS

• **PROBLEM 15-16**

Derive the Laurent series about $z_0 = 0$ for the function

$$f(z) = e^{\frac{u}{2}(z - \frac{1}{z})} \tag{1}$$

Show that the coefficients of the expansion are the Bessel functions of the first kind.

Solution: Since the function $e^{\frac{u}{2}(z - \frac{1}{z})}$ is analytic for $z \neq 0$, its Laurent expansion about $z_0 = 0$ is

$$f(z) = \sum_{n=-\infty}^{+\infty} c_n z^n \tag{2}$$

with coefficients given by

$$c_n = \frac{1}{2\pi i} \int_C \frac{f(\omega)}{\omega} d\omega \qquad (3)$$

and for the domain $0 < |z|$. Here, C is any simple closed path lying in the domain $|z| > 0$ and enclosing $z_0 = 0$. We can take $|z| = 1$, then

$$z = e^{i\theta} \qquad (4)$$

We have

$$e^{\frac{u}{2}(z - \frac{1}{z})} = e^{\frac{u}{2}(\cos\theta + i\sin\theta - \frac{1}{\cos\theta + i\sin\theta})}$$

$$= e^{iu\sin\theta} \qquad (5)$$

From (3), we obtain

$$c_n = \frac{1}{2\pi i} \int_C \frac{f(\omega) d\omega}{\omega^{n+1}} = \frac{1}{2\pi i} \int_C \frac{e^{\frac{u}{2}(z - \frac{1}{z})}}{z^{n+1}} dz$$

$$= \frac{1}{2\pi} \int_C \frac{e^{iu\sin\theta}}{e^{i\theta n}} d\theta = \frac{1}{2\pi} \int_0^{2\pi} e^{i(u\sin\theta - \theta n)} d\theta \qquad (6)$$

$$= \frac{1}{2\pi} \int_0^{2\pi} \cos\left[n\theta - u\sin\theta\right] d\theta$$

We see that the coefficients c_n are the Bessel functions of the first kind, usually denoted by J_n. Hence,

$$f(z) = \sum_{n=-\infty}^{+\infty} J_n(u) z^n \qquad (7)$$

where

$$J_n(u) = (-1)^n J_{-n}(u) = \frac{1}{2\pi} \int_0^{2\pi} \cos\left[n\theta - u\sin\theta\right] d\theta \qquad (8)$$

● **PROBLEM 15-17**

1. Show that the Laurent expansion of

$$f(z) = \frac{1}{e^z - 1} \qquad (1)$$

about $z_0 = 0$ is

$$f(z) = \sum_{n=0}^{\infty} \frac{B_n}{n!} z^{n-1} \qquad (2)$$

where B_n are Bernoulli's numbers. Compute $B_0, B_1, \ldots, B_7$.

2. Prove that

$$1 + \sum_{n=2}^{\infty} \frac{B_n}{n!} z^n = \frac{z}{e^z - 1} + \frac{z}{2} = \frac{z}{2} \coth \frac{z}{2} \qquad (3)$$

3. Utilizing the fact that $\frac{z}{2} \coth \frac{z}{2}$ is an even function, prove that

$$B_3 = B_5 = B_7 = \ldots = B_{2n+1} = 0 \qquad (4)$$

for $n = 1, 2, 3, \ldots$

<u>Solution</u>: 1. Since

$$e^z = 1 + z + \frac{z^2}{2!} + \frac{z^3}{3!} + \ldots \qquad (5)$$

we can denote

$$f(z) = \frac{1}{e^z - 1} = \frac{1}{z + \frac{z^2}{2!} + \frac{z^3}{3!} + \frac{z^4}{4!} + \ldots}$$

$$= \frac{1}{z} \cdot \frac{1}{1 + \frac{z}{2!} + \frac{z^2}{3!} + \frac{z^3}{4!} + \ldots} \qquad (6)$$

$$\frac{g(z)}{h(z)} = \frac{1}{1 + \frac{z}{2!} + \frac{z^2}{3!} + \frac{z^3}{4!} + \ldots} \qquad (7)$$

Thus,

and

$$a_0 = 1, \quad a_1 = a_2 = \ldots = 0$$

$$b_0 = 1, \quad b_1 = \frac{1}{2}, \quad b_2 = \frac{1}{6}, \quad b_3 = \frac{1}{24}, \quad b_4 = \frac{1}{120}, \qquad (8)$$

$$b_5 = \frac{1}{720}, \ldots$$

$$c_0 = \frac{a_0}{b_0} = 1$$

$$c_1 = a_1 - c_0 b_1 = -\frac{1}{2}$$

$$c_2 = -c_0 b_2 - c_1 b_1 = -\frac{1}{6} + \frac{1}{4} = \frac{1}{12} \qquad (9)$$

$$c_3 = -c_0 b_3 - c_1 b_2 - c_2 b_1 = -\frac{1}{24} + \frac{1}{12} - \frac{1}{24} = 0$$

$$c_4 = -c_0 b_4 - c_1 b_3 - c_2 b_2 - c_3 b_1 = -\frac{1}{120} + \frac{1}{2} \cdot \frac{1}{24} - \frac{1}{12} \cdot \frac{1}{6}$$

$$= -\frac{1}{720}$$

$$c_5 = 0$$

$$c_6 = \frac{1}{42} \cdot \frac{1}{6!}$$

Hence,

$$f(z) = \frac{1}{z}\left(1 - \frac{1}{2} z + \frac{1}{12} z^2 - \frac{1}{720} z^4 \ldots \right)$$

$$= z^{-1} - \frac{1}{2} + \frac{1}{12} z - \frac{1}{720} z^3 + \frac{1}{42 \cdot 6!} z^5 \ldots \quad (10)$$

$$= \sum_{n=0}^{\infty} \frac{B_n}{n!} z^{n-1}$$

$B_0 = 1$

$B_1 = -\frac{1}{2}$

$B_2 = \frac{1}{6}$

$B_3 = 0$

$B_4 = -\frac{1}{30}$ \quad (11)

$B_5 = 0$

$B_6 = \frac{1}{42}$

$B_7 = 0$

2. $$\frac{1}{e^z - 1} = \sum_{n=0}^{\infty} \frac{B_n}{n!} z^{n-1} \quad (12)$$

$$\frac{z}{e^z - 1} = \sum_{n=0}^{\infty} \frac{B_n}{n!} z^n$$

$$\frac{z}{e^z - 1} + \frac{z}{2} = B_0 + B_1 z + \frac{z}{2} + \sum_{n=2}^{\infty} \frac{B_n}{n!} z^n$$

$$= 1 + \sum_{n=2}^{\infty} \frac{B_n}{n!} z^n \quad (13)$$

Since $\cot hz = \frac{\cos hz}{\sin hz}$, we have

$$\frac{z}{2} \cot h \frac{z}{2} = \frac{z}{2} \left(\frac{e^{\frac{z}{2}} + e^{-\frac{z}{2}}}{e^{\frac{z}{2}} - e^{-\frac{z}{2}}} \right)$$

$$= \frac{z}{2} \left(\frac{e^{\frac{z}{2}}(e^{\frac{z}{2}} + e^{-\frac{z}{2}})}{e^{\frac{z}{2}}(e^{\frac{z}{2}} - e^{-\frac{z}{2}})} \right) \quad (14)$$

$$= \frac{z}{2} \left(\frac{e^z + 1}{e^z - 1} \right) = \frac{z}{e^z - 1} + \frac{z}{2}$$

3. The function $\frac{z}{2} \cot h \frac{z}{2}$ is an even function.

$$\frac{z}{2} \cot h \frac{z}{2} = \frac{(-z)}{2} \cot h \frac{(-z)}{2} \quad (15)$$

From

$$\frac{z}{2} \coth \frac{z}{2} = 1 + \sum_{n=2}^{\infty} \frac{B_n}{n!} z^n \qquad (16)$$

and the fact that the left-hand side is an even function, we conclude that

$$B_3 = B_5 = B_7 = \ldots = B_{2n+1} = 0 \qquad (17)$$

● **PROBLEM** 15-18

The Laurent expansion for

$$f(z) = \frac{1}{z^2(z+2)} \qquad (1)$$

about $z = -2$ is

$$f(z) = \sum_{n=-1}^{\infty} \frac{n+2}{2^{n+3}} (z+2)^n \quad \text{for } |z+2| < 2 \qquad (2)$$

$$f(z) = -\sum_{n=-3}^{-\infty} \frac{n+2}{2^{n+3}} (z+2)^n \quad \text{for } |z+2| > 2 \qquad (3)$$

Show what integrals can be easily evaluated from this information.

Solution: Let us recall the Laurent theorem:

If $f(z)$ is analytic in an annulus D with center at z_0, then

$$f(z) = \sum_{n=-\infty}^{\infty} c_n (z-z_0)^n \quad \text{for } z \in D \qquad (4)$$

where

$$c_n = \frac{1}{2\pi i} \int_C \frac{f(\omega) d\omega}{(\omega - z_0)^{n+1}} \qquad (5)$$

where C is any simple closed contour lying in D and enclosing z_0.

Comparing (2), (4) and (5), we find

$$\int_C \frac{f(\omega) d\omega}{(\omega - z_0)^{n+1}} = 2\pi i \, c_n = \begin{cases} \frac{n+2}{2^{n+3}} \cdot 2\pi i & \text{for } n \geq -1 \\ 0 & \text{for } n < -1 \end{cases} \qquad (6)$$

Thus, for any closed path lying in D_1 (see Fig. 1) and enclosing $z_0 = -2$,

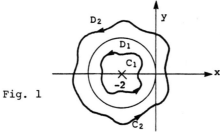

Fig. 1

we have

$$\int_C \frac{1}{z^2(z+2)(z+2)^{n+1}} dz \tag{7}$$

$$= \int_C \frac{1}{z^2(z+2)^{n+2}} dz = \frac{(n+2)\pi i}{2^{n+2}}$$

for $n = -1, 0, 1, 2, \ldots$

Similarly, from (3), (4) and (5) we find

$$\int_C \frac{f(\omega) d\omega}{(\omega - z_0)^{n+1}} = 2\pi i\, c_n = \begin{cases} -\dfrac{n+2}{2^{n+3}} \cdot 2\pi i & \text{for } n \leq -3 \\ 0 & \text{for } n > -3 \end{cases} \tag{8}$$

or

$$\int_C \frac{1}{z^2(z+2)^{n+2}} dz = \begin{cases} -\dfrac{(n+2)\pi i}{2^{n+2}} & \text{for } n \leq -3 \\ 0 & \text{for } n > -3 \end{cases} \tag{9}$$

where C is any closed simple path lying in D_2 and enclosing $z_0 = -2$.

• **PROBLEM** 15-19

Find the series expansion for

$$f(z) = \sum_{n=1}^{\infty} \frac{z^n}{1-z^n} \tag{1}$$

applying the Weierstrass double series theorem.

Solution: We denote

$$f_n(z) = \frac{z^n}{1-z^n}, \quad n = 1, 2, \ldots \tag{2}$$

Then,

$$f(z) = \sum_{n=1}^{\infty} f_n(z) \tag{3}$$

Each function f_n is analytic for $|z| < 1$. Let D be the unit disk. The series (1) is uniformly convergent on every compact subset of D. Let $A \subset D$ be a compact subset of D, and let $\delta > 0$ be the distance between A and $|z| = 1$ (see Fig. 1).

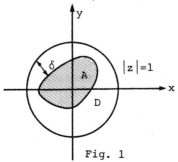

Fig. 1

Then, for all $z \in A$

$$|z| \leq 1 - \delta = \lambda < 1 \tag{4}$$

Hence,

$$\left|\frac{z^n}{1-z^n}\right| \leq \frac{\lambda^n}{1-\lambda^n} \leq \frac{\lambda^n}{1-\lambda} \tag{5}$$

The series $\sum_{n=1}^{\infty} \frac{\lambda^n}{1-\lambda}$ is a geometric series, and since $\lambda < 1$, it is convergent. From the Weierstrass M-test, it follows that series (1) is uniformly convergent on A.

We shall apply the following Weierstrass Double Series Theorem:

Suppose the series

$$f(z) = \sum_{n=1}^{\infty} f_n(z) \tag{6}$$

is uniformly convergent on every compact subset of the disk

$$D = \{z : |z-z_0| < r\}, \tag{7}$$

and suppose all f_n are analytic on D. Then $f(z)$ has the Taylor series expansion on D with the coefficients given by

$$c_k = \sum_{n=1}^{\infty} \frac{f_n^{(k)}(z_0)}{k!} \qquad k = 0,1,2,\ldots \tag{8}$$

Thus,

$$f(z) = \sum_{k=1}^{\infty} c_k z^k \qquad \text{for } |z| < 1 \tag{9}$$

To find the coefficients c_k of the Taylor expansion, we have to add all the coefficients of z^k in

$$f_n(z) = \frac{z^n}{1-z^n} = z^n + z^{2n} + \ldots \tag{10}$$

Thus in each expansion, the coefficient of z^k equals 1 if k is divisible by n, and zero otherwise.

We see that c_k equals the number of positive integers which divide k, including 1 and n.

$c_1 = 1$, $c_2 = 2$, $c_3 = 2$, $c_4 = 3$, $c_5 = 2$, $c_6 = 4$, $c_7 = 2$ etc.

● **PROBLEM 15-20**

Let

$$f(z) = \sum_{n=0}^{\infty} a_n z^n \tag{1}$$

be an entire function of z, and

$$g(z) = \sum_{n=0}^{\infty} \frac{b_n}{z^n} \quad (2)$$

be an entire function of $\frac{1}{z}$. Find the domain in which the function $f(z) \cdot g(z)$ is analytic, and compute its Laurent series expansion there.

Solution: The function $f(z) \cdot g(z)$ is analytic in

$$0 < |z| < \infty \quad (3)$$

Its Laurent series expansion in this domain is

$$f(z) \cdot g(z) = \sum_{n=0}^{\infty} c_n z^n + \sum_{m=-1}^{-\infty} d_m z^m \quad (4)$$

where

$$c_n = \sum_{k=0}^{\infty} a_{n+k} b_k \quad (5)$$

and

$$d_m = \sum_{k=0}^{\infty} b_{m+k} a_k \quad (6)$$

● **PROBLEM 15-21**

Let

$$f(z) = \sum_{n=-\infty}^{\infty} a_n z^n \quad (1)$$

and

$$g(z) = \sum_{n=-\infty}^{\infty} b_n z^n \quad (2)$$

be two Laurent series expansions which converge in the same annulus. Find the Laurent expansion of $f(z) \cdot g(z)$.

Solution: We shall use the following property of the series: If the two absolutely convergent series $\sum_{n=0}^{\infty} \alpha_n$ and $\sum_{n=0}^{\infty} \beta_n$ have the sums a and b, then the series

$$\sum_{n=0}^{\infty} c_n \quad (3)$$

$$c_n = \alpha_0 \beta_n + \alpha_1 \beta_{n-1} + \ldots + \alpha_n \beta_0 \quad (4)$$

also converges and has the sum ab.

It follows from the above that if $\sum_{n=0}^{\infty} \alpha_n$ and $\sum_{n=0}^{\infty} \beta_n$ converge absolutely, then every series,

$$\sum_{n=0}^{\infty} c_n$$

where c_n is any of the products $\alpha_k \beta_\ell$, with the restriction that to different n correspond different products, converges. If Σc_n in particular contains all products $\alpha_k \beta_\ell$, then

$$\Sigma c_n = \Sigma \alpha_k \Sigma \beta_\ell \tag{5}$$

Let z be any point for which the series (1) and (2) converge, then

$$c_n = \sum_{k=-\infty}^{\infty} a_k b_{n-k}$$

$$= z^{-n} \sum_{k=-\infty}^{\infty} (a_k z^k)(b_{n-k} z^{n-k}) \tag{6}$$

$$n = 0, \pm 1, \pm 2, \ldots$$

also converges. Hence, the series

$$\sum_{n=-\infty}^{\infty} c_n z^n \tag{7}$$

also converges. The series (7) represents the Laurent expansion of $f(z) \cdot g(z)$, where $f(z)$ and $g(z)$ are given by (1) and (2).

LAGRANGE'S EXPANSION

• **PROBLEM** 15-22

Consider the equation

$$z = a + \zeta f(z) \tag{1}$$

Let z be that root of (1) which has the value a when $\zeta = 0$. If $f(z)$ is analytic inside and on a circle C containing $z = 0$, then

$$z = a + \sum_{n=1}^{\infty} \frac{\zeta^n}{n!} \frac{d^{n-1}}{da^{n-1}} \left[f(a) \right]^n \tag{2}$$

This expansion, known as Lagrange's expansion, is very useful in finding approximate solutions of some equations.

Prove Lagrange's expansion.

Solution: We shall use the following theorem:

Let $\phi(z)$ and $\Psi(z)$ be analytic inside and on a simple closed curve C, except for the points $a_1, a_2, \ldots a_m$ which are poles of $\phi(z)$ of order $p_1, \ldots, p_m$ respectively. If $b_1, \ldots, b_n$ are the zeros of $\phi(z)$ of order $q_1, \ldots, q_n$ then

$$\frac{1}{2\pi i} \int_C \Psi(z) \frac{\phi'(z)}{\phi(z)} dz = \sum_{k=1}^{\infty} q_k \Psi(b_k) - \sum_{k=1}^{m} p_k \Psi(a_k) \qquad (3)$$

Let us take a simple closed curve C such that (1) has only one simple zero inside C.

We substitute

$$\Psi(z) = z$$
$$\phi(z) = z - a - \zeta f(z) \qquad (4)$$

into (3) to obtain

$$\frac{1}{2\pi i} \int_C \omega \frac{[1-\zeta f'(\omega)]}{[\omega-a-\zeta f(\omega)]} d\omega = z \qquad (5)$$

Eq.(5) can be written in the form

$$= \frac{1}{2\pi i} \int_C \frac{\omega}{\omega-a} \left[1-\zeta f'(\omega)\right] \left[\frac{1}{1-\frac{\zeta f(\omega)}{\omega-a}}\right] d\omega$$

$$= \frac{1}{2\pi i} \int_C \frac{\omega}{\omega-a} \left[1 - \zeta f'(\omega)\right] \left[\sum_{n=0}^{\infty} \frac{\zeta^n [f(\omega)]^n}{(\omega-a)^n}\right] d\omega \qquad (6)$$

We expressed $\dfrac{1}{1-\frac{\zeta f(\omega)}{\omega-a}}$ as a geometric series.

$$= \frac{1}{2\pi i} \int_C \frac{\omega}{\omega-a} d\omega$$

$$+ \sum_{n=1}^{\infty} \frac{\zeta^n}{2\pi i} \int_C \left[\frac{\omega [f(\omega)]^n}{(\omega-a)^{n+1}} - \frac{\omega f'(\omega)[f(\omega)]^{n-1}}{(\omega-a)^n}\right] d\omega \qquad (7)$$

$$= a - \sum_{n=1}^{\infty} \frac{\zeta^n}{2\pi i} \int_C \frac{\omega}{n} \frac{d}{d\omega} \frac{[f(\omega)]^n}{(\omega-a)^n} d\omega$$

$$= a + \sum_{n=1}^{\infty} \frac{\zeta^n}{2\pi i n} \int_C \frac{[f(\omega)]^n}{(\omega-a)^n} d\omega$$

$$= a + \sum_{n=1}^{\infty} \frac{\zeta^n}{n!} \frac{d^{n-1}}{da^{n-1}} \left\{ \left[f(a) \right]^n \right\}$$

That completes the proof.

• **PROBLEM 15-23**

Find the root of the equation
$$z = 1 + \zeta z^s \quad (1)$$
which is equal to 1 when $\zeta = 0$.

Solution: Applying Lagrange's expansion,
$$z = a + \zeta f(z) \quad (2)$$

$$z = a + \sum_{n=1}^{\infty} \frac{\zeta^n}{n!} \frac{d^{n-1}}{da^{n-1}} \left[f(a) \right]^n$$

we obtain for
$$f(z) = z^s \quad (3)$$

$$z = a + \sum_{n=1}^{\infty} \frac{\zeta^n}{n!} \frac{d^{n-1}}{da^{n-1}} \left[z^{ns} \right] \quad (4)$$

For $a = 1$, we have

$$\frac{d}{da}(a^{2s}) = 2s$$

$$\frac{d^2}{da^2}(a^{3s}) = 3s(3s-1) \quad (5)$$

$$\frac{d^3}{da^3}(a^{4s}) = 4s(4s-1)(4s-2)$$

... etc.

Hence, (4) becomes
$$z = 1 + \zeta + \frac{\zeta^2}{2!} 2s + \frac{\zeta^3}{3!} 3s(3s-1) + \frac{\zeta^4}{4!} 4s(4s-1)(4s-2) + \ldots \quad (6)$$

Lagrange's expansion, as given by equation (2), is a special case of a more general formula.

If $F(z)$ is analytic inside and on C, then
$$F(z) = F(a) + \sum_{n=1}^{\infty} \frac{\zeta^n}{n!} \frac{d^{n-1}}{da^{n-1}} \left[F'(a) [f(a)]^n \right] \quad (7)$$

• **PROBLEM 15-24**

The Legendre polynomials are given by Rodrigues' formula

$$P_n(t) = \frac{1}{2^n n!} \frac{d^n}{dt^n} (t^2-1)^n \qquad (1)$$

Prove that if C is any simple closed curve enclosing the point z = t, then

$$P_n(t) = \frac{1}{2\pi i} \cdot \frac{1}{2^n} \int_C \frac{(z^2-1)^n}{(z-t)^{n+1}} dz \qquad (2)$$

Eq.(1) is called Schlaefli's formula.

Also show that

$$P_n(t) = \frac{1}{2\pi} \int_0^{2\pi} (t+\sqrt{t^2-1}\cos\theta)^n d\theta \qquad (3)$$

Solution: Since C encloses z = t, we have

$$F^{(n)}(t) = \frac{n!}{2\pi i} \int_C \frac{F(z)}{(z-t)^{n+1}} dz \qquad (4)$$

Substituting

$$F(t) = (t^2-1)^n \qquad (5)$$

we obtain

$$P_n(t) = \frac{1}{2^n n!} \frac{d^n}{dt^n} (t^2-1)^n$$
$$= \frac{1}{2^n} \cdot \frac{1}{2\pi i} \int_C \frac{(z^2-1)^n}{(z-t)^{n+1}} dz \qquad (6)$$

To prove (3), let us choose a circle C with center at t, and radius $\sqrt{|t^2-1|}$ (see Fig. 1).

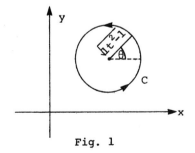

Fig. 1

The circle C is described by the equation

$$|z-t| = \sqrt{|t^2-1|} \qquad (7)$$

or

$$z = t + \sqrt{t^2-1}\, e^{i\theta} \qquad (8)$$

Thus,

$$0 \leq \theta < 2\pi$$

$$P_n(t) = \frac{1}{2^n} \cdot \frac{1}{2\pi i} \int_C \frac{(z^2-1)^n}{(z-t)^{n+1}} dz$$

$$= \frac{1}{2^n} \cdot \frac{1}{2\pi i} \int_0^{2\pi} \frac{[(t+\sqrt{t^2-1}\, e^{i\theta})^2 - 1]^n}{[\sqrt{t^2-1}\, e^{i\theta}]^{n+1}} \cdot \sqrt{t^2-1}\, ie^{i\theta} d\theta$$

$$= \frac{1}{2^n} \cdot \frac{1}{2\pi} \int_0^{2\pi} \frac{[(t^2-1)e^{-i\theta} + 2t\sqrt{t^2-1} + (t^2-1)e^{i\theta}]^n}{(t^2-1)^{\frac{n}{2}}} d\theta \quad (9)$$

$$= \frac{1}{2^n} \cdot \frac{1}{2\pi} \int_0^{2\pi} \frac{[2t\sqrt{t^2-1} + 2(t^2-1)\cos\theta]^n}{(t^2-1)^{\frac{n}{2}}} d\theta$$

$$= \frac{1}{2\pi} \int_0^{2\pi} (t + \sqrt{t^2-1}\, \cos\theta)^n d\theta$$

ANALYTIC CONTINUATION

● **PROBLEM 15-25**

The function $f(z)$ is defined by

$$f(z) = \frac{1}{1-z} \quad (1)$$

Its only singularity is a pole at $z_0 = 1$.

1. Find the Taylor series of $f(z)$ in powers of z.

2. Express $f(z)$ in terms of the Laurent series convergent in the annulus $0 < |z-1| < \infty$.

3. Express $f(z)$ in terms of the Laurent series convergent in the annulus

$$1 < |z| < \infty.$$

Solution: 1. The Taylor series of $f(z) = \frac{1}{1-z}$ is

$$f(z) = \sum_{n=0}^{\infty} z^n, \quad (2)$$

which is the geometric series.

2. The Laurent expansion of f(z) in the annulus $0 < |z-1| < \infty$ is the function itself

$$f(z) = \frac{1}{1-z} \qquad (3)$$

Note that (3) contains only powers of (1-z).

3. Here we have to find the Laurent expansion about $z_0 = 0$, thus

$$\frac{1}{1-z} = -\sum_{n=-1}^{-\infty} z^n = -\sum_{n=1}^{\infty} z^{-n} \qquad (4)$$

● **PROBLEM 15-26**

Show that the series

$$\sum_{n=0}^{\infty} \frac{z^n}{3^{n+1}} \qquad (1)$$

and

$$\sum_{n=0}^{\infty} \frac{(z-2i)^n}{(3-2i)^{n+1}} \qquad (2)$$

are analytic continuations of each other.

Solution: Series (1) is a geometric series with the first term $\frac{1}{3}$ and ratio $\frac{z}{3}$. Hence, it converges for

$$\left|\frac{z}{3}\right| < 1$$

or

$$|z| < 3 \qquad (3)$$

(see Fig. 1).

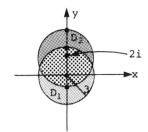

Fig. 1

$$\sum_{n=0}^{\infty} \frac{z^n}{3^{n+1}} = \frac{1}{3-z}$$

Series (1) in the domain

$$D_1 = \{z : |z| < 3\} \qquad (4)$$

represents function $\frac{1}{3-z}$.

Series (2) represents

$$\sum_{n=0}^{\infty} \frac{(z-2i)^n}{(3-2i)^{n+1}} = \frac{1}{3-z}, \tag{5}$$

which converges in the domain

$$D_2 = \{z : |z-2i| < |3-2i| = \sqrt{13}\} \tag{6}$$

Note that both series represent the same function, but have different domains of convergence.

We say that one series is the analytic continuation of another.

● **PROBLEM 15-27**

Prove that it is impossible to find the analytic continuation of the series

$$1 + z + z^2 + z^4 + z^8 + \ldots = 1 + \sum_{n=0}^{\infty} z^{2n} \tag{1}$$

beyond the domain $|z| < 1$.

Solution: Let us denote

$$f(z) = 1 + z + z^2 + z^4 + \ldots \tag{2}$$

Then, we have

$$f(z) = z + f(z^2) \tag{3}$$

$$f(z) = z + z^2 + f(z^4)$$

... etc.

Thus, the values of z such that

$$\begin{aligned} z &= 1 \\ z^2 &= 1 \\ z^4 &= 1 \\ z^8 &= 1 \\ &\vdots \end{aligned} \tag{4}$$

are singularities of $f(z)$.

All of them are located on the circle $|z| = 1$.

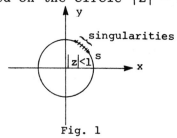

Fig. 1

Let s be any small arc on the circle $|z| = 1$ (see Fig. 1). It is obvious that this arc contains infinitely many singularities. These singularities constitute a barrier, and one cannot find an analytic continuation of (1) going beyond $z = 1$. A function cannot "leave" the domain $|z| < 1$ without passing through singularities.

CHAPTER 16

SINGULARITIES, RATIONAL AND MEROMORPHIC FUNCTIONS

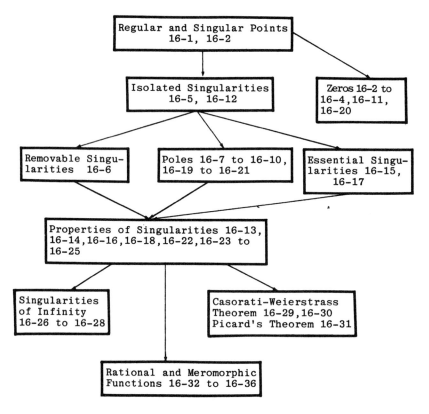

This chart is provided to facilitate rapid understanding of the interrelationships of the topics and subject matter in this chapter. Also shown are the problem numbers associated with the subject matter.

SINGULARITIES

● PROBLEM 16-1

Describe the concept of singularity.

Solution: Consider this situation:
 A function f is given, defined and analytic at all points

of D except the point z_0. The domain D is a subset of the complex plane.

Hence, the function f(z) is defined and analytic at all points of

$$D' = D - \{z_0\} \tag{1}$$

The question is what happens to the function f(z) at z_0. Sometimes, it is possible to define the value of f(z) at $z = z_0$

$$f(z_0) = a \tag{2}$$

in such a way that the new function g(z)

$$g(z) = \begin{cases} f(z), & z \neq z_0 \\ a, & \text{for } z = z_0 \end{cases} \tag{3}$$

is analytic and defined everywhere in D.

In other cases, like for example

$$f(z) = \frac{1}{z} \tag{4}$$

it is not possible to define f(0) in such a way that the new function is analytic everywhere. In general, a point at which f(z) fails to be analytic is called a singular point or singularity of f(z).

As you will learn, at the points where f(z) is not defined, the behavior of the function can vary. That will lead us into the classification of singularities and the appropriate treatment of them.

● **PROBLEM 16-2**

1. Find the regular points and the singular points (singularities) of the functions

    $\frac{1}{z}$

    $\frac{1}{1-z}$

    $\frac{1}{\sin \frac{1}{z}}$

2. Find the zeros and their order of the functions

    $\frac{1}{z}$

    $\cos z - 1$

    $z + 1$

Solution: 1. Let us repeat the definition:

If $f(z)$ is analytic at $z = z_0$, then z_0 is called a regular point of f. If $f(z)$ is not analytic at $z = z_0$, then z_0 is called a singular point of f.

The function $\frac{1}{z}$ is analytic everywhere except $z = 0$. Hence, it has singularity at $z = 0$. All points $z \neq 0$ are regular points.

Function $\frac{1}{1-z}$ has singularity at $z = 1$.

Function $\frac{1}{\sin \frac{1}{z}}$ has an infinite number of singular points,

$$z = \frac{1}{n\pi}, \quad n = \pm 1, \pm 2, \pm 3, \ldots$$

Definition

If $f(z)$ is analytic in $D' = D - \{z_0\}$, where z_0 is an interior point of D, then z_0 is an isolated singularity (we assume that f is not analytic at z_0).

2. Definition

A function $f(z)$ that is analytic at z_0, is said to have a zero of order n at z_0 if

$$f^{(k)}(z_0) = 0 \quad \text{for } k = 0, 1, 2, \ldots, n-1$$

but

$$f^{(n)}(z_0) \neq 0.$$

Function $f(z) = \frac{1}{z}$, has no zeros. Function $\cos z - 1$ has double zeros at $z = 2\pi n$ because

$$\cos(2\pi n) - 1 = 0$$

$$-\sin(2\pi n) = 0$$

but

$$\cos''(2\pi n) \neq 0.$$

At $z = -1$, function $f(z) = z + 1$ has a simple zero.

● **PROBLEM 16-3**

Theorem

Let $f(z)$ be analytic at z_0 and let z_0 be a zero of order m of $f(z)$. Then, in the neighborhood of $z = z_0$ the function $f(z)$ can be represented by

$$f(z) = (z-z_0)^m \sum_{n=0}^{\infty} a_n(z-z_0)^n \tag{1}$$

where $a_0 \neq 0$.

Prove this theorem

Solution: Since $f(z)$ is analytic at $z = z_0$, it has the Taylor expansion of the form

$$f(z) = \sum_{n=0}^{\infty} \frac{f^{(n)}(z_0)}{n!} (z-z_0)^n \qquad (2)$$

$$= f(z_0) + \frac{f'(z_0)}{1!}(z-z_0) + \frac{f''(z_0)}{2!}(z-z_0)^2 + \ldots$$

Since z_0 is a zero of order m

$$f(z_0) = 0, \quad f'(z_0) = 0, \ldots, f^{(m-1)}(z_0) = 0 \qquad (3)$$

and (2) reduces to

$$f(z) = \frac{f^{(m)}(z_0)}{m!}(z-z_0)^m + \frac{f^{(m+1)}(z_0)}{(m+1)!}(z-z_0)^{m+1} + \ldots$$

$$= (z-z_0)^m \left[\frac{f^{(m)}(z_0)}{m!} + \frac{f^{(m+1)}(z_0)}{(m+1)!}(z-z_0) + \ldots + \frac{f^{(m+k)}(z_0)}{(m+k)!}(z-z_0)^k + \ldots \right] \qquad (4)$$

$$= (z-z_0)^m \sum_{n=0}^{\infty} a_n (z-z_0)^n$$

● **PROBLEM 16-4**

Give an example of a function which is analytic everywhere, has a zero of order 2 at $z = 1$ and the value of its third derivative at $z = 1$ is 5.

Solution: Let $f(z)$ be an analytic function with a zero of order 2 at $z = 1$. Then, by the theorem proved in Problem 16-3, we can express function $f(z)$ as

$$f(z) = (z-1)^2 \sum_{n=0}^{\infty} a_n (z-1)^n \qquad (1)$$

$$= a_0(z-1)^2 + a_1(z-1)^3 + a_2(z-1)^4 + \ldots$$

Function $f(z)$ given by (1) has a zero at $z = 1$ of order two.

$$f'(z) = 2a_0(z-1) + 3a_1(z-1)^2 + 4a_2(z-1)^3 + \ldots \qquad (2)$$

$$f''(z) = 2a_0 + 2 \cdot 3a_1(z-1) + 3 \cdot 4a_2(z-1)^2 + \ldots \qquad (3)$$

$$f'''(z) = 6a_1 + 2 \cdot 3 \cdot 4a_2(z-1) + \ldots \qquad (4)$$

Since

$$f^{(3)}(1) = 5 \qquad (5)$$

from (4) we obtain

$$f^{(3)}(1) = 6a_1 = 5 \qquad (6)$$

$$a_1 = \frac{5}{6}$$

Hence, $f(z)$ is
$$f(z) = a_0(z-1)^2 + \frac{5}{6}(z-1)^3 \tag{7}$$
where a_0 is any complex constant.

• **PROBLEM 16-5**

Find the singularities of
$$f(z) = \frac{1}{\sin\frac{1}{z}} \tag{1}$$
Which of them are isolated, which are not?

Solution: Remember, that singularity is a point at which function $f(z)$ is not analytic. We shall limit our discussion to the fairly wide class of isolated singularities.

<u>Definition</u>

If $f(z)$ is analytic in a deleted neighborhood of z_0 ($0 < |z-z_0| < \varepsilon$, where $\varepsilon > 0$) but not at z_0 itself, we call z_0 an isolated singular point of $f(z)$.

Note that if z_0 is an isolated singular point of $f(z)$, then in the deleted neighborhood $0 < |z-z_0| < r$, there is a Laurent expansion of $f(z)$, where r is the distance from z_0 to the nearest other singular point of $f(z)$.

Function $f(z) = \frac{1}{\sin\frac{1}{z}}$ has an infinite number of isolated singularities

$$z = \pm\frac{1}{n\pi} \tag{2}$$

$n = \pm 1, \pm 2, \pm 3, \ldots$

The point $z = 0$ is a singular point of $f(z)$. It is not an isolated singular point, because every neighborhood of $z = 0$ contains an infinite number of singular points of $f(z) = \frac{1}{\sin\frac{1}{z}}$.

• **PROBLEM 16-6**

Consider the function,
$$f(z) = \begin{cases} \sum_{n=0}^{\infty} \frac{z^n}{n!}, & z \neq 0 \\ 5, & z = 0 \end{cases} \tag{1}$$

Show that $f(z)$ has a removable singularity at $z = 0$. Show that $\frac{\sin z}{z}$ has a removable singularity at 0.

Solution: Suppose z_0 is an isolated singular point of $f(z)$. Then,

$$f(z) = \sum_{n=0}^{\infty} a_n (z-z_0)^n + \sum_{n=1}^{\infty} b_n (z-z_0)^{-n}, \qquad (2)$$

$$0 < |z-z_0| < r$$

Definition

If the series (2) contains no negative powers of $(z-z_0)$, then z_0 is said to be a removable singular point of $f(z)$.

In such a case (2) reduces to

$$f(z) = \sum_{n=0}^{\infty} a_n (z-z_0)^n, \qquad (3)$$

$$0 < |z-z_0| < r$$

The right-hand side of (3) is an ordinary power series, which defines an analytic function

$$F(z) = \sum_{n=0}^{\infty} a_n (z-z_0)^n \qquad (4)$$

for $|z-z_0| < r$

Note that $f(z) = F(z)$ for $z \neq z_0$, and let

$$F(z_0) \equiv F_0 \qquad (5)$$

The function $f(z)$ can be made analytic by

$$f(z) = \begin{cases} \sum_{n=0}^{\infty} a_n (z-z_0)^n, & 0 < |z-z_0| < r \\ F_0 & \text{for } z = z_0 \end{cases} \qquad (6)$$

We see that if z_0 is a removable singular point of $f(z)$, then

$$\lim_{z \to z_0} f(z) = F_0 \qquad (7)$$

For the function (1) we can make $f(z)$ analytic by redefining $f(0)$ as

$$f(0) = \lim_{z \to 0} \sum_{n=0}^{\infty} \frac{z^n}{n!} = 1 \qquad (8)$$

The analytic function

$$f(z) = \frac{\sin z}{z} \qquad (9)$$

is defined in the domain

$$D = C - \{0\} \qquad (10)$$

The point $z_0 = 0$ is an isolated singularity of $\frac{\sin z}{z}$

$$\frac{\sin z}{z} = \frac{1}{z}\left(z - \frac{z^3}{3!} + \frac{z^5}{5!} - \frac{z^7}{7!} + \ldots \right)$$

$$= 1 - \frac{z^2}{3!} + \frac{z^4}{5!} - \ldots \qquad (11)$$

Thus,
$$\lim_{z \to 0} \frac{\sin z}{z} = 1$$

Point $z_0 = 0$ is a removable singularity of $\frac{\sin z}{z}$. The function

$$f(z) = \begin{cases} \frac{\sin z}{z} & \text{for } 0 < |z| \\ 1 & \text{for } z = 0 \end{cases} \tag{12}$$

is analytic everywhere.

POLES

• **PROBLEM 16-7**

1. Find the pole of
$$f(z) = \frac{\sinh z}{z^7} \tag{1}$$
and determine its order.

2. Find the principal part of
$$\frac{\sinh z}{z^7} \tag{2}$$

<u>Solution</u>: 1. Assume, that z_0 is an isolated singularity of a function $f(z)$.

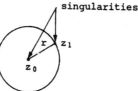

Fig. 1

The Laurent expansion of $f(z)$ exists for $0 < |z-z_0| < r$, where $r = |z_1-z_0|$ is the distance from z_0 to the nearest singularity of $f(z)$.

$$f(z) = \sum_{n=0}^{\infty} a_n(z-z_0)^n + \sum_{n=1}^{\infty} b_n(z-z_0)^{-n} \tag{3}$$

$$0 < |z-z_0| < r$$

<u>Definition</u>

The series in negative powers $\sum_{n=1}^{\infty} b_n(z-z_0)^{-n}$ is called the principal part of $f(z)$ at the isolated singularity z_0.

<u>Definition</u>

If all but a finite number of the coefficients b_n vanish

in (3), then z_0 is said to be a pole of $f(z)$. Furthermore, if

$$f(z) = \sum_{n=0}^{\infty} a_n(z-z_0)^n + \frac{b_1}{z-z_0} + \frac{b_2}{(z-z_0)^2} + \ldots$$

$$+ \frac{b_k}{(z-z_0)^k} \qquad (4)$$

and $\quad b_{k+1} = b_{k+2} = \ldots = 0$

then z_0 is said to be a pole of order k of $f(z)$. If $k = 1$, then z_0 is a simple pole of $f(z)$.

$$f(z) = \frac{\sinh z}{z^7} = \frac{1}{z^6} + \frac{1}{3!}\frac{1}{z^4} + \frac{1}{5!}\frac{1}{z^2} + \frac{1}{7!} + \frac{z^2}{9!} + \ldots \qquad (5)$$

The function $\frac{\sinh z}{z^7}$ has a pole of order 6 at $z = 0$.

2. The principal part of this function is

$$\frac{1}{z^6} + \frac{1}{3!}\frac{1}{z^4} + \frac{1}{5!}\frac{1}{z^2} \qquad (6)$$

● **PROBLEM 16-8**

Show that if z_0 is a pole of $f(z)$, then

$$\lim_{z \to z_0} f(z) = \infty \qquad (1)$$

Solution: The Laurent expansion of the function $f(z)$ in the deleted neighborhood of z_0 is

$$f(z) = \sum_{n=-\infty}^{+\infty} c_n(z-z_0)^n \qquad (2)$$

$0 < |z-z_0| < r$.

Since z_0 is a pole of, say order m, the series (2) contains only a finite number of negative powers of $(z-z_0)$. Therefore,

$$f(z) = \sum_{n=0}^{\infty} c_n(z-z_0)^n + \frac{c_{-1}}{z-z_0} + \ldots + \frac{c_{-m}}{(z-z_0)^m} \qquad (3)$$

where $c_{-m} \neq 0$.

Multiplying (3) by $(z-z_0)^m$ we obtain

$$(z-z_0)^m f(z) = \sum_{n=0}^{\infty} c_n(z-z_0)^{n+m} + c_{-1}(z-z_0)^{m-1}$$
$$+ c_{-2}(z-z_0)^{m-2} + \ldots + c_{-m} \qquad (4)$$

Note that the right-hand side of (4) is an ordinary power series, which for $z = z_0$ takes the value $c_{-m} \neq 0$. Thus, the point

z_0 is a removable singularity of the function $(z-z_0)^m f(z)$.

From (4) we conclude that

$$\lim_{z \to z_0} (z-z_0)^m f(z) = c_{-m} \tag{5}$$

and

$$\lim_{z \to z_0} f(z) = \lim_{z \to z_0} \frac{c_{-m}}{(z-z_0)^m} = \infty \tag{6}$$

● **PROBLEM 16-9**

The following theorem is very useful in determining the location of the poles and their order.

<u>Theorem</u>

Let the function $f(z)$ be analytic in a deleted neighborhood of z_0. Then

$$\begin{pmatrix} f(z) \text{ has a pole} \\ \text{of order } m \end{pmatrix} \iff \begin{pmatrix} \text{the function} \\ g(z)=(z-z_0)^m f(z) \\ \text{has a removable} \\ \text{singularity at } z_0 \\ \text{and} \\ \lim_{z \to z_0} g(z) \neq 0. \end{pmatrix} \tag{1}$$

<u>Solution</u>: $\Rightarrow$ In Problem 16-8 we proved that if $f(z)$ is analytic in a deleted neighborhood of $z = z_0$ and z_0 is a pole of order m of $f(z)$, then

$$g(z) = (z-z_0)^m f(z) \tag{2}$$

has a removable singularity at z_0 and

$$\lim_{z \to z_0} (z-z_0)^m f(z) \neq 0 \tag{3}$$

$\Leftarrow$

Suppose $g(z) = (z-z_0)^m f(z)$ has a removable singularity at $z = z_0$ and $\lim_{z \to z_0} g(z) \neq 0$, then $g(z)$ can be expanded in a Laurent series about $z = z_0$ with the coefficients of the negative powers vanishing. Therefore,

$$g(z) = (z-z_0)^m f(z) = \sum_{n=0}^{\infty} c_n (z-z_0)^n \tag{4}$$

Dividing both sides of (4) by $(z-z_0)^m$, we obtain

$$f(z) = \frac{g(z)}{(z-z_0)^m} = \frac{c_0}{(z-z_0)^m} + \frac{c_1}{(z-z_0)^{m-1}} + \ldots + \frac{c_{m-1}}{z-z_0}$$
$$+ \sum_{n=m}^{\infty} c_n (z-z_0)^n \tag{5}$$

Since,
$$c_0 = \lim_{z \to z_0} g(z) \neq 0, \tag{6}$$

we see that $f(z)$ has a pole of order m at $z = z_0$.

● **PROBLEM 16-10**

Find the poles of the function
$$f(z) = \frac{z^2 + 2}{z(z-3i)(z+1)^3} \tag{1}$$
and their order. Use theorem proved in Problem 16-9.

Solution: The function $f(z)$ has a simple pole at $z = 0$. Indeed
$$g_1(z) = zf(z) = \frac{z^2+2}{(z-3i)(z+1)^3}, \quad z \neq 0 \tag{2}$$
has a removable singularity at $z = 0$.
$$\lim_{z \to 0} g_1(z) = \frac{2}{-3i} \neq 0 \tag{3}$$

Therefore, $f(z)$ has a simple pole at $z = 3i$.
$$g_2(z) = (z-3i)f(z) = \frac{z^2 + 2}{z(z+1)^3}, \quad z \neq 3i \tag{4}$$

$$\lim_{z \to 3i} g_2(z) = \frac{(3i)^2+2}{3i(3i+1)^3} \neq 0 \tag{5}$$

At $z = -1$ the function $f(z)$ has a pole of order 3.
$$g_3(z) = (z+1)^3 f(z) = \frac{z^2 + 2}{z(z-3i)}$$
and
$$\lim_{z \to -1} g_3(z) = \frac{3}{1+3i} \neq 0 \tag{6}$$

Note that it would be much more difficult to find the poles of $f(z)$ by applying the definition of a pole. That is, finding the Laurent expansion of $f(z)$ (or rather three Laurent expansions).

The theorem proved in Problem 16-9 greatly simplifies the process of determining poles of a rational function.

● **PROBLEM 16-11**

Theorem
Suppose, $f(z)$ is an analytic function in a deleted neighborhood of $z = z_0$, and such that the coefficients of the Laurent expansion of $f(z)$ about $z = z_0$ do not all vanish. If

z_0 is a zero of $f(z)$, then a neighborhood of z_0 exists which contains no other zero of $f(z)$.

This theorem states that the zero of $f(z)$ is isolated. Prove it.

Solution: Let z_0 be a zero of order m. Then, by Problem 16-3, $f(z)$ can be written as

$$f(z) = (z-z_0)^m g(z) \tag{1}$$

where $g(z)$ is analytic and hence continuous in a neighborhood of z_0 with $g(z_0) \neq 0$.

Let us denote

$$|g(z_0)| = 2\alpha > 0 \tag{2}$$

Since $g(z)$ is continuous at $z = z_0$, there exists a neighborhood of z_0 such that

$$|z-z_0| < \delta, \tag{3}$$

in which

$$|g(z) - g(z_0)| < \alpha \tag{4}$$

Thus,

$$\begin{aligned}|g(z)| &= |g(z_0)-(g(z_0)-g(z))| \\ &\geq |g(z_0)| - |g(z)-g(z_0)| > 2\alpha - \alpha = \alpha\end{aligned} \tag{5}$$

for $|z-z_0| < \delta$.

Hence $g(z) \neq 0$ in the neighborhood $|z-z_0| < \delta$.

In this neighborhood $(z-z_0)^m \neq 0$ for $z \neq z_0$, hence

$$f(z) = (z-z_0)^m g(z) \neq 0 \tag{6}$$

for

$$0 < |z-z_0| < \delta \tag{7}$$

● **PROBLEM 16-12**

Prove that an analytic function cannot be bounded in the neighborhood of an isolated singularity.

Solution: Let $f(z)$ be analytic for $0 < |z-z_0| < R$,

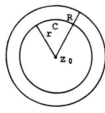

Fig. 1

and let z_0 be an isolated singularity of $f(z)$. Then $f(z)$ is analytic inside and on a circle C of radius r (r < R), except for $z = z_0$. The Laurent expansion of $f(z)$ is

$$f(z) = \sum_{n=-\infty}^{+\infty} C_n(z-z_0)^n \qquad (1)$$

where

$$C_{-n} = \frac{1}{2\pi i} \int_C \frac{f(z)dz}{(z-z_0)^{-n+1}} \qquad (2)$$

for $n = 1, 2, 3, \ldots$.

Suppose $f(z)$ is bounded, i.e., a number M exists such that

$$|f(z)| < M \qquad (3)$$

for z inside or on C.

From (2) we have

$$|C_{-n}| = \frac{1}{2\pi} \left| \int_C (z-z_0)^{n-1} f(z) dz \right|$$

$$\leq \frac{2\pi r}{2\pi} M r^{n-1} = Mr^n \qquad (4)$$

Since r can be made arbitrarily small, $a_{-n} = 0$ for $n = 1, 2, 3, \ldots$. That is all negative powers vanish and the Laurent series becomes a Taylor series about $z = z_0$. Hence $f(z)$ is analytic at $z = z_0$, and z_0 is not a singularity. Contradiction!

PROPERTIES OF SINGULARITIES • PROBLEM 16-13

Theorem

Let z_0 be a zero of order m of a function $f(z)$ analytic at $z = z_0$.

Then $\frac{1}{f(z)}$ is analytic in a deleted neighborhood of z_0 (i.e., with z_0 removed and z_0 is a pole of order m).

This theorem establishes relationship between zeros and poles. Prove it.

<u>Solution</u>: The function $f(z)$ is analytic at z_0 and z_0 is a zero of order m. Thus, its power series expansion is

$$f(z) = C_m(z-z_0)^m + C_{m+1}(z-z_0)^{m+1} + \ldots, \qquad (1)$$
$$C_m \neq 0$$

Equation (1) can be written in the form

$$f(z) = (z-z_0)^m \psi(z). \qquad (2)$$

Note, that function $\psi(z)$ is analytic at z_0 and $\psi(z_0) \neq 0$.

We can write

$$\frac{1}{f(z)} = \frac{1}{(z-z_0)^m} \cdot \frac{1}{\psi(z)} = \frac{\phi(z)}{(z-z_0)^m} \tag{3}$$

The function $\phi(z) = \frac{1}{\psi(z)}$ is analytic at z_0 and $\phi(z_0) \neq 0$.

Thus, $\frac{1}{f(z)}$ is analytic in a deleted neighborhood of z_0. We have

$$\phi(z) = \phi(z_0) + \phi'(z_0)(z-z_0) + \frac{\phi''(z_0)}{2!}(z-z_0)^2 + \ldots \tag{4}$$

then

$$\frac{1}{f(z)} = \frac{\phi(z_0)}{(z-z_0)^m} + \frac{\phi'(z_0)}{(z-z_0)^{m-1}} + \frac{\phi''(z_0)}{2!(z-z_0)^{m-2}} + \ldots \tag{5}$$

Thus, $\frac{1}{f(z)}$ has a pole of order m at z_0.

• **PROBLEM 16-14**

<u>Theorem</u>
  Let $f(z)$ be analytic in a deleted neighborhood of z_0 and let z_0 be a pole of order m. Then the function $\frac{1}{f(z)}$ is analytic at z_0 (provided that $\frac{1}{f(z_0)} \neq 0$), and z_0 is a zero of order m.

This is another theorem describing relationship between zeros and poles of a function. Prove it.

<u>Solution</u>: Since z_0 is a pole of order m of $f(z)$, we have

$$\lim_{z \to z_0} (z-z_0)^m f(z) = C_{-m} \neq 0 \tag{1}$$

The function $f(z)$ can be written as

$$f(z) = \frac{\psi(z)}{(z-z_0)^m} \tag{2}$$

where
$$\psi(z) \xrightarrow[z \to z_0]{} C_{-m} \tag{3}$$

Thus, $\psi(z)$ is analytic and non-zero at z_0, $\psi(z_0) = C_{-m} \neq 0$.

From (1) we conclude that

$$\frac{1}{f(z)} = (z-z_0)^m \frac{1}{\psi(z)} = (z-z_0)^m \phi(z) \tag{4}$$

The function $\phi(z) = \frac{1}{\psi(z)}$ is analytic at z_0 and $\phi(z_0) \neq 0$.

That proves that $\frac{1}{f(z)}$ is analytic at z_0 provided we set

$$\frac{1}{f(z_0)} = 0 \qquad (5)$$

Thus, from (4)

$$\frac{1}{f(z)} = (z-z_0)^m \phi(z_0) + (z-z_0)^{m+1} \phi'(z_0) \qquad (6)$$
$$+ (z-z_0)^{m+2} \frac{\phi''(z_0)}{2!} + \ldots$$

From (6) we conclude that $\frac{1}{f(z)}$ has a zero of order m at z_0.

• **PROBLEM 16-15**

Show that the function

$$f(z) = e^{\frac{1}{z^2}} \qquad (1)$$

has an essential singularity at $z = 0$.

Solution: Suppose that z_0 is an isolated singularity of the function $f(z)$. The Laurent series expansion of $f(z)$ is

$$f(z) = \sum_{n=0}^{\infty} a_n(z-z_0)^n + \sum_{n=1}^{\infty} b_n(z-z_0)^{-n} \qquad (2)$$

$$0 < |z-z_0| < r$$

Definition

If the Laurent expansion for $f(z)$ about z_0 contains infinitely many negative powers of $z - z_0$, then z_0 is said to be an essential singular point of $f(z)$.

For the function $f(z) = e^{\frac{1}{z^2}}$ we have

$$e^{\frac{1}{z^2}} = 1 + \frac{1}{z^2} + \frac{1}{2!} \cdot \frac{1}{z^4} + \frac{1}{3!} \cdot \frac{1}{z^6} + \ldots + \frac{1}{n!} \frac{1}{z^{2n}} + \ldots \qquad (3)$$

$$0 < |z|$$

The function $e^{\frac{1}{z^2}}$ has a singularity at $z = 0$. It is an essential singularity.

We can illustrate briefly the classification of points.

	Points			
		Singular		
Regular	Isolated			Non-isolated
	Removable	Poles	Essential	

• **PROBLEM 16-16**

In each case determine the kind of singularity:

1. $\dfrac{e^{2z}}{(z-1)^4}$, $z = 1$

2. $\dfrac{z}{(z+1)(z+2)}$, $z = -2$

3. $\dfrac{z - \sin z}{z^2}$, $z = 0$

4. $\dfrac{(\arcsin z)^2}{z^2}$, $z = 0$

5. $(z-2) \sin \dfrac{1}{z+2}$, $z = -2$

Solution: 1. Let $z - 1 = u$, then $z = u + 1$ and

$$\frac{e^{2z}}{(z-1)^4} = \frac{e^2}{u^4} \cdot e^{2u} = \frac{e^2}{u^4}\left[1 + 2u + \frac{(2u)^2}{2!} + \frac{(2u)^3}{3!} + \frac{(2u)^4}{4!} + \ldots\right]$$

$$= \frac{e^2}{(z-1)^4} + \frac{2e^2}{(z-1)^3} + \frac{2e^2}{(z-1)^2} + \frac{4e^2}{3(z-1)} + \frac{2^4 e^2}{4!} + \ldots \qquad (1)$$

This function has a pole of order 4 at $z = 1$.

2. Let $z + 2 = u$, then

$$\frac{z}{(z+1)(z+2)} = \frac{u-2}{(u-1)u} = \frac{2-u}{u} \cdot \frac{1}{1-u}$$

$$= \frac{2-u}{u}(1 + u + u^2 + u^3 + \ldots)$$

$$= \frac{2}{u} + 1 + u^2 + u^3 + \ldots \qquad (2)$$

$$= \frac{2}{z+2} + 1 + (z+2) + (z+2)^2 + (z+2)^3 + \ldots$$

At $z = -2$ the function has a pole of order 1.

3. $\dfrac{z - \sin z}{z^2} = \dfrac{1}{z^2}\left[z - \left(z - \dfrac{z^3}{3!} + \dfrac{z^5}{5!} - \dfrac{z^7}{7!} + \ldots\right)\right]$

$$= \frac{1}{z^2}\left[\frac{z^3}{3!} - \frac{z^5}{5!} + \frac{z^7}{7!} - \ldots\right] \qquad (3)$$

$$= \frac{z}{3!} - \frac{z^3}{5!} + \frac{z^5}{7!} - \frac{z^7}{9!} + \ldots$$

At $z = 0$ the function has a removable singularity.

4. $$\frac{(\arcsin z)^2}{z^2} = \frac{1}{z^2}\left[z^2 + \frac{2}{3}\cdot\frac{z^4}{2} + \frac{2}{3}\cdot\frac{4}{5}\cdot\frac{z^6}{3} + \frac{2}{3}\cdot\frac{4}{5}\cdot\frac{6}{7}\cdot\frac{z^8}{4} + \ldots\right] \quad (4)$$

$$= 1 + \frac{2}{3}\cdot\frac{z^2}{2} + \frac{2}{3}\cdot\frac{4}{5}\cdot\frac{z^4}{3} + \ldots$$

Hence at $z = 0$, the function has a removable singularity.

5. Let $z + 2 = u$ or $z = u - 2$,

then,
$$(z-2)\sin\frac{1}{z+2} = (u-4)\sin\frac{1}{u}$$

$$= (u-4)\cdot\left(\frac{1}{u} - \frac{1}{3!u^3} + \frac{1}{5!u^5} - \ldots\right) \quad (5)$$

$$= 1 - \frac{4}{u} - \frac{1}{3!u^2} + \frac{4}{3!u^3} + \frac{1}{5!u^4} - \ldots$$

$$= 1 - \frac{4}{z+2} - \frac{1}{3!(z+2)^2} + \frac{4}{3!(z+2)^3} + \ldots$$

This function at $z = -2$ has an essential singularity.

● **PROBLEM 16-17**

Prove the following useful property of the essential singular points.

$\begin{pmatrix} f(z) \text{ is non-vanishing in a} \\ \text{deleted neighborhood of } z_0, \\ z_0 \text{ is an essential singular} \\ \text{point of } f(z) \end{pmatrix} \Rightarrow \begin{pmatrix} \frac{1}{f(z)} \text{ has an essential} \\ \text{singular point at } z_0 \end{pmatrix}$

Solution: Consider the function
$$\psi(z) = \frac{1}{f(z)} \quad (1)$$

Assume that z_0 is not an essential singular point of $\psi(z)$.

Then, z_0 is either a pole or a removable singularity.

If z_0 is a pole of $\frac{1}{f(z)}$, then z_0 is a zero of $f(z)$. Contradiction.

If z_0 is a removable singularity, then
$$\lim_{z \to z_0} \psi(z) = 0 \quad (2)$$

or
$$\lim_{z \to z_0} \psi(z) \neq 0 \quad (3)$$

If $\lim_{z \to z_0} \psi(z) = 0$, then $f(z)$ has a pole at z_0. Contradiction.

If $\lim_{z \to z_0} \psi(z) \neq 0$, then $f(z)$ has a removable singular point at z_0. Contradiction.

● **PROBLEM 16-18**

Prove that if $f(z)$ has a pole of order k at the point z_0, then the mth derivative $f^{(m)}(z)$ has a pole of order $k+m$ at the point z_0.

Solution: The function $f(z)$ has a pole of order k at $z = z_0$, hence its Laurent series expansion around $z = z_0$ is

$$f(z) = \sum_{n=0}^{\infty} a_n (z-z_0)^n + \frac{b_1}{z-z_0} + \frac{b_2}{(z-z_0)^2} + \cdots + \frac{b_k}{(z-z_0)^k} \tag{1}$$

where $b_k \neq 0$.

Now, we should differentiate (1) m times and see what happens. Note that since we have to determine the order of the pole of $f^{(m)}(z)$ we are only interested in the highest negative power of $(z-z_0)$.

Differentiating the term with the highest negative power of $(z-z_0)$, that is the term

$$\frac{b_k}{(z-z_0)^k}, \tag{2}$$

we find the answer.

The mth derivative of (2) is

$$\left(\frac{b_k}{(z-z_0)^k}\right)^{(m)} = \frac{B}{(z-z_0)^{k+m}} \tag{3}$$

where B is a coefficient.

Thus, the function $f^{(m)}(z)$ has a pole of order $k+m$ at $z = z_0$.

● **PROBLEM 16-19**

At the point z_0 the function $f(z)$ has a pole of order k and the function $g(z)$ has a pole of order ℓ. Describe the following functions at the point z_0.

1. $f(z) \pm g(z)$

2. $f(z) \, g(z)$

3. $\dfrac{f(z)}{g(z)}$

Solution: Since $f(z)$ has a pole at z_0 of order k, we have

$$f(z) = \sum_{n=0}^{\infty} a_n(z-z_0)^n + \frac{b_1}{z-z_0} + \ldots + \frac{b_k}{(z-z_0)^k} \qquad (1)$$

and for $g(z)$

$$g(z) = \sum_{n=0}^{\infty} a_n'(z-z_0)^n + \frac{b_1'}{z-z_0} + \ldots + \frac{b_\ell'}{(z-z_0)^\ell} \qquad (2)$$

1. From (1) and (2) we conclude that $f(z) \pm g(z)$ has a pole of order equal to $\max(k,\ell)$, where

$$\max(k,\ell) = \begin{cases} k & \text{if } k \geq \ell \\ \ell & \text{if } k < \ell \end{cases} \qquad (3)$$

2. The function $f(z) \cdot g(z)$ has a pole of order $k + \ell$ at z_0 because

$$f(z) \cdot g(z) = \ldots + \frac{b_k \cdot b_\ell'}{(z-z_0)^{k+\ell}} \qquad (4)$$

3. If $k > \ell$ then,

$$\left[\ldots + \frac{b_k}{(z-z_0)^k} \right] = \left[\ldots + \frac{b_\ell'}{(z-z_0)^\ell} \right] \cdot \left[\ldots + \frac{B}{(z-z_0)^{k-\ell}} \right] \qquad (5)$$

and the function $\frac{f(z)}{g(z)}$ has a pole of order $k - \ell$ at the point $z = z_0$.

If $k < \ell$ then

$$\left[\ldots + \frac{b_k}{(z-z_0)^k} \right] = \left[\ldots + \frac{b_\ell'}{(z-z_0)^\ell} \right] \cdot \left[(z-z_0)^{\ell-k} + \ldots \right] \qquad (6)$$

and the function $\frac{f(z)}{g(z)}$ has a zero of order $\ell - k$.

• **PROBLEM 16-20**

At the point $z = z_0$ a function $f(z)$ has a zero of order k and a function $g(z)$ has a pole of order ℓ.

What kind of singularity or zero is point z_0 for:

1. $f(z) \pm g(z)$

2. $f(z) \cdot g(z)$

3. $\dfrac{f(z)}{g(z)}$

4. $\dfrac{g(z)}{f(z)}$

Solution: The function $f(z)$ can be represented by
$$f(z) = (z-z_0)^k \sum_{n=0}^{\infty} a_n (z-z_0)^n \tag{1}$$
(see Problem 16-3).

The function $g(z)$ can be written as
$$g(z) = \sum_{n=0}^{\infty} a'_n (z-z_0)^n + \frac{b_1}{z-z_0} + \ldots + \frac{b_\ell}{(z-z_0)^\ell} \tag{2}$$

1. From (1) and (2) we conclude that $f(z) \pm g(z)$ has a pole of order ℓ.

2. In the product $f(z) \cdot g(z)$ the term
$$\frac{a_0 b_\ell (z-z_0)^k}{(z-z_0)^\ell} \tag{3}$$
determines what kind of a point z_0 is.

For $k > \ell$, z_0 is a zero of order $k - \ell$.
For $k = \ell$, z_0 is a regular point.
For $k < \ell$, z_0 is a pole of order $\ell - k$.

3. Let
$$\frac{f}{g} = h \tag{4}$$
then,
$$(z-z_0)^k \sum_{n=0}^{\infty} a_n (z-z_0)^n = \left[\frac{b_1}{z-z_0} + \ldots + \frac{b_\ell}{(z-z_0)^\ell} + \sum_{n=0}^{\infty} a'_n (z-z_0)^n \right] h \tag{5}$$

From (5) we conclude that
$$a_0 (z-z_0)^k + \ldots = \frac{b_\ell}{(z-z_0)^\ell} \cdot (z-z_0)^{k+\ell} a''_0 + \ldots \tag{6}$$

Thus $\frac{f}{g}$ has a zero at $z = z_0$ of order $k + \ell$.

4.
$$\frac{g}{f} = \frac{\sum_{n=0}^{\infty} a'_n (z-z_0)^n + \frac{b_1}{z-z_0} + \ldots + \frac{b_\ell}{(z-z_0)^\ell}}{(z-z_0)^k \sum_{n=0}^{\infty} a_n (z-z_0)^n} \tag{7}$$

The function $\frac{g}{f}$ has a pole of order $k + \ell$ at $z = z_0$

• **PROBLEM 16-21**

Show that if the function $f(z)$ has a pole of order m at $z = z_0$, and if $P(u)$ is a polynomial of nth degree, then
$$P[f(z)] \tag{1}$$
has a pole of order mn at $z = z_0$.

Solution: Since $f(z)$ has a pole of order m at $z = z_0$, its Laurent series about $z = z_0$ is

$$f(z) = \sum_{n=0}^{\infty} a_n(z-z_0)^n + \frac{b_1}{z-z_0} + \ldots + \frac{b_m}{(z-z_0)^m} \quad (2)$$

In general, the polynomial of degree n can be written

$$P(u) = A_n u^n + A_{n-1} u^{n-1} + \ldots + A_1 u + A_0 \quad (3)$$

Substituting (2) into (3) we find function $P[f(z)]$.

Note, that the order of the pole at $z = z_0$ will be determined by the highest negative power of $(z-z_0)$.

Hence,

$$P[f(z)] = A_n \left[\ldots + \frac{b_1}{z-z_0} + \ldots + \frac{b_m}{(z-z_0)^m} \right]^n + \ldots \quad (4)$$

From (4) we conclude that the term

$$\frac{A_n b_m^n}{(z-z_0)^{mn}} \quad (5)$$

has the highest negative power of $(z-z_0)$. Therefore, the function $P[f(z)]$ has a pole of order mn at $z = z_0$.

REMOVABLE SINGULARITIES, SINGULARITIES AT INFINITY

• **PROBLEM 16-22**

In Problem 16.6 we defined a removable singularity. The following theorem explains why such singularities are called removable.

Theorem

Let z_0 be a removable singularity of a function $f(z)$ which is analytic in a domain $D' = D - \{z_0\}$ and let z_0 be an interior point of D. Then the limit exists

$$\lim_{z \to z_0} f(z) = \omega_0 \quad (1)$$

and the function

$$f(z_0) = \omega_0 \quad (2)$$

is analytic at z_0.

Prove this theorem.

Solution: Obviously, the function $f(z_0) = \omega_0$ is continuous at z_0. We shall prove that it is also analytic at z_0.

Let $C(z_0,R)$ be a circle contained in D' and let $F(z)$ be a function defined by

$$F(z) = \frac{1}{2\pi i} \int_{C(z_0,R)} \frac{f(\omega)}{\omega - z} d\omega \qquad (3)$$

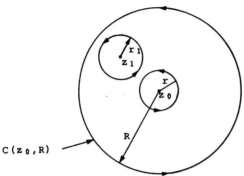

Fig. 1

The function $F(z)$ is analytic for $|z-z_0| < R$, because integral (3) is differentiable under the integral sign.

Let z_1 be such that

$$0 < |z_1-z_0| < R. \qquad (4)$$

We will show that $f(z_1) = F(z_1)$.

The circle $C(z_1,r_1)$ does not contain point z_0. We have

$$F(z_1) - f(z_1) = \frac{1}{2\pi i} \int_{C(z_0,R)} \frac{f(\omega)}{\omega-z_1} d\omega - \frac{1}{2\pi i} \int_{C(z_1,r_1)} \frac{f(\omega)}{\omega-z_1} d\omega \qquad (5)$$

Let $C(z_0,r)$ be a small circle which does not intersect $C(z_1,r_1)$, see Figure 1. By Cauchy integral theorem we have

$$\int_{C(z_0,R)} = \int_{C(z_1,r_1)} + \int_{C(z_0,r)} \qquad (6)$$

From (5) and (6) we obtain

$$F(z_1) - f(z_1) = \frac{1}{2\pi i} \int_{C(z_0,r)} \frac{f(\omega)}{\omega-z_1} d\omega \qquad (7)$$

Now, we shall show that integral in (7) is equal to zero.

Because $\lim_{z \to z_0} f(z) = \omega_0$ is finite, the function $|f(\omega)|$ is bounded for $|\omega-z_0| = r$. Also

$$|\omega-z_1| \geq |z_1-z_0| - r \qquad (8)$$

we have

$$\left|\frac{f(\omega)}{\omega - z_1}\right| < M \tag{9}$$

From (7) and (9) we conclude that

$$|F(z_1) - f(z_1)| < \frac{1}{2\pi} M \cdot 2\pi r = Mr \tag{10}$$

As $r \to 0$ we obtain

$$F(z_1) = f(z_1) \tag{11}$$

Hence, the new function

$$F(z) = \begin{cases} f(z) & \text{for } z \neq z_0 \\ \lim_{z \to z_0} f(z) & \text{for } z = z_0 \end{cases} \tag{12}$$

is analytic in D.

● **PROBLEM 16-23**

Let z_0 be an isolated singular point of $f(z)$. Consider the expression

$$\lim_{z \to z_0} f(z) \tag{1}$$

1. If z_0 is a removable singularity, then (1) exists and is finite, i.e.,

$$\lim_{z \to z_0} f(z) = C_0 \tag{2}$$

   (see Problem 16-6).

2. If z_0 is a pole, then

$$\lim_{z \to z_0} f(z) = \infty \tag{3}$$

   (see Problem 16-8).

3. If z_0 is an essential singularity then $\lim_{z \to z_0} f(z)$ does not exist.

Prove the converse statements.

Solution: 1. The converse statement to statement (1) is:

If $\lim_{z \to z_0} f(z)$ exists and is finite, then z_0 is a removable singularity of $f(z)$.

Suppose $f(z)$ is analytic in a deleted neighborhood of z_0, where z_0 is an interior point of this neighborhood. Suppose $\lim_{z \to z_0} f(z)$ exists and is finite. Then, if we define $f(z_0) = \omega_0 = $

$\lim_{z \to z_0} f(z)$, the function $f(z)$ is analytic at z_0 and hence in the neighborhood of z_0.

2. We have to prove that if

$$\lim_{z \to z_0} f(z) = \infty, \tag{4}$$

then z_0 is a pole.

Function $g(z) = \frac{1}{f(z)}$ has a zero at $z = z_0$ of order m. Hence,

$$g(z) = (z-z_0)^m g_1(z) \tag{5}$$

where $g_1(z)$ is analytic and non-zero at $z = z_0$.

We can write

$$f(z) = (z-z_0)^{-m} f_1(z), \tag{6}$$

where

$$f_1(z) = \frac{1}{g_1(z)} \tag{7}$$

is analytic and non-zero at z_0. Hence, $f_1(z)$ has a Taylor expansion

$$f_1(z) = a_0 + a_1(z-z_0) + a_2(z-z_0)^2 + \ldots \tag{8}$$

Multiplying (8) by $(z-z_0)^{-m}$ we get

$$f(z) = \frac{a_0}{(z-z_0)^m} + \frac{a_1}{(z-z_0)^{m-1}} + \ldots + \sum_{n=0}^{\infty} b_n (z-z_0)^n \tag{9}$$

Hence, $z = z_0$ is a pole of $f(z)$.

3. If $\lim_{z \to z_0} f(z)$ does not exist, then z_0 is an essential singularity. This becomes obvious when we eliminate cases 1 and 2.

• **PROBLEM 16-24**

Find and describe all singular points of each of the following functions:

1. $\dfrac{z}{\sin z}$

2. $e^{\cot\left(\frac{1}{z}\right)}$

3. $\dfrac{1}{\sin \frac{1}{z}}$

4. $\dfrac{z^3}{1+z^4}$

5. $\dfrac{1}{z^2(z^2+4)^3}$

6. $\dfrac{e^z+1}{e^{2z}}$

7. $\dfrac{1}{\sin z - \cos z}$

Solution: 1. The function $\dfrac{z}{\sin z}$ has simple poles at $z = n\pi$ $n = \pm 1, \pm 2, \pm 3, \ldots$, and removable singular point at $z = 0$.

2. The function $e^{\cot\left(\frac{1}{z}\right)}$ has essential singular points at $z = \dfrac{1}{n\pi}$, $n = \pm 1, \pm 2, \ldots$.

3. Singularities of this function are simple poles at $z = \dfrac{1}{n\pi}$, $n = \pm 1, \pm 2, \ldots$.

4. The function $\dfrac{z^3}{1+z^4}$ has simple poles at $\dfrac{1 \pm i}{\sqrt{2}}$ and $\dfrac{-1 \pm i}{\sqrt{2}}$ which are the solutions of the equation $z^4 + 1 = 0$.

5. This function has a double pole at $z = 0$ and two poles of order two at $z = \pm 2i$.

6. Function $\dfrac{e^z+1}{e^{2z}}$ has no singular points, it is an entire function.

7. Note that

$$\sin z - \cos z = \sqrt{2} \sin\left(z - \dfrac{\pi}{4}\right)$$

Hence, the function $\dfrac{1}{\sin z - \cos z}$ has simple poles at $z = n\pi + \dfrac{\pi}{4}$

• **PROBLEM 16-25**

Explain the differences between the function of the real variable

$$f(x) = \begin{cases} e^{-\frac{1}{x^2}} & \text{for } x \neq 0 \\ 0 & \text{for } x = 0 \end{cases} \quad (1)$$

and the function of the complex variable

$$f(z) = e^{-\frac{1}{z^2}} \quad (2)$$

in the neighborhood of $z = 0$.

Solution: First, consider the real function

$$\lim_{x \to 0} e^{-\frac{1}{x^2}} = 0 \tag{3}$$

The function $e^{-\frac{1}{x^2}}$ is differentiable infinitely many times at $x = 0$. All the derivatives at $x = 0$ are equal to zero. The graph of this function (symmetrical with respect to the y-axis) does not have any irregular features at $x = 0$.

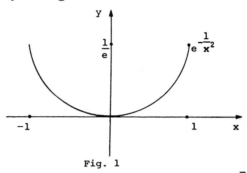

Fig. 1

At the point $z = 0$ the function $f(z) = e^{-\frac{1}{z^2}}$ has an essential singularity. Hence, in the neighborhood of $z = 0$, this function is not continuous and is not differentiable. It will be shown later in this chapter (see Casorati-Weierstrass theorem) that in every neighborhood of $z = 0$ the function comes arbitrarily close to every value.

● **PROBLEM 16-26**

Show that the nature of the singular point of $f(z)$ at $z = \infty$ is the same as that of the function

$$F(\omega) \equiv f\left(\frac{1}{\omega}\right) \quad \text{at } \omega = 0 \tag{1}$$

Solution: Suppose the function $f(z)$ is analytic for all values of z such that $|z| > R$, where $0 \le R < \infty$. Then, replacing z by $\frac{1}{\omega}$ we obtain a function $f\left(\frac{1}{\omega}\right)$ which is analytic for all values of ω such that

$$\left|\frac{1}{\omega}\right| > R \quad \text{or}$$

$$0 < |\omega| < \frac{1}{R} \tag{2}$$

Hence, the function

$$F(\omega) \equiv f\left(\frac{1}{\omega}\right) \tag{3}$$

is analytic for all values of ω such that $0 < |\omega| < \frac{1}{R}$.

For $F(\omega)$ we can compute the Laurent series expansion about $\omega = 0$.

$$f\left(\frac{1}{\omega}\right) = F(\omega) = \sum_{n=0}^{\infty} a_n \omega^n + \sum_{n=-1}^{\infty} b_n \omega^n \quad (4)$$

Having expansion (4) we can determine what kind of singularity is the point $\omega = 0$ for the function $F(\omega)$.

That information decides what kind of singularity the function $f(z)$ has at $z = \infty$.

Eq.(4) can be written in the form

$$f(z) = \sum_{n=0}^{\infty} a_n z^{-n} + \sum_{n=-1}^{\infty} b_n z^{-n} \quad (5)$$

$$= \sum_{n=0}^{\infty} a_n z^{-n} + \sum_{n=1}^{\infty} b_n z^n$$

where $b_{-n} = b_n$.

● **PROBLEM 16-27**

What kind of singularity do the functions

1. $\sqrt{(z-1)(z-2)}$

2. $e^{-\frac{1}{z^2}}$

have at $z = \infty$?

Solution: As usual we shall substitute

$$F(\omega) = f\left(\frac{1}{\omega}\right) \quad (1)$$

$$0 < |\omega| < \frac{1}{R}$$

and we will find the Laurent series expansion

$$f\left(\frac{1}{\omega}\right) = F(\omega) = \sum_{n=0}^{\infty} a_n \omega^n + \sum_{n=1}^{\infty} b_n \omega^{-n} \quad (2)$$

for $0 < |\omega| < \frac{1}{R}$

or

$$f(z) = \sum_{n=0}^{\infty} a_n z^{-n} + \sum_{n=1}^{\infty} b_n z^n \quad (3)$$

for $|z| > R$.

Definition
    If all but a finite number of the b_n vanish, then

$$f(z) = \sum_{n=0}^{\infty} a_n z^{-n} + b_1 z + b_2 z^2 + \ldots + b_k z^k \quad (4)$$

and the function $f(z)$ is said to have a pole of order k at $z=\infty$.

The expression

$$b_1 z + b_2 z^2 + \ldots + b_k z^k \tag{5}$$

is called the principal part of $f(z)$.

$$\sqrt{(z-1)(z-2)} = \sqrt{z^2 \left(1 - \frac{1}{z}\right)\left(1 - \frac{2}{z}\right)}$$

$$= \pm z \sqrt{1 - \frac{1}{z}} \cdot \sqrt{1 - \frac{2}{z}} \tag{6}$$

$$= \pm z \left(1 + \frac{a_1}{z} + \frac{a_2}{z^2} + \frac{a_3}{z^3} + \ldots\right)$$

This function has, on each of the two sheets a simple pole at $z = \infty$.

Definition

If all the b_n vanish then

$$f(z) = \sum_{n=0}^{\infty} a_n z^{-n}, \quad |z| > R \tag{7}$$

and the function $f(z)$ is said to have a removable singularity at $z = \infty$.

If we define $f(\infty) = a_0$, the function $f(z)$ is analytic at $z = \infty$.

$$e^{-\frac{1}{z^2}} = 1 - \frac{1}{z^2} + \frac{1}{2!}\frac{1}{z^4} - \ldots \tag{8}$$

Thus $e^{-\frac{1}{z^2}}$ has a removable singularity at $z = \infty$. It becomes analytic at $z = \infty$ if we define

$$\left. \left(e^{-\frac{1}{z^2}}\right)\right|_{z=\infty} = 1 \tag{9}$$

● **PROBLEM 16-28**

Show that the function

$$f(z) = \frac{z^3+4}{e^z} \tag{1}$$

has essential singularity at $z = \infty$.

Solution: The behavior of the function $f(z)$ at $z = \infty$ is the same as that of the function $F(\omega) = f\left(\frac{1}{\omega}\right)$ at $\omega = 0$.

The function $f(z)$ can be expanded in the Laurent series.

Definition

If in the Laurent series expansion

$$f(z) = \sum_{n=0}^{\infty} a_n z^{-n} + \sum_{n=1}^{\infty} b_n z^n, \quad |z| > R \tag{2}$$

an infinite number of b_n do not vanish then $f(z)$ has an essential singularity at $z = \infty$.

We have

$$\frac{z^3+4}{e^z} = (z^3+4) \cdot e^{-z}$$

$$= (z^3+4) \cdot \left[1 - z + \frac{z^2}{2!} - \frac{z^3}{3!} + \cdots\right] \tag{3}$$

Hence, the function $\frac{z^3+4}{e^z}$ has an essential singularity at $z = \infty$.

THEOREMS

• **PROBLEM 16-29**

The Casorati-Weierstrass theorem describes the behavior of a function at an essential singular point.

Theorem (Casorati-Weierstrass)

If $f(z)$ is analytic in a deleted neighborhood of z_0 and z_0 is an essential singular point of $f(z)$, then given any complex number W (finite or infinite), there exists a sequence (z_n) converging to z_0 such that

$$\lim_{n \to \infty} f(z_n) = W \tag{1}$$

where $z_n \to z_0$.

Prove this theorem.

Solution: Suppose $W = \infty$. We proved that an analytic function cannot be bounded in the neighborhood of an isolated singularity. Hence, for any positive integer n, there exists a point z_n such that

$$0 < |z_0 - z_n| < \frac{1}{n} \tag{2}$$

and

$$|f(z_n)| > n \tag{3}$$

Thus, the sequence (z_n) converges to z_0 and

$$\lim_{n \to \infty} f(z_n) = \infty = W \tag{4}$$

Suppose, now that W is any finite complex number. If every deleted neighborhood of z_0 contains a point z such that $f(z) = W$ the theorem is proved.

Suppose there is a deleted neighborhood of z_0 in which $f(z) \neq W$. The function

$$F(z) = \frac{1}{f(z)-W} \tag{5}$$

is analytic in this neighborhood and z_0 is an essential singularity. Hence, there exists a sequence (z_n) converging to z_0 such that

$$\lim_{n \to \infty} F(z_n) = \infty \tag{6}$$

From (6) we obtain $\quad \lim_{n \to \infty} F(z_n) = W \tag{7}$

That completes the proof.

● **PROBLEM 16-30**

Verify Casorati-Weierstrass theorem for the function

$$f(z) = e^{\frac{1}{z}} \tag{1}$$

at $z = 0$.

Solution: Indeed, from

$$f(z) = e^{\frac{1}{z}} = 1 + \frac{1}{z} + \frac{1}{2!}\frac{1}{z^2} + \frac{1}{3!}\frac{1}{z^3} + \ldots \tag{2}$$

we see that $z = 0$ is an essential singularity of $e^{\frac{1}{z}}$ (infinitely many terms with negative powers of z).

Now, we will show that for any complex number W there exists a sequence of points

$$z_n \xrightarrow[n \to \infty]{} z_0 = 0 \tag{3}$$

such that

$$f(z_n) \xrightarrow[n \to \infty]{} W \tag{4}$$

In our case if $W = 0$, then the sequence $z_n = -\frac{1}{n}$ satifies (3) and (4).

$$z_n \xrightarrow[n \to \infty]{} 0 \tag{5}$$

and

$$\lim_{n \to \infty} f(z_n) = \lim_{n \to \infty} e^{-n} = 0 \tag{6}$$

If $W = \infty$, then for $z_n = \frac{1}{n}$

$$z_n = \frac{1}{n} \xrightarrow[n \to \infty]{} 0 \tag{7}$$

and

$$\lim_{n \to \infty} f(z_n) = \lim_{n \to \infty} e^n = \infty \tag{8}$$

If $W \neq 0$ and $W \neq \infty$ then from

$$W = e^{\frac{1}{z}} \tag{9}$$

we get

$$z = \frac{1}{\ln W} \tag{10}$$

By $(\ln z)_0$ we can denote the branch of the logarithm such that

$$0 \leq \arg z < 2\pi \tag{11}$$

then (10) can be written as

$$z = \frac{1}{(\ln W)_0 + 2n\pi i}, \quad n = 1, 2, 3, \ldots \tag{12}$$

Eq. (12) defines a sequence

$$z_n = \frac{1}{2n\pi i + (\ln W)} \tag{13}$$

which

$$z_n \xrightarrow[n \to \infty]{} 0 \tag{14}$$

converges to zero and since

$$f(z_n) = e^{\frac{1}{z_n}} = W \tag{15}$$

$$\lim_{n \to \infty} f(z_n) = W \tag{16}$$

● **PROBLEM 16-31**

Verify Picard's Theorem for

$$f(z) = e^{\frac{1}{z}} \text{ at } z = 0.$$

Solution: Picard's theorem states that:

Theorem (Picard)

In every neighborhood of an analytic function where an essential singularity exists, there are an infinite number of z's (with one possible exception) satisfying the equation of the function.

The function $f(z) = e^{\frac{1}{z}}$ has an essential singularity at $z = 0$.

We will show that in every neighborhood of $z = 0$ there are an infinite number of z's satisfying the equation

$$e^{\frac{1}{z}} = \omega, \tag{1}$$

where $\omega \neq 0$ is any complex number

$$e^{\frac{1}{z}} = \omega = e^{\ln \omega} \tag{2}$$

Then,

$$\frac{1}{z} = \ln \omega + 2\pi i n, \tag{3}$$

$$n = 0, \pm 1, \pm 2, \ldots$$

Equation (3) defines a sequence

$$z_n = \frac{1}{\ln \omega + 2\pi i n} \quad n = 0, \pm 1, \pm 2 \ldots \tag{4}$$

such that

$$z_n \xrightarrow[n \to \infty]{} 0 \tag{5}$$

and

$$e^{\frac{1}{z_n}} = \omega \tag{6}$$

In this case the value excluded in Picard's theorem is zero, for $z \neq 0$, $e^{\frac{1}{z}} \neq 0$.

RATIONAL AND MEROMORPHIC FUNCTIONS

• **PROBLEM 16-32**

Let $f(z)$ be a rational function

$$f(z) = \frac{a_n z^n + \ldots + a_1 z + a_0}{b_m z^m + \ldots + b_1 z + b_0} \tag{1}$$

where $a_n \neq 0$ and $b_m \neq 0$.

Assuming that the numerator and denominator have no common zeros, describe the singular points of $f(z)$.

Solution: The fundamental theorem of algebra states that every polynomial of degree $m \geq 1$ has precisely m zeros.

In factored form, the factors of the polynomial are counted a number of times equal to the polynomial zeros. Hence, we can write

$$b_m z^m + \ldots + b_1 z + b_0 = b_m (z - z_1) \ldots (z - z_m) \tag{2}$$

It is possible that some of the elements of $z_1, z_2, \ldots, z_m$ are equal. Denoting by k_ℓ the order of zero z_ℓ, we have

$$b_m z^m + \ldots + b_1 z + b_0 = b_m (z-z_1)^{k_1} \ldots (z-z_r)^{k_r} \tag{3}$$

where

$$k_1 + \ldots + k_r = m \tag{4}$$

Function (1) can be written in the form

$$\frac{1}{f(z)} = \frac{b_m z^m + \ldots + b_1 z + b_0}{a_n z^n + \ldots + a_1 z + a_0}$$

$$= \frac{b_m (z-z_1)^{k_1} \ldots (z-z_r)^{k_r}}{a_n z^n + \ldots + a_1 z + a_0}, \quad b_m \neq 0 \tag{5}$$

By Problem 16-13 we know that a zero of order ℓ of a function $\frac{1}{f(z)}$ is a pole of order ℓ of $f(z)$.

Hence, the function given by (1) has a pole at each zero of the denominator $(b_m z^m + \ldots + b_1 z + z_0)$, of the same order as that of the zero.

• **PROBLEM 16-33**

1. Find the function $f(z)$ which has zero of the second order at $2i$, zero of the first order at 4 and two simple poles at i and $3i$.

2. Show that a rational function takes all values exactly the same number of times.

Solution: 1. We shall apply the following.

Theorem
 A function $f(z)$ is a rational function if, and only if, it has no essential singularities either in the finite complex plane or at infinity.

Since the function $f(z)$ we have to find, does not have any essential singularities it must be a rational function, i.e.,

$$f(z) = \frac{G(z)}{H(z)} \tag{1}$$

where $G(z)$ and $H(z)$ are relatively prime polynomials, that is polynomials whose only common divisors are constants. The function $f(z)$ has a zero of the second order at $2i$ and simple zero at 4. Therefore,

$$f(z) = \frac{(z-2i)^2(z-4)}{H(z)} \tag{2}$$

Since, its only singularities are two simple poles at i and 3i, we obtain

$$f(z) = \frac{(z-2i)^2(z-4)}{(z-i)(z-3i)} \tag{3}$$

2. Now we shall show that a rational function

$$f(z) = \frac{M(z)}{N(z)} \tag{4}$$

takes all values exactly the same number of times. Let

$$f(z) = a \tag{5}$$

then

$$M(z) - a N(z) = 0 \tag{6}$$

If the degree of M(z) is different from the degree of N(z) and $a \neq 0$, then the polynomial M(z)-aN(z) is always of the same degree which is independent of the value of a. Hence, the number of solutions of M(z)-aN(z) = 0 is independent of a. If the degree of M(z) is equal to the degree of N(z) then

$$\frac{M(z)}{N(z)} = \frac{M'(z)}{N(z)} + A \tag{7}$$

where A is a constant and degree of M'(z) = (degree of M(z)) -1. The reasoning can be repeated as for different degrees above.

• **PROBLEM 16-34**

1. Suppose you are a mathematician or a physicist handling some mathematical functions. If the function has singularities, what kind of singularities are easiest to deal with?

2. Which of the following functions are entire, which are meromorphic and which are neither?
   a) $z^3 e^{-z}$
   b) $\frac{1-\cos z}{z}$
   c) $z + \frac{1}{z^2}$
   d) $z \sin \frac{1}{z}$

Solution: 1. The most preferable situation is when the functions have no singularities, i.e., are analytic everywhere.

The second best would be the function with a removable singular point. In that case the series expansion of that function contains no negative powers and we can easily get rid of the singularity by redefining the function.

Next come the functions whose expansion contains a finite number of negative powers. These are the functions with poles.

The most difficult singularities to deal with are the essential singular points. Then the series expansion contains infinitely many negative powers.

2. Definition

A function which is analytic everywhere in the finite plane, i.e., everywhere except at ∞, is said to be an entire function or integral function.

Definition

A function is called meromorphic of its only singularities are isolated poles.

a) $z^3 e^{-z}$ is entire

b) $\dfrac{1-\cos z}{z}$ is entire

c) $z + \dfrac{1}{z^2}$ is meromorphic

d) $z \sin \dfrac{1}{z}$ is neither

• **PROBLEM 16-35**

The function f(z) is called meromorphic if its only singularities are isolated poles.

1. Determine which of the following functions are meromorphic, analytic, polynomials or rational.

    a) $e^{\frac{1}{z}}$

    b) $\dfrac{1}{\sin \frac{1}{z}}$

    c) $\dfrac{\sin z}{z}$

2. Prove that if the meromorphic function f(z) has finitely many poles, then f(z) can be expressed as

$$f(z) = \dfrac{h(z)}{P(z)} \qquad (1)$$

where h(z) is an analytic function and P(z) is a polynomial.

Solution: 1. An analytic function is differentiable everywhere, if it does not have any singularities, hence its meromorphic. Polynomial is an analytic function, hence it is meromorphic.

A rational function

$$f(z) = \frac{M(z)}{N(z)} \tag{2}$$

where $M(z)$ and $N(z)$ are polynomials has singularities which are poles, thus it is a meromorphic function.

a) $e^{\frac{1}{z}} = 1 + \frac{1}{z} + \frac{1}{2!z^2} + \ldots$ \hfill (3)

This function has an essential singularity at $z = 0$. It is not meromorphic.

b) $\dfrac{1}{\sin \frac{1}{z}}$

This function has simple poles at $z = \frac{1}{n\pi}$, $n = \pm 1, \pm 2, \ldots$. The pole at $z = 0$ is not isolated, $\dfrac{1}{\sin \frac{1}{z}}$ is not meromorphic

c) $\dfrac{\sin z}{z}$

This function has a removable singularity at $z = 0$, it is meromorphic.

2. Let $f(z)$ be a meromorphic function with finitely many poles $z_1, z_2, \ldots, z_k$. Their orders being $a_1, a_2, \ldots, a_k$, respectively. Then, $f(z)$ can be written in the form

$$f(z) = \frac{f_1(z)}{(z-z_1)^{a_1} \ldots (z-z_k)^{a_k}} \tag{4}$$

The denominator of (4) is a polynomial and $f_1(z)$ is an analytic function.

That completes the proof.

The above theorem is a special case of a more general.

Theorem

Every meromorphic function is the quotient of two analytic functions. Many mathematicians contributed to the solution of the problem of how to construct all meromorphic functions. Among them Hadamard, Weierstrass, Mittag-Leffler.

• **PROBLEM 16-36**

Prove that in each disk $|z| \leq r$ a meromorphic function can have only finitely many poles.

Solution: Let $f(z)$ be a meromorphic function. Then its only non-removable singularities are isolated poles.

Note that poles of a meromorphic function have to be isolated.

Suppose a meromorphic function $f(z)$ has infinitely many poles in a disk $|z| \leq r$.

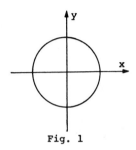

Fig. 1

We shall use the following:

<u>Theorem</u>

Every bounded set D of complex numbers containing an infinite number of points has an accumulation point.

Since the disk $|z| \leq r$ is a bounded set, the set of the poles of $f(z)$ in the disk is also a bounded set and therefore it has an accumulation point.

Such a point would be a non-isolated pole.   Contradiction.

CHAPTER 17

RESIDUE CALCULUS

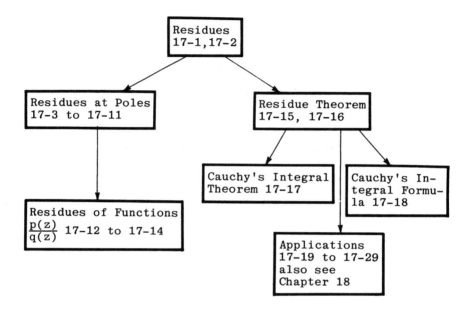

This chart is provided to facilitate rapid understanding of the interrelationships of the topics and subject matter in this chapter. Also shown are the problem numbers associated with the subject matter.

RESIDUES

● **PROBLEM 17-1**

Let $f(z)$ be analytic in an annulus D centered at $z = z_0$. Then $f(z)$ can be expanded in a Laurent series

$$f(z) = \sum_{n=0}^{\infty} a_n (z-z_0)^n + \sum_{n=-1}^{-\infty} b_n (z-z_0)^n \qquad (1)$$

in D. Eq.(1) can be written in the shorter form

$$f(z) = \sum_{n=-\infty}^{\infty} c_n (z-z_0)^n \qquad (2)$$

where

$$c_n = \frac{1}{2\pi i} \int_C \frac{f(z)}{(z-z_0)^{n+1}} dz \qquad (3)$$

$$n = 0, \pm 1, \pm 2, \pm 3, \ldots,$$

and C is a simple closed contour in D.

Note that by the Cauchy-Goursat theorem, if a function is analytic at all points interior to and on a simple closed contour C, then the value of the integral is zero, i.e.,

$$\int_C f(z) dz = 0 \qquad (4)$$

Show that

$$\int_C f(z) dz = 2\pi i \, c_{-1} \qquad (5)$$

Define the residue of $f(z)$ at $z = z_0$.

Solution: Setting $n = -1$ in (3), we obtain

$$\int_C f(z) dz = 2\pi i \, c_{-1} \qquad (6)$$

Definition

The coefficient c_{-1} of the term $\frac{1}{z-z_0}$ in the Laurent expansion of $f(z)$ in an annulus (or the deleted neighborhood of z_0) centered at z_0, is called the residue of $f(z)$ at $z = z_0$.

The residue of $f(z)$ at $z = z_0$ is denoted by $\operatorname{Res}_{z_0} f(z)$.

• **PROBLEM 17-2**

Determine the following residues:

1. $\operatorname{Res}_0(e^{\frac{1}{z}})$

2. $\operatorname{Res}_0(z^{-2} e^{\frac{1}{z}})$

3. $\operatorname{Res}_0(z^2 e^{\frac{1}{z}})$

What integrals can be evaluated from these residues?

Solution: Let us recall eq.(5) of Problem 17-1.

$$\int_C f(z) dz = 2\pi i \, c_{-1} \qquad (1)$$

which can be expressed

$$\text{Res}_{z_0} f(z) = c_{-1} = \frac{1}{2\pi i} \int_C f(z) dz \qquad (2)$$

1. The Laurent series for $e^{\frac{1}{z}}$ is

$$e^{\frac{1}{z}} = 1 + \frac{1}{z} + \frac{1}{2!z^2} + \frac{1}{3!z^3} + \cdots \qquad (3)$$

$$\text{for } 0 < |z|.$$

The coefficient of the term $\frac{1}{z-0}$ is 1, hence

$$c_{-1} = 1 \qquad (4)$$

or

$$\text{Res}_0(e^{\frac{1}{z}}) = 1 \qquad (5)$$

From (1),

$$\int_C e^{\frac{1}{z}} dz = 2\pi i \qquad (6)$$

2. $$\frac{1}{z^2} e^{\frac{1}{z}} = \frac{1}{z^2}(1 + \frac{1}{z} + \frac{1}{2!z^2} + \frac{1}{3!z^3} + \cdots)$$

$$= \frac{1}{z^2} + \frac{1}{z^3} + \frac{1}{2!z^4} + \cdots \qquad (7)$$

Hence,

$$\text{Res}_0(z^{-2} e^{\frac{1}{z}}) = 0 \qquad (8)$$

and

$$\int_C \frac{e^{\frac{1}{z}}}{z^2} dz = 0 \qquad (9)$$

where C can be the unit circle centered at $z_0 = 0$.

3. $$z^2 e^{\frac{1}{z}} = z^2(1 + \frac{1}{z} + \frac{1}{2!z^2} + \frac{1}{3!z^3} + \cdots)$$

$$= z^2 + z + \frac{1}{2!} + \frac{1}{3!z} + \cdots \qquad (10)$$

Hence,

$$\text{Res}_0(z^2 e^{\frac{1}{z}}) = \frac{1}{6} \qquad (11)$$

and

$$\int_C z^2 e^{\frac{1}{z}} dz = \frac{\pi i}{3} \qquad (12)$$

RESIDUES AT POLES

• **PROBLEM** 17-3

Show that if f(z) is analytic inside and on a simple closed curve C, except at a pole z_0 of order m, then the residue of f(z) at z_0 is given by

$$\text{Res}_{z_0} f(z) = \lim_{z \to z_0} \frac{1}{(m-1)!} \frac{d^{m-1}}{dz^{m-1}} \left[(z-z_0)^m f(z) \right] \quad (1)$$

Solution: Observe that eq.(1) enables us to evaluate the residue of the function f(z) without finding the Laurent series. If f(z) has a pole at $z = z_0$ of order m, then its Laurent series is

$$f(z) = c_0 + c_1(z-z_0) + c_2(z-z_0)^2 + \ldots + \frac{c_{-1}}{z-z_0} + \ldots$$
$$+ \frac{c_{-m}}{(z-z_0)^m} \quad (2)$$

Multiplying both sides by $(z-z_0)^m$, we obtain

$$f(z)(z-z_0)^m = c_{-m} + c_{-m+1}(z-z_0) + \ldots$$
$$+ c_{-1}(z-z_0)^{m-1} + c_0(z-z_0)^m + \ldots \quad (3)$$

The series (3) is the Taylor series of the analytic function $(z-z_0)^m f(z)$. Differentiating (3) (m-1) times, we obtain

$$\frac{d^{m-1}}{dz^{m-1}} \left[(z-z_0)^m f(z) \right] = (m-1)! c_{-1} + m(m-1)\ldots 2 \cdot c_0(z-z_0) + \ldots \quad (4)$$

Letting $z \to z_0$, we obtain

$$\lim_{z \to z_0} \frac{d^{m-1}}{dz^{m-1}} \left[(z-z_0)^m f(z) \right] = (m-1)! c_{-1} \quad (5)$$

Series (3) is the Taylor series. Coefficient c_{-1} of the term $(z-z_0)^{m-1}$ can be obtained directly from Taylor's theorem.

• **PROBLEM** 17-4

Calculate the residues of the following functions:

1. $\frac{3+5z}{z(z+1)}$ \hfill (1)

2. $\frac{1}{(z^2-1)^2}$ \hfill (2)

Solution: 1. Function (1) has simple poles (i.e., poles of order one) at $z = 0$ and $z = -1$.

In Problem 17-3, we derived the general formula for poles of order m,

$$\text{Res}_{z_0} f(z) = \frac{1}{(m-1)!} \lim_{z \to z_0} \frac{d^{m-1}}{dz^{m-1}} \left[(z-z_0)^m f(z) \right] \qquad (3)$$

which for simple poles reduces to

$$\text{Res}_{z_0} f(z) = \lim_{z \to z_0} \left[(z-z_0) f(z) \right] \qquad (4)$$

Thus, for function (1) we obtain

$$\text{Res}_0 \frac{3+5z}{z(z+1)} = \lim_{z \to 0} \frac{3+5z}{z+1} = 3 \qquad (5)$$

$$\text{Res}_{-1} \frac{3+5z}{z(z+1)} = \lim_{z \to -1} \frac{3+5z}{z} = 2 \qquad (6)$$

2. Function $\frac{1}{(z^2-1)^2}$ has two double poles at $z = -1$ and $z = 1$. For $m = 2$, equation (3) becomes

$$\text{Res}_{z_0} f(z) = \lim_{z \to z_0} \left[(z-z_0)^2 f(z) \right]' \qquad (7)$$

Since $\frac{1}{(z^2-1)^2} = \frac{1}{(z+1)^2 (z-1)^2}$

$$\text{Res}_{-1} \frac{1}{(z^2-1)^2} = \lim_{z \to -1} \left[\frac{1}{(z-1)^2} \right]' = \frac{1}{4} \qquad (8)$$

$$\text{Res}_1 \frac{1}{(z^2-1)^2} = \lim_{z \to 1} \left[\frac{1}{(z+1)^2} \right]' = -\frac{1}{4} \qquad (9)$$

● **PROBLEM 17-5**

Find the residues of the following functions:

1. $\dfrac{z^2+z}{(z-1)^2(z^2+4)}$ (1)

2. $\cot z$

Solution: 1. Function (1) has a double pole at $z = 1$, and two simple poles at $z = 2i$, $z = -2i$.

$$\text{Res}_1 f = \lim_{z \to 1} \frac{1}{1!} \frac{d}{dz} \left[(z-1)^2 \frac{z^2+z}{(z-1)^2(z^2+4)} \right]$$

$$= \lim_{z \to 1} \frac{d}{dz}\left(\frac{z^2+z}{z^2+4}\right) = \lim_{z \to 1} \frac{(2z+1)(z^2+4)-2z(z^2+z)}{(z^2+4)^2} \quad (2)$$

$$= \frac{11}{25}$$

$$\text{Res}_{2i} f = \lim_{z \to 2i}\left[(z-2i)\frac{z^2+z}{(z-1)^2(z^2+4)}\right]$$

$$= \lim_{z \to 2i}\left[\frac{z^2+z}{(z-1)^2(z+2i)}\right] = \frac{2i-4}{4i(2i-1)^2} = \frac{i-2}{8-6i} \quad (3)$$

$$\text{Res}_{-2i} f = \lim_{z \to -2i}\left[\frac{z^2+z}{(z-1)^2(z-2i)}\right] = \frac{i+2}{-8-6i} \quad (4)$$

2. Since

$$\cot z = \frac{\cos z}{\sin z}, \quad (5)$$

the singularities of cot z are simple poles which coincide with the zeros of sin z, i.e.,

$$z = k\pi, \quad \text{where k is an integer.}$$

Hence,

$$\text{Res}_{z_0} f(z) = \lim_{z \to z_0}\left[(z-z_0)f(z)\right]$$

$$= \lim_{z \to k\pi}\left[(z-k\pi)\frac{\cos z}{\sin z}\right] \quad (6)$$

We have

$$\sin z = (-1)^k \sin(z-k\pi) \quad (7)$$

and (6) leads to

$$= \lim_{z \to k\pi}\left[\frac{(z-k\pi)\cos z}{(-1)^k \sin(z-k\pi)}\right]$$

$$= \lim_{z \to k\pi}\left[\frac{\cos z}{(-1)^k}\right] \cdot \lim_{z \to k\pi}\left[\frac{z-k\pi}{\sin(2-k\pi)}\right] \quad (8)$$

$$= \frac{\cos k\pi}{(-1)^k} = \frac{(-1)^k}{(-1)^k} = 1$$

Hence, all residues are equal to one.

● **PROBLEM 17-6**

Find the residue of the function

$$f(z) = \frac{e^z}{\sin^2 z} \quad (1)$$

1. from the Laurent series of $f(z)$
2. from the formula

$$\operatorname{Res}_{z_0} f(z) = \frac{1}{(m-1)!} \lim_{z \to z_0} \frac{d^{m-1}}{dz^{m-1}}\left[(z-z_0)^m f(z)\right] \qquad (2)$$

Solution: 1. The function $\dfrac{e^z}{\sin^2 z}$ has double poles at $z = 0, \pm\pi, \pm 2\pi \ldots$ that is

$$z = k\pi, \quad k \text{ is an integer}$$

We shall expand $f(z)$ about $z = k\pi$. Substituting

$$\omega = z - k\pi, \qquad (3)$$

we obtain

$$\frac{e^z}{\sin^2 z} = \frac{e^{k\pi} e^\omega}{[\sin(\omega+k\pi)]^2} = e^{k\pi} \frac{e^\omega}{\sin^2 \omega}$$

$$= e^{k\pi} \frac{(1 + \omega + \frac{\omega^2}{2!} + \frac{\omega^3}{3!} + \ldots)}{(\omega - \frac{\omega^3}{3!} + \frac{\omega^5}{5!} - \ldots)^2}$$

$$= \frac{e^{k\pi}}{\omega^2} \frac{(1 + \omega + \frac{\omega^2}{2!} + \frac{\omega^3}{3!} + \ldots)}{(1 - \frac{\omega^2}{6} + \frac{\omega^4}{120} - \ldots)^2} \qquad (4)$$

$$= \frac{e^{k\pi}}{\omega^2} \cdot \frac{1 + \omega + \frac{\omega^2}{2!} + \frac{\omega^3}{3!} + \ldots}{1 - \frac{\omega^2}{3} + \frac{2\omega^4}{45} - \ldots}$$

$$= e^{k\pi} \left(\frac{1}{\omega^2} + \frac{1}{\omega} + \frac{5}{6} + \frac{\omega}{3} + \ldots\right)$$

$$= e^{k\pi} \cdot \frac{1}{z-k\pi} + \ldots$$

Hence,

$$\operatorname{Res}_{k\pi} \frac{e^z}{\sin^2 z} = e^{k\pi} \qquad (5)$$

2.
$$\lim_{z \to k\pi} \frac{1}{1!} \frac{d}{dz}\left[(z-k\pi)^2 \frac{e^z}{\sin^2 z}\right] \qquad (6)$$

$$= \lim_{z \to k\pi} \frac{[e^z(z-k\pi)^2 + 2(z-k\pi)e^z]\sin^2 z - 2\sin z \cos z \, e^z (z-k\pi)^2}{\sin^4 z}$$

$$= \lim_{z \to k\pi} \frac{e^z[(z-k\pi)^2 \sin z + 2(z-k\pi)\sin z - 2(z-k\pi)^2 \cos z]}{\sin^3 z}$$

We denote

$$\omega = z - k\pi,$$

and obtain from (6)

$$= \lim_{\omega \to 0} e^{k\pi} e^{\omega} \frac{\omega^2 \sin\omega + 2\omega\sin\omega - 2\omega^2\cos\omega}{\sin^3\omega} \qquad (7)$$

Since

$$\lim_{\omega \to 0} \frac{\omega^3}{\sin^3\omega} = \lim_{\omega \to 0} \left(\frac{\omega}{\sin\omega}\right)^3 = 1 \qquad (8)$$

Applying L'Hospital's rule to (7), we obtain

$$= e^{k\pi} \lim_{\omega \to 0} \frac{\omega^2 \sin\omega + 2\omega\sin\omega - 2\omega^2\cos\omega}{\omega^3} \cdot \frac{\omega^3}{\sin^3\omega}$$

$$= e^{k\pi} \lim_{\omega \to 0} \frac{\omega\sin\omega + 2\sin\omega - 2\omega\cos\omega}{\omega^2} \qquad (9)$$

$$= e^{k\pi} \lim_{\omega \to 0} \frac{2\cos\omega - \omega\sin\omega + 2\sin\omega + 2\omega\cos\omega}{2}$$

$$= e^{k\pi}$$

● **PROBLEM 17-7**

Show that the sum of the residues of the function

$$f(z) = \frac{1}{(z-z_1)(z-z_2)(z-z_3)} \qquad (1)$$

(where $\{z_1, z_2, z_3\}$ are distinct), is equal to zero.

Solution: Function $f(z)$ has three simple poles at z_1, z_2 and z_3.

$$\text{Res}_{z_1} f(z) = \lim_{z \to z_1} \left[(z-z_1)f(z)\right]$$

$$= \frac{1}{(z_1-z_2)(z_1-z_3)} \qquad (2)$$

$$\text{Res}_{z_2} f(z) = \frac{1}{(z_2-z_1)(z_2-z_3)} \qquad (3)$$

$$\text{Res}_{z_3} f(z) = \frac{1}{(z_3-z_1)(z_3-z_2)} \qquad (4)$$

$$\Sigma\text{Res} = \frac{(z_2-z_3) - (z_1-z_3) + (z_1-z_2)}{(z_1-z_2)(z_1-z_3)(z_2-z_3)} = 0 \qquad (5)$$

This result is true for the more general case

$$f(z) = \frac{1}{(z-z_1)(z-z_2)\ldots(z-z_k)}, \qquad (6)$$

where all $z_1,\ldots,z_k$ are distinct.

The sum of the residues of $f(z)$ is zero.

● **PROBLEM 17-8**

Compute the residue of the function

$$F(z) = \frac{\coth z \cot z}{z^3} \tag{1}$$

at $z = 0$.

Solution: The Laurent series of $F(z)$ is

$$F(z) = \frac{\cosh z \cos z}{z^3 \sinh z \sin z}$$

$$= \frac{1}{z^3} \cdot \frac{(1 + \frac{z^2}{2!} + \frac{z^4}{4!} + \ldots)}{(z + \frac{z^3}{3!} + \frac{z^5}{5!} + \ldots)} \cdot \frac{(1 - \frac{z^2}{2!} + \frac{z^4}{4!} - \ldots)}{(z - \frac{z^3}{3!} + \frac{z^5}{5!} - \ldots)}$$

$$= \frac{(1 - \frac{z^4}{6} + \ldots)}{z^5 (1 - \frac{z^4}{90} + \ldots)} \tag{2}$$

Dividing $(1 - \frac{z^4}{6} + \ldots)$ by $(1 - \frac{z^4}{90} + \ldots)$, we obtain $1 - \frac{7}{45} z^4 + \ldots$.

Thus the residue of $F(z)$ is the coefficient of the term $\frac{1}{z^5}(-\frac{7}{45} z^4)$, that is $-\frac{7}{45}$.

● **PROBLEM 17-9**

Show whether or not the following implication is true:

$$(\text{Res}_{z_0} f(z) = 0) \Rightarrow (f(z) \text{ is analytic at } z = z_0) \tag{1}$$

Solution: The information

$$\text{Res}_{z_0} f(z) = 0 \tag{2}$$

tells us merely that the coefficient c_{-1} in the Laurent expansion of $f(z)$ is zero. It does not mean that the other coefficients at the negative powers vanish. For example, the function

$$\frac{1}{z^2} \sin \frac{1}{z}$$

has residue zero at $z_0 = 0$,

$$\text{Res}_0 \frac{\sin \frac{1}{z}}{z^2} = 0 \tag{3}$$

but is not analytic at $z_0 = 0$. Hence, implication (1) is not true.

● **PROBLEM** 17-10

Evaluate
$$\text{Res}_0 \frac{1}{z^2(\sin z - \sinh z)} \qquad (1)$$

Solution: In this case, it is difficult to determine the order of the pole at $z = 0$. We shall expand the functions into Maclaurin series, and find the coefficient of the term $\frac{1}{z}$.

$$\frac{1}{z^2(\sin z - \sinh z)}$$

$$= \frac{1}{z^2\left[\left(z - \frac{z^3}{3!} + \frac{z^5}{5!} - \frac{z^7}{7!} + \ldots\right) - \left(z + \frac{z^3}{3!} + \frac{z^5}{5!} + \frac{z^7}{7!} + \ldots\right)\right]}$$

$$= \frac{1}{z^2\left[-\frac{2z^3}{3!} - \frac{2z^7}{7!} - \frac{2z^{11}}{11!} - \ldots\right]} \qquad (2)$$

$$= \frac{-3}{z^5\left[1 + \frac{6z^4}{7!} + \ldots\right]} = -\frac{3}{z^5}\left[1 - \frac{6z^4}{7!} - \ldots\right]$$

$$= -\frac{3}{z^5} + \frac{18}{7!}\frac{1}{z} + \ldots$$

Hence,
$$\text{Res}_0 \frac{1}{z^2(\sin z - \sinh z)} = \frac{18}{7!} \qquad (3)$$

Note that the application of the formula for a pole of order five

$$\text{Res}_0 f(z) = \frac{1}{4!} \lim_{z \to 0} \frac{d^4}{dz^4}\left[\frac{z^3}{\sin z - \sinh z}\right] \qquad (4)$$

leads to very complicated calculations.

● **PROBLEM** 17-11

Evaluate
$$\text{Res}_{z_0} f(z)g(z) \qquad (1)$$

if

1. $f(z)$ is analytic at z_0, and $g(z)$ has a simple pole at z_0.
2. $f(z)$ is analytic at z_0, and $g(z)$ has a pole of order m at z_0.
3. $f(z)$ has a pole of order k at z_0, and $g(z)$ has a pole of order ℓ at z_0.

Solution: 1. Since $f(z)$ is analytic at z_0,

$$f(z) = a_0 + a_1(z-z_0) + a_2(z-z_0)^2 + \ldots \qquad (2)$$

Function $g(z)$ has a simple pole at z_0.

$$g(z) = \frac{c_{-1}}{z-z_0} + c_0 + c_1(z-z_0) + c_2(z-z_0)^2 + \ldots \qquad (3)$$

In the product $f(z) \cdot g(z)$, there will be only one term of the type $\frac{A}{z-z_0}$. We have

$$f(z)g(z) = \ldots + \frac{a_0 c_{-1}}{z-z_0} + \ldots \qquad (4)$$

Thus,

$$\operatorname{Res}_{z_0} f(z)g(z) = a_0 c_{-1} \qquad (5)$$

2. Again, $f(z)$ is analytic, thus (2) is valid. Function $g(z)$ has a pole of order m.

$$\qquad (6)$$

$$g(z) = \frac{c_{-m}}{(z-z_0)^m} + \frac{c_{-m+1}}{(z-z_0)^{m-1}} + \ldots + \frac{c_{-1}}{z-z_0} + c_0 + c_1(z-z_0) + \ldots$$

From (2) and (6), we obtain

$$f(z)g(z) = \frac{a_0 c_{-1}}{z-z_0} + \frac{a_1 c_{-2}}{z-z_0} + \ldots + \frac{a_{m-1} c_{-m}}{z-z_0} + \ldots \qquad (7)$$

Hence,

$$\operatorname{Res}_{z_0} f(z)g(z) = a_0 c_{-1} + a_1 c_{-2} + \ldots + a_{m-1} c_{-m} \qquad (8)$$

3. Function $f(z)$ has a pole of order k at $z = z_0$.

$$f(z) = \frac{a_{-k}}{(z-z_0)^k} + \frac{a_{-k+1}}{(z-z_0)^{k-1}} + \ldots + \frac{a_{-1}}{z-z_0} + a_0 + a_1(z-z_0) + a_2(z-z_0)^2 + \ldots \qquad (9)$$

Function $g(z)$ has a pole of order ℓ at $z = z_0$.

$$g(z) = \frac{c_{-\ell}}{(z-z_0)^\ell} + \frac{c_{-\ell+1}}{(z-z_0)^{\ell-1}} + \ldots + \frac{c_{-1}}{z-z_0} + c_0 + c_1(z-z_0) + c_2(z-z_0)^2 + \ldots \qquad (10)$$

Multiplying (9) by (10), we obtain

$$f(z)g(z) = \frac{a_0 c_{-1}}{z-z_0} + \frac{a_1 c_{-2}}{z-z_0} + \ldots + \frac{a_{\ell-1} c_{-\ell}}{z-z_0} + \frac{a_{-1} c_0}{z-z_0} + \frac{a_{-2} c_1}{z-z_0} + \ldots + \frac{a_{-k} c_{k-1}}{z-z_0} + \ldots \qquad (11)$$

Thus,

$$\operatorname{Res}_{z_0} f(z)g(z) = a_0 c_{-1} + a_1 c_{-2} + \ldots + a_{\ell-1} c_{-\ell} + a_{-1} c_0 + a_{-2} c_1 + \ldots + a_{-k} c_{k-1} \qquad (12)$$

RESIDUES OF FUNCTIONS $\frac{p(z)}{q(z)}$

• **PROBLEM 17-12**

Suppose $f(z)$ is given by

$$f(z) = \frac{p(z)}{q(z)}, \qquad (1)$$

where both $p(z)$ and $q(z)$ are analytic in the neighborhood of z_0, and $q(z)$ has a simple zero at z_0, i.e.,

$$q(z_0) = 0 \quad \text{and} \quad q'(z_0) \neq 0 \qquad (2)$$

Show that

$$\operatorname{Res}_{z_0} \frac{p(z)}{q(z)} = \frac{p(z_0)}{q'(z_0)} \qquad (3)$$

Find the residues of

$$f(z) = \frac{5z-3}{z^2-z} \qquad (4)$$

Solution: Since $q(z)$ has a simple zero at z_0, the function $f(z)$ has a simple pole at z_0, thus

$$\operatorname{Res}_{z_0} f(z) = \lim_{z \to z_0} \left[(z-z_0)f(z)\right]$$

$$= \lim_{z \to z_0} \left[(z-z_0) \frac{p(z)}{q(z)}\right] \qquad (5)$$

$$= \lim_{z \to z_0} \left[\frac{p(z)}{\frac{q(z)}{z-z_0}}\right] = \frac{\lim_{z \to z_0} p(z)}{\lim_{z \to z_0} \frac{q(z)}{z-z_0}}$$

By L'Hospital's rule, we obtain

$$\lim_{z \to z_0} \frac{q(z)}{z-z_0} = q'(z_0) \qquad (6)$$

Hence,

$$\operatorname{Res}_{z_0} \frac{p(z)}{q(z)} = \frac{p(z_0)}{q'(z_0)} \qquad (7)$$

Function $f(z) = \frac{5z-3}{z(z-1)}$ has simple poles at $z = 0$ and $z = 1$.

$$\operatorname{Res}_{z_0} \frac{5z-3}{z(z-1)} = \left(\frac{5z-3}{2z-1}\right)\bigg|_{z=0} = 3 \qquad (8)$$

$$\operatorname{Res}_{z_0} \frac{5z-3}{z(z-1)} = \left(\frac{5z-3}{2z-1}\right)\bigg|_{z=1} = 2 \qquad (9)$$

• **PROBLEM 17-13**

Let $f(z)$ and $g(z)$ be regular at z_0. Compute the residue of

$$\frac{f(z)}{g(z)} \qquad (1)$$

assuming that $f(z_0) \neq 0$ and $g(z)$ has a zero of order 2. Carry out calculations for $g(z)$ with zero of order 3.

<u>Solution</u>: Function $f(z)$ is analytic at z_0 and $f(z_0) \neq 0$, thus

$$f(z) = a_0 + a_1(z-z_0) + a_2(z-z_0)^2 + \ldots \qquad (2)$$

$$a_0 \neq 0$$

Function $g(z)$ has a zero of order two at z_0

$$g(z) = b_2(z-z_0)^2 + b_3(z-z_0)^3 + \ldots \qquad (3)$$

$$b_2 \neq 0$$

Thus,

$$\frac{f(z)}{g(z)} = \frac{a_0+a_1(z-z_0)+a_2(z-z_0)^2+\ldots}{b_2(z-z_0)^2+b_3(z-z_0)^3+\ldots}$$

$$= \frac{1}{b_2(z-z_0)^2} \cdot \frac{a_0+a_1(z-z_0)+\ldots}{1+\frac{b_3}{b_2}(z-z_0)+\ldots}$$

$$= \frac{1}{b_2(z-z_0)^2} \cdot \left[a_0+a_1(z-z_0)+\ldots\right] \cdot \left[1-\frac{b_3}{b_2}(z-z_0)+\ldots\right]$$

$$= \frac{1}{(z-z_0)} \left[\frac{a_1}{b_2} - \frac{a_0 b_3}{b_2^2}\right] + \ldots \qquad (4)$$

Hence,

$$\text{Res}_{z_0} \frac{f(z)}{g(z)} = \frac{a_1 b_2 - a_0 b_3}{b_2^2} \qquad (5)$$

When $g(z)$ has a zero of order three,

$$g(z) = b_3(z-z_0)^3 + b_4(z-z_0)^4 + \ldots \qquad (6)$$

Thus,

$$\frac{f(z)}{g(z)} = \frac{a_0+a_1(z-z_0)+a_2(z-z_0)^2+\ldots}{b_3(z-z_0)^3+b_4(z-z_0)^4+\ldots}$$

$$= \frac{1}{b_3(z-z_0)^3} \cdot \frac{a_0+a_1(z-z_0)+a_2(z-z_0)^2+\ldots}{1+\frac{b_4}{b_3}(z-z_0)+\frac{b_5}{b_3}(z-z_0)^2+\ldots} \qquad (7)$$

$$= \frac{1}{b_3(z-z_0)^3} \cdot \left[a_0+a_1(z-z_0)+a_2(z-z_0)^2+\ldots\right]$$

$$\cdot \left[1-\frac{b_4}{b_3}(z-z_0) + \left(\frac{b_4^2}{b_3^2} - \frac{b_5}{b_3}\right)(z-z_0)^2+\ldots\right]$$

Thus,

$$\text{Res}_{z_0} \frac{f(z)}{g(z)} = \frac{a_0 b_4^2 - a_0 b_3 b_5 - a_1 b_3 b_4 + a_2 b_3^2}{b_3^3} \qquad (8)$$

• **PROBLEM 17-14**

Function $f(z)$ has a zero of order m at z_0. Compute the residues of the following at z_0.

1. $\dfrac{zf'(z)}{f(z)}$ (1)

2. $\dfrac{g(z)f'(z)}{f(z)}$ (2)

Function $g(z)$ is analytic at z_0.

Solution: 1. Since $f(z)$ has a zero of order m at z_0, we can denote

$$f(z) = a_m(z-z_0)^m + a_{m+1}(z-z_0)^{m+1} + a_{m+2}(z-z_0)^{m+2} + \ldots \quad (3)$$

$$f'(z) = ma_m(z-z_0)^{m-1} + (m+1)a_{m+1}(z-z_0)^m + \ldots \quad (4)$$

and

$$z = z_0 + (z-z_0) \quad (5)$$

Substituting (3), (4) and (5) into (1), we obtain

$$z\,\frac{f'(z)}{f(z)} = \frac{[z_0+(z-z_0)][ma_m(z-z_0)^{m-1}+\ldots]}{a_m(z-z_0)^m + \ldots} \quad (6)$$

Hence,

$$\operatorname{Res}_{z_0} z\,\frac{f'(z)}{f(z)} = \frac{z_0 m a_m}{a_m} = z_0 m \quad (7)$$

2. Function $g(z)$ is regular at z_0, therefore

$$g(z) = g(z_0) + g'(z_0)(z-z_0) + \ldots \quad (8)$$

We have

$$g(z)\,\frac{f'(z)}{f(z)} = \bigl[g(z_0) + g'(z_0)(z-z_0)\ldots\bigr]\,\frac{ma_m(z-z_0)^{m-1} + \ldots}{a_m(z-z_0)^m + \ldots} \quad (9)$$

Thus,

$$\operatorname{Res}_{z_0} g(z)\,\frac{f'(z)}{f(z)} = g(z_0)m \quad (10)$$

RESIDUE THEOREM

• **PROBLEM 17-15**

Prove the following Residue Theorem:

Let C be a positively oriented simple closed curve. If $f(z)$ is analytic inside and on C, except a finite number of points $z_1, z_2, \ldots z_n$ inside C, then

$$\int_C f(z)\,dz = 2\pi i \sum_{k=1}^{n} \operatorname{Res}_{z_k} f(z) \tag{1}$$

where $\operatorname{Res}_{z_k} f(z)$ denotes the residue of $f(z)$ at z_k.

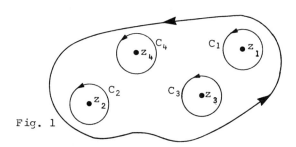

Fig. 1

Solution: Construct the circles $C_1, C_2, \ldots, C_n$ with centers at $z_1, z_2, \ldots, z_n$, respectively,

which do not overlap, and lie entirely inside C, as shown in Fig. 1. Points $z_1, \ldots, z_n$ are isolated singularities and interior points with respect to C. We have

$$\int_C f(z)\,dz = \int_{C_1} f(z)\,dz + \ldots + \int_{C_n} f(z)\,dz \tag{2}$$

On the other hand,

$$\int_{C_1} f(z)\,dz = 2\pi i \operatorname{Res}_{z_1} f(z)$$
$$\vdots \tag{3}$$
$$\int_{C_n} f(z)\,dz = 2\pi i \operatorname{Res}_{z_n} f(z)$$

From (2) and (3) we obtain

$$\int_C f(z)\,dz = 2\pi i \left[\operatorname{Res}_{z_1} f(z) + \ldots + \operatorname{Res}_{z_n} f(z) \right] \tag{4}$$

That completes the proof of the residue theorem for simply-connected domains containing a finite number of singularities of $f(z)$.

• PROBLEM 17-16

1. From the residue theorem, derive Cauchy's integral theorem.

2. Evaluate the integral

$$\int_C \frac{4z-2}{z(z-1)} dz \qquad (1)$$

where C is the circle $|z| = 2$.

Solution: 1. Cauchy's integral theorem states that if $f(z)$ is analytic within and on a simple closed path, then

$$\int_C f(z) dz = 0 \qquad (2)$$

In such a case, the Laurent series

$$\sum_{n=0}^{\infty} a_n (z-z_0)^n + \sum_{n=-1}^{-\infty} b_n (z-z_0)^n \qquad (3)$$

becomes a Taylor series

$$\sum_{n=0}^{\infty} a_n (z-z_0)^n \qquad (4)$$

All negative powers of $(z-z_0)$ vanish, hence

$$c_{-1} = 0 \qquad (5)$$

and from Problem 17-1, eq.(6), we obtain

$$\int_C f(z) dz = 2\pi i \, c_{-1} = 0 \qquad (6)$$

which is the Cauchy's integral theorem.

2. The integrand has two singularities at $z = 0$ and $z = 1$. They are both interior to C. We shall compute the residue at $z = 0$.

$$\frac{1}{1-z} = 1 + z + z^2 + z^3 + \ldots \qquad (7)$$

Hence,

$$\frac{4z-2}{z(z-1)} = \frac{2-4z}{z} \cdot (1+z+z^2+\ldots)$$

$$= (2-4z)\left(\frac{1}{z} + 1 + z + z^2 + \ldots\right) \tag{8}$$

$$= \frac{2}{z} + \ldots$$

The residue at $z = 0$ is 2.

$$\frac{4z-2}{z(z-1)} = \left[\frac{4(z-1)+2}{z-1}\right] \cdot \left[\frac{1}{1+(z-1)}\right] \tag{9}$$

$$= 4 + \frac{2}{z-1} \cdot \left[1-(z-1)+(z-1)^2 - \ldots\right]$$

$$= \frac{2}{z-1} + \ldots \qquad \text{for } 0 < |z-1| < 1.$$

We obtain

$$\int_C \frac{4z-2}{z(z-1)} dz = 2\pi i(2+2) = 8\pi i \tag{10}$$

• **PROBLEM 17-17**

Cauchy's integral formula can be formulated as follows:

If $f(z)$ is such that

$$f(z) = \frac{g(z)}{(z-z_0)^{m+1}} \tag{1}$$

where $g(z)$ is analytic inside and on the simple closed path C, and z_0 is inside C, then

$$\int_C \frac{g(z)}{(z-z_0)^{m+1}} dz = \frac{2\pi i}{m!} g^{(m)}(z_0) \tag{2}$$

Derive Cauchy's integral formula from the residue theorem. What is the relationship between

1. the residue theorem
2. Cauchy's integral theorem
3. Cauchy's integral formula

Solution: Suppose $f(z)$ has a pole of order m at $z = z_0$, but otherwise is analytic inside and on C. Then,

$$f(z) = \frac{g(z)}{(z-z_0)^m} \tag{3}$$

where $g(z)$ is analytic inside and on C.

By the residue theorem,

$$\int_C f(z) dz = 2\pi i \operatorname{Res}_{z_0} f(z) \tag{4}$$

Since g(z) is analytic everywhere,

$$g(z) = a_0 + a_1(z-z_0) + a_2(z-z_0)^2 + \ldots \tag{5}$$

and

$$f(z) = \frac{1}{(z-z_0)^m}\left[a_0 + a_1(z-z_0) + \ldots\right]$$

$$= \ldots + \frac{a_{m-1}(z-z_0)^{m-1}}{(z-z_0)^m} + \ldots = \ldots + \frac{a_{m-1}}{z-z_0} + \ldots \tag{6}$$

From (4) and (6),

$$\int_C f(z)dz = \int_C \frac{g(z)}{(z-z_0)^m} dz = 2\pi i \, a_{m-1} \tag{7}$$

Differentiating (5) (m-1) times,

$$g^{(m-1)}(z_0) = (m-1)! \, a_{m-1} \tag{8}$$

Thus, from (7) and (8), we obtain

$$\int_C f(z)dz = \int_C \frac{g(z)}{(z-z_0)^m} dz$$

$$= \frac{2\pi i}{(m-1)!} g^{(m-1)}(z_0) \tag{9}$$

From Problem 17-16 and the above considerations, we conclude that the residue theorem is the most general theorem concerning integrals around closed paths. Cauchy's integral theorem, and Cauchy's integral formula follow from the residue theorem.

• **PROBLEM 17-18**

The residue can be defined as that which is left, or the remainder. Keeping the properties of the Laurent series expansion in mind, explain why the theorem stated in Problem 17-15 is called the residue theorem.

Solution: Let C be a simple closed path, and f(z) be a function analytic everywhere, except $z = z_0$ and lying inside C.

From the residue theorem,

$$\int_C f(z)dz = 2\pi i \, \underset{z_0}{\mathrm{Res}} \, f(z) = 2\pi i \, c_{-1} \tag{1}$$

On the other hand, the Laurent series expansion of f(z) is

$$f(z) = \sum_{n=-\infty}^{\infty} c_n(z-z_0)^n \tag{2}$$

This series is uniformly convergent in the deleted neighborhood of z_0.

$$\int_C f(z)dz = \sum_{n=-\infty}^{\infty} c_n \int_C (z-z_0)^n dz \tag{3}$$

$$\int_C (z-z_0)^n dz = \begin{cases} 0 & \text{for } n \neq -1 \\ 2\pi i & \text{for } n = -1 \end{cases} \tag{4}$$

From (3) and (4),

$$\int_C f(z)dz = c_{-1} \cdot 2\pi i \tag{5}$$

We see that integration cancels all the terms of the Laurent series, except the term $c_{-1}(z-z_0)^{-1}$. Hence, the name "residue".

APPLICATIONS

• **PROBLEM** 17-19

Evaluate the integral of the function around the unit circle

1. $\dfrac{\sin \frac{1}{z}}{z}$

2. $z^k e^{\frac{1}{z^2}}$, k is any nonnegative integer

3. $z^2 \sin \frac{1}{z}$

4. $z^k \sin \frac{1}{z}$, k is any nonnegative integer

Solution: 1. The Laurent series of $\dfrac{\sin \frac{1}{z}}{z}$ is

$$\frac{1}{z}\left[\frac{1}{z} - \frac{1}{3!z^3} + \frac{1}{5!z^5} - \cdots\right] \tag{1}$$

The coefficient of the term $\frac{1}{z}$ is zero. Thus,

$$\int_{|z|=1} \frac{\sin \frac{1}{z}}{z} dz = 2\pi i \operatorname{Res}_0 \frac{\sin \frac{1}{z}}{z} = 0 \tag{2}$$

2. $z^k e^{\frac{1}{z^2}} = z^k \left[1 + \frac{1}{z^2} + \frac{1}{2z^4} + \frac{1}{3!z^6} + \cdots\right]$ \hfill (3)

We see that the residue depends on whether or not k is an even number.

$$\int_{|z|=1} z^k e^{\frac{1}{z^2}} dz = \begin{cases} 0 & \text{for k even} \\ \dfrac{2\pi i}{\left(\frac{m+1}{2}\right)!} & \text{for k odd} \end{cases} \qquad (4)$$

3. $$z^2 \sin \frac{1}{z} = z^2 \left[\frac{1}{z} - \frac{1}{3!z^3} + \cdots \right]$$

$$= z - \frac{1}{3!z} + \cdots \qquad (5)$$

Hence,

$$\int_{|z|=1} z^2 \sin \frac{1}{z} dz = -\frac{2\pi i}{3!} = -\frac{\pi i}{3} \qquad (6)$$

4. The Laurent series is

$$z^k \sin \frac{1}{z} = z^k \left[\frac{1}{z} - \frac{1}{3!z^3} + \frac{1}{5!z^5} - \cdots \right] \qquad (7)$$

For k odd the residue is zero.

$$\int_{|z|=1} z^k \sin \frac{1}{z} dz = \begin{cases} 0 & \text{for odd k} \\ 2\pi i (-1)^{\frac{m}{2}} \dfrac{1}{(m+1)!} & \text{for k even} \end{cases} \qquad (8)$$

To obtain (8), we set the following in (7):

$$z^k (-1)^{n-1} \frac{1}{(2n-1)! z^{2n-1}} = \alpha \frac{1}{z} \qquad (9)$$

where α is some coefficient. From (9), we obtain

$$k+1 = 2n-1 \qquad (10)$$

That leads to (8).

• **PROBLEM 17-20**

Evaluate the integrals by applying the residue theorem:

1. $$\int_C \frac{f(z)}{(z-z_1)(z-z_2)\cdots(z-z_k)} dz \qquad (1)$$

where C is the circle $|z| = r$, and all $z_1, z_2, \ldots, z_k$ are distinct and located inside the circle. Function $f(z)$ is regular in $|z| \leq r$.

2. $$\int_{|z|=n} \tan \pi z \, dz \qquad (2)$$

$n = 1, 2, 3, \ldots$

Solution: 1. The integrand has k simple poles at $z_1, z_2, \ldots, z_k$, located inside the circle $|z| = r$.

By the residue theorem,

$$\int_C \frac{f(z)}{(z-z_1)(z-z_2)\ldots(z-z_k)} dz$$

$$= 2\pi i \left[\text{Res}_{z_1} + \text{Res}_{z_2} + \ldots + \text{Res}_{z_k} \right] \quad (3)$$

$$\text{Res}_{z_1} = \lim_{z \to z_1} \frac{f(z)}{(z-z_2)(z-z_3)\ldots(z-z_k)}$$

$$= \frac{f(z_1)}{(z_1-z_2)(z_1-z_3)\ldots(z_1-z_k)} \quad (4)$$

$$\vdots$$

$$\text{Res}_{z_k} = \frac{f(z_k)}{(z_k-z_1)\ldots(z_k-z_{k-1})} \quad (5)$$

$$\int_C \frac{f(z)}{(z-z_1)(z-z_2)\ldots(z-z_k)} dz$$

$$= 2\pi i \left[\frac{f(z_1)}{(z_1-z_2)\ldots(z_1-z_k)} + \frac{f(z_2)}{(z_2-z_1)\ldots(z_2-z_k)} + \ldots \right.$$

$$\left. + \frac{f(z_k)}{(z_k-z_1)\ldots(z_k-z_{k-1})} \right] \quad (6)$$

2. We have

$$\tan \pi z = \frac{\sin \pi z}{\cos \pi z} \quad (7)$$

Function $\tan \pi z$ has simple poles at

$$z = n + \frac{1}{2}, \quad n = 0, \pm 1, \pm 2, \ldots \quad (8)$$

$$\text{Res}_{z \to n+\frac{1}{2}} \frac{\sin \pi z}{\cos \pi z} = \lim_{z \to n+\frac{1}{2}} \frac{(z-n-\frac{1}{2})\sin \pi z}{\cos \pi z}$$

$$= -\frac{1}{\pi} \quad (9)$$

Thus,

$$\int_{|z|=n} \tan \pi z \, dz = 2\pi i \left(-\frac{1}{\pi}\right) \cdot 2n = -4ni \quad (10)$$

• **PROBLEM 17-21**

Find the integral
$$\int_C e^{z+\frac{1}{z}} \, dz \tag{1}$$
where C is the circle $|z| = 1$, taken counterclockwise.

Solution: We shall apply the residue theorem.

$$\int_C e^{z+\frac{1}{z}} \, dz = \int_C e^z \cdot e^{\frac{1}{z}} \, dz$$

$$= \int_C \left(1 + z + \frac{z^2}{2!} + \frac{z^3}{3!} + \ldots\right)\left(1 + \frac{1}{z} + \frac{1}{2!z^2} + \ldots\right) dz \tag{2}$$

$$= 2\pi i \, \Sigma \, \text{Res}$$

where ΣRes indicates the sum of all the residues.

Multiplying term by term, we find

$$\Sigma \text{Res} = 1 + \frac{1}{1!2!} + \frac{1}{2!3!} + \frac{1}{3!4!} + \ldots$$

$$= \sum_{n=0}^{\infty} \frac{1}{n!(n+1)!} \tag{3}$$

Thus,
$$\int_C e^{z+\frac{1}{z}} \, dz = 2\pi i \sum_{n=0}^{\infty} \frac{1}{n!(n+1)!} \tag{4}$$

To solve this problem, we can also apply the following theorem:

Let C be any path interior to the circle of convergence of

$$f(z) = \sum_{n=0}^{\infty} a_n z^n$$

and let g(z) be any function continuous on C. Then,

$$\int_C f(z)g(z) \, dz = \sum_{n=0}^{\infty} a_n \int_C z^n g(z) \, dz \tag{5}$$

In our case,

$$f(z) = e^z = 1 + z + \frac{z^2}{2!} + \ldots \tag{6}$$

and

$$\int_C e^{z+\frac{1}{z}} \, dz = \sum_{n=0}^{\infty} \frac{1}{n!} \int_C z^n e^{\frac{1}{z}} \, dz \tag{7}$$

From the residue theorem,

$$\int_C z^n e^{\frac{1}{z}} dz = \int z^n (1 + \frac{1}{z} + \frac{1}{2!z^2} + \ldots) dz$$

$$= 2\pi i \; \text{Res}_0 (z^n e^{\frac{1}{z}}) = \frac{2\pi i}{(n+1)!} \tag{8}$$

From (7) and (8),

$$\int_C e^{z+\frac{1}{z}} dz = 2\pi i \sum_{n=0}^{\infty} \frac{1}{n!(n+1)!} \tag{9}$$

● **PROBLEM 17-22**

Evaluate

$$\int_C \frac{z^2 + 4}{z^3 + 2z^2 + 2z} dz \tag{1}$$

where C is the circle

$$|z+1-i| < \frac{3}{2}$$

taken counterclockwise.

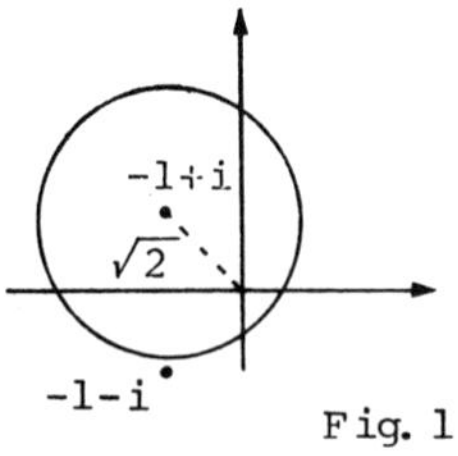

Fig. 1

Solution: Function

$$f(z) = \frac{z^2+4}{z(z^2+2z+2)} \tag{2}$$

has three simple poles at $z = 0$, $z = -1+i$ and $z = -1-i$, but only two of them lie inside C, see Fig. 1.

At $z = 0$, we have

$$\text{Res}_0 f(z) = \lim_{z \to 0} \left[\frac{z^2 + 4}{z^2+2z+2} \right] = 2 \tag{3}$$

At $z = -1+i$, we obtain

$$\text{Res}_{-1+i} f(z) = \lim_{z \to -1+i} \left[\frac{z^2+4}{z(z+1+i)} \right]$$

$$= \frac{3i-1}{2} \tag{4}$$

By the residue theorem

$$\int_C \frac{z^2 + 4}{z^3+2z^2+2z} dz = 2\pi i \left(2 + \frac{3i-1}{2} \right)$$

$$= 3\pi i - 3\pi \tag{5}$$

● **PROBLEM 17-23**

Evaluate the integral

$$\int_C \frac{e^z}{(z-1)^m} \, dz \qquad (1)$$

where C is the circle $|z| = 2$

1. directly from the definition
2. using formulas explained in this chapter.

Solution: 1. The only singular point of the integrand is located inside the circle $|z| = 2$. It is a pole of order m at $z = 1$. By the residue theorem,

$$\int_{|z|=2} \frac{e^z}{(z-1)^m} \, dz = 2\pi i \, \text{Res}_1 \frac{e^z}{(z-1)^m} \qquad (2)$$

Expanding $\dfrac{e^z}{(z-1)^m}$ in the Laurent series about $z = 1$, we obtain

$$\frac{e^z}{(z-1)^m} = e \frac{e^{z-1}}{(z-1)^m} \qquad (3)$$

$$= \frac{e}{(z-1)^m}\left[1+(z-1)+\ldots+\frac{(z-1)^{m-1}}{(m-1)!}+\frac{(z-1)^m}{m!}+\ldots\right]$$

$$= e\left[\frac{1}{(z-1)^m} + \frac{1}{(z-1)^{m-1}} + \ldots + \frac{1}{(m-1)!} \cdot \frac{1}{z-1} + \frac{1}{m!} + \ldots\right]$$

Hence,

$$\text{Res}_1 \frac{e^z}{(z-1)^m} = \frac{e}{(m-1)!} \qquad (4)$$

$$\int_{|z|=2} \frac{e^z}{(z-1)^m} \, dz = \frac{2\pi i e}{(m-1)!} \qquad (5)$$

2. The integrand has a pole of order m at $z = 1$.

$$\text{Res}_{z_0} f(z) = \frac{1}{(m-1)!} \lim_{z \to z_0} \frac{d^{m-1}}{dz^{m-1}}\left[(z-z_0)^m f(z)\right] \qquad (6)$$

$$= \frac{1}{(m-1)!} \lim_{z \to 1} \frac{d^{m-1}}{dz^{m-1}}\left[(z-1)^m \frac{e^z}{(z-1)^m}\right]$$

$$= \frac{1}{(m-1)!} \lim_{z \to 1} e^z = \frac{e}{(m-1)!}$$

● **PROBLEM 17-24**

Apply the residue theorem to show that

$$\int_0^\pi \cot(\theta-a)d\theta = \begin{matrix} \pi i & \text{if Im } a > 0 \\ -\pi i & \text{if Im } a < 0 \end{matrix} \qquad (1)$$

For Im a = 0 the integral diverges.

Solution: We shall substitute

$$e^{2i(\theta-a)} = z \qquad (2)$$

to obtain

$$dz = 2i\, e^{2i(\theta-a)}\, d\theta \qquad (3)$$

or

$$d\theta = \frac{dz}{2iz}$$

Since

$$\cot(\theta-a) = \frac{\cos(\theta-a)}{\sin(\theta-a)} \qquad (4)$$

we have

$$\cos(\theta-a) = \frac{e^{i(\theta-a)} + e^{-i(\theta-a)}}{2} \qquad (5)$$

$$\sin(\theta-a) = \frac{e^{i(\theta-a)} - e^{-i(\theta-a)}}{2i} \qquad (6)$$

and

$$\cot(\theta-a) = i\, \frac{e^{i(\theta-a)} + e^{-i(\theta-a)}}{e^{i(\theta-a)} - e^{-i(\theta-a)}} = i\, \frac{e^{2i(\theta-a)}+1}{e^{2i(\theta-a)}-1}$$

$$= i\, \frac{z+1}{z-1} \qquad (7)$$

Let

$$a = \alpha + i\beta \qquad (8)$$

then,

$$e^{2i(\theta-a)} = e^{2\beta} \cdot e^{2i(\theta-\alpha)} \qquad (9)$$

We obtain

$$\int_0^\pi \cot(\theta-a)d\theta = \int_C i\, \frac{z+1}{z-1}\, \frac{dz}{2iz} = \frac{1}{2} \int_C \frac{z+1}{z(z-1)}\, dz \qquad (10)$$

The integrand has two simple poles at z = 0 and z = 1. For β > 0, both poles are located inside C.

$$\int_0^\pi \cot(\theta-a)d\theta = \frac{1}{2}\int_C \frac{z+1}{z(z-1)}\,dz$$

$$= 2\pi i \cdot \frac{1}{2}\left[\text{Res}_0 + \text{Res}_1\right] \qquad (11)$$

$$= \pi i\left[-1 + 2\right] = \pi i$$

For β < 0, only one pole at z = 0 is located inside C. Hence,

$$\int_0^\pi \cot(\theta-a)d\theta = \pi i\,\text{Res}_0 = -\pi i \qquad (12)$$

For β = 0, the pole z = 1 is located on C and we cannot determine the value of the integral.

• **PROBLEM 17-25**

Evaluate the integral

$$\int_C \frac{e^{zt}}{\sinh z}\,dz \qquad (1)$$

where C is the positively oriented curve shown in Fig. 1.

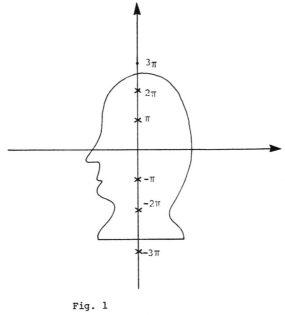

Fig. 1

Solution: We know that $\sinh z = 0$ if $z = n\pi i$. (2)

Thus, the integrand has simple poles at

$$z = 0, \quad z = \pi i, \quad z = -\pi i, \quad z = 2\pi i, \quad z = -2\pi i, \quad (3)$$

which are located inside contour C. We evaluate all necessary residues at $z = 0$.

$$\text{Res}_0 \frac{e^{zt}}{\sinh z} = \lim_{z \to 0} \frac{ze^{zt}}{\sinh z} \quad (4)$$

To find this limit, we shall apply L'Hospital's rule.

$$\lim_{z \to 0} \frac{ze^{zt}}{\sinh z} = \left(\frac{e^{zt}}{\cosh z}\right)\bigg|_{z=0} = 1 \quad (5)$$

$$\text{Res}_{\pi i} f(z) = \lim_{z \to \pi i} \frac{(z-\pi i)e^{zt}}{\sinh z} = \left(\frac{e^{zt}}{\cosh z}\right)\bigg|_{z=\pi i}$$

$$= \frac{e^{\pi i t}}{\cosh \pi i} \quad (6)$$

$$\text{Res}_{-\pi i} f(z) = \lim_{z \to -\pi i} \frac{(z+\pi i)e^{zt}}{\sinh z} = \frac{e^{-\pi i t}}{\cosh(-\pi i)} \quad (7)$$

$$\text{Res}_{2\pi i} f(z) = \frac{e^{2\pi i t}}{\cosh 2\pi i} \quad (8)$$

$$\text{Res}_{-2\pi i} f(z) = \frac{e^{-2\pi i t}}{\cosh(-2\pi i)} \quad (9)$$

By the residue theorem,

$$\int_C \frac{e^{zt}}{\sinh z} dz = 2\pi i \left[1 + \frac{e^{\pi i t} + e^{-\pi i t}}{\cosh \pi i} + \frac{e^{2\pi i t} + e^{-2\pi i t}}{\cosh 2\pi i}\right]$$

$$= 2\pi i \left[1 - 2\cos \pi t + 2\cos 2\pi t\right] \quad (10)$$

● **PROBLEM 17-26**

Let p and q be two parameters such that

$$p > q > 0 \quad (1)$$

Show that

$$\int_0^{2\pi} \frac{dx}{(p+q\cos x)^2} = \frac{2\pi p}{(p^2-q^2)^{\frac{3}{2}}} \quad (2)$$

Solution: We substitute
$$e^{ix} = z \tag{3}$$
to obtain
$$i\, e^{ix}\, dx = dz \tag{4}$$
and
$$\cos x = \frac{e^{ix} + e^{-ix}}{2}$$

Thus,
$$\int_0^{2\pi} \frac{dx}{(p+q\cos x)^2} = \int_{|z|=1} \frac{dz}{iz\left[p+q\left(\frac{z+z^{-1}}{2}\right)\right]^2} \tag{5}$$

$$= \int_{|z|=1} \frac{2\,dz}{i\left[pz+q\left(\frac{z^2+1}{1}\right)\right]^2} = -4i \int_{|z|=1} \frac{z\,dz}{(qz^2+2pz+q)^2}$$

The equation
$$qz^2 + 2pz + q = 0 \tag{6}$$
has two solutions
$$z_1 = \frac{-p + \sqrt{p^2-q^2}}{q}$$
$$z_2 = \frac{-p - \sqrt{p^2-q^2}}{q} \tag{7}$$

Thus, the integrand has two poles z_1 and z_2 of order 2. Because $p > q > 0$, only z_1 lies inside the circle $|z| = 1$.
$$|z_2| = \frac{p + \sqrt{p^2-q^2}}{q} > 1 \tag{8}$$

We obtain
$$-4i \int_{|z|=1} \frac{z\,dz}{(qz^2+2pz+q)^2} = \frac{-4i}{q^2} \int_{|z|=1} \frac{z\,dz}{(z-z_1)^2(z-z_2)^2}$$

$$= \frac{-4i}{q^2} \cdot 2\pi i\, \text{Res}_{z_1} \frac{z}{(z-z_1)^2(z-z_2)^2}$$

$$= \frac{8\pi}{q^2} \lim_{z \to z_1} \left[\frac{z}{(z-z_2)^2}\right]'$$

$$= \frac{8\pi}{q^2} \cdot \frac{(z_1-z_2)^2 - 2z_1(z_1-z_2)}{(z_1-z_2)^4} \tag{9}$$

569

$$= \frac{8\pi}{q^2} \cdot \frac{-z_1-z_2}{(z_1-z_2)^3} = \frac{-8\pi}{q^2} \cdot \frac{\frac{-2p}{q}}{\left[\frac{2\sqrt{p^2-q^2}}{q}\right]^3}$$

$$= \frac{2\pi p}{(p^2-q^2)^{\frac{3}{2}}}$$

● **PROBLEM 17-27**

Prove that

$$\int_0^{2\pi} \frac{dx}{1-2t\cos x + t^2} = \frac{2\pi}{1-t^2} \qquad (1)$$

$$0 < t < 1$$

Solution: We shall substitute

$$e^{ix} = z \qquad (2)$$

Then

$$i\, e^{ix}\, dx = dz$$

and

$$\cos x = \frac{e^{ix} + e^{-ix}}{2} \qquad (3)$$

We obtain

$$\int_0^{2\pi} \frac{dx}{1-2t\cos x + t^2} = \int_{|z|=1} \frac{dz}{i(t^2 z - t - tz^2 + z)}$$

$$= i \int_{|z|=1} \frac{dz}{tz^2 - (t^2+1)z + t} \qquad (4)$$

It is easy to see that $z = t$ is the solution of

$$tz^2 - (t^2+1)z + t = 0 \qquad (5)$$

The other solution is $z = \frac{1}{t}$. Hence,

$$i \int_{|z|=1} \frac{dz}{tz^2 - (t^2+1)z + t} = \frac{i}{t} \int_{|z|=1} \frac{dz}{(z-t)(z-\frac{1}{t})} \qquad (6)$$

The integrand has two singularities at $z = t$ and $z = \frac{1}{t}$. They are simple poles. Only one of them, $z = t$, is located inside the circle $|z| = 1$. We have

$$\frac{i}{t} \int_{|z|=1} \frac{dz}{(z-t)(z-\frac{1}{t})} = \frac{i}{t} \cdot 2\pi i \; \text{Res}_t \frac{1}{(z-t)(z-\frac{1}{t})}$$

$$= -\frac{2\pi}{t} \lim_{z \to t} \frac{1}{z - \frac{1}{t}} = \frac{-2\pi}{t(t-\frac{1}{t})} \quad (7)$$

$$= \frac{2\pi}{1 - t^2}$$

• **PROBLEM 17-28**

Let $f(z)$ be analytic on and inside a simple closed curve C. Except for $z = a$, the function $f(z)$ is different from zero ($f(z) \neq 0$), on and inside C. The point a is the simple zero of $f(z)$. Show that

$$a = \frac{f'(a)}{2\pi i} \int_C \frac{z}{f(z)} \, dz \quad (1)$$

Solution: Suppose $a \neq 0$. Then the function

$$\frac{z}{f(z)} \quad (2)$$

has a simple pole at $z = a$, because $f(z)$ has a simple zero at $z = a$. We have

$$\int_C \frac{z}{f(z)} \, dz = 2\pi i \; \text{Res}_a \frac{z}{f(z)} \quad (3)$$

From Problem 17-12, eq.(3), we obtain

$$\text{Res}_a \frac{z}{f(z)} = \frac{a}{f'(a)} \quad (4)$$

Substituting (4) into (3), we find

$$\frac{1}{2\pi i} f'(a) \int_C \frac{z}{f(z)} \, dz = a \quad (5)$$

For $a = 0$, we have

$$\int_C \frac{z}{f(z)} \, dz = 2\pi i \; \text{Res}_0 \frac{z}{f(z)} = 0 \quad (6)$$

• **PROBLEM 17-29**

Let D_n be the positively oriented boundary of the square whose edges are segments of the lines

$$x = \pm(n + \tfrac{1}{2})\pi$$

$$y = \pm(n + \tfrac{1}{2})\pi \qquad (1)$$

where n is a positive integer.

Show that

$$\int_{D_n} \frac{dz}{z^2 \sin z} = 2\pi i \left[\frac{1}{6} + 2 \sum_{k=1}^{n} \frac{(-1)^k}{k^2 \pi^2} \right] \qquad (2)$$

<u>Solution</u>: The integrand has a pole at $z = 0$ and simple poles at $z = \pm\pi$, $z = \pm 2\pi, \ldots, z = \pm n\pi$, (3)

which are located inside the square D_n. Hence,

$$\int_{D_n} \frac{dz}{z^2 \sin z} = 2\pi i \left[\text{Res}_0 + \text{Res}_\pi + \text{Res}_{-\pi} + \ldots + \text{Res}_{n\pi} + \text{Res}_{-n\pi} \right] \qquad (4)$$

At $z = 0$, we have

$$\frac{1}{z^2 \sin z} = \frac{1}{z^2 \left(z - \frac{z^3}{3!} + \frac{z^5}{5!} - \ldots \right)} = \frac{1}{z^3} \cdot \frac{1}{1 - \frac{z^2}{6} + \frac{z^4}{5!} - \ldots}$$

$$= \frac{1}{z^3} \cdot \left(1 + \frac{z^2}{6} + \ldots \right) \qquad (5)$$

Hence,

$$\text{Res}_0 \frac{1}{z^2 \sin z} = \frac{1}{6} \qquad (6)$$

For $z = k\pi$, we obtain

$$\text{Res}_{k\pi} \frac{1}{z^2 \sin z} = \lim_{z \to k\pi} \frac{z - k\pi}{z^2 \sin z}$$

$$= \left(\frac{1}{z^2 \cos z} \right) \Bigg|_{k\pi = z} = \frac{(-1)^k}{k^2 \pi^2} \qquad (7)$$

For $z = -k\pi$,

$$\text{Res}_{-k\pi} \frac{1}{z^2 \sin z} = \lim_{z \to -k\pi} \frac{z + k\pi}{z^2 \sin z} = \left(\frac{1}{z^2 \cos z} \right) \Bigg|_{z = -k\pi}$$

$$= \frac{(-1)^k}{k^2 \pi^2} \qquad (8)$$

Substituting (6), (7) and (8) into (4), we obtain

$$\int_{D_n} \frac{dz}{z^2 \sin z} = 2\pi i \left[\frac{1}{6} + 2 \sum_{k=1}^{n} \frac{(-1)^k}{k^2 \pi^2} \right] \qquad (9)$$

It can be shown that

$$\lim_{n \to \infty} \int_{D_n} \frac{dz}{z^2 \sin z} = 0 \qquad (10)$$

Thus,

$$\lim_{n \to \infty} \left[\frac{1}{6} + 2 \sum_{k=1}^{n} \frac{(-1)^k}{k^2 \pi^2} \right] = 0 \qquad (11)$$

or

$$\sum_{n=1}^{\infty} \frac{(-1)^{n+1}}{n^2} = \frac{\pi^2}{12} \qquad (12)$$

CHAPTER 18

APPLICATIONS OF RESIDUE CALCULUS, INTEGRATION OF REAL FUNCTIONS

● PROBLEM 18-1

Theorem

Let $R(\sin\theta,\cos\theta)$ be a rational function. Then integrals of the form

$$\int_0^{2\pi} R(\sin\theta,\cos\theta)d\theta \qquad (1)$$

can be transformed into integrals of the form (see Fig. 1)

$$\int_C f(z)dz \qquad (2)$$

where C is the unit circle positively oriented and

$$f(z) = \frac{1}{iz} R\left(\frac{z^2-1}{2iz}, \frac{z^2+1}{2z}\right) \qquad (3)$$

By the residue theorem

$$\int_0^{2\pi} R(\sin\theta,\cos\theta)d\theta = 2\pi i \sum_C \text{Res} f(z) \qquad (4)$$

The sum $\sum_C$ extends over all residues of $f(z)$ inside the unit circle.

Prove this theorem.

Solution: A rational function of two arguments $R(x,y)$ is a ratio of polynomials in x and y.

Substituting

$$z = e^{i\theta} \qquad (5)$$

we transform the interval $[0,2\pi]$ onto the unit circle.

$$\sin\theta = \frac{e^{i\theta} - e^{-i\theta}}{2i} = \frac{z - z^{-1}}{2i} = \frac{z^2 - 1}{2iz} \qquad (6)$$

$$\cos\theta = \frac{e^{i\theta} + e^{-i\theta}}{2} = \frac{z + z^{-1}}{2} = \frac{z^2 + 1}{2z} \qquad (7)$$

From (5) we obtain,

$$dz = ie^{i\theta} d\theta = iz\, d\theta \qquad (8)$$

Then,

$$\int_0^{2\pi} R(\sin\theta, \cos\theta) d\theta = \int_C R\left(\frac{z^2-1}{2iz}, \frac{z^2+1}{2z}\right) \frac{dz}{iz}$$

$$= \int_C f(z) dz \qquad (9)$$

Since $f(z)$ is a rational function of z, its singularities are poles.

This method enables us to replace the integral

$$\int_0^{2\pi} R(\sin\theta, \cos\theta) d\theta,$$

where $R(\sin\theta, \cos\theta)$ is a rational function, by the integral

$$\int_C f(z) dz,$$

where

$$f(z) = R\left(\frac{z - z^{-1}}{2i}, \frac{z + z^{-1}}{2}\right)$$

Function $f(z)$ is a rational function. Integral $\int_C f(z) dz$ can be calculated by the residue theorem.

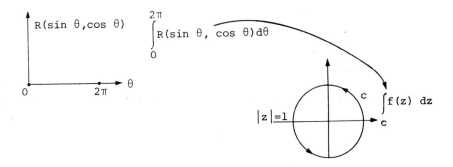

Fig. 1

• **PROBLEM 18-2**

Evaluate the integrals

1. $\displaystyle\int_0^{2\pi} \cos\theta \, d\theta$ (1)

2. $\displaystyle\int_0^{2\pi} \tan\theta \, d\theta$ (2)

<u>Solution</u>: 1. According to results of Problem 18-1 integral (1) can be transformed to

$$\int_0^{2\pi} \cos\theta \, d\theta = \int_C \frac{z^2+1}{2iz^2} \, dz \tag{3}$$

where C is the unit circle, $|z| = 1$. The integrand has a double pole at $z = 0$.

Thus,

$$\text{Res}_0 \frac{z^2+1}{2iz^2} = \lim_{z\to 0} \frac{d}{dz}\left(\frac{z^2+1}{2i}\right) = \lim_{z\to 0} \frac{z}{i} = 0 \tag{4}$$

We have

$$\int_0^{2\pi} \cos\theta \, d\theta = 2\pi i \, \text{Res}_0 = 0 \tag{5}$$

2. $\displaystyle\int_0^{2\pi} \tan\theta \, d\theta = \int_0^{2\pi} \frac{\sin\theta}{\cos\theta} \, d\theta$

$$= \int_C \frac{z^2-1}{2iz} \cdot \frac{2z}{z^2+1} \, dz \tag{6}$$

Note that the integrand has singularity for real θ, and the singular points are on the unit circle. This case cannot be computed using methods described in Problem 18-1.

• **PROBLEM 18-3**

Evaluate the integral

$$\int_0^{2\pi} \frac{\cos\theta}{3+2\cos\theta} \, d\theta \tag{1}$$

Solution: Substituting $e^{i\theta} = z$ we obtain

$$\cos\theta = \frac{z+z^{-1}}{2} \tag{2}$$

$$d\theta = \frac{dz}{iz} \tag{3}$$

Integral (1) is transformed to

$$\int_0^{2\pi} \frac{\cos\theta}{3+2\cos\theta} d\theta = \int_C \frac{z+z^{-1}}{2iz[3+z+z^{-1}]} dz$$

$$= \int_C \frac{z^2+1}{2iz(z^2+3z+1)} dz \tag{4}$$

To find the singularities of the integrand we solve the equation

$$z^2 + 3z + 1 = 0 \tag{5}$$

$$z_1 = \frac{-3+\sqrt{5}}{2}, \quad z_2 = \frac{-3-\sqrt{5}}{2} \tag{6}$$

The singularities of the integrand which are located inside the unit circle are

$$z = 0 \quad \text{and} \quad z = \frac{-3+\sqrt{5}}{2} \tag{7}$$

They are simple poles.

Therefore,

$$\int_C \frac{z^2+1}{2iz(z^2+3z+1)} = 2\pi i \left[\text{Res}_0 + \text{Res}_{\frac{-3+\sqrt{5}}{2}} \right]$$

$$= 2\pi i \left[\frac{1}{2i} - \frac{3}{2\sqrt{5}i} \right] = \pi - \frac{3\pi}{\sqrt{5}} \tag{8}$$

where

$$\text{Res}_0 = \lim_{z\to 0} \frac{z^2+1}{2iz(z^2+3z+1)} z = \frac{1}{2i} \tag{9}$$

and

$$\text{Res}_{\frac{-3+\sqrt{5}}{2}} = \lim_{z\to\frac{-3+\sqrt{5}}{2}} \left[\frac{z^2+1}{2iz(z-z_2)} \right]$$

$$= \frac{z_1^2+1}{2iz_1(z_1-z_2)} = \frac{-3}{2\sqrt{5}i} \tag{10}$$

• **PROBLEM 18-4**

Evaluate the integral

$$\int_0^{2\pi} \frac{d\theta}{p+\cos\theta}, \quad \text{where} \quad p^2 > 1. \tag{1}$$

Solution: Substituting
$$\cos\theta = \frac{z^2+1}{2z} \tag{2}$$

$$d\theta = \frac{dz}{iz} \tag{3}$$

we transform integral (1) into

$$\int_0^{2\pi} \frac{d\theta}{p+\cos\theta} = \int_C \frac{dz}{iz(p+\frac{z^2+1}{2z})}$$
$$= \int_C \frac{2dz}{i(z^2+2pz+1)} \tag{4}$$

The poles of the integrand are

$$z = -p \pm \sqrt{p^2-1} \tag{5}$$

For $p^2 > 1$ both poles are real and simple.

For $p < -1$, the pole $-p - \sqrt{p^2-1}$ is inside the unit circle and the other one is outside.

For $p > 1$, the pole $-p + \sqrt{p^2-1}$ is inside the circle.

In shorter form, for $p^2 > 1$, the enclosed simple pole may be written as

$$z_{enc} = -p + (\text{sign } p)\sqrt{p^2-1} \tag{6}$$

where

$$\text{sign } p = \frac{p}{|p|} = \begin{cases} 1 & \text{for } p > 0 \\ -1 & \text{for } p < 0 \end{cases} \tag{7}$$

Hence, the residue is

$$\text{Res}_{z_{enc}} f(z) = \frac{1}{i(p+z_{enc})} = \frac{\text{sign } p}{i\sqrt{p^2-1}} \tag{8}$$

and

$$\int_0^{2\pi} \frac{d\theta}{p+\cos\theta} = (\text{sign } p)\frac{2\pi}{\sqrt{p^2-1}} = \frac{p}{|p|}\frac{2\pi}{\sqrt{p^2-1}} \tag{9}$$

• **PROBLEM 18-5**

Show that

$$\int_0^{2\pi} \frac{d\theta}{a+b\sin\theta} = \frac{2\pi}{\sqrt{a^2-b^2}} \quad \text{where} \quad a > |b| \tag{1}$$

Solution:

$$\int_0^{2\pi} \frac{d\theta}{a+b\sin\theta} = \int_C \frac{2\,dz}{bz^2+2aiz-b} \tag{2}$$

where C is the unit circle.

Solving the equation

$$bz^2 + 2aiz - b = 0 \tag{3}$$

we find the singularities of the integrand of (2).

$$z = \frac{-ai \pm \sqrt{a^2-b^2}\, i}{b} \tag{4}$$

Because $a > |b|$, only the singularity $\frac{-ai + i\sqrt{a^2-b^2}}{b}$ lies inside C.

Therefore,

$$\left|\frac{-ai + i\sqrt{a^2-b^2}}{b}\right| = \left|\frac{\sqrt{a^2-b^2} - a}{b} \cdot \frac{\sqrt{a^2-b^2}+a}{\sqrt{a^2-b^2}+a}\right|$$

$$= \left|\frac{b}{a + \sqrt{a^2-b^2}}\right| < 1 \tag{5}$$

Since $z = \frac{-a + \sqrt{a^2-b^2}}{b} i$ is a simple pole the residue is

$$\text{Res}_{z_1} = \lim_{z \to z_1} \left[(z-z_1) \frac{2}{bz^2+2aiz-b}\right]$$

$$= \lim_{z \to z_1} \frac{2}{2bz+2ai} = \frac{1}{bz_1+ai} = \frac{1}{\sqrt{a^2-b^2}\, i} \tag{6}$$

Finally,

$$\int_0^{2\pi} \frac{d\theta}{a+b\sin\theta} = \frac{2\pi}{\sqrt{a^2-b^2}} \tag{7}$$

● **PROBLEM 18-6**

Evaluate the integral

$$\int_0^{2\pi} \frac{d\theta}{\sin\theta - 2\cos\theta + 3} \tag{1}$$

Solution: First we transform the integral to $\int_C f(z)\,dz$ and then apply the residue theorem. We have $z = e^{i\theta}$,

$$\sin\theta = \frac{z - z^{-1}}{2i} \tag{2}$$

and

$$\cos\theta = \frac{z + z^{-1}}{2} \tag{3}$$

$$\int_0^{2\pi} \frac{d\theta}{\sin\theta - 2\cos\theta + 3} = \int_C \frac{2dz}{(1-2i)z^2 + 6iz + (-2i-1)} \tag{4}$$

where C is the unit circle $|z| = 1$. Solving the equation,

$$(1-2i)z^2 + 6iz + (-2i-1) = 0 \tag{5}$$

we find the singularities of the integrand.

$$z = \frac{-6i \pm \sqrt{(6i)^2 - 4(1-2i)(-1-2i)}}{2(1-2i)}$$

$$= \frac{-6i \pm 4i}{2(1-2i)} \tag{6}$$

The singularities are two simple poles at

$$z_1 = 2 - i \quad \text{and} \quad z_2 = \frac{2-i}{5} \tag{7}$$

The simple pole z_2 lies inside the circle C.

$$\text{Res}_{\frac{2-i}{5}} = \lim_{z \to \frac{2-i}{5}} \left[\left(z - \frac{2-i}{5}\right) \cdot \frac{2}{(1-2i)z^2 + 6iz + (-1-2i)} \right]$$

$$= \lim_{z \to \frac{2-i}{5}} \frac{2}{2(1-2i)z + 6i} = \frac{1}{2i} \tag{8}$$

We obtain

$$\int_0^{2\pi} \frac{d\theta}{\sin\theta - 2\cos\theta + 3} = 2\pi i \cdot \frac{1}{2i} = \pi \tag{9}$$

• **PROBLEM 18-7**

Evaluate the integral

$$\int_0^{2\pi} \frac{d\theta}{(5 - 3\sin\theta)^2} \tag{1}$$

Solution: As usual, substituting

$$z = e^{i\theta} \tag{2}$$

we obtain
$$\sin\theta = \frac{z-z^{-1}}{2i}$$
$$d\theta = \frac{dz}{iz}$$
(3)

Integral (1) becomes

$$\int_0^{2\pi} \frac{d\theta}{(5-3\sin\theta)^2} = \int_C \frac{dz}{iz\left[5-3\frac{z-z^{-1}}{2i}\right]^2}$$

$$= -\frac{4}{i}\int_C \frac{z\,dz}{(3z^2-10iz-3)^2}$$
(4)

where C is the unit circle $|z|=1$ positively oriented. Solving

$$3z^2 - 10iz - 3 = 0$$

we find

$$z_1 = \frac{10i + \sqrt{-100+36}}{6} = \frac{18i}{6} = 3i$$
(5)

$$z_2 = \frac{10i - \sqrt{-100+36}}{6} = \frac{2i}{6} = \frac{i}{3}$$
(6)

Only z_2 lies inside the unit circle.

The integrand at z_2 has a pole of order two.

$$\operatorname{Res}_{z_2} = \lim_{z\to\frac{i}{3}} \frac{d}{dz}\left[\frac{z}{(3z^2-10iz-3)^2}\left(z-\frac{i}{3}\right)^2\right]$$

$$= \lim_{z\to\frac{i}{3}} \frac{d}{dz}\left[\frac{z}{\{3(z-3i)(z-\frac{i}{3})\}^2}\left(z-\frac{i}{3}\right)^2\right]$$
(7)

$$= \lim_{z\to\frac{i}{3}} \frac{d}{dz}\left[\frac{z}{9(z-3i)^2}\right] = -\frac{5}{256}$$

By the residue theorem

$$\int_0^{2\pi} \frac{d\theta}{(5-3\sin\theta)^2} = -\frac{4}{i}\cdot 2\pi i\left(-\frac{5}{256}\right) = \frac{5\pi}{32}$$
(8)

This result can be obtained from

$$\int_0^{2\pi} \frac{d\theta}{a+b\sin\theta} = \frac{2\pi}{\sqrt{a^2-b^2}}$$
(9)

which was proved in Problem 18-5.

Applying the Leibnitz rule to (9) and differentiating both sides with respect to a and considering b as constant, we obtain

$$\frac{d}{da}\int_0^{2\pi}\frac{d\theta}{a+b\sin\theta} = \int_0^{2\pi}\frac{\partial}{\partial a}\left(\frac{1}{a+b\sin\theta}\right)d\theta \qquad (10)$$

$$= \int_0^{2\pi}\frac{-d\theta}{(a+b\sin\theta)^2} = \frac{d}{da}\left(\frac{2\pi}{\sqrt{a^2-b^2}}\right)$$

$$= \frac{-2\pi a}{(a^2-b^2)^{\frac{3}{2}}}$$

or

$$\int_0^{2\pi}\frac{d\theta}{(a+b\sin\theta)^2} = \frac{2\pi a}{(a^2-b^2)^{\frac{3}{2}}} \qquad (11)$$

For a = 5 and b = -3 we get

$$\int_0^{2\pi}\frac{d\theta}{(5-3\sin\theta)^2} = \frac{10\pi}{16^{\frac{3}{2}}} = \frac{5\pi}{32} \qquad (12)$$

● **PROBLEM 18-8**

Evaluate the integral

$$I = \int_0^{2\pi}\frac{d\theta}{(a+b\cos\theta)^2} \qquad (1)$$

where $\quad 0 < b < a$

Solution: We substitute

$$z = e^{i\theta} \qquad (2)$$

Then,

$$\int_0^{2\pi}\frac{d\theta}{(a+b\cos\theta)^2} = \frac{4}{i}\int_C\frac{zdz}{[bz^2+2az+b]^2}$$

$$= \frac{4}{ib^2}\int_C\frac{zdz}{\left[z^2+\frac{2a}{b}z+1\right]^2} \qquad (3)$$

Let z_1 and z_2 denote the solutions of

$$z^2 + \frac{2a}{b} z + 1 = 0 \tag{4}$$

then

$$I = \frac{4}{ib^2} \int_C \frac{z\,dz}{(z-z_1)^2(z-z_2)^2} \tag{5}$$

where

$$z_1 = \frac{-a + \sqrt{a^2-b^2}}{b}$$
$$z_2 = \frac{-a - \sqrt{a^2-b^2}}{b} \tag{6}$$

The integrand has two poles of order 2. Only one pole z_1 is located inside the unit circle.

$$\text{Res}_{z_1} = \lim_{z \to z_1} \left[(z-z_1)^2 \frac{z}{(z-z_1)^2(z-z_2)^2} \right]^1$$
$$= \lim_{z \to z_1} \left[\frac{z}{(z-z_2)^2} \right]^1 = \frac{-(z_1+z_2)}{(z_1-z_2)^3} \tag{7}$$

From (6) we compute

$$z_1 + z_2 = -\frac{2a}{b}$$
$$z_1 - z_2 = \frac{2\sqrt{a^2-b^2}}{b} \tag{8}$$

$$I = 2\pi i \cdot \frac{4}{ib^2} \cdot \frac{2a}{b} \cdot \frac{b^3}{8(a^2-b^2)^{\frac{3}{2}}}$$

$$= \frac{2a\pi}{(a^2-b^2)^{\frac{3}{2}}}$$

● **PROBLEM 18-9**

Evaluate the integral

$$I = \int_0^{2\pi} \cos^{2n}\theta \, d\theta \tag{1}$$

Solution: In the usual manner we transform integral (1) to

$$I = \int_C \left(\frac{z+z^{-1}}{2} \right)^{2n} \frac{dz}{iz}$$
$$= \frac{1}{2^{2n}i} \int_C \frac{(z^2+1)^{2n}}{z^{2n+1}} dz \tag{2}$$

The integrand has a pole of order 2n+1 at z = 0. Thus,

$$I = 2\pi i \cdot \frac{1}{2^{2n}i} \operatorname{Res}_0 \frac{(z^2+1)^{2n}}{z^{2n+1}}$$

$$= \frac{\pi}{2^{2n-1}} \cdot \lim_{z \to 0} \frac{1}{(2n)!} \frac{d^{2n}}{dz^{2n}}(z^2+1)^{2n}$$

$$= \frac{\pi}{2^{2n-1}(2n)!} \lim_{z \to 0} \frac{d^{2n}}{dz^{2n}} \left[\binom{2n}{0}(z^2)^{2n} + \ldots + \binom{2n}{n}(z^2)^n + \ldots + \binom{2n}{2n} \right]$$

$$= \frac{\pi}{2^{2n-1}(2n)!} \cdot \binom{2n}{n}(2n)(2n-1)\ldots(2)(1)$$

$$= \frac{\pi}{2^{2n-1}} \binom{2n}{n}$$

(3)

Remember that

$$\binom{n}{m} = \frac{n!}{m!(n-m)!} \quad (4)$$

● **PROBLEM 18-10**

Evaluate the integral

$$I = \int_0^{2\pi} e^{2\cos\theta} d\theta \quad (1)$$

Solution:
$$\cos\theta = \frac{z+z^{-1}}{2}$$

$$d\theta = \frac{dz}{iz} \quad (2)$$

Thus,

$$I = \int_C \frac{e^{z+z^{-1}}}{iz} dz = \frac{1}{i} \int_C \frac{e^z}{z} \cdot e^{\frac{1}{z}} dz \quad (3)$$

We can replace e^z and e^{-z} with

$$e^z = 1 + z + \frac{z^2}{2!} + \frac{z^3}{3!} + \ldots \quad (4)$$

$$e^{\frac{1}{z}} = 1 + \frac{1}{z} + \frac{1}{2!z^2} + \frac{1}{3!z^3} + \ldots \quad (5)$$

to obtain

$$I = \frac{1}{i} \int_C \left[\frac{1}{z} + 1 + \frac{z}{2!} + \frac{z^2}{3!} + \ldots\right]\left[1 + \frac{1}{z} + \frac{1}{2!z^2} + \frac{1}{3!z^3} + \ldots\right] dz \quad (6)$$

$$= \frac{1}{i} 2\pi i \; \text{Res}\left[\left(\frac{1}{z} + 1 + \frac{z}{2!} + \frac{z^2}{3!} + \ldots\right)\left(1 + \frac{1}{z} + \frac{1}{2!z^2} + \frac{1}{3!z^3} + \ldots\right)\right]$$

Multiplying term by term we obtain a sequence whose terms with the residue different than zero will be

$$\frac{1}{z} \cdot 1, \quad 1 \cdot \frac{1}{z}, \quad \frac{z}{2!} \cdot \frac{1}{2!z^2}, \quad \frac{z^2}{3!} \cdot \frac{1}{3!z^3}, \ldots \quad (7)$$

Hence,

$$I = 2\pi\left[1 + 1 + \frac{1}{2!2!} + \frac{1}{3!3!} + \frac{1}{4!4!} + \ldots\right] \quad (8)$$

$$= 2\pi \sum_{n=0}^{\infty} \frac{1}{(n!)^2}$$

● **PROBLEM 18-11**

Let $f(z)$ and $g(z)$ be two analytic functions such that

$$f(z) = \sum_{n=0}^{\infty} a_n z^n, \quad |z| \leq r_1 \quad (1)$$

$$g(z) = \sum_{n=0}^{\infty} b_n z^n, \quad |z| \leq r_2 \quad (2)$$

Find the value of the integral

$$I = \int_{|t|=r_1} f(t) g\left(\frac{z}{t}\right) \frac{dt}{t}, \quad |z| < r_1 \cdot r_2 \quad (3)$$

Solution: Substituting (1) and (2) into (3) we obtain

$$I = \int_{|t|=r_1} \left[a_0 + a_1 t + a_2 t^2 + \ldots\right]\left[b_0 + b_1 \frac{z}{t} + b_2 \frac{z^2}{t^2} + \ldots\right]\frac{1}{t} dt$$

$$= \int_{|t|=r_1} \left[a_0 + a_1 t + a_2 t^2 + \ldots\right]\left[\frac{b_0}{t} + \frac{b_1 z}{t^2} + \frac{b_2 z^2}{t^3} + \ldots\right] dt \quad (4)$$

Multiplying term by term and omitting the terms whose integral is zero, we obtain

$$I = \int_{|t|=r_1} \left(\frac{a_0 b_0}{t} + \frac{a_1 b_1 z}{t} + \frac{a_2 b_2 z^2}{t} + \ldots\right) dt \quad (5)$$

Applying the theorem concerning the integration of the series, we find

$$I = \sum_{n=0}^{\infty} \int_{|t|=r_1} \frac{a_n b_n z^n}{t} \, dt \tag{6}$$

$$= 2\pi i \sum_{n=0}^{\infty} a_n b_n z^n$$

• **PROBLEM 18-12**

Evaluate the integral

$$I = \int_0^{2\pi} \frac{\cos 2\theta}{2 - \cos 2\theta} \, d\theta \tag{1}$$

<u>Solution</u>: Substituting

$$e^{i\theta} = z$$

leads to

$$\cos 2\theta = \frac{z^2 + z^{-2}}{2} \tag{2}$$

$$d\theta = \frac{dz}{iz}$$

Then,

$$I = \int_C \frac{z^2 + z^{-2}}{2\left(2 - \frac{z^2 + z^{-2}}{2}\right)} \frac{dz}{iz}$$

$$= -\frac{1}{i} \int_C \frac{z^4 + 1}{z(z^4 - 4z^2 + 1)} \, dz \tag{3}$$

$$z^4 - 4z^2 + 1 = 0 \tag{4}$$

For $z^2 = x$ we get

$$x^2 - 4x + 1 = 0 \quad \text{and}$$

$$x_1 = 2 + \sqrt{3}, \quad x_2 = 2 - \sqrt{3} \tag{5}$$

The integrand has simple poles inside the unit circle C at

$$z = 0, \quad z = \sqrt{2 - \sqrt{3}} \text{ and } z = -\sqrt{2 - \sqrt{3}} \tag{6}$$

$$\int \frac{z^4 + 1}{z(z^4 - 4z^2 + 1)} \, dz = 2\pi i \left[\text{Res}_0 + \text{Res}_{\sqrt{2-\sqrt{3}}} + \text{Res}_{-\sqrt{2-\sqrt{3}}} \right] \tag{7}$$

$$\text{Res}_0 = \lim_{z \to 0} \left[\frac{z^4 + 1}{z^4 - 4z^2 + 1} \right] = 1 \tag{8}$$

$$\text{Res}_{\sqrt{2-\sqrt{3}}} = \lim_{z \to \sqrt{2-\sqrt{3}}} \left[\frac{z^4 + 1}{z(z^2 - 2 - \sqrt{3})(z + \sqrt{2-\sqrt{3}})} \right] = -\frac{1}{\sqrt{3}} \tag{9}$$

$$\text{Res}_{-\sqrt{2-\sqrt{3}}} = \lim_{z \to -\sqrt{2-\sqrt{3}}} \left[\frac{z^4+1}{z(z^2-2-\sqrt{3})(z-\sqrt{2-\sqrt{3}})} \right] = -\frac{1}{\sqrt{3}} \quad (10)$$

We obtain

$$\int_0^{2\pi} \frac{\cos 2\theta}{2-\cos 2\theta} d\theta = -\frac{1}{i} \cdot 2\pi i \left(1 - \frac{1}{\sqrt{3}} - \frac{1}{\sqrt{3}} \right)$$

$$= 2\pi \left(\frac{2}{\sqrt{3}} - 1 \right) \quad (11)$$

● **PROBLEM 18-13**

Evaluate the integral

$$I = \int_0^{2\pi} \frac{\cos 3\theta}{10-8\cos\theta} d\theta \quad (1)$$

Solution: The standard substitution

$$z = e^{i\theta} \quad (2)$$

leads to

$$\cos 3\theta = \frac{e^{3i\theta}+e^{-3i\theta}}{2} = \frac{z^3+z^{-3}}{2} \quad (3)$$

$$d\theta = \frac{dz}{iz}$$

We obtain

$$I = \int_C \frac{(z^3+z^{-3})dz}{2iz\left[10-8\frac{z+z^{-1}}{2}\right]} = \int_C \frac{(z^3+z^{-3})dz}{2iz(10-4z-4z^{-1})}$$

$$= \int \frac{z^6+1}{4i(5z^4-2z^5-2z^3)} dz \quad (4)$$

$$= -\frac{1}{4i} \int \frac{z^6+1}{z^3(2z^2-5z+2)} dz = -\frac{1}{4i} \int \frac{z^6+1}{z^3(2z-1)(z-2)} dz$$

The integrand has a pole of order 3 at $z = 0$ and two simple poles at $z = 2$ and $z = \frac{1}{2}$. The pole at $z = 2$ lies outside of the unit circle.

$$\text{Res}_0 = \frac{1}{2!} \lim_{z \to 0} \frac{d^2}{dz^2} \left[\frac{z^6+1}{(2z-1)(z-2)} \right]$$

$$= \frac{1}{2} \lim_{z \to 0} \frac{6z^5(2z-1)(z-2) - (z^6+1)(4z-5)}{(2z-1)^2(z-2)^2} \qquad (5)$$

$$= \frac{21}{8}$$

$$\text{Res}_{\frac{1}{2}} = \lim_{z \to \frac{1}{2}} \left[\frac{z^6+1}{z^3(2z-1)(z-2)} (z-\tfrac{1}{2}) \right]$$

$$= \frac{1}{2} \lim_{z \to \frac{1}{2}} \frac{z^6+1}{(z-2)z^3} = -\frac{65}{24} \qquad (6)$$

Thus,

$$\int_0^{2\pi} \frac{\cos 3\theta}{10 - 8\cos\theta}\, d\theta = -\frac{1}{4i} 2\pi i (\text{Res}_0 + \text{Res}_{\frac{1}{2}})$$

$$= -\frac{\pi}{2}\left(\frac{21}{8} - \frac{65}{24}\right) = -\frac{\pi}{2}\left(-\frac{1}{12}\right) = \frac{\pi}{24} \qquad (7)$$

• **PROBLEM 18-14**

Find all possible values of the integral

$$I = \int_0^{2\pi} \frac{\cos\theta}{a + b\cos\theta}\, d\theta \qquad (1)$$

Solution: Substituting

$$z = e^{i\theta} \qquad (2)$$

we obtain $\cos\theta = \frac{z + z^{-1}}{2}$ and

$$I = \int_C \frac{z + z^{-1}}{2iz\left(a + b\frac{z+z^{-1}}{2}\right)}\, dz$$

$$= \int_C \frac{z^2 + 1}{iz(2az + bz^2 + b)}\, dz \qquad (3)$$

$$= \int_C \frac{(z^2+1)\, dz}{biz(z^2 + 2\frac{a}{b}z + 1)}$$

where C is the unit circle $z = 1$. The integrand has simple poles at

$$z_1 = 0$$

$$z_2 = \frac{-a+\sqrt{a^2-b^2}}{b} \quad (4)$$

$$z_3 = \frac{-a-\sqrt{a^2-b^2}}{b}$$

If only z_1 is located inside the unit circle, then

$$I = 2\pi i \, \text{Res}_0 = 2\pi i \lim_{z \to 0} \frac{z^2+1}{bi(z-z_1)(z-z_2)} \quad (5)$$

$$= 2\pi i \cdot \frac{1}{biz_1 z_2} = \frac{2\pi}{b}$$

But this situation is not possible. Since

$$z_2 z_3 = 1$$

we have

$$|z_2||z_3| = 1 \quad (6)$$

Hence, the only possibilities are

1. z_1 and z_2 lie inside C and z_3 outside.
2. z_1 and z_3 lie inside C and z_2 outside.

We have for case 1:

$$\text{Res}_{z_1} = \lim_{z \to z_1} \frac{z^2+1}{biz(z-z_2)} = \frac{z_1^2+1}{biz_1(z_1-z_2)}$$

$$= \frac{z_1^2+z_1 z_2}{biz_1(z_1-z_2)} = \frac{z_1+z_2}{bi(z_1-z_2)} = \frac{-a}{bi\sqrt{a^2-b^2}} \quad (7)$$

and

$$I = 2\pi i \left[\frac{1}{bi} - \frac{a}{bi\sqrt{a^2-b^2}}\right] = \frac{2\pi}{b}\left(1 - \frac{a}{\sqrt{a^2-b^2}}\right) \quad (8)$$

For case 2 we get

$$\text{Res}_{z_2} = \lim_{z \to z_2} \frac{z^2+1}{biz(z-z_1)} = \frac{z_2^2+1}{biz_2(z_2-z_1)} = \frac{z_2^2+z_1 z_2}{biz_2(z_2-z_1)}$$

$$= \frac{z_1+z_2}{bi(z_2-z_1)} = \frac{a}{bi\sqrt{a^2-b^2}} \quad (9)$$

Thus,

$$I = 2\pi i \left[\frac{1}{bi} + \frac{a}{bi\sqrt{a^2-b^2}}\right] = \frac{2\pi}{b}\left(1 + \frac{a}{\sqrt{a^2-b^2}}\right) \quad (10)$$

● **PROBLEM 18-15**

Show that for real a and b

$$I = \int_0^{2\pi} \frac{d\theta}{a^2\cos^2\theta + b^2\sin^2\theta} = \frac{2\pi}{ab} \quad (1)$$

Solution: Since $\sin^2\theta = 1 - \cos^2\theta$ we get,

$$I = \int_0^{2\pi} \frac{d\theta}{(a^2-b^2)\cos^2\theta + b^2} \qquad (2)$$

Substituting

$$e^{i\theta} = z \qquad (3)$$

we find

$$I = \int_C \frac{dz}{iz\left[(a^2-b^2) \cdot \frac{z^2+2+z^{-2}}{4} + b^2\right]}$$

$$= \frac{4}{i} \int_C \frac{z\,dz}{4b^2z^2 + (a^2-b^2)z^4 + 2(a^2-b^2)z^2 + (a^2-b^2)} \qquad (4)$$

$$= \frac{4}{i(a^2-b^2)} \int_C \frac{z\,dz}{z^4 + 2\frac{a^2+b^2}{a^2-b^2}z^2 + 1}$$

where C is the unit circle, $|z| = 1$. Let us denote

$$A = \frac{a^2+b^2}{a^2-b^2} \qquad (5)$$

then,

$$z^4 + 2Az^2 + 1 = 0 \qquad (6)$$

Setting $z^2 = x$ we find

$$x_1 = -A + \sqrt{A^2-1}, \qquad x_2 = -A - \sqrt{A^2-1} \qquad (7)$$

$$\operatorname*{Res}_{\sqrt{-A+\sqrt{A^2-1}}} = \lim_{z \to \sqrt{-A+\sqrt{A^2-1}}} \frac{z}{(z^2-x_2)(z+\sqrt{x_1})}$$

$$= \frac{1}{4\sqrt{A^2-1}} \qquad (8)$$

$$\operatorname*{Res}_{-\sqrt{-A+\sqrt{A^2-1}}} = \frac{1}{4\sqrt{A^2-1}} \qquad (9)$$

We obtain

$$I = \int_0^{2\pi} \frac{d\theta}{a^2\cos^2\theta + b^2\sin^2\theta}$$

$$= \frac{4}{i(a^2-b^2)} \cdot 2\pi i \cdot \frac{2}{4\sqrt{\frac{(a^2+b^2)^2}{(a^2-b^2)^2} - 1}} = \frac{2\pi}{ab} \qquad (10)$$

• **PROBLEM 18-16**

1. Which of the following functions are even and which are odd?

   $\cos\theta$

   $\sin 3\theta$

   $\cos\theta \sin\theta$

   $\dfrac{2\sin\theta}{3+4\cos\theta}$

2. Evaluate the integral

$$\int_{-\pi}^{\pi} \frac{\sin\theta + \sin 2\theta}{3 + \cos^2\theta}\, d\theta$$

Solution: 1. A function $f(\alpha)$ is an even function of α if

$$f(-\alpha) = f(\alpha) \tag{1}$$

and an odd function if

$$f(-\alpha) = -f(\alpha) \tag{2}$$

The function $\cos m\theta$ is even and $\sin m\theta$ is an odd function.

The product or quotient of an odd function and an even function is odd. Hence, $\cos\theta\sin\theta$ is an odd function.

$$\frac{2\sin(-\theta)}{3+4\cos(-\theta)} = -\frac{2\sin\theta}{3+4\cos\theta} \tag{3}$$

Hence, it is an odd function.

2. Observe that the integrand is an odd function:

$$\frac{\sin(-\theta)+\sin 2(-\theta)}{3+\cos^2(-\theta)} = -\frac{\sin\theta+\sin 2\theta}{3+\cos^2\theta} \tag{4}$$

We shall apply the following theorem which is easy to prove.

Theorem

If $f(\theta)$ is an odd function, then

$$\int_{-a}^{a} f(\theta)\, d\theta = 0 \tag{5}$$

We obtain

$$\int_{-\pi}^{\pi} \frac{\sin\theta + \sin 2\theta}{3 + \cos^2\theta}\, d\theta = 0 \tag{6}$$

A similar theorem exists for even functions.

Theorem

If $f(\theta)$ is an even function, then

$$\int_{-a}^{a} f(\theta)d\theta = 2\int_{0}^{a} f(\theta)d\theta \qquad (7)$$

• **PROBLEM 18-17**

1. Show that any function can be written as the sum of an even and an odd function.

2. Evaluate the integral

$$\int_{0}^{2\pi} \frac{\sin 3\theta}{\cos\theta + 2\cos 2\theta} d\theta \qquad (1)$$

Solution: 1. Let $f(\alpha)$ be any function. Then we denote

$$f(\alpha) = \frac{1}{2}\left[f(\alpha)+f(-\alpha)\right] + \frac{1}{2}\left[f(\alpha)-f(-\alpha)\right] \qquad (2)$$

The first term $\frac{1}{2}\left[f(\alpha)+f(-\alpha)\right]$ is an even function while $\frac{1}{2}\left[f(\alpha)-f(-\alpha)\right]$ is an odd function.

2. It can be proved that if $R(\cos\theta, \sin\theta)$ is a rational function of $\cos\theta$ and $\sin\theta$, and a is any real constant, then

$$\int_{-\pi}^{\pi} R d\theta = \int_{0}^{2\pi} R d\theta = \int_{a}^{2\pi+a} R d\theta \qquad (3)$$

Integral (1) can be written

$$\int_{0}^{2\pi} \frac{\sin 3\theta}{\cos\theta + 2\cos 2\theta} d\theta = \int_{-\pi}^{\pi} \frac{-\sin 3\theta}{\cos\theta + 2\cos 2\theta} d\theta \qquad (4)$$

The integrand is an odd function:

$$\frac{\sin 3(-\theta)}{\cos(-\theta)+2\cos 2(-\theta)} = \frac{-\sin 3\theta}{\cos\theta+2\cos 2\theta} \qquad (5)$$

For any odd function $f(\theta)$ we have

$$\int_{-a}^{a} f(\theta)d\theta = 0 \qquad (6)$$

Therefore,

$$\int_0^{2\pi} \frac{\sin 3\theta}{\cos\theta + 2\cos 2\theta} d\theta = 0 \tag{7}$$

• **PROBLEM 18-18**

1. Show that:
   a. sums, products and quotients of even functions are even.
   b. the product or quotient of an odd function and an even function is odd.

2. Evaluate

$$\int_0^\pi \frac{\cos 2\theta}{2 + \cos\theta} d\theta \tag{1}$$

<u>Solution</u>: 1. a. Let $f(z)$ and $g(z)$ be two even functions. Then

$$f(-z) = f(z), \quad g(-z) = g(z) \tag{2}$$

and

$$f(z) + g(z) = (f+g)(z) = f(-z) + g(-z) = (f+g)(-z) \tag{3}$$

$$(f \cdot g)(-z) = f(-z)g(-z) = (f \cdot g)(z) \tag{4}$$

Similarly,

$$\frac{f}{g}(-z) = \frac{f(-z)}{g(-z)} = \frac{f(z)}{g(z)} = \frac{f}{g}(z) \tag{5}$$

b. Let $f(z) = f(-z)$ be an even function and $g(-z) = -g(z)$ be an odd function. Then,

$$(f \cdot g)(-z) = f(-z)g(-z) = -f(z) \cdot g(z) = -(f \cdot g)(z)$$

and

$$\frac{f}{g}(-z) = \frac{f(-z)}{g(-z)} = \frac{f(z)}{-g(z)} = -\frac{f}{g}(z) \tag{6}$$

2. The integrand is an even function of θ. Hence,

$$\int_0^\pi \frac{\cos 2\theta}{2 + \cos\theta} d\theta = \frac{1}{2} \int_0^{2\pi} \frac{\cos 2\theta}{2 + \cos\theta} d\theta$$

$$= \frac{1}{2} \int_C \frac{z^2 + z^{-2}}{2\left(2 + \frac{z + z^{-1}}{2}\right)} \cdot \frac{dz}{iz} \tag{7}$$

$$= \frac{1}{2i} \int_C \frac{z^2 + z^{-2}}{z^2 + 4z + 1} dz = \frac{1}{2i} \int_C \frac{z^4 + 1}{z^2(z^2 + 4z + 1)} dz$$

$$= \frac{1}{2i} \int_C \frac{z^4+1}{z^2(z-z_1)(z-z_2)} dz$$

where
$$z_1 = -2 + \sqrt{3}$$
$$z_2 = -2 - \sqrt{3}$$
(8)

The integrand has a double pole at the origin, and a simple pole at z_1 located inside the unit circle.

$$\text{Res}_0 = \lim_{z \to 0} \frac{d}{dz}\left(\frac{z^4+1}{z^2+4z+1}\right) = -4 \qquad (9)$$

$$\text{Res}_{-2+\sqrt{3}} = \frac{(\sqrt{3}-2)^4 + 1}{2\sqrt{3}(\sqrt{3}-2)^2}$$

We obtain,

$$\int_0^\pi \frac{\cos 2\theta}{2+\cos\theta} = \frac{1}{2i} \cdot 2\pi i (\text{Res}_0 + \text{Res}_{-2+\sqrt{3}})$$

$$= \pi\left[-4 + \frac{1+(\sqrt{3}-2)^4}{2\sqrt{3}(\sqrt{3}-2)^2}\right] \qquad (10)$$

• **PROBLEM 18-19**

Evaluate the integral

$$\int_{-\infty}^{\infty} \frac{\cos nx}{x^4+1} dx, \quad n \geq 0 \qquad (1)$$

Solution: We shall describe the general method of evaluating the integrals of the form

$$\int_{-\infty}^{\infty} e^{inx} R(x) dx \qquad (2)$$

where $R(x)$ is a real-valued rational function

$$R(x) = \frac{p(x)}{q(x)} \qquad (3)$$

such that

$$\lim_{x \to \pm\infty} xR(x) = 0 \qquad (4)$$

The integrals with one or two infinite limits are called improper integrals. They can be interpreted as

$$\lim_{a \to \infty} \int_{-a}^{a} e^{inx} R(x) dx = \int_{-\infty}^{\infty} e^{inx} R(x) dx \qquad (5)$$

If $R(x)$ has no real poles the limit exists.

When the constant n is different from zero, integral (2) is called the Fourier transform (or inverse Fourier transform).

Taking the real and imaginary parts of (2) we obtain Fourier cosine and sine integrals

$$\int_{-\infty}^{\infty} (\cos nx) \; R(x) dx \qquad (6)$$

$$\int_{-\infty}^{\infty} (\sin nx) \cdot R(x) dx \qquad (7)$$

These integrals have numerous applications in physics, partial differential equations, etc.

We shall apply the following:

Theorem

Let $p(x)$ and $q(x)$ be real polynomials such that

$$\text{degree of } q(x) \geq 2 + \text{degree of } p(x) \qquad (8)$$

and $q(x)$ has no real zeros.

Then,

$$\int_{-\infty}^{\infty} e^{inx} \frac{p(x)}{q(x)} dx = \begin{cases} 2\pi i \sum_{C_+} \text{Res } \frac{p(x)}{q(x)} e^{inx}, & \text{if } n \geq 0 \\ -2\pi i \sum_{C_-} \text{Res } \frac{p(x)}{q(x)} e^{inx}, & \text{if } n < 0 \end{cases} \qquad (9)$$

Here $\sum_{C_+}$ denotes the sum of all residues in the upper half-plane, C_+ indicates the upper half-plane

$$C_+ \equiv \{z : \text{Im } z > 0\}$$

and C_- indicates the lower half-plane

$$C_- \equiv \{z : \text{Im } z < 0\}$$

We shall evaluate

$$\int_{-\infty}^{\infty} \frac{e^{inx}}{x^4+1} dx \qquad (10)$$

and then take the real part of it.

The integrand

$$\frac{e^{inx}}{x^4+1} \tag{11}$$

has four distinct simple poles

$$x_1 = \sqrt{-i}$$
$$x_2 = -\sqrt{-i} \tag{12}$$
$$x_3 = \sqrt{i}$$
$$x_4 = -\sqrt{i}$$

Two of the poles are located in the upper-half plane

$$z_1 = \sqrt{-i} = -\frac{1}{\sqrt{2}} + i\frac{1}{\sqrt{2}}$$
$$z_3 = \sqrt{i} = \frac{1}{\sqrt{2}} + i\frac{1}{\sqrt{2}} \tag{13}$$

The assumptions of the theorem are fulfilled. From (9) we obtain

$$\int_{-\infty}^{\infty} \frac{e^{inx}}{x^4+1}\,dx = 2\pi i\left[\text{Res}_{z_1} + \text{Res}_{z_3}\right]$$

$$= 2\pi i\left(\frac{e^{inz_1}}{4z_1^3} + \frac{e^{inz_3}}{4z_3^3}\right) \tag{14}$$

$$= \frac{\pi e^{\frac{n}{\sqrt{2}}}}{\sqrt{2}}\left[\cos\frac{n}{\sqrt{2}} + \sin\frac{n}{\sqrt{2}}\right]$$

$$= \int_{-\infty}^{\infty} \frac{\cos nx}{x^4+1}\,dx$$

Note that for n = 0, (14) reduces to

$$\int_{-\infty}^{\infty} \frac{1}{x^4+1}\,dx = \frac{\sqrt{2}\pi}{2} \tag{15}$$

• **PROBLEM 18-20**

Evaluate the integral

$$\int_{-\infty}^{\infty} \frac{\cos nx}{1+x^2}\,dx, \quad n \geq 0 \tag{1}$$

Find

$$\int_{-\infty}^{\infty} \frac{dx}{1+x^2} \qquad (2)$$

as a special case of (1).

Solution: We shall consider the integral

$$\int_{-\infty}^{\infty} \frac{e^{inx}}{1+x^2} dx \qquad (3)$$

The function $\frac{1}{1+z^2}$ has simple poles at $z = i$ and $z = -i$.

Therefore (see Problem 18-19),

$$\int_{-\infty}^{\infty} \frac{e^{inx}}{1+x^2} dx = 2\pi i \cdot \text{Res}_i \left(\frac{e^{inz}}{1+z^2} \right)$$

$$= 2\pi i \lim_{z \to i} \left[\frac{e^{inz}}{(z+i)(z-i)} (z-i) \right] = 2\pi i \frac{e^{-n}}{2i} \qquad (4)$$

$$= \pi e^{-n}$$

Comparing the real and imaginary parts we obtain

$$\text{Re} \int_{-\infty}^{\infty} \frac{e^{inx}}{1+x^2} dx = \int_{-\infty}^{\infty} \frac{\cos nx}{1+x^2} dx$$

$$= \text{Re}(\pi e^{-n}) = \pi e^{-n} \qquad (5)$$

Note that for the imaginary part we obtain

$$\text{Im} \int_{-\infty}^{\infty} \frac{e^{inx}}{1+x^2} dx = \int_{-\infty}^{\infty} \frac{\sin nx}{1+x^2} dx$$

$$= \text{Im}(\pi e^{-n}) = 0 \qquad (6)$$

Setting $n = 0$ in (5) we find

$$\int_{-\infty}^{\infty} \frac{1}{1+x^2} dx = \pi \qquad (7)$$

● **PROBLEM 18-21**

Evaluate

1. $\int_{-\infty}^{\infty} \frac{dx}{x^2+a^2}$, $a > 0$ (1)

2. $\int_{-\infty}^{\infty} \frac{dx}{(x^2+a^2)^2}$, $a > 0$ (2)

<u>Solution</u>: 1. Since $a > 0$,

$$\frac{1}{x^2+a^2} = \frac{1}{(x+ia)(x-ia)} \tag{3}$$

the integrand has one simple pole located in the upper-half plane at $z = ia$.

$$\underset{ia}{\text{Res}} \frac{1}{z^2+a^2} = \lim_{z \to ia} \frac{1}{z+ia} = \frac{1}{2ia} \tag{4}$$

Hence,

$$\int_{-\infty}^{\infty} \frac{dx}{x^2+a^2} = 2\pi i \cdot \frac{1}{2ia} = \frac{\pi}{a} \tag{5}$$

2. By the Leibnitz's rule we obtain from (5)

$$\frac{d}{da} \int_{-\infty}^{\infty} \frac{dx}{x^2+a^2} = \int_{-\infty}^{\infty} \frac{\partial}{\partial a}\left(\frac{1}{x^2+a^2}\right) dx$$

$$= \int_{-\infty}^{\infty} \frac{-2a}{(x^2+a^2)^2} dx = \frac{d}{da}\left(\frac{\pi}{a}\right) \tag{6}$$

$$= \frac{-\pi}{a^2}$$

Hence, $\int_{-\infty}^{\infty} \frac{dx}{(x^2+a^2)^2} = \frac{\pi}{2a^3}$ (7)

● **PROBLEM 18-22**

1. Find the integral

$$\int_{-\infty}^{\infty} \frac{\cos nx}{x^2+a^2} dx, \quad a > 0, \quad n > 0 \tag{1}$$

2. Evaluate

$$\int_{-\infty}^{\infty} \frac{\cos nx}{(x^2+a^2)^2}\, dx, \quad n > 0, \quad a > 0 \tag{2}$$

Solution: 1. First we shall evaluate

$$\int_{-\infty}^{\infty} \frac{e^{inx}}{x^2+a^2}\, dx \tag{3}$$

$$\frac{1}{x^2+a^2} = \frac{1}{(x+ia)(x-ia)} \tag{4}$$

Since $a > 0$, the integrand has one simple pole at $z = ia$ located in the upper-half plane.

$$\operatorname*{Res}_{ia} \frac{e^{inz}}{z^2+a^2} = \lim_{z \to ia} \frac{e^{inz}}{z+ia} = \frac{e^{-na}}{2ia} \tag{5}$$

Hence,

$$\int_{-\infty}^{\infty} \frac{e^{inx}}{x^2+a^2}\, dx = 2\pi i \frac{e^{-na}}{2ia} = \frac{\pi}{a} e^{-na} \tag{6}$$

By taking real and imaginary parts, we obtain

$$\int_{-\infty}^{\infty} \frac{e^{inx}}{x^2+a^2}\, dx = \int_{-\infty}^{\infty} \frac{\cos nx}{x^2+a^2}\, dx + i \int_{-\infty}^{\infty} \frac{\sin nx}{x^2+a^2}\, dx = \frac{\pi}{a} e^{-na} \tag{7}$$

(the $\sin nx$ integral equals 0)

Therefore,

$$\int_{-\infty}^{\infty} \frac{\cos nx}{x^2+a^2}\, dx = \frac{\pi}{a} e^{-na} \tag{8}$$

2. Here we shall apply Leibnitz's rule for differentiation under the integral sign.

$$\frac{d}{d\alpha} \int_a^b F(x,\alpha)\, dx = \int_a^b \frac{\partial F}{\partial \alpha}\, dx \tag{9}$$

where a and b are constants and α is a real parameter such that $\alpha_1 \leq \alpha \leq \alpha_2$ where α_1 and α_2 are constants. The function $F(x,\alpha)$ is continuous and has a continuous partial derivative with respect to α. The limits of integral (9) can be infinite.

From (7) differentiating with respect to a we obtain

$$\frac{d}{da} \int_{-\infty}^{\infty} \frac{\cos nx}{x^2+a^2} \, dx = \int_{-\infty}^{\infty} \frac{\partial}{\partial a}\left(\frac{\cos nx}{x^2+a^2}\right) dx$$

$$= \int_{-\infty}^{\infty} \frac{-2a\cos nx}{(x^2+a^2)^2} \, dx = \frac{\partial}{\partial a}\left(\frac{\pi}{a} e^{-na}\right) \quad (10)$$

$$= -\frac{\pi}{a^2} e^{-na} + \frac{\pi}{a}(-n)e^{-na}$$

or

$$\int_{-\infty}^{\infty} \frac{\cos nx}{(x^2+a^2)^2} \, dx = \frac{\pi e^{-na}}{2a^2}\left(\frac{1}{a} + n\right) \quad (11)$$

Compare the results with Problems 18-20 and 18-21.

● **PROBLEM** 18-23

Evaluate the integral

$$\int_{-\infty}^{\infty} \frac{dx}{ax^2+bx+c} \quad (1)$$

where $4ac > b^2$.

Solution: Since $4ac > b^2$, the equation $ax^2 + bx + c = 0$ does not have any real roots.

We shall use the formula

$$\int_{-\infty}^{\infty} \frac{\cos nx}{x^2+2ax+b} \, dx = \frac{\pi}{\sqrt{b-a^2}} e^{-n\sqrt{b-a^2}} \cos na \quad (2)$$

which for $n = 0$ becomes

$$\int_{-\infty}^{\infty} \frac{1}{x^2+2ax+b} \, dx = \frac{\pi}{\sqrt{b-a^2}} \quad (3)$$

Hence,

$$\int_{-\infty}^{\infty} \frac{dx}{ax^2+bx+c} = \frac{1}{a} \int_{-\infty}^{\infty} \frac{dx}{x^2 + 2\frac{b}{2a}x + \frac{c}{a}}$$

$$= \frac{1}{a} \frac{\pi}{\sqrt{\frac{c}{a} - \frac{b^2}{4a^2}}} = \frac{2\pi}{\sqrt{4ac-b^2}} \quad (4)$$

• **PROBLEM 18-24**

Evaluate the integral

$$\int_{-\infty}^{\infty} \frac{\sin 3x}{x^2+2x+3} \, dx \qquad (1)$$

Solution: First we evaluate

$$\int_{-\infty}^{\infty} \frac{e^{3ix}}{x^2+2x+3} \, dx \qquad (2)$$

and then take the imaginary part of it.

Solving

$$z^2 + 2z + 3 = 0 \qquad (3)$$

we find

$$z_1 = \frac{-2+\sqrt{4-12}}{2} = -1 + \sqrt{2}\, i$$

$$z_2 = -1 - \sqrt{2}\, i \qquad (4)$$

The integrand has only one simple pole in the upper-half plane. Therefore,

$$\int_{-\infty}^{\infty} \frac{e^{3ix}}{x^2+2x+3} \, dx = 2\pi i \operatorname*{Res}_{-1+\sqrt{2}\, i}$$

$$= 2\pi i \left[\frac{e^{3iz}}{(z-z_2)} \right]_{z=z_1}$$

$$= 2\pi i \, \frac{e^{3i(-1+\sqrt{2}\, i)}}{2\sqrt{2}\, i} \qquad (5)$$

$$= \frac{\pi}{\sqrt{2}} \cdot e^{-3\sqrt{2}} \cdot e^{-3i}$$

We obtain

$$\int_{-\infty}^{\infty} \frac{\sin 3x}{x^2+2x+3} \, dx = \operatorname{Im}\left[\int_{-\infty}^{\infty} \frac{e^{3ix}}{x^2+2x+3} \, dx \right]$$

$$= \operatorname{Im}\left[\frac{\pi}{\sqrt{2}} \cdot e^{-3\sqrt{2}} \cdot e^{-3i} \right] \qquad (6)$$

$$= \frac{\pi}{2} e^{-3\sqrt{2}} \cdot \sin(-3) = -\frac{\pi}{2} e^{-3\sqrt{2}} \sin 3$$

• **PROBLEM 18-25**

Evaluate

$$\int_{-\infty}^{\infty} \frac{\cos nx}{x^2 + 2ax + b} \, dx, \qquad n > 0 \qquad (1)$$

What restrictions should be imposed on parameters a and b if one wants to apply the Fourier transform theorem?

Solution: We shall evaluate

$$\int_{-\infty}^{\infty} \frac{e^{inx}}{x^2 + 2ax + b} \, dx \qquad (2)$$

The Fourier transform theorem can be applied when the equation

$$x^2 + 2ax + b = 0 \qquad (3)$$

does not have real solutions.

$$\begin{aligned} x_1 &= -a + \sqrt{a^2 - b} \\ x_2 &= -a - \sqrt{a^2 - b} \end{aligned} \qquad (4)$$

The roots are complex when $a^2 < b$.

The integrand of (2) has one simple pole located in the upper-half plane.

$$\operatorname{Res}_{z_1} \frac{e^{inz}}{z^2 + 2az + b}$$

$$= \lim_{z \to z_1} \frac{e^{inz}}{z - z_2} = \frac{e^{inz_1}}{z_1 - z_2} \qquad (5)$$

$$= \frac{e^{in[-a + \sqrt{a^2 - b}]}}{2\sqrt{a^2 - b}}$$

Hence,

$$\int_{-\infty}^{\infty} \frac{e^{inx}}{x^2 + 2ax + b} \, dx = 2\pi i \operatorname{Res}_{z_1}$$

$$= \frac{2\pi i}{2i\sqrt{b - a^2}} e^{-ina} e^{in\sqrt{b - a^2}} \qquad (6)$$

602

$$= \frac{\pi}{\sqrt{b-a^2}} e^{-n\sqrt{b-a^2}} e^{-ina}$$

$$\int_{-\infty}^{\infty} \frac{\cos nx}{x^2+2ax+b} dx = \frac{\pi}{\sqrt{b-a^2}} e^{-n\sqrt{b-a^2}} \cos na \qquad (7)$$

As a by-product we obtain

$$\int_{-\infty}^{\infty} \frac{\sin nx}{x^2+2ax+b} dx = \frac{-\pi}{\sqrt{b-a^2}} e^{-n\sqrt{b-a^2}} \sin na \qquad (8)$$

● **PROBLEM 18-26**

Evaluate the integral

$$\int_{-\infty}^{\infty} \frac{e^{ix}}{x^2-2x+2} dx \qquad (1)$$

and find corresponding integrals with cosx and sinx.

Solution: The singularities of the function

$$\frac{1}{z^2-2z+2} \qquad (2)$$

are simple poles located at

$$z_1 = 1 + i$$
$$z_2 = 1 - i \qquad (3)$$

Therefore,

$$\int_{-\infty}^{\infty} \frac{e^{ix}}{x^2-2x+2} dx = 2\pi i \sum_{C_+} \text{Res}$$

$$= 2\pi i \operatorname*{Res}_{1+i} \frac{e^{iz}}{z^2-2z+2}$$

$$= 2\pi i \lim_{z \to 1+i} \left[\frac{e^{iz}}{(z-z_1)(z-z_2)} (z-z_1) \right] \qquad (4)$$

$$= 2\pi i \frac{e^{iz_1}}{z_1-z_1} = 2\pi i \frac{e^{i(1+i)}}{2i} = \frac{\pi}{e} e^{i}$$

Here C_+ denotes the upper half-plane

$$C_+ \equiv \{z : \text{Im} z > 0\} \qquad (5)$$

Taking the real and imaginary parts of (4) we obtain

$$\int_{-\infty}^{\infty} \frac{\cos x}{x^2-2x+2} \, dx = \frac{\pi}{e} \cos 1 \qquad (6)$$

$$\int_{-\infty}^{\infty} \frac{\sin x}{x^2-2x+2} \, dx = \frac{\pi}{e} \sin 1 \qquad (7)$$

● **PROBLEM 18-27**

Evaluate the integrals

$$\int_{-\infty}^{\infty} \frac{\cos nx}{x^4+a^4} \, dx, \qquad a > 0, \quad n > 0 \qquad (1)$$

$$\int_{-\infty}^{\infty} \frac{\sin nx}{x^4+a^4} \, dx, \qquad a > 0, \quad n > 0 \qquad (2)$$

Solution: We compute

$$\int_{-\infty}^{\infty} \frac{e^{inx}}{x^4+a^4} \, dx \qquad (3)$$

The integrand

$$\frac{e^{inz}}{z^4+a^4} \qquad (4)$$

has four distinct simple poles:

$$\begin{aligned} z_1 &= a\sqrt{-i} \\ z_2 &= -a\sqrt{-i} \\ z_3 &= a\sqrt{i} \\ z_4 &= -a\sqrt{i} \end{aligned} \qquad (5)$$

Only two of them z_1 and z_3 are located in the upper-half plane.

$$\begin{aligned} z_1 = a\sqrt{-i} &= a\, e^{i\frac{3\pi}{4}} \\ &= a\left(\cos \frac{3\pi}{4} + i \sin \frac{3\pi}{4}\right) \\ &= a\left(-\frac{1}{\sqrt{2}} + i\frac{1}{\sqrt{2}}\right) \end{aligned} \qquad (6)$$

$$z_3 = a\sqrt{i} = ae^{i\frac{\pi}{4}} = a\left(\cos\frac{\pi}{4} + i\sin\frac{\pi}{4}\right)$$
$$= a\left(\frac{1}{\sqrt{2}} + i\frac{1}{\sqrt{2}}\right) \tag{7}$$

We obtain

$$\int_{-\infty}^{\infty} \frac{e^{inx}}{x^4+a^4}\,dx = 2\pi i(\text{Res}_{z_1} + \text{Res}_{z_3}) \tag{8}$$

$$\text{Res}_{z_1} = \frac{e^{inz_1}}{4z_1^3} \tag{9}$$

$$\text{Res}_{z_3} = \frac{e^{inz_3}}{4z_3^3}$$

Hence,

$$\int_{-\infty}^{\infty} \frac{e^{inx}}{x^4+a^4}\,dx = 2\pi i\left(\frac{e^{inz_1}}{4z_1^3} + \frac{e^{inz_3}}{4z_3^3}\right)$$

$$= \frac{\pi}{2a^3} e^{-\frac{na}{\sqrt{2}}} \left[\frac{e^{-\frac{na}{\sqrt{2}}i}}{\frac{1}{\sqrt{2}} - i\frac{1}{\sqrt{2}}} + \frac{e^{i\frac{na}{\sqrt{2}}}}{\frac{1}{\sqrt{2}} + i\frac{1}{\sqrt{2}}}\right]$$

$$= \frac{\pi e^{-\frac{na}{\sqrt{2}}}}{\sqrt{2}\,a^3} \left[\cos\frac{na}{\sqrt{2}} + \sin\frac{na}{\sqrt{2}}\right] \tag{10}$$

● **PROBLEM 18-28**

Evaluate the integral

$$\int_{-\infty}^{\infty} \frac{\cos nx}{(1+x^2)^2}\,dx \tag{1}$$

where $n > 0$.

Solution: First we evaluate

$$\int_{-\infty}^{\infty} \frac{e^{inx}}{(1+x^2)^2}\,dx \tag{2}$$

The function

$$\frac{1}{(1+z^2)^2} \tag{3}$$

has two double poles at $z = i$ and $z = -i$.

Because $n > 0$ we will take into account poles located in the upper half-plane, i.e., the pole at $z = i$.

$$\int_{-\infty}^{\infty} \frac{e^{inx}}{(1+x^2)^2} dx = 2\pi i \, \underset{i}{\text{Res}} \tag{4}$$

where

$$\underset{i}{\text{Res}} \frac{e^{inz}}{(1+z^2)^2} = \lim_{z \to i} \frac{d}{dz}\left[\frac{e^{inz}}{(z-i)^2(z+i)^2}(z-i)^2\right]$$

$$= \lim_{z \to i} \frac{d}{dz}\left[\frac{e^{inz}}{(z+i)^2}\right] = e^{-n} \frac{n+1}{4i} \tag{5}$$

From (4) and (5), we obtain

$$\int_{-\infty}^{\infty} \frac{e^{inx}}{(1+x^2)^2} dx = 2\pi i \cdot e^{-n} \frac{n+1}{4i}$$

$$= \frac{\pi e^{-n}(n+1)}{2} \tag{6}$$

Hence,

$$\int_{-\infty}^{\infty} \frac{\cos nx}{(1+x^2)^2} dx = \frac{\pi e^{-n}(n+1)}{2} \tag{7}$$

For $n = 0$, we obtain

$$\int_{-\infty}^{\infty} \frac{1}{(1+x^2)^2} dx = \frac{\pi}{2} \tag{8}$$

• **PROBLEM 18-29**

Evaluate the integral

$$\int_0^{\infty} \frac{x^2}{x^8+1} dx \tag{1}$$

Solution: The integrand is an even function. Hence,

$$\int_0^{\infty} \frac{x^2}{x^8+1} dx = \frac{1}{2} \int_{-\infty}^{\infty} \frac{x^2}{x^8+1} dx \tag{2}$$

We shall apply the theorem about the Fourier transform.

The poles of the function

$$f(z) = \frac{z^2}{1+z^8} \tag{3}$$

are the roots of

$$1 + z^8 = 0 \tag{4}$$

There are eight roots ω_k of (4). All of them are distinct. Function $f(z)$ has eight simple poles.

$$\operatorname*{Res}_{\omega_k} f(z) = \lim_{z \to \omega_k}\left[f(z)(z-\omega_k)\right]$$

$$= \left(\frac{z^2}{8z^7}\right)\bigg|_{z=\omega_k} = \frac{\omega_k^2}{8\omega_k^7} = \frac{\omega_k^3}{8\omega_k^8} = \frac{\omega_k^3}{-8} \tag{5}$$

The poles ω_k are

$$\omega_k = e^{\pi i \frac{1+2k}{8}}, \quad k = 0, 1, \ldots 7 \tag{6}$$

Only four of them are located in the upper-half plane, $\omega_0, \omega_1, \omega_2, \omega_3$.

We have

$$\omega_k^3 = e^{3\pi i \frac{1+2k}{8}} \tag{7}$$

$$\int_0^\infty \frac{x^2}{1+x^8}\, dx = \frac{1}{2} \int_{-\infty}^\infty \frac{x^2}{1+x^8}\, dx$$

$$= \frac{1}{2} \cdot 2\pi i \cdot \left(-\frac{1}{8}\right)\left[\omega_0^3 + \omega_1^3 + \omega_2^3 + \omega_3^3\right]$$

$$= -\frac{\pi i}{8}\left[e^{\frac{3}{8}\pi i} + e^{\frac{9}{8}\pi i} + e^{\frac{15}{8}\pi i} + e^{\frac{21}{8}\pi i}\right] \tag{8}$$

$$= -\frac{\pi i}{8}\left[2\cos\frac{\pi}{4} - 1 + i\right]$$

● **PROBLEM 18-30**

Evalutate

$$\int_0^\infty \frac{dx}{x^4 + x^2 + 1} \tag{1}$$

Solution: Since the integrand is an even function,

$$\int_0^\infty \frac{dx}{x^4+x^2+1} = \frac{1}{2} \int_{-\infty}^\infty \frac{dx}{x^4+x^2+1} \tag{2}$$

We shall find the singularities of the integrand.

$$z^4 + z^2 + 1 = 0$$
$$z^2 = \omega \tag{3}$$
$$\omega^2 + \omega + 1 = 0$$

Hence,

$$\omega_1 = -\frac{1}{2} + i\frac{\sqrt{3}}{2} = e^{i\frac{2\pi}{3}}$$
$$\omega_2 = -\frac{1}{2} - i\frac{\sqrt{3}}{2} = e^{i\frac{4\pi}{3}} \tag{4}$$

The integrand $\frac{1}{z^4+z^2+1}$ has four simple poles:

$$z_1 = \sqrt{\omega_1} = e^{i\frac{\pi}{3}}$$
$$z_2 = -\sqrt{\omega_1} = -e^{i\frac{\pi}{3}}$$
$$z_3 = \sqrt{\omega_2} = e^{i\frac{2\pi}{3}} \tag{5}$$
$$z_4 = -\sqrt{\omega_2} = -e^{i\frac{2\pi}{3}}$$

Only two poles lie in the upper half-plane: z_1 and z_3.

$$\operatorname*{Res}_{z_1} = \lim_{z \to e^{i\frac{\pi}{3}}} \frac{z - e^{i\frac{\pi}{3}}}{z^4+z^2+1} \tag{6}$$

$$= \frac{1}{2e^{i\frac{\pi}{3}}\left(e^{i\frac{2\pi}{3}} - e^{i\frac{4\pi}{3}}\right)}$$

$$\operatorname*{Res}_{z_3} = \lim_{z \to e^{i\frac{2\pi}{3}}} \frac{1}{\left(z^2 - e^{i\frac{2\pi}{3}}\right)\left(z + e^{i\frac{2\pi}{3}}\right)} \tag{7}$$

$$= \frac{1}{2e^{i\frac{2\pi}{3}}\left(e^{i\frac{4\pi}{3}} - e^{i\frac{2\pi}{3}}\right)}$$

$$\int_0^\infty \frac{dx}{x^4+x^2+1} = \frac{1}{2} \cdot 2\pi i \, \frac{1}{2\left(e^{i\frac{4\pi}{3}} - e^{i\frac{2\pi}{3}}\right)} \left[\frac{1}{e^{i\frac{2\pi}{3}}} - \frac{1}{e^{i\frac{\pi}{3}}}\right]$$

$$= \frac{\pi i}{2e^{i\frac{\pi}{3}}\left(e^{i\frac{\pi}{3}}+1\right)} = \frac{\pi i}{(1+i\sqrt{3})\left(\frac{3}{2}+i\frac{\sqrt{3}}{2}\right)} \quad (8)$$

$$= \frac{\pi}{2\sqrt{3}}$$

● **PROBLEM** 18-31

Evaluate the integrals

1. $\int_{-\infty}^{\infty} \frac{\cos nx}{(x^2+a^2)(x^2+b^2)}\, dx,$ $\quad n \geq 0$
   $a > b > 0$ or $\quad$ (1)
   $b > a > 0$

2. $\int_{-\infty}^{\infty} \frac{x \sin nx}{(x^2+a^2)(x^2+b^2)}\, dx \quad (2)$

3. $\int_{-\infty}^{\infty} \frac{dx}{(x^2+a^2)(x^2+b^2)} \quad (3)$

Solution: 1. First we compute

$$\int_{-\infty}^{\infty} \frac{e^{inz}}{(z^2+a^2)(z^2+b^2)}\, dz \quad (4)$$

$$\frac{e^{inz}}{(z^2+a^2)(z^2+b^2)} = \frac{e^{inz}}{(z+ia)(z-ia)(z+ib)(z-ib)} \quad (5)$$

In the upper half-plane the integrand has two simple poles at $z = ia$ and $z = ib$.

$$\operatorname{Res}_{ia} = \lim_{z \to ia} \frac{e^{inz}}{(z+ia)(z+ib)(z-ib)} = \frac{e^{-na}}{2ia(a+b)(b-a)} \quad (6)$$

$$\operatorname{Res}_{ib} = \lim_{z \to ib} \frac{e^{inz}}{(z+ia)(z-ia)(z+ib)} = \frac{e^{-nb}}{2ib(a+b)(a-b)} \quad (7)$$

Hence,

$$\int_{-\infty}^{\infty} \frac{e^{inz}}{(z^2+a^2)(z^2+b^2)}\, dz = 2\pi i \left[\frac{e^{-na}}{2ia(a+b)(b-a)} - \frac{e^{-nb}}{2ib(a+b)(b-a)}\right] \quad (8)$$

$$= \frac{\pi}{(a+b)(a-b)}\left[\frac{e^{-nb}}{b} - \frac{e^{-na}}{a}\right]$$

We obtain

$$\int_{-\infty}^{\infty} \frac{\cos nx}{(x^2+a^2)(x^2+b^2)} \, dx = \frac{\pi}{(a+b)(a-b)} \left(\frac{e^{-nb}}{b} - \frac{e^{-na}}{a} \right) \tag{9}$$

2. By the Leibnitz's rule from (9), we obtain

$$\frac{d}{dn} \int_{-\infty}^{\infty} \frac{\cos nx}{(x^2+a^2)(x^2+b^2)} \, dx = \int_{-\infty}^{\infty} \frac{-x \sin nx}{(x^2+a^2)(x^2+b^2)} \, dx$$

$$= \frac{\partial}{\partial n} \left[\frac{\pi}{a^2-b^2} \left(\frac{e^{-nb}}{b} - \frac{e^{-na}}{a} \right) \right] \tag{10}$$

$$= \frac{\pi}{a^2-b^2} (e^{-na} - e^{-nb})$$

Thus,

$$\int_{-\infty}^{\infty} \frac{x \sin nx}{(x^2+a^2)(x^2+b^2)} \, dx = \frac{\pi}{(a^2-b^2)} (e^{-nb} - e^{-na}) \tag{11}$$

3. This integral can be found from eq.(9) setting $n = 0$

$$\int_{-\infty}^{\infty} \frac{1}{(x^2+a^2)(x^2+b^2)} \, dx = \frac{\pi}{(a+b)(a-b)} \left(\frac{1}{b} - \frac{1}{a} \right) \tag{12}$$

$$= \frac{\pi}{ab(a+b)}$$

CHAPTER 19

APPLICATIONS OF RESIDUE CALCULUS PART II

● **PROBLEM 19-1**

Evaluate the integral:

$$\int_{-\infty}^{\infty} \frac{dx}{(x^2+a^2)^3}, \qquad a > 0$$

Solution: An important application of the theory of residues is the evaluation of certain types of integrals. An improper integral of a continuous function f(x) taken over an infinite interval $[0, \infty)$ is defined by

$$\int_0^{\infty} f(x)dx = \lim_{r \to \infty} \int_0^r f(x)dx \qquad (1)$$

When the limit exists the improper integral is said to converge. Similarly, for a function continuous for all x, we define

$$\int_{-\infty}^{\infty} f(x)dx = \lim_{r_1 \to \infty} \int_0^{r_1} f(x)dx + \lim_{r_2 \to \infty} \int_{-r_2}^{0} f(x)dx \qquad (2)$$

When both limits exist, integral $\int_{-\infty}^{\infty} f(x)dx$ converges and its value is given by (2). We can assign to integral $\int_{-\infty}^{\infty} f(x)dx$ another number, namely the Cauchy principal value.

$$\int_{-\infty}^{\infty} f(x)dx = \lim_{r \to \infty} \int_{-r}^{r} f(x)dx \qquad (3)$$

When the integral in (2) converges, its value is the same as the Cauchy principal value. The opposite is not true. For example, for f(x) = x, eq.(3) yields

$$\int_{-\infty}^{\infty} xdx = 0$$

and according to (2) the integral does not converge.

To evaluate our integral consider the function

$$f(z) = \frac{1}{(z^2+a^2)^3} \qquad (4)$$

and the contour C shown in Fig. 1,

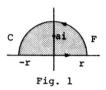

Fig. 1

consisting of the segment of the x axis $[-r,r]$ and a semicircular arc F of radius r such that $r > a$.

The function (4) has a pole of order 3 at $z = ai$.

$$\underset{ai}{\text{Res}} = \frac{1}{(3-1)!} \lim_{z \to ai} \frac{d^2}{dz^2} \frac{(z-ai)^3}{(z^2+a^2)^3}$$

$$= \frac{1}{2} \frac{d^2}{dz^2} \frac{1}{(z+ai)^3} \bigg|_{z=ai} = \frac{1}{2} \cdot \frac{(-3)(-4)}{(2ai)^5} \qquad (5)$$

$$= \frac{3}{16ia^5}$$

Hence, by the residue theorem

$$\int_C f(z)dz = \int_F f(z)dz + \int_{-r}^{r} f(x)dx$$

$$= 2\pi i \underset{ai}{\text{Res}} f(z) = \frac{3\pi}{8a^5} \qquad (6)$$

What happens with $\int_F f(z)dz$ as $r \to \infty$? For $z \in F$ we have

$$\frac{1}{|z^2+a^2|} = \frac{1}{|z^2-(-a^2)|} \leq \frac{1}{||z^2|-|a^2||} = \frac{1}{r^2-a^2} \qquad (7)$$

Hence,

$$\left| \int_F f(z)dz \right| \leq \frac{\pi r}{(r^2-a^2)^3} \qquad (8)$$

and

$$\lim_{r \to \infty} \int_F f(z)dz = 0 \qquad (9)$$

We obtain,

$$\lim_{r \to \infty} \int_{-r}^{r} f(x)dx = \int_{-\infty}^{\infty} f(x)dx$$

$$= \int_{-\infty}^{\infty} \frac{dx}{(x^2+a^2)^3} = \frac{3\pi}{8a^5} \qquad (10)$$

● **PROBLEM 19-2**

Evaluate the integral:

$$\int_0^\infty \frac{\cos x}{x^2+a^2} dx, \quad a > 0 \qquad (1)$$

Solution: Let C be the contour of integration.

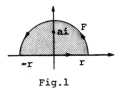

Fig.1

We shall evaluate the integral

$$\int_C \frac{e^{iz}}{z^2+a^2} dz = \int_F \frac{e^{iz}}{z^2+a^2} dz + \int_{-r}^r \frac{e^{ix}}{x^2+a^2} dx \qquad (2)$$

For the points on F we have

$$\left| e^{iz} \right| \leq 1, \qquad (3)$$

thus

$$\left| \int_F \frac{e^{iz}}{z^2+a^2} dz \right| \leq \frac{\pi r}{r^2-a^2} \qquad (4)$$

and

$$\lim_{r \to \infty} \int_F \frac{e^{iz}}{z^2+a^2} dz = 0 \qquad (5)$$

The integrand has one singular point inside the contour C. It is a simple pole at $z = ai$.

$$\operatorname*{Res}_{ai} \frac{e^{iz}}{z^2+a^2} = \left[\frac{e^{iz}}{2z} \right]_{z=ai} = \frac{e^{-a}}{2ai} \qquad (6)$$

By the residue theorem

$$\int_C f(z)dz = \int_F f(z)dz + \int_{-r}^r \frac{e^{ix}}{x^2+a^2} dx$$

$$= 2\pi i \cdot \frac{e^{-a}}{2ai} = \frac{\pi e^{-a}}{a} \qquad (7)$$

Taking the limit as $r \to \infty$ we obtain from (7)

$$\int_{-\infty}^{\infty} \frac{e^{ix}}{x^2+a^2} dx = \frac{\pi e^{-a}}{a} \qquad (8)$$

Taking the real part of (8) we find

$$\int_{-\infty}^{\infty} \frac{\cos x}{x^2+a^2} dx = \frac{\pi e^{-a}}{a} \qquad (9)$$

Since the integrand is an even function,

$$\frac{1}{2} \int_{-\infty}^{\infty} \frac{\cos x}{x^2+a^2} dx = \int_{0}^{\infty} \frac{\cos x}{x^2+a^2} dx = \frac{\pi e^{-a}}{2a} \qquad (10)$$

● **PROBLEM 19-3**

Show that, if $f(z)$ is such that

$$|f(z)| \leq \frac{M}{r^k} \qquad (1)$$

for $z = re^{i\theta}$, (2)

where M and $k > 1$ are constants, then

$$\lim_{r \to \infty} \int_F f(z)dz = 0, \qquad (3)$$

where F is the semicircle of radius r shown in Fig. 1.

Solution: We shall use the following property of the complex functions:

If M is an upper bound of $|f(z)|$ on C,

$$|f(z)| \leq M, \qquad (4)$$

and L is the length of C, then

$$\left| \int_C f(z)dz \right| \leq ML \qquad (5)$$

Thus

$$\left| \int_F f(z)dz \right| \leq \frac{M}{r^k} \cdot r\pi = \frac{\pi M}{r^{k-1}} \qquad (6)$$

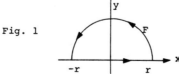

Fig. 1

Taking the limit of both sides of (6) as $r \to \infty$ we obtain,

$$\lim_{r \to \infty} \frac{\pi M}{r^{k-1}} = 0, \qquad k-1 > 0 \tag{7}$$

Thus,

$$\lim_{r \to \infty} \left| \int_F f(z) dz \right| = 0 \tag{8}$$

and

$$\lim_{r \to \infty} \int_F f(z) dz = 0 \tag{9}$$

● **PROBLEM 19-4**

Evaluate:

$$\int_0^\infty \frac{dx}{x^6+1} \tag{1}$$

Solution: Consider the contour C consisting of the line along the x axis from -r to r and the semicircle (see Fig. 1).

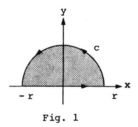

Fig. 1

The contour C is positively oriented. We shall compute

$$\int_C \frac{dz}{z^6+1} \tag{2}$$

The integrand has poles at

$$\alpha_k = e^{i \frac{\pi(2k+1)}{6}}, \qquad k = 0,1,2,3,4,5 \tag{3}$$

which are simple poles.

Only three poles

$$e^{i \frac{\pi}{6}}, \quad e^{i \frac{3\pi}{6}}, \quad e^{i \frac{5\pi}{6}} \tag{4}$$

are located inside C.

To compute the residues we shall use l'Hospital's rule.

$$\text{Res}_{e^{i\frac{\pi}{6}}} = \lim_{z \to e^{i\frac{\pi}{6}}} \left[\frac{z - e^{i\frac{\pi}{6}}}{z^6 + 1} \right] = \lim_{z \to e^{i\frac{\pi}{6}}} \frac{1}{6z^5} \qquad (5)$$

$$= \frac{1}{6} e^{-i\frac{5\pi}{6}}$$

$$\text{Res}_{e^{i\frac{3\pi}{6}}} = \lim_{z \to e^{i\frac{3\pi}{6}}} \frac{1}{6z^5} = \frac{1}{6} e^{-i\frac{5\pi}{2}} \qquad (6)$$

$$\text{Res}_{e^{i\frac{5\pi}{6}}} = \lim_{z \to e^{i\frac{5\pi}{6}}} \frac{1}{6z^5} = \frac{1}{6} e^{-i\frac{25\pi}{6}} \qquad (7)$$

Thus,

$$\int_C \frac{dz}{z^6 + 1} = 2\pi i \cdot \frac{1}{6} \left[e^{-i\frac{5\pi}{6}} + e^{-i\frac{5\pi}{6}} + e^{-i\frac{25\pi}{6}} \right]$$

$$= \frac{2\pi}{3} \qquad (8)$$

$$\int_C \frac{dz}{z^6 + 1} = \int_{-r}^{r} \frac{dz}{z^6 + 1} + \int_{\text{semicircle}} \frac{dz}{z^6 + 1} = \frac{2\pi}{3} \qquad (9)$$

Taking the limit of both sides of (9) we find (see Problem 19-3)

$$\lim_{r \to \infty} \int_{-r}^{r} \frac{dz}{z^6 + 1} + \lim_{r \to \infty} \int_{\text{semicircle}} \frac{dz}{z^6 + 1}$$

$$= \lim_{r \to \infty} \int_{-r}^{r} \frac{dz}{z^6 + 1} = \int_{-\infty}^{\infty} \frac{dx}{x^6 + 1} = \frac{2\pi}{3} \qquad (10)$$

Since the integrand in (1) is an even function,

$$\int_0^{\infty} \frac{dx}{x^6 + 1} = \frac{1}{2} \int_{-\infty}^{\infty} \frac{dx}{x^6 + 1} = \frac{\pi}{3} \qquad (11)$$

● **PROBLEM 19-5**

Evaluate:

$$\int_0^{\infty} \frac{2x^2 + 1}{x^4 + 4x^2 + 3} \, dx \qquad (1)$$

Solution: Since the integrand is an even function we have,

$$\int_0^\infty \frac{2x^2+1}{x^4+4x^2+3} dx = \frac{1}{2} \int_{-\infty}^\infty \frac{2x^2+1}{x^4+4x^2+3} dx \tag{2}$$

Consider the function

$$f(z) = \frac{2z^2+1}{z^4+4z^2+3} = \frac{2z^2+1}{(z^2+1)(z^2+3)} \tag{3}$$

and the contour C shown in Fig. 1.

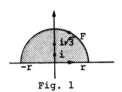

Fig. 1

Then,

$$\int_C f(z)dz = \int_F f(z)dz + \int_{-r}^r f(z)dz$$
$$= 2\pi i \; \Sigma \; \text{Res} \; f(z) \tag{4}$$

The integrand has two simple poles inside the contour C, at $z = i$ and $z = i\sqrt{3}$.

Hence,

$$\underset{i}{\text{Res}} \frac{2z^2+1}{(z^2+1)(z^2+3)} = \lim_{z \to i} \frac{2z^2+1}{(z+i)(z^2+3)}$$
$$= -\frac{1}{4i} \tag{5}$$

$$\underset{i\sqrt{3}}{\text{Res}} f(z) = \lim_{z \to i\sqrt{3}} \frac{2z^2+1}{(z^2+1)(z+i\sqrt{3})} = \frac{5}{4\sqrt{3} \; i} \tag{6}$$

Hence,

$$\int_F f(z)dz + \int_{-r}^r f(z)dz = 2\pi i \left(-\frac{1}{4i} + \frac{5}{4\sqrt{3}\;i} \right)$$
$$= \frac{5\pi}{2\sqrt{3}} - \frac{\pi}{2} \tag{7}$$

Taking the limit as $r \to \infty$ we obtain,

$$\int_{-\infty}^\infty f(z)dz = \frac{5\pi}{2\sqrt{3}} - \frac{\pi}{2} \tag{8}$$

and

$$\int_0^\infty \frac{2x^2+1}{x^4+4x^2+3} dx = \frac{5\pi}{4\sqrt{3}} - \frac{\pi}{4} \tag{9}$$

● **PROBLEM 19-6**

Evaluate:
$$\int_0^\infty \frac{\sin x}{x} dx \tag{1}$$

Solution: We shall consider the complex function

$$f(z) = \frac{e^{iz}}{z} \tag{2}$$

whose imaginary part coincides with the integrand (1) on the real axis.

Note, that we cannot employ the contour used in Problem 19-5. The function f(z) becomes infinite at the origin. We shall use the contour C shown in Fig. 1.

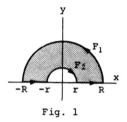

Fig. 1

The function f(z) is analytic inside C, thus

$$\int_C f(z)dz = \int_{-R}^{-r} \frac{e^{ix}}{x} dx + \int_{F_2} \frac{e^{iz}}{z} dz$$
$$+ \int_r^R \frac{e^{ix}}{x} dx + \int_{F_1} \frac{e^{iz}}{z} dz = 0 \tag{3}$$

Taking the limit of (3) as both

$$\begin{aligned} R &\to \infty \\ r &\to 0 \end{aligned} \tag{4}$$

we obtain,

$$\int_{-\infty}^0 \frac{e^{ix}}{x} dx + \lim_{r \to 0} \int_{F_2} \frac{e^{iz}}{z} dz + \int_0^\infty \frac{e^{ix}}{x} dx$$

$$+ \lim_{R\to\infty} \int_{F_1} \frac{e^{iz}}{z} dz = 0 \tag{5}$$

Integrating by parts we find

$$\int_{F_1} \frac{e^{iz}}{z} dz = \int_{F_1} \frac{d(e^{iz})}{iz} = \int_{F_1} \frac{e^{iz}}{iz^2} dz + \left.\frac{e^{iz}}{iz}\right|_{-R}^{R}$$

$$= \frac{1}{i} \int_{F_1} \frac{e^{iz}}{z^2} dz + \frac{e^{iR} + e^{-iR}}{iR} \tag{6}$$

and

$$\left|\int_{F_1} \frac{e^{iz}}{z} dz\right| \leq \left|\int_{F_1} \frac{e^{iz}}{z^2} dz\right| + \left|\frac{e^{iR} + e^{-iR}}{iR}\right|$$

$$\leq \frac{\pi R}{R^2} + \frac{2}{R} \xrightarrow[R\to\infty]{} 0 \tag{7}$$

Hence,

$$\lim_{R\to\infty} \int_{F_1} \frac{e^{iz}}{z} dz = 0 \tag{8}$$

This result can be obtained directly from the theorem proved in Problem 19-3.

For $z = re^{i\theta}$ we have

$$\left|\frac{e^{iz}}{z}\right| = \left|\frac{e^{ir(\cos\theta + i\sin\theta)}}{r(\cos\theta + i\sin\theta)}\right|$$

$$= \frac{1}{re^{r\sin\theta}} \leq \frac{M}{r^k} \tag{9}$$

and that leads to (8).

To evaluate

$$\lim_{r\to 0} \int_{F_2} \frac{e^{iz}}{z} dz \tag{10}$$

we take the Laurent expansion of the integrand, or

$$\frac{e^{iz}}{z} = \frac{1 + iz + \frac{(iz)^2}{2!} + \cdots}{z} = \frac{1}{z} + F(z), \tag{11}$$

where $F(z)$ is analytic at $z = 0$.

Thus,

$$\lim_{r\to 0} \int_{F_2} \frac{e^{iz}}{z} dz = \lim_{r\to 0} \int_{F_2} \frac{dz}{z} + \lim_{r\to 0} \int_{F_2} F(z) dz$$

$$= \lim_{r \to 0} \int_{F_2} \frac{dz}{z} \tag{12}$$

$$\int_{F_2} \frac{dz}{z} = \int_{\pi}^{0} \frac{ire^{i\theta}}{re^{i\theta}} d\theta = -i\pi \tag{13}$$

Hence,

$$\lim_{r \to 0} \int_{F_2} \frac{e^{iz}}{z} dz = -\pi i \tag{14}$$

Substituting (8), (14) into (5) we get

$$\int_{-\infty}^{0} \frac{e^{ix}}{x} dx + \int_{0}^{\infty} \frac{e^{ix}}{x} dx = \pi i \tag{15}$$

Taking the imaginary part

$$\int_{-\infty}^{0} \frac{\sin x}{x} dx + \int_{0}^{\infty} \frac{\sin x}{x} dx = \pi \tag{16}$$

or

$$\int_{0}^{\infty} \frac{\sin x}{x} dx = \frac{\pi}{2} \tag{17}$$

• **PROBLEM 19-7**

Show that:

$$\int_{0}^{\infty} \frac{x^{\alpha-1}}{x+1} dx = \frac{\pi}{\sin \alpha \pi}, \qquad 0 < \alpha < 1 \tag{1}$$

Solution: Consider the integral

$$\int_{C} \frac{z^{\alpha-1}}{z+1} dz. \tag{2}$$

where C is the contour shown in Fig. 1.

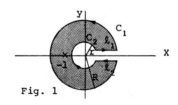

Fig. 1

Point $z = 0$ is a branch point and lines ℓ_1 and ℓ_2 coincide with the x axis. The integrand has one simple pole at $z = -1$ inside C. Thus,

$$\operatorname*{Res}_{-1} \frac{z^{\alpha-1}}{z+1} = \lim_{z \to -1} \frac{z^{\alpha-1}}{z+1} (z+1) = (e^{\pi i})^{\alpha-1}$$
$$= e^{(\alpha-1)\pi i} \tag{3}$$

and

$$\int_C \frac{z^{\alpha-1}}{z+1} dz = 2\pi i \, e^{(\alpha-1)\pi i} \tag{4}$$

On the other hand,

$$\int_{C_1} \frac{z^{\alpha-1}}{z+1} dz + \int_{C_2} \frac{z^{\alpha-1}}{z+1} dz + \int_{\ell_1} \frac{z^{\alpha-1}}{z+1} dz + \int_{\ell_2} \frac{z^{\alpha-1}}{z+1} dz$$
$$= 2\pi i \, e^{(\alpha-1)\pi i} \tag{5}$$

$$\int_0^{2\pi} \frac{(Re^{i\theta})^{\alpha-1}}{Re^{i\theta}+1} iR\, e^{i\theta} d\theta + \int_{2\pi}^0 \frac{(re^{i\theta})^{\alpha-1}}{re^{i\theta}+1} ir\, e^{i\theta} d\theta$$
$$+ \int_r^R \frac{x^{\alpha-1}}{x+1} dx + \int_R^r \frac{(xe^{2\pi i})^{\alpha-1}}{xe^{2\pi i}+1} dx = 2\pi i \, e^{(\alpha-1)\pi i} \tag{6}$$

Taking the limit as

$$r \to 0$$
$$R \to \infty \tag{7}$$

we get,

$$\int_0^{\infty} \frac{x^{\alpha-1}}{x+1} dx + \int_{\infty}^0 \frac{e^{2\pi i (\alpha-1)} x^{\alpha-1}}{x+1} dx$$
$$= 2\pi i \, e^{(\alpha-1)\pi i} \tag{8}$$

or

$$\left[1 - e^{2\pi i(\alpha-1)}\right] \int_0^{\infty} \frac{x^{\alpha-1}}{x+1} dx = 2\pi i \, e^{(\alpha-1)\pi i} \tag{9}$$

Hence,

$$\int_0^{\infty} \frac{x^{\alpha-1}}{x+1} dx = \frac{2\pi i \, e^{(\alpha-1)\pi i}}{1 - e^{(\alpha-1)2\pi i}} = \frac{\pi}{\sin \alpha \pi} \tag{10}$$

• **PROBLEM 19-8**

Evaluate the Fresnel integrals:

$$\int_0^\infty \sin x^2 dx, \quad \int_0^\infty \cos x^2 dx \tag{1}$$

<u>Solution</u>: We shall choose the integrand

$$f(z) = e^{iz^2}$$

and the contour shown in Fig. 1.

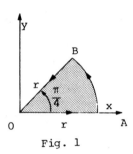

Fig. 1

By Cauchy's theorem

$$\int_C e^{iz^2} dz = 0 \tag{2}$$

or

$$\int_0^r e^{ix^2} dx + \int_{AB} e^{iz^2} dz + \int_r^0 e^{-r^2} \sqrt{i}\, dr = 0 \tag{3}$$

Note that along BO line $z = r\left(\dfrac{1}{\sqrt{2}} + i\dfrac{1}{\sqrt{2}}\right) = r\sqrt{i}$.

Therefore

$$e^{iz^2} = e^{-r^2}. \tag{4}$$

Taking the limit as $r \to \infty$ we find,

$$\int_0^\infty e^{ix^2} dx + \lim_{r\to\infty} \int_{AB} e^{iz^2} dz - \sqrt{i} \int_0^\infty e^{-r^2} dr = 0 \tag{5}$$

But

$$\int_{AB} e^{iz^2} dz = \int_{AB} \frac{d(e^{iz^2})}{2iz}$$

$$= \frac{e^{iz^2}}{2iz}\Bigg|_r^{\sqrt{i}\,r} + \frac{1}{2i}\int_{AB}\frac{e^{iz^2}}{z^2}dz \qquad (6)$$

Observe that, for $z \in AB$

$$\left|\frac{e^{-r^2}}{2i\sqrt{i}\,r} - \frac{e^{ir^2}}{2ir}\right| \leq \frac{e^{-r^2}+1}{2r} \xrightarrow[r\to\infty]{} 0 \qquad (7)$$

and

$$\left|\frac{e^{iz^2}}{z^2}\right| = \left|\frac{e^{ir^2(\cos 2\theta + i\sin 2\theta)}}{z^2}\right| = \frac{e^{-r^2\sin 2\theta}}{r^2} \qquad (8)$$

For $z \in AB$, we have

$$z = r(\cos\theta + i\sin\theta) \text{ and}$$

$$0 \leq \sin 2\theta \quad e^{-r^2\sin 2\theta} \leq 1 \qquad (9)$$

Hence,

$$\left|\frac{e^{iz^2}}{z^2}\right| \leq \frac{1}{r^2} \qquad (10)$$

and

$$\left|\int_{AB}\frac{e^{iz^2}}{z}dz\right| \leq \frac{1}{r^2}\frac{\pi r}{4} = \frac{\pi}{4r} \xrightarrow[r\to\infty]{} 0 \qquad (11)$$

Hence,

$$\lim_{r\to\infty}\int_{AB} e^{iz^2}dz = 0 \qquad (12)$$

From (5) and (12) we obtain,

$$\int_0^\infty e^{ix^2}dx = \sqrt{i}\int_0^\infty e^{-r^2}dr \qquad (13)$$

Since,

$$\int_0^\infty e^{-r^2}dr = \frac{\sqrt{\pi}}{2} \qquad (14)$$

we get

$$\int_0^\infty e^{ix^2}dx = \sqrt{i}\,\frac{\sqrt{\pi}}{2} = \frac{\sqrt{\pi}}{2}\frac{1+i}{\sqrt{2}} \qquad (15)$$

Taking the real and imaginary parts of (15) we get

$$\int_0^\infty \cos x^2\, dx = \frac{1}{2}\sqrt{\frac{\pi}{2}}$$

$$\int_0^\infty \sin x^2 \, dx = \frac{1}{2}\sqrt{\frac{\pi}{2}} \tag{16}$$

• **PROBLEM 19-9**

Show that:

$$\int_0^\infty \frac{\cosh \alpha x}{\cosh x} \, dx = \frac{\pi}{2\cos\left(\frac{\pi\alpha}{2}\right)} \tag{1}$$

where $-1 < \alpha < 1$.

Solution: We shall take the function

$$f(z) = \frac{e^{\alpha z}}{\cosh z} \tag{2}$$

Its singularities are simple poles at

$$z = \left(k + \frac{1}{2}\right)\pi i, \qquad k = 0, \pm 1, \pm 2, \ldots \tag{3}$$

We shall choose a contour which contains one singularity at $z = \frac{\pi i}{2}$, as shown in Fig. 1.

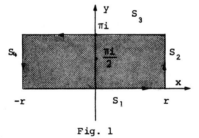

Fig. 1

We have,

$$\int_C \frac{e^{\alpha z}}{\cosh z} \, dz = 2\pi i \operatorname*{Res}_{\frac{\pi i}{2}} \frac{e^{\alpha z}}{\cosh z}$$

$$= 2\pi i \left[\frac{e^{\alpha z}}{(\cosh z)'}\right]\bigg|_{z=\frac{\pi i}{2}} = 2\pi i \frac{e^{\frac{\alpha \pi i}{2}}}{\sinh \frac{\pi i}{2}} \tag{4}$$

$$= \int_{S_1} + \int_{S_2} + \int_{S_3} + \int_{S_4}$$

$$= 2\pi i \frac{e^{\frac{\alpha \pi i}{2}}}{i \sin \frac{\pi}{2}} = 2\pi e^{\frac{\alpha \pi i}{2}}$$

or

$$\int_{-r}^{r} \frac{e^{\alpha x}}{\cosh x} dx + \int_{0}^{\pi} \frac{e^{\alpha(r+iy)}}{\cosh(r+iy)} i\,dy + \int_{r}^{-r} \frac{e^{\alpha(x+\pi i)}}{\cosh(x+\pi i)} dx$$

$$+ \int_{\pi}^{0} \frac{e^{\alpha(-r+iy)}}{\cosh(-r+iy)} i\,dy = 2\pi e^{\frac{\alpha \pi i}{2}} \quad (5)$$

We shall show that

$$\lim_{r \to \infty} \int_{S_2} = \lim_{r \to \infty} \int_{S_4} = 0 \quad (6)$$

Indeed,

$$|\cosh(r+iy)| = \left| \frac{e^{r+iy} + e^{-r-iy}}{2} \right|$$

$$\geq \frac{1}{2} \left(\left| e^{r+iy} \right| - \left| e^{-r-iy} \right| \right) = \frac{1}{2}(e^r - e^{-r}) \geq \frac{1}{4} e^r \quad (7)$$

and

$$\left| \int_{0}^{\pi} \frac{e^{\alpha(r+iy)}}{\cosh(r+iy)} i\,dy \right| \leq \int_{0}^{\pi} \frac{e^{\alpha r}}{\frac{1}{4} e^r} dy$$

$$= 4\pi e^{r(\alpha-1)} \xrightarrow[r \to \infty]{} 0 \quad (8)$$

In a similar manner we prove the second part of (6).

Taking the limit of (5) as $r \to \infty$ we get,

$$\int_{-\infty}^{\infty} \frac{e^{\alpha x}}{\cosh x} dx + e^{\alpha \pi i} \int_{-\infty}^{\infty} \frac{e^{\alpha x}}{\cosh x} dx = 2\pi e^{\frac{\alpha \pi i}{2}} \quad (9)$$

Note that in (9) we used

$$\cosh(x+\pi i) = -\cosh x \quad (10)$$

$$\int_{0}^{\infty} \frac{e^{\alpha x}}{\cosh x} dx = \frac{2\pi e^{\frac{\alpha \pi i}{2}}}{1 + e^{\alpha \pi i}}$$

$$= \frac{2\pi}{e^{\frac{\alpha \pi i}{2}} + e^{\frac{-\alpha \pi i}{2}}} = \frac{\pi}{\cos\left(\frac{\pi \alpha}{2}\right)} \quad (11)$$

Eq.(11) can be written as

$$\int_{-\infty}^{0} \frac{e^{\alpha x}}{\cosh x} dx + \int_{0}^{\infty} \frac{e^{\alpha x}}{\cosh x} dx$$

$$= \int_0^\infty \frac{e^{-\alpha x}}{\cosh x} dx + \int_0^\infty \frac{e^{\alpha x}}{\cosh x} dx \qquad (12)$$

$$= 2 \int_0^\infty \frac{\cosh \alpha x}{\cosh x} dx = \frac{\pi}{\cos\left(\frac{\pi \alpha}{2}\right)}$$

In the first step we replaced x by -x in $\int_{-\infty}^0 \frac{e^{\alpha x}}{\cosh x} dx$.

● **PROBLEM 19-10**

Show that:

$$\int_{-\infty}^\infty \frac{e^{\alpha x}}{e^x + 1} dx = \frac{\pi}{\sin \alpha \pi}, \qquad 0 < \alpha < 1 \qquad (1)$$

Solution: Consider the function

$$f(z) = \frac{e^{\alpha z}}{e^z + 1} \qquad (2)$$

Now, the problem is to find the appropriate contour.

The function $f(z)$ has singularities at $z = \pi i(2k+1)$ where $k = 0, \pm 1, \pm 2, \ldots$.

The best contour, for our purpose, containing one singularity, for example, $z = \pi i$ is the rectangle shown in Fig. 1.

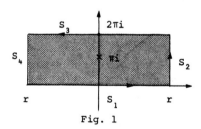

Fig. 1

It consists of four segments S_1, S_2, S_3, S_4. By the residue theorem

$$\int_{S_1} + \int_{S_2} + \int_{S_3} + \int_{S_4} = 2\pi i \operatorname*{Res}_{\pi i} f(z) \qquad (3)$$

$$= 2\pi i \lim_{z \to \pi i} \frac{e^{\alpha z}}{(e^z + 1)'} = -2\pi i \, e^{\alpha \pi i}$$

$$\int_{S_1} f(z)dz = \int_{-r}^{r} \frac{e^{\alpha x}}{e^x+1} dx \qquad (4)$$

$$\int_{S_3} f(z)dz = \int_{r}^{-r} \frac{e^{\alpha(x+2\pi i)}}{e^{x+2\pi i}+1} dx = -e^{2\alpha\pi i} \int_{-r}^{r} \frac{e^{\alpha x}}{e^x+1} dx \qquad (5)$$

For $z \in S_2$ we have

$$|f(z)| = \left|\frac{e^{\alpha(r+iy)}}{e^{r+iy}+1}\right| \leq \frac{e^{\alpha r}}{e^r-1} = \frac{e^{(\alpha-1)r}}{1-e^{-r}} \qquad (6)$$

and for $z \in S_4$

$$|f(z)| = \left|\frac{e^{\alpha(-r+iy)}}{e^{-r+iy}+1}\right| \leq \frac{e^{-\alpha r}}{1-e^{-r}} \qquad (7)$$

Thus,

$$\left|\int_{S_2} f(z)dz\right| \leq \frac{e^{(\alpha-1)r}}{1-e^{-r}} 2\pi \xrightarrow[r\to\infty]{} 0 \qquad (8)$$

and

$$\left|\int_{S_4} f(z)dz\right| \leq \frac{e^{-\alpha r}}{1-e^{-r}} 2\pi \xrightarrow[r\to\infty]{} 0 \qquad (9)$$

where $0 < \alpha < 1$

Taking the limit of (3) as $r \to \infty$ we obtain

$$\int_{-\infty}^{\infty} \frac{e^{\alpha x}}{e^x+1} dx - e^{2\alpha\pi i} \int_{-\infty}^{\infty} \frac{e^{\alpha x}}{e^x+1} dx = -2\pi i\, e^{\alpha\pi i} \qquad (10)$$

or

$$\int_{-\infty}^{\infty} \frac{e^{\alpha x}}{e^x+1} dx = \frac{-2\pi i\, e^{\alpha\pi i}}{1-e^{2\alpha\pi i}} = \frac{\pi}{\sin\alpha\pi}, \qquad 0 < \alpha < 1 \qquad (11)$$

● **PROBLEM 19-11**

Show that:

$$\int_0^{\infty} \frac{\ln(x^2+1)}{x^2+1} dx = \pi \ln 2 \qquad (1)$$

<u>Solution</u>: At first glance it looks like we should consider the function

$$\frac{\ln(z^2+1)}{z^2+1}$$

That leads to some problems in evaluation of the residue at the simple pole z = i (we obtain $\ln(i^2+1)$).

Instead, we shall try the function

$$\frac{\ln(z+i)}{z^2+1} \qquad (2)$$

computed around the contour C shown in Fig. 1.

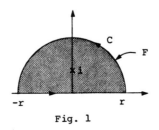

Fig. 1

The only pole inside C is a simple pole at z = i.

$$\int_C \frac{\ln(z+i)}{z^2+1} \, dz = 2\pi i \operatorname*{Res}_i$$

$$= 2\pi i \left[\frac{\ln(z+i)}{(z^2+1)'} \right]_{z=i} \qquad (3)$$

$$= 2\pi i \, \frac{\ln 2i}{2i} = \pi \ln(2i) = \pi \ln 2 + \pi \ln i$$

$$= \pi \ln 2 + \pi \ln e^{\frac{\pi i}{2}} = \pi \ln 2 + \frac{1}{2}\pi^2 i$$

Eq.(3) can be written

$$\int_{-r}^{r} \frac{\ln(x+i)}{x^2+1} \, dx + \int_F \frac{\ln(z+i)}{z^2+1} \, dz = \pi \ln 2 + \frac{1}{2}\pi^2 i \qquad (4)$$

or

$$\int_{-r}^{0} \frac{\ln(x+i)}{x^2+1} \, dx + \int_0^r \frac{\ln(x+i)}{x^2+1} \, dx + \int_F \frac{\ln(z+i)}{z^2+1} \, dz$$

$$\qquad (5)$$

$$= \int_0^r \frac{\ln(i-x)}{x^2+1} + \int_0^r \frac{\ln(i+x)}{x^2+1} \, dx + \int_F \frac{\ln(z+i)}{z^2+1}$$

$$= \pi \ln 2 + \frac{1}{2}\pi^2 i$$

Since,
$$\ln(i-x) + \ln(i+x) = \ln(i^2-x^2) = \ln(1+x^2) + \pi i \qquad (6)$$
we obtain
$$\int_0^r \frac{\ln(x^2+1)}{x^2+1} dx + \int_0^r \frac{\pi i}{x^2+1} dx + \int_F \frac{\ln(z+i)}{z^2+1} dz = \pi \ln 2 + \frac{1}{2}\pi^2 i \qquad (7)$$

Taking the limit of (7) as $r \to \infty$ we find

$$\int_0^\infty \frac{\ln(x^2+1)}{x^2+1} dx + \int_0^\infty \frac{\pi i}{x^2+1} dx = \pi \ln 2 + \frac{1}{2}\pi^2 i \qquad (8)$$

It can be shown that
$$\lim_{r \to \infty} \int_F \frac{\ln(z+i)}{z^2+1} dz = 0 \qquad (9)$$

Taking the real part of (8)
$$\int_0^\infty \frac{\ln(x^2+1)}{x^2+1} dx = \pi \ln 2 \qquad (10)$$

As a by-product from (8) we get
$$\int_0^\infty \frac{1}{x^2+1} dx = \frac{\pi}{2} \qquad (11)$$

● **PROBLEM 19-12**

Show that:
$$\int_0^{\pi/2} \ln \sin x \, dx = \int_0^{\pi/2} \ln \cos x \, dx \qquad (1)$$
$$= -\frac{1}{2} \pi \ln 2$$

Solution: We shall use eq.(1) from Problem 19-10

$$\int_0^\infty \frac{\ln(x^2+1)}{x^2+1} dx = \pi \ln 2 \qquad (2)$$

Setting
$$x = \tan\alpha \tag{3}$$

in (2) and noting that as x goes from 0 to ∞, α changes from 0 to $\frac{\pi}{2}$, we get

$$\int_0^{\frac{\pi}{2}} \frac{\ln(\tan^2\alpha+1)}{\tan^2\alpha+1} \cdot d(\tan\alpha)d\alpha \tag{4}$$

$$= \int_0^{\frac{\pi}{2}} \frac{\ln(\tan^2\alpha+1)}{\tan^2\alpha+1} \sec^2\alpha \, d\alpha$$

$$= -2 \int_0^{\frac{\pi}{2}} \ln\cos\alpha \, d\alpha = \pi\ln 2$$

or

$$\int_0^{\frac{\pi}{2}} \ln\cos\alpha \, d\alpha = -\frac{1}{2}\pi\ln 2 \tag{5}$$

In the same way setting

$$\alpha = \frac{\pi}{2} - \theta \tag{6}$$

in (5) we obtain

$$\int_0^{\frac{\pi}{2}} \ln\sin\theta \, d\theta = -\frac{1}{2}\pi\ln 2 \tag{7}$$

● **PROBLEM 19-13**

Show that:

$$\int_0^\infty \frac{(\ln x)^2}{x^2+1} dx = \frac{\pi^3}{8} \tag{1}$$

Solution: We shall consider the function

$$f(z) = \frac{(\ln z)^2}{z^2+1} \tag{2}$$

evaluated along the contour C shown in Fig. 1. Note that F_1 and F_2 are semicircles.

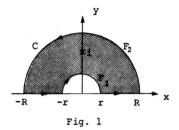

Fig. 1

The integrand of (2) has a simple pole at $z = i$ inside C. By the residue theorem

$$\int_C \frac{(\ln z)^2}{z^2+1} dz = 2\pi i \operatorname{Res}_i f(z)$$

$$= 2\pi i \left[\frac{(\ln z)^2}{2z}\right]_{z=i} = 2\pi i \frac{(\ln i)^2}{2i} \quad (3)$$

$$= 2\pi i \cdot \frac{\left(\frac{\pi i}{2}\right)^2}{2i} = -\frac{\pi^3}{4}$$

Eq.(3) can be written as

$$\int_{F_2} f(z)dz + \int_{-R}^{-r} f(z)dz + \int_{F_1} f(z)dz + \int_r^R f(z)dz$$
$$= -\frac{\pi^3}{4} \quad (4)$$

Replacing z by -u in the second integral we get

$$\ln z = \ln(-u) = \ln u + \ln(-1) = \pi i + \ln u \quad (5)$$

We have

$$\int_{F_2} \frac{(\ln z)^2}{z^2+1} dz + \int_r^R \frac{(\pi i + \ln u)^2}{u^2+1} du + \int_{F_1} \frac{(\ln z)^2}{z^2+1} dz$$

$$+ \int_r^R \frac{(\ln u)^2}{u^2+1} du = -\frac{\pi^3}{4} \quad (6)$$

For clarity in the last integral we replaced z by u.

Taking the limit of (6) as $r \to 0$ and $R \to \infty$ and observing that

$$\lim_{r \to 0} \int_{F_1} f(z)dz = \lim_{R \to \infty} \int_{F_2} f(z)dz = 0 \quad (7)$$

we obtain

$$\int_0^\infty \frac{(\pi i + \ln u)^2}{u^2+1} du + \int_0^\infty \frac{(\ln u)^2}{u^2+1} du =$$

$$2\int_0^\infty \frac{(\ln u)^2}{u^2+1} du + 2\pi i \int_0^\infty \frac{\ln u}{u^2+1} du \qquad (8)$$

$$-\pi^2 \int_0^\infty \frac{du}{u^2+1} = -\frac{\pi^3}{4}$$

$$\int_0^\infty \frac{du}{u^2+1} = \arc\tan u \Big|_0^\infty = \frac{\pi}{2} \qquad (9)$$

Finally, we obtain

$$\int_0^\infty \frac{(\ln u)^2}{u^2+1} du + \pi i \int_0^\infty \frac{\ln u}{u^2+1} du = \frac{\pi^3}{8} \qquad (10)$$

Comparing the real and imaginary parts we obtain:

$$\int_0^\infty \frac{(\ln x)^2}{x^2+1} dx = \frac{\pi^3}{8} \quad \text{and}$$

$$\int_0^\infty \frac{\ln x}{x^2+1} dx = 0 \qquad (11)$$

● **PROBLEM 19-14**

Show that:

$$\int_0^\infty \frac{\sin ax}{x(x^2+b^2)^2} dx = \frac{\pi}{2b^4}\left[1 - \frac{e^{-ab}}{2}(ab+2)\right] \qquad (1)$$

$$a > 0, \quad b > 0.$$

Solution: We shall consider the function

$$f(z) = \frac{e^{iaz}}{z(z^2+b^2)^2} \qquad (2)$$

This function has a simple pole on the real axis at $z = 0$ and one double pole at $z = bi$ in the upper half-plane.

The residues are,

$$\operatorname{Res}_0 f(z) = \lim_{z \to 0} \frac{e^{iaz}}{(z^2+b^2)^2} = \frac{1}{b^4} \qquad (3)$$

and

$$\text{Res } f(z) \bigg|_{bi} = \lim_{z \to bi} \frac{d}{dz}\left[\frac{e^{iaz}}{z(z^2+b^2)^2}(z-ib)^2\right] \quad (4)$$

$$= \lim_{z \to bi} \frac{d}{dz}\left[\frac{e^{iaz}}{z(z+ib)^2}\right] = \frac{-e^{-ab}}{4b^4}(ab+2)$$

Applying the formula

$$\int_{-\infty}^{\infty} f(z)dz = 2\pi i \sum_{C_t} \text{Res } f(z_n) + \pi i \sum_{x\text{-axis}} \text{Res } f(x), \quad (5)$$

we obtain

$$\int_{-\infty}^{\infty} \frac{e^{iaz}}{z(z^2+b^2)^2} dz = 2\pi i \frac{-e^{-ab}}{4b^4}(ab+2) + \pi i \cdot \frac{1}{b^4}$$

$$= \frac{i\pi}{b^4}\left[1 - \frac{e^{-ab}}{2}(ab+2)\right] \quad (6)$$

Comparing the imaginary parts of (6) we get

$$\int_{-\infty}^{\infty} \frac{\sin az}{z(z^2+b^2)^2} dz = \frac{\pi}{b^4}\left[1 - \frac{e^{-ab}}{2}(ab+2)\right] \quad (7)$$

Since the integrand is an even function,

$$\int_{0}^{\infty} \frac{\sin ax}{x(x^2+b^2)^2} dx = \frac{\pi}{2b^4}\left[1 - \frac{e^{-ab}}{2}(ab+2)\right] \quad (8)$$

• **PROBLEM 19-15**

Evaluate the integral:

$$\int_{0}^{\infty} \frac{\ln x}{(x^2+1)^2} dx \quad (1)$$

Solution: We shall take the function

$$f(z) = \frac{\ln z}{(z^2+1)^2} \quad (2)$$

and the contour C shown in Fig. 1. We choose ln z to be the branch which satisfies

$$-\pi < \text{Im ln } z = \arg z \leq \pi \quad (3)$$

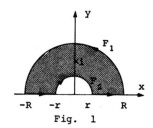

Fig. 1

The function $f(z)$ is analytic everywhere in and on C except $z=i$, where $f(z)$ has a pole of order 2.

$$\operatorname*{Res}_{i} f(z) = \lim_{z \to i} \frac{d}{dz}\left[\frac{\ln z}{(z+i)^2}\right] = \frac{\pi+2i}{8} \qquad (4)$$

By the residue theorem

$$\int_{F_1} f(z)dz + \int_{-R}^{-r} f(z)dz + \int_{F_2} f(z)dz + \int_{r}^{R} f(z)dz$$

$$= 2\pi i\left(\frac{\pi+2i}{8}\right) = \frac{\pi^2 i}{4} - \frac{\pi}{2} \qquad (5)$$

For the points on F_1

$$z = Re^{i\theta}, \quad 0 \leq \theta \leq \pi \quad \text{and}$$

$$|\ln z| = |\ln|z| + i \arg z| = \sqrt{\ln^2 R + \theta^2}$$
$$\leq \sqrt{\ln^2 R + \pi^2} \leq 2 \ln R \qquad (6)$$

As $R \to \infty$

$$\left|\int_{F_1} f(z)dz\right| \leq \frac{2 \ln R}{(R^2-1)^2} \pi R \xrightarrow[R \to \infty]{} 0 \qquad (7)$$

For the points on F_2

$$z = re^{i\theta}, \quad 0 \leq \theta \leq \pi \qquad (8)$$

and

$$|\ln z| \leq 2 \ln \frac{1}{r} \qquad (9)$$

For $r \to 0$

$$\left|\int_{F_2} f(z)dz\right| \leq \frac{2 \ln \frac{1}{r}}{(1-r^2)^2} \pi r \to 0 \qquad (10)$$

Taking the limit of (5) as $r \to 0$ and $R \to \infty$ we obtain

$$\int_{-\infty}^{0} \frac{\ln x}{(x^2+1)^2} dx + \int_{0}^{\infty} \frac{\ln x}{(x^2+1)^2} dx = \frac{\pi^2 i}{4} - \frac{\pi}{2} \qquad (11)$$

Since,
$$\ln(-x) = \ln x + \pi i \qquad (12)$$
we get

$$\int_{-\infty}^{0} \frac{\ln x}{(x^2+1)^2} dx = \int_{0}^{\infty} \frac{\ln x}{(x^2+1)^2} dx + \pi i \int_{0}^{\infty} \frac{dx}{(x^2+1)^2} \qquad (13)$$

and

$$2 \int_{0}^{\infty} \frac{\ln x}{(x^2+1)^2} dx + \pi i \int_{0}^{\infty} \frac{dx}{(x^2+1)^2}$$
$$= \frac{\pi^2 i}{4} - \frac{\pi}{2} \qquad (14)$$

Comparing the real and imaginary parts we find

$$\int_{0}^{\infty} \frac{\ln x}{(x^2+1)^2} dx = -\frac{\pi}{4}$$

$$\int_{0}^{\infty} \frac{dx}{(x^2+1)^2} = \frac{\pi}{4} \qquad (15)$$

• **PROBLEM 19-16**

Evaluate the integral

$$\int_{\alpha-i\infty}^{\alpha+i\infty} \frac{e^{zt}}{\sqrt{z+1}} dt, \qquad (1)$$

where α and t are positive constants.

Solution: At $z = -1$ the integrand has a branch point. As a branch line we choose the part of the real axis

$$x < -1 \qquad (2)$$

The contour C cannot cross the branch line, see Fig. 1.

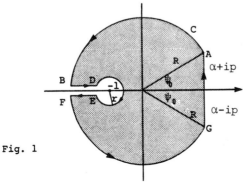

Fig. 1

Inside and on C the function $\dfrac{e^{zt}}{\sqrt{z+1}}$ is analytic; thus by Cauchy's theorem

$$\int_C \frac{e^{zt}}{\sqrt{z+1}} \, dz = 0 \tag{3}$$

or

$$\int_{AB} + \int_{BD} + \int_{DE} + \int_{EF} + \int_{FG} + \int_{GA} = 0 \tag{4}$$

For the points on arcs AB and FG

$$z = Re^{i\psi}, \quad dz = iRe^{i\psi} \tag{5}$$

Thus,

$$\int_{AB} \frac{e^{zt}}{\sqrt{z+1}} \, dz = \int_{\psi_0}^{\pi} \frac{e^{Re^{i\psi}t}}{\sqrt{Re^{i\psi}+1}} \, iRe^{i\psi} d\psi \tag{6}$$

$$\int_{FG} \frac{e^{zt}}{\sqrt{z+1}} \, dz = \int_{\pi}^{2\pi-\psi_0} \frac{e^{Re^{i\psi}t}}{\sqrt{Re^{i\psi}+1}} \, iRe^{i\psi} d\psi \tag{7}$$

Along DE $z+1 = re^{i\phi}$ \hfill (8)

and

$$\int_{DE} \frac{e^{zt}}{\sqrt{z+1}} \, dz = \int_{\pi}^{-\pi} \frac{e^{(re^{i\phi}-1)t}}{\sqrt{re^{i\phi}}} \cdot ire^{i\phi} d\phi \tag{9}$$

Along BD

$$z+1 = u\,e^{\pi i}$$

and

$$\sqrt{z+1} = \sqrt{u}\, e^{\frac{\pi}{2}i} = i\sqrt{u} \tag{10}$$

$$\int_{BD} \frac{e^{zt}}{\sqrt{z+1}} \, dz = -\int_{R-1}^{r} \frac{e^{(u+1)(-t)}}{i\sqrt{u}} \, du \tag{11}$$

Along EF

$$z+1 = u\,e^{-\pi i}$$

and

$$\sqrt{z+1} = \sqrt{u}\, e^{-\frac{\pi}{2}i} = -i\sqrt{u} \tag{12}$$

$$\int_{EF} \frac{e^{zt}}{\sqrt{z+1}} \, dz = -\int_{r}^{R-1} \frac{e^{(u+1)(-t)}}{-i\sqrt{u}} \, du \tag{13}$$

Taking the limit of (u) as

$$R \to \infty$$

and (14)
$$r \to 0$$

we find that integrals along AB, DE and FG approach zero.

Hence,
$$\lim_{\substack{R\to\infty \\ r\to 0}} \int_{BD} + \int_{EF} + \int_{G}^{A} = 0 \tag{15}$$

or

$$\int_{\alpha-i\infty}^{\alpha+i\infty} \frac{e^{zt}}{\sqrt{z+1}} dz = \lim_{\substack{R\to\infty \\ r\to 0}} 2i \int_{r}^{R-1} \frac{e^{-(u+1)t}}{\sqrt{u}} du$$
$$= 2i \int_{0}^{\infty} \frac{e^{-(u+1)t}}{\sqrt{u}} du \tag{16}$$

Substituting $u = \omega^2$

we get

$$\int_{\alpha-i\infty}^{\alpha+i\infty} \frac{e^{zt}}{\sqrt{z+1}} dz = 2ie^{-t} \int_{0}^{\infty} \frac{e^{-\omega^2 t}}{\omega} \cdot 2\omega d\omega$$
$$= \frac{e^{-t}}{\sqrt{\pi t}} 2\pi i = \frac{2i\sqrt{\pi}}{\sqrt{t}} e^{-t} \tag{17}$$

• **PROBLEM 19-17**

Let S_n be a square with vertices as indicated in Fig. 1.

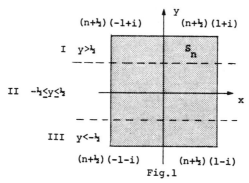

Fig.1

Show that on S_n

$$|\cot \pi z| < M, \tag{1}$$

where M is a constant.

Solution: The relation $|\cot \pi z| < M$ will be proved separately for the three regions indicated in Fig. 1.

I. For $y > \frac{1}{2}$ and $z = x + iy$, we have

$$|\cot \pi z| = \left|\frac{\cos \pi z}{\sin \pi z}\right| = \left|\frac{e^{i\pi z} + e^{-i\pi z}}{e^{i\pi z} - e^{-i\pi z}}\right|$$

$$\leq \frac{|e^{i\pi z}| + |e^{-i\pi z}|}{|e^{-i\pi z}| - |e^{i\pi z}|} = \frac{e^{-\pi y} + e^{\pi y}}{e^{\pi y} - e^{-\pi y}} \quad (2)$$

$$= \frac{1 + e^{-2\pi y}}{1 - e^{-2\pi y}} \leq \frac{1 + e^{-\pi}}{1 - e^{-\pi}} = m_1$$

II. For $-\frac{1}{2} \leq y \leq \frac{1}{2}$, first take $z = n + \frac{1}{2} + iy$, then

$$|\cot \pi z| = |\cot \pi(n + \frac{1}{2} + iy)|$$

$$= \left|\cot\left(\frac{\pi}{2} + i\pi y\right)\right| = |\tanh \pi y| \leq \tanh \frac{\pi}{2} = m_2 \quad (3)$$

For $z = -n - \frac{1}{2} + iy$ we obtain

$$|\cot \pi z| = |\cot \pi(-n - \frac{1}{2} + iy)|$$

$$= \left|\cot\left(-\frac{\pi}{2} + i\pi y\right)\right| = |\tanh \pi y| \leq \tanh \frac{\pi}{2} = m_2 \quad (4)$$

III. For $y < -\frac{1}{2}$ we have from (2)

$$|\cot \pi z| \leq \frac{|e^{i\pi x - \pi y}| + |e^{-i\pi x + \pi y}|}{|e^{i\pi x - \pi y}| - |e^{-i\pi x + \pi y}|} \quad (5)$$

$$= \frac{e^{-\pi y} + e^{\pi y}}{e^{-\pi y} - e^{\pi y}} = \frac{1 + e^{2\pi y}}{1 - e^{2\pi y}} \leq \frac{1 + e^{-\pi}}{1 - e^{-\pi}} = m_1$$

Setting

$$M = \max(m_1, m_2) \quad (6)$$

we complete the proof.

• **PROBLEM 19-18**

A square S_N is shown in Fig. 1.

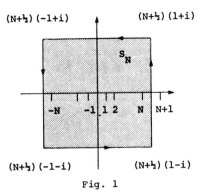

Fig. 1

Show that if $f(z)$ has a finite number of poles and along S_N

$$|f(z)| \le \frac{M}{|z|^\alpha}, \qquad (1)$$

where M and $\alpha > 1$ are constants independent of N, then

$$\sum_{n=-\infty}^{\infty} f(n) = - \sum_{\text{poles of } f(z)} \text{Res } \pi \cot \pi z \cdot f(z) \qquad (2)$$

Solution: Since $f(z)$ has a finite number of poles, N can be chosen so large that S_N encloses all poles of $f(z)$.

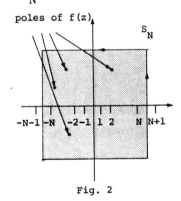

Fig. 2

Function $\cot \pi z$ has simple poles at $z = 0, \pm 1, \pm 2, \ldots, \pm n$. Assuming that $f(z)$ is continuous at $z = n$ we obtain

$$\text{Res}_n \pi \cot \pi z \cdot f(z) = \lim_{z \to n} (z-n) \pi \cot \pi z \cdot f(z)$$

$$= \lim_{z \to n} \pi \left(\frac{z-n}{\sin \pi z}\right) \cos \pi z \cdot f(z) = \left[\frac{\pi}{\pi \cos \pi z} \cdot \cos \pi z \cdot f(z)\right]\bigg|_{z=n} \qquad (3)$$

$$= f(n)$$

By the residue theorem

$$\frac{1}{2\pi i} \int_{S_N} \pi \cot \pi z \cdot f(z) dz = \sum_{n=-N}^{N} f(n) + R, \qquad (4)$$

where R is the sum of the residues of $\pi \cot \pi z \cdot f(z)$ evaluated at the poles of $f(z)$.

$$\left| \int_{S_N} \pi \cot \pi z \cdot f(z) dz \right| \leq (\text{Length of } S_N) \qquad (5)$$
$\cdot$ (upper bound of $|\pi \cot \pi z \cdot f(z)|$ on S_N)

$$= 4 \cdot (2N+1) \cdot \pi A \cdot \frac{M}{(N+\tfrac{1}{2})^\alpha}$$

Since $\alpha > 1$

$$\lim_{N \to \infty} \int_{S_N} \pi \cot \pi z \cdot f(z) dz = 0 \qquad (6)$$

Then,

$$\sum_{n=-\infty}^{\infty} f(n) = -R \qquad (7)$$

Note that in (5) we used the inequality

$$|\cot \pi z| < A \qquad (8)$$

proved in Problem 19-17.

● **PROBLEM 19-19**

Prove that:

$$\sum_{n=1}^{\infty} \frac{1}{n^2+a^2} = \frac{\pi}{2a} \coth a - \frac{1}{2a^2}, \quad a > 0 \qquad (1)$$

<u>Solution</u>: First we shall evaluate

$$\sum_{n=-\infty}^{\infty} \frac{1}{n^2+a^2} \qquad (2)$$

Let

$$f(z) = \frac{1}{z^2+a^2} \qquad (3)$$

This function has poles at $z = ia$ and $z = -ia$.

We obtain

$$\operatorname*{Res}_{ai} \frac{\pi \cot \pi z}{z^2+a^2} = \lim_{z \to ai} (z-ai) \frac{\pi \cot \pi z}{(z+ai)(z-ai)}$$

$$= \frac{\pi \cot \pi ai}{2ai} = \frac{-\pi}{2a} \coth \pi a \qquad (4)$$

$$\operatorname*{Res}_{-ai} \frac{\pi \cot \pi z}{z^2+a^2} = \frac{-\pi}{2a} \coth \pi a \qquad (5)$$

Thus,

$$\sum_{n=-\infty}^{\infty} \frac{1}{n^2+a^2} = -\left(\text{Res}_{ai} + \text{Res}_{-ai}\right) = \frac{\pi}{a}\coth\pi a \tag{6}$$

Eq.(6) can be written in the form

$$\sum_{n=-\infty}^{-1} \frac{1}{n^2+a^2} + \frac{1}{a^2} + \sum_{n=1}^{\infty} \frac{1}{n^2+a^2}$$

$$= 2\sum_{n=1}^{\infty} \frac{1}{n^2+a^2} + \frac{1}{a^2} = \frac{\pi}{a}\coth\pi a \tag{7}$$

or

$$\sum_{n=1}^{\infty} \frac{1}{n^2+a^2} = \frac{\pi}{2a}\coth\pi a - \frac{1}{2a^2} \tag{8}$$

● **PROBLEM 19-20**

Show that:

$$\frac{1}{1^2} + \frac{1}{2^2} + \frac{1}{3^2} + \ldots = \frac{\pi^2}{6} \tag{1}$$

Solution: In this case it is easy to find the proper function $f(z)$ which is

$$f(z) = \frac{1}{z^2} \tag{2}$$

Then,

$$\sum_{-\infty}^{\infty} f(n) = -(\text{sum of residues of } \pi\cot\pi z\, f(z) \text{ at the poles of } f(z)) \tag{3}$$

$$\pi\cot\pi z \cdot f(z) = \frac{\pi\cot\pi z}{z^2} = \frac{\pi\cos\pi z}{z^2\sin\pi z}$$

$$= \frac{\pi\left(1 - \frac{\pi^2 z^2}{2!} + \frac{\pi^4 z^4}{4!} - \ldots\right)}{z^2\left(\pi z - \frac{\pi^3 z^3}{3!} + \frac{\pi^5 z^5}{5!} - \ldots\right)}$$

$$= \frac{1 - \frac{\pi^2 z^2}{2!} + \frac{\pi^4 z^4}{4!} - \ldots}{z^3\left(1 - \frac{\pi^2 z^2}{3!} + \frac{\pi^4 z^4}{5!} - \ldots\right)} \tag{4}$$

$$= \frac{1}{z^3} \cdot \left(1 - \frac{\pi^2 z^2}{3} + \ldots\right) = -\frac{\pi^2}{3z} + \ldots$$

The residue at $z = 0$ is $-\frac{\pi^2}{3}$.

Thus,

$$\frac{1}{2\pi i} \int_{S_N} \frac{\pi \cot \pi z}{z^2} dz = \sum_{n=-N}^{-1} \frac{1}{n^2} + \sum_{n=1}^{N} \frac{1}{n^2} - \frac{\pi^2}{3} \qquad (5)$$

$$= 2 \sum_{n=1}^{N} \frac{1}{n^2} - \frac{\pi^2}{3}$$

As $N \to \infty$ the integral (5) approaches zero, then

$$2 \sum_{n=1}^{\infty} \frac{1}{n^2} = \frac{\pi^2}{3} \qquad (6)$$

or

$$\sum_{n=1}^{\infty} \frac{1}{n^2} = \frac{\pi^2}{6} \qquad (7)$$

● **PROBLEM 19-21**

Show that:

$$\sum_{n=1}^{\infty} \frac{1}{n^6} = \frac{\pi^6}{945} \qquad (1)$$

Solution: We shall choose the function

$$f(z) = \frac{1}{z^6} \qquad (2)$$

Then,

$$\frac{\pi \cot \pi z}{z^6} = \frac{\pi \cos \pi z}{z^6 \sin \pi z} = \frac{\pi (1 - \frac{\pi^2 z^2}{2!} + \frac{\pi^4 z^4}{4!} - \cdots)}{z^6 (\pi z - \frac{\pi^3 z^3}{3!} + \frac{\pi^5 z^5}{5!} - \cdots)}$$

$$= \frac{1}{z^7} (1 - \frac{\pi^2 z^2}{2!} + \frac{\pi^4 z^4}{4!} - \frac{\pi^6 z^6}{6!} + \cdots) \frac{1}{(1 - \frac{\pi^2 z^2}{3!} + \frac{\pi^4 z^4}{5!} - \cdots)} \qquad (3)$$

Carrying out the division we find

$$1 \Big/ (1 - \frac{\pi^2 z^2}{3!} + \frac{\pi^4 z^4}{5!} - \frac{\pi^6 z^6}{7!} + \cdots)$$

$$= 1 + \frac{\pi^2 z^2}{6} + \frac{\pi^4 z^4 \cdot 7}{360} + \pi^6 z^6 \left(\frac{1}{7!} - \frac{1}{6 \cdot 120} + \frac{7}{6 \cdot 360} \right) + \cdots \qquad (4)$$

Thus,

$$\cdots = \left(\frac{1}{z^7} - \frac{\pi^2}{2! z^5} + \frac{\pi^4}{4! z^3} - \frac{\pi^6}{6! z} + \cdots \right)$$

$$\cdot \left[1 + \frac{\pi^2 z^2}{6} + \frac{\pi^4 z^4 \cdot 7}{360} + \pi^6 z^6 \left(\frac{1}{7!} - \frac{1}{6 \cdot 120} + \frac{7}{6 \cdot 360} \right) + \cdots \right]$$

$$= \frac{\pi^6}{z} \left(\frac{1}{120 \cdot 6 \cdot 7} - \frac{1}{6 \cdot 120} + \frac{7}{6 \cdot 360} \right) - \frac{\pi^6}{z} \cdot \frac{7}{2 \cdot 360} \qquad (5)$$

$$+ \frac{\pi^6}{z} \frac{1}{6 \cdot 24} - \frac{\pi^6}{z} \frac{1}{120 \cdot 6} + \ldots$$

$$= \frac{\pi^6}{z} \left[\frac{1}{120} \left(\frac{1}{6 \cdot 7} + \frac{7}{6 \cdot 3} - \frac{9}{6} \right) + \frac{1}{6 \cdot 24} \right] + \ldots$$

$$= -\frac{2\pi^6}{945z} + \ldots$$

Hence,

$$\int_{S_N} \frac{\pi \cot \pi z}{z^6} dz = \sum_{n=-N}^{-1} \frac{1}{n^6} + \sum_{n=1}^{N} \frac{1}{n^6} - \frac{2\pi^6}{945} \tag{6}$$

As $N \to \infty$ the integral (6) tends to zero and

$$\sum_{n=1}^{\infty} \frac{1}{n^6} = \frac{\pi^6}{945} \tag{7}$$

● **PROBLEM 19-22**

Show that:

$$\frac{\coth \pi}{1^3} + \frac{\coth 2\pi}{2^3} + \frac{\coth 3\pi}{3^3} + \ldots = \frac{7\pi^3}{180} \tag{1}$$

Solution: The corresponding function is

$$f(z) = \frac{\coth \pi z}{z^3} \tag{2}$$

We shall integrate the function

$$F(z) = \frac{\pi \cot \pi z \coth \pi z}{z^3} \tag{3}$$

around the square S_N shown in Fig. 1.

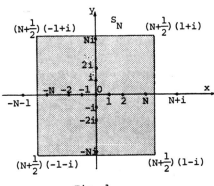

Fig. 1

Inside the square S_N the function $F(z)$ has the following singularities

at $z = 0$ a pole

at z = n   simple poles
at z = ni   simple poles
where n = ±1, ±2, ..., ±N

The residues are:

At z = 0

$$F(z) = \frac{\pi \cos \pi z}{z^3 \sin \pi z} \frac{\cosh \pi z}{\cosh \pi z}$$

$$= \frac{\pi \cdot (1 - \frac{\pi^2 z^2}{2!} + \frac{\pi^4 z^4}{4!} - \ldots)(1 + \frac{\pi^2 z^2}{2!} + \frac{\pi^4 z^4}{4!} + \ldots)}{z^3 (\pi z - \frac{\pi^3 z^3}{3!} + \frac{\pi^5 z^5}{5!} - \ldots)(\pi z + \frac{\pi^3 z^3}{3!} + \frac{\pi^5 z^5}{5!} + \ldots)} \quad (4)$$

$$= \frac{\pi}{z^5 \pi^2} \cdot \frac{(1 - \frac{\pi^4 z^4}{6} + \ldots)}{(1 - \frac{\pi^4 z^4}{90} + \ldots)} = \frac{1}{\pi z^5} \left[1 - \frac{7 \pi^4 z^4}{45} + \ldots \right]$$

$$\underset{o}{\text{Res}} F(z) = -\frac{7\pi^3}{45} \quad (5)$$

At z = n

$$\underset{n}{\text{Res}} F(z) = \lim_{z \to n} \left[\frac{(z-n)}{\sin \pi z} \frac{\pi \cos \pi z \coth \pi z}{z^3} \right]$$

$$= \frac{\coth n\pi}{n^3} \quad (6)$$

At z = ni

$$\underset{ni}{\text{Res}} F(z) = \lim_{z \to ni} \left[\frac{(z-ni)}{\sinh \pi z} \frac{\pi \cot \pi z \cosh \pi z}{z^3} \right]$$

$$= \frac{\coth n\pi}{n^3} \quad (7)$$

By the residue theorem

$$\frac{1}{2\pi i} \int_{S_N} \frac{\pi \cot \pi z \coth \pi z}{z^3} dz$$

$$= -\frac{7\pi^3}{45} + 2 \sum_{n=1}^{N} \frac{\coth n\pi}{n^3} + 2 \sum_{n=-1}^{-N} \frac{\coth n\pi}{n^3} \quad (8)$$

$$= -\frac{7\pi^3}{45} + 4 \sum_{n=1}^{N} \frac{\coth n\pi}{n^3}$$

Taking the limit of (8) as $N \to \infty$ we get

$$\sum_{n=1}^{\infty} \frac{\coth n\pi}{n^3} = \frac{7\pi^3}{180} \quad (9)$$

Note that as $N \to \infty$ the integral in (8) approaches zero.

• PROBLEM 19-23

Let $f(z)$ be a function such that along S_N

$$|f(z)| \leq \frac{M}{|z|^\alpha} \qquad (1)$$

where $\alpha > 1$ and M are constant independent of N, and $f(z)$ has a finite number of poles.

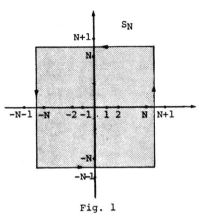

Fig. 1

Show that

$$\sum_{n=-\infty}^{\infty} (-1)^n f(n) = -\Sigma \text{ Res } \pi \csc \pi z f(z) \qquad (2)$$

Sum is taken over all poles of $f(z)$.

Solution: The function $\csc \pi z$ has simple poles at $z = 0, \pm 1, \pm 2, \ldots$.

At $z = n$,

$$\text{Res}_n \pi \csc \pi z \cdot f(z) = \lim_{z \to n} (z-n) \pi \csc \pi z \cdot f(z)$$

$$= \lim_{z \to n} \pi \left(\frac{z-n}{\sin \pi z} \right) f(z) = (-1)^n f(n) \qquad (3)$$

By the residue theorem

$$\frac{1}{2\pi i} \int_{S_N} \pi \csc \pi z \cdot f(z) dz = \sum_{n=-N}^{N} (-1)^n f(n) + R \qquad (4)$$

where R is the sum of the residues of $\pi \csc \pi z \cdot f(z)$ at the poles of $f(z)$. We assume that $f(z)$ has a finite number of poles, none of which coincides with the poles of $\csc \pi z$.

As $N \to \infty$ the integral in (4) approaches zero.

Thus,

$$\sum_{-\infty}^{\infty} (-1)^n f(n) = R \qquad (5)$$

• **PROBLEM** 19-24

Show that:

$$\sum_{n=-\infty}^{\infty} \frac{(-1)^n}{(n+a)^2} = \frac{\pi^2 \cos \pi a}{\sin^2 \pi a}, \tag{1}$$

where a is a real number different from any integer.

<u>Solution</u>: We shall consider

$$f(z) = \frac{1}{(z+a)^2} \tag{2}$$

which has a double pole at z = -a. The residue is

$$\operatorname*{Res}_{-a} \frac{\pi \csc \pi z}{(z+a)^2} = \lim_{z \to -a} \frac{d}{dz}\left[(z+a)^2 \frac{\pi \csc \pi z}{(z+a)^2}\right]$$

$$= \lim_{z \to -a} \frac{d}{dz}\left(\frac{\pi}{\sin \pi z}\right) = \left.\frac{-\pi^2 \cos \pi z}{\sin^2 \pi z}\right|_{z=-a} \tag{3}$$

$$= \frac{-\pi^2 \cos \pi a}{\sin^2 \pi a}$$

Thus,

$$\sum_{n=-\infty}^{\infty} \frac{(-1)^n}{(n+a)^2} = -\left(\text{sum of the residues of } \frac{\pi \csc \pi z}{(z+a)^2} \text{ at the poles of } \frac{1}{(z+a)^2}\right) \tag{4}$$

$$= \frac{\pi^2 \cos \pi a}{\sin^2 \pi a}$$

• **PROBLEM** 19-25

Compute the value of the series:

$$\frac{1}{1^2} - \frac{1}{2^2} + \frac{1}{3^2} - \frac{1}{4^2} + \ldots \tag{1}$$

<u>Solution</u>: Consider the function

$$F(z) = \frac{\pi \csc \pi z}{z^2} \tag{2}$$

and a square S_N as shown in Problem 19-23.

This function has a singularity at z = 0. The Laurent series at z = 0 is

$$\frac{\pi \csc \pi z}{z^2} = \frac{\pi}{z^2 \left(\pi z - \frac{\pi^3 z^3}{3!} + \ldots\right)}$$

646

$$= \frac{1}{z^3}\left(1 + \frac{\pi^2 z^2}{3!} + \ldots\right) = \frac{\pi^2}{6}\frac{1}{z} + \ldots \qquad (3)$$

The residue of F(z) at z = 0 is $\frac{\pi^2}{6}$. We have

$$\int_{S_N} \frac{\pi \csc \pi z}{z^2} dz = \sum_{n=-1}^{-N} \frac{(-1)^n}{n^2} + \sum_{n=1}^{N} \frac{(-1)^n}{n^2} + \frac{\pi^2}{6} \qquad (4)$$

Taking the limit of (4) as $N \to \infty$ and observing that

$$\lim_{N \to \infty} \int_{S_N} \frac{\pi \csc \pi z}{z^2} dz = 0, \qquad (5)$$

$$2 \sum_{n=1}^{\infty} \frac{(-1)^n}{n^2} = -\frac{\pi^2}{6} \qquad (6)$$

Hence,

$$\frac{1}{1^2} - \frac{1}{2^2} + \frac{1}{3^2} - \frac{1}{4^2} + \ldots = \frac{\pi^2}{12} \qquad (7)$$

● **PROBLEM 19-26**

Prove that:

$$\sum_{m=-\infty}^{\infty} \sum_{n=-\infty}^{\infty} \frac{1}{(m^2+a^2)(n^2+b^2)} = \frac{\pi^2}{ab} \coth \pi a \coth \pi b \qquad (1)$$

where $a > 0$ and $b > 0$.

Solution: In Problem 19-19 we proved the equation

$$\sum_{n=-\infty}^{\infty} \frac{1}{n^2+b^2} = \frac{\pi}{b} \coth \pi b \qquad (2)$$

Let m_1 be any integer then,

$$\frac{1}{m_1^2 + a^2} \sum_{n=-\infty}^{\infty} \frac{1}{n^2 + b^2} = \frac{1}{m_1^2 + a^2} \frac{\pi}{b} \coth \pi b \qquad (3)$$

Thus,

$$\sum_{m=-\infty}^{\infty} \sum_{n=-\infty}^{\infty} \frac{1}{m^2+a^2} \cdot \frac{1}{n^2+b^2}$$

$$= \sum_{m=-\infty}^{\infty} \frac{1}{m^2+a^2} \cdot \frac{\pi}{b} \coth \pi b$$

$$= \frac{\pi}{b} \coth \pi b \sum_{m=-\infty}^{\infty} \frac{1}{m^2+a^2} \qquad (4)$$

$$= \frac{\pi}{ab} \coth\pi a \coth\pi b$$

● **PROBLEM 19-27**

Prove that:

1. $\sum_{n=-\infty}^{\infty} f\left(\frac{2n+1}{2}\right) = \Sigma \, \text{Res}\,\pi\tan\pi z \cdot f(z)$ (1)

2. $\sum_{n=-\infty}^{\infty} (-1)^n f\left(\frac{2n+1}{2}\right) = \Sigma \, \text{Res}\,\pi\sec\pi z \cdot f(z)$ (2)

Both sums in (1) and (2) are computed over all the poles of f(z). We assume that f(z) has a finite number of poles.

Solution: 1. The function f(z) has a finite number of poles. We choose N so large that a square S_N contains all poles of f(z).

The function $\tan\pi z$ has poles at

$$z = \frac{2n+1}{2} \quad \text{where } n = 0, \pm 1, \tag{3}$$

$$\text{Res}\,\pi \atop \frac{2n+1}{2}} \tan\pi z \cdot f(z) = \lim_{z \to \frac{2n+1}{2}} \frac{\pi \sin\pi z}{\cos\pi z}\left(z - \frac{2n+1}{2}\right) \cdot f(z)$$

$$= -f\left(\frac{2n+1}{2}\right) \tag{4}$$

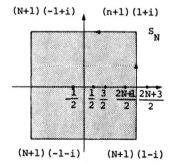

Fig.1

By the residue theorem

$$\frac{1}{2\pi i} \int_{S_N} \pi \tan\pi z \cdot f(z) dz = -\sum_{n=-N}^{N} f\left(\frac{2n+1}{2}\right) + R \tag{5}$$

where R is the sum of the residues of $\pi\tan\pi z \cdot f(z)$ at the poles of f(z). We can show (see Problems 19-17 and 19-18) that

$$\lim_{N \to \infty} \int_{S_N} \pi \tan\pi z \cdot f(z) dz = 0 \tag{6}$$

Taking the limit of (5), we obtain

$$\sum_{n=-\infty}^{\infty} f\left(\frac{2n+1}{2}\right) = \text{sum of residues of } \pi \tan \pi z \cdot f(z) \text{ at all the poles of } f(z) \quad (7)$$

2. The function

$$\sec \pi z = \frac{1}{\cos \pi z} \quad (8)$$

has poles at

$$z = \frac{2n+1}{2} \quad \text{where } n = 0, \pm 1, \ldots \quad (9)$$

$$\text{Res}_{\frac{2n+1}{2}} \pi \cdot \sec \pi z \cdot f(z) = \lim_{z \to \frac{2n+1}{2}} \frac{\pi \left(z - \frac{2n+1}{2}\right)}{\cos \pi z} \cdot f(z)$$

$$= \frac{\pi}{-\pi \sin \pi z} \cdot f(z) \bigg|_{z=\frac{2n+1}{2}} = \frac{-1}{\sin \frac{2n+1}{2}\pi} \cdot f\left(\frac{2n+1}{2}\right) \quad (10)$$

$$= -(-1)^n f\left(\frac{2n+1}{2}\right)$$

Applying the same reasoning as in part 1, we obtain

$$\sum_{n=-\infty}^{\infty} (-1)^n f\left(\frac{2n+1}{2}\right) = \text{sum of residues of } \pi \sec \pi z \cdot f(z) \text{ at all poles of } f(z) \quad (11)$$

• **PROBLEM 19-28**

Show that:

$$\frac{1}{1^3} - \frac{1}{3^3} + \frac{1}{5^3} - \ldots = \frac{\pi^3}{32} \quad (1)$$

Solution: The function

$$f(z) = \frac{\pi \sec \pi z}{z^3} \quad (2)$$

is useful in this case.

Its series expansion is

$$f(z) = \frac{\pi}{z^3 \cos \pi z} = \frac{\pi}{z^3 \left(1 - \frac{\pi^2 z^2}{2!} + \frac{\pi^4 z^4}{4!} - \ldots\right)}$$

$$= \frac{\pi}{z^3} \cdot \left(1 + \frac{\pi^2 z^2}{2} + \ldots\right) = \frac{\pi^3}{2z} + \ldots \quad (3)$$

Hence, the residue of $f(z)$ at $z = 0$ is $\frac{\pi^3}{2}$.

The function $f(z)$ has simple poles at

$$z = n + \frac{1}{2}, \quad n = 0, \pm 1, \pm 2, \quad (4)$$

The residue of $f(z)$ at $z = n + \frac{1}{2}$ is

$$\operatorname*{Res}_{z=n+\frac{1}{2}} \frac{\pi \sec \pi z}{z^3} = \lim_{z \to n+\frac{1}{2}} \left[z - \left(n+\frac{1}{2}\right)\right] \frac{\pi}{z^3 \cos \pi z}$$

$$= \frac{\pi}{(n+\frac{1}{2})^3} \lim_{z \to n+\frac{1}{2}} \frac{z-(n+\frac{1}{2})}{\cos \pi z}$$

$$= \frac{\pi}{(n+\frac{1}{2})^3} \left[\frac{1}{-\pi \sin \pi z}\right]\bigg|_{z=n+\frac{1}{2}} \quad (5)$$

$$= \frac{-(-1)^n}{(n+\frac{1}{2})^3} = \frac{-8(-1)^n}{(2n+1)^3}$$

Let S_N be a square with vertices as shown in Fig. 1.

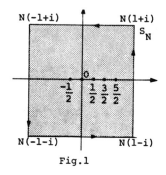

Fig.1

Then, evaluating the integral

$$\int_{S_N} \frac{\pi \sec \pi z}{z^3} dz$$

and substituting (3) and (5) we obtain (1).

● **PROBLEM 19-29**

Show that:

$$\frac{1}{1^5} - \frac{1}{3^5} + \frac{1}{5^5} - \frac{1}{7^5} + \ldots = \frac{5\pi^5}{1536} \quad (1)$$

Solution: We shall use eq.(2) from Problem 19-27.

$$f(z) = \frac{1}{z^5} \quad (2)$$

$$F(z) = \pi \sec \pi z \cdot \frac{1}{z^5} = \frac{\pi}{z^5 \cos \pi z} \quad (3)$$

Function f(z) has a pole of order five at $z = 0$. To evaluate the residue we take the Laurent expansion

$$F(z) = \frac{\pi}{z^5 \cos \pi z} = \frac{\pi}{z^5 \left(1 - \frac{\pi^2 z^2}{2!} + \frac{\pi^4 z^4}{4!} - \frac{\pi^6 z^6}{6!} + \ldots \right)} \quad (4)$$

$$= \frac{\pi}{z^5} \left(1 + \frac{\pi^2 z^2}{2} + \frac{5}{24} \pi^4 z^4 + \ldots \right)$$

Hence,

$$\operatorname*{Res}_{0} \frac{\pi}{z^5 \cos \pi z} = \frac{5 \pi^5}{24} \quad \text{and} \quad (5)$$

$$\sum_{n=-\infty}^{\infty} (-1)^n \frac{2^5}{(2n+1)^5} = \frac{5 \pi^5}{24} \quad (6)$$

or

$$\frac{1}{1^5} - \frac{1}{3^5} + \frac{1}{5^5} - \frac{1}{7^5} + \ldots = \frac{5 \pi^5}{1536} \quad (7)$$

• **PROBLEM 19-30**

Let s be different than any integer,

$$s \ne 0, \pm 1, \pm 2, \pm 3, \ldots \quad (1)$$

Show that

$$\frac{s^2 + 1}{(s^2-1)^2} - \frac{s^2 + 4}{(s^2-4)^2} + \frac{s^2 + 9}{(s^2-9)^2} - \ldots$$

$$= \frac{1}{2s^2} - \frac{\pi^2 \cos \pi s}{2 \sin^2 \pi s} \quad (2)$$

Solution: In Problem 19-24 we proved the formula

$$\sum_{n=-\infty}^{\infty} \frac{(-1)^n}{(n+s)^2} = \frac{\pi^2 \cos \pi s}{\sin^2 \pi s} \quad (3)$$

Eq.(3) can be written in the form

$$\frac{1}{s^2} - \left[\frac{1}{(s+1)^2} + \frac{1}{(s-1)^2}\right] + \left[\frac{1}{(s+2)^2} + \frac{1}{(s-2)^2}\right] - \left[\frac{1}{(s+3)^2} + \frac{1}{(s-3)^2}\right] \quad (4)$$

$$+ \ldots = \frac{\pi^2 \cos \pi s}{\sin^2 \pi s} = \ldots$$

Note, that since the series is absolutely convergent the grouping of terms is permissible

$$\ldots = \frac{1}{s^2} - \frac{2(s^2+1)}{(s^2-1)^2} + \frac{2(s^2+4)}{(s^2-4)^2} - \frac{2(s^2+9)}{(s^2-9)^2} + \ldots \quad (5)$$

$$= \frac{\pi^2 \cos \pi s}{\sin^2 \pi s}$$

or

$$\frac{s^2+1}{(s^2-1)^2} - \frac{s^2+4}{(s^2-4)^2} + \frac{s^2+9}{(s^2-9)^2} - \cdots$$
$$= \frac{1}{2s^2} - \frac{\pi^2 \cos \pi s}{2 \sin^2 \pi s} \tag{6}$$

● **PROBLEM 19-31**

Show that:

$$\frac{1}{1^3 \sinh \pi} - \frac{1}{2^3 \sinh 2\pi} + \frac{1}{3^3 \sinh 3\pi} - \cdots = \frac{\pi^3}{360} \tag{1}$$

<u>Solution</u>: We shall investigate the function

$$F(z) = \frac{\pi}{z^3 \sin \pi z \cdot \sinh \pi z} \tag{2}$$

and the integral

$$\int_{S_N} F(z) dz \tag{3}$$

taken around the square S_N

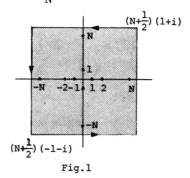

Fig.1

The function $F(z)$ has poles at

$$z = 0, \quad z = n, \quad z = ni, \tag{4}$$

where $n = \pm 1, \pm 2, \ldots, \pm N$.

The residues are at $z = 0$, at $z = n$ and at $z = ni$.

At $z = 0$ we have

$$\frac{\pi}{z^3 \sin \pi z \cdot \sinh \pi z} = \frac{\pi}{z^3 [\pi z - \frac{\pi^3 z^3}{3!} + \frac{\pi^5 z^5}{5!} - \cdots][\pi z + \frac{\pi^3 z^3}{3!} + \frac{\pi^5 z^5}{5!} + \cdots]}$$

$$= \frac{1}{\pi z^5 (1 - \frac{\pi^4 z^4}{90} + \cdots)} = \frac{\pi^3}{90} \cdot \frac{1}{z} + \cdots \tag{5}$$

Hence,
$$\operatorname*{Res}_{0} F(z) = \frac{\pi^3}{90} \tag{6}$$

At $z = n$ we have
$$\operatorname*{Res}_{n} F(z) = \lim_{z \to n} \frac{(z-n)}{\sin\pi z} \cdot \frac{\pi}{z^3 \sinh\pi z}$$
$$= \frac{1}{z^3 \cos\pi z \cdot \sinh\pi z}\bigg|_{z=n} = (-1)^n \frac{1}{n^3 \sinh\pi n} \tag{7}$$

At $z = ni$ we have
$$\operatorname*{Res}_{ni} F(z) = \lim_{z \to ni} \frac{(z-ni)}{\sinh\pi z} \cdot \frac{\pi}{z^3 \sin\pi z}$$
$$= \frac{1}{\cosh\pi z \cdot z^3 \sin\pi z}\bigg|_{z=ni} = \frac{1}{(ni)^3 \sin\pi ni \cdot \cosh\pi ni}$$
$$= \frac{1}{n^3 \sinh n\pi \cdot \cos n\pi} = \frac{(-1)^n}{n^3 \sinh n\pi} \tag{8}$$

Hence,
$$\frac{1}{2\pi i} \int_{S_N} \frac{\pi}{z^3 \sin\pi z \cdot \sinh\pi z}\, dz = \frac{\pi^3}{90} + 2 \sum_{n=1}^{N} \frac{(-1)^n}{n^3 \sinh n\pi} \tag{9}$$
$$+ 2 \sum_{n=-1}^{-N} \frac{(-1)^n}{n^3 \sinh n\pi}$$

Taking the limit of (9) as $N \to \infty$ and observing that
$$\lim_{N \to \infty} \int_{S_N} F(z)\, dz = 0, \tag{10}$$

we find,
$$\frac{1}{1^3 \sinh\pi} - \frac{1}{2^3 \sinh 2\pi} + \frac{1}{3^3 \sinh 3\pi} - \cdots = \frac{\pi^3}{360} \tag{11}$$

● **PROBLEM** 19-32

Let $f(z)$ be analytic in the finite z-plane except points
$$\alpha_1, \alpha_2, \alpha_3, \cdots \tag{1}$$
which are simple poles. A sequence $\{\alpha_n\}$ is such that
$$|\alpha_1| \leq |\alpha_2| \leq |\alpha_3| \leq \cdots \tag{2}$$
The family of circles C_N of radius r_N is such that none of C_N passes through any poles α_n. Furthermore,

$$|f(z)| < M \tag{3}$$

and

$$\lim_{N \to \infty} r_N = \infty \tag{4}$$

Prove Mittag-Leffler expansion theorem:

$$f(z) = f(0) + \sum_{n=1}^{\infty} a_n \left[\frac{1}{z-\alpha_n} + \frac{1}{\alpha_n} \right], \tag{5}$$

where $a_1, a_2, \ldots, a_n$ are residues of $f(z)$ at $\alpha_1, \alpha_2, \ldots, \alpha_n$.

Solution: Let ω be any point of the z-plane which is not a pole of $f(z)$. The function

$$\frac{f(z)}{z - \omega} \tag{6}$$

has poles at $\alpha_1, \alpha_2, \ldots$ and ω.

$$\operatorname*{Res}_{\alpha_n} \frac{f(z)}{z - \omega} = \lim_{z \to \alpha_n} (z-\alpha_n) \frac{f(z)}{z - \omega} = \frac{a_n}{\alpha_n - \omega} \tag{7}$$

$$\operatorname*{Res}_{\omega} \frac{f(z)}{z - \omega} = \lim_{z \to \omega} \frac{f(z)}{z - \omega} (z-\omega) = f(\omega) \tag{8}$$

By the residue theorem

$$\frac{1}{2\pi i} \int_{C_N} \frac{f(z)}{z - \omega} dz = f(\omega) + \sum_n \frac{a_n}{\alpha_n - \omega} \tag{9}$$

The sum is taken over all poles inside the circle C_N of radius r_N such that

$$|\omega| < r_N \tag{10}$$

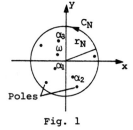

Fig. 1

If $f(z)$ is analytic at $z = 0$, then setting $\omega = 0$ in (9) we obtain,

$$\frac{1}{2\pi i} \int_{C_N} \frac{f(z)}{z} dz = f(0) + \sum_n \frac{a_n}{\alpha_n} \tag{11}$$

Subtracting (11) from (9) we find

$$\frac{1}{2\pi i} \int_{C_N} \left[\frac{1}{z-\omega} - \frac{1}{z} \right] f(z)dz = \frac{\omega}{2\pi i} \int_{C_N} \frac{f(z)}{z(z-\omega)} dz$$

$$= f(\omega) - f(0) + \sum_n a_n \left[\frac{1}{\alpha_n - \omega} - \frac{1}{\alpha_n} \right] \tag{12}$$

For $z \in C_N$

$$|z-\omega| \geq |z| - |\omega| = r_N - |\omega| \tag{13}$$

and

$$|f(z)| < M$$

Hence,

$$\left| \int_{C_N} \frac{f(z)}{z(z-\omega)} dz \right| \leq 2\pi r_N \cdot \frac{M}{r_N(r_N - |\omega|)} \tag{14}$$

and since

$$\lim_{N \to \infty} r_N = \infty$$

we get

$$\lim_{N \to \infty} \int_{C_N} \frac{f(z)}{z(z-\omega)} dz = 0 \tag{15}$$

Taking the limit of (12) as $N \to \infty$ we find

$$f(\omega) = f(0) + \sum_{n=1}^{\infty} a_n \left[\frac{1}{\alpha_n} + \frac{1}{\omega - \alpha_n} \right] \tag{16}$$

● **PROBLEM 19-33**

Applying Mittag-Leffler theorem show that:

$$\cot z = \frac{1}{z} + \sum_{n=1}^{\infty} \left(\frac{1}{z-n\pi} + \frac{1}{n\pi} \right) + \sum_{n=-1}^{-\infty} \left(\frac{1}{z-n\pi} + \frac{1}{n\pi} \right) \tag{1}$$

Solution: We shall consider the function

$$f(z) = \cot z - \frac{1}{z} = \frac{z\cos z - \sin z}{z \sin z} \tag{2}$$

Note that at $z = 0$, the function $f(z)$ has a removable singularity. Therefore,

$$\lim_{z \to 0} \frac{z\cos z - \sin z}{z \sin z} = 0 \tag{3}$$

Thus, we can define

$$f(0) = 0 \tag{4}$$

Function f(z) has simple poles at $z = n\pi$, $n = \pm 1, \pm 2, \pm 3, \ldots$ \hfill (5)

$$\operatorname*{Res}_{n\pi} f(z) = \lim_{z \to n\pi} (z-n\pi) \frac{z\cos z - \sin z}{z \sin z}$$

$$= \lim_{z \to n\pi} \left(\frac{z-n\pi}{\sin z}\right) \lim_{z \to n\pi} \left(\frac{z\cos z - \sin z}{z}\right) = 1 \hfill (6)$$

The function $f(z) = \cot z - \frac{1}{z}$ is bounded on circles C_N having center at the origin and radius

$$r_N = (N+\tfrac{1}{2})\pi \hfill (7)$$

By Problem 19-32 we get

$$\cot z = \frac{1}{z} + \sum_n \left(\frac{1}{z-n\pi} + \frac{1}{n\pi}\right) \hfill (8)$$

where $\quad n = \pm 1, \pm 2, \pm 3, \ldots$

Remember that $f(0) = 0$.

● **PROBLEM 19-34**

Show that:

$$\csc z = \frac{1}{z} - 2z\left(\frac{1}{z^2-\pi^2} - \frac{1}{z^2-4\pi^2} + \frac{1}{z^2-9\pi^2} - \cdots\right) \hfill (1)$$

<u>Solution</u>: Consider the function

$$f(z) = \frac{1}{\sin z} - \frac{1}{z} \hfill (2)$$

which has simple poles at $z = n\pi$, $n = \pm 1, \pm 2, \pm 3, \ldots$.

The residue is

$$\operatorname*{Res}_{n\pi} f(z) = \lim_{z \to n\pi} \frac{(z-n\pi)}{\sin z} \cdot \frac{(z-\sin z)}{z}$$

$$= \lim_{z \to n\pi} \frac{z-\sin z}{z \cos z} = (-1)^n \hfill (3)$$

The assumption of the Mittag-Leffler expansion theorem are fulfilled.

At $z = 0$

$$\lim_{z \to 0} \frac{z-\sin z}{z \sin z} = \lim_{z \to 0} \frac{1-\cos z}{\sin z + z \cos z} = 0 \hfill (4)$$

The function $f(z)$ has a removable singularity and

$$f(0) = 0 \hfill (5)$$

Thus,

$$\frac{1}{\sin z} = \frac{1}{z} + \sum_n (-1)^n \left[\frac{1}{z-n\pi} + \frac{1}{n\pi} \right] \tag{6}$$

where $n = \pm 1, \pm 2, \pm 3, \ldots$

Eq. (6) can be written in the form

$$\frac{1}{\sin z} = \frac{1}{z} - \left(\frac{1}{z-\pi} + \frac{1}{\pi} \right) - \left(\frac{1}{z+\pi} - \frac{1}{\pi} \right)$$

$$+ \left(\frac{1}{z-2\pi} + \frac{1}{2\pi} \right) + \left(\frac{1}{z+2\pi} - \frac{1}{2\pi} \right) - \left(\frac{1}{z-3\pi} + \frac{1}{3\pi} \right) - \left(\frac{1}{z+3\pi} - \frac{1}{3\pi} \right) \tag{7}$$

$$+ \ldots = \frac{1}{z} - 2z \left(\frac{1}{z^2 - \pi^2} - \frac{1}{z^2 - 4\pi^2} + \frac{1}{z^2 - 9\pi^2} - \ldots \right)$$

● **PROBLEM 19-35**

Show that:
$$\tan z = 2z \left[\frac{1}{\left(\frac{\pi}{2}\right)^2 - z^2} + \frac{1}{\left(\frac{3\pi}{2}\right)^2 - z^2} + \frac{1}{\left(\frac{5\pi}{2}\right)^2 - z^2} + \ldots \right] \tag{1}$$

Solution: The function

$$\tan z = \frac{\sin z}{\cos z} \tag{2}$$

has simple poles at

$$z = \frac{2n+1}{2} \pi \tag{3}$$

where $n = 0, \pm 1, \pm 2, \ldots$

The residue of $f(z)$ at $z = \frac{2n+1}{2}\pi$, is

$$\operatorname*{Res}_{\frac{2n+1}{2}\pi} \frac{\sin z}{\cos z} = \lim_{z \to \frac{2n+1}{2}\pi} \frac{\sin z}{\cos z} \left(z - \frac{2n+1}{2}\pi \right)$$

$$= \frac{\sin z}{-\sin z} = -1 \tag{4}$$

By Mittag-Leffler theorem

$$\tan z = \sum_n (-1) \left[\frac{1}{z - \frac{2n+1}{2}\pi} + \frac{1}{\frac{2n+1}{2}\pi} \right] \tag{5}$$

Eq. (5) can be written in the form

$$\tan z = -\left[\frac{1}{z - \frac{\pi}{2}} + \frac{1}{\frac{\pi}{2}} \right] - \left[\frac{1}{z - \frac{3}{2}\pi} + \frac{1}{\frac{3}{2}\pi} \right]$$

$$-\left[\frac{1}{z+\frac{\pi}{2}} - \frac{1}{\frac{\pi}{2}}\right] - \left[\frac{1}{z-\frac{5}{2}\pi} + \frac{1}{\frac{5}{2}\pi}\right] - \ldots$$

$$= \left[\frac{1}{\frac{\pi}{2}-z} - \frac{1}{\frac{\pi}{2}+z}\right] + \left[\frac{1}{\frac{3}{2}\pi-z} - \frac{1}{\frac{3}{2}\pi+z}\right] \tag{6}$$

$$+ \left[\frac{1}{\frac{5}{2}\pi-z} - \frac{1}{\frac{5}{2}\pi+z}\right] + \ldots$$

$$= 2z\left[\frac{1}{\left(\frac{\pi}{2}\right)^2-z^2} + \frac{1}{\left(\frac{3\pi}{2}\right)^2-z^2} + \frac{1}{\left(\frac{5\pi}{2}\right)^2-z^2} + \ldots\right]$$

• **PROBLEM 19-36**

Show that:

$$\frac{\pi^2}{(\sin\pi z)^2} = \sum_{n=-\infty}^{\infty} \frac{1}{(z-n)^2} \tag{1}$$

<u>Solution</u>: Consider the function

$$f(z) = \cot\pi z = \frac{\cos\pi z}{\sin\pi z} \tag{2}$$

which has simple poles at

$$z = n, \quad n = 0, \pm 1, \pm 2, \ldots \tag{3}$$

The residue of $f(z)$ at $z = n$ is

$$\operatorname*{Res}_{z=n} \cot\pi z = \lim_{z \to n} \frac{\cos\pi z}{\sin\pi z}(z-n) = \frac{1}{\pi} \tag{4}$$

From the definition of the residue we conclude that the principal part of $\cot\pi z$ at $z_n = n$ is

$$\frac{1}{\pi} \cdot \frac{1}{z-z_n} \tag{5}$$

Thus, in a neighborhood of z_n

$$0 < |z-z_n| < 1 \tag{6}$$

we have

$$\cot\pi z = \frac{\cos\pi z}{\sin\pi z} = \frac{1}{\pi} \cdot \frac{1}{z-z_n} + F_n(z) \tag{7}$$

where $F_n(z)$ is analytic for $|z-z_n| < 1$.

Differentiating (7) we obtain

$$\frac{\pi^2}{(\sin\pi z)^2} = \frac{1}{(z-z_n)^2} - \pi F_n'(z) \tag{8}$$

Since $F_n'(z)$ is analytic the principal part of $\frac{\pi^2}{(\sin\pi z)^2}$ at each z_n is $\frac{1}{(z-n)^2}$.

We shall form the series

$$u(z) = \sum_{n=-\infty}^{\infty} \frac{1}{(z-n)^2} \tag{9}$$

which for each non-integer z, it converges uniformly in a neighborhood of z. Then, by the Weierstrass theorem, u(z) given by (9) is analytic for any z which is not an integer.

Comparing (8) and (9), we see that both functions have the same principal parts $\frac{1}{(z-n)^2}$ at z = n. The function

$$v(z) = \frac{\pi^2}{(\sin\pi z)^2} - u(z) \tag{10}$$

is therefore analytic everywhere. We shall prove that

$$v(z) = 0 \tag{11}$$

From (10) we obtain,

$$v\left(\frac{z}{2}\right) = \frac{\pi^2}{\left(\sin\frac{\pi z}{2}\right)^2} - \sum_{n=-\infty}^{\infty} \frac{4}{(z-2n)^2} \tag{12}$$

and

$$v\left(\frac{z+1}{2}\right) = \frac{\pi^2}{\left(\cos\frac{\pi z}{2}\right)^2} - \sum_{n=-\infty}^{\infty} \frac{4}{(z-2n+1)^2} \tag{13}$$

Adding (12) and (13) we find,

$$v\left(\frac{z}{2}\right) + v\left(\frac{z+1}{2}\right) = \frac{4\pi^2}{(\sin\pi z)^2} - \sum_{n=-\infty}^{\infty} \frac{4}{(z-n)^2} \tag{14}$$

$$= 4v(z)$$

Thus,

$$v(z) = \frac{1}{4}\left[v\left(\frac{z}{2}\right) + v\left(\frac{z+1}{2}\right)\right] \tag{15}$$

Let

$$a = \max|v(z)| \quad \text{for } |z| \leq 1 \tag{16}$$

then

$$|v(z)| = \frac{1}{4}\left|v\left(\frac{z}{2}\right) + v\left(\frac{z+1}{2}\right)\right| \leq \frac{1}{4}\left[\left|v\left(\frac{z}{2}\right)\right| + \left|v\left(\frac{z+1}{2}\right)\right|\right] \tag{17}$$

$$\leq \frac{1}{4}(a+a) = \frac{a}{2}$$

and

$$a \leq \frac{a}{2}$$

or

$$a = 0 \tag{18}$$

For $|z| \leq 1$, $v(z) = 0$ and

$$\frac{\pi^2}{(\sin \pi z)^2} = \sum_{n=-\infty}^{\infty} \frac{1}{(z-n)^2} \tag{19}$$

● **PROBLEM 19-37**

Show that:

$$\frac{\pi^2}{(\cos \pi z)^2} = \sum_{n=-\infty}^{\infty} \frac{1}{(z-\frac{1}{2}-n)^2} \tag{1}$$

Solution: To prove (1) we can follow the method described in Problem 19-36. However, there is a shorter way. Replace z by $z - \frac{1}{2}$ in (1) in Problem 19-36. Then,

$$\frac{\pi^2}{[\sin \pi (z-\frac{1}{2})]^2} = \sum_{n=-\infty}^{\infty} \frac{1}{(z-\frac{1}{2}-n)^2} \tag{2}$$

The simple trigonometric transformation of (2) leads to

$$\frac{\pi^2}{[\sin \pi (z-\frac{1}{2})]^2} = \frac{\pi^2}{[\cos \pi z]^2} = \sum_{n=-\infty}^{\infty} \frac{1}{(z-\frac{1}{2}-n)^2} \tag{3}$$

The proof is completed.

● **PROBLEM 19-38**

Show that:

$$\pi \tan \pi z = -\sum_{n=0}^{\infty} \left[\frac{1}{z-\frac{1}{2}-n} + \frac{1}{z+\frac{1}{2}+n} \right] \tag{1}$$

$$= \sum_{n=0}^{\infty} \frac{2z}{(n+\frac{1}{2})^2 - z^2}$$

Solution: Observe that

$$\frac{d}{dz}(\pi \tan \pi z) = \frac{\pi^2}{(\cos \pi z)^2} \tag{2}$$

From Problem 19-36 eq.(1) we have

$$\frac{d}{dz}(\pi\tan\pi z) = \sum_{n=-\infty}^{\infty} \frac{1}{(z-\tfrac{1}{2}-n)^2} \tag{3}$$

Differentiate (1)

$$\frac{d}{dz}(\pi\tan\pi z) = -\sum_{n=0}^{\infty} \frac{d}{dz}\left[\frac{1}{z-\tfrac{1}{2}-n} + \frac{1}{z+\tfrac{1}{2}+n}\right]$$

$$= \sum_{n=0}^{\infty} \left[\frac{1}{(z-\tfrac{1}{2}-n)^2} + \frac{1}{(z+\tfrac{1}{2}+n)^2}\right] \tag{4}$$

Now we will show that

$$\sum_{n=-\infty}^{\infty} \frac{1}{(z-\tfrac{1}{2}-n)^2} = \sum_{n=0}^{\infty} \left[\frac{1}{(z-\tfrac{1}{2}-n)^2} + \frac{1}{(z+\tfrac{1}{2}+n)^2}\right] \tag{5}$$

Indeed,

$$\sum_{n=-\infty}^{\infty} \frac{1}{(z-\tfrac{1}{2}-n)^2} = \sum_{n=-1}^{-\infty} \frac{1}{(z-\tfrac{1}{2}-n)^2} + \sum_{n=0}^{\infty} \frac{1}{(z-\tfrac{1}{2}-n)^2}$$

$$= \sum_{n=0}^{\infty} \frac{1}{(z-\tfrac{1}{2}-n)^2} + \sum_{n=0}^{\infty} \frac{1}{(z+\tfrac{1}{2}+n)^2} \tag{6}$$

It is easy to see that

$$\sum_{n=-1}^{-\infty} \frac{1}{(z-\tfrac{1}{2}-n)^2} = \sum_{n=0}^{\infty} \frac{1}{(z+\tfrac{1}{2}+n)^2} \tag{7}$$

CHAPTER 20

MAPPING BY ELEMENTARY FUNCTIONS AND LINEAR FRACTIONAL TRANSFORMATIONS

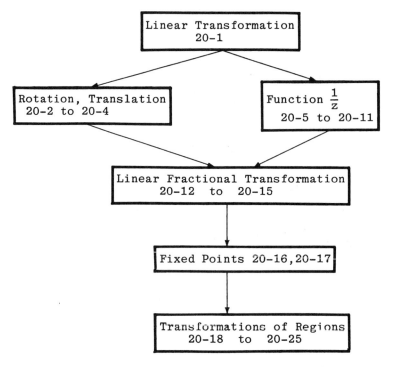

This chart is provided to facilitate rapid understanding of the interrelationships of the topics and subject matter in this chapter. Also shown are the problem numbers associated with the subject matter.

LINEAR TRANSFORMATION

● **PROBLEM 20-1**

Write the equation for and explain a general linear transformation. Give an example.

<u>Solution</u>: We shall describe two transformations: translation and rotation, and then combine them into one equation. Let $A = a + ib$ be a complex constant, then the mapping of the z plane onto the ω plane

$$\omega = z + A \tag{1}$$

is called a translation. If

$$\omega = u + iv \tag{2}$$
$$z = x + iy$$

then

$$u + iv = (x+a) + i(y+b) \tag{3}$$

The image of any point (x,y) in the z plane is the point (u,v) in the ω plane.

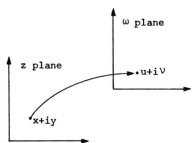

Fig. 1

The mapping

$$\omega = Bz \tag{4}$$

where B is a non-zero complex constant, consists of an expansion or contraction and a rotation. Indeed, let

$$B = be^{i\beta} \tag{5}$$
$$z = re^{i\theta} \tag{6}$$

then

$$\omega = bre^{i(\beta+\theta)} \tag{7}$$

Transformation (7) maps the non-zero point z with polar coordinates (r,θ) on to the non-zero point ω with coordinates (rb, θ+β).

A linear transformation

$$\omega = Bz + A \tag{8}$$

is a composition of the transformations

$$\begin{aligned} z' &= Bz \\ \omega &= z' + A \end{aligned} \tag{9}$$

For example

$$\omega = (2+3i)z + (1-i) \tag{10}$$

is a linear transformation.

ROTATION, TRANSLATION

• **PROBLEM** 20-2

The linear transformation is given by

$$\omega = (1+i)z + (1-i) \quad (1)$$

What is the image of the rectangular region shown in Fig. 1 under this transformation?

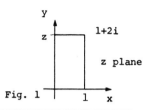

Fig. 1

Solution: Transformation (1) is a composition of the transformations

$$z' = (1+i)z \quad (2)$$
$$\omega = z' + (1-i)$$

Since

$$1+i = \sqrt{2}\, e^{i\frac{\pi}{4}} \quad (3)$$

transformation $z' = (1+i)z = \sqrt{2}\, e^{i\frac{\pi}{4}} z$ is an expansion by the factor $\sqrt{2}$, and a rotation through the angle $\frac{\pi}{4}$ as shown in Fig. 2.

Transformation $\omega = z' + (1-i)$ is a translation (see Fig. 2).

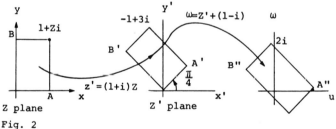

Fig. 2

To get a clear picture, follow the points A, B, 1+2i from the first part.

$$A \longrightarrow A' \longrightarrow A''$$
$$A=1 \quad A'=(1+i)A = 1+i \quad A''=A'+(1-i)=2$$

$$B \longrightarrow B' \longrightarrow B'' \quad (4)$$
$$B=2i \quad B'=(1+i)2i=-2+2i \quad B''=(-2+2i)+(1-i)=-1+i$$

$$1+2i \longrightarrow \quad \longrightarrow$$
$$\quad (1+i)(1+2i)=-1+3i \quad (-1+3i)+(1-i)=2i$$

• **PROBLEM** 20-3

Discuss the transformation

$$\omega = iz \quad (1)$$

and find the image of the infinite strip

$$0 < x < 1 \tag{2}$$

Solution: The complex number i in the polar coordinates is

$$i = e^{i\frac{\pi}{2}} \tag{3}$$

Hence, there is no contraction or expansion.

Transformation (1) is a rotation of the z plane through the angle $\frac{\pi}{2}$.

$$\omega = u + iv = iz = i(x+iy) = ix - y \tag{4}$$

The infinite strip $0 < x < 1$ in the z plane becomes an infinite strip $0 < v < 1$ in the ω plane.

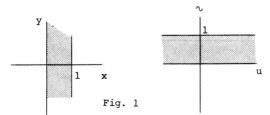

Fig. 1

• **PROBLEM 20-4**

Find the image of the semi-infinite strip

$$x > 0 \tag{1}$$

$$0 < y < 4$$

under the transformation

$$\omega = iz + 2 \tag{2}$$

Solution: The transformation (2) can be written in the form

$$\omega = iz + 2 = e^{i\frac{\pi}{2}} z + 2 \tag{3}$$

It consists of the rotation through the angle $\frac{\pi}{2}$ and the translation.

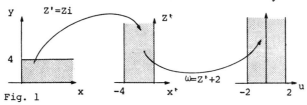

Fig. 1

The successive transformations are shown in Fig. 1.

The strip $x > 0$

$0 < y < 4$

under the transformation

$$\omega = iz + 2$$

is transformed into

$$v > 0$$
$$-2 < u < 2 \tag{4}$$

FUNCTION $\frac{1}{z}$

● **PROBLEM 20-5**

The transformation

$$\omega = \frac{1}{z} \tag{1}$$

determines a one-to-one correspondence between the points $z \neq 0$ of the z plane and points of the ω plane. It can be viewed as two successive transformations

$$z' = \frac{z}{|z|^2} \tag{2}$$

and

$$\omega = \overline{z'} \tag{3}$$

Indeed

$$\omega = \overline{z'} = \overline{\left(\frac{z}{|z|^2}\right)} = \frac{\overline{z}}{|z|^2} = \frac{\overline{z}}{z\overline{z}}$$
$$= \frac{1}{z} \tag{4}$$

Transformation (2) is an inversion with respect to the unit circle $|z| = 1$.

$$|z'| = \frac{1}{|z|} \qquad \arg z' = \arg z \tag{5}$$

The points exterior to the circle $|z| = 1$ are mapped onto the points $z \neq 0$ interior to it, and conversely (see Fig. 1).

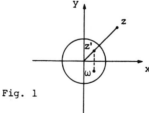

Fig. 1

Transformation (3) is a reflection in the real axis.

Since

$$\lim_{z \to 0} \frac{1}{z} = \infty \quad \text{and} \quad \lim_{z \to \infty} \frac{1}{z} = 0, \tag{6}$$

eq.(1) can be extended in such a way that

$$\omega = T(z) = \begin{cases} \dfrac{1}{z} & \text{for } z \neq 0 \text{ and } z \neq \infty \\ 0 & \text{for } z = \infty \\ \infty & \text{for } z = 0 \end{cases} \quad (7)$$

Transformation $T(z) = \omega$, defined in (7), is continuous throughout the extended z plane.

Show that the mapping $\omega = \dfrac{1}{z}$ transforms circles and lines into circles and lines.

Solution: Let $z = x+iy$ be a non-zero point transformed into $\omega = u + iv$ under the transformation

$$\omega = \frac{1}{z} \quad (8)$$

Then

$$u+iv = \frac{\bar{z}}{|z|^2} = \frac{x - iy}{x^2+y^2} \quad (9)$$

and

$$u = \frac{x}{x^2+y^2} \qquad v = \frac{-y}{x^2+y^2} \quad (10)$$

On the other hand, since

$$z = \frac{1}{\omega} = \frac{\bar{\omega}}{|\omega|^2}$$

$$x = \frac{u}{u^2+v^2} \qquad y = \frac{-v}{u^2+v^2} \quad (11)$$

Let a,b,c,d be real numbers such that

$$b^2 + c^2 > 4ad \quad (12)$$

Then,

$$a(x^2+y^2) + bx + cy + d = 0 \quad (13)$$

represents a circle for $a \neq 0$ and a line for $a = 0$.

If x and y satisfy eq.(13), then by substituting (11) into (13), we obtain

$$a(x^2+y^2) + bx + cy + d$$

$$= a\left[\frac{u^2}{(u^2+v^2)^2} + \frac{v^2}{(u^2+v^2)^2}\right] + \frac{bu}{u^2+v^2} - \frac{cv}{u^2+v^2} + d \quad (14)$$

$$= \frac{a}{u^2+v^2} + \frac{bu}{u^2+v^2} - \frac{cv}{u^2+v^2} + d = 0$$

or

$$d(u^2+v^2) + bu - cv + a = 0 \quad (15)$$

which also represents a circle or a line. Conversely, if u and v satisfy (13), it follows that x and y also satisfy (13).

Thus, a circle not passing through the origin (d≠0) in the z plane is transformed into a circle not passing through the origin (a≠0) in the ω plane.

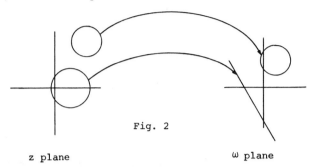

Fig. 2

z plane   ω plane

A circle through the origin in the z plane is transformed into a line which does not pass through the origin in the ω plane.

● **PROBLEM 20-6**

Find the image of a vertical line

$$x = a_1 \quad (b \neq 0) \qquad (1)$$

under the transformation

$$\omega = \frac{1}{z} \qquad (2)$$

Consider different values of b (b > 0, b < 0).

Solution: The line is given by

$$z = x + iy = a_1 \qquad (3)$$

Then, eq.(13) of Problem 20-5 reduces to

$$x - a_1 = 0 \qquad (4)$$

and eq.(15) of Problem 20-5 becomes

$$-a_1(u^2+v^2) + u = 0 \qquad (5)$$

which after some simplifications leads to

$$\left(u - \frac{1}{2a_1}\right)^2 + v^2 = \left(\frac{1}{2a_1}\right)^2 \qquad (6)$$

This is the equation of a circle tangent to the v axis and centered on the u axis. A point on the line (a_1,y) is transformed to

$$(a_1,y) \rightarrow (u,v) = \left(\frac{a_1}{a_1^2 + y^2}, \frac{-y}{a_1^2 + y^2}\right) \qquad (7)$$

For $a_1 > 0$, the circle is located on the right-hand side of the v axis. For $a_1 < 0$, the circle is located on the left.

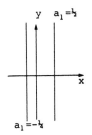
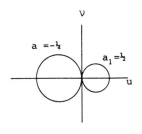

Fig. 1

● **PROBLEM 20-7**

Find the image of a horizontal line

$$y = a_2 \quad (a_2 \neq 0) \qquad (1)$$

under the transformation

$$\omega = \frac{1}{z}. \qquad (2)$$

Solution: From Problem 20-5 we see that the line

$$y - a_2 = 0 \qquad (3)$$

is transformed into

$$\left(v + \frac{1}{2a_2}\right)^2 + u^2 = \left(\frac{1}{2a_2}\right)^2, \qquad (4)$$

which is a circle tangent to the u axis and centered on the v axis. For $a_2 > 0$, the circle is located below the u axis, for $a_2 < 0$ the circle is located above the u axis.

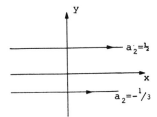
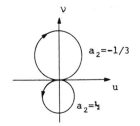

Fig. 1

For $a_2 > 0$, as the point on the line moves to the right, its image traverses the circle counterclockwise. The point at infinity in the extended z plane corresponds to the origin in the ω plane.

● **PROBLEM 20-8**

Find the image of a half plane $x > a_1$ $(a_1 > 0)$ under the transformation

$$\omega = \frac{1}{z} \qquad (1)$$

Solution: We have

$$z = \frac{1}{\omega} = \frac{\bar{\omega}}{|\omega|^2} \qquad (2)$$

or

$$x = \frac{u}{u^2+v^2}, \quad y = \frac{-v}{u^2+v^2} \quad (3)$$

Hence,

$$\frac{u}{u^2+v^2} > a_1 \quad (4)$$

or

$$\left(u - \frac{1}{2a_1}\right)^2 + v^2 < \left(\frac{1}{2a_1}\right)^2 \quad (5)$$

The image of any point (x,y) in the half-plane $x > a_1$ lies inside the circle

$$\left(u - \frac{1}{2a_1}\right)^2 + v^2 = \left(\frac{1}{2a_1}\right)^2 \quad (6)$$

Conversely, every point inside the circle given by (6) satisfies inequality (4) and is the image of a point in the half-plane.

Thus, the image of the half-plane is the entire circular domain.

● **PROBLEM 20-9**

Find the image of the infinite strip

$$0 < y < \frac{1}{2a}, \quad a > 0 \quad (1)$$

under the transformation $\omega = \frac{1}{z}$.

Solution: Transformation $\omega = \frac{1}{z}$ can be expressed in terms of transformations of its components.

$$x = \frac{u}{u^2+v^2} \quad (2)$$

$$y = \frac{-v}{u^2+v^2} \quad (3)$$

Combining (1) and (3), we obtain

$$0 < \frac{-v}{u^2+v^2} < \frac{1}{2a} \quad (4)$$

which after some calculations leads to

$$v < 0$$
$$a^2 < u^2 + (v+a)^2 \quad (5)$$

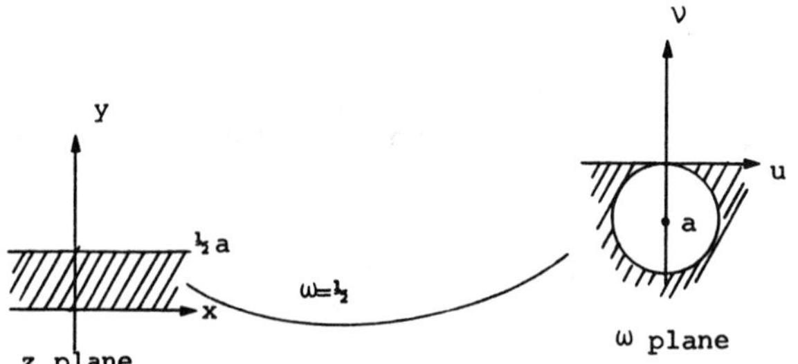

Fig. 1

The shaded area in the z plane (Fig. 1) is mapped under $\omega = \frac{1}{z}$ onto the shaded area in the ω plane.

• **PROBLEM** 20-10

Show that the transformation $\omega = \frac{1}{z}$ maps the shaded areas of the z plane onto the shaded areas of the ω plane.

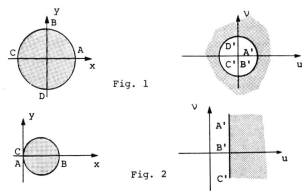

Fig. 1

Fig. 2

Solution: I. The shaded area (Fig. 1) in the z plane is described by

$$x^2 + y^2 \leq a^2 \tag{1}$$

The transformation $\omega = \frac{1}{z}$ is given by

$$x = \frac{u}{u^2+v^2}$$
$$y = \frac{-v}{u^2+v^2} \tag{2}$$

Hence,

$$x^2+y^2 = \frac{u^2+v^2}{(u^2+v^2)^2} \leq a^2 \tag{3}$$

or

$$\frac{1}{a^2} \leq u^2 + v^2 \tag{4}$$

which describes the shaded area in the ω plane.

II. The area is described by

$$(x-a)^2 + y^2 \leq a^2 \tag{5}$$

Substituting (2) into (5), we find

$$\left[\frac{u}{u^2+v^2} - a\right]^2 + \frac{v^2}{(u^2+v^2)^2} \leq a^2 \tag{6}$$

or

$$\frac{1}{2a} \leq u \tag{7}$$

which is the shaded area in the ω plane.

• **PROBLEM 20-11**

Show that the transformation

$$\omega = \frac{1}{z} \tag{1}$$

maps the hyperbola

$$x^2 - y^2 = 1 \tag{2}$$

onto the lemniscate

$$\rho^2 = \cos 2\phi \tag{3}$$

Solution: Since

$$x = \frac{z + \bar{z}}{2}$$
$$y = \frac{z - \bar{z}}{2i} \tag{4}$$

we can denote

$$x^2 - y^2 = \left(\frac{z+\bar{z}}{2}\right)^2 - \left(\frac{z-\bar{z}}{2i}\right)^2 = \frac{z^2 + \bar{z}^2}{2} \tag{5}$$

Equation (2) becomes

$$z^2 + \bar{z}^2 = 2 \tag{6}$$

Under the transformation $\omega = \frac{1}{z}$, (6) is mapped into

$$\frac{1}{\omega^2} + \frac{1}{\bar{\omega}^2} = 2 \tag{7}$$

Expressing ω in the polar form

$$\omega = \rho e^{i\phi} \tag{8}$$

we transform (7) into

$$\frac{1}{\rho^2 e^{2i\phi}} + \frac{1}{\rho^2 e^{-2i\phi}} = 2 \tag{9}$$

or

$$\rho^2 = \cos 2\phi \tag{10}$$

which is the equation of the lemniscate.

LINEAR FRACTIONAL TRANSFORMATION

• **PROBLEM 20-12**

I. If a, b, c, d are complex constants such that

$$ad - bc \neq 0 \tag{1}$$

then the transformation

$$T(z) = \frac{az+b}{cz+d} \tag{2}$$

is called a linear fractional transformation, or Möbius

transformation.

Explain eq.(1).

II. Define the linear fractional transformation T given by (2) on the extended z plane.

Solution: I. Eq.(2) can be written in the form

$$\omega = T(z) = \frac{a}{c} + \frac{bc-ad}{c} \cdot \frac{1}{cz+d} \tag{3}$$

Condition (1) ensures that a linear fractional transformation is never a constant function. Eq.(3) can be written as

$$Az\omega + Bz + C\omega + D = 0 \tag{4}$$

which is a bilinear transformation, or

$$z = \frac{-d\omega+b}{c\omega - a} \tag{5}$$

II. The linear fractional transformation

$$T(z) = \frac{az+b}{cz+d}, \quad ad - bc \neq 0 \tag{6}$$

can be defined on the extended z plane. If $c = 0$

$$T(\infty) = \infty \tag{7}$$

if $c \neq 0$

$$T\left(-\frac{d}{c}\right) = \infty$$

$$T(\infty) = \frac{a}{c} \tag{8}$$

This "new" definition of T(z) ensures that the linear fractional transformation is a one-to-one mapping of the extended z plane onto the extended ω plane. If $z_1 \neq z_2$, then $T(z_1) \neq T(z_2)$, and for each ω there exists z in the z plane such that

$$T(z) = \omega. \tag{9}$$

We can define the inverse transformation

$$T^{-1}(\omega) = z \quad \text{if} \quad T(z) = \omega.$$

● **PROBLEM 20-13**

Show that a linear fractional transformation transforms circles and lines into circles and lines.

Solution: A linear fractional transformation

$$\omega = \frac{az+b}{cz+d}, \quad ad - bc \neq 0 \tag{1}$$

when $c \neq 0$, can be expressed

$$\omega = \frac{a}{c} + \frac{bc-ad}{c} \cdot \frac{1}{cz+d} \tag{2}$$

This last transformation can be written as a succession of transformations.

$$z' = cz + d \tag{3}$$

$$z'' = \frac{1}{z'} \tag{4}$$

$$\omega = \frac{a}{c} + \frac{bc-ad}{c} \cdot z'' \tag{5}$$

Transformation (3) is a linear transformation which transforms lines and circles into lines and circles.

Transformation $z'' = \frac{1}{z'}$ transforms lines and circles into lines and circles (see Problem 20-5).

Transformation (5) is also a linear transformation. That completes the proof.

● **PROBLEM 20-14**

Find a linear fractional transformation which maps points z_1, z_2, z_3 of the z plane into points $\omega_1, \omega_2, \omega_3$ of the ω plane respectively.

Solution: We have

$$\omega - \omega_k = \frac{az+b}{cz+d} - \frac{az_k+b}{cz_k+d}$$

$$= \frac{(ad-bc)(z-z_k)}{(cz+d)(cz_k+d)} \tag{1}$$

for $k = 1, 2, 3$.

Then

$$\omega - \omega_1 = \frac{(ad-bc)(z-z_1)}{(cz+d)(cz_1+d)} \tag{2}$$

and

$$\omega - \omega_3 = \frac{(ad-bc)(z-z_3)}{(cz+d)(cz_3+d)} \tag{3}$$

Replacing ω by ω_2 and z by z_2, we obtain

$$\omega_2 - \omega_1 = \frac{(ad-bc)(z_2-z_1)}{(cz_2+d)(cz_1+d)} \tag{4}$$

$$\omega_2 - \omega_3 = \frac{(ad-bc)(z_2-z_3)}{(cz_2+d)(cz_3+d)} \tag{5}$$

Dividing (2) by (3) and (5) by (4) and multiplying the corresponding results, we find

$$\frac{(\omega-\omega_1)(\omega_2-\omega_3)}{(\omega-\omega_3)(\omega_2-\omega_1)} = \frac{(z-z_1)(z_2-z_3)}{(z-z_3)(z_2-z_1)} \tag{6}$$

Solving (6) for ω in terms of z gives the required transformation.

• **PROBLEM 20-15**

Find a bilinear transformation which maps points

$z_1 = 2$, $z_2 = 0$, $z_3 = -1$ of the z plane into

$\omega_1 = 1$, $\omega_2 = \infty$, $\omega_3 = i$ of the ω plane.

Solution: We shall apply formula (6) of Problem 20-14.

$$\frac{(\omega-\omega_1)(\omega_2-\omega_3)}{(\omega-\omega_3)(\omega_2-\omega_1)} = \frac{(z-z_1)(z_2-z_3)}{(z-z_3)(z_2-z_1)} \qquad (1)$$

The left-hand side of (1) can be expressed as

$$\frac{(\omega-\omega_1)(1-\frac{\omega_3}{\omega_2})}{(\omega-\omega_3)(1-\frac{\omega_1}{\omega_2})} = \frac{\omega-\omega_1}{\omega-\omega_3} \qquad (2)$$

because $\omega_2 = \infty$ and

$$\frac{\omega-\omega_1}{\omega-\omega_3} = \frac{(z-z_1)(z_2-z_3)}{(z-z_3)(z_2-z_1)} \qquad (3)$$

Substituting the corresponding numbers into (3), we find

$$\frac{\omega-1}{\omega-i} = \frac{z-2}{-2(z+1)} \qquad (4)$$

or

$$\omega = \frac{2(z+1)+i(z-2)}{3z} \qquad (5)$$

FIXED POINTS

• **PROBLEM 20-16**

Show that every linear fractional transformation which is not the identity transformation $\omega = z$, has at most two fixed points in the extended z plane.

A fixed point of a transformation $\omega = f(z)$ is a point z_0, such that

$$f(z_0) = z_0 \qquad (1)$$

Solution: We shall utilize the formula proved in Problem 20-14.

$$\frac{(\omega-\omega_1)(\omega_2-\omega_3)}{(\omega-\omega_3)(\omega_2-\omega_1)} = \frac{(z-z_1)(z_2-z_3)}{(z-z_3)(z_2-z_1)} \qquad (2)$$

Let us assume that a linear fractional transformation which is not the identity transformation has three fixed points: z_1, z_2 and z_3. Then in eq.(2) we substitute

$$\frac{(\omega-z_1)(z_2-z_3)}{(\omega-z_3)(z_2-z_1)} = \frac{(z-z_1)(z_2-z_3)}{(z-z_3)(z_2-z_1)} \qquad (3)$$

Multiplying and simplifying (3) we obtain

$$\omega = z \qquad (4)$$

which is the contradiction.

Hence, a linear fractional transformation (not the identity) has at most two fixed points.

• **PROBLEM 20-17**

Show that if the origin is a fixed point of a linear fractional transformation, then the transformation can be written in the form

$$\omega = \frac{z}{cz+d}, \quad d \neq 0 \tag{1}$$

<u>Solution</u>: In general, a linear fractional transformation can be written in the form

$$\omega = \frac{az+b}{cz+d}, \quad ad - bc \neq 0 \tag{2}$$

Since the origin is a fixed point of the transformation

$$0 = \frac{a \cdot 0 + b}{c \cdot 0 + d} \tag{3}$$

or

$$b = 0,$$

Eq.(2) becomes

$$\omega = \frac{az}{cz+d}, \quad ad \neq 0 \tag{4}$$

or

$$\omega = \frac{z}{\frac{c}{a}z + \frac{d}{a}}, \quad \frac{d}{a} \neq 0 \tag{5}$$

which is eq.(1).

TRANSFORMATIONS OF REGIONS

• **PROBLEM 20-18**

Find the family of all linear fractional transformations that map the upper half plane Im z > 0 onto the open disk

$$|\omega| < 1 \tag{1}$$

and the boundary Imz = 0 onto the boundary $|\omega| = 1$.

<u>Solution</u>: We shall start with the equation

$$\omega = \frac{az+b}{cz+d}, \quad ad - bc \neq 0 \tag{2}$$

The line Imz = 0 is transformed into $|\omega| = 1$.

Since the image of z = ∞ is ω = ∞ when c = 0, we obtain

$$c \neq 0 \tag{3}$$

When $c \neq 0$, the image of $z = \infty$ is $\omega = \dfrac{a}{c}$ and the condition $|\omega| = 1$ for $z = \infty$ leads to

$$|a| = |c| \neq 0 \tag{4}$$

Since $|\omega| = 1$ for $z = 0$, we obtain

$$|b| = |d| \neq 0 \tag{5}$$

Eq.(4) is

$$\omega = \left(\frac{a}{c}\right) \frac{z + \frac{b}{a}}{z + \frac{d}{c}} \tag{6}$$

Since

$$\left|\frac{a}{c}\right| = 1 \tag{7}$$

we can denote

$$\omega = e^{i\alpha} \frac{z-z_0}{z-z_1} \tag{8}$$

where α is a real constant and z_0, z_1 are non-zero complex constants.

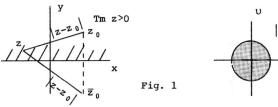

Fig. 1

Since

$$\left|\frac{d}{c}\right| = \left|\frac{b}{a}\right| \tag{9}$$

we get

$$|z_1| = |z_0| \tag{10}$$

When $z = 1$, $|\omega| = 1$, and (8) leads to

$$|1-z_1| = |1-z_0| \tag{11}$$

or

$$(1-z_1)(1-\overline{z}_1) = (1-z_0)(1-\overline{z}_0) \tag{12}$$

or

$$z_1 + \overline{z}_1 = z_0 + \overline{z}_0 \tag{13}$$

because $|z_1| = |z_0|$ and $z_1\overline{z}_1 = z_0\overline{z}_0$.

$$\mathrm{Re}\, z_0 = \mathrm{Re}\, z_1$$

Thus, since $|z_0| = |z_1|$, we have either $z_0 = z_1$ or $z_1 = \overline{z}_0$.

For $z_0 = z_1$, transformation (8) becomes a constant function. We are left with

$$z_1 = \overline{z}_0 \tag{14}$$

677

and (8) becomes

$$\omega = e^{i\alpha} \frac{z-z_0}{z-\bar{z}_0} \tag{15}$$

Transformation (15) maps z_0 onto the origin $\omega = 0$. Since points $|\omega| < 1$ are the images of points above the real axis, we have

$$\mathrm{Im}\, z_0 > 0 \tag{16}$$

Thus, the answer is

$$\omega = e^{i\alpha} \frac{z-z_0}{z-\bar{z}_0}, \qquad \mathrm{Im}\, z_0 > 0 \tag{17}$$

Fig. 1 illustrates that the converse is also true. Any linear fractional transformation (17) has the desired mapping properties (see Problem 20-19).

• **PROBLEM 20-19**

Show that the linear fractional transformation

$$\omega = e^{i\alpha} \frac{z-z_0}{z-\bar{z}_0}, \qquad \mathrm{Im}\, z_0 > 0 \tag{1}$$

maps the upper half of the z plane onto $|\omega| < 1$.

Solution: From (1) we have

$$|\omega| = \left| e^{i\alpha} \frac{z-z_0}{z-\bar{z}_0} \right| = \left| \frac{z-z_0}{z-\bar{z}_0} \right| \tag{2}$$

Points z_0 and z are located in the upper half of the z plane (see Fig. 1).

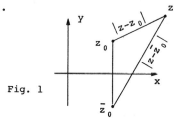

Fig. 1

From the figure we see that

$$|z-z_0| \leq |z-\bar{z}_0| \tag{3}$$

Hence, $|\omega| \leq 1$.

• **PROBLEM 20-20**

Consider the transformation

$$\omega = z^2 \tag{1}$$

1. Find the image of the first quadrant in the z plane under the transformation (1).

2. Show that the straight lines $x = c$ in the z plane are mapped into a family of parabolas in the ω plane.

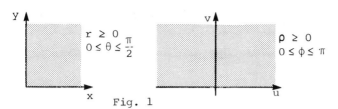

Fig. 1

Solution: 1. Transformation (1) can be described in terms of polar coordinates.

If $z = re^{i\theta}$ and $\omega = \rho e^{i\phi}$, then $\rho e^{i\phi} = r^2 e^{2i\theta}$ (2)

Hence, $|\omega| = |z|^2$ and $\arg \omega = 2 \arg z$ (3)

The image of the first quadrant $r \geq 0$, $0 \leq \theta \leq \frac{\pi}{2}$ in the z plane is the upper half $\rho \geq 0$, $0 \leq \phi \leq \pi$ of the ω plane (Fig. 1).

2. The transformation $\omega = z^2$ can be represented by

$$\omega = u + iv = z^2 = x^2 - y^2 + 2xyi \quad (4)$$

or

$$u = x^2 - y^2$$
$$v = 2xy \quad (5)$$

The straight line $x = c$ is transformed into a parabola in the ω plane

$$u = c^2 - y^2$$
$$v = 2cy \quad -\infty < y < \infty \quad (6)$$

or

$$v^2 = -4c^2(u-c^2) \quad (7)$$

Note that the lines $y = a$ in the z plane are mapped into a family of parabolas in the ω plane.

$$u = x^2 - a^2$$
$$v = 2ax \quad -\infty < x < \infty \quad (8)$$

or

$$v^2 = 4a^2(u+a^2) \quad (9)$$

● PROBLEM 20-21

Find the image of the region D bounded by

$$x = 2, \quad y = 0, \quad x^2 - y^2 = 1, \quad x \geq 0, \quad y \geq 0 \quad (1)$$

under the transformation

$$\omega = z^2 \quad (2)$$

Solution: Transformation (2) can be written in the component form

$$u = x^2 - y^2$$
$$v = 2xy \quad (3)$$

The image of the region D is the region D' in the ω plane bounded by

$$v = 0, \quad v^2 = -16(u-4), \quad v \geq 0, \quad u = 1 \tag{4}$$

as shown in Fig. 1.

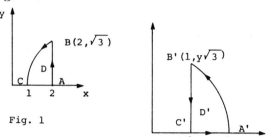

Fig. 1

The corresponding points are shown with arrows indicating corresponding directions.

● **PROBLEM** 20-22

Verify that the transformation

$$\omega = z^2 \tag{1}$$

maps the regions shown in Fig. 1.

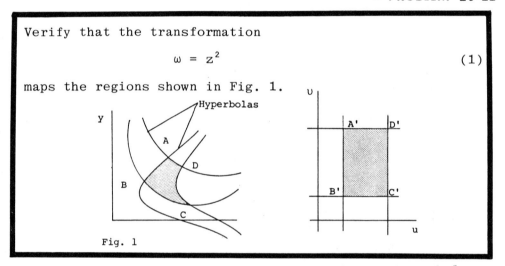

Fig. 1

Solution: We shall show that each branch of a hyperbola

$$x^2 - y^2 = c, \quad c > 0 \tag{2}$$

is mapped one-to-one onto the vertical line $u = c$.

We have

$$x^2 - y^2 = u = c$$
$$v = 2y\sqrt{y^2+c} \quad -\infty < y < \infty \tag{3}$$

Hence, the hyperbola is mapped onto the vertical line.

In a similar manner, we can show that each branch of a hyperbola

$$2xy = c \quad (c > 0) \tag{4}$$

is transformed into the line

$$v = c \quad (c > 0) \tag{5}$$

as shown in Fig. 2.

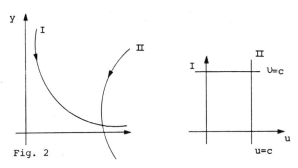

Fig. 2

That verifies Fig. 1.

● **PROBLEM 20-23**

Find the transformation (analytic) which maps the entire z plane onto the open unit disk $|\omega| < 1$ (see Fig. 1), and which is one-to-one.

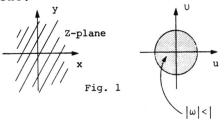

Fig. 1

Solution: We will show that no such analytic mapping exists. Assume the contrary. Then the mapping is bounded. By Liouville's theorem (a bounded entire function is a constant), we conclude that the mapping is a constant. Furthermore, no such linear fractional transformation exists.

A linear fractional transformation defined on the set $C \cup \{\infty\}$ (C is the set of complex numbers), is one-to-one and onto. The transformation we seek is not onto $C \cup \{\infty\}$.

Obviously, no linear fractional transformation exists which maps the open unit disk onto the entire z plane.

● **PROBLEM 20-24**

Find a linear fractional transformation which maps the open disk $|z| < 1$ onto the upper half of the ω plane.

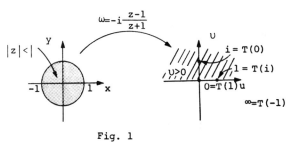

Fig. 1

Solution: The transformation we look for maps the circle $|z| = 1$ onto the real u axis.

The linear fractional transformation is determined by three values (see Problem 20-14). We choose the points on the unit circle $|z| = 1$;

$$z_0 = 1, \quad z_1 = i, \quad z_2 = -1, \tag{1}$$

and their images on the u axis

$$\omega_0 = 0, \quad \omega_1 = 1, \quad \omega_2 = \infty$$

respectively.

Thus, the transformation

$$\omega = T(z) = \frac{i+1}{i-1} \frac{z-1}{z+1} = -i \frac{z-1}{z+1} \tag{2}$$

maps z_0, z_1, z_2 into $\omega_0, \omega_1, \omega_2$. Transformation (2) maps $|z| < 1$ onto the upper half of the ω plane ($T(0) = i$).

• **PROBLEM 20-25**

Find all linear fractional transformations that map the unit open disk onto itself.

Solution: Let us use the auxiliary complex η plane (see Fig. 1). By $P(z)$ we denote any linear fractional transformation of the disk $|z| < 1$ onto the upper half of the η plane

$$P(z) = \eta \tag{1}$$

Every linear fractional transformation

$$\omega = T(z) \tag{2}$$

of the disk $|z| < 1$ onto the disk $|\omega| < 1$ has a unique composition

$$\omega = T(z) = T'(P(z)) \tag{3}$$

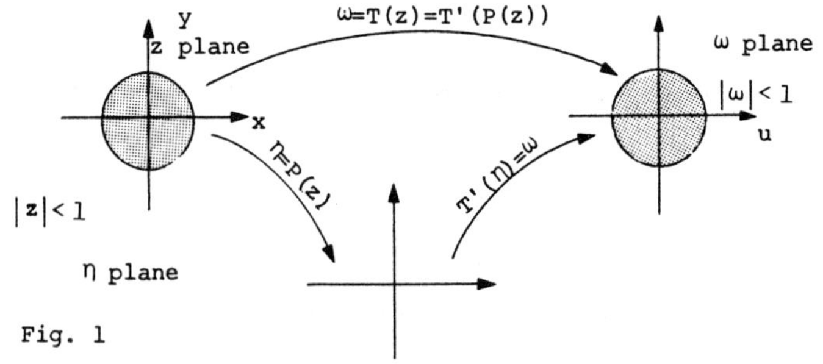

Fig. 1

where $\omega = T'(\eta)$ is any mapping of the upper half plane onto the unit disk.

$$T'(\eta) = T(P^{-1}(\eta)) \tag{4}$$

Every T' has the form

$$\omega = T'(\eta) = e^{i\phi} \frac{\eta - \eta_0}{\eta - \bar{\eta}_0} \tag{5}$$

(see Problem 20-18).

From Problem 20-24, we obtain transformation $P(z)$

$$\eta = P(z) = -i\frac{z-1}{z+1} \quad (6)$$

Substituting (5) and (6) into

$$\omega = T(z) = T'(P(z)), \quad (7)$$

we find

$$\omega = T(z) = e^{i\alpha}\frac{z-z_1}{z\bar{z}_1-1} \quad (8)$$

where α is real and $|z_1| < 1$.

SOME SPECIAL FUNCTIONS

• **PROBLEM** 20-26

Define the principal branch of the double-valued function $\sqrt{z}$.

Solution: When $z \neq 0$, the values of $z^{\frac{1}{2}}$ are two square roots of z. In polar coordinates

$$z = re^{i\theta}, \quad -\pi < \theta \leq \pi \quad (1)$$

Then,

$$z^{\frac{1}{2}} = \sqrt{r}\, e^{\frac{i(\theta+2k\pi)}{2}} \quad (2)$$

where $k = 0, 1$.

For $k = 0$, eq.(2) yields the principal root.

The principal branch $F_0(z)$ of the double-valued function $z^{\frac{1}{2}}$ is obtained by taking $k = 0$

$$F_0(z) = \sqrt{r}\, e^{\frac{i\theta}{2}} \quad (3)$$

$$r \geq 0, \quad -\pi < \theta < \pi$$

Another way is to express $z^{\frac{1}{2}}$ by

$$z^{\frac{1}{2}} = e^{\frac{1}{2}\log z} \quad (4)$$

and to take the principal branch of $\log z$, which is sometimes indicated by $\text{Log } z$, then

$$F_0(z) = e^{\frac{1}{2}\text{Log } z} \quad (5)$$

$$|z| > 0, \quad -\pi < \text{Arg } z < \pi$$

The ray $\theta = \pi$ is the branch cut for $F_0(z)$, and the point $z = 0$ is the branch point.

● **PROBLEM 20-27**

Show that the transformation

$$\omega = F_0(z) \tag{1}$$

where $F_0(z)$ denotes the principal branch of $z^{\frac{1}{2}}$, maps the region

$$0 < r \le r_0, \quad -\pi < \theta < \pi \tag{2}$$

onto the half disk

$$0 < \rho \le \sqrt{r_0}, \quad -\frac{\pi}{2} < \phi < \frac{\pi}{2} \tag{3}$$

in the ω plane (see Fig. 1).

Solution: Transformation $z^{\frac{1}{2}}$ can be written in the form

$$z^{\frac{1}{2}} = \sqrt{r}\, e^{\frac{i(\theta+2k\pi)}{2}} \tag{4}$$

where $k = 0, 1$.

The principal branch $F_0(z)$ is obtained from (4) by setting $k = 0$

$$F_0(z) = \sqrt{r}\, e^{\frac{i\theta}{2}} \tag{5}$$

Eq.(1) becomes

$$\omega = \sqrt{r}\, e^{\frac{i\theta}{2}} \tag{6}$$

Under transformation 6, region $0 < r \le r_0$, $-\pi < \theta < \pi$ is mapped onto $0 < \rho \le \sqrt{r_0}$, $-\frac{\pi}{2} < \phi < \frac{\pi}{2}$.

From (6), we obtain

$$\rho = \sqrt{r}, \quad \phi = \frac{\theta}{2} \tag{7}$$

$0 < r \le r_0$
$-\pi < \theta < \pi$

Fig. 1

$0 < \rho \le \sqrt{r_0}$
$-\frac{\pi}{2} < \phi < \frac{\pi}{2}$

● **PROBLEM 20-28**

Find the branches of the double-valued function

$$f(z) = (z - z_0)^{\frac{1}{2}} \tag{1}$$

Solution: Function (1) can be decomposed into translation
$$z' = z - z_0 \quad \text{and function } z'^{\frac{1}{2}}.$$

In polar coordinates
$$z' = R e^{i\theta} \qquad (2)$$

and the branches of $z'^{\frac{1}{2}}$ are

$$z'^{\frac{1}{2}} = \sqrt{R}\, e^{\frac{i\theta}{2}}, \quad 0 < R,\ \alpha < \theta < \alpha + 2\pi \qquad (3)$$

Since
$$R = |z - z_0| \quad \text{and}$$
$$\theta = \arg(z - z_0), \qquad (4)$$

denoting
$$H = \text{Arg}(z - z_0) \qquad (5)$$

we obtain two branches of $(z - z_0)^{\frac{1}{2}}$

$$H_0(z) = \sqrt{R}\, e^{\frac{iH}{2}}, \quad R > 0,\ -\pi < H < \pi$$
and
$$h_0(z) = \sqrt{R}\, e^{\frac{i\theta}{2}}, \quad R > 0,\ 0 < \theta < 2\pi \qquad (6)$$

The branch of $z'^{\frac{1}{2}}$ in $H_0(z)$ is defined at all points in the z' plane, except for the origin and the points on the ray $\text{Arg } z' = \pi$.

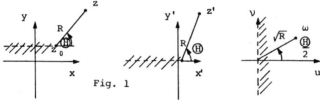

Fig. 1

The transformation $\omega = H_0(z)$ is a one-to-one mapping of the domain $|z - z_0| > 0$, $-\pi < \text{Arg}(z - z_0) < \pi$ onto the right half of the plane $\text{Re}\,\omega > 0$.

CHAPTER 21

CONFORMAL MAPPINGS, BOUNDARY VALUE PROBLEM

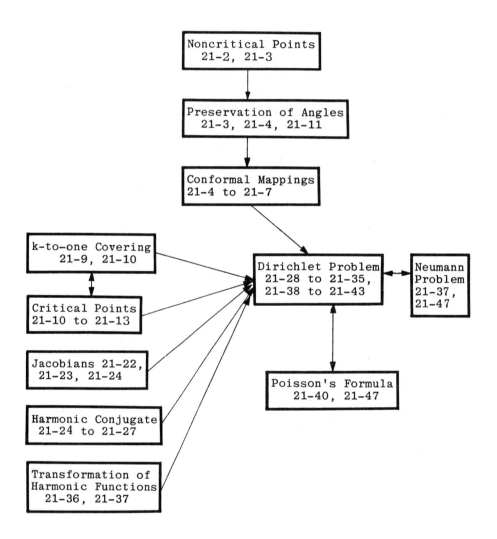

This chart is provided to facilitate rapid understanding of the interrelationships of the topics and subject matter in this chapter. Also shown are the problem numbers associated with the subject matter.

CRITICAL AND NON CRITICAL POINTS

● **PROBLEM** 21-1

Locate all critical points of the following functions and determine the order:

1. z^2
2. e^z
3. z^k
4. $\cos z$
5. $z + \dfrac{1}{z}$

Solution: We shall start with the following definition:

If $f'(z_0) = 0$, then z_0 is a critical point for $f(z)$. If the first $k - 1$ derivatives of $f(z)$ vanish at z_0,

$$f'(z_0) = \ldots = f^{(k-1)}(z_0) = 0, \qquad (1)$$

but

$$f^{(k)}(z_0) \neq 0 \qquad (2)$$

then $f(z)$ has order k at z_0.

1. The derivatives are

$$z^2, \quad \frac{d}{dz}(z^2) = 2z, \quad \frac{d}{dz}(2z) = 2, \quad \frac{d}{dz}(2) = 0$$

Hence, $z = 0$ is a critical point for z^2 of order 2.

2. Function e^z has no critical points.

3. Function z^k has order k at $z = 0$, and order 1 at all other points.

4. The critical points of $\cos z$ are zeros of $-\sin z$, i.e.

$$z = 0, \quad \pm\pi, \quad \pm 2\pi, \ldots \qquad (3)$$

They are critical points of order 2.

5. The critical points are of order 2 at

$$z = 1, \quad z = -1.$$

● **PROBLEM** 21-2

The following theorem describes an interesting property of analytic functions:

If f is an analytic function in a domain D containing z_0, which is not a critical point for f, i.e., $f'(z_0) \neq 0$, then f is locally one-to-one near z_0 (see Fig. 1). An open neighborhood D_0 of z_0 exists such that $D_0 \subset D$. If $z_1 \neq z_2$ in D_0, then

$$f(z_1) \neq f(z_2) \tag{1}$$

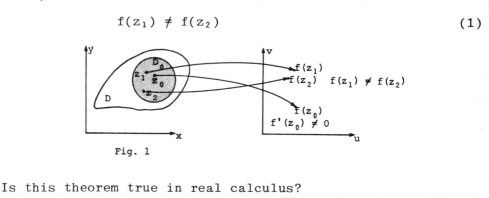

Fig. 1

Is this theorem true in real calculus?

Solution: If the function $f(x)$ at x_0 is such that $f'(x_0) \neq 0$, then its graph $y = f(x)$ is decreasing or increasing and the function is locally one-to-one. The theorem holds for real functions.

The difference between real and imaginary functions lies in the conditions $f'(z_0) \neq 0$ (or $f'(x_0) \neq 0$). Real function $f(x)$ can be one-to-one in the neighborhood of a critical point. For example, $f(x) = x^3$ is one-to-one near $x_0 = 0$.

This is not true for analytic functions.

• **PROBLEM 21-3**

Let C be a smooth arc, defined by

$$z = z(t), \qquad a \leq t \leq b, \tag{1}$$

and let $w = f(z) = f(z(t))$ be a function defined at all points of C.

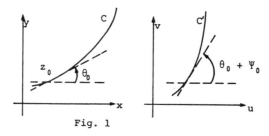

Fig. 1

The image of C under the transformation $w = f(z)$ is C', represented by

$$w = f(z(t)), \qquad a \leq t \leq b \tag{2}$$

Let z_0 be any point on C and let $f(z)$ be analytic at z_0 and $f'(z_0) \neq 0$ (z_0 is the noncritical point). (3)

Show that under this transformation, the tangent at z_0 to

C in the z plane is rotated through the angle arg $f'(z_0)$, i.e., show that

$$\psi_0 = \arg f'(z_0) \tag{4}$$

See Fig. 1.

Solution: Applying the chain rule to the function $\omega = f(z) = f(z(t))$, we obtain

$$\omega'(t_0) = f'[z(t_0)] z'(t_0) \tag{5}$$

We recall an important identity involving arguments:

$$\arg(z_1 z_2) = \arg z_1 + \arg z_2, \tag{6}$$

where z_1 and z_2 are any complex numbers. From (5) we find

$$\arg \omega'(t_0) = \arg f'[z(t_0)] + \arg z'(t_0) \tag{7}$$

From Fig. 1 we see that if T denotes the unit tangent vector at z_0,

$$T = \frac{z'(t_0)}{|z'(t_0)|} \tag{8}$$

its angle of inclination is $\arg z'(t_0)$, and

$$\arg z'(t_0) = \theta_0 \tag{9}$$

Similarly,

$$\arg \omega'(t_0) = \theta_0 + \psi_0 \tag{10}$$

From (7), (9) and (10), we obtain

$$\arg f'(z_0) = \psi_0 \tag{11}$$

CONFORMAL MAPPING

● **PROBLEM 21-4**

Define conformal and isogonal transformations and prove the following theorem:

If $f(z)$ is analytic and $f'(z) \neq 0$ in a region D, then the mapping $\omega = f(z)$ is conformal at all points of D.

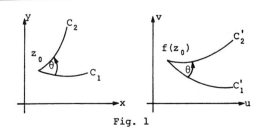

Fig. 1

Solution: Let

$$\omega = f(z) \tag{1}$$

be a transformation between the z plane and the ω plane.

Let C_1 and C_2 be two smooth arcs passing through z_0 which are mapped into C_1' and C_2', and passing through $\omega_0 = f(z_0)$. If the transformation is such that the angle at z_0 between C_1 and C_2 is equal to the angle at $\omega_0 = f(z_0)$ between C_1' and C_2', both in magnitude and sense, the transformation is called conformal at z_0. If the transformation preserves the magnitudes of angles, but not the sense, it is called isogonal.

Now we shall prove the theorem. Two smooth curves C_1 and C_2 are passing through z_0. Let θ_1 and θ_2 be angles of inclination of directed lines tangent to C_1 and C_2, respectively, at z_0 (see Fig. 1). The angles of inclination of directed lines tangent to the image curves C_1' and C_2' at the point $\omega_0 = f(z_0)$ are

$$\phi_1 = \theta_1 + \arg f'(z_0)$$
$$\phi_2 = \theta_2 + \arg f'(z_0) \tag{2}$$

(see Problem 21-3).

Hence,
$$\phi_2 - \phi_1 = \theta_2 - \theta_1 \tag{3}$$

The angle $\phi_2 - \phi_1$ is the same in magnitude and sense from C_1' to C_2' as the angle $\theta_2 - \theta_1$. Note that if the transformation $f(z)$ is conformal at z_0, then it is conformal at each point in a neighborhood of z_0.

● **PROBLEM 21-5**

Show that the mapping

$$f(z) = \bar{z} \tag{1}$$

is not conformal.

Solution: Let L_1 and L_2 be two straight lines in the z plane, as shown in Fig. 1. The transformation

$$f(z) = \bar{z} = \overline{x + iy} = x - iy \tag{2}$$

is a reflection across the x axis.

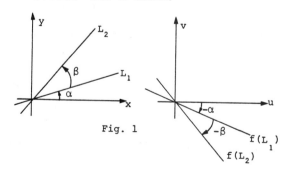

Fig. 1

A conformal transformation preserves the magnitude and sense of the angles. The angle from L_1 to L_2 is β, while the angle from $f(L_1)$ to $f(L_2)$ is $-\beta$.

Therefore, $f(z) = \bar{z}$ is not a conformal transformation.

• **PROBLEM 21-6**

Find the image of the region bounded by $x = 1$, $y = 1$ and $x + y = 1$ under the transformation

$$f(z) = z^2 \tag{1}$$

Using the boundaries of the region, verify that transformation (1) is conformal.

Solution: In the component form, $z^2 = w$ becomes

$$u + iv = x^2 - y^2 + 2ixy \tag{2}$$

$$u = x^2 - y^2$$

$$v = 2xy$$

Line $x = 1$ maps into

$$u = 1 - y^2, \quad v = 2y \quad \text{or}$$

$$u = 1 - \frac{v^2}{4} \tag{3}$$

Line $y = 1$ maps into

$$u = x^2 - 1, \quad v = 2x \quad \text{or}$$

$$u = \frac{v^2}{4} - 1 \tag{4}$$

and line $x + y = 1$ maps into

$$u = 2x - 1, \quad v = 2x - 2x^2 \quad \text{or}$$

$$v = \frac{1}{2}(1 - u^2), \tag{5}$$

as shown in Fig. 1.

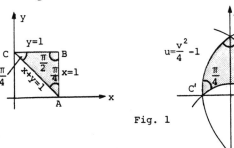

Fig. 1

Calculating the angles between C'B' and B'A', C'B' and C'A', C'A' and A'B', we find that they are equal to corresponding angles of the triangles ABC. This result can be obtained much faster from the theorem of Problem 21-4. The transformation $f(z) = z^2$ is analytic in the domain ABC (see Fig. 1), and $f(z) \neq 0$ everywhere in ABC. Therefore $f(z) = z^2$ is conformal in ABC.

• **PROBLEM** 21-7

In the previous problems, we have investigated the behavior of an analytic function at noncritical points, i.e., points z_0, such that $f'(z_0) \neq 0$.

The following theorem summarizes the basic facts:

Let $f : C \rightarrow W$ be an analytic function and $z_0 \in C$ a noncritical point, $f'(z_0) \neq 0$.

Then,

1. f is conformal at z_0;

2. f rotates angles at z_0 through constant arg $f'(z_0)$;

3. the distances near z_0 are magnified by a factor approximately equal to $|f'(z_0)|$.

Prove this theorem.

Solution: Part 1 has been proved in Problem 21-4.

From Problem 21-3 and 21-4, we conclude that a conformal mapping rotates angles at a noncritical point z_0 through arg $f'(z_0)$. Suppose z is near z_0. Then

$$|f'(z_0)| = \lim_{z \to z_0} \frac{|f(z) - f(z_0)|}{|z - z_0|} \tag{1}$$

Thus, for z close to z_0, $|f'(z_0)|$ is approximately equal to

$$|f'(z_0)| \approx \frac{|f(z) - f(z_0)|}{|z - z_0|} \tag{2}$$

or

$$|f(z) - f(z_0)| \approx |f'(z_0)||z - z_0| \tag{3}$$

Hence, the distances near z_0 are magnified by the factor $|f'(z_0)|$, which is called the local magnification factor.

Remember,

arg $f'(z_0)$:	rotates
$|f'(z_0)|$:	magnifies.

• **PROBLEM 21-8**

The fact that an analytic mapping has no critical points in D does not guarantee that it is globally one-to-one.

Discuss the mapping

$$f(z) = e^z \qquad (1)$$

and its one-to-one property around the origin.

Solution: Since

$$\frac{d}{dz}(e^z) = e^z \neq 0 \qquad (2)$$

mapping e^z has no critical points in C. If $w \neq 0$, then the equation $e^z = w$ has infinite solutions z. We see that even though $w = e^z$ is locally one-to-one, it is not globally one-to-one.

For example, $z_0 = 0$ is not a critical point of e^z. Hence, e^z is locally one-to-one around $z_0 = 0$. It is one-to-one in the disk $D(0,r)$.

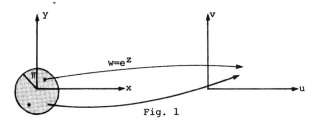

Fig. 1

The function e^z is periodic with period $2\pi i$.

$$e^{z+2k\pi i} = e^z \qquad (3)$$

The largest radius r of $D(0,r)$ for which e^z is one-to-one is

$$r = \pi \qquad (4)$$

• **PROBLEM 21-9**

A ray $x = y$ in the z plane from the origin is mapped by

$$f(z) = z^2 \qquad (1)$$

into the curve in the ω plane. Sketch the image curve and compare the angles at the origin.

For this case check the veracity of the theorem of Problem 20-11.

Solution: A ray $x = y$ consists of the points

$$z = x + ix \qquad (2)$$

or
$$z = t + it, \quad t \geq 0$$

It can be parametrized by

$$z = se^{i\frac{\pi}{4}}, \quad s \geq 0 \tag{3}$$

Its image in the plane is

$$f(z) = z^2 = s^2 e^{i\frac{\pi}{2}} \tag{4}$$

as shown in Fig. 1.

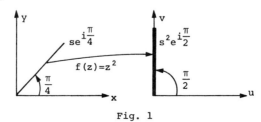

Fig. 1

The function $f(z) = z^2$ doubles angles at the origin.

According to the theorem of Problem 20-11 the angle θ at the origin in the z plane, becomes in the w plane

$$\arg f^{(k)}(z_0) + k\theta \tag{5}$$

In our case
$$f(z) = z^2$$
$$f'(z) = 2z \tag{6}$$
$$f''(z) = 2$$

and
$$f(0) = f'(0) = 0 \text{ while}$$
$$f''(0) = 2 \tag{7}$$

Hence, $f(z) = z^2$ has order 2 at $z_0 = 0$. Eq.(5) leads to

$$\arg f''(0) + 2\theta = 2\theta \tag{8}$$

The result is the same as in (4).

To summarize the results we can remark that if $f(z)$ has order k at z_0, i.e.,

$$f(z) = f(z_0) + \frac{f^{(k)}(z_0)}{k!}(z - z_0)^k + \text{ higher powers}$$
$$\text{of } (z - z_0) \tag{9}$$

$$f'(z_0) = f''(z_0) = \ldots = f^{(k-1)}(z_0) = 0 \tag{10}$$

then for z near z_0, $f(z)$ behaves like a polynomial of the form

$$P(z) = f(z_0) + \frac{f^{(k)}(z_0)}{k!}(z - z_0)^k \tag{11}$$

In particular if $k = 1$, $f'(z_0) \neq 0$ and $f(z)$ can be replaced near z_0, by a linear function

$$f(z_0) + f'(z_0)(z - z_0)$$

● **PROBLEM 21-10**

The theorem of Problem 21-2 described function's behavior near a noncritical point z_0. The following theorem deals with the critical points:

If $f : D \to C$ is analytic and f has order k at the point $z_0 \in D$, then f yields a local k-to-one covering near z_0.

Applying this theorem, describe the local covering property of

$$f(z) = (z - 2i)^3 e^z \qquad (1)$$

at the point $z_0 = 2i$.

Solution: The derivatives of $f(z)$ are

$$f'(z) = 3(z - 2i)^2 e^z + e^z(z - 2i)^3,$$
$$f''(z) = 6(z - 2i)e^z + f'(z) + 3e^z(z - 2i)^3, \qquad (2)$$
$$f^{(3)}(z) = 6e^z + \text{terms with } (z - 2i).$$

Thus, at $z = 2i$,

$$f'(2i) = f''(2i) = 0$$
$$f^{(3)}(2i) \neq 0 \qquad (3)$$

The function $f(z) = (z - 2i)^3 e^z$ has order 3 at $2i$. Since $f(z)$ is analytic, it gives a local 3-to-1 covering near $z_0 = 2i$.

● **PROBLEM 21-11**

Let $f : D \to C$ be an analytic function and $z_0 \in D$ be a critical point. f has order k at z_0.

A ray of the form

$$z = z_0 + se^{i\theta}, \quad s \geq 0, \qquad (1)$$

is mapped by f to the curve

$$f(z) = f(z_0 + se^{i\theta}) \qquad (2)$$

Find the tangent to $f(z)$ at the point $\omega_0 = f(z_0)$.

Solution: Since $f(z)$ has order k at z_0, its Taylor series is

$$f(z) = f(z_0) + \frac{f^{(k)}(z_0)}{k!}(z - z_0)^k$$
$$+ \frac{f^{(k+1)}(z_0)}{(k+1)!}(z - z_0)^{k+1} + \ldots \qquad (3)$$

From (1) we obtain

$$(z - z_0)^k = s^k e^{ik\theta} \qquad (4)$$

and (3) becomes

$$f(z) = \omega = \omega_0 + \frac{f^{(k)}(z_0)}{k!} s^k e^{ik\theta} + \frac{f^{(k+1)}(z_0)}{(k+1)!} s^{k+1} e^{i(k+1)\theta} + \ldots \qquad (5)$$

Substituting

$$s^k = t \qquad (6)$$

we obtain

$$\omega = \omega_0 + \frac{f^{(k)}(z_0)}{k!} t e^{ik\theta} + \frac{f^{(k+1)}(z_0)}{(k+1)!} t^{\frac{k+1}{k}} e^{i(k+1)\theta} + \ldots \qquad (7)$$

Differentiating (7) with respect to t and setting t = 0, we obtain a vector

$$\frac{f^{(k)}(z_0)}{k!} e^{ik\theta} \qquad (8)$$

The angle between this vector and the horizontal is

$$\arg\left[\frac{f^{(k)}(z_0)}{k!} e^{ik\theta}\right] = \arg f^{(k)}(z_0) + k\theta \qquad (9)$$

This result can be expressed in the form of the following theorem:

If $f : D \to C$ is analytic and has order $K \geq 2$ at a point $z_0 \in D$, then a ray $z_0 + se^{i\theta}$ extending from z_0 at an angle θ to the horizontal is mapped by f to a curve extending from $f(z_0) = w_0$ and at w_0, making an angle

$$\arg f^{(k)}(z_0) + k\theta$$

with the horizontal.

● **PROBLEM 21-12**

Find the image of a circle D under the transformation

$$\omega = f(z) = \frac{1}{2}\left(z + \frac{1}{z}\right) \qquad (1)$$

The circle D in the z plane has its center on the x axis. Point z = -1 is its interior point and it passes through z = 1.

Solution: First we must find the critical points of f(z).

$$\frac{df}{dz} = \frac{1}{2}\left(1 - \frac{1}{z^2}\right) = 0 \qquad (2)$$

Hence, z = 1 is a critical point of f(z).

At z = 1, the Taylor series of f(z) is

$$\omega = 1 + \frac{1}{2}\left[(z-1)^2 - (z-1)^3 + (z-1)^4 - \ldots\right] \qquad (3)$$

Transformation (1) doubles the angles at z = 1.

The angle at z = 1 exterior to C is π, therefore the angle at ω = 1 exterior to the image of C is 2π.

The image of D has a sharp pointed tail at ω = 1.

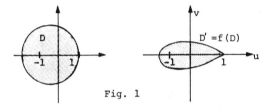

Fig. 1

Note that under the transformation

$$f(z) = \frac{1}{2}\left(z + \frac{1}{z}\right) \qquad (4)$$

the circle $|z|$ = 1 is mapped into the slit from ω = 1 to ω = -1.

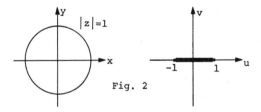

Fig. 2

When circle D changes its shape approaching $|z|$ = 1, the tear-like shape of D' approaches the straight line [-1,1].

● **PROBLEM 21-13**

Consider the transformation and the circle as described in Problem 20-13. What happens to the image if the circle is moved so its center is in the upper half plane, still passing through z = 1 and containing z = -1?

Solution: Point z = 1 is a critical point of the transformation

$$f(z) = \frac{1}{2}\left(z + \frac{1}{z}\right) \qquad (1)$$

Hence, as before, we obtain a sharp tail at $\omega = 1$.

When the circle does not entirely enclose the circle $|z| = 1$, its image does not enclose the image of $|z| = 1$.

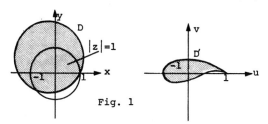

Fig. 1

The circle $|z| = 1$ is mapped onto $[-1,1]$. Since D does not entirely enclose $|z| = 1$, its image D' will enclose that portion of $[-1,1]$ which corresponds to the part of $|z| = 1$ inside D.

Moving the circle D, we obtain different shapes D'. The shape of D' resembles the cross-section of the wing of an airplane. This is an important fact in aerodynamic theory. The wing-like shapes are called Joukowski airfoils or profiles and the transformation

$$f(z) = \frac{1}{2}\left(z + \frac{1}{z}\right) \qquad (2)$$

is called a Joukowski transformation.

● **PROBLEM 21-14**

1. Is the image of a domain also a domain under a nonconstant analytic function?

2. Show that an analytic function assuming real boundary values must be a real constant.

Solution: 1. The image of a connected set under a continuous mapping is a connected set.

The image of an open set under a continuous function is an open set. Therefore, a nonconstant analytic mapping maps a domain into a domain.

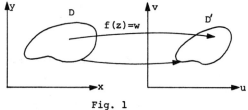

Fig. 1

Function $f(z)$ assumes real boundary values. Therefore, D' is located on the real u axis and cannot be an open subset of the plane (it does not contain any open subset).

THEOREMS, JACOBIAN OF THE TRANSFORMATION

• **PROBLEM 21-15**

Prove that an analytic function one-to-one mapping f(z) of the z plane into the ω plane is linear.

$$f(z) = az + b, \quad a \neq 0 \tag{1}$$

Solution: We shall examine the behavior of f(z) at infinity. Suppose f is an entire function with a removable singularity at z = ∞, that is $\lim_{z \to \infty} f(z)$ exists and is finite. It follows that f(z) is bounded in the entire z-plane.

By Liouville's theorem, a bounded entire function is a constant.

We conclude that f(z) is a constant, and therefore cannot be one-to-one. We conclude that at infinity

$$\lim_{z \to \infty} f(z) = \infty \tag{2}$$

and that the function f(z) has a pole.

On the other hand, an entire function f(z) that is not a polynomial has an essential singularity at z = ∞. Thus, f(z) is a polynomial. Since f(z) is one-to-one, we are left with the following possibility:

$$f(z) = az + b, \quad a \neq 0 \tag{3}$$

• **PROBLEM 21-16**

Prove Rouche's theorem:

Let D be a domain and L a Jordan curve in D, with interior contained in D. Let f(z) be non-zero on L and analytic in D.

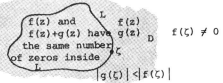

Fig. 1

If g(z) is analytic in D and

$$|g(\zeta)| < |f(\zeta)| \quad \text{for all } \zeta \in L, \tag{1}$$

then

$$f(z) \quad \text{and} \quad f(z) + g(z) = F(z) \tag{2}$$

have the same number of zeros inside L.

Solution:
$$F(z) = f(z)\left[1 + \frac{g(z)}{f(z)}\right] = f(z)h(z) \qquad (3)$$

Function $h(z)$ is analytic except possibly for the zeros of $f(z)$.

For $\zeta \in L$,
$$\left|\frac{g(\zeta)}{f(\zeta)}\right| < 1, \quad \text{hence,} \qquad (4)$$
$$h(\zeta) \neq 0.$$

Function $F(z)$, therefore, has no zeros on L.

By the Argument Principle, the number of zeros of $F(z)$ inside L is
$$\frac{1}{2\pi i}\int_L \frac{F'(\zeta)}{F(\zeta)}\, d\zeta = \frac{1}{2\pi i}\int_L \frac{f'(\zeta)h(\zeta) + h'(\zeta)f(\zeta)}{f(\zeta)h(\zeta)}\, d\zeta \qquad (5)$$
$$= \frac{1}{2\pi i}\int_L \frac{f'(\zeta)}{f(\zeta)}\, d\zeta + \frac{1}{2\pi i}\int_L \frac{h'(\zeta)}{h(\zeta)}\, d\zeta$$

The integral $\dfrac{1}{2\pi i}\displaystyle\int_L \dfrac{f'(\zeta)}{f(\zeta)}\, d\zeta$ equals the number of zeros of $f(z)$ inside L. To complete the proof, we show that
$$\frac{1}{2\pi i}\int_L \frac{h'(\zeta)}{h(\zeta)}\, d\zeta = 0 \qquad (6)$$

This integral equals the winding number
$$n(h(L); 0) \qquad (7)$$
where
$$h(z) = 1 + \frac{g(z)}{f(z)} \qquad (8)$$

Since $\left|\dfrac{g(\zeta)}{f(\zeta)}\right| < 1$, the curve $h(L)$ remains inside an open disk $D(1,1)$, centered at $\omega = 1$ of radius 1. Thus, $h(L)$ cannot wind around the origin in the ω plane and
$$n(h(L); 0) = 0 \qquad (9)$$

That completes the proof.

● **PROBLEM 21-17**

Prove the fundamental theorem of algebra by applying Rouché's theorem.

Solution: Consider a polynomial
$$F(z) = \underbrace{a_n z^n}_{f(z)} + \underbrace{a_{n-1}z^{n-1} + \ldots + a_1 z + a_0}_{g(z)} \qquad (1)$$

or

$$F(z) = f(z) + g(z) \tag{2}$$

The function $g(z)$ has degree $n - 1$ (or less). We can always find a sufficiently large circle C that

$$|g(\zeta)| < |a_n \zeta^n| \tag{3}$$

for all $\zeta \in C$

Hence, by Rouché's theorem, $f(z) = a_n z^n$ and $F(z)$ have the same number of zeros inside C. Thus, inside C, $F(z)$ given by (1) has n complex zeros.

● **PROBLEM 21-18**

Prove that all of the roots of

$$z^6 - 4z^2 + 10 = 0 \tag{1}$$

lie between the circles $|z| = 1$ and $|z| = 2$.

Solution: Let

$$f(z) = 10 \quad \text{and} \quad g(z) = z^6 - 4z^2 \tag{2}$$

Then, for the points on the circle $|z| = 1$,

$$|g(z)| = |z^6 - 4z^2| \leq |z^6| + |4z^2| = 5 < 10 = |f(z)| \tag{3}$$

Function $f(z) = 10$ has no zeros inside $|z| = 1$, hence,

$$f(z) + g(z) = z^6 - 4z^2 + 10 \tag{4}$$

has no zeros inside $|z| = 1$.

Now, let

$$f(z) = z^6, \quad g(z) = 10 - 4z^2 \tag{5}$$

For the points on $|z| = 2$, we have

$$|g(z)| = |10 - 4z^2| \leq 10 + 4|z^2| = 50 < 2^6 = |f(z)| \tag{6}$$

Function $f(z) = z^6$ has 6 zeros inside circle $|z| = 2$. Hence,

$$z^6 - 4z^2 + 10 = 0 \tag{7}$$

also has six zeros inside this circle.

● **PROBLEM 21-19**

A well known theorem of topology states:

If $f : \overline{K} \to \overline{K}$ is a continuous function mapping a closed disk into itself, then $f(z)$ has at least one fixed point z_0,

$$f(z_0) = z_0, \quad z_0 \in \overline{K} \tag{1}$$

For example, rotation of a disk about its center (angle $\neq 2\pi$) changes all points except the center, which remains fixed.

This theorem is known as the Brouwer fixed-point theorem. Prove the following simplified version, dealing with the analytic functions:

If $f : \overline{K(0,r)} \to K(0,r)$ is an analytic function, then f has exactly one fixed point in $K(0,r)$.

Solution: We shall consider $g(z) = -z$ as the main function and $f(z)$ its perturbation and apply Rouché's theorem.

On the circle $C(0,r)$, we have

$$|f(z)| < |z|, \quad z \in C(0,r) \tag{2}$$

also $g(z) \neq 0$ for $z \in C(0,r)$.

By Rouché's theorem, $g(z) = -z$ and $f(z) + g(z) = f(z) - z$ have the same number of zeros in $D(0,r)$.

Function $g(z) = -z$ has only one zero, at $z = 0$. Hence, the only fixed point of f is the solution

$$f(z) - z = 0 \tag{3}$$

• **PROBLEM 21-20**

Verify the Brouwer fixed-point theorem for the function

$$f(z) = az^k, \quad |a| < 1 \tag{1}$$

defined on the closed unit disk $|z| \le 1$.

Solution: Transformation

$$az^k = f(z) : \overline{K(0,1)} \to K(0,1) \tag{2}$$

maps the closed unit disk into an open disk $|z| < 1$, since for any $|z| \le 1$, the value of $|az^k|$ is

$$|az^k| = |a||z|^k \le |a| < 1 \tag{3}$$

The origin $z = 0$ is the fixed point of the mapping $f(z)$

$$f(0) = a0^k = 0 \tag{4}$$

The point $z = 0$ is the unique fixed point in $K(0,1)$. Equation

$$f(z) = az^k = z \quad \text{or}$$

$$az^k - z = z(az^{k-1} - 1) = 0 \tag{5}$$

has only one solution in $K(0,1)$.

Mapping $f(z)$ has other fixed points in $C(0,1)$ (circle of radius one) which are the roots of equation

$$az^{k-1} - 1 = 0 \qquad (6)$$

• **PROBLEM 21-21**

Transformation

$$w = f(z) = u + iv \qquad (1)$$

is analytic in a region D. Show that

$$\frac{\partial(u,v)}{\partial(x,y)} = |f'(z)|^2 \qquad (2)$$

Solution: The Jacobian of the transformation (1) is defined as follows:

$$\frac{\partial(u,v)}{\partial(x,y)} = \begin{vmatrix} \frac{\partial u}{\partial x} & \frac{\partial u}{\partial y} \\ \frac{\partial v}{\partial x} & \frac{\partial v}{\partial y} \end{vmatrix} \qquad (3)$$

Since $f(z)$ is analytic in D, it satisfies the Cauchy-Riemann equations

$$\frac{\partial u}{\partial x} = \frac{\partial v}{\partial y}$$

and $\qquad (4)$

$$\frac{\partial v}{\partial x} = -\frac{\partial u}{\partial y}$$

Then

$$\frac{\partial(u,v)}{\partial(x,y)} = \begin{vmatrix} \frac{\partial u}{\partial x} & \frac{\partial u}{\partial y} \\ \frac{\partial v}{\partial x} & \frac{\partial v}{\partial y} \end{vmatrix} = \frac{\partial z}{\partial x}\frac{\partial v}{\partial y} - \frac{\partial u}{\partial y}\frac{\partial v}{\partial x}$$

$$= \left(\frac{\partial u}{\partial x}\right)^2 + \left(\frac{\partial u}{\partial y}\right)^2 = \left|\frac{\partial u}{\partial x} + i\frac{\partial u}{\partial y}\right|^2 \qquad (5)$$

$$= |f'(z)|^2$$

• **PROBLEM 21-22**

Find the Jacobian of the transformation

$$f(z) = \sqrt{3}\, e^{i\frac{\pi}{4}}\, z + 2 - i \qquad (1)$$

and its geometrical interpretation.

Solution: The Jacobian of a transformation

$$f(z) = w = u + iv$$

is given by

$$\frac{\partial(u,v)}{\partial(x,y)} = |f'(z)|^2 \qquad (2)$$

Hence,

$$\frac{\partial(u,v)}{\partial(x,y)} = \left|\sqrt{3}\, e^{i\frac{\pi}{4}}\right|^2 = 3 \qquad (3)$$

The factor $|f'(z)|^2$ is called the magnification factor. Any region in the z plane is mapped under transformation (1) into a region in the w plane in such a way that the area of the image is three times larger than the area of the region in the z plane.

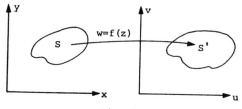

Fig. 1

If S is the area of the region and S' the area of the image of this region, then

$$S' = S|f'(z)|^2 \qquad (4)$$

Another way to compute the Jacobian is to write $f(z)$ in the form

$$f(z) = u + iv = \left(2 + \sqrt{\frac{3}{2}}\, x - \sqrt{\frac{3}{2}}\, y\right) + i\left(y\sqrt{\frac{3}{2}} + x\sqrt{\frac{3}{2}} - 1\right) \qquad (5)$$

and apply the definition

$$\frac{\partial(u,v)}{\partial(x,y)} = \begin{vmatrix} \frac{\partial u}{\partial x} & \frac{\partial u}{\partial y} \\ \frac{\partial v}{\partial x} & \frac{\partial v}{\partial y} \end{vmatrix} = \begin{vmatrix} \sqrt{\frac{3}{2}} & -\sqrt{\frac{3}{2}} \\ \sqrt{\frac{3}{2}} & \sqrt{\frac{3}{2}} \end{vmatrix}$$

$$= \frac{3}{2} + \frac{3}{2} = 3 \qquad (6)$$

● **PROBLEM 21-23**

Prove that

$$\frac{\partial(u,v)}{\partial(x,y)} \cdot \frac{\partial(x,y)}{\partial(u,v)} = 1 \qquad (1)$$

Solution: The Jacobian of the transformation

$$u = u(x,y)$$
$$v = v(x,y) \tag{2}$$

is

$$\frac{\partial(u,v)}{\partial(x,y)}$$

Solving (2) for x and y in terms of u and v, we obtain the inverse transformation

$$x = x(u,v)$$
$$y = y(u,v) \tag{3}$$

with Jacobian

$$\frac{\partial(x,y)}{\partial(u,v)}$$

We obtain from (2) and (3)

$$du = \frac{\partial u}{\partial x} dx + \frac{\partial u}{\partial y} dy$$
$$dv = \frac{\partial v}{\partial x} dx + \frac{\partial v}{\partial y} dy \tag{4}$$

$$dx = \frac{\partial x}{\partial u} du + \frac{\partial x}{\partial v} dv$$
$$dy = \frac{\partial y}{\partial u} du + \frac{\partial y}{\partial v} dv \tag{5}$$

Hence,

$$du = \frac{\partial u}{\partial x}\left[\frac{\partial x}{\partial u} du + \frac{\partial x}{\partial v} dv\right] + \frac{\partial u}{\partial y}\left[\frac{\partial y}{\partial u} du + \frac{\partial y}{\partial v} dv\right]$$
$$= du\left[\frac{\partial u}{\partial x}\frac{\partial x}{\partial u} + \frac{\partial u}{\partial y}\frac{\partial y}{\partial u}\right] + dv\left[\frac{\partial u}{\partial x}\frac{\partial x}{\partial v} + \frac{\partial u}{\partial y}\frac{\partial y}{\partial v}\right] \tag{6}$$

Thus,

$$\frac{\partial u}{\partial x}\frac{\partial x}{\partial u} + \frac{\partial u}{\partial y}\frac{\partial y}{\partial u} = 1$$
$$\frac{\partial u}{\partial x}\frac{\partial x}{\partial v} + \frac{\partial u}{\partial y}\frac{\partial y}{\partial v} = 0 \tag{7}$$

Similarly, we find

$$\frac{\partial v}{\partial x}\frac{\partial x}{\partial v} + \frac{\partial v}{\partial y}\frac{\partial y}{\partial v} = 1$$
$$\frac{\partial v}{\partial x}\frac{\partial x}{\partial u} + \frac{\partial v}{\partial y}\frac{\partial y}{\partial u} = 0 \tag{8}$$

Finally, we obtain

$$\frac{\partial(u,v)}{\partial(x,y)} \cdot \frac{\partial(x,y)}{\partial(u,v)} = \begin{vmatrix} \frac{\partial u}{\partial x} & \frac{\partial u}{\partial y} \\ \frac{\partial v}{\partial x} & \frac{\partial v}{\partial y} \end{vmatrix} \cdot \begin{vmatrix} \frac{\partial x}{\partial u} & \frac{\partial x}{\partial v} \\ \frac{\partial y}{\partial u} & \frac{\partial y}{\partial v} \end{vmatrix}$$

$$= \begin{vmatrix} \frac{\partial u}{\partial x}\frac{\partial x}{\partial u} + \frac{\partial u}{\partial y}\frac{\partial y}{\partial u} & \frac{\partial u}{\partial x}\frac{\partial x}{\partial v} + \frac{\partial u}{\partial y}\frac{\partial y}{\partial v} \\ \frac{\partial v}{\partial x}\frac{\partial x}{\partial u} + \frac{\partial v}{\partial y}\frac{\partial y}{\partial u} & \frac{\partial v}{\partial x}\frac{\partial x}{\partial v} + \frac{\partial v}{\partial y}\frac{\partial y}{\partial v} \end{vmatrix}$$

$$= \begin{vmatrix} 1 & 0 \\ 0 & 1 \end{vmatrix} = 1 \tag{9}$$

HARMONIC CONJUGATE, DIRICHLET PROBLEM

• **PROBLEM 21-24**

Let $u(x,y)$ by any given harmonic function defined on a simply connected domain D. Show that $u(x,y)$ has a harmonic conjugate $v(x,y)$ in D, given by

$$v(x,y) = \int_{(x_0,y_0)}^{(x,y)} -\frac{\partial u(X,Y)}{\partial Y} dX = \frac{\partial u(X,Y)}{\partial X} dY \tag{1}$$

where (x_0, y_0) is a fixed point in D and $(x,y) \in D$.

Solution: Two functions $u(x,y)$ and $v(x,y)$ with the first-order partial derivatives satisfying the Cauchy-Riemann equations

$$\frac{\partial u}{\partial x} = \frac{\partial v}{\partial y} \tag{2}$$

$$\frac{\partial u}{\partial y} = -\frac{\partial v}{\partial x} \tag{3}$$

are called harmonic conjugates, i.e., v is a harmonic conjugate of u and u is a harmonic conjugate of v. Let us recall some facts from advanced calculus.

If $f(x,y)$ and $g(x,y)$ have continuous first-order partial derivatives in a simply connected domain D and

$$\frac{\partial f}{\partial y} = \frac{\partial g}{\partial x}, \tag{4}$$

then,

$$\int_{(x_0,y_0)}^{(x,y)} f(X,Y)dX + g(X,Y)dY, \tag{5}$$

the line integral from $(x_0, y_0) \in D$ to $(x,y) \in D$ along the contour C in D, is independent of C. Furthermore, if

$$F(x,y) = \int_{(x_0,y_0)}^{(x,y)} f(X,Y)dX + g(X,Y)dY \qquad (6)$$

then

$$\frac{\partial F(x,y)}{\partial x} = f(x,y) \qquad (7)$$

$$\frac{\partial F(x,y)}{\partial y} = g(x,y) \qquad (8)$$

Hence, from (1), (7) and (8) we obtain

$$\frac{\partial v(x,y)}{\partial x} = -\frac{\partial u(x,y)}{\partial y} \qquad (9)$$

$$\frac{\partial v(x,y)}{\partial y} = \frac{\partial u(x,y)}{\partial x} \qquad (10)$$

Note that since the first-order partial derivatives of $u(x,y)$ are continuous, the derivatives of $v(x,y)$ are also continuous. The function

$$u(x,y) + iv(x,y) \qquad (11)$$

is analytic in D.

• **PROBLEM 21-25**

Find the harmonic conjugate of

$$u(x,y) = xy \qquad (1)$$

Solution: It is easy to verify that (1) is a harmonic function throughout the entire plane.

$$\nabla^2(xy) = \frac{\partial^2}{\partial x^2}(xy) + \frac{\partial^2}{\partial y^2}(xy) = 0 \qquad (2)$$

According to eq.(1) of Problem 21-25, the harmonic conjugate of $u(x,y)$ is

$$v(x,y) = \int_{(x_0,y_0)}^{(x,y)} -XdX + YdY \qquad (3)$$

We can choose (x_0, y_0) to be $(0,0)$

$$v = \int_{(0,0)}^{(x,y)} -XdX + YdY \qquad (4)$$

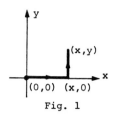

Fig. 1

Integrating along the contour shown in Fig. 1 we obtain

$$v(x,y) = -\frac{1}{2}x^2 + \frac{1}{2}y^2 \qquad (5)$$

The corresponding analytic function is

$$f(z) = xy - \frac{i}{2}(x^2 - y^2) = -\frac{i}{2}z^2 \qquad (6)$$

● **PROBLEM 21-26**

Show that if a function $u(x,y)$ is harmonic in a simply connected domain D then its partial derivatives of all orders are continuous in D.

Solution: Since $u(x,y)$ is harmonic in a simply connected domain D, it has a harmonic conjugate $v(x,y)$ in D. The function $f(z)$ defined by

$$f(z) = u(x,y) + i\,v(x,y) \qquad (1)$$

is analytic in D. We shall use the following theorem:

If a function f is analytic in D, then its derivatives of all orders are also analytic functions in D.

It follows that the partial derivatives of u and v of all orders of $f(z) = u + iv$ are continuous at any point where f is analytic.

Thus, we conclude that a harmonic function $u(x,y)$ in a simply connected domain D has continuous partial derivatives of all orders in D.

● **PROBLEM 21-27**

Frequently, especially in applied mathematics, one has to find a function which is harmonic in a specified domain and which satisfies certain conditions on the boundary of that domain. This problem is known as a Dirichlet problem or a boundary value problem of the first kind.

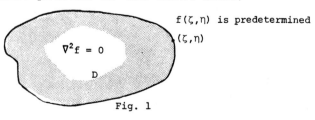

Fig. 1

Show that

$$f(x,y) = e^{-y} \sin x \qquad (1)$$

is the solution of the following Dirichlet problem

$$f(0,y) = 0, \quad f(\pi,y) = 0$$
$$f(x,0) = \sin x, \quad \lim_{y \to \infty} f(x,y) = 0 \qquad (2)$$

Solution: First, we have to show that (1) is a harmonic function, or that

$$\nabla^2 f = \frac{\partial^2}{\partial x^2}(e^{-y} \sin x) + \frac{\partial^2}{\partial y^2}(e^{-y} \sin x) = 0 \qquad (3)$$

Substituting (1) into (2) we quickly show that $e^{-y} \sin x$ satisfies (2). We can show that $e^{-y} \sin x$ is a harmonic function without any calculations.

Note, that since

$$-ie^{iz} = e^{-y} \sin x - ie^{-y} \cos x \qquad (4)$$

the function $e^{-y} \sin x$ is the real part of an analytic function $-ie^{iz}$. If a function $f(z) = u(x,y) + iv(x,y)$ is analytic in a domain D, its component functions u and v are harmonic in D.

Hence, $e^{-y} \sin x$ is harmonic.

● **PROBLEM 21-28**

Very often in applications the solution to a given boundary problem can be found by identifying it as a real or imaginary part of an analytic function. The following theorem is helpful:

If an analytic function

$$w = f(z) = u(x,y) + iv(x,y) \qquad (1)$$

transforms a domain D in the z plane onto a domain D' in the w plane and if g(u,v) is a harmonic function on D', then the function

$$G(x,y) = g\left[u(x,y), v(x,y)\right] \qquad (2)$$

is harmonic in D.

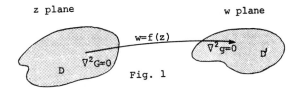

Fig. 1

In Problem 21-27 we have shown that
$$g(u,v) = e^{-v} \sin u \qquad (3)$$
is harmonic for all $v > 0$.

Taking the transformation
$$f(z) = z^2 \qquad (4)$$
show that
$$G(x,y) = e^{-2xy} \sin(x^2 - y^2) \qquad (5)$$
is harmonic in
$$D = \{(x,y) : x > 0, y > 0\} \qquad (6)$$

<u>Solution</u>: The function $g(u,v) = e^{-v} \sin u$ is harmonic in the upper half of the w plane, $v > 0$.

The transformation
$$w = z^2 = (x+iy)^2 = (x^2-y^2) + i\,2xy = u + iv \qquad (7)$$
maps the domain
$$D = \{(x,y) : x > 0, y > 0\} \qquad (8)$$
in the z plane onto the domain.
$$D' = \{(u,v) : v > 0\} \qquad (9)$$
in the w plane. Hence,
$$\begin{aligned} G(x,y) &= g\big[u(x,y), v(x,y)\big] \\ &= g\big[x^2 - y^2, 2xy\big] = e^{-2xy} \sin(x^2-y^2) \end{aligned} \qquad (10)$$
is harmonic in D.

• **PROBLEM 21-29**

Transformation
$$w = e^z \qquad (1)$$
maps the horizontal strip $0 < y < \pi$ onto the upper half plane $v > 0$.

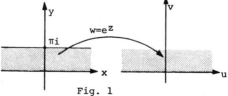

Fig. 1

The function
$$f(u,v) = \text{Re}(w^2) = u^2 - v^2 \qquad (2)$$

is harmonic, in the upper half plane, $v > 0$. Show that

$$F(x,y) = e^{2x} \cos 2y \qquad (3)$$

is harmonic in the horizontal strip of the z plane.

Solution: We will apply the theorem of Problem 21-28.
From (1)

$$w = e^z = e^{x+iy} = e^x(\cos y + i \sin y) \qquad (4)$$
$$= e^x \cos y + i e^x \sin y$$

Then, from (2), (4) and eq.(2) of Problem 21-29 we have

$$F(x,y) = f\left[u(x,y), v(x,y)\right] \qquad (5)$$
$$= f\left[e^x \cos y, e^x \sin y\right] = (e^x \cos y)^2 - (e^x \sin y)^2$$
$$= e^{2x} \cos 2y$$

• **PROBLEM 21-30**

Let us recall the Dirichlet problem. The domain is D and its boundary is C. A continuous (or harmonic) function f is defined on the boundary C. The Dirichlet problem comprises of three parts:

1. Existence. Does a function u defined and continuous (or harmonic) on D and such that on the boundary C it agrees with f exist?

2. Uniqueness. Is the function u uniquely determined by f?

3. Representation. Find a formula for u which involves only f.

Answer question 2 for the case of continuous functions.

Solution: We can apply the following theorem about continuous functions.

Theorem

Let D be a bounded domain and let u, v be functions continuous on $\overline{D}$ and such that for the boundary points of $\overline{D}$, u = v. Then

$$u(z) = v(z)$$

for all $z \in \overline{D}$.

Thus a continuous function u is uniquely determined by f.

• **PROBLEM** 21-31

Prove the uniqueness of the solution to Dirichlet's problem for harmonic functions.

Solution: As usual, D is a domain and C its boundary. Function f is defined and harmonic on C. Assume that Dirichlet's problem has two solutions u_1 and u_2 and

$$u = u_1 - u_2 \tag{1}$$

Then,

$$\nabla^2 u_1 = \frac{\partial^2 u_1}{\partial x^2} + \frac{\partial^2 u_1}{\partial y^2} = 0 \quad \text{in D}$$

$$\nabla^2 u_2 = 0 \quad \text{in D} \tag{2}$$

$$u_1 = u_2 = f(x,y) \quad \text{on C} \tag{3}$$

and

$$\nabla^2 u = 0 \quad \text{in D} \quad \text{and} \quad u = 0 \quad \text{on C} \tag{4}$$

We will show that $u = 0$ identically in D.

Applying Green's theorem we obtain

$$\int_C u\left(\frac{\partial u}{\partial x} dx - \frac{\partial u}{\partial y} dy\right) = -\iint_D \left[u\left(\frac{\partial^2 u}{\partial x^2} + \frac{\partial^2 u}{\partial y^2}\right) + \left(\frac{\partial u}{\partial x}\right)^2 + \left(\frac{\partial u}{\partial y}\right)^2\right] dxdy \tag{5}$$

Since $u = 0$ on C and $\nabla^2 u = 0$ in D, (5) reduces to

$$\iint_D \left[\left(\frac{\partial u}{\partial x}\right)^2 + \left(\frac{\partial u}{\partial y}\right)^2\right] dxdy = 0 \tag{6}$$

Hence u is identically equal to a constant in D. By continuity we must have

$$u = 0 \quad \text{in D} \tag{7}$$

That proves that the solution to the Dirichlet's problem is unique.

• **PROBLEM** 21-32

Here is a simple physical interpretation of the Dirichlet's problem. Let D be the disk made of some metal and C its rim. The flame heats up the rim and after a while the steady-state flow of heat is established. The temperature along the rim is $f(z)$. Heat flows from the rim inward.

Using this physical model explain Dirichlet's problem. Apply the maximum principle and mean-value property to this model and draw physical conclusions.

Solution: Using a thermometer we can measure the temperature at any point z inside D. The temperature u(z) is a continuous function in D ∪ C and a harmonic function in D. On the boundary C, u(z) is equal to f(z). The sources and sinks of heat are all located on the boundary C. Hence, we cannot have a hot interior point surrounded by cooler points, or a cold point surrounded by hotter points. The hottest and coldest points occur on the rim C.

This illustrates the maximum principle (see Fig. 1).

The temperature at center of disk is some kind of average of the temperatures around the rim C. This states the mean-value property.

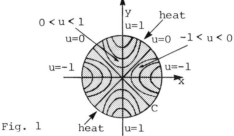

Fig. 1

APPLICATIONS

• **PROBLEM 21-33**

Let D be the semi-infinite strip

$$D = \{(x,y); \; x < 0, \; -\pi < y < \pi\} \quad (1)$$

with a boundary temperature f(x,y)

$$f(x,\pi) = f(x,-\pi) = e^{-x}$$
$$f(0,y) = -\cos y \quad (2)$$

Find the solution of this boundary-value problem.

Solution: The semi-infinite strip is shown in Fig. 1.

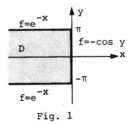

Fig. 1

From the boundary conditions we can guess that the suitable analytic function is $-e^{-z}$ or

$$-e^{-z} = -e^{-x} \cos y + i e^{-x} \sin y \quad (3)$$

Its real part is $-e^{-x} \cos y$.

The boundary conditions are fulfilled.

$$\left(-e^{-x} \cos y\right)\bigg|_{y=\pi} = \left(-e^{-x} \cos y\right)\bigg|_{y=-\pi} = e^{-x}$$

$$\left(-e^{-x} \cos y\right)\bigg|_{x=0} = -\cos y \tag{4}$$

The steady-state temperature on D is given by

$$T(x,y) = \operatorname{Re}\left[-e^{-z}\right] = -e^{-x} \cos y \tag{5}$$

● **PROBLEM 21-34**

Find the solution of the Dirichlet's problem for an infinite strip

$$D = \{(x,y);\ 0 < x < \tfrac{\pi}{2}\} \tag{1}$$

with the boundary conditions

$$f\left(\tfrac{\pi}{2}, y\right) = 0$$
$$f(0,y) = \cosh y \tag{2}$$

Hint:
$$\cos z = \cos x \cosh y - i \sin x \sinh y \tag{3}$$

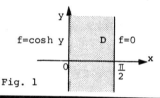

Fig. 1

Solution: From (2) and (3) we see that the real part of $\cos z$, i.e., $\cos x \cos hy$ at the boundaries of D has the values

$$(\cos x \cosh y)\big|_{x=0} = \cosh y$$
$$(\cos x \cosh y)\big|_{x=\tfrac{\pi}{2}} = 0 \tag{4}$$

Since $\cos x \cos hy$ is the real part of an analytic function, it is a harmonic function, which is the solution of the problem

$$T(x,y) = \cos x \cosh y \tag{5}$$

● **PROBLEM 21-35**

Let $f(z) = w$ be analytic and $f'(z) \neq 0$. Transformation $f(z)$ transforms $\phi(x,y)$ into $\phi(u,v)$. Show that

$$\nabla_z^2 \phi = |f'(z)|^2 \nabla_w^2 \phi \tag{1}$$

where

$$\nabla_z^2 = \frac{\partial^2}{\partial x^2} + \frac{\partial^2}{\partial y^2}, \quad \nabla_w^2 = \frac{\partial^2}{\partial u^2} + \frac{\partial^2}{\partial v^2} \tag{2}$$

Solution: Since
$$u = u(x,y) \tag{3}$$
$$v = v(x,y)$$

we have

$$\frac{\partial \phi}{\partial x} = \frac{\partial \phi}{\partial u}\frac{\partial u}{\partial x} + \frac{\partial \phi}{\partial v}\frac{\partial v}{\partial x} \tag{4}$$

$$\frac{\partial \phi}{\partial y} = \frac{\partial \phi}{\partial u}\frac{\partial u}{\partial y} + \frac{\partial \phi}{\partial v}\frac{\partial v}{\partial y} \tag{5}$$

$$\frac{\partial^2 \phi}{\partial x^2} = \frac{\partial \phi}{\partial u}\frac{\partial^2 u}{\partial x^2} + \frac{\partial u}{\partial x}\left[\frac{\partial^2 \phi}{\partial u^2}\frac{\partial u}{\partial x} + \frac{\partial^2 \phi}{\partial u \partial v}\frac{\partial v}{\partial x}\right] + \frac{\partial \phi}{\partial v}\frac{\partial^2 v}{\partial x^2}$$
$$+ \frac{\partial v}{\partial x}\left[\frac{\partial^2 \phi}{\partial u \partial v}\frac{\partial u}{\partial x} + \frac{\partial^2 \phi}{\partial v^2}\frac{\partial v}{\partial x}\right] \tag{6}$$

Similarly

$$\frac{\partial^2 \phi}{\partial y^2} = \frac{\partial \phi}{\partial u}\frac{\partial^2 u}{\partial y^2} + \frac{\partial u}{\partial y}\left[\frac{\partial^2 \phi}{\partial u^2}\frac{\partial u}{\partial y} + \frac{\partial^2 \phi}{\partial u \partial v}\frac{\partial v}{\partial y}\right] + \frac{\partial \phi}{\partial v}\frac{\partial^2 v}{\partial y^2}$$
$$+ \frac{\partial v}{\partial y}\left[\frac{\partial^2 \phi}{\partial u \partial v}\frac{\partial u}{\partial y} + \frac{\partial^2 \phi}{\partial v^2}\frac{\partial v}{\partial y}\right] \tag{7}$$

Adding (6) to (7) we obtain

$$\frac{\partial^2 \phi}{\partial x^2} + \frac{\partial^2 \phi}{\partial y^2} = \frac{\partial \phi}{\partial u}\left(\frac{\partial^2 u}{\partial x^2} + \frac{\partial^2 u}{\partial y^2}\right) + \frac{\partial \phi}{\partial v}\left(\frac{\partial^2 v}{\partial x^2} + \frac{\partial^2 v}{\partial y^2}\right)$$
$$+ \frac{\partial^2 \phi}{\partial u^2}\left[\left(\frac{\partial u}{\partial x}\right)^2 + \left(\frac{\partial u}{\partial y}\right)^2\right] + 2\frac{\partial^2 \phi}{\partial u \partial v}\left[\frac{\partial u}{\partial x}\frac{\partial v}{\partial x} + \frac{\partial u}{\partial y}\frac{\partial v}{\partial y}\right] \tag{8}$$
$$+ \frac{\partial^2 \phi}{\partial v^2}\left[\left(\frac{\partial v}{\partial x}\right)^2 + \left(\frac{\partial v}{\partial y}\right)^2\right]$$

Functions $u(x,y)$ and $v(x,y)$ are harmonic, they also satisfy the Cauchy-Riemann equations. Thus,

$$\frac{\partial^2 \phi}{\partial x^2} + \frac{\partial^2 \phi}{\partial y^2} = \frac{\partial^2 \phi}{\partial u^2}\left[\left(\frac{\partial u}{\partial x}\right)^2 + \left(\frac{\partial u}{\partial y}\right)^2\right] + \frac{\partial^2 \phi}{\partial v^2}\left[\left(\frac{\partial v}{\partial x}\right)^2 + \left(\frac{\partial v}{\partial y}\right)^2\right]$$
$$= |f'(z)|^2 \left(\frac{\partial^2 \phi}{\partial u^2} + \frac{\partial^2 \phi}{\partial v^2}\right) \tag{9}$$

Here we used

$$\left(\frac{\partial u}{\partial x}\right)^2 + \left(\frac{\partial u}{\partial y}\right)^2 = \left(\frac{\partial v}{\partial x}\right)^2 + \left(\frac{\partial v}{\partial y}\right)^2 = \left(\frac{\partial u}{\partial x}\right)^2 + \left(\frac{\partial v}{\partial x}\right)^2$$
$$= \left|\frac{\partial u}{\partial x} + i\frac{\partial v}{\partial x}\right|^2 = |f'(z)|^2 \tag{10}$$

● **PROBLEM 21-36**

Prove that an analytic function $f(z)$ where $f'(z) \neq 0$, transforms a harmonic function into a harmonic function.

Solution: Let $\phi(x,y)$ be a harmonic function. Then, by Problem 21-36

$$\frac{\partial^2 \phi}{\partial x^2} + \frac{\partial^2 \phi}{\partial y^2} = |f'(z)|^2 \left(\frac{\partial^2 \phi}{\partial u^2} + \frac{\partial^2 \phi}{\partial v^2} \right) \tag{1}$$

Since $\phi(x,y)$ is harmonic it follows from (1) that, since $|f'(z)|^2 \neq 0$

$$\frac{\partial^2 \phi}{\partial u^2} + \frac{\partial^2 \phi}{\partial v^2} = 0 \tag{2}$$

Thus, $\phi(u,v)$ is harmonic.

• **PROBLEM 21-37**

The Dirichlet and Neumann problems are essentially different in three dimensions and require different techniques. The situation is simpler in two dimensions.

Show that in two-dimensional space every Neumann problem can be reduced to a corresponding Dirichlet problem.

Solution: We denote by $\frac{\partial}{\partial n}$ the directional derivative at a boundary point in the direction determined by the outwards pointing normal of C at this point. By $\frac{\partial}{\partial s}$ we denote the derivative along the positively oriented boundary C. If

$$f(z) = u + iv \tag{1}$$

is an analytic function, then

$$\frac{\partial f}{\partial s} = i \frac{\partial f}{\partial n} \tag{2}$$

and

$$\frac{\partial u}{\partial n} = \frac{\partial v}{\partial s} \tag{3}$$

Let s_0 be the value of the length parameter s at a given point of C, then

$$v\left[x(s),y(s)\right] = v\left[x(s_0),y(s_0)\right] + \int_{s_0}^{s} \frac{\partial u}{\partial n} ds \tag{4}$$

Thus, the boundary values of $v(x,y)$ can be obtained from the prescribed values of $\frac{\partial u}{\partial n}$.

The Neumann problem for $u(x,y)$ is equivalent to a Dirichlet problem for the conjugate harmonic function $v(x,y)$. Note, that if the integration is extended over the entire closed contour C, then from (4) we obtain

$$\int_C \frac{\partial u}{\partial n} ds = 0 \tag{5}$$

• **PROBLEM** 21-38

Solve the Dirichlet problem for a function f(z) in the upper half of the z plane which on the x axis takes the values

$$F(x) = \begin{cases} 1, & x > 0 \\ 0, & x < 0 \end{cases} \qquad (1)$$

Solution: We have to find a function f(x,y) such that

$$\frac{\partial^2 f}{\partial x^2} + \frac{\partial^2 f}{\partial y^2} = 0 \quad \text{for} \quad y > 0 \qquad (2)$$

and

$$\lim_{y \to 0^+} f(x,y) = F(x) \qquad (3)$$

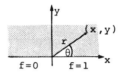

Fig. 1

Consider a simple function

$$a\theta + b \qquad (4)$$

where a, b are real constants. This is a harmonic function, because it is the imaginary part of an analytic function

$$a \ln z + b \qquad (5)$$

Applying the boundary conditions we determine a and b.

For $\theta = 0$, f = 1

For $\theta = \pi$, f = 0 $\qquad (6)$

Thus,

$$a \cdot 0 + b = 1$$

$$a \cdot \pi + b = 0 \qquad (7)$$

and

$$b = 1, \quad a = -\frac{1}{\pi} \qquad (8)$$

The solution is

$$f = a\theta + b = 1 - \frac{\theta}{\pi} = 1 - \frac{1}{\pi} \arctan\left(\frac{y}{x}\right) \qquad (9)$$

• **PROBLEM** 21-39

Solve Problem 21-38 using Poisson's formula for the half plane.

Solution: In this case the domain is the upper half of the z plane with the x axis as the boundary C. Hence, $f(x,y)$ has to be harmonic for $y > 0$.

Therefore,

$$\frac{\partial^2 f(x,y)}{\partial x^2} + \frac{\partial^2 f(x,y)}{\partial y^2} = 0 \quad \text{for} \quad y > 0 \tag{1}$$

with the prescribed value $F(x)$ on the x axis

$$f(x,0) = F(x), \quad -\infty < x < \infty \tag{2}$$

The solution to this problem is given by Poisson's formula for the half plane

$$f(x,y) = \frac{1}{\pi} \int_{-\infty}^{\infty} \frac{yF(\zeta)}{y^2 + (x-\zeta)^2} d\zeta \tag{3}$$

Substituting into (3) the boundary values (1) from Problem 21-39 we obtain,

$$f(x,y) = \frac{1}{\pi} \int_{-\infty}^{\infty} \frac{yF(\zeta)}{y^2 + (x-\zeta)^2} d\zeta$$

$$= \frac{1}{\pi} \int_{-\infty}^{0} \frac{y \cdot 0}{y^2 + (x-\zeta)^2} d\zeta + \frac{1}{\pi} \int_{0}^{\infty} \frac{y \cdot 1}{y^3 + (x-\zeta)^2} d\zeta$$

$$= \frac{1}{\pi} \left[\arctan\left(\frac{\zeta - x}{y}\right) \right]\Big|_{\zeta=0}^{\infty} \tag{4}$$

$$= \frac{1}{2} + \frac{1}{\pi} \arctan \frac{x}{y} = 1 - \frac{1}{\pi} \arctan \frac{y}{x}$$

• **PROBLEM 21-40**

Applying Poisson's formula for the half plane solve the Dirichlet problem for the upper half plane $y > 0$ with the boundary conditions

$$F(x) = \begin{cases} a_0, & x < -1 \\ a_1, & -1 < x < 1 \\ a_2, & x > 1 \end{cases} \tag{1}$$

where a_0, a_1, a_2 are constants.

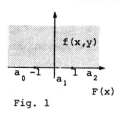

Fig. 1

Solution:

From Poisson's formula we get

$$f(x,y) = \frac{1}{\pi} \int_{-\infty}^{\infty} \frac{yF(\zeta)}{y^2+(x-\zeta)^2} d\zeta$$

$$= \frac{1}{\pi} \int_{-\infty}^{-1} \frac{ya_0}{y^2+(x-\zeta)^2} d\zeta + \frac{1}{\pi} \int_{-1}^{1} \frac{ya_1}{y^2+(x-\zeta)^2} d\zeta \quad (2)$$

$$+ \frac{1}{\pi} \int_{1}^{\infty} \frac{ya_2}{y^2+(x-\zeta)^2} d\zeta$$

$$= \frac{a_0}{\pi} \arctan\left(\frac{\zeta-x}{y}\right)\Big|_{\zeta=-\infty}^{-1}$$

$$+ \frac{a_1}{\pi} \arctan\left(\frac{\zeta-x}{y}\right)\Big|_{\zeta=-1}^{1} + \frac{a_2}{\pi} \arctan\left(\frac{\zeta-x}{y}\right)\Big|_{\zeta=1}^{\infty}$$

$$= \frac{a_0-a_1}{\pi} \arctan\left(\frac{y}{x+1}\right) + \frac{a_1-a_2}{\pi} \arctan\left(\frac{y}{x-1}\right) + a_2$$

● **PROBLEM** 21-41

Solve the Dirichlet problem for the unit circle $|z| = 1$ with the boundary condition

$$F(\theta) = \begin{cases} 1 & \text{for } 0 < \theta < \pi \\ 0 & \text{for } \pi < \theta < 2\pi \end{cases} \quad (1)$$

on the circumference of the circle.

Solution: The Dirichlet and Neumann problems can be solved for any simply connected domain D which can be mapped conformally by an analytic function onto a half plane or a unit circle. That conclusion can be drawn from Riemann's mapping theorem, which we shall discuss later.

The solution consists of three steps.

1. The domain D is mapped by an analytic function into the unit circle or half plane.

2. The problem is solved for the unit circle or half plane with or without the use of Poisson's formulas.

3. The inverse mapping function transforms the solution found in 2 into the solution of Dirichlet problem for domain D.

We shall use the transformation

$$z = \frac{i - w}{i + w} \tag{2}$$

to transform the interior of the circle $|z| = 1$ onto the upper half of the w plane (see Fig. 1).

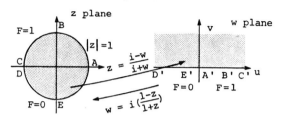

Fig. 1

Arc ABC is mapped onto the positive real axis and DEF onto the negative. We shall use the following theorem:

If $F = a$ is a constant on the boundary C or part of the boundary of a domain D in the z plane, then $F' = a$ on the image C' in the w plane.

Now, we solve the Dirichlet problem for the upper half plane with the boundary conditions

$$F'(u) = \begin{cases} 0 & \text{for } u < 0 \\ 1 & \text{for } u > 0 \end{cases} \tag{3}$$

From Problem 21-40 we have

$$f' = 1 - \frac{1}{\pi} \arctan \frac{v}{u} \tag{4}$$

From

$$w = i\left(\frac{1 - z}{1 + z}\right) \tag{5}$$

we find

$$u = \frac{2y}{(1+x)^2 + y^2} \tag{6}$$

$$v = \frac{1 - (x^2+y^2)}{(1+x)^2 + y^2} \tag{7}$$

Using (6) and (7) we express (4) in (x,y) coordinates

$$f = 1 - \frac{1}{\pi} \arctan \frac{2y}{1 - (x^2+y^2)} \tag{8}$$

or in polar coordinates

$$f = 1 - \frac{1}{\pi} \arctan \frac{2r \sin \theta}{1 - r^2} \tag{9}$$

• **PROBLEM 21-42**

Solve the Dirichlet problem for the unit circle $|z| = 1$ with the boundary condition

$$F(\theta) = \begin{cases} 1 & \text{for } \alpha < \theta < \beta \\ 0 & \text{on the rest of the circle} \end{cases} \quad (1)$$

Solution: The function

$$f(z) = e^{\frac{1}{2}(\alpha-\beta)i} \frac{z - e^{i\beta}}{z - e^{i\alpha}} \quad (2)$$

maps the domain $|z| < 1$ on the upper half plane $\text{Im } w > 0$. The point $e^{i\beta}$ has the image 0 and the point $e^{i\alpha}$ has the image ∞. Hence, the arc $\alpha < \theta < \beta$ on $|z| = 1$ is mapped on the negative real axis u. Therefore the solution is given by

$$u = \frac{1}{\pi} \arg f(z)$$

$$= \frac{1}{\pi} \arg \left[e^{\frac{1}{2}(\alpha-\beta)i} \frac{z - e^{i\beta}}{z - e^{i\alpha}} \right] \quad (3)$$

This problem can be solved directly by Poisson's formula for a circle

$$u(r,\theta) = \frac{1}{2\pi} \int_0^{2\pi} \frac{(1-r^2)F(\phi)d\phi}{1 + r^2 - 2r\cos(\theta-\phi)} \quad (4)$$

which in this case reduces to

$$u(r,\theta) = \frac{1}{2\pi} \int_\alpha^\beta \frac{1 - r^2}{1 + r^2 - 2r\cos(\theta-\phi)} d\phi \quad (5)$$

To compute integral (5), the following integral is used

$$\int \frac{dx}{a + \cos x} = \frac{2}{\sqrt{a^2-1}} \arctan\left(\frac{\sqrt{a^2-1} \tan \frac{x}{2}}{a+1}\right) \quad (6)$$

$$a > 1$$

Comparing the results, and writing (3) in the real form, we have

$$u = \frac{2}{\pi} \arctan \frac{\text{Im}\left[e^{\frac{1}{2}(\alpha-\beta)i} \dfrac{z - e^{i\beta}}{z - e^{i\alpha}}\right]}{\text{Re}\left[e^{\frac{1}{2}(\alpha-\beta)i} \dfrac{z - e^{i\beta}}{z - e^{i\alpha}}\right]} \quad (7)$$

BOUNDARY VALUE PROBLEM, NEUMANN PROBLEM, POISSON'S FORMULA

● **PROBLEM 21-43**

The following theorem is frequently used in solving boundary value problems.

Theorem

Let

$$w = f(z) = u(x,y) + iv(x,y) \qquad (1)$$

be a conformal transformation on a smooth curve C and let C' be the image of C under f(z). If a function h(u,v) satisfies along C' either of the conditions

$$h = h_0 \quad \text{or} \quad \frac{\partial h}{\partial n} = 0 \qquad (2)$$

then along C the function

$$H(x,y) = h\left[u(x,y), v(x,y)\right] \qquad (3)$$

satisfies the corresponding conditions

$$H = h_0 \quad \text{or} \quad \frac{\partial H}{\partial N} = 0 \qquad (4)$$

Here h_0 is a real constant, $\frac{\partial h}{\partial n}$ denotes derivative normal to C' and $\frac{\partial H}{\partial N}$ denotes derivative normal to C.

Prove this theorem.

Solution: We shall start with the easier part

$$(h = h_0 \text{ on } C') \Rightarrow (H = h_0 \text{ on } C) \qquad (5)$$

From (3) we conclude that the value of H at any point (x,y) on C is the same as the value of h at the image (u,v) of (x,y) under the transformation f(z).

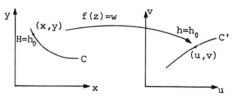

Fig. 1

Since the image (u,v) lies on C' and since $h = h_0$ along C', we conclude that $H = h_0$ along C.

Now, we will prove that

$$\left(\frac{\partial h}{\partial n} = 0 \text{ on } C'\right) \Rightarrow \left(\frac{\partial H}{\partial N} = 0 \text{ along } C\right) \qquad (6)$$

The derivative $\frac{\partial h}{\partial n}$ is equal to

$$\frac{\partial h}{\partial n} = (\nabla h) \cdot \bar{n} \quad \text{(we assume } \nabla h \neq 0\text{)} \tag{7}$$

where ∇h is the gradient of h and $\bar{n}$ is a unit vector normal to C'. Since,

$$\frac{\partial h}{\partial n} = 0 = (\nabla h) \cdot \bar{n} \tag{8}$$

we conclude that ∇h is normal to $\bar{n}$ and ∇h is tangent to C' (see Fig. 2).

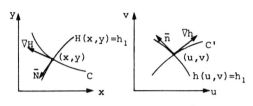

Fig. 2

On the other hand, ∇h is orthogonal to a level curve

$$h(u,v) = h_1 \tag{9}$$

passing through (u,v).

The level curve $H(x,y) = h_1$ passing through (x,y) can be written

$$H(x,y) = h_1 = h\left[u(x,y), v(x,y)\right] \tag{10}$$

Hence, the level curve $H(x,y) = h_1$ is transformed into $h(u,v) = h_1$ and the curve C is transformed into C' by a conformal transformation $f(z)$ (preserves angles). Since C' is orthogonal to $h(u,v) = h_1$, C is orthogonal to $H(x,y) = h_1$.

Therefore ∇H is tangent to C at (x,y), and ∇H and $\bar{N}$ are orthogonal, where $\bar{N}$ denotes a unit vector normal to C at (x,y).

$$(\nabla H) \cdot \bar{N} = 0 \tag{11}$$

Finally

$$\frac{\partial H}{\partial N} = (\nabla H) \cdot \bar{N} = 0 \tag{12}$$

If $\nabla h = 0$ then from

$$|\nabla H(x,y)| = |f'(z)| \cdot |\nabla h(u,v)| \tag{13}$$

immediately follows

$$\nabla H = 0 \tag{14}$$

and

$$\frac{\partial H}{\partial N} = 0 \tag{15}$$

• PROBLEM 21-44

Verify theorem of Problem 21-44 for the function

$$h(u,v) = 2v + 3 \qquad (1)$$

and the transformation

$$w = f(z) = iz^2 = -2xy + i(x^2 - y^2) \qquad (2)$$

The contour C in the z plane is the line

$$y = x, \quad x > 0 \qquad (3)$$

Solution: The line $y = x (x > 0)$ is mapped by $f(z) = iz^2$ onto

$$w = iz^2 = -2xy + i(x^2 - y^2) = -2x^2 \qquad (4)$$

the negative u axis.

The value of $h(u,v)$ on the negative u axis is

$$h(u,v) = 3 \qquad (5)$$

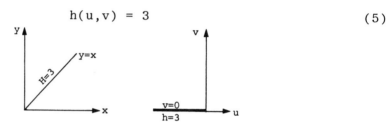

Fig. 1

$$H(x,y) = h\left[u(x,y), v(x,y)\right]$$
$$= 2(x^2 - y^2) + 3 \qquad (6)$$

On the line $y = x$ we get

$$H(x,y) = 2(x^2 - x^2) + 3 = 3 \qquad (7)$$

That verifies the theorem.

• PROBLEM 21-45

Find the image of the segment $0 \le y \le \pi$ of the y axis under the transformation

$$w = e^z \qquad (1)$$

Show that the function

$$h(u,v) = \text{Re}\left(3 - w + \frac{1}{w}\right) \qquad (2)$$

is harmonic everywhere in the w plane except for the origin. Illustrate and verify the theorem of Problem 21-43.

Solution: The transformation $w = e^z$ maps the segment $0 \leq y \leq \pi$ of the y axis on the semicircle

$$u^2 + v^2 = 1, \quad v \geq 0 \tag{3}$$

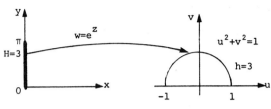

Fig. 1

On the semicircle the function $h(u,v)$ takes the value

$$h(u,v) = \text{Re}\left[3 - w + \frac{1}{w}\right] = 3 - u + \frac{u}{u^2 + v^2} \tag{4}$$

$$= 3 - u + u = 3$$

The corresponding function

$$H(x,y) = h\left[u(x,y), v(x,y)\right] \tag{5}$$

can be evaluated with the use of

$$e^z = e^{x+iy} = e^x \cos y + ie^x \sin y \tag{6}$$

On the segment $0 \leq y \leq \pi$ at $x = 0$, we have

$$H(x,y) = 3 - e^x \cos y + \frac{e^x \cos y}{e^{2x}\cos^2 y + e^{2x}\sin^2 y}$$

$$= 3 - e^x \cos y + \frac{\cos y}{e^x} = 3 \tag{7}$$

• **PROBLEM 21-46**

Let

$$w = f(z) = u(x,y) + iv(x,y) \tag{1}$$

be an analytic function which maps a domain D in the z plane onto the domain D' in the w plane. Let $\rho(u,v)$ be a function with continuous first- and second-order partial derivatives, which satisfies Poisson's equation

$$\frac{\partial^2 \rho}{\partial u^2} + \frac{\partial^2 \rho}{\partial v^2} = G(u,v) \tag{2}$$

in a domain D'. Here $G(u,v)$ is a prescribed function.

Show that

$$\Gamma(x,y) = \rho\left[u(x,y), v(x,y)\right] \tag{3}$$

satisfies Poisson's equation

$$\frac{\partial^2 \Gamma}{\partial x^2} + \frac{\partial^2 \Gamma}{\partial y^2} = |f'(z)|^2 \, G\!\left[u(x,y),\, v(x,y)\right] \qquad (4)$$

Solution: From Problem 21-36 we have

$$\nabla_z^2 \Gamma = |f'(z)|^2 \, \nabla_w^2 \rho \qquad (5)$$

Substituting

$$\nabla_w^2 \rho = \frac{\partial^2 \rho}{\partial u^2} + \frac{\partial^2 \rho}{\partial v^2} = G(u,v) \qquad (6)$$

into (5) we obtain

$$\frac{\partial^2 \Gamma}{\partial x^2} + \frac{\partial^2 \Gamma}{\partial y^2} = |f'(z)|^2 \left(\frac{\partial^2 \rho}{\partial u^2} + \frac{\partial^2 \rho}{\partial v^2}\right)$$
$$= |f'(z)|^2 \, G\!\left[u(x,y),\, v(x,y)\right] \qquad (7)$$

● **PROBLEM 21-47**

Show that if a function $u(x,y)$ is a solution of a Neumann problem, then

$$u(x,y) + a \qquad (1)$$

where a is any real constant, is also a solution of that problem.

Solution: Neumann problem, or the boundary value problem of the second kind can be formulated as follows:

A specified domain is given and one has to find a function harmonic in this domain with the prescribed values of the normal derivative along the boundary of that domain.

Let $u(x,y)$ be the solution of a Neumann problem.

$$\nabla^2 u = \frac{\partial^2 u}{\partial x^2} + \frac{\partial^2 u}{\partial y^2} = 0 \quad \text{in D} \qquad (2)$$

and

$$\frac{\partial u}{\partial n} = (\text{grad } u) \cdot \bar{n} = f(x_1, y_1) \qquad (3)$$

where $f(x_1,y_1)$ is a prescribed function along the boundary C of the domain D. Then, the function $u(x,y) + a$ is also harmonic in D.

$$\nabla^2(u + a) = \nabla^2 u = 0 \quad \text{in D} \qquad (4)$$

and satisfies the same boundary conditions

$$\frac{\partial(u+a)}{\partial n} = \left[\text{grad}(u+a)\right] \cdot \bar{n}$$
$$= (\text{grad } u) \cdot \bar{n} = f(x_1, y_1) \qquad (5)$$

CHAPTER 22

APPLICATIONS IN PHYSICS

HEAT FLOW

• PROBLEM 22-1

In many applications of physical theories we have to find a function u = u(x,y,z) which is a solution of the Laplace equation,

$$\frac{\partial^2 u}{\partial x^2} + \frac{\partial^2 u}{\partial y^2} + \frac{\partial^2 u}{\partial z^2} = 0 \qquad (1)$$

in a three-dimensional region V.

The function u(x,y,z) has to satisfy certain conditions which depend on the nature of the problem. In many cases, the physical data describe the behavior of u(x,y,z) on the boundary S of the region V. That physical problem leads to what is called in mathematics the boundary value problem. If the physical situation is such that u does not depend on the coordinate z, (1) reduces to

$$\frac{\partial^2 u}{\partial x^2} + \frac{\partial^2 u}{\partial y^2} = 0 \qquad (2)$$

Explain how the theory of conformal mappings and analytic functions can be applied to solve (2).

Solution: Any solution of (2) is the real part of an analytic function of a complex variable. The Dirichlet and Neumann problems can be solved for any simply-connected region which can be mapped conformally onto the region with the known solution of Dirichlet problem. This procedure consists of three steps.

1. Using an analytic function which transforms the boundary-value problem for D into a corresponding problem for an "easier" domain D'.
2. Solve the problem for the "easier" domain D'.
3. Using the inverse function transform the solution for D' into the solution for D.

• PROBLEM 22-2

Consider the following mathematical model of heat conduction. The temperature in a solid body is described by the tempera-ture function

$$T = T(x,y,z,t) \tag{1}$$

such that all its partial derivatives of the first and second order are continuous at each point interior to the solid. For simplicity assume that the flow is steady, that is, it does not vary with time,

$$T = T(x,y,z) \tag{2}$$

and that the temperature depends on only x and y coordinates

$$T = T(x,y) \tag{3}$$

The flux across a surface is defined by

$$\phi = -k \frac{dT}{dN}, \quad k > 0 \tag{4}$$

where k is the thermal conductivity and $\frac{dT}{dN}$ is the normal derivative. Assuming that there are no sinks or heat sources (i.e., no thermal energy is destroyed or created) within the solid body, show that T(x,y) is a harmonic function, that is

$$\frac{\partial^2 T(x,y)}{\partial x^2} + \frac{\partial^2 T(x,y)}{\partial y^2} = 0 \tag{5}$$

Solution: Consider a rectangular prism of unit height with the base Δx and Δy.

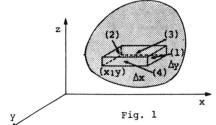

Fig. 1

The time rate of flow of heat toward the right across face 1 is

$$-k \frac{\partial T}{\partial x} (x + \Delta x, y) \Delta y \tag{6}$$

across face (2) is

$$-k \frac{\partial T}{\partial x} (x,y) \Delta y \tag{7}$$

Subtracting, we obtain the rate of heat loss through two faces (1) and (2)

$$-k \left[\frac{\frac{\partial T(x+\Delta x, y)}{\partial x} - \frac{\partial T(x,y)}{\partial x}}{\Delta x} \right] \Delta x \Delta y = -k \frac{\partial^2 T}{\partial x^2}(x,y) \Delta x \Delta y \tag{8}$$

Similarly for the remaining two faces

$$-k \frac{\partial^2 T}{\partial y^2} (x,y) \Delta x \Delta y \tag{9}$$

Since the temperature is steady and there are no sinks or heat sources the sum of (8) and (9) is zero. Thus,

$$\frac{\partial^2 T(x,y)}{\partial x^2} + \frac{\partial^2 T(x,y)}{\partial y^2} = 0 \tag{10}$$

The function $T(x,y)$ satisfies Laplace's equation at each interior point of the solid and its derivatives are continuous. Thus $T(x,y)$ is a harmonic function of x and y in the interior of the solid.

• **PROBLEM 22-3**

Find the steady-state temperature of the semi-infinite strip ($0 \leq x \leq \pi$, $y \geq 0$) with the boundary conditions given in Fig. 1.

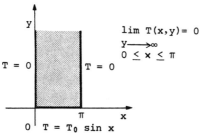

Fig. 1

Try to guess the answer checking familiar analytic functions.

Solution: Consider the analytic function

$$f(z) = e^{iz} \tag{1}$$

Its real and imaginary parts are

$$e^{iz} = e^{-y} \cos x + i e^{-y} \sin x \tag{2}$$

We have two functions

$$u(x,y) = e^{-y} \cos x, \quad v(x,y) = e^{-y} \sin x \tag{3}$$

which are harmonic, that is, they are solutions of Laplace's equation. Now, it is easy to show that

$$v(x,y) = e^{-y} \sin x \tag{4}$$

satisfies the boundary conditions.

$$v(0,y) = 0$$

$$v(\pi,y) = 0$$

$$v(x,0) = \sin x \tag{5}$$

$$\lim_{y \to \infty} v(x,y) = 0$$

Hence, the solution is

$$T(x,y) = T_0 e^{-y} \sin x \tag{6}$$

• **PROBLEM 22-4**

Find the expression for the steady-state temperature T(x,y) of a thin semi-infinite plate (y ≥ 0) whose edge is kept at temperature

$$T(x,0) = \begin{cases} 1 & \text{for } |x| < 1 \\ 0 & \text{for } |x| > 1 \end{cases} \quad (1)$$

Solution: We have to solve the boundary value problem

$$\frac{\partial^2 T}{\partial x^2} + \frac{\partial^2 T}{\partial y^2} = 0, \quad -\infty < x < \infty, \quad 0 < y$$

$$T(x,0) = \begin{cases} 1, & |x| < 1 \\ 0, & |x| > 1 \end{cases} \quad (2)$$

Furthermore we assume that T(x,y) is bounded

$$|T(x,y)| < M \quad \text{for all x and y and}$$

$$\lim_{y \to \infty} T(x,y) = 0 \quad (3)$$

We will find an analytic function f(z) = w in the domain y > 0 which is conformal along y = 0 except at the points x = ±1.

The function

$$w = \log \frac{z-1}{z+1} \quad (4)$$

maps the upper half plane y ≥ 0, except for the points z = ±1 onto the strip 0 < v < π in the w plane (see Appendix, Figure 21).

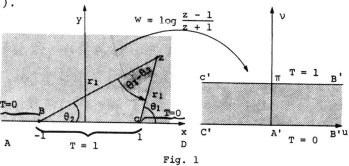

Fig. 1

From Fig. 1 we have

$$z - 1 = r_1 e^{i\theta_1}$$

$$z + 1 = r_2 e^{i\theta_2}$$

where

$$-\frac{\pi}{2} < \theta_1 < \frac{3\pi}{2} \quad (5)$$

$$-\frac{\pi}{2} < \theta_2 < \frac{3\pi}{2}$$

Then
$$w = \log \frac{z-1}{z+1} = \log \frac{r_1}{r_2} + i(\theta_1 - \theta_2) \tag{6}$$

Now we find a bounded harmonic function of u and v that satisfies the boundary conditions
$$T(u,0) = 0$$
$$T(u,\pi) = 1 \tag{7}$$

This function is
$$T = \frac{1}{\pi} v \tag{8}$$

Using
$$w = \log\left|\frac{z-1}{z+1}\right| + i \arg \frac{z-1}{z+1} \tag{9}$$

we find
$$v = \arg\left[\frac{(z-1)(\bar{z}+1)}{(z+1)(\bar{z}+1)}\right] = \arg\left[\frac{x^2+y^2-1+2iy}{(x+1)^2+y^2}\right]$$
$$= \arctan\left(\frac{2y}{x^2+y^2-1}\right) \tag{10}$$

Function (4) in x,y coordinates becomes
$$T = \frac{1}{\pi} \arctan \frac{2y}{x^2+y^2-1} \tag{11}$$

which is the solution of the problem.

● **PROBLEM 22-5**

Find the temperature distribution of a semi-infinite slab bounded by the planes $y = 0$ and $x = \pm \frac{\pi}{2}$, with the boundary conditions
$$T(x,0) = 1$$
$$T\left(\frac{\pi}{2}, y\right) = T\left(-\frac{\pi}{2}, y\right) = 0 \tag{1}$$
$$y > 0, \quad -\frac{\pi}{2} < x < \frac{\pi}{2}$$

Solution: The slab is shown in Fig. 1.

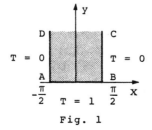

Fig. 1

We shall find a function T(x,y) such that

$$\frac{\partial^2 T(x,y)}{\partial x^2} + \frac{\partial^2 T(x,y)}{\partial y^2} = 0 \quad \begin{pmatrix} y > 0 \\ -\frac{\pi}{2} < x < \frac{\pi}{2} \end{pmatrix} \quad (2)$$

with the boundary conditions (1). It is easy to observe that the mapping

$$w = \sin z \quad (3)$$

transforms this boundary problem to the one solved in Problem 22-4 (also see Fig. 6 of Problem 21-49).

Rewriting eq.(11) of Problem 22-4 in terms of u and v we get

$$T = \frac{1}{\pi} \arctan \frac{2v}{u^2+v^2-1} \quad (4)$$

From (3)

$$\begin{aligned} u &= \sin x \cosh y \\ v &= \cos x \sinh y \end{aligned} \quad (5)$$

Substituting (5) into (4) we find

$$T = \frac{1}{\pi} \arctan \left[\frac{2\cos x \sinh y}{\sin^2 x \cosh^2 y + \cos^2 x \sinh^2 y - 1} \right]$$

$$= \frac{1}{\pi} \arctan \left[\frac{2\cos x \sinh y}{(1-\cos^2 x)\cosh^2 y + \cos^2 x \sinh^2 y - \cosh^2 y + \sin^2 y} \right]$$

$$= \frac{1}{\pi} \arctan \frac{2 \cos x \sinh y}{\sinh^2 y - \cos^2 x}$$

$$= \frac{1}{\pi} \arctan \left[\frac{2 \frac{\cos x}{\sinh y}}{1 - \left(\frac{\cos x}{\sinh y}\right)^2} \right] \quad (6)$$

$$= \frac{2}{\pi} \arctan \left[\frac{\cos x}{\sinh y} \right]$$

Since the argument of arc tan is non-negative, the range of arc tan is 0 to $\frac{\pi}{2}$ and

$$0 \le T(x,y) \le 1 \quad (7)$$

• **PROBLEM 22-6**

Find the isotherms T(x,y) = const for the temperature functions in Problem 22-4 and Problem 22-5.

<u>Solution</u>: The surfaces

$$T(x,y) = a \quad (1)$$

where a is any real constant, are called the isotherms within the solid.

For the temperature function

$$T = \frac{1}{\pi} \arctan \frac{2y}{x^2 + y^2 - 1} \tag{2}$$

the isotherms are

$$T(x,y) = a = \frac{1}{\pi} \arctan \frac{2y}{x^2 + y^2 - 1} \tag{3}$$

Thus,

$$\tan \pi a = \frac{2y}{x^2 + y^2 - 1} \tag{4}$$

$$x^2 + (y^2 - 2y \cot \pi a - 1) = 0$$

$$x^2 + (y - \cot \pi a)^2 = \cot^2 \pi a + 1 = \frac{\sin^2 \pi a + \cos^2 \pi a}{\sin^2 \pi a} \tag{5}$$

Hence the isotherms

$$x^2 + (y - \cot \pi a)^2 = \csc^2 \pi a \tag{6}$$

are arcs of the circles (6) with centers on the y axis and passing through the points (1,0) and (-1,0).

For the temperature function

$$T = \frac{2}{\pi} \arctan \left[\frac{\cos x}{\sinh y} \right] \tag{7}$$

the isotherms are

$$a = \frac{2}{\pi} \arctan \left[\frac{\cos x}{\sinh y} \right] \tag{8}$$

or

$$(\sinh y) \cdot \left(\tan \frac{\pi a}{2} \right) = \cos x \tag{9}$$

Hence, the isotherms are parts of the surfaces (9) within the slab.

• **PROBLEM 22-7**

The temperature distribution in a solid is steady-state and two-dimensional,

$$T = T(x,y) \tag{1}$$

The heat flux is

$$F = -K \left(\frac{\partial T}{\partial x} + i \frac{\partial T}{\partial y} \right) \tag{2}$$

where K is the constant thermal conductivity.

Define the complex temperature $\Omega(z)$ and show that it is an analytic function.

Solution: Let us denote in (2)

$$F = -K\left(\frac{\partial T}{\partial x} + i\frac{\partial T}{\partial y}\right) = P_x + i P_y \qquad (3)$$

where

$$P_x = -K\frac{\partial T}{\partial x}$$
$$P_y = -K\frac{\partial T}{\partial y} \qquad (4)$$

If C is any simple closed curve in the z plane and P_n and P_t are normal and tangential components of the heat flux F, then

$$\int_C P_t ds = \int_C P_x dx + P_y dy = 0 \qquad (5)$$

$$\int_C P_n ds = \int_C P_x dy - P_y dx = 0 \qquad (6)$$

Since the process is steady there are no sinks or sources inside C. Eq.(6) yields

$$\frac{\partial P_x}{\partial x} + \frac{\partial P_y}{\partial y} = 0 \qquad (7)$$

From (4) and (7) we obtain

$$\frac{\partial^2 T}{\partial x^2} + \frac{\partial^2 T}{\partial y^2} = 0 \qquad (8)$$

Function $T = T(x,y)$ is harmonic, therefore it has a harmonic conjugate $R(x,y)$ such that

$$\Omega(z) = T(x,y) + i R(x,y) \qquad (9)$$

is an analytic function.

The curves $T(x,y) = \alpha$ are called isothermal lines, the curves $R(x,y) = \beta$ are called flux lines and $\Omega(z)$ is called the complex temperature.

● **PROBLEM 22-8**

Find the temperature distribution of a semi-infinite slab (see Fig. 1). The boundary conditions are as shown and the process is steady-state.

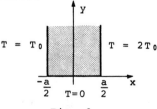

Fig. 1

T_0 is a real positive constant.

Solution: The transformation
$$w = \sin \frac{\pi z}{a} \qquad (1)$$

maps the slab of Fig. 1 into the upper-half of the w plane (see Fig. 2).

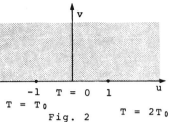

Fig. 2

This Dirichlet problem was solved in Problem 21-41. From eq.(2) of this problem we obtain

$$T = \frac{T_0}{\pi} \arctan\left(\frac{v}{u+1}\right) - \frac{2T_0}{\pi} \arctan\left(\frac{v}{u-1}\right) + 2T_0 \qquad (2)$$

The real and imaginary parts of (1) are

$$u = \sin\left(\frac{\pi x}{a}\right) \cosh\left(\frac{\pi y}{a}\right)$$
$$v = \cos\left(\frac{\pi x}{a}\right) \sinh\left(\frac{\pi y}{a}\right) \qquad (3)$$

Eq.(2) can be expressed in terms of x and y coordinates.

$$T(x,y) = \frac{T_0}{\pi} \arctan\left[\frac{\cos\frac{\pi x}{a} \sinh\frac{\pi y}{a}}{\sin\frac{\pi x}{a} \cosh\frac{\pi y}{a} + 1}\right]$$
$$- \frac{2T_0}{\pi} \arctan\left[\frac{\cos\frac{\pi x}{a} \sinh\frac{\pi y}{a}}{\sin\frac{\pi x}{a} \cosh\frac{\pi y}{a} - 1}\right] + 2T_0 \qquad (4)$$

● **PROBLEM 22-9**

Find the steady-state temperature of the shaded region with the boundary conditions as indicated.

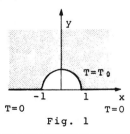

Fig. 1

Solution: As usual we shall find a conformal mapping which transforms a difficult Dirichlet problem into one that can be solved easily.

In this case, the appropriate function is

$$w = z + \frac{1}{z} \tag{1}$$

It maps the shaded area into the upper half of the w plane (see Appendix, Fig. 9). The new boundary conditions are shown in Fig.2.

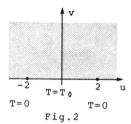

Fig.2

This Dirichlet problem was solved in 21-41. Thus,

$$T(u,v) = \frac{T_0}{\pi} \arctan\left(\frac{v}{u-2}\right) - \frac{T_0}{\pi} \arctan\left(\frac{v}{u+2}\right) \tag{2}$$

To return to the x,y coordinates we compute from (1)

$$u = x + \frac{x}{x^2 + y^2}$$

$$v = y - \frac{y}{x^2 + y^2} \tag{3}$$

and substitute into (2), we have

$$T(x,y) = \frac{T_0}{\pi} \arctan\left[\frac{y(x^2+y^2-1)}{x(x^2+y^2+1) - 2(x^2+y^2)}\right]$$

$$- \frac{T_0}{\pi} \arctan\left[\frac{y(x^2+y^2-1)}{x(x^2+y^2+1) + 2(x^2+y^2)}\right] \tag{4}$$

● **PROBLEM 22-10**

Find the steady-state temperature distribution of the two-dimensional problem shown in Fig. 1.

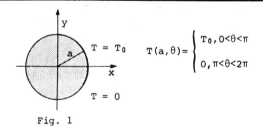

Fig. 1

Solution: The solution can be found directly from the Poisson's integral formula for a circle

$$T(r,\theta) = \frac{1}{2\pi} \int_0^\pi \frac{T_0(a^2-r^2)d\phi}{(a^2+r^2) - 2ar\cos(\phi-\theta)} \tag{1}$$

Now, let

$$\psi = \phi - \theta \qquad (2)$$

then

$$T(r,\psi) = \frac{T_0(a^2-r^2)}{2\pi} \int_{-\theta}^{\pi-\theta} \frac{d\psi}{(a^2+r^2)-2ar\cos\psi}$$

$$= \frac{T_0(a^2-r^2)}{2\pi} \frac{2}{a^2-r^2} \arctan\left[\frac{a+r}{a-r} \tan\frac{\psi}{2}\right]\Bigg|_{\psi=-\theta}^{\psi=\pi-\theta} \qquad (3)$$

Hence,

$$T(r,\theta) = -\frac{T_0}{\pi} \arctan \frac{a^2-r^2}{2ar\sin\theta} \qquad (4)$$

We shall show another method of solving this Dirichlet problem by using the Fourier series expansion.

For $|z| < a$ the converging MacLaurin series expansion is

$$\frac{a+z}{a-z} = \frac{1+\frac{z}{a}}{1-\frac{z}{a}} = \left(1+\frac{z}{a}\right) \sum_{n=0}^{\infty}\left(\frac{z}{a}\right)^n$$

$$= 1 + 2\sum_{n=1}^{\infty}\left(\frac{z}{a}\right)^n = 1 + 2\sum_{n=1}^{\infty}\left(\frac{r}{a}\right)^n e^{in(\theta-\phi)} \qquad (5)$$

Hence,

$$u(r,\theta) = \text{Re} \frac{1}{2\pi} \int_0^{2\pi} \left[1 + 2\sum_{n=1}^{\infty}\left(\frac{r}{a}\right)^n e^{in(\theta-\phi)}\right] T(\phi) d\phi \qquad (6)$$

Note that (6) is a Maclaurin expansion of an analytic function for $|z| < a$, hence integrating term by term

$$u(r,\theta) = \frac{1}{2\pi} \int_0^{2\pi} T(\phi) d\phi$$

$$+ \sum_{n=1}^{\infty}\left(\frac{r}{a}\right)^n \left[\frac{1}{\pi}\left\{\int_0^{2\pi} T(\phi)\cos n\phi \, d\phi\right\} \cos n\theta \right.$$

$$\left. + \frac{1}{\pi}\left\{\int_0^{2\pi} T(\phi)\sin n\phi \, d\phi\right\} \sin n\theta \right] \qquad (7)$$

$$= \frac{a_0}{2} + \sum_{n=1}^{\infty}\left(\frac{r}{a}\right)^n (a_n\cos n\theta + b_n\sin n\theta)$$

where

$$a_n = \frac{1}{\pi} \int_0^{2\pi} T(\phi)\cos n\phi \, d\phi$$

$$b_n = \frac{1}{\pi} \int_0^{2\pi} T(\phi)\sin n\phi \, d\phi \tag{8}$$

Formula (7) is the Fourier series expansion for the Dirichlet problem for the circle.

For the boundary values of Fig. 1 we obtain

$$a_0 = \frac{1}{\pi} \int_0^\pi d\phi = 1$$

$$a_n = \frac{1}{\pi} \int_0^\pi \cos n\phi \, d\phi = 0 \quad n = 1, 2, \ldots \tag{9}$$

$$b_n = \frac{1}{\pi} \int_0^\pi \sin n\phi \, d\phi = \frac{1}{n\pi}\left[1 - (-1)^n\right]$$

and

$$u(r,\theta) = \frac{1}{2} + \frac{2}{\pi} \sum_{n=0}^\infty \left(\frac{r}{a}\right)^{2n+1} \frac{\sin(2n+1)\theta}{2n+1} \tag{10}$$

● **PROBLEM 22-11**

Compute the complex temperature for the process described in Problem 22-10.

Solution: The mapping used from the z plane to the w plane was

$$w = \ln z = \ln r + i\theta \tag{1}$$

The temperature distribution of the corresponding region in the w plane was

$$T(u,v) = \frac{T_1}{\pi} v \tag{2}$$

Applying the Cauchy-Riemann equations we find the harmonic conjugate $F(u,v)$ of $T(u,v)$.

$$F(u,v) = -\frac{T_1}{\pi}(u + C) \tag{3}$$

where C is a real constant.

The function

$$\Omega(w) = T + iF = \frac{T_1}{\pi} v - i\frac{T_1}{\pi}(u+C)$$

$$= -i \frac{T_1}{\pi}\left[u+iv\right] + iC_1 \qquad (4)$$

$$= -i \frac{T_1}{\pi} w + iC_1$$

is analytic.

The complex temperature of the region of the z plane is

$$\Omega(z) = -i \frac{T_1}{\pi} w + iC_1 = -i \frac{T_1}{\pi}\left[\ln r + i\theta\right] + iC_1 \qquad (5)$$

$$= \frac{T_1}{\pi} \theta - \frac{T_1}{\pi} i \ln r + iC_1$$

● **PROBLEM 22-12**

Compute the temperature function for the steady-state heat transfer in the upper half-plane with the boundary conditions indicated in Fig. 1.

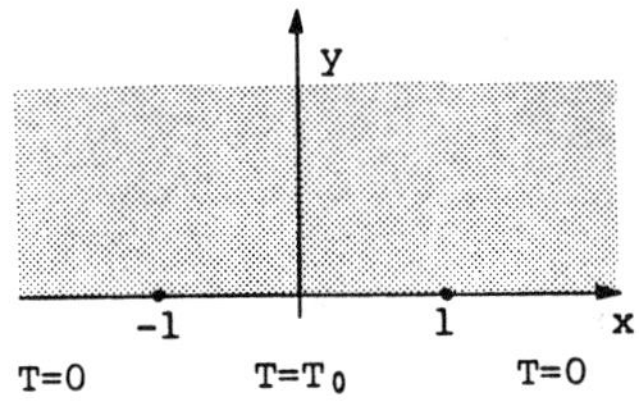

$$T = \begin{cases} T_0 & \text{for } |x| < 1 \\ 0 & \text{for } |x| > 1 \end{cases} \qquad (1)$$

T=0   T=T$_0$ T=0

Fig. 1

Solution: In this case the Dirichlet problem can be solved directly from Poisson's formula for the half plane

$$T(x,y) = \frac{1}{\pi} \int_{-\infty}^{\infty} \frac{T_0 y}{(x-\eta)^2 + y^2} d\eta$$

$$= \frac{T_0}{\pi} \arctan \frac{2y}{x^2 + y^2 - 1} \qquad (2)$$

Note that

$$\arctan x + \arctan y = \arctan \frac{x+y}{1-xy} \qquad (3)$$

Compare this method with the one used in Problem 22-4.

● **PROBLEM 22-13**

Find the steady-state temperature distribution for the semi-circular disk with the boundary conditions as indicated in Fig. 1.

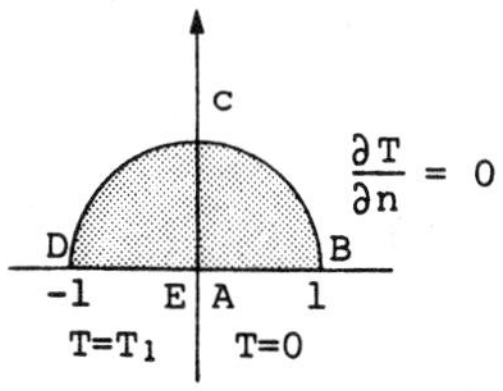

$\frac{\partial T}{\partial n} = 0$

-1 E | A   1
T=T$_1$ | T=0

Fig. 1

Solution: The transformation

$$w = \ln z = \ln r + i\theta, \quad 0 \le \theta < \pi \qquad (1)$$

maps the semicircular disk into the semi-infinite strip shown in Fig. 2.

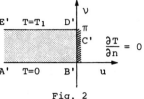

Fig. 2

Indeed, for AB in Fig. 1, $0 \le r \le 1$, and $\theta = 0$. Hence, $u = \ln r$, $v = 0$ and u changes from $-\infty$ to 0. For the arc BCD, $r = 1$ and $0 \le \theta \le \pi$. Hence, $u = \ln 1 = 0$, $0 \le v \le \pi$, that yields segment B'D'. Finally for DE, $u = \ln r$, $0 \le r \le 1$, $v = \pi$, that yields segment D'E'.

Note that there is no heat flux across D'B', because $\frac{\partial T}{\partial n} = 0$. The temperature distribution in the stripe A'B'C'D'E' is

$$T = \frac{T_1}{\pi} v \qquad (2)$$

In x,y coordinates (2) becomes

$$T(x,y) = \frac{T_1}{\pi} \theta = \frac{T_1}{\pi} \arctan \frac{y}{x} \qquad (3)$$

● **PROBLEM 22-14**

Find the steady-state temperature of a plate of the form of a quadrant ($x \ge 0$, $y \ge 0$).

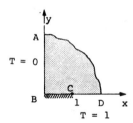

Fig. 1

The boundary temperatures are as shown and the segment BC is insulated.

Solution: This boundary value problem can be formulated as follows:

$$\frac{\partial^2 T(x,y)}{\partial x^2} + \frac{\partial^2 T(x,y)}{\partial y^2} = 0 \qquad x > 0, \ y > 0$$

$$T(x,0) = 1 \quad \text{for} \quad x > 1$$

$$\frac{\partial T(x,0)}{\partial y} = 0 \quad \text{for} \quad 0 < x < 1 \qquad (1)$$

$$T(0,y) = 0 \quad \text{for} \quad y > 0$$

The function

$$z = \sin w \qquad (2)$$

is a one-to-one and onto mapping of the strip $v \ge 0$, $0 \le u \le \frac{\pi}{2}$ onto the quadrant $x \ge 0$, $y \ge 0$.

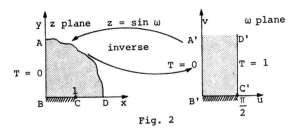

Fig. 2

Since (2) is one-to-one and onto, the inverse transformation exists. Transformation (2) is conformal throughout the strip except at the point $w = \frac{\pi}{2}$. Hence, the inverse transformation is conformal in the first quadrant except $z = 1$. We shall solve the boundary value problem for the region of the w plane shown in Fig. 2. The temperature function is

$$T(u,v) = \frac{2}{\pi} u \qquad (3)$$

To express T in terms of x and y note that from (2)

$$x = \sin u \cosh v$$
$$y = \cos u \sinh v \qquad (4)$$

For $0 < u < \frac{\pi}{2}$ both $\sin u \neq 0$ and $\cos u \neq 0$, thus

$$\frac{x^2}{\sin^2 u} - \frac{y^2}{\cos^2 u} = 1 \qquad (5)$$

For any fixed u the foci of hyperbola (5) are

$$z = \pm\sqrt{\sin^2 u + \cos^2 u} = \pm 1 \qquad (6)$$

The line segment joining the two vertices is $2 \sin u$. Hence, the absolute value of the difference of the distances between the foci and a point (x,y) of the hyperbola in the first quadrant is

$$2 \sin u = \sqrt{(x+1)^2+y^2} - \sqrt{(x-1)^2+y^2} \qquad (7)$$

Computing u from (7) and substituting into (3) we get

$$T = \frac{2}{\pi} \arcsin \frac{1}{2}\left[\sqrt{(x+1)^2+y^2} - \sqrt{(x-1)^2+y^2}\right] \qquad (8)$$

• **PROBLEM 22-15**

Find the steady-state temperature distribution of a semi-infinite slab shown in Fig. 1.

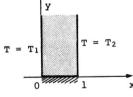

Fig. 1

The segment [0,1] on the x-axis is insulated.

Solution: No heat transfer takes place along the insulated segment. Therefore the temperature problem shown in Fig. 1 is equivalent to the one shown in Fig. 2.

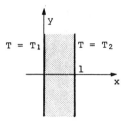

Fig. 2

The temperature between two parallel planes kept at constant temperatures T_1 and T_2, respectively, is

$$T(x,y) = T_1 + (T_2 - T_1)x$$

That is also the solution of problem posed in Fig. 1.

• **PROBLEM** 22-16

Find the steady-state temperature of an infinite wedge of angle $\frac{\pi}{4}$ with the boundary conditions shown in Fig. 1.

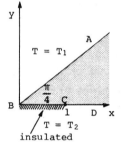

Fig. 1

Consider the following consecutive transformations:

1. A wedge into the first quadrant

$$\zeta = z^2 \tag{1}$$

2. The first quadrant into a semi-infinite slab

$$\zeta = \sin \frac{\pi w}{2} \tag{2}$$

Solution: The transformation $\zeta = z^2$ maps the wedge in the z plane into the first quadrant of the ζ plane with the boundary conditions as shown (Fig. 2b).

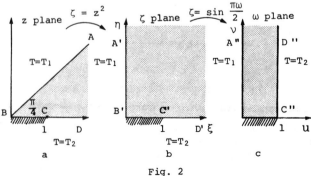

Fig. 2

The transformation $\zeta = \sin\frac{\pi w}{2}$ maps the first quadrant of the ζ plane into the slab in the w plane with the boundary conditions shown in Fig. 2c. The solution of the last problem is

$$T(u,v) = T_1 + (T_2 - T_1)u \tag{3}$$

If we express (3) in terms of z, we have

$$w = \frac{2}{\pi} \arcsin \zeta \tag{4}$$

$$w = \frac{2}{\pi} \arcsin z^2 \tag{5}$$

$$u = \frac{2}{\pi} \operatorname{Re}\left[\arcsin z^2\right] \tag{6}$$

Substituting (4), (5) and (6) into (3) we have,

$$T(z) = T_1 + \frac{2(T_2 - T_1)}{\pi} \operatorname{Re}\left[\arcsin z^2\right] \tag{7}$$

which is the solution of the problem.

• **PROBLEM 22-17**

Compute the temperature distribution of an infinite rod. Its initial temperature

$$T(x,0) = f(x) \tag{1}$$

vanishes outside a finite interval.

Solution: We must solve the following equation

$$\frac{\partial^2 T}{\partial x^2} = \frac{1}{a^2}\frac{\partial T}{\partial t}, \quad -\infty < x < \infty, \quad t > 0$$

with the boundary conditions

$$T(x,0) = f(x) \tag{2}$$

$$\lim_{|x|\to\infty} T(x,t) = 0$$

Assume that $T(x,t)$ and $f(x)$ have Fourier transforms in the variable x, then

$$U(z,t) = \frac{1}{\sqrt{2\pi}} \int_{-\infty}^{\infty} T(x,t) e^{-izx} dx \tag{3}$$

$$(iz)^2 U(z,t) = \frac{1}{a^2}\frac{\partial U}{\partial t} \tag{4}$$

$$U(z,0) = \frac{1}{\sqrt{2\pi}} \int_{-\infty}^{\infty} f(x)e^{-izx} \, dx = H(z) \tag{5}$$

Hence,

$$\frac{\partial U}{\partial t} + a^2 z^2 U = 0 \tag{6}$$

$$U(z,t) = U(z,0)e^{-z^2 a^2 t} = H(z)e^{-z^2 a^2 t} \tag{7}$$

Applying the formula

$$\frac{1}{\sqrt{2\pi}} \int_{-\infty}^{\infty} e^{-\frac{x^2}{2}} e^{-i\eta x} \, dx = e^{-\frac{\eta^2}{2}} \tag{8}$$

and setting

$$\frac{\eta^2}{2} = z^2 a^2 t \tag{9}$$

we obtain

$$e^{-z^2 a^2 t} = \frac{1}{\sqrt{2\pi}} \int_{-\infty}^{\infty} e^{-\frac{x^2}{2}} e^{-iza\sqrt{2t} \, x} \, dx$$

$$= \frac{1}{2a\sqrt{\pi t}} \int_{-\infty}^{\infty} e^{-\frac{T^2}{4a^2 t}} e^{-izT} \, dT \tag{10}$$

$$= F\left[\frac{1}{a\sqrt{2t}} e^{-\frac{T^2}{4a^2 t}}\right]$$

By the convolution theorem

$$T(x,t) = \frac{1}{2a\sqrt{\pi t}} \int_{-\infty}^{\infty} f(y) e^{-\frac{(x-y)^2}{4a^2 t}} \, dy \tag{11}$$

• **PROBLEM 22-18**

Find the temperature distribution of a semi-infinite heat conducting rod. The initial temperature is

$$T(x,0) = f(x), \quad 0 < x < \infty$$

and

$$f(0) = 0, \quad T(0,t) = 0 \tag{1}$$

Solution: The mathematical description of the problem is

$$\frac{\partial^2 T}{\partial x^2} = \frac{1}{a^2}\frac{\partial T}{\partial t}, \quad 0 < x < \infty, \quad t > 0$$

$$T(x,0) = f(x), \quad T(0,t) = 0 \qquad (2)$$

$$\lim_{x \to \infty} T(x,t) = 0$$

We cannot use one-sided Fourier transform with respect to x because we don't know

$$\left.\frac{\partial T(x,t)}{\partial x}\right|_{x=0}$$

The solution to this problem can be obtained directly from the results of Problem 22-17. If $f(x)$ is an odd function

$$T(x,t) = \frac{1}{2a\sqrt{\pi t}} \int_{-\infty}^{\infty} f(y)\, e^{-\frac{(x-y)^2}{4a^2 t}}\, dy$$

Substituting $y = -p$, we have

$$T(x,t) = \frac{1}{2a\sqrt{\pi t}} \int_{-\infty}^{\infty} f(-p)\, e^{-\frac{(x+p)^2}{4a^2 t}}\, dp \qquad (3)$$

$$= -\frac{1}{2a\sqrt{\pi t}} \int_{-\infty}^{\infty} f(p)\, e^{-\frac{(-x-p)^2}{4a^2 t}}\, dp$$

$$= -T(-x,t)$$

Similarly,

$$T(0,t) = \frac{1}{2a\sqrt{\pi t}} \int_{-\infty}^{\infty} f(y)\, e^{-\frac{y^2}{4a^2 t}}\, dy$$

$$= -\frac{1}{2a\sqrt{\pi t}} \int_{-\infty}^{\infty} f(p)\, e^{-\frac{p^2}{4a^2 t}}\, dp = -T(0,t) \qquad (4)$$

Thus $T(0,t) = 0$, and the solution is given by

$$T(x,t) = \frac{1}{2a\sqrt{\pi t}} \int_{-\infty}^{\infty} f(y)\, e^{-\frac{(x-y)^2}{4a^2 t}}\, dy \qquad (5)$$

where $f(y)$ is an odd function.

ELASTICITY

• **PROBLEM** 22-19

In the theory of elasticity two-dimensional problems can be reduced to the solution of the biharmonic equation

$$\nabla^2(\nabla^2 U) = \nabla^4 U = 0 \tag{1}$$

where

$$\nabla^2 = \frac{\partial^2}{\partial x^2} + \frac{\partial^2}{\partial y^2}$$

Eq.(1) can be written

$$\frac{\partial^4 U}{\partial x^4} + 2\frac{\partial^4 U}{\partial x^2 \partial y^2} + \frac{\partial^4 U}{\partial y^4} = 0 \tag{2}$$

Function U is called Airy's stress function. Its second derivatives are the components of the stress tensor.

Show that if f(z) and g(z) are analytic in a domain D, then

$$U = \operatorname{Re}\left[\bar{z}f(z) + g(z)\right] \tag{3}$$

is biharmonic in D.

Solution: Eq.(1) is equivalent to the equations

$$\nabla^2 T = 0 \quad \text{and} \quad \nabla^2 U = T \tag{4}$$

Function T is harmonic. Let U_1 and U_2 be the solutions for given T, then

$$\nabla^2(U_1 - U_2) = T - T = 0 \tag{5}$$

Hence the solutions of $\nabla^2 U = T$ are of the form

$$U' + W \tag{6}$$

where U' is a particular solution and W is harmonic.

Assume that two harmonic functions u and v exist such that

$$T = \frac{\partial u}{\partial x} = \frac{\partial v}{\partial y} \tag{7}$$

Then,

$$\nabla^2(xu+yv) = x\nabla^2 u + y\nabla^2 v + 2\left(\frac{\partial u}{\partial x} + \frac{\partial v}{\partial y}\right) = 4T \tag{8}$$

and

$$U' = \frac{1}{4}(xu + yv) \tag{9}$$

If D is simply connected we can choose S so that

$$T + iS = F(z) \tag{10}$$

is analytic in D.

Then, equation

$$u + iv = f(z) = \int F(z)dz \tag{11}$$

defines harmonic functions u and v such that

$$T = \frac{\partial u}{\partial x} = \frac{\partial v}{\partial y} \tag{12}$$

and

$$U' = \frac{1}{4}(xu + iv) = \frac{1}{4} \operatorname{Re}\left[\overline{z}f(z)\right] \tag{13}$$

Hence the general solution U can be written

$$U = U' + W = \operatorname{Re}\left[\frac{\overline{z}f(z)}{4} + g(z)\right] \tag{14}$$

The factor $\frac{1}{4}$ can be included in f(z). We also proved that if D is simply connected, then all harmonic functions in D can be represented by (3).

• **PROBLEM 22-20**

The ends of an elastic string are fixed at x = 0 and x = π. The initial velocity and displacement are zero. Find the displacement of the string when it moves under the influence of a force k sin x.

Solution: The problem is formulated as follows

$$\frac{\partial^2 u}{\partial x^2} = \frac{1}{a^2} \frac{\partial^2 u}{\partial t^2} - k \sin x \quad \begin{array}{c} 0 < x < \pi \\ t > 0 \end{array} \tag{1}$$

$$u(x,0) = u_t(x,0) = 0$$

$$u(0,t) = u(\pi,t) = 0$$

Taking the one-sided Fourier transform with respect to t,

$$U(x,z) = F[u]$$

we obtain

$$U_{xx} + \frac{z^2}{a^2} U = \frac{-k \sin x}{\sqrt{2\pi} \, iz} \tag{2}$$

and

$$U(x,z) = A \sin \frac{zx}{a} + B \cos \frac{zx}{a} - \frac{a^2 k \sin x}{\sqrt{2\pi} \, iz(z^2 - a^2)} \tag{3}$$

From the boundary conditions we evaluate the coefficients
$$A = B = 0 \tag{4}$$

and (3) reduces to
$$U(x,z) = \frac{-a^2 k \sin x}{\sqrt{2\pi}\, iz(z^2-a^2)} \tag{5}$$

Applying the inversion integral to (5) we obtain

$$u(x,t) = \frac{-a^2 k \sin x}{2\pi i} \int_{-\infty+i\delta}^{\infty+i\delta} \frac{e^{izt}}{z(z^2-a^2)}\, dz \tag{6}$$

where $\delta < -a$.

The poles are at $z = 0$ and $z = \pm a$. For $t < 0$ the integral is zero. For $t > 0$ we can choose the contour to be the semi-circle in the upper half-plane. Thus,

$$u(x,t) = \begin{cases} 0 & \text{for } t < 0 \\ k \sin x(1 - \cos at) & \text{for } t > 0 \end{cases} \tag{7}$$

● **PROBLEM 22-21**

Compute the displacement of an infinitely long elastic string. Its initial displacement is

$$u(x,0) = f(x) \tag{1}$$

which is zero outside a finite interval. Its initial velocity is

$$\left.\frac{\partial u(x,t)}{\partial t}\right|_{t=0} = u_t(x,0) = h(x) \tag{2}$$

which is zero outside of a finite interval.

Solution: The problem can be formulated as follows

$$\frac{\partial^2 u}{\partial x^2} = \frac{1}{a^2}\frac{\partial^2 u}{\partial t^2} \quad t > 0,\ -\infty < x < \infty$$

$$\begin{aligned} u(x,0) &= f(x) \\ u_t(x,0) &= h(x) \\ \lim_{x \to \pm\infty} u(x,t) &= 0, \quad t > 0 \end{aligned} \tag{3}$$

Assuming $u(x,t)$, $f(x)$, $h(x)$ have Fourier transforms (and considering only small displacement), we have

$$U(z,t) = \frac{1}{\sqrt{2\pi}} \int_{-\infty}^{\infty} u(x,t) e^{-izx}\, dx \tag{4}$$

and

$$(iz)^2 U(z,t) = \frac{1}{a^2}\frac{\partial^2 U}{\partial t^2} \tag{5}$$

$$U(z,0) = \frac{1}{\sqrt{2\pi}} \int_{-\infty}^{\infty} f(x) e^{-izx} \, dx = F(z) \tag{6}$$

$$U_t(z,0) = \frac{1}{\sqrt{2\pi}} \int_{-\infty}^{\infty} h(x) e^{-izx} \, dx = H(z) \tag{7}$$

Then

$$U_{tt} + a^2 z^2 U = 0 \tag{8}$$

$$U(z,t) = A \sin azt + B \cos azt \tag{9}$$

$$U(z,0) = F(z) = B, \quad U_t(z,0) = H(z) = azA \tag{10}$$

Hence,

$$U(z,t) = \frac{H(z)}{az} \sin azt + F(z) \cos azt$$

$$= \frac{H(z)}{2iaz}(e^{iazt} - e^{-iazt}) \tag{11}$$

$$+ \frac{F(z)}{2}(e^{iazt} + e^{-iazt})$$

Applying the inversion integral we find

$$u(x,t) = \frac{1}{2\sqrt{2\pi}} \int_{-\infty+i\beta}^{\infty+i\beta} F(z) \left[e^{iz(x+at)} + e^{iz(x-at)} \right] dz$$

$$+ \frac{1}{2a\sqrt{2\pi}} \int_{-\infty+i\beta}^{\infty+i\beta} \frac{H(z)}{iz} \left[e^{iz(x+at)} - e^{iz(x-at)} \right] dz \tag{12}$$

$$= \frac{1}{2}\left[f(x+at) + f(x-at) \right] + \frac{1}{2a} \int_{x-at}^{x+at} g(y) \, dy$$

HYDRODYNAMICS and AERODYNAMICS

• **PROBLEM 22-22**

1. Consider the motion of an incompressible fluid (that is a liquid or gas) at velocities much less than the velocity of sound. Define the steady-state two-dimensional velocity field.

2. Define the circulation and the flux.

Solution: 1. A vector function giving the velocity of the fluid at every point of a given region and at every time is called a velocity field.

The flow which is independent of time is called stationary or steady-state. The velocity field is two-dimensional if it depends on two coordinates. Thus, steady-state two-dimensional flow is described by the complex velocity

$$w(x,y) = u(x,y) + iv(x,y) \tag{1}$$

where $u(x,y)$ is the x component and $v(x,y)$ the y component of the flow.

2. Flow $w = u + iv$ is defined in a domain D. Let C be a piecewise smooth closed curve in D, parametrized by S or

$$z = z(S), \quad 0 \leq S \leq \ell \tag{2}$$

where ℓ is the length.

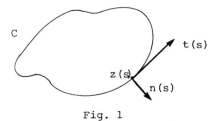

Fig. 1

Let w_t be the tangent component of w and w_n the normal component to C. The integral

$$\int_C w_t \, ds = \int_C (u+iv)_t \, ds = \int_C u \, dx + v \, dy \tag{3}$$

is called the circulation around C, and the integral

$$\int_C w_n \, ds = \int_C (u+iv)_n \, ds = \int_C -v \, dx + u \, dy \tag{4}$$

is called the flux through C.

• **PROBLEM 22-23**

Prove the following theorem:

A continuously differentiable flow $w = u + iv$ defined in a simply-connected domain D is irrotational and solenoidal if and only if

$$u + iv = \overline{\frac{df}{dz}} \tag{1}$$

where $f(z)$, called the complex potential of the flow, is analytic in D.

Solution: Suppose the field $w = u + iv$ is solenoidal (i.e., the flux through any piecewise smooth closed curve is zero) and irrotational (i.e., the circulation around C is zero). Then, both $u\,dx + v\,dy$ and $-v\,dx + u\,dy$ are exact differentials (see Problem 22-22) of two real functions.

$$u\,dx + v\,dy = d\phi \tag{2}$$

$$-v\,dx + u\,dy = d\psi \tag{3}$$

Then,

$$u = \frac{\partial \phi}{\partial x} = \frac{\partial \psi}{\partial y}$$
$$v = \frac{\partial \phi}{\partial y} = -\frac{\partial \psi}{\partial x} \tag{4}$$

The functions ϕ and ψ satisfy the Cauchy-Riemann equations in D. Hence the function

$$f(z) = \phi + i\psi \tag{5}$$

is analytic in D. From (4) we get

$$f'(z) = \frac{\partial \phi}{\partial x} + i\frac{\partial \psi}{\partial x} = u - iv \tag{6}$$

and

$$\overline{f'(z)} = u + iv \tag{7}$$

Assuming that $w = u + iv$ satisfies (1), and because $f(z)$ and $f'(z)$ are analytic, then

$$\int_C \overline{w}\,dz = \int_C f'(z)\,dz = 0 \tag{8}$$

On the other hand

$$\int_C \overline{w}\,dz = \int_C \overline{(u+iv)}(dx+i\,dy) \tag{9}$$

$$= \int_C u\,dx + v\,dy + i\int_C -v\,dx + u\,dy$$

or

$$\mathrm{Re}\int_C \overline{w}\,dz = \int_C u\,dx + v\,dy \tag{10}$$

$$\mathrm{Im}\int_C \overline{w}\,dz = \int_C -v\,dx + u\,dy \tag{11}$$

Combining (8), (10) and (11) we conclude that the field w is irrotational and solenoidal.

● PROBLEM 22-24

A fluid is moving with a constant velocity $\bar{V}_0 = (V_0 \cos\alpha, V_0 \sin\alpha)$. Find,

1. the complex potential
2. velocity potential and stream function
3. the equations for equipotential lines and for the streamlines.

Solution: 1. The complex velocity w is given by

$$w = V_0 \cos\alpha + iV_0 \sin\alpha = V_0(\cos\alpha + i\sin\alpha)$$
$$= V_0 e^{i\alpha} \tag{1}$$

The complex potential is defined by

$$\overline{f'(z)} = w \tag{2}$$

Thus,

$$\overline{f'(z)} = V_0 e^{i\alpha} \tag{3}$$

and, integrating (3), we obtain

$$f(z) = V_0 z e^{-i\alpha} \tag{4}$$

2. We have

$$f(z) = \text{Re } f(z) + i \text{ Im } f(z) = \phi + i\psi \tag{5}$$

where ϕ is the velocity potential and ψ is the stream function. Thus,

$$f(z) = V_0 z e^{-i\alpha} = V_0(x\cos\alpha + y\sin\alpha) + iV_0(y\cos\alpha - x\sin\alpha) \tag{6}$$

and

$$\phi = V_0(x\cos\alpha + y\sin\alpha) \tag{7}$$

$$\psi = V_0(y\cos\alpha - x\sin\alpha) \tag{8}$$

3. The equipotential lines are

$$\phi = V_0(x\cos\alpha + y\sin\alpha) = a = \text{const} \tag{9}$$

The points on an equipotential line are at equal potential.

The streamlines

$$\psi = V_0(y\cos\alpha - x\sin\alpha) = b = \text{const} \tag{10}$$

represent the path of fluid particles.

● **PROBLEM 22-25**

The complex potential of a flow defined in the whole plane is

$$f(z) = \alpha z \qquad (1)$$

Find its complex velocity, velocity potential, stream function, equipotentials and streamlines.

Solution: The velocity is

$$w = \overline{f'(z)} = \overline{\alpha} \qquad (2)$$

If $\alpha = a + ib$ then

$$f(z) = \alpha z = (a+ib)(x+iy) = (ax-by) + i(ay+bx) \qquad (3)$$

The velocity potential is

$$\theta(x,y) = ax - by \qquad (4)$$

and the stream function is

$$\psi(x,y) = ay + bx \qquad (5)$$

The equipotentials are

$$ax - by = \text{const} \qquad (6)$$

and the streamlines are

$$bx + ay = \text{const} \qquad (7)$$

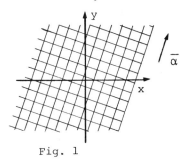

Fig. 1

● **PROBLEM 22-26**

The complex potential defined for the whole plane is

$$f(z) = z^2 \qquad (1)$$

Find its velocity, velocity potential, stream function, equipotentials and streamlines.

Solution: The velocity of the flow is

$$w = u + iv = \overline{f'(z)} = 2\bar{z} \quad (2)$$

From

$$f(z) = \phi + i\psi = (x+iy)^2 = (x^2-y^2) + i2xy \quad (3)$$

we obtain the velocity potential

$$\phi(x,y) = x^2 - y^2 \quad (4)$$

and the stream function

$$\psi(x,y) = 2xy \quad (5)$$

The equipotentials are

$$x^2 - y^2 = \text{const} \quad (6)$$

and the streamlines are

$$2xy = \text{const} \quad (7)$$

Eq.(6) and (7) represent two families of equilateral hyperbolas. The coordinate axes $x = 0$ and $y = 0$ are streamlines $xy = 0$.

Assuming the flow takes place in the first quadrant ($x \geq 0$, $y \geq 0$), we obtain a model of a flow around a corner as shown in Fig. 1.

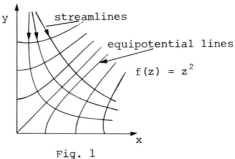

Fig. 1

● **PROBLEM 22-27**

Prove that the equipotentials and the streamlines form an orthogonal system.

Solution: The complex potential is

$$f(z) = \phi + i\psi \quad (1)$$

where ϕ is the velocity potential and ψ the stream function. We define

$$\phi(x,y) = \text{const}, \quad \psi(x,y) = \text{const} \quad (2)$$

to be equipotentials and streamlines respectively. The mapping

$$f(z) = w = \zeta + i\eta \tag{3}$$

where $f(z)$ is the complex potential, is conformal everywhere except at the points $f'(z) = 0$. Function $f(z)$ maps (2) into

$$\zeta = \text{const}, \quad \eta = \text{const} \tag{4}$$

Curves (4) in the w plane form an orthogonal system. Hence, curves (2) also form an orthogonal system except at the points where $f'(z) = 0$. Function $f(z)$ is analytic, hence

$$\frac{\partial \phi}{\partial x} = \frac{\partial \psi}{\partial y}, \quad \frac{\partial \phi}{\partial y} = -\frac{\partial \psi}{\partial x} \tag{5}$$

From (2) and (5) we get

$$\frac{\partial \phi}{\partial x} dx + \frac{\partial \phi}{\partial y} dy = u\, dx + v\, dy = 0 \tag{6}$$

and

$$\frac{\partial \psi}{\partial x} dx + \frac{\partial \psi}{\partial y} dy = -v\, dx + u\, dy = 0 \tag{7}$$

because

$$f'(z) = \frac{\partial \phi}{\partial x} + i \frac{\partial \psi}{\partial x} = u - iv \tag{8}$$

We see that at every point (x,y) except where $f'(z) = 0$, the velocity is normal to the equipotential line through (x,y) and tangential to the streamline through (x,y). The streamlines are the actual trajectories of the flowing particles of the fluid.

• **PROBLEM** 22-28

The complex potential of a flow

$$f(z) = A \ln(z - \alpha) \tag{1}$$

where $A > 0$, is defined for the whole complex plane except $z = \alpha$. Find its complex velocity, equipotentials and streamlines.

Solution: Note that function (1) is not single-valued, but it has a single-valued derivative

$$f'(z) = \frac{A}{z - \alpha} \tag{2}$$

Let

$$z - \alpha = re^{i\theta} \tag{3}$$

then up to a multiple of $2\pi A i$

$$f(z) = A \ln r + A\theta i \tag{4}$$

Equipotential lines of the form

$$A \ln r = \text{const} \quad \text{or}$$
$$r = \text{const} \tag{5}$$

are circles with center at $z = \alpha$ (see Fig. 1). The streamlines are

$$\theta = \text{const} \tag{6}$$

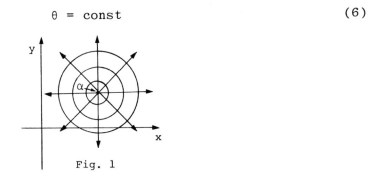

Fig. 1

The velocity is

$$v = \overline{f'(z)} = \frac{A}{\overline{z - \alpha}} = \frac{A}{r} e^{i\theta} \tag{7}$$

$$= \frac{A}{r}(\cos\theta + i \sin\theta) = \frac{A}{r}\cos\theta + i \frac{A}{r}\sin\theta$$

and the speed is

$$|v| = \frac{A}{r} \tag{8}$$

● **PROBLEM 22-29**

Show that the flow described in Problem 22-28 is locally source and sink free but globally does not have this property.

Solution: Let C be any closed path containing the point $z = \alpha$ in its interior, then

$$\int_C f'(z) dz = \int_C \frac{A}{z - \alpha} dz = 2\pi A i \tag{1}$$

On the other hand

$$\int_C f'(z) dz = \text{circulation} + i \cdot \text{flux} \tag{2}$$

Thus,

$$\text{Re}\left[\int_C f'(z) dz\right] = \text{circulation} = 0 \tag{3}$$

$$\text{Im}\left[\int_C f'(z)dz\right] = \text{flux} = 2\pi A \tag{4}$$

Point $z = \alpha$ is the source of strength $M = 2\pi A$. The strength M of the source is the rate at which a fluid of unit density flows across any closed path containing $z = \alpha$ in its interior. If the strength is negative, there is a sink at $z = \alpha$ and the flow is inward. The flow is not globally source free. It is, however, locally source free. Taking any closed contour C which does not contain $z = \alpha$, we find circulation and flux to be zero

$$\int_C f'(z)dz = 0 \tag{5}$$

● **PROBLEM 22-30**

For the complex potential
$$f(z) = iA \ln z, \quad A > 0 \tag{1}$$
find the streamlines, equipotentials, velocity and speed.

Solution: Let
$$z = re^{i\theta} \tag{2}$$

then
$$f(z) = \phi + i\psi = iA(\ln r + i\theta) \tag{3}$$
$$= -A\theta + iA \ln r$$

The streamlines are
$$\psi = \text{const} \quad \text{or} \quad r = \text{const} \tag{4}$$

and the equipotential lines are
$$\phi = \text{const} \quad \text{or} \quad \theta = \text{const} \tag{5}$$

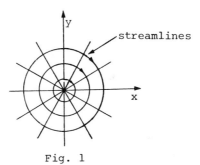

Fig. 1

$$f'(z) = \frac{iA}{r} = \frac{A \sin\theta}{r} + \frac{iA \cos\theta}{r} \tag{6}$$

The complex velocity is

$$v = \overline{f'(z)} = \frac{A \sin\theta}{r} - \frac{iA \cos\theta}{r} \tag{7}$$

Hence, the direction of flow is clockwise, with a speed of magnitude

$$|v| = \frac{A}{r} \tag{8}$$

• **PROBLEM 22-31**

Find the circulation about the vortex of the flow described in Problem 22-30.

Solution: We have

$$\int_C f'(z)dz = \int_C \frac{iA}{z} dz = 2\pi A \tag{1}$$

where C is any closed contour containing the point $z = 0$ in its interior. The flux across C is

$$\text{Im}\left[\int_C f'(z)dz\right] = 0 \tag{2}$$

and the circulation about C is

$$\text{Re}\left[\int_C f'(z)dz\right] = 2\pi A \tag{3}$$

The point $z = 0$ is called a vortex of strength $N = 2\pi A$. For the contour not containing $z = 0$ the integral

$$\int_C f'(z) = 0 \tag{4}$$

Hence, locally the flow is circulation and source free.

• **PROBLEM 22-32**

An infinite line source emanates fluid at a constant rate. The fluid density is α. Find the speed of the fluid and the complex potential of the flow.

Solution: We choose the system of coordinates in such a way that the line source occupies the z axis. Consider a cylinder of radius r and unit length, see Fig. 1. Let V_r be the radial velocity of the fluid at a distance r.

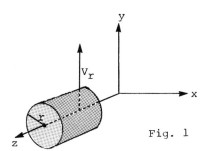
Fig. 1

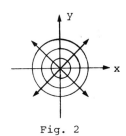
Fig. 2

The total mass of fluid leaving the cylinder per unit time = Radial velocity × Fluid Density × Surface area

$= V_r \cdot \alpha \cdot 2\pi r.$

Denoting the total mass, which is constant by k we get

$$V_r = \frac{k}{2\pi\alpha} \cdot \frac{1}{r} = A \cdot \frac{1}{r} \quad (1)$$

Coefficient $A = \frac{k}{2\pi\alpha}$ is the strength of the source.

From

$$V_r = \frac{\partial \phi}{\partial r} = \frac{A}{r} \quad (2)$$

we obtain

$$\phi = A \ln r \quad (3)$$

But,

$$f(z) = \phi + i\psi \quad (4)$$

Therefore,

$$f(z) = A \ln z \quad (5)$$

which is the complex potential.

● **PROBLEM 22-33**

A source of strength A is located at z = -a and the sink of the same strength at z = a.

Find the complex potential, equipotentials and streamlines.

Solution: The complex potential due to source at z = -a is $A \ln(z + a)$. The complex potential due to sink at z = a is $-A \ln(z - a)$.

Thus, the complex potential of the system is

$$f(z) = A \ln(z + a) - A \ln(z - a) \quad (1)$$

$$= A \ln\left(\frac{z+a}{z-a}\right)$$

Let
$$z + a = r_1 e^{i\theta_1}$$
$$z - a = r_2 e^{i\theta_2} \tag{2}$$

then
$$f(z) = \phi + i\psi = A \ln\left(\frac{r_1 e^{i\theta_1}}{r_2 e^{i\theta_2}}\right) \tag{3}$$

$$= A \ln\left(\frac{r_1}{r_2}\right) + iA(\theta_1 - \theta_2)$$

The equipotential lines are
$$\phi = A \ln\left(\frac{r_1}{r_2}\right) = \alpha \tag{4}$$

and the streamlines are
$$\psi = A(\theta_1 - \theta_2) = \beta \tag{5}$$

Since
$$r_1 = \sqrt{(x+a)^2 + y^2}$$
$$r_2 = \sqrt{(x-a)^2 + y^2} \tag{6}$$

and
$$\theta_1 = \arctan\left(\frac{y}{x+a}\right)$$
$$\theta_2 = \arctan\left(\frac{y}{x-a}\right) \tag{7}$$

eq.(4) can be written in the form
$$\sqrt{\frac{(x+a)^2 + y^2}{(x-a)^2 + y^2}} = e^{\frac{\alpha}{A}} \tag{8}$$

or
$$\left[x - a \coth\left(\frac{\alpha}{A}\right)\right]^2 + y^2 = a^2 \operatorname{csch}^2\left(\frac{\alpha}{A}\right) \tag{9}$$

The last equation represents the family of circles with centers at $a \coth\left(\frac{\alpha}{A}\right)$ and radii equal to $a\left|\operatorname{csch}\left(\frac{\alpha}{A}\right)\right|$, see Fig. 1.

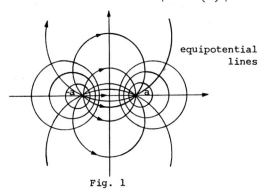

Fig. 1

equipotential lines

The streamlines are

$$\arctan\left(\frac{y}{x+a}\right) - \arctan\left(\frac{y}{x-a}\right) = \frac{\beta}{A} \quad (10)$$

or

$$\left[y + a\cot\left(\frac{\beta}{A}\right)\right]^2 + x^2 = a^2\csc^2\left(\frac{\beta}{A}\right) \quad (11)$$

which represents the family of circles with centers at $-a\cot\left(\frac{\beta}{A}\right)$ and radii $a\left|\csc\left(\frac{\beta}{A}\right)\right|$.

● **PROBLEM 22-34**

Find the complex potential, velocity and the stream function for a flow in the angular region

$$r \geq 0, \quad 0 \leq \theta \leq \frac{\pi}{4}.$$

<u>Solution</u>: The angular region is mapped by

$$f(z) = Az^4 \quad (1)$$

onto the upper half-plane. Function (1) is analytic in the whole plane, thus it is a complex potential of the flow.

$$z = re^{i\theta}$$

$$f(z) = Az^4 = Ar^4(\cos 4\theta + i\sin 4\theta)$$
$$= Ar^4\cos 4\theta + i\, Ar^4\sin 4\theta = \phi + i\psi \quad (2)$$

Hence, the stream function is

$$\psi = Ar^4\sin 4\theta \quad (3)$$

The streamlines are given by

$$r^4\sin 4\theta = \text{const} \quad (4)$$

as shown in Fig. 1.

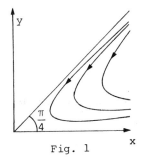

Fig. 1

• **PROBLEM 22-35**

For the flow in the semi-infinite well $\left(-\frac{\pi}{2} \leq x \leq \frac{\pi}{2},\ y \geq 0\right)$ find the complex potential, velocity, speed and streamlines.

Solution: The transformation
$$f(z) = \sin z \qquad (1)$$

maps the semi-infinite strip onto the upper half-plane. Thus, for the complex potential we choose
$$f(z) = A \sin z \qquad (2)$$

The velocity vector is given by
$$V = \overline{f'(z)} \qquad (3)$$

Hence,
$$V = \overline{\frac{d}{dz} A \sin z} = \overline{A \cos z}$$
$$= \overline{A(\cos x \cosh y - i \sin x \sinh y)} \qquad (4)$$
$$= A(\cos x \cosh y + i \sin x \sinh y)$$

The speed is given by
$$|V| = |f'(z)| \qquad (5)$$
$$|V| = |A \cos z| = A|\cos z| = A\sqrt{\cos^2 x + \sinh^2 y} \qquad (6)$$

The velocity potential and the stream function are
$$f(z) = A \sin z = A(\sin x \cosh y + i \cos x \sinh y) \qquad (7)$$
$$= \phi + i\psi$$

The streamlines are
$$\psi = A \cos x \sinh y = \text{const} \qquad (8)$$

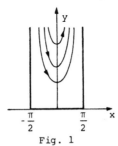

Fig. 1

• **PROBLEM 22-36**

Find the complex potential for the flow in the region $r \geq 1$, $0 \leq \theta \leq \frac{\pi}{2}$. Find the velocity potential and the stream function. Show that along the boundary $\psi(x,y) = 0$.

Solution: We shall show that

$$f(z) = z^2 + \frac{1}{z^2} \tag{1}$$

maps the region of flow onto the upper half-plane, see Fig. 1.

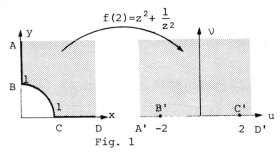

Fig. 1

Hence, the complex potential is

$$f(z) = A\left(z^2 + \frac{1}{z^2}\right) \tag{2}$$

Let

$$z = re^{i\theta}$$

then

$$\begin{aligned} f(z) &= A\left(r^2 e^{2i\theta} + r^{-2} e^{-2i\theta}\right) \\ &= A\left[r^2(\cos 2\theta + i\sin 2\theta) + r^{-2}(\cos 2\theta - i\sin 2\theta)\right] \\ &= A\cos 2\theta (r^2 + r^{-2}) + iA\sin 2\theta (r^2 - r^{-2}) \end{aligned} \tag{3}$$

Thus,

$$\begin{aligned} \phi &= A \cos 2\theta (r^2 + r^{-2}) \\ \psi &= A \sin 2\theta (r^2 - r^{-2}) \end{aligned} \tag{4}$$

Line ABCD (see Fig. 1) is mapped by $f(z) = z^2 + \frac{1}{z^2}$ onto the real u axis. Therefore if (x,y) is a point on the boundary then

$$\text{Im } f(z) = 0; \quad z = x + iy \tag{5}$$

$$\psi(x,y) = 0$$

Note that for any physical flow the surface of the boundary confining this flow must be part of a streamline $\psi(x,y) = \text{const}$. The component of the flow normal to such a surface is zero.

● **PROBLEM 22-37**

The sink of strength $m > 0$ is located at $z = 0$ and a source of the same strength is located at $z = \alpha$. This flow has the complex potential

$$f(z) = \frac{m}{2\pi} \ln(z - \alpha) - \frac{m}{2\pi} \ln z \qquad (1)$$

What happens to the complex potential when the distance between the source and sink approaches zero and the strength increases according to

$$m|\alpha| = k = \text{const}? \qquad (2)$$

<u>Solution:</u> Let

$$\alpha = |\alpha|e^{i\lambda} \qquad (3)$$

where λ is constant, then (1) can be expressed in the form

$$f(z) = -\frac{k}{2\pi} e^{i\lambda} \left[\frac{\ln(z-\alpha) - \ln z}{-z} \right] \qquad (4)$$

Taking the limit as $\alpha \to 0$ we obtain

$$f_0(z) = \lim_{\alpha \to 0} f(z) = \frac{-ke^{i\lambda}}{2\pi} \cdot \frac{d}{dz}(\ln z) = \frac{-ke^{i\lambda}}{2\pi} \cdot \frac{1}{z} \qquad (5)$$

This system has a doublet at $z = 0$ of strength k. The vector $e^{i\lambda}$ is called the axis of the doublet.

Let $z = re^{i\theta}$ then

$$\begin{aligned} f_0(z) &= \frac{-k}{2\pi r} e^{-i(\theta-\lambda)} \\ &= \frac{-k}{2\pi r}\left[\cos(\theta-\lambda) - i\sin(\theta-\lambda)\right] \end{aligned} \qquad (6)$$

and

$$\begin{aligned} \phi &= \frac{-k}{2\pi r} \cos(\theta - \lambda) \\ \psi &= \frac{k}{2\pi r} \sin(\theta - \lambda) \end{aligned} \qquad (7)$$

The streamlines are circles passing through the origin and tangent to the axis of the doublet. The equipotential lines are circles passing through the origin and perpendicular to the axis of the doublet.

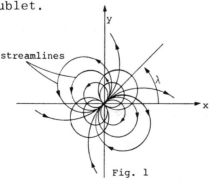

Fig. 1

From (5) we compute the velocity

$$V = \overline{f_0'(z)} = \frac{ke^{-i\lambda}}{2\pi} \cdot \frac{1}{z^2}$$

$$= \frac{k}{2\pi r^2}\left[\cos(2\theta-\lambda) + i\sin(2\theta-\lambda)\right] \tag{8}$$

The speed is given by

$$|V| = \frac{k}{2\pi r^2} \tag{9}$$

• **PROBLEM 22-38**

The complex potential of a flow is

$$f(z) = A\left(\frac{z}{e^{i\lambda}} + \frac{k^2}{z}e^{i\lambda}\right) \quad A > 0, \; k > 0 \tag{1}$$

Compute the velocity vector and find the equipotential lines and streamlines.

Solution: Let

$$z = re^{i\theta} \tag{2}$$

then

$$f(z) = A\left[re^{i\theta}e^{-i\lambda} + k^2 r^{-1} e^{-i\theta} e^{i\lambda}\right]$$

$$= Ar\left[\cos\theta\cos\lambda + \sin\theta\sin\lambda\right] + Ak^2 r^{-1}\left[\cos\theta\cos\lambda + \sin\theta\sin\lambda\right]$$

$$+ Ari\left[\sin\theta\cos\lambda - \cos\theta\sin\lambda\right] + Ak^2 r^{-1} i\left[\cos\theta\sin\lambda - \sin\theta\cos\lambda\right]$$

$$= A\left(r + \frac{k^2}{r}\right)\cos(\theta-\lambda) + iA\left(r - \frac{k^2}{r}\right)\sin(\theta-\lambda) \tag{3}$$

The equipotential lines are

$$\left(r + \frac{k^2}{r}\right)\cos(\theta - \lambda) = a_1 \tag{4}$$

and the streamlines are

$$\left(r - \frac{k^2}{r}\right)\sin(\theta - \lambda) = a_2 \tag{5}$$

where a_1 and a_2 are arbitrary constants. The velocity vector is given by

$$v = \overline{f'(z)} \tag{6}$$

From (1)

$$f'(z) = A\left(e^{-i\lambda} - \frac{k^2}{z^2}e^{i\lambda}\right) \tag{7}$$

and
$$v = A\left(e^{i\lambda} - k^2 r^{-2} e^{2i\theta} e^{-i\lambda}\right) \tag{8}$$

where
$$z = re^{i\theta}$$

FLOW AROUND AN OBJECT

• **PROBLEM 22-39**

Consider the uniform flow around a long circular cylinder of unit radius, which is perpendicular to the direction of flow.

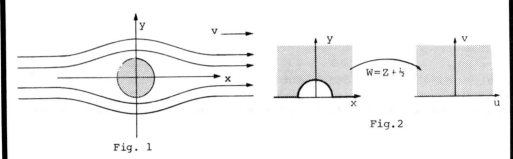

Fig. 1

Fig. 2

The flow far away from the cylinder is parallel to the x axis. Find the complex potential and the streamlines of this flow.

Solution: The equation of the cylinder is $x^2 + y^2 = 1$. The flow is symmetric for $y > 0$ and $y < 0$, hence we need only to consider the upper half-plane. Transformation

$$w = z + \frac{1}{z} \tag{1}$$

maps the boundary of the region of flow, that is the upper semi-circle and the parts of the x axis exterior to the circle onto the u axis.

The complex potential of the uniform flow in the upper half of the w plane is

$$f(w) = aw, \quad a > 0 \tag{2}$$

Hence, the complex potential of the flow in the z plane is

$$f(z) = a\left(z + \frac{1}{z}\right) \tag{3}$$

The velocity

$$v = \overline{f'(z)} = a\left(1 - \frac{1}{z^2}\right) \tag{4}$$

tends to

$$\lim_{|z|\to\infty} v = a \qquad (5)$$

In polar coordinates the stream function is

$$\psi = \text{Im } f(z) = a\left(r - \frac{1}{r}\right)\sin\theta \qquad (6)$$

and the streamlines are

$$a\left(r - \frac{1}{r}\right)\sin\theta = \text{const} \qquad (7)$$

In particular for const = 0

$$\left(r - \frac{1}{r}\right)\sin\theta = 0 \qquad (8)$$

the streamline consists of the circle $r = 1$ and the parts of the x axis outside of the circle.

● **PROBLEM 22-40**

The complex potential of the flow around an infinite cylinder of unit radius is

$$f(z) = A\left(z + \frac{1}{z}\right) \qquad (1)$$

Compute the speed of the fluid on the surface of the cylinder. Show that the fluid pressure on the cylinder is greatest at the points $z = \pm 1$ and least at the points $z = \pm i$. Use Bernoulli's law: $P + \frac{1}{2}\rho|V|^2 = \text{const}$.

<u>Solution</u>: The velocity field is

$$V = A\left(1 - \frac{1}{z^2}\right) \qquad (2)$$

The points on the surface of the cylinder are

$$z = e^{i\theta}$$

Then,

$$V = A\left(1 - \frac{1}{e^{-2i\theta}}\right) \qquad (3)$$

and the speed on the cylinder is

$$|V| = A|1 - \cos 2\theta - i\sin 2\theta| = A\sqrt{2 - 2\cos 2\theta}$$
$$= 2A|\sin\theta| \qquad (4)$$

By Bernoulli's law

$$\frac{P}{\rho} + \frac{1}{2}|v|^2 = \frac{P}{\rho} + 2A\sin^2\theta = \text{const} \tag{5}$$

The pressure is greatest when $\sin\theta = 0$, that is for $z = \pm 1$.

The least pressure, for $\sin^2\theta = 1$ exists for the points $z = \pm i$.

• **PROBLEM 22-41**

Consider the flow past a circular cylinder of radius R. The flow is solenoidal in the flow domain D and irrotational in every simply connected subdomain of D. Far away from the cylinder the velocity is $w = u + iv$. Find the Laurent expansion of the complex potential $f(z)$ and $f'(z)$. From the condition that the surface of the cylinder is one of the streamlines, simplify the results as much as possible.

Solution: If $f(z)$ is the complex potential of the flow, then $f'(z)$ is an analytic function for $|z| > R$. Its Laurent expansion at infinity is

$$f'(z) = \bar{w} + \frac{c_1}{z} + \frac{c_2}{z^2} + \frac{c_3}{z^3} + \ldots \tag{1}$$

and

$$f(z) = \bar{w}z + c_1 \ln z - \frac{c_2}{z} - \frac{c_3}{2z^2} - \ldots \tag{2}$$

The stream function is given by

$$\psi(x,y) = \operatorname{Im} f(z) \tag{3}$$

or in polar coordinates

$$z = re^{i\theta}$$

Setting

$$c_k = a_k + ib_k, \quad k = 1, 2, \ldots \tag{4}$$

we find

$$\psi(r,\theta) = a_1\theta + b_1 \ln r + \frac{a_2 + r^2 u}{r}\sin\theta - \frac{b_2 + r^2 v}{r}\cos\theta \tag{5}$$
$$+ \frac{a_3}{2r^2}\sin 2\theta - \frac{b_3}{2r^2}\cos 2\theta + \ldots$$

The surface of the cylinder $|z| = R$ must be one of the streamlines

$$\psi(r,\theta)\Big|_{|z|=R} = \text{const} \tag{6}$$

From (5) and (6) we obtain

$$a_1 = a_2 + R^2 u = a_3 = \ldots = 0 \tag{7}$$

$$b_2 + R^2 v = b_3 = \ldots = 0$$

Hence, (1) and (2) reduce to

$$f'(z) = \bar{w} + \frac{ib_1}{z} - \frac{R^2 w}{z^2} \tag{8}$$

$$f(z) = \bar{w} z + i b_1 \ln z + \frac{R^2 w}{z} \tag{9}$$

• PROBLEM 22-42

Consider the flow described in Problem 22-41. Let A be the circulation around the circle $|z| = R$. In eqs.(8) and (9) of Problem 22-41 replace the real constant b by A.

Solution: Let C be the circle $|z| = R$. Then the circulation A around C is

$$A = \operatorname{Re} \int_C f'(z)\,dz = \operatorname{Re} \int_{|z|=r} f'(z)\,dz \tag{1}$$

for $r > R$.

$$A = \operatorname{Re} \int_{|z|=r} \left[\bar{w} + \frac{ib_1}{z} - \frac{R^2 \bar{w}}{z^2} \right] dz = -2\pi b_1 \tag{2}$$

Then

$$f'(z) = \bar{w} - \frac{iA}{z 2\pi} - \frac{R^2 w}{z^2} \tag{3}$$

and

$$f(z) = \bar{w} z - \frac{iA}{2\pi} \ln z + \frac{R^2 w}{z} \tag{4}$$

• PROBLEM 22-43

Find the stagnation points of the flow with the complex potential

$$f(z) = \bar{w} z - \frac{iA}{2\pi} \ln z + \frac{R^2 w}{z} \tag{1}$$

where $w = u + iv$ is the velocity of the fluid at infinity.

Solution: Rotating the coordinate axes we can obtain

$$w = u > 0 \tag{2}$$

Then,

$$f(z) = uz - \frac{iA}{2\pi} \ln z + \frac{R^2 u}{z} \quad (3)$$

and

$$f'(z) = u - \frac{iA}{2\pi z} - \frac{R^2 u}{z^2} \quad (4)$$

The stagnation points are points where the velocity vanishes. Setting $f'(z) = 0$ we obtain from (4)

$$z^2 - \frac{iA}{2\pi u} z - R^2 = 0 \quad (5)$$

$$z_{1,2} = \frac{iA}{4\pi u} \pm \sqrt{R^2 - \left(\frac{A}{4\pi u}\right)^2} \quad (6)$$

For $|A| = 4\pi Ru$, there is only one stagnation point (see Fig. 1a).

If $|A| < 4\pi Ru$ there are two stagnation points z_1 and z_2 located on the circle $|z| = R$. The points are symmetric with respect to the imaginary axis (see Fig. 1b).

If $|A| > 4\pi Ru$, stagnation points are imaginary. Only one of them lies outside the circle $|z| = R$ (i.e., in the domain of flow) because $z_1 z_2 = -R^2$ (see Fig. 1c).

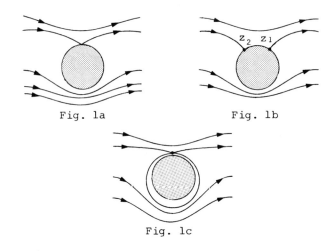

Fig. 1a   Fig. 1b

Fig. 1c

If the circulation vanishes, $A = 0$, we have two stagnation points

$$z_{1,2} = \pm R \quad (7)$$

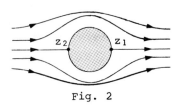

Fig. 2

• **PROBLEM 22-44**

Find the complex potential of the flow past a cylinder of cross section C, where C is a piecewise smooth closed curve. At infinity the velocity is w = u + iv. The flow is solenoidal in the domain of flow D and irrotational in every simply connected subdomain of D.

Solution: A unique conformal mapping $h(z)$ of the exterior of C onto the exterior of the unit circle $|\xi| = 1$ exists such that

$$h(\infty) = \infty \quad \text{and} \quad h'(\infty) > 0 \tag{1}$$

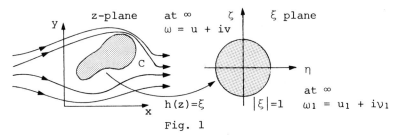

Fig. 1

The Laurent expansion of $h(z)$ at infinity is

$$\xi = h(z) = c_0 + cz + \frac{c_1}{z} + \frac{c_2}{z^2} + \ldots \tag{2}$$

From Problem 22-43 the complex potential of the flow past the cylinder $|\zeta| = 1$ is

$$F(\zeta) = \overline{w}_1 \zeta - \frac{iA}{2\pi} \ln \zeta + \frac{w_1}{\zeta} \tag{3}$$

or substituting $\xi = h(z)$

$$f(z) = F(h(z)) = \overline{w}_1 h(z) - \frac{iA}{2\pi} \ln h(z) + \frac{w_1}{h(z)} \tag{4}$$

Since the velocity of the flow at infinity is

$$w = u + iv$$

we have

$$\overline{w} = f'(\infty) = F'(\infty) \cdot h'(\infty) = \overline{w}_1 c \tag{5}$$

Hence,

$$w_1 = \frac{w}{c} = \frac{w}{h'(\infty)} \tag{6}$$

The complex potential of the flow past C with the circulation A around C and velocity w at infinity is

$$f(z) = \frac{\overline{w}}{\overline{h'(\infty)}} \cdot h(z) - \frac{iA}{2\pi} \ln h(z) + \frac{w}{h'(\infty) h(z)} \tag{7}$$

• **PROBLEM 22-45**

Compute the force exerted by the flow past a cylinder on a unit length of the cylinder of cross section C. Apply Bernoulli's law.

Solution: Let ρ be the density of the fluid and $p(x,y)$ the pressure. Bernoulli's law states that the expression

$$p + \frac{1}{2}\rho|v|^2 \qquad (1)$$

is constant along streamlines. In particular, since C is a streamline we have

$$p + \frac{1}{2}\rho|v|^2 = A \quad \text{along C.} \qquad (2)$$

Let $dz = dx + idy$ be an element of C, then the pressure acting on C is normal to C, see Fig. 1.

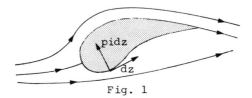

Fig. 1

The force on dz is $pidz$.

$$pidz = Aidz - \frac{1}{2}\rho i|v|^2 dz \qquad (3)$$

Integrating (3) along C we obtain the total force acting on C

$$X + iY = \int_C pidz = Ai\int_C dz - \frac{1}{2}\rho i\int_C |v|^2 dz$$

$$= -\frac{1}{2}\rho i\int_C |v|^2 dz \qquad (4)$$

Since C is a streamline, the velocity v must be tangent to C and

$$v = \overline{f'(z)} = |v|e^{i\theta} \qquad (5)$$

where $\arg dz = \theta$.

$$|v| = \overline{f'(z)}e^{-i\theta} \qquad (6)$$

$f(z)$ is the complex potential of the flow.

Substituting (6) into (4)

$$X + iY = -\frac{1}{2}\rho i\int_C \left[\overline{f'(z)}\right]^2 e^{-2i\theta} dz \qquad (7)$$

Since,
$$e^{-2i\theta}dz = e^{-i\theta}|dz| = d\bar{z} \tag{8}$$

(7) can be written as

$$X + iY = -\frac{1}{2}\rho i \int_C \left[\overline{f'(z)}\right]^2 d\bar{z} \tag{9}$$

Taking the complex conjugate of (9)

$$X - iY = \frac{1}{2}\rho i \int_C \left[f'(z)\right]^2 dz \tag{10}$$

Contour C can be replaced by any large enough circle $|z| = r$ surrounding C. Then

$$X - iY = \frac{1}{2}\rho i \int_{|z|=r} \left[f'(z)\right]^2 dz \tag{11}$$

This result is known as the Blasius theorem.

● **PROBLEM 22-46**

From eq.(11) of Problem 22-45 derive the Joukowski theorem (sometimes called Kutta-Joukowski theorem).

$$X + iY = -\rho A v i \tag{1}$$

where v is the uniform wind (or liquid) of velocity v and A is the circulation around the airfoil.

Solution:
$$X - iY = \frac{1}{2}\rho i \int_{|z|=r} \left[f'(z)\right]^2 dz \tag{2}$$

The complex potential $f(z)$ can be expressed by

$$f'(z) = v - \frac{iA}{2\pi z} - \frac{r^2 v}{z^2} \tag{3}$$

(see Problem 22-43).

Then,
$$X - iY = \frac{1}{2}\rho i \int_{|z|=r} \left[v - \frac{Ai}{2\pi z} + \ldots\right]^2 dz$$

$$= \frac{1}{2}\rho i \cdot \frac{2A\overline{v}}{2\pi i} \int_{|z|=r} \frac{dz}{z} = \frac{1}{2}\rho i \cdot \frac{2A\overline{v}}{2\pi i} \cdot 2\pi i \tag{4}$$

$$= i\rho A\overline{v}$$

Hence,

$$X + iY = -i\rho Av \qquad (5)$$

which is the Joukowski theorem.

Applications of this theorem are numerous in aerodynamics and hydrodynamics.

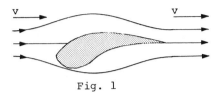

Fig. 1

The theorem states that if an airfoil is at rest in a wind of velocity v, then the force exerted on the airfoil is

$$\rho |Av|,$$

where A is the circulation around the airfoil. The force is perpendicular to the wind. The direction of the force is obtained by rotating v through 90° in the direction opposite to the circulation.

● **PROBLEM 22-47**

The force exerted by the flow past a cylinder on a unit length of the cylinder of cross section C is

$$X - iY = \frac{1}{2}\rho i \int_C \left[f'(z)\right]^2 dz \qquad (1)$$

(see Problem 22-45).

Show that the total moment about the origin is

$$M = \text{Re}\left\{-\frac{1}{2}\rho \int_C z \left[\frac{df}{dz}\right]^2 dz\right\} \qquad (2)$$

Solution: Let ds be an element of length on C

$$dz = ds\, e^{i\theta} \qquad (3)$$

Then the moment about the origin of the force acting on element ds is

$$dM = (p\, ds\, \cos\theta)x + (p\, ds\, \sin\theta)y$$
$$= p(y\,dy + x\,dx) \qquad (4)$$

Integrating (4) and applying Bernoulli's law, we find

$$M = \int_C p(y\,dy + x\,dx) = \int_C (A - \tfrac{1}{2}\rho v^2)(y\,dy + x\,dx)$$

$$= A \int_C (y\,dy + x\,dx) - \frac{1}{2}\rho \int_C v^2(y\,dy + x\,dx)$$

$$= -\frac{1}{2}\rho \int_C v^2(x\cos\theta + y\sin\theta)ds \tag{5}$$

Since $y\,dy + x\,dx$ is an exact differential, the integral $\int_C y\,dy + x\,dx$ vanishes.

Expression (5) for M can be further simplified

$$M = \mathrm{Re}\left[-\frac{1}{2}\rho \int_C v^2(x+iy)(\cos\theta - i\sin\theta)ds\right]$$

$$= \mathrm{Re}\left[-\frac{1}{2}\rho \int_C v^2 z\, e^{-i\theta} ds\right] \tag{6}$$

$$= \mathrm{Re}\left[-\frac{1}{2}\rho \int_C z(v^2 e^{-2i\theta})e^{i\theta} ds\right]$$

$$= \mathrm{Re}\left[-\frac{1}{2}\rho \int_C z\left(\frac{df}{dz}\right)^2 dz\right]$$

The counterclockwise moments are positive.

• **PROBLEM 22-48**

The complex potential of a flow is

$$f(z) = v_0\left(z + \frac{a^2}{z}\right) - \frac{iA}{2\pi}\ln z \tag{1}$$

where v_0 is the speed of the fluid at infinity and A is the circulation. Find the force acting on the obstacle.

Solution: The force acting on the obstacle is

$$X - iY = \frac{1}{2}i\rho \int_C \left(\frac{df}{dz}\right)^2 dz \tag{2}$$

(see Problem 22-45).

Substituting (1) into (2) we find

$$X - iY = \frac{1}{2}i\rho \int_C \left[v_0\left(1 - \frac{a^2}{z^2}\right) - \frac{iA}{2\pi z}\right]^2 dz$$

$$= \frac{1}{2} i\rho \int_C \left[v_0^2 \left(1 - \frac{a^2}{z^2}\right)^2 - \frac{2iv_0 A}{2\pi z}\left(1 - \frac{a^2}{z^2}\right) - \frac{A^2}{4\pi^2 z^2}\right] dz \quad (3)$$

$$= \rho v_0 A$$

ELECTROSTATIC FIELDS

• **PROBLEM** 22-49

An electric field is a vector field giving the force exerted on a unit positive charge. The electric field is usually defined for a given region and for every instant of time,

$$\overline{E} = \overline{E}(x,y,z,t)$$

If the field is independent of time it is called stationary. The stationary electric field is described by the Maxwell equations

$$\operatorname{curl} \overline{E} = \overline{0} \quad (1)$$

$$\operatorname{div} \overline{E} = 4\pi\rho$$

where ρ is continuous and $\overline{E}$ is continuously differentiable.

Assuming that $\overline{E}$ is independent of the z-coordinate obtain the two-dimensional version of the Maxwell equations.

Solution:
$$\operatorname{curl} \overline{E} = \nabla \times \overline{E} = \left(\frac{\partial}{\partial x}, \frac{\partial}{\partial y}, \frac{\partial}{\partial z}\right) \times (E_x, E_y, E_z)$$

$$= \left(\frac{\partial E_z}{\partial y} - \frac{\partial E_y}{\partial z}, \frac{\partial E_x}{\partial z} - \frac{\partial E_z}{\partial x}, \frac{\partial E_y}{\partial x} - \frac{\partial E_x}{\partial y}\right) \quad (2)$$

$$= \left(\frac{\partial E_z}{\partial y}, \frac{\partial E_z}{\partial x}, \frac{\partial E_y}{\partial x} - \frac{\partial E_x}{\partial y}\right) = \overline{0}$$

Thus

$$\frac{\partial E_z}{\partial x} = \frac{\partial E_z}{\partial y} = 0 \quad (3)$$

and we can choose $E_z = 0$.

Hence, (2) leads to

$$\frac{\partial E_y}{\partial x} - \frac{\partial E_x}{\partial y} = 0 \quad (4)$$

$$\text{div } \bar{E} = \left(\frac{\partial}{\partial x}, \frac{\partial}{\partial y}, \frac{\partial}{\partial z}\right) \cdot (E_x, E_y, E_z) \quad (5)$$

$$= \frac{\partial E_x}{\partial x} + \frac{\partial E_y}{\partial y} = 4\pi\rho$$

Thus, the stationary two-dimensional electric field is described by

$$\frac{\partial E_y}{\partial x} - \frac{\partial E_x}{\partial y} = 0$$

$$\frac{\partial E_x}{\partial x} + \frac{\partial E_y}{\partial y} = 4\pi\rho \quad (6)$$

where ρ is the surface charge density.

• **PROBLEM 22-50**

Applying Green's theorem to

$$\frac{\partial E_y}{\partial x} - \frac{\partial E_x}{\partial y} = 0 \quad (1)$$

$$\frac{\partial E_x}{\partial x} + \frac{\partial E_y}{\partial y} = 4\pi\rho \quad (2)$$

show that two real functions $\phi = \phi(x,y)$ and $\psi = \psi(x,y)$ exist such that

$$E_x = -\frac{\partial \phi}{\partial x} \qquad E_y = -\frac{\partial \phi}{\partial y}$$

$$E_x = \frac{\partial \psi}{\partial y} \qquad E_y = -\frac{\partial \psi}{\partial x} \quad (3)$$

Solution: According to Green's thoerem, if the functions $f = f(x,y)$, $g = g(x,y)$ and their first partial derivatives are continuous on a piecewise smooth closed curve C and in its interior, then

$$\iint_{\text{int C}} \left(\frac{\partial g}{\partial x} - \frac{\partial f}{\partial y}\right) dx\,dy = \int_C f\,dx + g\,dy \quad (4)$$

Eq.(1) leads to

$$\int_C E_x\,dx + E_y\,dy = 0 \quad (5)$$

and (2) leads to

$$\int_C -E_y\,dx + E_x\,dy = 0 \quad (6)$$

777

Thus, $E_x dx + E_y dy$ and $-E_y dx + E_x dy$ are exact differentials and we can define two real functions $\phi(x,y)$ and $\psi(x,y)$ such that

$$-E_x dx - E_y dy = d\phi \tag{7}$$

$$-E_y dx + E_x dy = d\psi \tag{8}$$

The minus sign on the left-hand side of (7) is there for historical reasons. Finally, from (7) and (8) we get

$$E_x = -\frac{\partial \phi}{\partial x}, \quad E_y = -\frac{\partial \phi}{\partial y} \tag{9}$$

$$E_y = -\frac{\partial \psi}{\partial x}, \quad E_x = \frac{\partial \psi}{\partial y} \tag{10}$$

Note that the electrostatic field is irrotational and solenoidal.

• **PROBLEM 22-51**

Show that the functions $\phi(x,y)$ and $\psi(x,y)$ satisfy the Cauchy-Riemann equations in a simply connected domain D. Define the complex potential, stream function, electrostatic potential, lines of force, and equipotentials.

Solution: From eq.(9) and (10) of Problem 22-50 we get

$$\frac{\partial \phi}{\partial y} = \frac{\partial \psi}{\partial x} \tag{1}$$

and

$$-\frac{\partial \phi}{\partial x} = \frac{\partial \psi}{\partial y} \tag{2}$$

Hence, ϕ and ψ satisfy the Cauchy-Riemann equations in D. The function

$$f(z) = \psi(x,y) + i\,\phi(x,y) \tag{3}$$

called the complex potential, is therefore analytic in D.

$$f'(z) = \frac{\partial \psi}{\partial x} + i\frac{\partial \phi}{\partial x} = -E_y - iE_x = -i(E_x - iE_y) \tag{4}$$

or

$$-i\,\overline{f'(z)} = E_x + iE_y \tag{5}$$

The function ψ is called the stream function and the function ϕ the electrostatic potential.

The lines of force are defined by

$$\psi(x,y) = \text{const} \tag{6}$$

and equipotentials are defined by

$$\phi(x,y) = \text{const} \tag{7}$$

Note that curves (6) and (7) form an orthogonal system.

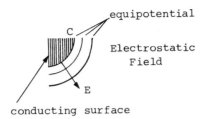

Let C be a conducting surface set in an electrostatic field. Then C is a part of an equipotential, that is the tangential component of the electric field is zero. Otherwise the charges would move along C.

• **PROBLEM** 22-52

The complex potential of an electric field is

$$f(z) = az, \quad a > 0 \tag{1}$$

Find all elements of the field defined in Problem 22-51.

Solution: The electric field is

$$-i \,\overline{f'(z)} = E_x + i E_y = -ia \tag{2}$$

$$E_x = 0 \tag{3}$$
$$E_y = -a$$

Since

$$f(z) = az = a(x+iy) = ax + i\,ay = \psi + i\phi \tag{4}$$

the stream function is

$$\psi = ax \tag{5}$$

and the electrostatic potential is

$$\phi = ay \tag{6}$$

Hence the lines of force are

$$x = \text{const} \tag{7}$$

and the equipotentials are

$$y = \text{const} \tag{8}$$

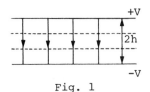

Fig. 1

For example, for a condenser shown in Fig. 1., consisting of two infinite parallel-plates with one plate at potential V and the other -V, we choose $a = \frac{V}{h}$ to obtain

$$\psi = \frac{V}{h} x, \quad \phi = \frac{V}{h} y \quad (9)$$

$$E = -i \frac{V}{h}$$

• **PROBLEM** 22-53

Find the electric field, stream function, electrostatic potential, lines of force and equipotentials due to the complex potential

$$f(z) = \frac{1}{z^2} \quad (1)$$

Solution:

$$f(z) = \frac{1}{z^2} = \frac{1}{(x+iy)^2} = \frac{x^2 - y^2}{(x^2+y^2)^2} + i \frac{-2xy}{(x^2+y^2)^2} \quad (2)$$

$$= \psi + i\phi$$

Hence, the stream function is

$$\psi = \frac{x^2 - y^2}{(x^2+y^2)^2} \quad (3)$$

and the electrostatic potential is

$$\phi = \frac{-2xy}{(x^2+y^2)^2} \quad (4)$$

Lines of force are given by

$$\psi(x,y) = \frac{x^2 - y^2}{(x^2+y^2)^2} = a_1 = \text{const} \quad (5)$$

while equipotentials are

$$\phi(x,y) = \frac{-2xy}{(x^2+y^2)^2} = a_2 = \text{const} \quad (6)$$

The electrostatic field $E = E_x + i E_y$ can be evaluated from

$$E_x + i E_y = -i \overline{f'(z)} \quad (7)$$

Thus,

$$E_x + i E_y = -i \left[\frac{-2}{z^3}\right] = 2i \frac{1}{z^3} \tag{8}$$

• **PROBLEM 22-54**

The equipotentials of an electric field are the circles

$$(x - a)^2 + y^2 = a^2 \tag{1}$$

Find the ratio between the magnitude of the field at the point $(4a,0)$ and its magnitude at (a,a).

Solution: From (1) we obtain the equation for equipotentials in the form

$$\frac{x^2 + y^2}{x} = 2a \tag{2}$$

Hence, the electrostatic potential is

$$\phi(x,y) = \frac{x^2 + y^2}{x} \tag{3}$$

Since, $f(z) = \psi + i\phi$ is an analytic function we have

$$f'(z) = \frac{\partial \phi}{\partial y} + i \frac{\partial \phi}{\partial x} \tag{4}$$

$$\frac{\partial \phi}{\partial y} = \frac{2y}{x}, \quad \frac{\partial \phi}{\partial x} = \frac{x^2 - y^2}{x^2} \tag{5}$$

and

$$f'(z) = \frac{2y}{x} + i \frac{x^2 - y^2}{x^2} \tag{6}$$

The electrostatic field is related to the complex potential by

$$E_x + i E_y = -i \overline{f'(z)} \tag{7}$$

Combining (6) and (7) we find

$$E_x + i E_y = -\frac{x^2 - y^2}{x^2} - i \frac{2y}{x} \tag{8}$$

Furthermore

$$(E_x + i E_y)\big|_{(4a,0)} = -1$$

$$(E_x + i E_y)\big|_{(a,a)} = -2i \tag{9}$$

Hence, the ratio of magnitudes is

$$\frac{\left|(E_x + i E_y)\right|_{(4a,0)}}{\left|(E_x + i E_y)\right|_{(a,a)}} = \frac{|-1|}{|-2i|} = \frac{1}{2} \qquad (10)$$

● **PROBLEM 22-55**

Find the electrostatic field generated by a semi-infinite condenser. The condenser consists of two parallel plates at a distance 2a from each other, kept at steady potentials. The upper plate has a steady potential V_0 and the lower plate $-V_0$.

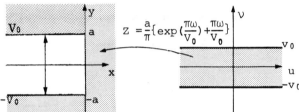

Fig. 1

The transformation

$$z = \frac{a}{\pi}\left(e^{\frac{\pi w}{V_0}} + \frac{\pi w}{V_0}\right) \qquad (1)$$

maps the region between two infinite planes onto the exterior of a semi-infinite condenser, see Fig. 1.

Solution: Taking the real and imaginary parts of (1) we get

$$x = \frac{a}{\pi}\left(e^{\frac{\pi u}{V_0}} \cos\frac{\pi v}{V_0} + \frac{\pi u}{V_0}\right) \qquad (2)$$

$$y = \frac{a}{\pi}\left(e^{\frac{\pi u}{V_0}} \sin\frac{\pi v}{V_0} + \frac{\pi v}{V_0}\right) \qquad (3)$$

Setting $u = c_1 = $ const in (2) and (3)

$$x = \frac{a}{\pi}\left(e^{\frac{\pi c_1}{V_0}} \cos\frac{\pi v}{V_0} + \frac{\pi c_1}{V_0}\right)$$

$$y = \frac{a}{\pi}\left(e^{\frac{\pi c_1}{V_0}} \sin\frac{\pi v}{V_0} + \frac{\pi v}{V_0}\right) \qquad (4)$$

and then eliminating v from (4) we obtain the lines of force. Similarly setting $v = c_2 = $ const we obtain the equipotentials.

The electric field can be derived from the complex potential

$$E_x + i E_y = -i \overline{f'(z)} \tag{5}$$

or

$$E = -i \overline{\frac{dw}{dz}} = -i \frac{1}{\overline{\frac{dz}{dw}}} = -i \frac{V}{a}(1 + e^{\overline{\frac{\pi w}{V_0}}})^{-1} \tag{6}$$

• **PROBLEM 22-56**

Find the complex potential due to a line of charge q per unit length. The line is perpendicular to the xy plane at z = 0.

Solution: The tangential component of the electric field due to a line charge is zero. Let C be a cylinder of radius r with axis at z = 0. By Gauss's theorem,

$$\int_C E_n ds = E_r \int_C ds = E_r \cdot 2\pi r = 4\pi q \tag{1}$$

Then,

$$E_r = \frac{2q}{r} \tag{2}$$

The electric field is radial and the normal component of the field is

$$E_n = E_r \tag{3}$$

Since,

$$E_r = -\frac{\partial \psi}{\partial r} \tag{4}$$

we have

$$\psi = -2q \ln r \tag{5}$$

and the complex potential is

$$f(z) = -2q \ln z = -2q \ln r - 2q i\theta \tag{6}$$

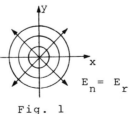

Fig. 1

Note that in this problem the value of the complex potential was multiplied by i in comparison to the definition given in Problem 22-51. We shall from now on use the "new" value of the complex potential.

$$f(z)_{now} = \phi(x,y) + i\psi(x,y); \quad f(z)_{before} = \psi(x,y) + i\phi(x,y)$$

where ϕ is electrostatic potential and ψ the stream function

$$i\, f(z)_{now} = f(z)_{before}$$

• **PROBLEM 22-57**

Find the potential due to a line charge q per unit length at $z = z_0$ and a line charge -q per unit length located at $z = \bar{z}_0$.

Solution: The complex potential due to a line charge q per unit length located at z_0 is

$$f_1(z) = -2q \ln(z - z_0) \tag{1}$$

and the complex potential due to a line charge -q at $\bar{z}_0$ is

$$f_2(z) = -2(-q)\ln(z - \bar{z}_0) \tag{2}$$

Hence, the total complex potential is

$$f(z) = -2q \ln(z-z_0) + 2q \ln(z-\bar{z}_0)$$
$$= 2q \ln\left(\frac{z - \bar{z}_0}{z - z_0}\right) \tag{3}$$

The electrostatic potential is the real part of f(z), hence

$$\text{Re } f(z) = 2q \text{ Re}\left[\ln\left(\frac{z - \bar{z}_0}{z - z_0}\right)\right] \tag{4}$$

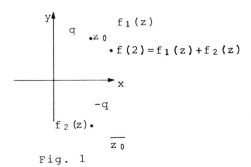

Fig. 1

• **PROBLEM 22-58**

Find the electrostatic potential inside a long hollow cylinder shown in Fig. 1. The cross section of the cylinder is a circle $x^2 + y^2 = 1$. The potential of the upper half is $V = 0$ and of the lower half is $V = 1$.

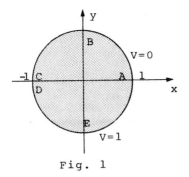

Fig. 1

Solution: The transformation

$$z = \frac{i - w}{i + w} \tag{1}$$

maps the upper half plane w > 0 onto the interior of the unit circle (see Fig. 2).

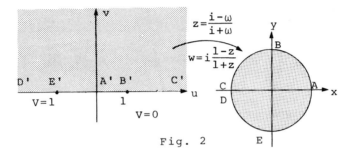

Fig. 2

Now, we have to solve the potential problem for the half-space. Taking the real and imaginary parts of the function $\frac{1}{\pi} \ln w$ we get

$$\frac{1}{\pi} \ln w = \frac{1}{\pi} \ln \rho + \frac{1}{\pi} i\phi \tag{2}$$

where $\rho > 0$, $0 \le \phi \le \pi$

and $w = \rho e^{i\phi}$. The imaginary part $\frac{\phi}{\pi}$ is harmonic in the upper half-plane and takes the required boundary values, V = 0 for ϕ = 0 and V = 1 for $\phi = \pi$.

$$V = \frac{\phi}{\pi} = \frac{1}{\pi} \arctan\left(\frac{v}{u}\right) \tag{3}$$

Taking the inverse transformation of (1) we get

$$w = u + iv = i\left(\frac{1 - z}{1 + z}\right) \tag{4}$$

and

$$V = \frac{1}{\pi} \arctan\left(\frac{1 - x^2 - y^2}{2y}\right) \tag{5}$$

The equipotential curves are

or

$$\frac{1}{\pi} \arctan\left(\frac{1 - x^2 - y^2}{2y}\right) = a \quad (6)$$

$$x^2 + (y + \tan \pi a)^2 = \sec^2 \pi a \quad (7)$$

● **PROBLEM 22-59**

Verify that
$$V = \frac{1}{\pi} \arctan\left(\frac{1 - x^2 - y^2}{2y}\right) \quad (1)$$

is the solution of Problem 22-58. Find the equipotentials and streamlines (sometimes called the flux lines) of equation (1).

Solution: Function (1) is harmonic inside the circle $x^2 + y^2 = 1$

$$\frac{\partial^2 V}{\partial x^2} + \frac{\partial^2 V}{\partial y^2} = 0 \quad (2)$$

We will show that V assumes the required values on the semicircles. Since,

$$\lim_{\substack{\alpha \to 0 \\ \alpha > 0}} \arctan \alpha = 0 \quad (3)$$

and

$$\lim_{\substack{\alpha \to 0 \\ \alpha < 0}} \arctan \alpha = \pi \quad (4)$$

let us take any sequence of points inside the cylinder that converges to the upper semicircle. Then

$$\lim V = \lim_{\substack{x^2 + y^2 < 1 \\ y > 0 \\ x^2 + y^2 \to 1}} \frac{1}{\pi} \arctan\left[\frac{1 - (x^2 + y^2)}{2y}\right] \quad (5)$$

$$= \frac{1}{\pi} \cdot 0 = 0$$

For the sequence converging to the lower semicircle

$$\lim V = \frac{1}{\pi} \lim_{\substack{x^2 + y^2 \to 1 \\ x^2 + y^2 < 1 \\ y < 0}} \arctan\left[\frac{1 - (x^2 + y^2)}{2y}\right] \quad (6)$$

$$= \frac{1}{\pi} \cdot \pi = 1$$

From From (1) setting $V(x,y) = c_1 = $ const ($0 < c_1 < 1$) we obtain equipotential curves of the form

$$x^2 + (y + \tan \pi c_1)^2 = \sec^2 \pi c_1 \tag{7}$$

To find the streamlines, note that a harmonic conjugate $\psi(x,y)$ of $V(x,y)$ is

$$-\frac{1}{\pi} \ln \rho = \text{Im}\left[-\frac{i}{\pi} \ln w\right] \tag{8}$$

Hence,

$$\psi = -\frac{1}{\pi} \ln \left|\frac{1-z}{1+z}\right| \tag{9}$$

• **PROBLEM 22-60**

Find the electrostatic potential V in the space enclosed by the half cylinder $x^2 + y^2 = 1$, $y \geq 0$ and the plane $y = 0$ with the boundary conditions shown in Fig. 1.

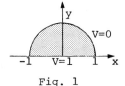

Fig. 1

Hint: Verify that the transformation

$$z = \frac{1-w}{i+w} \tag{1}$$

maps the first quadrant into the shaded region in Fig. 1.

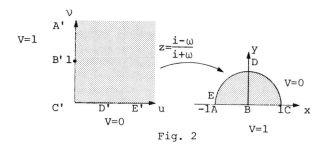

Fig. 2

Solution: The imaginary part of

$$\frac{2}{\pi} \ln w = \frac{2}{\pi} \ln \rho + i \frac{2}{\pi} \phi \tag{2}$$

where $w = \rho e^{i\phi}$, satisfies the boundary conditions in the first quadrant. Hence, the desired harmonic function is

$$V = \frac{2}{\pi} \phi = \frac{2}{\pi} \arctan\left(\frac{v}{u}\right) \tag{3}$$

The inverse mapping to (1) is

$$w = i\frac{1-z}{1+z} \tag{4}$$

Expressing u and v in terms of x,y and substituting into (3) we find

$$V = \frac{2}{\pi} \arctan\left(\frac{1 - x^2 - y^2}{2y}\right) \tag{5}$$

($0 \leq \arctan \alpha \leq \pi$).

• **PROBLEM 22-61**

Find the potential due to a line charge q per unit length located at $z = z_0$. The plane $y = 0$ has potential $V_0 = 0$.

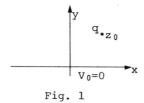

Fig. 1

Solution: Consider the system described in Problem 22-57. Its complex potential is

$$f(z) = 2q \ln\left(\frac{z - \bar{z}_0}{z - z_0}\right) \tag{1}$$

On the x axis, $z = x$ and (1) reduces to

$$f(x) = 2q \ln\left(\frac{x - \bar{z}_0}{x - z_0}\right) \tag{2}$$

But

$$\overline{f(x)} = 2q \ln\left(\frac{x - z_0}{x - \bar{z}_0}\right) = -f(x) \tag{3}$$

hence

$$\text{Re}\left[f(x)\right] = 0 \tag{4}$$

The charge $-q$ located at $z = \bar{z}_0$ can be replaced by a plane $y = 0$ at zero potential.

• **PROBLEM 22-62**

Two infinite parallel planes at a distance a from each other have potential zero. Determine the potential between the planes, when a line charge q per unit length is placed between the planes at a distance between the planes at a distance b from one of them, as shown in Fig. 1a.

Solution: We choose the coordinate system in such a way that the x axis is located on one of the planes and the charge is on the imaginary axis.

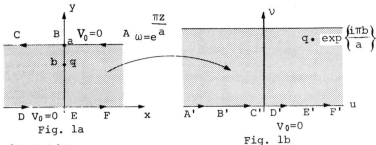

Fig. 1a Fig. 1b

The transformation

$$w = e^{\frac{\pi z}{a}} \tag{1}$$

maps an infinite strip of width a onto the upper half-plane. The line charge q located at $z = bi$ is mapped into the line charge at $w = e^{i\frac{\pi b}{a}}$.

The potential in the upper half of the w plane due to a line charge q at $w = e^{i\frac{\pi b}{a}}$ is (see Problem 22-61)

$$V(w) = 2q \, \mathrm{Re}\left[\frac{w - e^{-i\frac{\pi b}{a}}}{w - e^{i\frac{\pi b}{a}}}\right] \tag{2}$$

Substituting (1) into (2) we obtain the potential of the infinite stripe in Fig. 1a.

$$V(z) = 2q \, \mathrm{Re}\left[\frac{e^{\frac{\pi z}{a}} - e^{-i\frac{\pi b}{a}}}{e^{\frac{\pi z}{a}} - e^{i\frac{\pi b}{a}}}\right] \tag{3}$$

• **PROBLEM** 22-63

Find the potential everywhere in the z-space if the potential on the real axis is

$$V = \begin{cases} V_0 & \text{for } x > 0 \text{ and} \\ -V_0 & \text{for } x < 0 \end{cases} \tag{1}$$

Determine the lines of force and equipotentials.

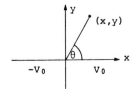

Fig. 1

Solution: Function V has to be harmonic in the whole plane and take the prescribed values on the x axis. Let a and b be real constants, then

$$a\theta + b \tag{2}$$

is harmonic. Hence,

$$a \cdot 0 + b = V_0$$
$$a \cdot \pi + b = -V_0 \tag{3}$$

and

$$a = \frac{-2V_0}{\pi}, \quad b = V_0 \tag{4}$$

The potential is

$$V_0 \left(1 - \frac{2}{\pi}\theta\right) = V_0 \left(1 - \frac{2}{\pi} \arctan \frac{y}{x}\right) \tag{5}$$

The equipotential lines are

$$V_0 \left(1 - \frac{2}{\pi}\theta\right) = \alpha = \text{const} \tag{6}$$

that is

$$\theta = \text{const} \quad \text{or}$$
$$y = kx \tag{7}$$

which are straight lines passing through the origin.

The lines of force are orthogonal to equipotentials

$$x^2 + y^2 = \beta = \text{const}$$

which are circles with center at the origin.

● **PROBLEM 22-64**

Find the electrostatic potential in the first quadrant, when the potentials on the positive x and y axes are constant and equal to V_0 and $-V_0$, respectively.

Solution: The transformation

$$w = z^2 \tag{1}$$

maps the first quadrant onto the upper half-plane, as shown in Fig. 1.

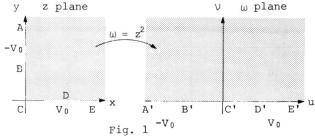

Fig. 1

From Problem 22-63 we obtain the potential in the w plane

$$V(w) = V_0 \left[1 - \frac{2}{\pi} \arctan \frac{v}{u}\right] \tag{2}$$

From (1)

$$w = z^2 = (x+iy)^2 = x^2 - y^2 + i2xy \tag{3}$$

and

$$u = x^2 - y^2$$
$$v = 2xy \tag{4}$$

The electrostatic potential in the first quadrant of the z plane is

$$V(z) = V_0 \left[1 - \frac{2}{\pi} \arctan \frac{2xy}{x^2 - y^2}\right] \tag{5}$$

● **PROBLEM 22-65**

Find the electrostatic potential in the region $0 < r < 1$, $0 < \phi < \frac{\pi}{4}$ bounded by the surfaces $\phi = 0$, $\phi = \frac{\pi}{4}$ and $r = 1$. The boundary conditions are as indicated in Fig. 1.

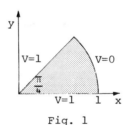

Fig. 1

Solution: The transformation

$$w = z^4 \tag{1}$$

maps the shaded region of Fig. 2a into the shaded region in Fig. 2b.

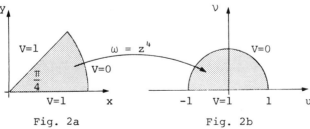

Fig. 2a   Fig. 2b

From Problem 22-60 we obtain the electrostatic potential for the region in Fig. 2b

$$V(u,v) = \frac{2}{\pi} \arctan \left[\frac{1 - u^2 - v^2}{2v}\right] \tag{2}$$

Let

$$z = re^{i\theta} \quad \text{and} \quad w = \rho e^{i\theta} \tag{3}$$

then

$$w = u + iv = \rho e^{i\theta} = \rho(\cos\theta + i\sin\theta) \tag{4}$$

and

$$V(\rho,\theta) = \frac{2}{\pi} \arctan\left[\frac{1-\rho^2}{2\rho\sin\theta}\right] \tag{5}$$

$$w = z^4 = r^4(\cos 4\phi + i\sin 4\phi) \tag{6}$$

Equation (5) can be expressed in terms of r and ϕ as

$$V(r,\phi) = \frac{2}{\pi} \arctan\left[\frac{1-r^8}{2r^4 \sin 4\phi}\right] \tag{7}$$

• **PROBLEM 22-66**

Evaluate the electrostatic potential $V(x,y)$ in the region $y > 0$ bounded by an infinite plane $y = 0$. The steady potential of the plane $y = 0$ is

$$V(x,0) = \begin{cases} 1 & \text{for } |x| < a \\ 0 & \text{for } |x| > a \end{cases} \tag{1}$$

Solution: The transformation

$$w = \ln\frac{z-a}{z+a} \tag{2}$$

maps the upper half-plane into an infinite strip (see Fig. 1).

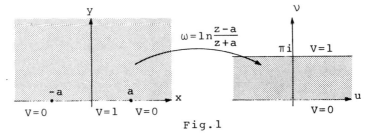

Fig.1

A bounded harmonic function, $V(u,v)$, that is unity for $v = \pi$ and zero for $v = 0$ is

$$V(u,v) = \frac{1}{\pi} v \tag{3}$$

To express (3) in x,y coordinates note that

$$w = \ln\left|\frac{z-a}{z+a}\right| + i \arg\frac{z-a}{z+a} \tag{4}$$

792

and
$$v = \arg\left(\frac{z - a}{z + a}\right) \tag{5}$$

Since, $z = x + iy$ we get

$$v = \arg\left[\frac{(z-a)(\bar{z}+a)}{(z+a)(\bar{z}+a)}\right] = \arg\left[\frac{x^2+y^2-a^2+2iay}{(x+a)^2+y^2}\right] \tag{6}$$

or
$$v = \arctan\left[\frac{2ay}{x^2+y^2-a^2}\right] \tag{7}$$

and
$$V(x,y) = \frac{1}{\pi} \arctan\left[\frac{2ay}{x^2+y^2-a^2}\right] \tag{8}$$

• PROBLEM 22-67

Find the electrostatic potential in the space between the planes $y = 0$ and $y = \pi$. For $x > 0$ the potential on the planes is $V = 0$ and for $x < 0$ the potential on the planes is $V = 1$.

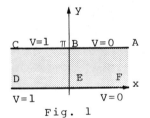

Fig. 1

Solution: The transformation

$$w = e^z \tag{1}$$

maps the strip in Fig. 1 into the upper half of the w plane, see Fig. 2.

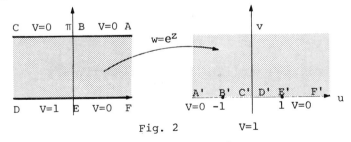

Fig. 2

In the w plane the potential is given by

$$V(u,v) = \frac{1}{\pi} \arctan\left[\frac{2v}{u^2 + v^2 - 1}\right] \tag{2}$$

(see Problem 22-66).

Then, from (1)

$$w = u + iv = e^z = e^{x+iy} = e^x(\cos y + i \sin y) \qquad (3)$$

Substituting (3) into (2) we obtain

$$V(x,y) = \frac{1}{\pi} \arctan\left[\frac{2e^x \sin y}{e^{2x}\cos^2 y + e^{2x}\sin^2 y - 1}\right] \qquad (4)$$

$$= \frac{1}{\pi} \arctan\left[\frac{\sin y}{\sinh x}\right]$$

● **PROBLEM** 22-68

Consider an infinite cylinder of radius $r = 1$. On the surface of the cylinder the potential is

$$V(r,\theta) = \begin{cases} 0 & r=1,\ 0 < \theta < \frac{\pi}{2} \\ 1 & r=1,\ \frac{\pi}{2} < \theta < 2\pi \end{cases} \qquad (1)$$

Find the potential inside the cylinder.

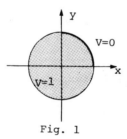

Fig. 1

Hint: Consider the transformation

$$w = e^{i\frac{\pi}{4}} \frac{1-z}{z-i} \qquad (2)$$

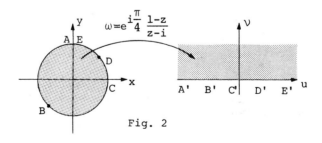

Fig. 2

Solution: Transformation (2) maps the disk ABCDE into the upper half of the w plane. We have to solve the boundary problem shown in Fig. 3.

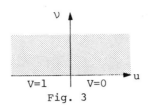

Fig. 3

The solution is

$$V(u,v) = \frac{1}{\pi} \arctan\left(\frac{v}{u}\right) \qquad (3)$$

Function (3) should be expressed in x,y coordinates

$$w = e^{i\frac{\pi}{4}} \frac{1-z}{z-i} = \left(\frac{1}{\sqrt{2}} + i\frac{1}{\sqrt{2}}\right) \frac{(1-x-iy)(x-iy+i)}{(x+iy-i)(x-iy+i)}$$

$$= \frac{2x + 2y - x^2 - y^2 - 1}{\sqrt{2}(x^2+y^2+1)} + i\frac{1 - x^2 - y^2}{\sqrt{2}(x^2+y^2+1)} \qquad (4)$$

Hence,

$$V = \frac{1}{\pi} \arctan\left[\frac{1 - x^2 - y^2}{2x + 2y - x^2 - y^2 - 1}\right] \qquad (5)$$

● PROBLEM 22-69

Find the electrostatic potential in the semi-infinite space bounded by two half-planes and a half cylinder. The boundary potentials are as shown in Fig. 1.

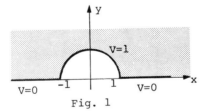

Fig. 1

Solution: The function

$$w = z + \frac{1}{z} \qquad (1)$$

maps the shaded area of Fig. 1 into the upper half of the w plane with the boundary conditions transformed as shown in Fig. 2.

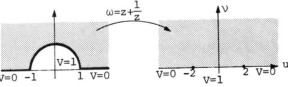

Fig. 2

The solution of the boundary value problem for the upper half-plane is

$$V(u,v) = \frac{1}{\pi} \arctan\left[\frac{uv}{u^2 + v^2 - 4}\right] \tag{2}$$

(see Problem 22-66 eq.(8)).

Now, we shall express $V(u,v)$ in terms of x and y coordinates. From (1)

$$w = u + iv = z + \frac{1}{z} = x + \frac{x}{x^2 + y^2} + i\left(y - \frac{y}{x^2 + y^2}\right) \tag{3}$$

$$u = x + \frac{x}{x^2 + y^2}$$
$$v = y - \frac{y}{x^2 + y^2} \tag{4}$$

Substituting (4) into (2) we obtain

$$V = \frac{1}{\pi} \arctan\left[\frac{4y - \frac{4y}{x^2+y^2}}{\left(x + \frac{x}{x^2+y^2}\right)^2 + \left(y - \frac{y}{x^2+y^2}\right)^2 - 4}\right]$$

$$= \frac{1}{\pi}\left[\frac{4y(x^2+y^2)(x^2+y^2-1)}{x^2(x^2+y^2+1)^2+y^2(x^2+y^2-1)^2-4(x^2+y^2)^2}\right] \tag{5}$$

$$= \frac{1}{\pi} \arctan\left[\frac{4y(x^2+y^2-1)}{x^4+2x^2y^2-2x^2+y^4-6y^2+1}\right]$$

Since,

$$2 \arctan x = \arctan \frac{2x}{1 - x^2} \tag{6}$$

we obtain

$$V = \frac{2}{\pi} \arctan\left[\frac{2y}{x^2 + y^2 - 1}\right] \tag{7}$$

• **PROBLEM** 22-70

Find the electrostatic potential in the lens-shaped region bounded by two circular arcs. The potential on the arc ABC is $-V_0$ and on the arc CDA is V_0. The angle of the lens is $\frac{\pi}{2}$.

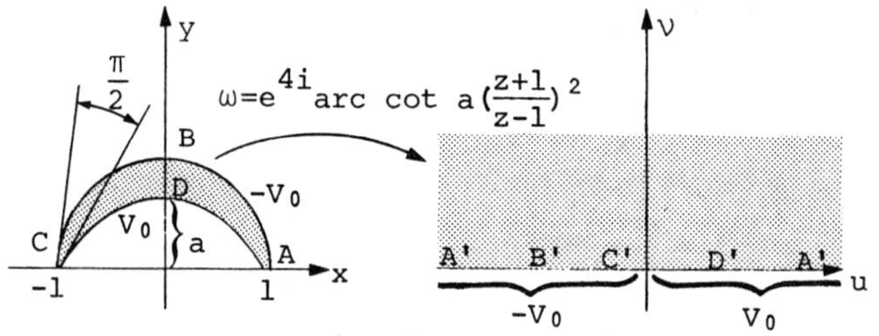

$$\omega = e^{4i} \text{arc cot } a\left(\frac{z+1}{z-1}\right)^2$$

Fig. 1

Hint: The transformation

$$w = e^{4i} \text{ arc cot } a\left(\frac{z+1}{z-1}\right)^2 \qquad (1)$$

maps the lens-shaped region onto the upper half of the w plane.

Solution: For the upper half-plane with the boundary conditions as depicted in Fig. 1., the solution is

$$V(u,v) = V_0\left(1 - \frac{2}{\pi} \text{ arc tan } \frac{v}{u}\right) \qquad (2)$$

(see Problem 22-63 eq.(5)).

From (1) we obtain

$$w = u + iv = (\cos 4 + i\sin 4)\text{arc cot } a\left(\frac{x+1+iy}{x-1+iy}\right)^2$$

$$= \text{arc cot } a\left[\cos 4 + i\sin 4\right] \cdot \left[\frac{x^2+y^2-1}{(x-1)^2+y^2} + i\frac{-2y}{(x-1)^2+y^2}\right] \qquad (3)$$

$$= \frac{\text{arc cot } a}{(x-1)^2+y^2}\left[\cos 4 \cdot (x^2+y^2-1) + 2y\sin 4\right]$$

$$+ i\frac{\text{arc cot } a}{(x-1)^2+y^2}\left[(x^2+y^2-1)\sin 4 - 2y\cos 4\right]$$

From (2) and (3) we obtain

$$V(x,y) = V_0 - \frac{2V_0}{\pi} \text{ arc tan}\left[\frac{(x^2+y^2-1)\sin 4 - 2y\cos 4}{(x^2+y^2-1)\cos 4 + 2y\sin 4}\right] \qquad (4)$$

• **PROBLEM** 22-71

Using the method of separation of variables one finds the solution of the Dirichlet problem

$$\frac{\partial^2 V}{\partial x^2} + \frac{\partial^2 V}{\partial y^2} = 0$$

$$V(0,y) = V(a,y) = 0 \quad 0 < y < b \qquad (1)$$

$$V(x,0) = 0, \quad v(x,b) = 1 \quad 0 < x < a$$

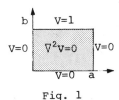

Fig. 1

to be

$$V = \frac{4}{\pi} \sum_{n=1}^{\infty} \frac{\sinh\left(\frac{k\pi y}{a}\right)}{k \sinh\left(\frac{k\pi b}{a}\right)} \sin\left(\frac{k\pi x}{a}\right) \qquad (2)$$

$$k = 2n - 1$$

Using this result and the transformation $w = \ln z$ find the potential $V(r,\theta)$ in the space

$$\begin{aligned} 1 < r < r_0 \\ 0 < \theta < \pi \end{aligned} \qquad (3)$$

with the boundary conditions shown in Fig. 2a.

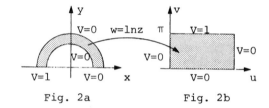

Fig. 2a   Fig. 2b

Solution: Let $z = re^{i\theta}$ then

$$w = u + iv = \ln z = \ln r + i\theta \qquad (4)$$

Substituting $u = \ln r$ and $v = \theta$ into (2), we obtain

$$V = \frac{4}{\pi} \sum_{n=1}^{\infty} \frac{\sinh\left[\frac{k\pi \theta}{\ln r_0}\right]}{k \sinh\left[\frac{k\pi^2}{\ln r_0}\right]} \sin\left[\frac{k\pi}{\ln r_0} \ln r\right] \qquad (5)$$

$$k = 2n - 1.$$

● **PROBLEM 22-72**

Two infinitely long coaxial conducting cylinders of radii r_1 and r_2 ($r_2 > r_1$) are kept at constant potentials V_1 and V_2, respectively. Find the electrostatic potential in the space between the cylinders.

Fig. 1

Solution: It is reasonable to assume that the solution is of the form

$$f(z) = \alpha \ln z + \beta \tag{1}$$

where α and β are real constants. The electrostatic potential is Re $f(z) = V$.

If $z = re^{i\theta}$, then

$$f(z) = V + i\psi = \alpha \ln r + \alpha i\theta + \beta \tag{2}$$

and

$$V = \alpha \ln r + \beta \tag{3}$$

$$\psi = \alpha\theta$$

The function $V(r,\theta)$ is harmonic everywhere in the region $r_1 < r < r_2$. From the boundary conditions we compute

$$V_1 = \alpha \ln r_1 + \beta$$
$$V_2 = \alpha \ln r_2 + \beta \tag{4}$$

Hence,

$$\alpha = \frac{V_2 - V_1}{\ln\left(\frac{r_2}{r_1}\right)} \tag{5}$$

$$\beta = \frac{V_1 \ln r_2 - V_2 \ln r_1}{\ln\left(\frac{r_2}{r_1}\right)} \tag{6}$$

Substituting (5) and (6) into the expression for the electrostatic potential we find

$$V(r,\theta) = \frac{V_2 - V_1}{\ln\left(\frac{r_2}{r_1}\right)} \ln r + \frac{V_1 \ln r_2 - V_2 \ln r_1}{\ln\left(\frac{r_2}{r_1}\right)} \tag{7}$$

Note that the electric field is

$$E = -\text{grad } V = -\frac{\partial V}{\partial r} = \frac{V_1 - V_2}{\ln\left(\frac{r_2}{r_1}\right)} \cdot \frac{1}{r} \tag{8}$$

● PROBLEM 22-73

Compute the capacitance of the condenser consisting of the two cylinders described in Problem 22-72.

Solution: The quality of a conductor, called the capacitance, depends only on the geometry of a conductor and is defined by

$$C = \frac{q}{V} \quad (1)$$

where q is the charge and V the potential.

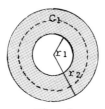

Fig. 1

Let C_1 be a simple closed curve containing the cylinder of radius r_1, but no part of the cylinder of radius r_2. Then, denoting by q the charge of cylinder r_1 and applying Gauss's theorem, we get

$$\int_{C_1} E_n ds = \int_{\theta=0}^{2\pi} \frac{V_1 - V_2}{\ln\left(\frac{r_2}{r_1}\right)} \cdot \frac{1}{r} \cdot r d\theta$$

$$= \frac{2\pi(V_1 - V_2)}{\ln\left(\frac{r_2}{r_1}\right)} = 4\pi q \quad (2)$$

Hence,

$$q = \frac{V_1 - V_2}{2\ln\left(\frac{r_2}{r_1}\right)} \quad (3)$$

The capacitance of the system is

$$C = \frac{q}{\Delta V} = \frac{\text{charge}}{\text{difference in potentials}}$$

$$= \frac{V_1 - V_2}{2\ln\left(\frac{r_2}{r_1}\right) \cdot (V_1 - V_2)} = \frac{1}{2\ln\left(\frac{r_2}{r_1}\right)} \quad (4)$$

Here, we assumed that there is a vacuum between the conductors. Otherwise the charge q should be replaced by $\frac{q}{k}$, where k is the dielectric constant of the material between the cylinders. The result (4) should then be replaced by

$$C = \frac{1}{2k \ln\left(\frac{r_2}{r_1}\right)} \quad (5)$$

CHAPTER 23

APPLICATIONS OF CONFORMAL MAPPINGS THE SCHWARZ-CHRISTOFFEL TRANSFORMATION

MAPPING OF POLYGONS ONTO THE REAL AXIS

• **PROBLEM** 23-1

Consider a polygon in the w plane with vertices at w_1, w_2, ..., w_n and corresponding interior angles α_1, α_2, ..., α_n. The vertices of the polygon are mapped into points x_1, x_2, ..., x_n on the real axis in the z plane, see Fig. 1.

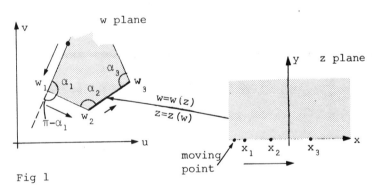

Fig 1

Show that the mapping function

$$\frac{dw}{dz} = A(z - x_1)^{\frac{\alpha_1}{\pi} - 1} (z - x_2)^{\frac{\alpha_2}{\pi} - 1} \cdots (z - x_n)^{\frac{\alpha_n}{\pi} - 1} \quad (1)$$

maps the polygon of the w plane into the real axis of the z plane.

Solution: From (1) we have

$$\arg dw = \arg dz + \arg A + \left(\frac{\alpha_1}{\pi} - 1\right)\arg(z - x_1) \quad (2)$$

801

$$+ \left[\frac{\alpha_2}{\pi} - 1\right] \arg(z - x_2) + \ldots + \left[\frac{\alpha_n}{\pi} - 1\right] \arg(z - x_n)$$

A point moves along the real x axis from $-\infty$ to x_1. Assume that its image moves along a side of the polygon toward w_1. When the point passes x_1, arg $(z - x_1)$ changes from π to 0. All other terms in (2) stay the same. Thus, arg dw decreases by

$$\left[\frac{\alpha_1}{\pi} - 1\right] \arg(z - x_1) = \left[\frac{\alpha_1}{\pi} - 1\right] \pi = \alpha_1 - \pi \tag{3}$$

The image was moving along the $w_n w_1$ side and at w_1 the direction changed by $\alpha_1 - \pi$. Hence, the motion takes place along the $w_1 w_2$ side.

In the same way we can show that as the point passes x_2, $\arg(z - x_1)$ and $\arg(z - x_2)$ change from π to 0, while all other terms stay constant. Hence, a turn through angle $\alpha_2 - \pi$ occurs.

Carrying out, we see that as the point moves along the x axis its image moves along the sides of the polygon.

• **PROBLEM** 23-2

Show that for closed polygons the sum of the exponents in the Schwarz-Christoffel transformation

$$\frac{dw}{dz} = A(z - x_1)^{\frac{\alpha_1}{\pi} - 1} \ldots (z - x_n)^{\frac{\alpha_n}{\pi} - 1} \tag{1}$$

is equal to -2.

Solution: The sum of the exterior angles of any closed polygon is 2π.

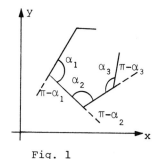

Fig. 1

Thus, from Figure 1 we get

$$(\pi - \alpha_1) + (\pi - \alpha_2) + \ldots + (\pi - \alpha_n) = 2\pi \tag{2}$$

Dividing by $-\pi$ we have

$$\left[\frac{\alpha_1}{\pi} - 1\right] + \left[\frac{\alpha_2}{\pi} - 1\right] + \ldots + \left[\frac{\alpha_n}{\pi} - 1\right] = -2 \tag{3}$$

● **PROBLEM 23-3**

Show that the Schwarz-Christoffel transformation

$$w = A \int (z - x_1)^{\frac{\alpha_1}{\pi} - 1} (z - x_2)^{\frac{\alpha_2}{\pi} - 1} \ldots (z - x_n)^{\frac{\alpha_n}{\pi} - 1} dz + B \tag{1}$$

maps the interior of the polygon into the upper half plane.

Solution: First, we shall show that (1) maps the interior of the polygon onto the unit circle.

Let $w = f(z)$ be the function mapping the polygon P in the w plane onto the unit circle C in the z plane and let $f(z)$ be analytic inside C.

We shall show that $f(z)$ is one-to-one and onto. If w_0 is a point inside P, then by Cauchy's integral formula

$$\frac{1}{2\pi i} \int_P \frac{dw}{w - w_0} = 1 \tag{2}$$

Since, $w - w_0 = f(z) - w_0$ we get

$$\frac{1}{2\pi i} \int_C \frac{f'(z) dz}{f(z) - w_0} = 1 \tag{3}$$

The function $f(z) - w_0$ is analytic inside C. Hence, there is only one zero of $f(z) - w_0$ inside C, that is

$$f(z_0) = w_0 \tag{4}$$

Therefore, to each point w_0 inside P there corresponds one and only one point z_0 inside C, such that $f(z_0) = w_0$. Since, the unit circle can be mapped onto the upper half plane, the proof is completed.

● **PROBLEM 23-4**

Show that if in the Schwarz-Christoffel transformation

$$\frac{dw}{dz} = A(z - x_1)^{\frac{\alpha_1}{\pi} - 1} \ldots (z - x_n)^{\frac{\alpha_n}{\pi} - 1} \tag{1}$$

one of the points $x_1, x_2, \ldots, x_n$, say x_n is chosen at infinity, then the factor in (1) corresponding to this point is equal to one.

Solution: Let us denote the factor A as

$$A = \frac{M}{(-x_n)^{\frac{\alpha_n}{\pi} - 1}} \qquad (2)$$

Then, (1) becomes

$$\frac{dw}{dz} = M(z - x_1)^{\frac{\alpha_1}{\pi} - 1} \cdots (z - x_{n-1})^{\frac{\alpha_{n-1}}{\pi} - 1} \left(\frac{x_n - z}{x_n}\right)^{\frac{\alpha_n}{\pi} - 1} \qquad (3)$$

Taking the limit

$$\lim_{x_n \to \infty} \left(\frac{x_n - z}{x_n}\right)^{\frac{\alpha_n}{\pi} - 1} = 1 \qquad (4)$$

Thus,

$$\frac{dw}{dz} = M(z - x_1)^{\frac{\alpha_1}{\pi} - 1} \cdots (z - x_{n-1})^{\frac{\alpha_{n-1}}{\pi} - 1} \qquad (5)$$

TRIANGLES, RECTANGLES AND DEGENERATE POLYGONS

• **PROBLEM 23-5**

Find the transformation which maps the shaded region in the w plane onto the upper half of the z plane, see Figure 1.

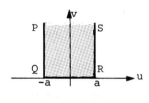
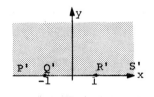

Fig. 1

Solution: Suppose the points P,Q,R,S of the polygon are mapped into P',Q',R',S', respectively. Points P' and S' are at infinity. By the Schwarz-Christoffel transformation we get

$$\frac{dw}{dz} = A(z + 1)^{\frac{\pi}{2\pi} - 1} (z - 1)^{\frac{\pi}{2\pi} - 1}$$

$$= \frac{A}{\sqrt{z^2 - 1}} = \frac{M}{\sqrt{1 - z^2}} \tag{1}$$

Integrating

$$w = M \int \frac{dz}{\sqrt{1 - z^2}} + B = M \text{ arc sin } z + B \tag{2}$$

Since, for $z = -1$, $w = -a$

$$-a = M \text{ arc sin}(-1) + B \tag{3}$$

and for $z = 1$, $w = a$

$$a = M \text{ arc sin}(1) + B \tag{4}$$

Solving (3) and (4) we find

$$M = \frac{2a}{\pi}, \quad B = 0 \tag{5}$$

Hence,

$$w = \frac{2a}{\pi} \text{ arc sin } z \tag{6}$$

or

$$z = \sin \frac{\pi w}{2a} \tag{7}$$

● **PROBLEM 23-6**

Find the Schwarz-Christoffel transformation which maps the triangle ABC in the w plane onto the upper half of the z plane.

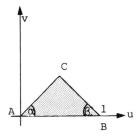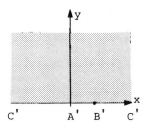

Fig. 1

Solution: Let the vertices ABC map into points A'B'C' as shown in Figure 1. The Schwarz-Christoffel transformation is

$$\frac{dw}{dz} = A z^{\frac{\alpha}{\pi} - 1} (z - 1)^{\frac{\beta}{\pi} - 1} = M z^{\frac{\alpha}{\pi} - 1} (1 - z)^{\frac{\beta}{\pi} - 1} \tag{1}$$

Integrating (1) we obtain

$$w = M \int_0^z \eta^{\frac{\alpha}{\pi} - 1} (1 - \eta)^{\frac{\beta}{\pi} - 1} d\eta + B \tag{2}$$

When z = 0, then w = 0 and B = 0. When z = 1, then w = 1 and

$$1 = M \int_0^1 \eta^{\frac{\alpha}{\pi} - 1} (1 - \eta)^{\frac{\beta}{\pi} - 1} d\eta = \frac{\Gamma\left(\frac{\alpha}{\pi}\right) \Gamma\left(\frac{\beta}{\pi}\right)}{\Gamma\left(\frac{\alpha+\beta}{\pi}\right)} \tag{3}$$

where $\Gamma(z)$ is the gamma function defined by

$$\Gamma(z) = \int_0^\infty t^{z-1} e^{-t} dt \tag{4}$$

for Re z > 0.

Hence

$$M = \frac{\Gamma\left(\frac{\alpha+\beta}{\pi}\right)}{\Gamma\left(\frac{\alpha}{\pi}\right) \Gamma\left(\frac{\beta}{\pi}\right)} \tag{5}$$

and

$$w = \frac{\Gamma\left(\frac{\alpha+\beta}{\pi}\right)}{\Gamma\left(\frac{\alpha}{\pi}\right) \Gamma\left(\frac{\beta}{\pi}\right)} \int_0^z \eta^{\frac{\alpha}{\pi} - 1} (1 - \eta)^{\frac{\alpha}{\pi} - 1} d\eta \tag{6}$$

● **PROBLEM 23-7**

Find the transformation which maps the shaded region in the w plane onto the upper half of the z plane.

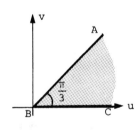
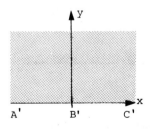

Fig. 1

Solution: Applying the formula for the Schwarz-Christoffel transformation we find

$$\frac{dw}{dz} = Az^{\frac{\pi}{3\pi} - 1} \tag{1}$$

Integrating (1) we obtain

$$w = 3A\, z^{\frac{1}{3}} + B \qquad (2)$$

From the condition that z = 0 when w = 0, we get

$$B = 0 \qquad (3)$$

The required transformation is

$$w = 3A\, z^{\frac{1}{3}} = M z^{\frac{1}{3}} \qquad (4)$$

where M is a constant.

In order to determine the value of M we should have one more point. For example, if B = 1 in the w plane is mapped to B' = 1 in the z plane. Then,

$$M = 1 \qquad (5)$$

and (4) becomes

$$w = z^{\frac{1}{3}}$$
or,
$$z = w^3 \qquad (6)$$

• **PROBLEM** 23-8

Apply the Schwarz-Christoffel transformation to find a function which maps the shaded region in the w plane onto the upper half of the z plane.

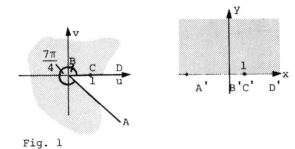

Fig. 1

Solution: By the Schwarz-Christoffel transformation we obtain

$$\frac{dw}{dz} = A z^{\frac{7\pi/4}{\pi} - 1} (z - 1)^{\frac{\pi/\pi} - 1} = A z^{\frac{3}{4}} \qquad (1)$$

Integrating (1) we obtain

$$w = A \frac{z^{\frac{3}{4} + 1}}{\frac{3}{4} + 1} + B = \frac{4A}{7} z^{\frac{7}{4}} + B \qquad (2)$$

For w = 0 and z = 0, B = 0   (3)

For w = 1 and z = 1, $\frac{4A}{7} = 1$ (4)

Thus,
$$w = z^{\frac{7}{4}}$$
or,
$$z = w^{\frac{4}{7}}$$
    (5)

● **PROBLEM** 23-9

Using the Schwarz-Christoffel transformation determine a function which maps the shaded region in the w plane onto the upper half of the z plane.

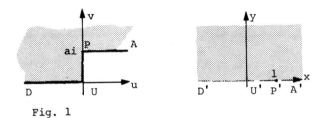

Fig. 1

<u>Solution</u>: Note that the interior angles at U and P are $\frac{\pi}{2}$ and $\frac{3\pi}{2}$, respectively. Then,

$$\frac{dw}{dz} = Az^{\frac{\pi}{2\pi} - 1}(z - 1)^{\frac{3\pi}{2\pi} - 1} = A\sqrt{\frac{z-1}{z}}$$

$$= M\sqrt{\frac{1-z}{z}}$$
    (1)

Let us substitute
$$z = \sin^2\alpha$$
    (2)

Then,
$$w = 2M\int \cos^2\alpha \, d\alpha = M\int(1 + \cos 2\alpha)d\alpha$$
$$= M\left[\alpha + \frac{1}{2}\sin 2\alpha\right] + B$$
    (3)
$$= M\left[\alpha + \sin\alpha\cos\alpha\right] + B = M\left[\arcsin\sqrt{z} + \sqrt{z(1-z)}\right] + B$$

When w = 0, z = 0 and B = 0.

For w = ai, z = 1 and
$$M = \frac{2ai}{\pi}$$
    (4)

The required transformation is

$$w = \frac{2ai}{\pi}\left[\arcsin\sqrt{z} + \sqrt{z(1-z)}\right] \qquad (5)$$

• **PROBLEM** 23-10

Applying the Schwarz-Christoffel transformation find the function which maps the shaded region in the w plane onto the upper half of the z plane.

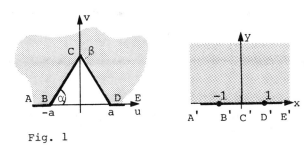

Fig. 1

Solution: The Schwarz-Christoffel transformation yields

$$\frac{dw}{dz} = A(z+1)^{\frac{\pi-\alpha}{\pi}-1} z^{\frac{\pi+2\alpha}{\pi}-1} (z-1)^{\frac{\pi-\alpha}{\pi}-1} \qquad (1)$$

Note that the interior angles at B and D are $\pi - \alpha$ and at C $2\pi - (\pi - 2\alpha) = \pi + 2\alpha$. From (1) we obtain

$$\frac{dw}{dz} = A \frac{z^{\frac{2\alpha}{\pi}}}{(z^2-1)^{\frac{\alpha}{\pi}}} = A' \frac{z^{\frac{2\alpha}{\pi}}}{(1-z^2)^{\frac{\alpha}{\pi}}} \qquad (2)$$

Integrating (2) we find

$$w = A' \int_0^z \frac{\eta^{\frac{2\alpha}{\pi}}}{(1-\eta^2)^{\frac{\alpha}{\pi}}} d\eta + B \qquad (3)$$

For $z = 0$, $w = bi$, hence

$$w = A' \int_0^z \frac{\eta^{\frac{2\alpha}{\pi}}}{(1-\eta^2)^{\frac{\alpha}{\pi}}} d\eta + bi \qquad (4)$$

when $z = 1$, $w = a$ therefore

$$a = A' \int_0^1 \frac{\eta^{\frac{2\alpha}{\pi}}}{(1 - \eta^2)^{\frac{\alpha}{\pi}}} d\eta + bi \qquad (5)$$

and

$$A' = \frac{\sqrt{\pi}(a - bi)}{\Gamma\left(\frac{\alpha}{\pi} + \frac{1}{2}\right) \Gamma\left(1 - \frac{\alpha}{\pi}\right)} \qquad (6)$$

• **PROBLEM 23-11**

Find a function which maps the shaded region in the w plane onto the upper half of the z plane.

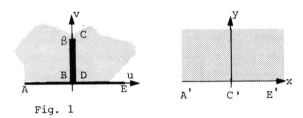

Fig. 1

Solution: The shaded region represents the upper half of the w plane with the cut along the CD segment.

The mapping can be found from the Schwarz-Christoffel transformation. The interior angles are at B and D, $\frac{\pi}{2}$, and at C 2π. Another way to solve this problem is to use the results of Problem 23-10.

$$w = bi - bi \int_0^z \frac{\eta d\eta}{\sqrt{1 - \eta^2}}$$

$$= bi\sqrt{1 - z^2} \qquad (1)$$

• **PROBLEM 23-12**

Find the transformation which maps a polygon P in the w plane onto the unit circle in the η plane.

Solution: The Schwarz-Christoffel transformation maps the polygon P in the w plane, onto the x axis of the z plane.

$$w = A \int (z - x_1)^{\frac{\alpha_1}{\pi} - 1} \cdots (z - x_n)^{\frac{\alpha_n}{\pi} - 1} dz + B \qquad (1)$$

The transformation

$$\eta = \frac{i-z}{i+z} \quad \text{or} \quad z = i\frac{1-\eta}{1+\eta} \qquad (2)$$

maps the upper half of the z plane onto the unit circle in the η plane.

Let $x_1, x_2, \ldots, x_n$ map into $\eta_1, \eta_2, \ldots, \eta_n$, respectively on the unit circle.

Then,

$$z - z_k = i\left(\frac{1-\eta}{1+\eta}\right) - i\left(\frac{1-\eta_k}{1+\eta_k}\right)$$

$$= -2i\frac{\eta - \eta_k}{(1+\eta)(1+\eta_k)} \qquad (3)$$

From (2) we have

$$dz = -2i\frac{d\eta}{(1+\eta)^2} \qquad (4)$$

Substituting into (1) we find

$$w = D\int (\eta - \eta_1)^{\frac{\alpha_1}{\pi}-1}(\eta - \eta_2)^{\frac{\alpha_2}{\pi}-1}\ldots(\eta - \eta_n)^{\frac{\alpha_n}{\pi}-1}d\eta + B \qquad (5)$$

where D is an arbitrary constant.

• **PROBLEM 23-13**

Write the transformation which maps the real x axis onto a triangle with vertices w_1, w_2, and w_3. What are the changes when w_3 is the image of the point of infinity?

Solution: Note that the transformation given by the Schwarz-Christoffel formula is one-to-one. Therefore for a triangle with vertices w_1, w_2, and w_3 we have

$$w = A\int_{z_0}^{z}(\zeta - x_1)^{\frac{\alpha_1}{\pi}-1}(\zeta - x_2)^{\frac{\alpha_2}{\pi}-2}(\zeta - x_3)^{\frac{\alpha_3}{\pi}-1}d\zeta + B \qquad (1)$$

where

$$\alpha_1 + \alpha_2 + \alpha_3 = \pi \qquad (2)$$

Angles α_1, α_2, and α_3 are internal angles of the triangle. We can assign arbitrary values to the points x_1, x_2, x_3. When one of these points, say x_3, is at infinity, then the corresponding term in (1) is equal to one.

Transformation (1) is reduced to

$$w = A \int_{z_0}^{z} (\zeta - x_1)^{\frac{\alpha_1}{\pi} - 1} (\zeta - x_2)^{\frac{\alpha_2}{\pi} - 1} d\zeta + B \tag{3}$$

Arbitrary real values can be assigned to x_1 and x_2.

● **PROBLEM 23-14**

Find the transformation which maps the real x axis onto an equilateral triangle.

Solution: The internal angles of the equilateral triangle are

$$\alpha_1 = \alpha_2 = \alpha_3 = \frac{\pi}{3} \tag{1}$$

Hence,

$$w = A \int_{z_0}^{z} (\zeta - x_1)^{-\frac{2}{3}} (\zeta - x_2)^{-\frac{2}{3}} (\zeta - x_3)^{-\frac{2}{3}} d\zeta + B \tag{2}$$

The points x_1, x_2 and x_3 are arbitrary real constants. Let us choose

$$x_1 = -1, \quad x_2 = 1, \quad x_3 = \infty \quad \text{and} \quad z_0 = 1.$$

Then, (2) becomes

$$w = A \int_{1}^{z} (\zeta + 1)^{-\frac{2}{3}} (\zeta - 1)^{-\frac{2}{3}} d\zeta + B \tag{3}$$

Since A and B are arbitrary constants we can set $A = 1$ and $B = 0$. Therefore,

$$w = \int_{1}^{z} (\zeta + 1)^{-\frac{2}{3}} (\zeta - 1)^{-\frac{2}{3}} d\zeta \tag{4}$$

From (4) we get

$$w(1) = 0 \quad \text{or} \quad w_2 = 0$$

When $z = -1$ we set in (4) $\zeta = x \; (-1 < x < 1)$.

We have

$$\begin{aligned} & x + 1 > 0 \\ & \arg(x + 1) = 0 \\ & |x - 1| = 1 - x \\ & \arg(x - 1) = \pi \end{aligned} \tag{5}$$

Thus,

$$w = \int_1^{-1} (x+1)^{-\frac{2}{3}} (1-x)^{-\frac{2}{3}} e^{-\frac{2\pi i}{3}} dx \qquad (6)$$

$$= e^{\frac{\pi i}{3}} \int_0^1 \frac{2dx}{(1-x^2)^{\frac{2}{3}}}$$

Denoting the last integral by a, we get

$$a = \int_0^1 \frac{2dx}{(1-x^2)^{\frac{2}{3}}} = \int_0^1 s^{-\frac{1}{2}} (1-s)^{-\frac{2}{3}} ds = B\left(\frac{1}{2}, \frac{1}{3}\right) \qquad (7)$$

where $x = \sqrt{s}$ and $B(.,.)$ is the beta function.

$$w_1 = w(z_1) = a\, e^{\frac{\pi i}{3}} \qquad (8)$$

The image of $x_3 = \infty$ is

$$w(x_3) = w_3 = \int_1^\infty \frac{dx}{(x^2-1)^{\frac{2}{3}}} \qquad (9)$$

Hence, w_3 is located on the positive u axis. On the other hand

$$w_3 = \int_1^{-1} (|x+1||x-1|)^{-\frac{2}{3}} e^{-\frac{2\pi i}{3}} dx$$

$$+ \int_{-1}^{-\infty} (|x+1||x-1|)^{-\frac{2}{3}} e^{-\frac{4\pi i}{3}} dx \qquad (10)$$

$$= w_1 + e^{-\frac{4\pi i}{3}} \int_{-1}^{-\infty} (|x+1||x-1|)^{-\frac{2}{3}} dx$$

$$= a\, e^{\frac{\pi i}{3}} + e^{-\frac{\pi i}{3}} \int_1^\infty \frac{dx}{(x^2-1)^{\frac{2}{3}}}$$

$$= a\, e^{\frac{\pi i}{3}} + w_3\, e^{-\frac{\pi i}{3}}$$

Thus,
$$w_3 = a \tag{11}$$
Transformation (4) maps the x axis onto the equilateral triangle.

● **PROBLEM 23-15**

Find the transformation which maps the real x axis onto a rectangle.

Solution: The internal angles of a rectangle are
$$\alpha_1 = \alpha_2 = \alpha_3 = \alpha_4 = \frac{\pi}{2} \tag{1}$$

Hence,
$$w = A \int_{z_0}^{z} (\zeta-x_1)^{-\frac{1}{2}}(\zeta-x_2)^{-\frac{1}{2}}(\zeta-x_3)^{-\frac{1}{2}}(\zeta-x_4)^{-\frac{1}{2}} d\zeta + B \tag{2}$$

Points x_1, x_2, x_3, and x_4 are arbitrary, therefore we can choose
$$x_1 = -a, \quad x_2 = -1$$
$$x_3 = 1, \quad x_4 = a \tag{3}$$

Then
$$0 \le \arg(z - x_k) \le \pi, \quad k = 1, 2, 3, 4 \tag{4}$$

and setting $z_0 = 0$ we find
$$w = -\int_{0}^{z} (\zeta+a)^{-\frac{1}{2}}(\zeta+1)^{-\frac{1}{2}}(\zeta-1)^{-\frac{1}{2}}(\zeta-a)^{-\frac{1}{2}} d\zeta \tag{5}$$

In (5) we set $A = -1$ and $B = 0$.

● **PROBLEM 23-16**

Verify that transformation (5) of Problem 23-15 maps the x axis onto a rectangle.

Solution: We shall show that
$$w = -\int_{0}^{z} (\zeta+a)^{-\frac{1}{2}}(\zeta+1)^{-\frac{1}{2}}(\zeta-1)^{-\frac{1}{2}}(\zeta-a)^{-\frac{1}{2}} d\zeta \tag{1}$$

maps the x axis onto a rectangle. Let us denote

$$F(\zeta) = (\zeta + a)^{-\frac{1}{2}}(\zeta + 1)^{-\frac{1}{2}}(\zeta - 1)^{-\frac{1}{2}}(\zeta - a)^{-\frac{1}{2}} \qquad (2)$$

If $-a < x < -1$, then

$$F(x) = \left[e^{-\frac{\pi i}{2}}\right]^3 |F(x)| = i|F(x)| \qquad (3)$$

and

$$w(x_1) = w_1 = -\int_0^{-a} F(x)dx = -\int_0^{-1} F(x)dx$$
$$-\int_{-1}^{-a} F(x)dx = \int_0^{-1} |F(x)|dx - i\int_{-1}^{-a} |F(x)|dx \qquad (4)$$

Note that for $-1 < x < 0$

$$F(x) = \left[e^{-\frac{\pi i}{2}}\right]^2 |F(x)| = -|F(x)| \qquad (5)$$

and

$$\arg(x + a) = \arg(x + 1) = 0$$
$$\arg(x - 1) = \arg(x - a) = \pi \qquad (6)$$

Let us denote

$$b = \int_0^1 |F(x)|dx = \int_0^1 \frac{dx}{\sqrt{(1-x^2)(a^2-x^2)}} \qquad (7)$$

and

$$c = \int_1^a |F(x)|dx = \int_1^a \frac{dx}{\sqrt{(x^2-1)(a^2-x^2)}} \qquad (8)$$

Then from (4), (7) and (8) we find

$$w_1 = -b + ic \qquad (9)$$

In the same manner we can show that $w_4 = b + ic$.

$$w_2 = -\int_0^{-1} F(x)dx = \int_0^{-1} |F(x)|dx = -b \qquad (10)$$

Similarly we can show that $w_3 = b$.

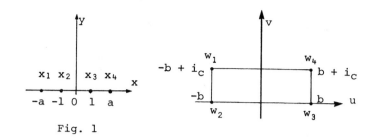

Fig. 1

• **PROBLEM 23-1**

Find the transformation which maps the x axis onto the edges of an infinite strip

$$0 < v < \pi \qquad (1)$$

Solution: Let $w_1 = \pi i$, w_2, $w_3 = 0$, and w_4 be the vertices of a rhombus.

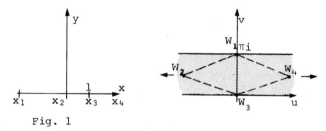

Fig. 1

As points w_1 and w_4 move apart, the rhombus "approaches" an infinite strip. The angles are

$$\alpha_1 = \pi, \quad \alpha_2 = 0, \quad \alpha_3 = \pi, \quad \alpha_4 = 0 \qquad (2)$$

Since points x_1, x_2, x_3, and x_4 are arbitrary let

$$x_2 = 0, \quad x_3 = 1 \quad \text{and} \quad x_4 = \infty \qquad (3)$$

Point x_1 will be determined later, then

$$\frac{dw}{dz} = A(z - x_1)^0 \, z^{-1}(z - 1)^0 = \frac{A}{z} \qquad (4)$$

Hence,

$$w = A \ln z + B \qquad (5)$$

For $z = 1$, w becomes zero, therefore, the constant B must be zer

When $z = x > 0$, point w lies on the real axis therefore A is a real constant. The image of $z = x_1$ is $w_1 = \pi i$, thus

$$\pi i = A \ln x_1 = A \ln|x_1| + A\pi i \qquad (6)$$

and $|x_1| = 1$, $A = 1$.

We obtain

$$w = \ln z \tag{7}$$

• **PROBLEM** 23-18

The transformation

$$w = A \int_{z_0}^{z} (\zeta-x_1)^{\frac{\alpha_1}{\pi}-1} (\zeta-x_2)^{\frac{\alpha_2}{\pi}-1} (\zeta-x)^{\frac{\alpha_3}{\pi}-1} d\zeta + B \tag{1}$$

maps the positive x axis onto a triangle with vertices w_1, w_2, and w_3.

Using (1) and setting

$$A = e^{\frac{3\pi i}{4}}, \quad B = 0 \tag{2}$$

find the transformation which maps the x axis onto an isosceles right triangle.

Solution: Since x_1, x_2, and x_3 are arbitrary points let us choose

$$x_1 = -1, \quad x_2 = 0, \quad x_3 = 1 \tag{3}$$

The vertices of the triangle can be chosen in such a way that

$$\alpha_1 = \frac{\pi}{4}, \quad \alpha_2 = \frac{\pi}{2}, \quad \alpha_3 = \frac{\pi}{4} \tag{4}$$

Setting $z_0 = 0$ from (1) we obtain

$$w = e^{\frac{3\pi i}{4}} \int_{0}^{z} (\zeta+1)^{-\frac{3}{4}} \zeta^{-\frac{1}{2}} (\zeta-1)^{-\frac{3}{4}} d\zeta \tag{5}$$

The point $x_2 = 0$ is mapped to $w_2 = 0$ and the point $x_3 = 1$ is mapped to

$$w_3 = e^{\frac{3\pi i}{4}} \int_{0}^{1} (\zeta^2-1)^{-\frac{3}{4}} \zeta^{-\frac{1}{2}} d\zeta \tag{6}$$

Denoting by a the integral

$$\int_{0}^{1} (1-x^2)^{-\frac{3}{4}} x^{-\frac{1}{2}} dx \tag{7}$$

we obtain $w_3 = a$.

The image of $x_1 = -1$ is $w_1 = ai$.

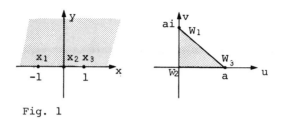

Fig. 1

• **PROBLEM 23-19**

Find the Schwarz-Christoffel transformation which maps the x axis onto a square.

In the formula set $A = i$ and $B = 0$ and find the vertices of the square.

Solution: The angles of the square are

$$\alpha_1 = \alpha_2 = \alpha_3 = \alpha_4 = \frac{\pi}{2} \tag{1}$$

Assume that

$$x_4 = \infty \tag{2}$$

and $x_1 = -1$, $x_2 = 0$, $x_3 = 1$.

Hence,

$$\begin{aligned} w &= A \int_{z_0}^{z} (\zeta+1)^{-\frac{1}{2}} \zeta^{-\frac{1}{2}} (\zeta-1)^{-\frac{1}{2}} d\zeta + B \\ &= i \int_{0}^{z} (\zeta+1)^{-\frac{1}{2}} (\zeta-1)^{-\frac{1}{2}} \zeta^{-\frac{1}{2}} d\zeta \end{aligned} \tag{3}$$

From (3) we can find the vertices of the square.

$$w_2 = w(0) = 0$$

$$w_3 = i \int_0^1 (x^2 - 1)^{-\frac{1}{2}} x^{-\frac{1}{2}} dx \tag{4}$$

Substituting

$$x^2 = t \tag{5}$$

into (4), we obtain

$$w_3 = i \int_0^1 (t-1)^{-\frac{1}{2}} t^{-\frac{1}{4}} \frac{1}{2} t^{-\frac{1}{2}} dt$$

$$= \frac{1}{2} i \int_0^1 (t-1)^{-\frac{1}{2}} t^{-\frac{3}{4}} dt \tag{6}$$

$$= \frac{1}{2} \int_0^1 (1-t)^{-\frac{1}{2}} t^{-\frac{3}{4}} dt = \frac{1}{2} B(\tfrac{1}{4}, \tfrac{1}{2})$$

where $B(.,.)$ is the beta function defined by

$$B(m,n) = \int_0^1 t^{m-1} (1 - t)^{n-1} dt \tag{7}$$

In the same manner we find

$$w_1 = w(-1) = \frac{1}{2} i B\left(\frac{1}{4}, \frac{1}{2}\right) \tag{8}$$

and

$$w_4 = \frac{1}{2} B\left(\frac{1}{4}, \frac{1}{2}\right) \left[1 + i\right] \tag{9}$$

• **PROBLEM** 23-20

The transformation

$$w = \pi i + z - \ln z \tag{1}$$

maps the x axis with the points ABCDE onto the subset of the w plane as shown in Figure 1.

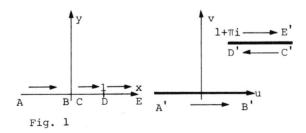

Fig. 1

Applying the Schwarz-Christoffel transformation derive (1) from Figure 1.

Solution: Consider a point P moving along the x axis from $-\infty$ to $+\infty$. As P moves from A to B its image P' moves from A' to B'. As P moves from 0 to 1 (that is from C to D) its image moves from C' to D' = $1 + \pi i$ and finally as P moves from D to E, P' moves from D' to E'. At the images of the points z = 0 and z = 1, P' changes the direction of its motion.

Thus,

$$\frac{dw}{dz} = A(z - 0)^{\frac{0}{\pi} - 1} (z - 1)^{\frac{2\pi}{\pi} - 1}$$

$$= A(z - 0)^{-1}(z - 1) = A\frac{z-1}{z}$$

$$= A\left(1 - \frac{1}{z}\right)$$

(2)

and

$$w = A(z - \ln z) + B \tag{3}$$

From Figure 1 we get

$$A = 1 \quad \text{and} \quad B = \pi i \tag{4}$$

Then

$$w = z - \ln z + \pi i \tag{5}$$

• **PROBLEM 23-21**

Find the transformation which maps the upper half plane onto a semi-infinite strip shown in Figure 1.

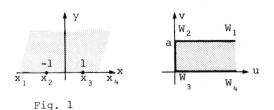

Fig. 1

Solution: We choose the points x_1 and x_4 at infinity. The Schwarz-Christoffel transformation yields

$$w = A \int_0^z (\zeta+1)^{-\frac{1}{2}}(\zeta-1)^{-\frac{1}{2}} d\zeta + B$$

$$= A \int_0^z \frac{d\zeta}{\sqrt{\zeta^2 - 1}} + B = A \text{ arc cosh } z + B$$

(1)

From Figure 1 we get the following conditions:

for z = -1   the value of w is $w_2 = ia$

for z = 1   the value of w is $w_3 = 0$

Therefore,

$$B = 0 \quad \text{and} \quad A = \frac{a}{\pi} \tag{2}$$

Substituting (2) into (1) we find

$$w = \frac{a}{\pi} \text{ arc cosh } z \tag{3}$$

$$0 \leq \text{Im}(\text{arc cosh } z) \leq \pi$$

VARIOUS POLYGONS

• **PROBLEM 23-22**

Find the transformation which maps the upper half of the z plane onto the shaded region shown in Figure 1b.

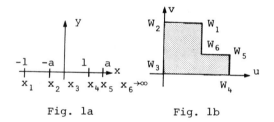

Fig. 1a   Fig. 1b

Solution: We choose the points on the real axis

$$x_1 = -1, \quad x_2 = -a, \quad x_3 = 0, \quad x_4 = 1, \quad x_5 = a, \quad x_6 = \infty \tag{1}$$

The Schwarz-Christoffel transformation is

$$w = A \int_0^z (\zeta+1)^{-\frac{1}{2}}(\zeta+a)^{-\frac{1}{2}}\zeta^{-\frac{1}{2}}(\zeta-a)^{-\frac{1}{2}}(\zeta-1)^{-\frac{1}{2}} d\zeta + B$$

$$= A \int_0^z \frac{d\zeta}{\sqrt{(\zeta^2-1)(\zeta^2-a^2)\zeta}} + B \tag{2}$$

$$= \frac{A}{a} \int_0^z \frac{d\zeta}{\sqrt{(1-z^2)(1-p^2\zeta^2)\zeta}} + B$$

821

where
$$p^2 = \frac{1}{a^2}$$

Setting $\frac{A}{a} = 1$ and $B = 0$ we find

$$w = \int_0^z \frac{d\zeta}{\sqrt{(1-z^2)(1-p^2\zeta^2)\zeta}}, \quad p > 1 \tag{3}$$

• **PROBLEM** 23-23

Applying the Schwarz-Christoffel transformation find the mapping of the upper half of the z plane onto the shaded region in Figure 1b.

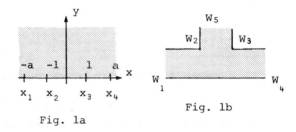

Fig. 1a

Fig. 1b

Solution: Since the choice of the points $x_1, x_2, \ldots, x_5$ is arbitrary let

$$x_1 = -a, \; x_2 = -1, \; x_3 = 1, \; x_4 = a, \; x_5 = \infty \tag{1}$$

Then, by the Schwarz-Christoffel transformation we get

$$w = A \int_0^z (\zeta+a)^{-1}(\zeta+1)^{\frac{1}{2}}(\zeta-1)^{\frac{1}{2}}(\zeta-a)^{-1} d\zeta + B$$

$$= A \int_0^z \frac{\sqrt{\zeta^2 - 1}}{\zeta^2 - a^2} d\zeta + B \tag{2}$$

$$= -\frac{Ai}{a^2} \int_0^z \frac{\sqrt{1 - \zeta^2}}{1 - p^2\zeta^2} d\zeta + B$$

where
$$\frac{1}{a^2} = p^2$$

Setting $B = 0$ in (2) we find

$$w = A' \int_0^z \frac{\sqrt{1 - \zeta^2}}{1 - p^2\zeta^2} d\zeta \tag{3}$$

where A' is an arbitrary constant.

• **PROBLEM 23-24**

Find the conformal transformation which maps the upper half of the z plane onto the shaded region shown in Figure 1.

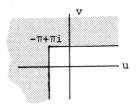

Fig. 1

Solution: Let us first find the transformation which maps the upper half of the z plane onto the "negative" of the shaded region, see Figure 2.

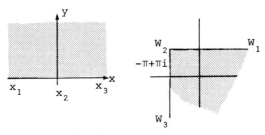

Fig. 2

The Schwarz-Christoffel transformation yields

$$w = A \int_0^z \zeta^{-\frac{1}{2}} d\zeta + B = A' z^{\frac{1}{2}} + B \qquad (1)$$

Since for $z = 0$, $w = -\pi + \pi i$ we have

$$B = -\pi + \pi i \qquad (2)$$

Thus,

$$w = A' z^{\frac{1}{2}} - \pi + \pi i \qquad (3)$$

Setting $A' = 1$ we obtain the transformation of the upper half of the z plane onto the shaded region of Figure 1.

$$w = z^{\frac{1}{2}} - \pi + \pi i \qquad (4)$$

APPLICATIONS IN PHYSICS

• **PROBLEM 23-25**

Apply the Schwarz-Christoffel transformation to find a complex potential F(w) of a flow of a fluid over a step shown in Figure 1b. The velocity of a fluid at infinity is $V_0 > 0$.

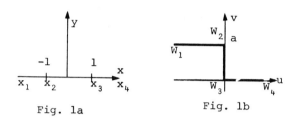

Fig. 1a   Fig. 1b

Solution: Points x_1 and x_4 are chosen at infinity. Hence, the Schwarz-Christoffel transformation yields

$$w = A \int_0^z (\zeta+1)^{\frac{1}{2}} (\zeta-1)^{-\frac{1}{2}} d\zeta + B$$

$$= A\left[\sqrt{z^2 - 1} + \text{arc cosh } z\right] + B \tag{1}$$

For $z = -1$, $w = w_2 = ia$

and for $z = 1$, $w = w_3 = 0$ \tag{2}

Therefore,

$$B = 0 \quad \text{and} \quad A = \frac{a}{\pi} \tag{3}$$

$$f(z) = w = \frac{a}{\pi}\left[\sqrt{z^2 - 1} + \text{arc cosh } z\right] \tag{4}$$

Let $F'(z)$ be a complex potential for the half plane $\text{Im } z \geq 0$ with the real axis x as a streamline. Then,

$$F'(z) = F(w) = F[f(z)] \tag{5}$$

$$\frac{dF'}{dz} = \frac{dF}{dw}\frac{dw}{dz} = \frac{dF}{dw} \cdot \frac{a}{\pi}\sqrt{\frac{z+1}{z-1}} \tag{6}$$

Note that

$$\lim_{z \to \infty} \frac{dF'}{dz} = \frac{aV_0}{\pi} \tag{7}$$

Therefore,

$$F'(z) = \frac{aV_0}{\pi} z \tag{8}$$

and
$$F(w) = \frac{aV_0}{\pi} f^{-1}(w) = \frac{aV_0}{\pi} z \qquad (9)$$

• **PROBLEM** 23-26

Find the electrostatic potential due to a line charge of density q located between two semi-infinite grounded planes intersecting at an angle α, see Figure 1.

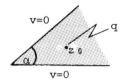

Fig. 1

Solution: The electrostatic potential $V(x,y)$ is a harmonic function in a domain not containing the charges. Let

$$F(z) = U(x,y) + iV(x,y) \qquad (1)$$

The electrostatic potential due to a line charge of density q located at the origin (see Figure 2) is

$$V = -2q \ln r, \quad |r| = z \qquad (2)$$

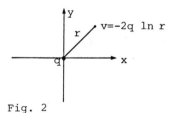

Fig. 2

From Coulomb's law we obtain the coefficient $-2q$ in (2).

$$F(z) = U + iV = -2qi \ln z \qquad (3)$$

Consider the system shown in Figure 3.

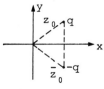

Fig. 3

Charge q is located at z_0 and charge $-q$ is located at $\bar{z}_0$. The potential of this system is

$$F(z) = U + iV = -2qi \ln \frac{z - z_0}{z - \bar{z}_0} \tag{4}$$

Hence, eq.(4) determines the potential due to an infinite grounded plane located in the x plane and a line charge q situated at z_0, see Figure 4.

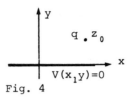

Fig. 4

Let us find the mapping of the upper half of the w plane onto the interior of the angle α.

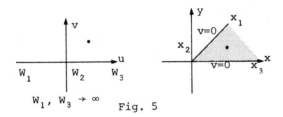

Fig. 5

By the Schwarz-Christoffel transformation

$$z = A \int_0^w \zeta^{\frac{\alpha}{\pi} - 1} d\zeta + B = A_1 w^{\frac{\alpha}{\pi}} + B \tag{5}$$

Since for $z = x_2 = 0$, $w = w_2 = 0$ we have $B = 0$.

$$z = A_1 w^{\frac{\alpha}{\pi}} \tag{6}$$

The inverse mapping is

$$w = f(z) = A_2 z^{\frac{\pi}{\theta}}, \quad A_2 > 0 \tag{7}$$

Point z_0 is mapped to

$$w_0 = A_2 z_0^{\frac{\pi}{\alpha}} \tag{8}$$

Denoting by V_1 the electrostatic potential due to an infinite grounded plate and a line charge q we get

$$F_1(w) = U_1 + iV_1 = -2qi \ln \frac{w - w_0}{w - \overline{w_0}} \tag{9}$$

and

$$U + iV = F(z) = F_1(f(z)) = -2qi \ln \frac{z^{\frac{\pi}{\alpha}} - z_0^{\frac{\pi}{\alpha}}}{z^{\frac{\pi}{\alpha}} - \overline{z_0}^{\frac{\pi}{\alpha}}} \tag{10}$$

From (10) we obtain the electrostatic potential

$$V(x,y) = \operatorname{Im}\left[-2qi \ln \frac{z^{\frac{\pi}{\alpha}} - z_0^{\frac{\pi}{\alpha}}}{z^{\frac{\pi}{\alpha}} - \overline{z_0}^{\frac{\pi}{\alpha}}}\right] \tag{11}$$

CHAPTER 24

SPECIAL TOPICS OF COMPLEX ANALYSIS

ANALYTIC CONTINUATION

● **PROBLEM** 24-1

Explain the principle of analytic continuation. What boundaries are called natural?

Solution: Consider a function $f_1(z)$ analytic in D_1 and suppose $f_2(z)$ is analytic in D_2 and such that in $D_1 \cap D_2$

$$f_1(z) = f_2(z) \quad \text{for } z \in D_1 \cap D_2 \tag{1}$$

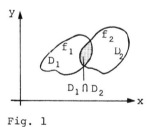

Fig. 1

We call $f_2(z)$ an analytic continuation of $f_1(z)$.

The function $f(z)$ can be defined as analytic in $D_1 \cup D_2$ and such that

$$\begin{aligned} f(z) &= f_1(z) \quad \text{for } z \in D_1 \\ f(z) &= f_2(z) \quad \text{for } z \in D_2 \end{aligned} \tag{2}$$

Analytic continuation can be carried out to domains $D_2, D_3, D_4, \ldots$ The functions $f_2, f_3, f_4, \ldots$ defined in $D_2, D_3, D_4, \ldots$ are called function elements.

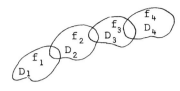

Fig. 2

Sometimes it is impossible to extend a function analytically beyond the boundary of a domain. Then such a boundary is called a natural boundary. A function $f_1(z)$ defined in D_1 can be continued analytically to D_n along two different paths.

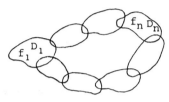

Fig. 3

Two analytic continuations will be identical if there is no singularity between the paths. This fact is known as the uniqueness theorem for analytic continuation.

● **PROBLEM** 24-2

Show that if $f(z)$ is analytic in D and $f(z) = 0$ for all points along an arc AB inside D then

$$f(z) = 0 \quad \text{for all } z \in D. \qquad (1)$$

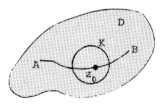

Fig. 1

Solution: Let z_0 be any point on AB.

Then a circle K exists with a center at z_0 such that $K \subset D$ and for any $z \in K$, $f(z)$ has a Taylor series expansion

$$f(z) = f(z_0) + f'(z_0)(z - z_0) + \frac{1}{2} f''(z_0)(z - z_0) + \ldots \qquad (2)$$

But

$$f(z_0) = f'(z_0) = f''(z_0) = \ldots = 0 \qquad (3)$$

Therefore, for any $z \in K$

$$f(z) = 0, \quad z \in K \quad (4)$$

Choosing another arc in K we can continue this procedure and show that

$$f(z) = 0 \quad \text{for all } z \in D \quad (5)$$

● **PROBLEM 24-3**

The identity

$$\sin^2 x + \cos^2 x = 1 \quad (1)$$

holds for all real values of x. Applying the principle of analytic continuation, show that (1) holds for all complex values.

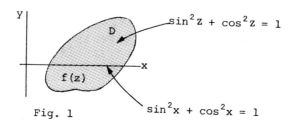

Fig. 1

Solution: Let D be a domain in the z plane containing a part of the x axis.

Because sin z and cos z are analytic in D, the function

$$f(z) = \sin^2 z + \cos^2 z - 1 \quad (2)$$

is also analytic in D.

Furthermore, $f(z) = 0$ along the x axis. Hence, by Problem 24-2

$f(z) = 0$ everywhere in D or

$$\sin^2 z + \cos^2 z = 1, \quad z \in D \quad (3)$$

Because (3) holds for an arbitrary domain, it must be true for the whole complex space.

● **PROBLEM 24-4**

Let $f_1(z)$ be analytic in a domain D_1 and on the boundary AB. Prove that if a function $f_2(z)$ exists, analytic in D_2 and on the boundary AB such that

$$f_1(z) = f_2(z) \quad \text{for all } z \in AB \tag{1}$$

then, the function $f(z)$ defined by

$$f(z) = \begin{cases} f_1(z) & \text{for } z \in D_1 \\ f_2(z) & \text{for } z \in D_2 \end{cases} \tag{2}$$

is analytic in $D_1 \cup D_2$ ($D_1 \cup D_2$ denoted the sum of sets D_1 and D_2).

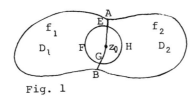

Fig. 1

Solution: Figure 1 illustrates the situation.

Let z_0 be any point on AB. To show that $f(z)$ is analytic at z_0 we shall apply Morera's theorem.

$$\int_{EFGH} f(z)dz = \int_{EFG} f(z)dz + \int_{GE} f(z)dz + \int_{EG} f(z)dz + \int_{GHE} f(z)dz \tag{3}$$

By Cauchy's theorem any integral along any simple closed path in D is zero. Therefore,

$$\int_{EFGH} f(z)dz = \int_{EFGE} f_1(z)dz + \int_{EGHE} f_2(z)dz = 0$$

By Morera's theorem we conclude that $f(z)$ defined in (2) is analytic in $D = D_1 \cup D_2$.

The function $f_2(z)$ is called an analytic continuation of $f_1(z)$.

• **PROBLEM 24-5**

Let f_1, f_2 be analytic in D_1, D_2, respectively. Show that the analytic continuation of f_1 to D_2 is unique.

Solution: Suppose f_2 and f_2' are analytic continuations of f_1 to D_2. Then,

$$f_1(z) = f_2(z) = f_2'(z) \quad \text{for } z \in D_1 \cap D_2 \tag{1}$$

or

$$(f_2 - f_2')(z) = 0 \text{ for } z \in D_1 \cap D_2 \tag{2}$$

Let us choose any arc AB in $D_1 \cap D_2$. Then for all $z \in AB$

$$(f_2 - f_2')(z) = 0 \text{ for } z \in AB \tag{3}$$

By Problem 24-2

$$(f_2 - f_2')(z) = 0 \text{ for all } z \in D_2 \tag{4}$$

or

$$f_2(z) = f_2'(z) \text{ for all } z \in D_2.$$

Hence, the analytic continuation of f_1 to D_2 is unique.

• **PROBLEM 24-6**

Let $f_1(z)$ and $f_2(z)$ be two functions analytic in D and such that on an arc AB in D

$$f_1(z) = f_2(z), \quad z \in AB \tag{1}$$

Prove that

$$f_1(z) = f_2(z) \text{ for all } z \in D \tag{2}$$

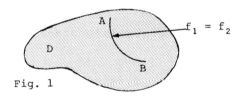

Fig. 1

Solution: Let

$$f(z) = f_1(z) - f_2(z) \tag{3}$$

Then along AB

$$f(z) = 0 \tag{4}$$

By Problem 24-2

$$f(z) = 0 \text{ for all } z \in D \tag{5}$$

or

$$f_1(z) = f_2(z) \text{ for all } z \in D \tag{6}$$

• **PROBLEM 24-7**

Find the domain for which
$$f_1(z) = z - z^2 + z^3 - z^4 + \ldots \quad (1)$$
is analytic and determine the analytic continuation of $f_1(z)$.

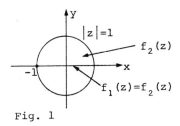

Fig. 1

Solution: The ratio test states that if

$$\lim_{n \to \infty} \left| \frac{a_{n+1}}{a_n} \right| < 1 \quad (2)$$

then $\sum_{1}^{\infty} a_n$ converges absolutely. Hence,

$$\lim_{n \to \infty} \left| \frac{z^{n+1}}{z^n} \right| = |z| \quad (3)$$

The series $f_1(z)$ converges for $|z| < 1$. Thus, the series represents an analytic function in this domain.

The sum of the series for $|z| < 1$ is

$$f_2(z) = \frac{z}{1+z} \quad (4)$$

Function $f_2(z)$ is analytic in the whole complex plane except $z = -1$. For $|z| < 1$

$$f_1(z) = f_2(z) \quad (5)$$

Function $f_2(z)$ is therefore an analytic continuation of $f_1(z)$.

• **PROBLEM 24-8**

Consider the function $f_1(z)$,

$$f_1(z) = \int_0^\infty e^{-zt} dt \quad (1)$$

Find the domain for which $f_1(z)$ is analytic and show that

$$f_2(z) = i \sum_{n=0}^{\infty} \left(\frac{z+i}{i}\right)^n \qquad (2)$$

is the analytic continuation of $f_1(z)$.

Solution: Integral (1) exists for $\operatorname{Re} z > 0$

$$f_1(z) = \int_0^{\infty} e^{-zt} \, dt = \frac{1}{z}, \quad \operatorname{Re} z > 0 \qquad (3)$$

The function $f_1(z)$ is analytic in the domain D_1

$$D_1 = \{z : \operatorname{Re} z > 0\} \qquad (4)$$

The series (2) is convergent for $|z + i| < 1$, which is the unit circle around the point $z = -i$.

$$f_2(z) = i \sum_{n=0}^{\infty} \left(\frac{z+i}{i}\right)^n = i \frac{1}{1 - \frac{z+i}{i}} = \frac{1}{z} \qquad (5)$$

$$\text{for } |z + i| < 1$$

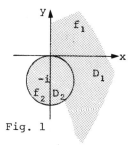

Fig. 1

$$D_2 = \{z : |z + i| < 1\} \qquad (6)$$

For $z \in D_1 \cap D_2$

$$f_1(z) = f_2(z)$$

thus, f_2 is the analytic continuation of f_1 into D_2.

Let us denote

$$f(z) = \frac{1}{z} \qquad (7)$$

which is analytic in the whole complex plane except the origin. The functions f_1 and f_2 are elements of f.

• **PROBLEM** 24-9

Show that

$$f_1(z) = \int_0^\infty (1 + t)e^{-zt}\, dt \tag{1}$$

is convergent for Re z > 0. Find the analytic continuation of $f_1(z)$ into the left half plane Re z < 0.

Solution: Integrating by parts we find

$$f_1(z) = \int_0^\infty (1 + t)e^{-zt}\, dt$$

$$= \left(\frac{e^{-zt}}{-z}\right)\Bigg|_{t=0}^{\infty} + \int_0^\infty t\left(\frac{e^{-zt}}{-z}\right)'\, dt$$

$$= \left[\frac{e^{-zt}}{-z} - \frac{te^{-zt}}{z} - \frac{e^{-zt}}{z^2}\right]\Bigg|_{t=0}^{\infty} \tag{2}$$

$$= \frac{1}{z} + \frac{1}{z^2} = \frac{1 + z}{z^2}$$

For Re z > 0 the function

$$f_2(z) = \frac{1 + z}{z^2}, \quad z \neq 0 \tag{3}$$

is analytic in the whole complex plane except the origin. Thus $f_2(z)$ is the analytic continuation of $f_1(z)$ into the left half-plane, Re z < 0.

• **PROBLEM** 24-10

In Problem 24-6 we showed that if two analytic functions coincide in a disk (or on an arc) then the two functions are identical.

Prove the following stronger statement. Let (z_n) be a sequence of points in D converging to a point $a \neq z_k$, and let $f_1(z)$ and $f_2(z)$ be two functions analytic in D. If

$$f_1(z_k) = f_2(z_k), \quad k = 1, 2, \ldots \tag{1}$$

then
$$f_1(z) = f_2(z), \quad z \in D \tag{2}$$

<u>Solution</u>: Let
$$f(z) = f_1(z) - f_2(z) \tag{3}$$

The function $f(z)$ is analytic at a. Its Taylor expansion around $z = a$ is
$$f(z) = a_0 + a_1(z-a) + a_2(z-a)^2 + \ldots \tag{4}$$

which converges for $|z - a| < r$. We have
$$f(z_k) = f_1(z_k) - f_2(z_k) = 0, \text{ for } k = 1, 2, \ldots \tag{5}$$

Since, $f(z)$ is continuous at a and
$$\lim_{k \to \infty} z_k = a \tag{6}$$

we conclude that
$$a_0 = f(a) = 0 \tag{7}$$

We shall show that all other coefficients in (4) are equal to zero. Indeed, assume that
$$a_1 = a_2 = \ldots = a_{\ell-1} = 0, \; a_\ell \neq 0 \tag{8}$$

then
$$\begin{aligned} f(z) &= a_\ell (z-a)^\ell + a_{\ell+1}(z-a)^{\ell+1} + \ldots \\ &= (z-a)^\ell F(z) \end{aligned} \tag{9}$$

But
$$f(z_k) = 0 \tag{10}$$

therefore $F(z_k) = 0$.

Function $F(z)$ is continuous at a, hence
$$a_\ell = F(a) = 0 \tag{11}$$

All coefficients in (4) vanish. Therefore, $f(z) = 0$ and
$$f_1(z) = f_2(z) \quad \text{for } z \in D.$$

PRINCIPLE OF REFLECTION

• **PROBLEM** 24-11

The following theorem is known as Schwarz's principle of reflection.

Theorem

Let $f_1(z)$ be analytic in some domain D_1 which includes a segment of the x-axis, and which assumes real values on this segment. Then the analytic continuation of $f_1(z)$ into a domain D_2 which is the mirror image of D_1 with respect to the x-axis, is given by

$$f_2(z) = \overline{f_1(\bar{z})} \tag{1}$$

Prove this theorem.

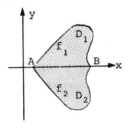

Fig. 1

Solution: On the segment AB we have

$$f_1(z) = f_1(x) = \overline{f_1(x)} = \overline{f_1(\bar{z})} \tag{2}$$

We shall prove that $f_2(z) = \overline{f_1(\bar{z})}$ is analytic in D_2. Let

$$f_1(z) = u(x,y) + iv(x,y) \tag{3}$$

The function $f_1(z)$ is analytic in D_1. Hence, by the Cauchy-Riemann equations

$$\frac{\partial u}{\partial x} = \frac{\partial v}{\partial y}, \quad \frac{\partial v}{\partial x} = -\frac{\partial u}{\partial y} \tag{4}$$

The partial derivatives in (4) are continuous. We have

$$f_1(\bar{z}) = f_1(x - iy) = u(x, -y) + iv(x, -y) \tag{5}$$

and

$$\overline{f_1(\bar{z})} = u(x, -y) - iv(x, -y) = f_2(z) \tag{6}$$

From (4) we obtain

$$\frac{\partial(-v)}{\partial(-y)} = \frac{\partial v}{\partial y}, \quad \frac{\partial(-v)}{\partial x} = -\frac{\partial v}{\partial x}, \quad \frac{\partial u}{\partial(-y)} = -\frac{\partial u}{\partial y} \tag{7}$$

and

$$\frac{\partial u}{\partial x} = \frac{\partial(-v)}{\partial(-y)}, \quad \frac{\partial(-v)}{\partial x} = -\frac{\partial u}{\partial(-y)} \tag{8}$$

Therefore, $f_2(z)$ is analytic.

Since, on AB $f_1(z) = f_2(z)$ and $f_2(z)$ is analytic in D_2, $f_2(z)$ is the analytic continuation of $f_1(z)$ into D_2.

• **PROBLEM 24-12**

Prove the following version of the reflection principle.

<u>Theorem</u>

Let D be a symmetric domain with respect to the x-axis and containing a segment AB of the x-axis. If $f(z)$ is analytic in D and assuming the real values on AB then

$$f(\bar{z}) = \overline{f(z)}, \quad z \in D \tag{1}$$

Conversely, if (1) holds then

$$f(x) \text{ for } x \in AB \text{ is real.}$$

<u>Solution</u>: Eq.(1) can be written as

$$f(z) = \overline{f(\bar{z})} \tag{2}$$

where

$$f(z) = u(x,y) + iv(x,y) \tag{3}$$

and

$$\overline{f(\bar{z})} = u(x,-y) - iv(x,-y) \tag{4}$$

At a point $(x,0)$ of the real axis

$$u(x,0) + iv(x,0) = u(x,0) - iv(x,0) \tag{5}$$

Hence,

$$v(x,0) = 0 \tag{6}$$

and

$$f(x) \text{ is a real function.}$$

The converse statement, i.e., that if $f(x)$ is real then $f(\bar{z}) = \overline{f(z)}$ has been proved in problem 24-11.

GAMMA FUNCTION

• **PROBLEM** 24-13

Define the gamma function $\Gamma(z)$ and show that

$$\Gamma(z + 1) = z\Gamma(z) \tag{1}$$

Solution: The gamma function $\Gamma(z)$ is defined for all z such that for Re $z > 0$.

$$\Gamma(z) \equiv \int_0^\infty t^{z-1} e^{-t} \, dt \tag{2}$$

Eq.(1) is called the recursion formula.

Integrating by parts we obtain

$$\Gamma(z+1) = \int_0^\infty t^z e^{-t} \, dt = \lim_{p \to \infty} \int_0^P t^z e^{-t} \, dt$$

$$= \lim_{p \to \infty} \left[t^z \cdot (-e^{-t}) \Big|_{t=0}^P - \int_0^P z t^{z-1}(-e^{-t}) \, dt \right] \tag{3}$$

$$= z \int_0^\infty t^{z-1} e^{-t} \, dt = z\Gamma(z); \quad \text{Re } z > 0$$

Note that

$$\lim_{p \to \infty} \frac{p^z}{e^p} = 0 \tag{4}$$

• **PROBLEM** 24-14

Show that if Re $p > 0$ then

$$\Gamma(p) = 2 \int_0^\infty x^{2p-1} e^{-x^2} \, dx \tag{1}$$

Solution: Substituting $t = x^2$ and $dt = 2x\,dx$ into the definition of the gamma function (see Problem 24-13) we obtain

$$\Gamma(p) = \int_0^\infty t^{p-1} e^{-t} dt$$

$$= \int_0^\infty (x^2)^{p-1} e^{-x^2} \cdot 2x\,dx \qquad (2)$$

$$= 2 \int_0^\infty x^{2p-1} e^{-x^2} dx$$

Function gamma appears in many branches of pure and applied analysis. In the forthcoming problems equation (2) of Problem 24-13 and (1) of this problem will be used.

• **PROBLEM 24-15**

Prove that $\Gamma(z)$ is an analytic function of z in the half-plane Re $z > 0$.

Solution: We shall show that the limit

$$\lim_{h \to 0} \frac{\Gamma(z+h) - \Gamma(z)}{h} \qquad (1)$$

exists. Indeed

$$\frac{\Gamma(z+h) - \Gamma(z)}{h} = \frac{1}{h} \int_0^\infty t^{z+h-1} e^{-t} dt - \frac{1}{h} \int_0^\infty t^{z-1} e^{-t} dt$$

$$= \int_0^\infty t^{z-1} e^{-t} \left[\frac{t^h - 1}{h} \right] dt \qquad (2)$$

The limit is equal to

$$\lim_{h \to 0} \frac{\Gamma(z+h) - \Gamma(z)}{h} = \lim_{h \to 0} \int_0^\infty t^{z-1} e^{-t} \left[\frac{t^h - 1}{h} \right] dt$$

$$= \int_0^\infty t^{z-1} e^{-t} \ln t\, dt \qquad (3)$$

To evaluate the limit

$$\lim_{h \to 0} \frac{t^h - 1}{h} = \left[\frac{d}{dh} \left(\frac{t^h - 1}{h} \right) \right]_{h=0} = \ln t \qquad (4)$$

we applied l'Hospital's rule. Hence, $\Gamma(z)$ is an analytic function in the half-plane Re z > 0.

• **PROBLEM 24-16**

In Problem 24-13 we defined the gamma function $\Gamma(z)$ and in Problem 24-15 we showed that $\Gamma(z)$ is analytic in the half-plane Re z > 0.

With the help of the functional equation

$$z\Gamma(z) = \Gamma(z + 1) \tag{1}$$

find the analytic continuation beyond the original domain Re z > 0.

Solution: Since (1) holds for Re z > 0 it must remain valid for all analytic continuations of the functions appearing in (1). The function $\Gamma(z + 1)$ can be continued analytically to all points z such that Re(z + 1) > 0, that is

$$\text{Re } z > -1 \tag{2}$$

The same is true for the function $z\Gamma(z)$. Eq.(1) holds for the half-plane Re z > -1. The values of $\Gamma(z)$ for the strip -1 < Re z ≤ 0 can be evaluated from

$$\Gamma(z) = \frac{\Gamma(z + 1)}{z} \tag{3}$$

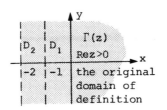

Fig. 1

Originally $\Gamma(z)$ was defined for the domain Re z > 0. Eq.(3) extends the domain of definition by D_1, -1 < Re z ≤ 0. From (3) we conclude that $\Gamma(z)$ is analytic for all points of the strip, except at z = 0 where it has a simple pole.

In the same manner we can add the additional stripes to the domain of definition to obtain Re z > -n, where n is an arbitrary positive integer. The function $\Gamma(z)$ is analytic for all points of Re z > -n except for the simple poles located at z = 0, -1, -2, -3, ..., -(n - 1),....

● **PROBLEM** 24-17

Show that $\Gamma(z)$ satisfies the functional equation
$$\Gamma(z) = \frac{\Gamma(z+n)}{z(z+1)(z+2)\ldots(z+n-1)} \quad (1)$$
where n is a positive integer.

Solution: We have
$$\Gamma(z+1) = z\Gamma(z) \quad (2)$$
and
$$\Gamma(z+2) = (z+1)\Gamma(z+1) \quad (3)$$
Substituting (2) into (3) we get
$$\Gamma(z+2) = (z+1)z\,\Gamma(z) \quad (4)$$
Similarly for $z+3$
$$\Gamma(z+3) = (z+2)\Gamma(z+2) = (z+2)(z+1)z\Gamma(z) \quad (5)$$
and, in general
$$\Gamma(z+n) = (z+n-1)(z+n-2)\ldots(z+1)z\Gamma(z) \quad (6)$$

● **PROBLEM** 24-18

Show that
$$\Gamma(z) \cdot \Gamma(1-z) = \frac{\pi}{\sin \pi z} \quad (1)$$

Solution: First, we shall prove (1) for z such that Im z = 0 and $0 < z < 1$, and then by analytic continuation we will extend the result to other values of z.

From Problem 24-14 we obtain
$$\Gamma(m)\Gamma(1-m) = \left[2\int_0^\infty x^{2m-1} e^{-x^2} dx\right]\left[2\int_0^\infty y^{1-2m} e^{-y^2} dy\right]$$
$$= 4\int_0^\infty \int_0^\infty x^{2m-1} y^{1-2m} e^{-(x^2+y^2)} dx\,dy \quad (2)$$

In polar coordinates,
$$x = r\cos\theta$$
$$y = r\sin\theta \quad (3)$$

therefore, (2) becomes

$$4 \int_{\theta=0}^{\frac{\pi}{2}} \int_{r=0}^{\infty} (\tan^{1-2m}\theta) r e^{-r^2} \, dr \, d\theta \qquad (4)$$

$$= 2 \int_0^{\frac{\pi}{2}} \tan^{1-2m}\theta \, d\theta = \frac{\pi}{\sin m\pi}$$

Hence,

$$\Gamma(z)\Gamma(1-z) = \frac{\pi}{\sin \pi z} \qquad (5)$$

Here, we used the formula

$$\int_0^\infty \frac{x^{\alpha-1}}{1+x} \, dx = \frac{\pi}{\sin \alpha\pi} \qquad (6)$$

where $0 < \alpha < 1$.

● **PROBLEM 24-19**

Find the value of $\Gamma(z)$ at the point $z = \frac{1}{2}$.

Solution: We shall apply the following formula

$$\Gamma(p) = 2 \int_0^\infty x^{2p-1} e^{-x^2} \, dx, \quad p > 0 \qquad (1)$$

Setting $p = \frac{1}{2}$ we obtain

$$\Gamma\left(\frac{1}{2}\right) = 2 \int_0^\infty e^{-x^2} \, dx = \sqrt{\pi} \qquad (2)$$

because using (5) from Problem 24-18 and setting $z = \frac{1}{2}$

$$\Gamma\left(\frac{1}{2}\right)\Gamma\left(\frac{1}{2}\right) = \Gamma^2\left(\frac{1}{2}\right) = \pi \qquad (3)$$

or

$$\Gamma\left(\frac{1}{2}\right) = \sqrt{\pi}$$

Another way of finding the solution is to follow the method shown in Problem 24-18.

$$\left[\Gamma\left(\frac{1}{2}\right)\right]^2 = \left[2\int_0^\infty e^{-x^2}dx\right]\left[2\int_0^\infty e^{-y^2}dy\right]$$

$$= 4\int_0^\infty\int_0^\infty e^{-(x^2+y^2)}dxdy = 4\int_{\theta=0}^{\frac{\pi}{2}}\int_{r=0}^\infty e^{-r^2}rdrd\theta \qquad (4)$$

$$= \pi$$

Hence,

$$\Gamma\left(\frac{1}{2}\right) = \sqrt{\pi} \qquad (5)$$

• **PROBLEM 24-20**

Evaluate:

1. $\Gamma\left(\frac{5}{2}\right)$

2. $\Gamma\left(-\frac{1}{2}\right)$

Solution: We shall use the formula

$$\Gamma(z + 1) = z\Gamma(z) \qquad (1)$$

From Problem 24-19 we get

$$\Gamma\left(\frac{1}{2}\right) = \sqrt{\pi} \qquad (2)$$

Then

$$\Gamma\left(\frac{5}{2}\right) = \Gamma\left(\frac{3}{2} + 1\right) = \frac{3}{2}\Gamma\left(\frac{3}{2}\right)$$

$$= \frac{3}{2}\Gamma\left(\frac{1}{2} + 1\right) = \frac{3}{2}\cdot\frac{1}{2}\Gamma\left(\frac{1}{2}\right) = \frac{3\sqrt{\pi}}{4} \qquad (3)$$

Similarly,

$$\Gamma\left(\frac{1}{2}\right) = \Gamma\left(-\frac{1}{2} + 1\right) = -\frac{1}{2}\Gamma\left(-\frac{1}{2}\right) \qquad (4)$$

and

$$\Gamma\left(-\frac{1}{2}\right) = -2\sqrt{\pi} \qquad (5)$$

• **PROBLEM 24-21**

The gamma function has the following important property

$$\Gamma(n) = n^{n-\frac{1}{2}} e^{-n} \sqrt{2\pi}\, e^{\frac{\theta(n)}{12n}} \quad \text{for } n > 0, \quad (1)$$

where $\theta(n)$ denotes a function of n such that $0 < \theta(n) < 1$. From (1) derive the Stirling approximation to n!

$$n! \sim n^{n+\frac{1}{2}} e^{-n} \sqrt{2\pi} \quad (2)$$

Solution: If n is a positive integer then

$$\Gamma(n+1) = n\Gamma(n) = n(n-1)\Gamma(n-1) = \ldots$$
$$= n(n-1) \cdot \ldots \cdot 2 \cdot 1 \cdot \Gamma(1) \quad (3)$$

But

$$\Gamma(1) = \int_0^\infty e^{-t}\, dt = 1 \quad (4)$$

Hence,

$$\Gamma(n+1) = n! \quad (5)$$

From (1) we have

$$\frac{n\Gamma(n)}{n^{n+\frac{1}{2}} e^{-n} \sqrt{2\pi}} = e^{\frac{\theta(n)}{12n}} \quad (6)$$

Since for all n, $0 < \theta(n) < 1$, it follows that

$$\lim_{n \to \infty} \frac{\Gamma(n+1)}{n^{n+\frac{1}{2}} e^{-n} \sqrt{2\pi}} = 1 \quad (7)$$

From (5) and (7) we obtain

$$n! \sim n^{n+\frac{1}{2}} e^{-n} \sqrt{2\pi} \quad (8)$$

It is a fairly good approximation.

For example for n = 10

$$n! = 3{,}628{,}800 \quad (9)$$

while the right-hand side of (8) is

$$\sim 3{,}610{,}000 \quad (10)$$

• **PROBLEM 24-22**

Prove that for k = 1,2,3,...

$$\Gamma\left(\frac{1}{k}\right)\Gamma\left(\frac{2}{k}\right) \cdots \Gamma\left(\frac{k-1}{k}\right) = \frac{(2\pi)^{\frac{k-1}{2}}}{\sqrt{k}} \quad (1)$$

Solution: Let us denote the left-hand side of (1) by A then

$$A = \Gamma\left(\frac{1}{k}\right)\Gamma\left(\frac{2}{k}\right) \cdots \Gamma\left(\frac{k-1}{k}\right)$$
$$= \Gamma\left(1-\frac{1}{k}\right)\Gamma\left(1-\frac{2}{k}\right) \cdots \Gamma\left(1-\frac{k-1}{k}\right) \quad (2)$$

Multiplying term by term and applying

$$\Gamma(k)\,\Gamma(1-k) = \frac{\pi}{\sin \pi k} \quad (3)$$

we find

$$A^2 = \left[\Gamma\left(\frac{1}{k}\right)\Gamma\left(1-\frac{1}{k}\right)\right]\left[\Gamma\left(\frac{2}{k}\right)\Gamma\left(1-\frac{2}{k}\right)\right] \cdots \left[\Gamma\left(1-\frac{1}{k}\right)\Gamma\left(\frac{1}{k}\right)\right]$$
$$= \frac{\pi}{\sin \frac{\pi}{k}} \cdot \frac{\pi}{\sin \frac{2\pi}{k}} \cdots \frac{\pi}{\sin \frac{(k-1)\pi}{k}} \quad (4)$$

But, for k = 1,2,3,...

$$\sin \frac{\pi}{k} \cdot \sin \frac{2\pi}{k} \cdot \sin \frac{3\pi}{k} \cdot \ldots \cdot \sin \frac{(k-1)\pi}{k}$$
$$= \frac{k}{2^{k-1}} \quad (5)$$

Substituting (5) into (4) we find

$$\Gamma\left(\frac{1}{k}\right)\Gamma\left(\frac{2}{k}\right) \cdots \Gamma\left(\frac{k-1}{k}\right) = \sqrt{\frac{\pi^{k-1}}{k} 2^{k-1}}$$
$$= \frac{(2\pi)^{\frac{k-1}{2}}}{\sqrt{k}} \quad (6)$$

• **PROBLEM 24-23**

Consider the path shown in Figure 1.

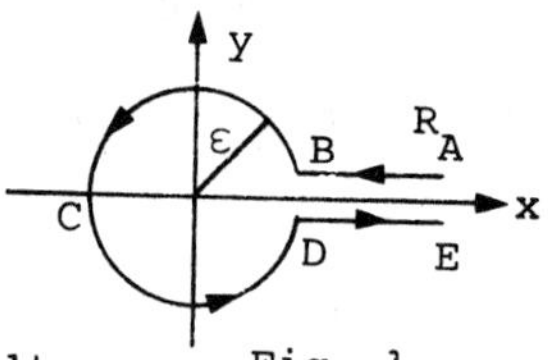

Show that

$$\Gamma(z) = \frac{1}{e^{2\pi i z}-1} \int_{ABCDE} t^{z-1} e^{-t}\, dt \quad (1)$$

Fig. 1

Solution: Note that along AB, $t = x$; along BCD, $t = e^{i\theta}$; and along DE, $t = xe^{2\pi i}$.

Then

$$\int_{ABCDE} t^{z-1} e^{-t} dt = \int_R^\varepsilon x^{z-1} e^{-x} dx$$

$$+ \int_0^{2\pi} (\varepsilon e^{i\theta})^{z-1} e^{-\varepsilon e^{i\theta}} i\varepsilon e^{i\theta} d\theta \qquad (2)$$

$$+ \int_\varepsilon^R x^{z-1} e^{2\pi i(z-1)} e^{-x} dx$$

$$= (e^{2\pi i z} - 1) \int_\varepsilon^R x^{z-1} e^{-x} dx + i \int_0^{2\pi} \varepsilon^z e^{i\theta z} e^{-\varepsilon e^{i\theta}} d\theta$$

For Re $z > 0$, taking the limit as $\varepsilon \to 0$ and $R \to \infty$ we obtain

$$\int_{ABCDE} t^{z-1} e^{-t} dt = (e^{2\pi i z} - 1) \int_0^\infty x^{z-1} e^{-x} dx \qquad (3)$$

$$= (e^{2\pi i z} - 1) \Gamma(z)$$

Since the functions on both sides of (3) are analytic, we have

$$\Gamma(z) = \frac{1}{e^{2\pi i z} - 1} \oint t^{z-1} e^{-t} dt \qquad (4)$$

• **PROBLEM 24-24**

Evaluate the integrals:

a) $\int_0^\infty y^5 e^{-2y} dy$ \qquad (1)

b) $\int_0^\infty y^{\frac{3}{2}} e^{-3y} dy$ \qquad (2)

Solution: a) Substituting $2y = t$ into (1) we find

$$\int_0^\infty y^5 e^{-2y} \, dy = \int_0^\infty \left(\frac{t}{2}\right)^5 e^{-t} \frac{1}{2} \, dt$$

$2y = t$
$y = t/2$
$\frac{dy}{dt} = \frac{1}{2}$. (3)
$dy = \frac{1}{2} dt$

$$= \frac{1}{2^6} \int_0^\infty t^5 e^{-t} \, dt$$

But,

$$\Gamma(z) = \int_0^\infty t^{z-1} e^{-t} \, dt \tag{4}$$

Therefore, (3) is equal to $\frac{1}{2^6} \Gamma(6)$. (5)

Applying $\Gamma(z + 1) = z \, \Gamma(z)$ we get

$$\int_0^\infty y^5 e^{-2y} \, dy = \frac{1}{2^6} \cdot 5 \cdot 4 \cdot 3 \cdot 2 = \frac{15}{8} \tag{6}$$

b) To evaluate integral (2) substitute

$3y = t$ $y = \frac{t}{3}$ $\frac{dy}{dt} = \frac{1}{3}$ $dy = \frac{1}{3} dt$ (7)

$$\int_0^\infty y^{\frac{3}{2}} e^{-3y} \, dy = \int_0^\infty \left(\frac{t}{3}\right)^{\frac{3}{2}} e^{-t} \frac{1}{3} \, dt$$

$$= 3^{-\frac{5}{2}} \int_0^\infty t^{\frac{3}{2}} e^{-t} \, dt = 3^{-\frac{5}{2}} \Gamma\left(\frac{5}{2}\right) \tag{8}$$

$$= 3^{-\frac{5}{2}} \cdot \frac{3}{2} \cdot \frac{1}{2} \Gamma\left(\frac{1}{2}\right) = \frac{\sqrt{3\pi}}{36}$$

● **PROBLEM 24-25**

Prove the functional identity

$$\int_0^\infty x^m e^{-ax^n} \, dx = \frac{1}{n} a^{-\frac{m+1}{n}} \Gamma\left(\frac{m+1}{n}\right) \tag{1}$$

where m, n, and a are positive constants.

Solution: Substituting

$$ax^n = t \tag{2}$$

we find
$$x = \left(\frac{t}{a}\right)^{\frac{1}{n}}, \quad anx^{n-1} dx = dt \qquad (3)$$

and
$$\int_0^\infty x^m e^{-ax^n} dx = \int_0^\infty \left(\frac{t}{a}\right)^{\frac{m}{n}} e^{-t} \frac{1}{an} \cdot \frac{1}{\left(\frac{t}{a}\right)^{\frac{n-1}{n}}} dt$$

$$= \frac{1}{n} \cdot a^{-\frac{m+1}{n}} \int_0^\infty t^{\frac{m-n+1}{n}} e^{-t} dt \qquad (4)$$

$$= \frac{1}{n} a^{-\frac{m+1}{n}} \Gamma\left(\frac{m+1}{n}\right)$$

BETA FUNCTION

● **PROBLEM 24-26**

Define the beta function and show that
$$B(n,m) = B(m,n) \qquad (1)$$

<u>Solution</u>: We define the beta function by
$$B(n,m) = \int_0^1 t^{n-1} (1-t)^{m-1} dt \qquad (2)$$

for Re $n > 0$ and Re $m > 0$.

Setting
$$t = 1 - u, \qquad (3)$$

we obtain
$$B(n,m) = \int_0^1 t^{n-1} (1-t)^{m-1} dt$$

$$= \int_0^1 (1-u)^{n-1} u^{m-1} du = B(m,n) \qquad (4)$$

• **PROBLEM** 24-27

Show that

$$B(m,n) = 2\int_0^{\pi/2} \sin^{2m-1}\theta \cos^{2n-1}\theta \, d\theta$$

$$= 2\int_0^{\pi/2} \cos^{2m-1}\theta \sin^{2n-1}\theta \, d\theta \qquad (1)$$

Solution: Let

$$t = \sin^2\theta \qquad (2)$$

then,

$$dt = 2\sin\theta\cos\theta \, d\theta \qquad (3)$$

and

$$B(m,n) = \int_0^1 t^{m-1}(1-t)^{n-1} \, dt$$

$$= \int_0^{\pi/2} (\sin^2\theta)^{m-1}(\cos^2\theta)^{n-1} \, 2\sin\theta\cos\theta \, d\theta \qquad (4)$$

$$= 2\int_0^{\pi/2} \sin^{2m-1}\theta \cos^{2n-1}\theta \, d\theta$$

Since $B(m,n) = B(n,m)$ from (4) we get

$$B(m,n) = 2\int_0^{\pi/2} \cos^{2m-1}\theta \sin^{2n-1}\theta \, d\theta \qquad (5)$$

• **PROBLEM** 24-28

Show that

$$B(m,n) = \frac{\Gamma(m)\Gamma(n)}{\Gamma(m+n)} \qquad (1)$$

Solution: We shall use the functional identity

$$\Gamma(m) = 2 \int_0^\infty x^{2m-1} e^{-x^2} dx, \quad m > 0 \tag{2}$$

Transforming to polar coordinates we have

$$\Gamma(m)\Gamma(n) = \left[2 \int_0^\infty x^{2m-1} e^{-x^2} dx \right] \left[2 \int_0^\infty y^{2n-1} e^{-y^2} dy \right]$$

$$= 4 \int_0^\infty \int_0^\infty x^{2m-1} y^{2n-1} e^{-(x^2+y^2)} dxdy \tag{3}$$

$$= 4 \int_{\theta=0}^{\frac{\pi}{2}} \int_{r=0}^\infty (\cos^{2m-1}\theta \sin^{2n-1}\theta)(r^{2m+2n-1} e^{-r^2}) drd\theta$$

$$= \left[2 \int_0^{\frac{\pi}{2}} \cos^{2m-1}\theta \sin^{2n-1}\theta d\theta \right] \left[2 \int_0^\infty r^{2(m+n)-1} e^{-r^2} dr \right]$$

$$= B(m,n) \, \Gamma(m+n)$$

Hence,

$$B(m,n) = \frac{\Gamma(m)\Gamma(n)}{\Gamma(m+n)} \tag{4}$$

● **PROBLEM 24-29**

Show that

$$\frac{B(m,n+1)}{B(m+1,n)} = \frac{n}{m} \tag{1}$$

Solution: We shall utilize the functional identity proved in Problem 24-28.

$$B(m,n) = \frac{\Gamma(m)\Gamma(n)}{\Gamma(m+n)} \tag{2}$$

Then,

$$\frac{B(m,n+1)}{B(m+1,n)} = \frac{\frac{\Gamma(m)\Gamma(n+1)}{\Gamma(m+n+1)}}{\frac{\Gamma(m+1)\Gamma(n)}{\Gamma(m+n+1)}}$$

$$= \frac{\Gamma(m)\Gamma(n+1)}{\Gamma(m+1)\Gamma(n)} = \frac{n\Gamma(n)\Gamma(m)}{m\Gamma(m)\Gamma(n)} = \frac{n}{m} \tag{3}$$

We used the identity

$$\Gamma(z + 1) = z\Gamma(z) \qquad (4)$$

● **PROBLEM 24-30**

Prove that

$$\frac{B\left(\frac{q+1}{2}, \frac{1}{2}\right)}{B\left(\frac{q+1}{2}, \frac{q+1}{2}\right)} = 2^q \qquad (1)$$

Solution: Since,

$$B(m,n) = \frac{\Gamma(m)\Gamma(n)}{\Gamma(m+n)} \qquad (2)$$

the left-hand side of (1) can be written in the form

$$\frac{B\left(\frac{q+1}{2}, \frac{1}{2}\right)}{B\left(\frac{q+1}{2}, \frac{q+1}{2}\right)} = \frac{\Gamma\left(\frac{q+1}{2}\right)\Gamma\left(\frac{1}{2}\right)\Gamma(q+1)}{\Gamma\left(\frac{q+2}{2}\right)\Gamma\left(\frac{q+1}{2}\right)\Gamma\left(\frac{q+1}{2}\right)}$$

$$= \frac{\Gamma\left(\frac{1}{2}\right)q\Gamma(q)}{\Gamma\left(\frac{q}{2}+1\right)\Gamma\left(\frac{q}{2}+\frac{1}{2}\right)} = \frac{2\Gamma\left(\frac{1}{2}\right)\Gamma(q)}{\Gamma\left(\frac{q}{2}\right)\Gamma\left(\frac{q}{2}+\frac{1}{2}\right)} \qquad (3)$$

We shall apply the duplication formula

$$\Gamma(z)\,\Gamma\left(z + \frac{1}{2}\right) = \frac{\sqrt{\pi}\,\Gamma(2z)}{2^{2z-1}} \qquad (4)$$

to obtain from (3)

$$\frac{2\Gamma\left(\frac{1}{2}\right)\Gamma(q)}{\Gamma\left(\frac{q}{2}\right)\Gamma\left(\frac{q}{2}+\frac{1}{2}\right)} = \frac{2\sqrt{\pi}\,\Gamma(q)}{\sqrt{\pi}\,\Gamma(q)} \cdot 2^{q-1}$$

$$= 2^q \qquad (5)$$

● **PROBLEM 24-31**

Evaluate:

1. $B\left(3, \frac{7}{2}\right)$

2. $B(k,\ell)$, where k, ℓ are positive integers

3. $B\left(\frac{1}{3}, \frac{2}{3}\right)$

Solution: 1.
$$B\left(3, \frac{7}{2}\right) = \frac{\Gamma(3)\Gamma\left(\frac{7}{2}\right)}{\Gamma\left(3 + \frac{7}{2}\right)}$$

$$= \frac{2 \cdot \frac{5}{2} \cdot \frac{3}{2} \cdot \frac{1}{2} \Gamma\left(\frac{1}{2}\right)}{\frac{11}{2} \cdot \frac{9}{2} \cdot \frac{7}{2} \cdot \frac{5}{2} \cdot \frac{3}{2} \cdot \frac{1}{2} \cdot \Gamma\left(\frac{1}{2}\right)} \qquad (1)$$

$$= \frac{2 \cdot 8}{11 \cdot 9 \cdot 7} = \frac{16}{693}$$

2.
$$B(k, \ell) = \frac{\Gamma(k)\Gamma(\ell)}{\Gamma(k + \ell)} = \frac{(k-1)! \ (\ell-1)!}{(k+\ell-1)!}$$

$$= \frac{(\ell-1)!}{k \cdot (k+1) \ldots (k+\ell-1)} \qquad (2)$$

3.
$$B\left(\frac{1}{3}, \frac{2}{3}\right) = \frac{\Gamma\left(\frac{1}{3}\right)\Gamma\left(\frac{2}{3}\right)}{\Gamma\left(\frac{1}{3} + \frac{2}{3}\right)} = \Gamma\left(\frac{1}{3}\right)\Gamma\left(\frac{2}{3}\right) \qquad (3)$$

Utilizing the formula

$$\Gamma(z)\Gamma(1-z) = \frac{\pi}{\sin \pi z} \qquad (4)$$

we obtain

$$B\left(\frac{1}{3}, \frac{2}{3}\right) = \frac{\pi}{\sin \frac{\pi}{3}} = \frac{2\pi}{\sqrt{3}} \qquad (5)$$

• **PROBLEM 24-32**

Evaluate the integral

$$\int_0^a \frac{dy}{\sqrt{a^4 - y^4}} \qquad (1)$$

Solution: Substituting

$$t = \frac{y^4}{a^4} \qquad (2)$$

where

$$dt = \frac{4y^3 dy}{a^4} \qquad (3)$$

we obtain

$$\int_0^a \frac{dy}{\sqrt{a^4 - y^4}} = \int_0^1 \frac{1}{a^2\sqrt{1-t}} \cdot \frac{a^4}{4a^3 t^{\frac{3}{4}}} dt$$

$$= \frac{1}{4a} \int_0^1 t^{-\frac{3}{4}} (1-t)^{-\frac{1}{2}} dt$$

$$= \frac{1}{4a} B\left(\frac{1}{4}, \frac{1}{2}\right) \tag{4}$$

We can express the answer in terms of the gamma function

$$\frac{1}{4a} B\left(\frac{1}{4}, \frac{1}{2}\right) = \frac{1}{4a} \frac{\Gamma(\frac{1}{4})\Gamma(\frac{1}{2})}{\Gamma(\frac{1}{4}+\frac{1}{2})} \tag{5}$$

Since,

$$2^{2z-1} \Gamma(z)\Gamma(z + \tfrac{1}{2}) = \sqrt{\pi}\, \Gamma(2z) \tag{6}$$

we have

$$\Gamma\left(\frac{1}{4} + \frac{1}{2}\right) = \frac{\pi\sqrt{2}}{\Gamma(\frac{1}{4})} \tag{7}$$

and

$$\int_0^a \frac{dy}{\sqrt{a^4 - y^4}} = \frac{1}{4a} \frac{\Gamma(\frac{1}{4}) \sqrt{\pi} \cdot \Gamma(\frac{1}{4})}{\pi\sqrt{2}}$$

$$= \frac{1}{4a} \frac{[\Gamma(\frac{1}{4})]^2}{\sqrt{2\pi}} \tag{8}$$

● **PROBLEM** 24-33

Evaluate the integrals

1. $\int_0^3 \sqrt{5x(3-x)}\, dx$

2. $\int_0^{\frac{\pi}{2}} \sqrt{\tan\theta}\, d\theta$

3. $\int_0^{\frac{\pi}{2}} \sin^6\theta \cos^4\theta\, d\theta$

Solution: 1. Substituting $x = 3t$ we find

$$\int_0^3 \sqrt{5x(3-x)}\, dx = \int_0^1 \sqrt{5 \cdot 3t \cdot 3(1-t)}\, 3\, dt$$

$$= 9\sqrt{5} \int_0^1 t^{\frac{1}{2}}(1-t)^{\frac{1}{2}}\, dt \qquad (1)$$

$$= 9\sqrt{5}\, B\left(\frac{3}{2}, \frac{3}{2}\right) = 9\sqrt{5}\, \frac{\Gamma\left(\frac{3}{2}\right)\Gamma\left(\frac{3}{2}\right)}{\Gamma(3)}$$

$$= \frac{9\sqrt{5} \cdot \pi}{8}$$

2. $$\int_0^{\frac{\pi}{2}} \sqrt{\tan\theta}\, d\theta = \int_0^{\frac{\pi}{2}} \sin^{\frac{1}{2}}\theta\, \cos^{-\frac{1}{2}}\theta\, d\theta \qquad (2)$$

Using the formula

$$B(m,n) = 2 \int_0^{\frac{\pi}{2}} \sin^{2m-1}\theta\, \cos^{2n-1}\theta\, d\theta \qquad (3)$$

and because $\Gamma(z)\Gamma(1-z) = \dfrac{\pi}{\sin \pi z}$, we find

$$\int_0^{\frac{\pi}{2}} \sqrt{\tan\theta}\, d\theta = \frac{1}{2} B\left(\frac{3}{4}, \frac{1}{4}\right) = \frac{1}{2} \Gamma\left(\frac{3}{4}\right)\Gamma\left(\frac{1}{4}\right)$$

$$= \frac{1}{2} \frac{\pi}{\sin \frac{\pi}{4}} = \frac{\pi\sqrt{2}}{2} \qquad (4)$$

3. $$\int_0^{\frac{\pi}{2}} \sin^6\theta\, \cos^4\theta\, d\theta = \frac{1}{2} B\left(\frac{7}{2}, \frac{5}{2}\right)$$

$$= \frac{1}{2} \frac{\Gamma\left(\frac{5}{2}\right)\Gamma\left(\frac{7}{2}\right)}{\Gamma(6)} = \frac{\frac{3}{2} \cdot \frac{1}{2} \cdot \frac{5}{2} \cdot \frac{3}{2} \cdot \frac{1}{2} \sqrt{\pi} \cdot \sqrt{\pi}}{2 \cdot 2 \cdot 3 \cdot 4 \cdot 5} \qquad (5)$$

$$= \frac{3\pi}{512}$$

INFINITE PRODUCTS

• **PROBLEM 24-34**

Prove that

$$\left[\prod_{n=1}^{\infty} (1 + |w_n|) \text{ converges}\right] \iff \left[\sum_{n=1}^{\infty} |w_n| \text{ converges}\right] \qquad (1)$$

Solution: The symbol $\prod_{n=1}^{k} (1 + w_n)$ denotes the product

$$\prod_{n=1}^{k} (1 + w_n) = (1 + w_1)(1 + w_2)\ldots(1 + w_k) \qquad (2)$$

If there exists a value $A \neq 0$ such that

$$\lim_{k \to \infty} \prod_{n=1}^{k} (1 + w_n) = A \qquad (3)$$

we say that the infinite product converges to A. Otherwise the product diverges.

($\Leftarrow$) Suppose $\sum_{n=1}^{\infty} |w_n|$ converges.

If $x > 0$, then $1 + x \leq e^x$. Hence,

$$\prod_{n=1}^{k} (1 + |w_n|) = (1 + |w_1|)(1 + |w_2|)\ldots(1 + |w_k|) \qquad (4)$$

$$\leq \exp(|w_1|) \cdot \exp(|w_2|) \cdot \ldots \cdot \exp(|w_k|)$$

$$= \exp(|w_1| + \ldots + |w_k|)$$

Since $\sum_{n=1}^{\infty} |w_n|$ converges, the sequence $\prod_{n=1}^{k} (1 + |w_n|)$ is a bounded, monotonically increasing sequence. Therefore it has a limit, and the infinite product

$$\prod_{n=1}^{\infty} (1 + |w_n|) \text{ converges.}$$

($\Rightarrow$) Suppose now that $\prod_{n=1}^{\infty} (1 + |w_n|)$ converges. Then,

$$(1 + |w_1|)(1 + |w_2|)\ldots(1 + |w_n|) \geq 1 + |w_1| + \ldots + |w_n| \qquad (5)$$

$$= 1 + \sum_{k=1}^{n} |w_k| \geq 1$$

If $\lim_{n\to\infty} \Pi(1 + |w_n|)$ exists then $\Sigma |w_n|$ is a bounded monotonically increasing sequence. Therefore it has a limit and the sequence

$$\sum_{n=1}^{\infty} |w_n| \text{ converges.}$$

● **PROBLEM 24-35**

Prove that $\prod_{n=1}^{\infty} \left(1 - \frac{z^2}{n^2 + 1}\right)$ converges.

Solution: Let us denote

$$w_n = -\frac{z^2}{n^2 + 1} \tag{1}$$

Then

$$|w_n| = \frac{|z|^2}{n^2 + 1} \tag{2}$$

Let

$$w'_n = -\frac{z^2}{n^2}$$

then

$$|w'_n| = \frac{|z|^2}{n^2} \tag{3}$$

The series

$$\sum_{n=1}^{\infty} |w'_n| = |z|^2 \sum_{n=1}^{\infty} \frac{1}{n^2} \tag{4}$$

converges.

Therefore, the series $\sum_{n=1}^{\infty} |w_n| = |z|^2 \sum_{n=1}^{\infty} \frac{1}{n^2 + 1}$

also converges, because

$$|w_n| < |w'_n| \tag{5}$$

Hence the product

$$\prod_{n=1}^{\infty} \left(1 - \frac{z^2}{n^2 + 1}\right) \tag{6}$$

converges.

● **PROBLEM 24-36**

Prove that

$$\sin z = z\left(1 - \frac{z^2}{\pi^2}\right)\left(1 - \frac{z^2}{4\pi^2}\right)\cdots = z\prod_{n=1}^{\infty}\left(1 - \frac{z^2}{n^2\pi^2}\right) \quad (1)$$

Hint: Use

$$\cot z = \frac{1}{z} + 2z\left[\frac{1}{z^2 - \pi^2} + \frac{1}{z^2 - 4\pi^2} + \cdots\right] \quad (2)$$

Solution: Since,

$$\int\left(\cot t - \frac{1}{t}\right)dt = \ln\left(\frac{\sin t}{t}\right) + C \quad (3)$$

we can write

$$\int_0^z \left(\cot t - \frac{1}{t}\right)dt = \left[\ln\left(\frac{\sin t}{t}\right)\right]\Big|_{t=0}^z$$

$$= \ln\left(\frac{\sin z}{z}\right) \quad (4)$$

Substituting (2) into the integral we get

$$\int_0^z \left(\cot t - \frac{1}{t}\right)dt$$

$$= \int_0^z \left[\frac{2t}{t^2 - \pi^2} + \frac{2t}{t^2 - 4\pi^2} + \cdots\right]dt$$

$$= \sum_{n=1}^{\infty} \ln\left(1 - \frac{z^2}{n^2\pi^2}\right) = \ln\prod_{n=1}^{\infty}\left(1 - \frac{z^2}{n^2\pi^2}\right) \quad (5)$$

Comparing (4) and (5) we obtain

$$\sin z = z\prod_{n=1}^{\infty}\left(1 - \frac{z^2}{n^2\pi^2}\right) \quad (6)$$

● **PROBLEM 24-37**

Determine which of the products is convergent and evaluate them:

1. $\prod_{n=1}^{\infty}\left[1 + \frac{1}{n(n+2)}\right]$ \quad (1)

2. $\dfrac{2}{1} \cdot \dfrac{5}{4} \cdot \ldots \cdot \dfrac{n^2 + 1}{n^2} \cdot \ldots$ \hfill (2)

Solution: 1. <u>Theorem</u>

The product $\prod_{n=1}^{\infty} (1 + w_n)$ converges absolutely, if and only if the series $\sum_{n=1}^{\infty} w_n$ converges absolutely.

By virtue of the theorem, (1) converges absolutely. We get

$$\prod_{n=1}^{\infty} \left[1 + \dfrac{1}{n(n+2)} \right] = \prod_{n=1}^{\infty} \dfrac{(n+1)^2}{n(n+2)}$$

$$= \lim_{n \to \infty} \dfrac{2 \cdot 2}{1 \cdot 3} \cdot \dfrac{3 \cdot 3}{2 \cdot 4} \cdot \dfrac{4 \cdot 4}{3 \cdot 5} \cdot \ldots \cdot \dfrac{(n+1)(n+1)}{n(n+2)} \quad (3)$$

$$= \lim_{n \to \infty} 2 \dfrac{n+1}{n+2} = 2$$

2. In the formula

$$\dfrac{\sin \pi z}{\pi z} = \prod_{n=1}^{\infty} \left(1 - \dfrac{z^2}{n^2} \right) \hfill (4)$$

we set $z = i$ to obtain

$$\dfrac{\sin \pi i}{\pi i} = \prod_{n=1}^{\infty} \left(1 + \dfrac{1}{n^2} \right) \hfill (5)$$

Hence,

$$\dfrac{\sin \pi i}{\pi i} = \dfrac{e^{-\pi} - e^{\pi}}{-2\pi} = \dfrac{e^{\pi} - e^{-\pi}}{2\pi}$$

$$= \prod_{n=1}^{\infty} \dfrac{n^2 + 1}{n^2} = \dfrac{2}{1} \cdot \dfrac{5}{4} \cdot \ldots \cdot \dfrac{n^2 + 1}{n^2} \cdot \ldots \hfill (6)$$

● **PROBLEM 24-38**

The product expansion of the function $\sin \pi z$ is

$$\sin \pi z = \pi z \prod_{n=1}^{\infty} \left(1 - \dfrac{z^2}{n^2} \right) \hfill (1)$$

Compare this result with the power series for $\sin \pi z$.

Solution: The power series is

$$\dfrac{\sin \pi z}{\pi z} = 1 - \dfrac{\pi^2 z^2}{6} + \dfrac{\pi^4 z^4}{120} - \ldots \hfill (2)$$

Eq.(1) can be written

$$\frac{\sin \pi z}{\pi z} = \prod_{n=1}^{\infty}\left(1 - \frac{z^2}{n^2}\right) = (1 - z^2)\left(1 - \frac{z^2}{4}\right)\left(1 - \frac{z^2}{9}\right)\ldots \quad (3)$$

$$= 1 - \left[\sum_{n=1}^{\infty}\frac{1}{n^2}\right]z^2 + \left[\sum_{m=1}^{\infty}\left(\sum_{n=m+1}^{\infty}\frac{1}{m^2n^2}\right)\right]z^4 - \ldots$$

Comparing (2) and (3) we obtain

$$\frac{\pi^2}{6} = \sum_{n=1}^{\infty}\frac{1}{n^2} \quad (4)$$

and

$$\frac{\pi^4}{120} = \sum_{m=1}^{\infty}\frac{1}{m^2}\left(\sum_{n=m+1}^{\infty}\frac{1}{n^2}\right) = \sum_{n=2}^{\infty}\frac{1}{n^2}\left(\sum_{m=1}^{n-1}\frac{1}{m^2}\right) \quad (5)$$

Hence, from (4) and (5) we compute

$$\left(1 + \frac{1}{2^2} + \frac{1}{3^2} + \ldots\right)^2 - \left(1 + \frac{1}{2^4} + \frac{1}{3^4} + \ldots\right) = \frac{\pi^4}{60} \quad (6)$$

and

$$\sum_{n=1}^{\infty}\frac{1}{n^4} = \frac{\pi^4}{36} - \frac{\pi^4}{60} = \frac{\pi^4}{90} \quad (7)$$

• **PROBLEM 24-39**

Applying the product expansion

$$\frac{\sin \pi z}{\pi z} = \prod \left(1 - \frac{z^2}{n^2}\right) \quad (1)$$

show that

$$\sqrt{2} = \frac{2}{1} \cdot \frac{2}{3} \cdot \frac{6}{5} \cdot \frac{6}{7} \cdot \frac{10}{9} \cdot \frac{10}{11} \ldots \quad (2)$$

Solution: By substituting $z = \frac{1}{2}$ and $z = \frac{1}{4}$ into (1) we have

$$\frac{\sin \frac{\pi}{2}}{\frac{\pi}{2}} = \left(1 - \frac{1}{(2n)^2}\right) \quad (3)$$

and

$$\frac{\sin \frac{\pi}{4}}{\frac{\pi}{4}} = \left(1 - \frac{1}{(4n)^2}\right) \quad (4)$$

Dividing (4) by (3) we obtain

$$\frac{\sin\frac{\pi}{4}}{\frac{\pi}{4}} \cdot \frac{\frac{\pi}{2}}{\sin\frac{\pi}{2}} = \sqrt{2} = \frac{\Pi\left(1 - \frac{1}{(4n)^2}\right)}{\Pi\left(1 - \frac{1}{(2n)^2}\right)} \qquad (5)$$

But,

$$\Pi\left(1 - \frac{1}{(2n)^2}\right) = \Pi\left(1 - \frac{1}{(4m)^2}\right)\left(1 - \frac{1}{(4m-2)^2}\right) \qquad (6)$$

Substituting (6) into (5) we find

$$\sqrt{2} = \frac{1}{\Pi\left(1 - \frac{1}{(4m-2)^2}\right)} = \frac{2\cdot 2}{1\cdot 3} \cdot \frac{6\cdot 6}{5\cdot 7} \cdot \frac{10\cdot 10}{9\cdot 11} \cdot \ldots \qquad (7)$$

• **PROBLEM 24-40**

Applying the Weierstrass factor theorem for infinite products find the infinite product for the gamma function.

Solution: We shall quote the Weierstrass theorem for infinite products:

Theorem

Let $f(z)$ be an entire function with simple zeroes at $a_1, a_2, \ldots$ where

$$0 < |a_1| < |a_2| < \ldots \qquad (1)$$

and

$$\lim_{n \to \infty} |a_n| = \infty \qquad (2)$$

Then, the function $f(z)$ can be expressed as an infinite product

$$f(z) = f(0)\, e^{\frac{f'(0)z}{f(0)}} \prod_{n=1}^{\infty}\left[\left(1 - \frac{z}{a_n}\right)e^{\frac{z}{a_n}}\right] \qquad (3)$$

Let

$$f(z) = \frac{1}{\Gamma(z+1)} \qquad (4)$$

The function $f(z)$ is analytic everywhere (entire) and has simple zeros at $-1, -2, -3, \ldots$. From (3) we obtain

$$f(z) = \frac{1}{\Gamma(z+1)} = e^{f'(0)z} \prod_{n=1}^{\infty}\left(1 + \frac{z}{n}\right)e^{-\frac{z}{n}} \qquad (5)$$

To evaluate $f'(0)$ set $z = 1$ in (5)

$$1 = e^{f'(0)} \prod_{n=1}^{\infty}\left(1 + \frac{1}{n}\right)e^{-\frac{1}{n}}$$

$$= e^{f'(0)} \lim_{N \to \infty} \prod_{n=1}^{N} \left(1 + \frac{1}{n}\right) e^{-\frac{1}{n}} \qquad (6)$$

Taking the logarithm of (6) we find

$$f'(0) = \left\{ \lim_{N \to \infty} \frac{1}{1} + \frac{1}{2} + \frac{1}{3} + \ldots + \frac{1}{N} - \ln\left[\left(1 + \frac{1}{1}\right)\left(1 + \frac{1}{2}\right) \cdots \left(1 + \frac{1}{N}\right)\right] \right\}$$

$$= \lim_{N \to \infty} \left[1 + \frac{1}{2} + \frac{1}{3} + \ldots + \frac{1}{N} - \ln N \right] = \gamma \qquad (7)$$

where γ is Euler's constant.

Substituting (7) into (5) and noting that $\Gamma(z + 1) = z\Gamma(z)$ we obtain the solution.

● **PROBLEM 24-41**

Find the product expansion of the following entire functions:

1. $e^z - 1$
2. $e^z - e^{z_0}$
3. $\cos \pi z$

Solution: 1.

$$e^z - 1 = e^{\frac{z}{2}} \left(e^{\frac{z}{2}} - e^{-\frac{z}{2}} \right)$$

$$= 2i \, e^{\frac{z}{2}} \frac{e^{\frac{z}{2}} - e^{-\frac{z}{2}}}{2i} = 2i \, e^{\frac{z}{2}} \frac{e^{i\frac{z}{2i}} - e^{-i\frac{z}{2i}}}{2i} \qquad (1)$$

$$= 2i \, e^{\frac{z}{2}} \sin \frac{z}{2i}$$

From the formula

$$\frac{\sin \pi z}{\pi z} = \prod \left(1 - \frac{z^2}{n^2}\right) \qquad (2)$$

we obtain

$$2i \, e^{\frac{z}{2}} \sin \pi \frac{z}{2i\pi} = 2i \, e^{\frac{z}{2}} \cdot \pi \frac{z}{2i\pi} \prod_{n=1}^{\infty} \left(1 + \frac{z^2}{4\pi^2 n^2}\right)$$

$$= e^{\frac{z}{2}} \cdot z \cdot \prod_{n=1}^{\infty} \left(1 + \frac{z^2}{4\pi^2 n^2}\right) \qquad (3)$$

2. Assuming that

$$z_0 \neq 2k\pi i, \quad k = 0, \pm 1, \pm 2, \ldots \tag{4}$$

we obtain

$$e^z - e^{z_0} = e^{z_0}(e^{z-z_0} - 1) \tag{5}$$

$$= e^{z_0} e^{\frac{z-z_0}{2}} \cdot (z-z_0) \prod_{n=1}^{\infty} \left(1 + \frac{(z-z_0)^2}{4\pi^2 n^2}\right)$$

3.
$$\cos \pi z = \frac{\sin 2\pi z}{2 \sin \pi z} \tag{6}$$

$$= \frac{2\pi z \cdot \prod \left(1 - \frac{4z^2}{n^2}\right)}{2 \cdot \pi z \prod \left(1 - \frac{z^2}{n^2}\right)} = \prod_{n=1}^{\infty} \left(1 - \frac{4z^2}{(2n-1)^2}\right)$$

DIFFERENTIAL EQUATIONS

• **PROBLEM** 24-42

For the following linear differential equations determine which points are ordinary points, which are regular singular points and which are irregular singular points:

a) $(1 - z^2)Y'' + 3zY' + 5Y = 0$ (1)

b) $z^3 Y'' + (2 - z^3)Y' + (z + 1)Y = 0$ (2)

Solution: Consider a linear differential equation

$$Y'' + p(z)Y' + q(z)Y = 0 \tag{3}$$

The point, $z = a$, is called an ordinary point of the differential equation if $p(z)$ and $q(z)$ are analytic at $z = a$. All other points are called singular points. If $z = a$ is a singular point but $(z - a)p(z)$ and $(z - a)^2 q(z)$ are analytic at $z = a$, then a is called a regular singular point. All other singular points are called irregular singular points.

Ordinary Points	Singular Points	
	Regular Singular Points	Irregular Singular Points

Let us write (1) and (2) in the form

$$Y'' + \frac{3z}{1-z^2} Y' + \frac{5}{1-z^2} Y = 0 \tag{4}$$

$$Y'' + \frac{2-z^3}{z^3} Y' + \frac{z+1}{z^3} Y = 0 \tag{5}$$

The singular points of eq.(4) are $z = \pm 1$. Since,

$$\frac{3z}{1-z^2} (z+1) = \frac{3z}{1-z}$$

is analytic at $z = -1$ and

$$\frac{5}{1-z^2} (z+1)^2$$

is also analytic at $z = -1$, point $z = -1$ is a regular singular point of equation (4).

Eq.(5) has a singular point at $z = 0$. It is an irregular singular point, because $\frac{2-z^3}{z^3} \cdot z$ is not analytic at $z = 0$.

● **PROBLEM 24-43**

Find a power series solution of the differential system

$$y'' + y = 0 \tag{1}$$

$$y(0) = 0 \quad y'(0) = 1 \tag{2}$$

<u>Solution</u>: Assume that the solution has the form of a power series

$$y = \sum_{n=0}^{\infty} a_n x^n \tag{3}$$

Substituting (3) into (1) we obtain

$$\sum_{n=2}^{\infty} n(n-1) a_n x^{n-2} + \sum_{n=0}^{\infty} a_n x^n = 0 \tag{4}$$

or

$$\sum_{n=0}^{\infty} \left[(n+2)(n+1) a_{n+2} + a_n \right] x^n = 0 \tag{5}$$

For a power series (5) to be identically zero each of its coefficents must be zero, hence

$$(n+2)(n+1) a_{n+2} + a_n = 0 \quad n = 0,1,2,\ldots \tag{6}$$

Utilizing the boundary conditions (2) we find

$$a_0 = 0, \quad a_1 = 1 \tag{7}$$

Hence,

$$a_0 = a_2 = a_4 = \ldots = 0 \tag{8}$$

and

$$a_{2n+1} = \frac{-a_{2n-1}}{2n(2n+1)} \tag{9}$$

$$a_1 = 1, \quad a_3 = \frac{-1}{2 \cdot 3}, \quad a_5 = \frac{1}{2 \cdot 3 \cdot 4 \cdot 5}, \ldots \tag{10}$$

The power series solution of $y'' + y = 0$ is

$$y = x - \frac{x^3}{3!} + \frac{x^5}{5!} - \frac{x^7}{7!} + \ldots \tag{11}$$

Power series (11) represents the function

$$y = \sin x \tag{12}$$

● **PROBLEM 24-44**

Applying the power series method solve the differential equation

$$w'' + zw = 0 \tag{1}$$

Solution: Suppose the solution has the form

$$w(z) = \sum_{n=0}^{\infty} a_n z^n \tag{2}$$

Then, substituting (2) into (1) we find

$$\sum_{n=2}^{\infty} a_n n(n-1) z^{n-2} + \sum_{n=0}^{\infty} a_n z^{n+1} = 0 \tag{3}$$

Setting $z = 0$ in (3) we find that

$$a_2 = 0 \tag{4}$$

Then (3) can be written as

$$\sum_{n=0}^{\infty} \left[a_{n+3}(n+3)(n+2) + a_n \right] z^{n+1} = 0 \tag{5}$$

Since, (5) has to hold for all values of z we obtain

$$a_2 = 0, \quad a_{n+3} = \frac{-a_n}{(n+3)(n+2)} \tag{6}$$

Let $a_0 = A$, then

$$a_0 = A, \quad a_3 = \frac{-A}{2 \cdot 3}, \quad a_6 = \frac{A}{6 \cdot 5 \cdot 3 \cdot 2}, \ldots \tag{7}$$

and let $a_1 = B$, then

$$a_1 = B, \quad a_4 = \frac{-B}{4 \cdot 3}, \quad a_7 = \frac{B}{7 \cdot 6 \cdot 4 \cdot 3} \quad (8)$$

Substituting (7) and (8) into (2) we find

$$w(z) = A\left[1 - \frac{z^3}{2 \cdot 3} + \frac{z^6}{2 \cdot 3 \cdot 5 \cdot 6} - \frac{z^9}{2 \cdot 3 \cdot 5 \cdot 6 \cdot 8 \cdot 9}\right]$$
$$+ B\left[z - \frac{z^4}{3 \cdot 4} + \frac{z^7}{3 \cdot 4 \cdot 6 \cdot 7} - \cdots\right] \quad (9)$$

where A and B are arbitrary complex numbers.

• **PROBLEM 24-45**

Find the power series which is the solution of

$$w'' + z^2 w = 0$$
$$w(0) = 1, \quad w'(0) = -1 \quad (1)$$

and determine its region of convergence.

Solution: Suppose

$$w(z) = \sum_{n=0}^{\infty} a_n z^n \quad (2)$$

is the solution of (1). Then,

$$\sum_{n=2}^{\infty} a_n \cdot n \cdot (n-1) z^{n-2} + \sum_{n=0}^{\infty} a_n z^{n+2} = 0 \quad (3)$$

or

$$\sum_{n=0}^{\infty} a_{n+2}(n+2)(n+1) z^n + \sum_{n=0}^{\infty} a_n z^{n+2} = 0 \quad (4)$$

Since (4) holds for all values of z, it must be that

$$a_2 = a_3 = 0 \quad (5)$$

Then, (4) becomes

$$\sum_{n=2}^{\infty} \left[a_{n+2}(n+2)(n+1) + a_{n-2}\right] z^n = 0 \quad (6)$$

and, since (6) holds for all values of (6)

$$a_{n+2} = \frac{-a_{n-2}}{(n+2)(n+1)} \quad (7)$$

$$a_0 = 1, \quad a_1 = -1, \quad a_2 = 0, \quad a_3 = 0,$$

$$a_4 = \frac{1}{3 \cdot 4}, \quad a_5 = \frac{1}{4 \cdot 5}, \quad a_6 = 0, \quad a_7 = 0,$$ (8)

$$a_7 = \frac{1}{3 \cdot 4 \cdot 7 \cdot 8}, \quad a_9 = \frac{-1}{4 \cdot 5 \cdot 9 \cdot 8}$$

Hence,

$$w(z) = 1 - z - \frac{z^4}{3 \cdot 4} + \frac{z^5}{4 \cdot 5} + \frac{z^8}{3 \cdot 4 \cdot 7 \cdot 8} - \frac{z^9}{4 \cdot 5 \cdot 8 \cdot 9}$$ (9)

Series (9) is convergent for

$$|z| < \infty$$ (10)

• **PROBLEM 24-46**

Solve the equation

$$2z \frac{d^2w}{dz^2} + \frac{dw}{dz} - w = 0$$ (1)

<u>Solution</u>: Let us rewrite (1) in the form

$$\frac{d^2w}{dz^2} + \frac{1}{2z} \frac{dw}{dz} - \frac{1}{2z} w = 0$$ (2)

Thus, $z = 0$ is a regular singular point and the indicial equation is

$$r^2 - \frac{1}{2} r = 0$$ (3)

with the roots

$$r_1 = \frac{1}{2}, \quad r_2 = 0$$ (4)

Substituting

$$w_1(z) = \sum_{n=0}^{\infty} a_n z^{n+\frac{1}{2}}$$ (5)

into the equation (1) we obtain

$$\sum_{n=0}^{\infty} 2\left(n + \frac{1}{2}\right)\left(n - \frac{1}{2}\right) a_n z^{n-\frac{1}{2}} + \sum_{n=0}^{\infty} \left(n + \frac{1}{2}\right) a_n z^{n-\frac{1}{2}} - \sum_{n=0}^{\infty} a_n z^{n+\frac{1}{2}} = 0$$ (6)

or

$$\sum_{n=1}^{\infty} \left[n(2n+1)a_n - a_{n-1}\right] z^{n-\frac{1}{2}} = 0$$ (7)

Thus,

$$a_n = \frac{2a_{n-1}}{(2n+1)2n} \quad \text{or} \quad a_n = \frac{2^n a_0}{(2n+1)!} \tag{8}$$

Substituting (8) into (5) we find

$$w_1(z) = a_0 z^{\frac{1}{2}} \sum_{n=0}^{\infty} \frac{2^n}{(2n+1)!} z^n \tag{9}$$

For $r_2 = 0$ we find

$$w_2(z) = b_0 \sum_{n=0}^{\infty} \frac{2^n}{(2n)!} z^n \tag{10}$$

a_0 and b_0 are arbitrary constants. The series (9) and (10) are two linearly independent solutions of (1).

● **PROBLEM 24-47**

Find the solution of the equation

$$z \frac{d^2w}{dz^2} + (2n+1) \frac{dw}{dz} + zw = 0 \tag{1}$$

in the form

$$w = \int_C e^{zt} F(t) dt \tag{2}$$

Solution: Let

$$w = \int_C e^{zt} F(t) dt$$

then,

$$\frac{dw}{dz} = \int_C t e^{zt} F(t) dt \tag{3}$$

and

$$\frac{d^2w}{dz^2} = \int_C t^2 e^{zt} F(t) dt \tag{4}$$

If we choose C in such a way that the functional values at the initial and final point are the same, then

$$zw = \int_C z e^{zt} F(t) dt = e^{zt} F(t) \Big|_A^A - \int_C e^{zt} \frac{dF}{dt} dt \tag{5}$$

$$= -\int_C e^{zt} \frac{dF}{dt} dt$$

$$(2n+1)\frac{dw}{dz} = \int_C (2n+1)t e^{zt} F(t) dt \qquad (6)$$

$$z\frac{d^2w}{dz^2} = \int_C zt^2 e^{zt} F(t)dt = e^{zt} t^2 F(t)\bigg|_A^A - \int_C e^{zt} \frac{d}{dt}\left[t^2 F(t)\right] dt \qquad (7)$$

$$= -\int_C e^{zt} \frac{d}{dt}\left[t^2 F(t)\right] dt$$

Thus,

$$0 = z\frac{d^2w}{dz^2} + (2n+1)\frac{dw}{dz} + zw$$

$$= \int_C e^{zt}\left\{-\frac{dF}{dt} + (2n+1)tF(t) - \frac{d}{dt}\left[t^2 F(t)\right]\right\} dt \qquad (8)$$

The integral is zero when we choose the integrand to be zero, then

$$-\frac{dF}{dt} + (2n+1)t F(t) - \frac{d}{dt}\left[t^2 F\right] = 0$$

or

$$\frac{dF}{dt} = \frac{(2n-1)t}{t^2+1} F(t) \qquad (9)$$

Integrating (9) we find

$$F(t) = a(t^2 + 1)^{n-\frac{1}{2}} \qquad (10)$$

where a is an arbitrary constant.

From (2) and (10) we obtain

$$w(z) = a \int_C e^{zt} (t^2 + 1)^{n-\frac{1}{2}} dt \qquad (11)$$

• **PROBLEM 24-48**

Solve the equation

$$\frac{d^2w}{dz^2} - \frac{dw}{dz} - 2w = 0 \qquad (1)$$

Solution: Suppose the solution can be written in the form

$$w = \int_C e^{zt} F(t) dt \qquad (2)$$

Then,

$$\frac{dw}{dz} = \int_C t\, e^{zt} F(t) dt \qquad (3)$$

$$\frac{d^2w}{dz^2} = \int_C t^2\, e^{zt} F(t) dt \qquad (4)$$

Substituting (2), (3), and (4) into (1) we obtain

$$\int_C e^{zt}(t^2 - t - 2) F(t) dt = 0 \qquad (5)$$

Eq.(5) is satisfied if we choose

$$F(t) = \frac{1}{t^2 - t - 2} \qquad (6)$$

then

$$w(z) = \int_C \frac{e^{zt}}{t^2 - t - 2} dt \qquad (7)$$

The integrand has two simple poles at $t = 2$ and $t = -1$.

Choosing C in such a way that the pole at $t = 2$ is inside C and the pole at $t = -1$ is outside we find

$$w_1(z) = 2\pi i\, e^{2z} \qquad (8)$$

Now, choosing C so that $t = -1$ is inside and $t = 2$ outside, we get

$$w_2(z) = 2\pi i\, e^{-z} \qquad (9)$$

Hence, the solution is

$$w(z) = A e^{2z} + B e^{-z} \qquad (10)$$

where A and B are arbitrary constants.

BESSEL FUNCTION

• **PROBLEM** 24-49

Find the Laurent series expansion of the function

$$e^{\frac{u}{2}(z-\frac{1}{z})} = \sum_{n=-\infty}^{\infty} J_n(u) z^n \qquad (1)$$

where

$$J_n(u) = \frac{1}{2\pi} \int_0^{2\pi} \cos(n\theta - u\sin\theta) d\theta \qquad (2)$$

The function $e^{\frac{u}{2}(z-\frac{1}{z})}$ is called the generating function for the Bessel functions of the first kind and the coefficients $J_n(u)$ are known as the Bessel functions of the first kind.

Solution: Function (1) has a simple pole at $z = 0$. The coefficients of the Laurent expansion are

$$J_n(u) = \frac{1}{2\pi i} \int_C \frac{e^{\frac{u}{2}(w-\frac{1}{w})}}{w^{n+1}} dw \qquad (3)$$

Changing (3) to polar coordinates and choosing the radius of the circle C to be one, we find

$$J_n(u) = \frac{1}{2\pi i} \int_0^{2\pi} \frac{e^{\frac{u}{2}(e^{i\theta} - e^{-i\theta})}}{e^{i(n+1)\theta}} \cdot i \cdot e^{i\theta} d\theta$$

$$= \frac{1}{2\pi} \int_0^{2\pi} e^{i[u\sin\theta - n\theta]} d\theta \qquad (4)$$

$$= \frac{1}{2\pi} \int_0^{2\pi} \cos(n\theta - u\sin\theta) d\theta$$

• **PROBLEM** 24-50

Find the general solution of Bessel's differential equation

$$z^2 w'' + zw' + (z^2 - n^2)w = 0 \qquad (1)$$

where n is not an integer.

Solution: Since $z = 0$ is a regular singular point the solution of (1) can be written in the form

$$w = \sum_{k=0}^{\infty} a_k z^{k+\ell} \qquad (2)$$

Then,

$$w' = \Sigma(k + \ell)a_k z^{k+\ell-1} \qquad (3)$$

$$w'' = \Sigma(k + \ell)(k + \ell - 1)a_k z^{k+\ell-2} \qquad (4)$$

Substituting (2), (3), and (4) into (1) we find

$$z^2 w'' + zw' + (z^2 - n^2)w$$
$$= \Sigma\{[(k+\ell)^2 - n^2]a_k + a_{k-2}\}z^{k+\ell} = 0 \qquad (5)$$

Hence,

$$[(k + \ell)^2 - n^2]a_k + a_{k-2} = 0 \qquad (6)$$

If $k = 0$ and $a_{-2} = 0$, then

$$(\ell^2 - n^2)a_0 = 0 \qquad (7)$$

If $a_0 \neq 0$, then

$$\ell^2 - n^2 = 0 \qquad (8)$$

which is the indicial equation with roots $\ell = \pm n$.

For $\ell = n$, (6) yields

$$k(2n + k)a_k + a_{k-2} = 0 \qquad (9)$$

Since $a_{-1} = 0$, for $k = 1$ we have $a_1 = 0$ and

$$a_1 = a_3 = a_5 = \ldots = 0 \qquad (10)$$

Similarly from (9) if $a_0 \neq 0$

$$a_2 = \frac{-a_0}{2(2n+2)}$$

$$a_4 = \frac{-a_2}{4(2n+4)} = \frac{a_0}{2 \cdot 4(2n+2)(2n+4)} \qquad (11)$$

and

$$w_1 = \Sigma a_k z^{k+\ell} = a_0 z^n \left[1 - \frac{z^2}{2(2n+2)} + \frac{z^4}{2 \cdot 4(2n+2)(2n+4)} - \cdots\right] \qquad (12)$$

For $\ell = -n$ we get

$$w_2 = a_0 z^{-n}\left[1 - \frac{z^2}{2(2-2n)} + \frac{z^4}{2 \cdot 4(2n+2)(2n+4)} - \cdots\right] \qquad (13)$$

The general solution is

$$w = Aw_1 + Bw_2 \tag{14}$$

where A and B are arbitrary constants.

• **PROBLEM** 24-51

Show that

$$z J_{n-1}(z) - 2n J_n(z) + z J_{n+1}(z) = 0 \tag{1}$$

Solution: Differentiating the identity

$$e^{\frac{1}{2}z(u - \frac{1}{u})} = \sum_{n=-\infty}^{\infty} J_n(z) u^n \tag{2}$$

with respect to u we get

$$e^{\frac{1}{2}z(u - \frac{1}{u})} \cdot \frac{1}{2} z \left(1 + \frac{1}{u^2}\right) = \sum_{n=-\infty}^{\infty} \frac{z}{2}\left(1 + \frac{1}{u^2}\right) J_n(z) u^n$$

$$= \sum_{n=-\infty}^{\infty} n J_n(z) u^{n-1} \tag{3}$$

Then,

$$\sum_{n=-\infty}^{\infty} z J_n(z) u^n + \sum_{n=-\infty}^{\infty} z J_n(z) u^{n-2}$$

$$= \sum_{n=-\infty}^{\infty} 2n J_n(z) u^{n-1} \tag{4}$$

Equating coefficients of u^n on both sides we have

$$z J_n(z) + z J_{n+2}(z) = 2(n + 1) J_{n+1}(z) \tag{5}$$

Replacing n by n - 1 we get

$$z J_{n-1}(z) + z J_{n+1}(z) = 2n J_n(z) \tag{6}$$

This result also holds for n which is not an integer.

Eq.(6) is known as the recursion formula for Bessel functions.

• **PROBLEM** 24-52

Let C be a simple closed contour enclosing t = 0. Show that

$$J_n(z) = \frac{1}{2\pi i} \int_C u^{-n-1} e^{\frac{1}{2}z(u - \frac{1}{u})} du \tag{1}$$

Solution: Multiplying

$$e^{\frac{1}{2}z(u-\frac{1}{u})} = \sum_{m=-\infty}^{\infty} J_m(z) u^m \qquad (2)$$

by u^{-n-1} we obtain

$$e^{\frac{1}{2}z(u-\frac{1}{u})} u^{-n-1} = \sum_{m=-\infty}^{\infty} J_m(z) u^{m-n-1} \qquad (3)$$

Hence,

$$\int_C e^{\frac{1}{2}z(u-\frac{1}{u})} u^{-n-1} \, du = \sum_{m=-\infty}^{\infty} J_m(z) \int_C u^{m-n-1} \, du \qquad (4)$$

But

$$\int_C u^{m-n-1} \, du = \begin{cases} 0 & \text{if } m \neq n \\ 2\pi i & \text{if } m = n \end{cases} \qquad (5)$$

and (4) leads to

$$\int_C e^{\frac{1}{2}z(u-\frac{1}{u})} u^{-n-1} \, du = 2\pi i \, J_n(z) \qquad (6)$$

• **PROBLEM 24-53**

Show that if $\alpha \neq \beta$

$$\int_0^z u J_n(\alpha u) J_n(\beta u) \, du = z \, \frac{\alpha J_n(\beta z) J_n'(\alpha z) - \beta J_n(\alpha z) J_n'(\beta z)}{\beta^2 - \alpha^2} \qquad (1)$$

Solution: Both functions $J_n(\alpha u)$ and $J_n(\beta u)$ satisfy the Bessel's differential equation

$$u^2 J_n''(\alpha u) + u J_n'(\alpha u) + (\alpha^2 u^2 - n^2) J_n(\alpha u) = 0 \qquad (2)$$

$$u^2 J_n''(\beta u) + u J_n'(\beta u) + (\beta^2 u^2 - n^2) J_n(\beta u) = 0 \qquad (3)$$

Multiplying (2) by $J_n(\beta u)$ and (3) by $J_n(\alpha u)$ and subtracting we obtain

$$u^2 \left[J_n(\beta u) J_n''(\alpha u) - J_n(\alpha u) J_n''(\beta u) \right]$$
$$+ u \left[J_n'(\alpha u) J_n(\beta u) - J_n'(\beta u) J_n(\alpha u) \right] = (\beta^2 - \alpha^2) u^2 \, J_n(\alpha u) J_n(\beta u) \qquad (4)$$

Eq. (4) can be written in the form

$$\frac{d}{du}\left\{u\left[J_n(\beta u)J_n'(\alpha u) - J_n(\alpha u)J_n'(\beta u)\right]\right\}$$
$$= (\beta^2 - \alpha^2)u\, J_n(\alpha u)J_n(\beta u) \tag{5}$$

Integrating (5) with respect to u from 0 to z we find

$$\int_0^z uJ_n(\alpha u)J_n(\beta u)\,du = \left.\frac{u[J_n(\beta u)J_n'(\alpha u)-J_n(\alpha u)J_n'(\beta u)]}{\beta^2-\alpha^2}\right|_{u=0}^z$$
$$= \frac{z[\alpha J_n(\beta z)J_n'(\alpha z)-\beta J_n(\alpha z)J_n'(\beta z)]}{\beta^2-\alpha^2} \tag{6}$$

LEGENDRE POLYNOMIALS

● **PROBLEM** 24-54

Legendre's differential equation of order n is given by

$$(1 - z^2)w'' - 2zw' + n(n+1)w = 0 \tag{1}$$

If n is zero or a positive integer, the solutions of (1) are polynomials of degree n called Legendre polynomials. They are given by Rodrigues' formula

$$P_n(1) = 1$$

$$P_n(z) = \frac{1}{2^n n!}\frac{d^n}{dz^n}(z^2-1)^n \tag{2}$$

The following interesting property of Legendre polynomials can be established with the help of the generating function

$$\frac{1}{\sqrt{w^2-2zw+1}} = \sum_{n=0}^{\infty} P_n(z)w^n \tag{3}$$

Prove that

$$\int_{-1}^1 P_m(z)P_n(z)\,dz = 0 \quad \text{if} \quad m \neq n \tag{4}$$

<u>Solution</u>: Since both $P_m(z)$ and $P_n(z)$ are solutions of (1) we have

$$(1 - z^2)P_m'' - 2zP_m' + m(m+1)P_m = 0 \tag{5}$$

$$(1 - z^2)P_n'' - 2zP_n' + n(n + 1)P_n = 0 \tag{6}$$

Multiplying (5) by P_n and (6) by P_m and subtracting, we obtain

$$\frac{d}{dz}\left[(1 - z^2)(P_n P_m' - P_m P_n')\right]$$
$$= P_m P_n \left[n(n + 1) - m(m + 1)\right] \tag{7}$$

Integrating (7) from -1 to 1, we get

$$(1 - z^2)(P_n P_m' - P_m P_n')\Big|_{-1}^{1} = 0$$
$$= \left[n(n + 1) - m(m + 1)\right] \int_{-1}^{1} P_m(z)P_n(z)dz \tag{8}$$

Since, $m \neq n$

$$\int_{-1}^{1} P_m(z)P_n(z)dz = 0 \tag{9}$$

Hence, the Legendre polynomials form an orthogonal set.

● **PROBLEM 24-55**

Show that

$$\int_{-1}^{1} P_n(z)P_n(z)dz = \frac{2}{2n + 1} \tag{1}$$

Solution: From eq.(3) of Problem 24-54 we obtain

$$\frac{1}{w^2 - 2zw + 1} = \sum_{m=0}^{\infty} \sum_{n=0}^{\infty} P_m(z)P_n(z)w^{m+n} \tag{2}$$

Integrating (2) from -1 to 1 and applying eq.(9) of Problem 24-54 we find

$$\int_{-1}^{1} \frac{dz}{w^2 - 2zw + 1} = \sum_{m=0}^{\infty} \sum_{n=0}^{\infty} \int_{-1}^{1} P_m(z)P_n(z)w^{m+n} dz$$

$$\tag{3}$$

$$= \sum_{n=0}^{\infty} \left[\int_{-1}^{1} P_n(z)P_n(z)\,dz \right] w^{2n}$$

The left side of (3) is equal to

$$-\frac{1}{2w} \ln(w^2 - 2zw + 1)\Big|_{z=-1}^{1} = \frac{1}{w} \ln\left(\frac{1+w}{1-w}\right)$$

$$= \sum_{n=0}^{\infty} \frac{2}{2n+1} w^{2n} \quad (4)$$

Comparing the coefficients of w^{2n} in (3) and (4), we find

$$\int_{-1}^{1} P_n(z)P_n(z)\,dz = \frac{2}{2n+1} \quad (5)$$

• **PROBLEM** 24-56

Prove the recursion formula for Legendre polynomials

$$z(2n+1)P_n(z) = nP_{n-1}(z) + (n+1)P_{n+1}(z) \quad (1)$$

Solution: Differentiating

$$\frac{1}{\sqrt{w^2 - 2zw + 1}} = \sum_{n=0}^{\infty} P_n(z) w^n \quad (2)$$

with respect to w, we find

$$\frac{z-w}{(w^2 - 2zw + 1)^{\frac{3}{2}}} = \sum_{n=0}^{\infty} n\, P_n(z) w^{n-1} \quad (3)$$

Multiplying (3) by $w^2 - 2zw + 1$, we get

$$(z-w) \sum_{n=0}^{\infty} P_n(z) w^n = (w^2 - 2zw + 1) \sum_{n=0}^{\infty} n\, P_n(z) w^{n-1} \quad (4)$$

or

$$\sum_{n=0}^{\infty} P_n(z) zw^n - \sum_{n=0}^{\infty} P_n(z) w^{n+1}$$
$$= \sum_{n=0}^{\infty} n\, P_n(z) w^{n+1} - \sum_{n=0}^{\infty} 2nz\, P_n(z) w^n + \sum_{n=0}^{\infty} nP_n(z) w^{n-1} \quad (5)$$

Comparing the coefficients at w^n, we have

$$z P_n(z) - P_{n-1}(z) = (n-1)P_{n-1}(z) - 2nzP_n(z)$$
$$+ (n+1)P_{n+1}(z) \quad (6)$$

or

$$z(2n+1)P_n(z) = n P_{n-1}(z) + (n+1)P_{n+1}(z) \quad (7)$$

HYPERGEOMETRIC FUNCTION

• **PROBLEM** 24-57

The solution of Gauss' differential equation or the hypergeometric equation

$$z(1-z)w'' + \left[c - z(a+b+1)\right]w' - abw = 0 \quad (1)$$

is the hypergeometric function given by

$$F(a,b,c,z) = 1 + \frac{a \cdot b}{1 \cdot c} z + \frac{a(a+1)b(b+1)}{1 \cdot 2 \cdot c \cdot (c+1)} z^2 + \ldots \quad (2)$$

The series (2) is absolutely convergent for $|z| < 1$ and divergent for $|z| > 1$. For $|z| > 1$ the function is defined by analytic continuation. Show that

$$F\left(\frac{1}{2}, \frac{1}{2}, \frac{3}{2}, z^2\right) = \frac{\arcsin z}{z} \quad (3)$$

Solution: The power series expansion of $\dfrac{\arcsin z}{z}$ is

$$\frac{\arcsin z}{z} = 1 + \frac{1}{2}\frac{z^2}{3} + \frac{1 \cdot 3}{2 \cdot 4}\frac{z^4}{5} + \frac{1 \cdot 3 \cdot 5}{2 \cdot 4 \cdot 6}\frac{z^6}{7} + \ldots \quad (4)$$

for $|z| < 1$

Substituting the appropriate values of a, b, c, and z into (2) we get

$$F\left(\frac{1}{2}, \frac{1}{2}, \frac{3}{2}, z^2\right) = 1 + \frac{\frac{1}{2} \cdot \frac{1}{2}}{1 \cdot \frac{3}{2}} z^2 + \frac{\frac{1}{2} \cdot \frac{3}{2} \cdot \frac{1}{2} \cdot \frac{3}{2}}{1 \cdot 2 \cdot \frac{3}{2} \cdot \frac{5}{2}} z^4$$

$$+ \frac{\frac{1}{2} \cdot \frac{3}{2} \cdot \frac{5}{2} \cdot \frac{1}{2} \cdot \frac{3}{2} \cdot \frac{5}{2}}{1 \cdot 2 \cdot 3 \cdot \frac{3}{2} \cdot \frac{5}{2} \cdot \frac{7}{2}} \cdot z^6 + \ldots \quad (5)$$

$$= 1 + \frac{1}{2} \cdot \frac{z^2}{3} + \frac{1 \cdot 3}{2 \cdot 4}\frac{z^4}{5} + \frac{1 \cdot 3 \cdot 5}{2 \cdot 4 \cdot 6}\frac{z^6}{7} + \ldots$$

$$= \frac{\arcsin z}{z}$$

● **PROBLEM** 24-58

Prove that

$$\ln(1 + z) = z\, F(1,1,2,-z) \tag{1}$$

Solution: The left side of eq.(1) can be expressed by the power series

$$\ln(1 + z) = z - \frac{z^2}{2} + \frac{z^3}{3} - \frac{z^4}{4} + \ldots + (-1)^{n-1}\frac{z^n}{n} + \ldots \tag{2}$$

Substituting the appropriate numbers for the hypergeometric function and multiplying by z we find

$$z\, F(1,1,2,-z) = z\left[1 + \frac{1 \cdot 1}{1 \cdot 2}(-z) + \frac{1 \cdot 2 \cdot 1 \cdot 2}{1 \cdot 2 \cdot 2 \cdot 3}(-z)^2 \right.$$
$$\left. + \frac{1 \cdot 2 \cdot 3 \cdot 1 \cdot 2 \cdot 3}{1 \cdot 2 \cdot 3 \cdot 2 \cdot 3 \cdot 4}(-z)^3 + \ldots \right] \tag{3}$$

$$= z - \frac{z^2}{2} + \frac{z^3}{3} - \frac{z^4}{4} + \ldots$$

Comparing (2) and (3) we obtain (1).

ZETA FUNCTION

● **PROBLEM** 24-59

The zeta function defined by

$$\zeta(z) = \sum_{n=1}^{\infty} \frac{1}{n^z} \tag{1}$$

plays a prominent part in the analytic theory of numbers. Series (1) converges uniformly in any closed subdomain of the half-plane Re z > 1. Find the analytic continuation of $\zeta(z)$ to points outside this half-plane.

Solution: Since

$$\int_0^{\infty} t^{z-1} e^{-nt}\, dt = \frac{1}{n^z} \int_0^{\infty} t^{z-1} e^{-t}\, dt = \frac{\Gamma(z)}{n^z} \tag{2}$$

we have

$$\Gamma(z) \sum_{n=1}^{m} \frac{1}{n^z} = \int_0^\infty t^{z-1} \left(\sum_{n=1}^{m} e^{-nt} \right) dt$$

$$= \int_0^\infty t^{z-1} e^{-t} \left(\frac{1 - e^{-mt}}{1 - e^{-t}} \right) dt \qquad (3)$$

We shall show that

$$P = \int_0^\infty \frac{t^{z-1} e^{-mt}}{1 - e^{-t}} e^{-t} \, dt \qquad (4)$$

tends to zero as $m \to \infty$.

Since

$$\frac{e^{-t}}{1 - e^{-t}} = \frac{1}{e^t - 1} \leq \frac{1}{t} \qquad (5)$$

and

$$\left| t^{z-1} \right| = t^{x-1} \qquad (6)$$

we have

$$|P| \leq \int_0^\infty t^{x-2} e^{-mt} \, dt = \frac{1}{m^{x-1}} \int_0^\infty t^{x-2} e^{-t} \, dt$$

$$= \frac{\Gamma(x-1)}{m^{x-1}} \qquad (7)$$

Since Re $z = x > 1$

$$\lim_{m \to \infty} \frac{\Gamma(x-1)}{m^{x-1}} = 0 \qquad (8)$$

Taking the limit of (3) as $m \to \infty$ we get

$$\zeta(z) = \frac{1}{\Gamma(z)} \int_0^\infty \frac{t^{z-1}}{e^t - 1} \, dt \qquad (9)$$

● **PROBLEM 24-60**

Show that the zeta function

$$\zeta(z) = \sum_{n=1}^{\infty} \frac{1}{n^z} \qquad (1)$$

is analytic in the domain Re z > 1.

Solution: Note that each term $\frac{1}{n^z}$ of (1) is an analytic function. Let ε be any fixed positive number. Then, if

$$x = \text{Re } z \geq 1 + \varepsilon \tag{2}$$

we have

$$\left|\frac{1}{n^z}\right| = \left|\frac{1}{e^{z \ln n}}\right| = \frac{1}{e^{x \ln n}} = \frac{1}{n^x} \leq \frac{1}{n^{1+\varepsilon}} \tag{3}$$

Because $\sum_{n=1}^{\infty} \frac{1}{n^{1+\varepsilon}}$ converges, by the Weierstrass M we conclude that $\sum_{n=1}^{\infty} \frac{1}{n^z}$ converges uniformly for Re z > 1.

Applying the theorem: if $f_n(z)$; n = 0,1,2,... are analytic and $\Sigma f_n(z)$ is uniformly convergent in D; then $\Sigma f_n(z)$ is analytic in D; we conclude that $\zeta(z)$ is analytic for Re z > 1.

DOUBLY PERIODIC FUNCTIONS, ELLIPTIC FINCTIONS

• **PROBLEM 24-61**

Prove that

$$\text{sn}(-z) = -\text{sn } z \tag{1}$$

$$\text{cn}(-z) = \text{cn } z \tag{2}$$

$$\text{dn}(-z) = \text{dn } z \tag{3}$$

Solution: The elliptic integral of the first kind is defined by

$$z = \int_0^w \frac{dt}{\sqrt{(1-t^2)(1-k^2 t^2)}}, \quad |k| < 1 \tag{4}$$

Eq.(4) is sometimes written briefly as

$$z = \text{sn}^{-1} w \tag{5}$$

That leads to the definition of an elliptic function

$$w = \operatorname{sn} z \tag{6}$$

Replacing in (4), t by -t we obtain

$$z = \int_0^{-w} \frac{-dt}{\sqrt{(1-t^2)(1-k^2t^2)}} \tag{7}$$

That is

$$-z = \operatorname{sn}^{-1}(-w) \tag{8}$$

or

$$\operatorname{sn}(-z) = -w = -\operatorname{sn} z \tag{9}$$

Relating the elliptic function, $w = \operatorname{sn} z$, with the trigonometric functions, we define

$$\operatorname{cn} z = \sqrt{1 - \operatorname{sn}^2 z} \tag{10}$$

and

$$\operatorname{tn} z = \frac{\operatorname{sn} z}{\operatorname{cn} z} \tag{11}$$

$$\operatorname{dn} z = \sqrt{1 - k^2 \operatorname{sn}^2 z} \tag{12}$$

We have

$$\operatorname{cn}(-z) = \sqrt{1 - \sin^2(-z)} = \sqrt{1 - \operatorname{sn}^2 z} = \operatorname{cn} z \tag{13}$$

$$\operatorname{dn}(-z) = \sqrt{1 - k^2 \operatorname{sn}^2(-z)} = \sqrt{1 - k^2 \operatorname{sn}^2 z} = \operatorname{dn} z \tag{14}$$

● **PROBLEM 24-62**

Prove that

$$\frac{d}{dz} \operatorname{sn} z = \operatorname{cn} z \operatorname{dn} z \tag{1}$$

$$\frac{d}{dz} \operatorname{cn} z = -\operatorname{sn} z \operatorname{dn} z \tag{2}$$

<u>Solution</u>: If

$$z = \int_0^w \frac{dt}{\sqrt{(1-t^2)(1-k^2t^2)}} \tag{3}$$

then

$$w = \operatorname{sn} z \tag{4}$$

Thus,

$$\frac{d}{dz} \operatorname{sn} z = \frac{dw}{dz} = \frac{1}{\frac{dz}{dw}} = \sqrt{(1-w^2)(1-k^2w^2)} \qquad (5)$$

$$= \operatorname{cn} z \, \operatorname{dn} z$$

$$\frac{d}{dz} \operatorname{cn} z = \frac{d}{dz} \sqrt{1 - \operatorname{sn}^2 z} = \frac{1}{2\sqrt{1 - \operatorname{sn}^2 z}} \frac{d}{dz}(-\operatorname{sn}^2 z) \qquad (6)$$

$$= \frac{1}{2\sqrt{1 - \operatorname{sn}^2 z}} (-2\operatorname{sn} z)\frac{d}{dz} \operatorname{sn} z$$

$$= \frac{-\operatorname{cn} z \, \operatorname{dn} z \, \operatorname{sn} z}{\operatorname{cn} z} = -\operatorname{sn} z \, \operatorname{dn} z$$

● **PROBLEM 24-63**

Prove that
$$\operatorname{sn}(z + 2P) = -\operatorname{sn} z \qquad (1)$$
$$\operatorname{cn}(z + 2P) = -\operatorname{cn} z \qquad (2)$$

Solution: By definition

$$z = \int_0^w \frac{dt}{\sqrt{(1-t^2)(1-k^2t^2)}}, \quad |k| < 1 \qquad (3)$$

$$w = \operatorname{sn} z \qquad (4)$$

Substituting $t = \sin\theta$ and $w = \sin\phi$ we can transform (3) to

$$z = \int_0^\phi \frac{d\theta}{\sqrt{1 - k^2\sin^2\theta}} \qquad (5)$$

which is often written in the form
$$\phi = \operatorname{am} z \qquad (6)$$

We have
$$w = \sin\phi = \operatorname{sn} z \qquad (7)$$
and
$$\cos\phi = \operatorname{cn} z \qquad (8)$$

Thus,

$$\int_0^{\phi+\pi} \frac{d\theta}{\sqrt{1-k^2\sin^2\theta}} = \int_0^\pi \frac{d\theta}{\sqrt{1-k^2\sin^2\theta}} + \int_\pi^{\phi+\pi} \frac{d\theta}{\sqrt{1-k^2\sin^2\theta}}$$

$$= 2 \int_0^{\frac{\pi}{2}} \frac{d\theta}{\sqrt{1-k^2\sin^2\theta}} + \int_0^{\phi} \frac{d\lambda}{\sqrt{1-k^2\sin^2\lambda}} \qquad (9)$$

where

$$\lambda = \theta - \pi$$

Therefore, denoting

$$P = \int_0^{\frac{\pi}{2}} \frac{d\theta}{\sqrt{1-k^2\sin^2\theta}} \qquad (10)$$

we get

$$\phi + \pi = \text{am}(z + 2P) \qquad (11)$$

and

$$\text{sn}(z + 2P) = \sin\left[\text{am}(z + 2P)\right] = \sin(\phi + \pi)$$
$$= -\sin\phi = -\text{sn } z \qquad (12)$$

$$\text{cn}(z + 2P) = \cos\left[\text{am}(z + 2P)\right] = \cos(\phi + \pi)$$
$$= -\cos\phi = -\text{cn } z \qquad (13)$$

• **PROBLEM 24-64**

Show that the functions sn z and cn z are periodic functions with period 4P.

Show that dn z has period 2P.

Solution: A function $f(z)$ defined in the complex plane is periodic with period p if

$$f(z + p) = f(z) \quad \text{for every } p.$$

Applying eq.(1) of Problem 24-63 we obtain

$$\text{sn}(z + 4P) = \text{sn}(z + 2P + 2P) = -\text{sn}(z + 2P) = \text{sn } z \qquad (1)$$

$$\text{cn}(z + 4P) = -\text{cn}(z + 2P) = \text{cn } z \qquad (2)$$

$$\text{dn}(z + 2P) = \sqrt{1 - k^2\text{sn}^2(z + 2P)} = \sqrt{1 - k^2\text{sn}^2 z} = \text{dn } z \qquad (3)$$

● **PROBLEM 24-65**

Let us denote
$$k' = \sqrt{1 - k^2} \tag{1}$$

$$P' = \int_0^1 \frac{dt}{\sqrt{(1-t^2)(1-k'^2 t^2)}} \tag{2}$$

Prove that
$$\operatorname{sn}(P + iP') = \frac{1}{k} \tag{3}$$

$$\operatorname{cn}(P + iP') = -\frac{ik'}{k} \tag{4}$$

Solution: We denote
$$s = \frac{1}{\sqrt{1 - k'^2 t^2}} \tag{5}$$

Then, when $t = 0$, $s = 1$, and when $t = 1$, $s = \frac{1}{k}$.

$$\sqrt{1 - t^2} = \frac{-ik's}{\sqrt{1 - k'^2 s^2}} \tag{6}$$

Then,
$$P' = -i \int_1^{\frac{1}{k}} \frac{ds}{\sqrt{(1-s^2)(1-k^2 s^2)}} \tag{7}$$

and we obtain
$$P + iP' = \int_0^1 \frac{ds}{\sqrt{(1-s^2)(1-k^2 s^2)}} + \int_1^{\frac{1}{k}} \frac{ds}{\sqrt{(1-s^2)(1-k^2 s^2)}} \tag{8}$$

$$= \int_0^{\frac{1}{k}} \frac{ds}{\sqrt{(1-s^2)(1-k^2 s^2)}}$$

Therefore,
$$\operatorname{sn}(P + iP') = \frac{1}{k} \tag{9}$$

and
$$\operatorname{cn}(P + iP') = \sqrt{1 - \operatorname{sn}^2(P + iP')}$$

$$= \sqrt{1 - \frac{1}{k^2}} = \frac{-i\sqrt{1-k^2}}{k} = \frac{-ik'}{k} \qquad (10)$$

From (3) we also obtain

$$dn(P + iP') = \sqrt{1 - k^2 sn^2(P + iP')} = \sqrt{1 - k^2 \cdot \frac{1}{k^2}} = 0 \qquad (11)$$

● **PROBLEM** 24-66

Prove that

$$sn(z_1+z_2) = \frac{sn z_1 cn z_2 dn z_2 + sn z_2 cn z_1 dn z_1}{1 - k^2 sn^2 z_1 sn^2 z_2} \qquad (1)$$

Solution: Let us denote

$$z_1 + z_2 = a \qquad (2)$$

where a is an arbitrary constant,

$$A = sn\ z_1$$
$$B = sn\ z_2 \qquad (3)$$

Then,

$$\frac{dz_2}{dz_1} = -1 \qquad (4)$$

and

$$\frac{dA}{dz_1} = A' = cn\ z_1\ dn\ z_1 \qquad (5)$$

$$\frac{dB}{dz_1} = B' = \frac{dB}{dz_2}\frac{dz_2}{dz_1} = -cn\ z_2\ dn\ z_2 \qquad (6)$$

Then,

$$A'^2 = (1 - A^2)(1 - k^2 A^2) \qquad (7)$$

$$B'^2 = (1 - B^2)(1 - k^2 B^2) \qquad (8)$$

Differentiating with respect to z_1 we get

$$A'' = 2k^2 A^3 - (1 + k^2)A \qquad (9)$$

$$B'' = 2k^2 B^3 - (1 + k^2)B \qquad (10)$$

Multiplying (9) by B and (10) by A and subtracting we find

$$A''B - B''A = 2k^2 AB(A^2 - B^2) \qquad (11)$$

and

$$A'B - AB' = \frac{(1-k^2 A^2 B^2)(B^2 - A^2)}{A'B + AB'} \qquad (12)$$

886

Dividing (11) by (12) we get

$$\frac{A''B - B''A}{A'B - AB'} = \frac{-2k^2 AB(A'B + AB')}{1 - k^2 A^2 B^2} \tag{13}$$

or

$$\frac{(A'B - AB')'}{A'B - AB'} = \frac{(1 - k^2 A^2 B^2)'}{1 - k^2 A^2 B^2} \tag{14}$$

Integrating (14) we find

$$\frac{A'B - AB'}{1 - k^2 A^2 B^2} = \alpha_1 = \text{const} \tag{15}$$

Thus,

$$\frac{sn z_1 cn z_2 dn z_2 + cn z_1 sn z_2 dn z_1}{1 - k^2 sn^2 z_1 sn^2 z_2} = \alpha_1 \tag{16}$$

Since $z_1 + z_2 = a$, both constants must be related by

$$\alpha_1 = F(z_1 + z_2) \tag{17}$$

Setting $z_2 = 0$, we have $F(z_1) = sn\ z_1$. Thus

$$F(z_1 + z_2) = sn(z_1 + z_2) \tag{18}$$

From (16), (17), and (18) we obtain (1).

● **PROBLEM** 24-67

Prove that

$$sn(2P + 2iP') = 0 \tag{1}$$

$$cn(2P + 2iP') = 1 \tag{2}$$

$$dn(2P + 2iP') = -1 \tag{3}$$

Solution: In formula (1) of Problem 24-66 we substitute

$$z_1 = z_2 = P + iP' \tag{4}$$

Then, because $dn(P + iP') = 0$

$$sn(2P + 2iP') = \frac{2 sn(P + iP') cn(P + iP') dn(P + iP')}{1 - k^2 sn^4(P + iP')} = 0$$

$$cn(2P + 2iP') = \sqrt{1 - sn^2(2P + 2iP')} = 1 \tag{5}$$

$$dn(2P + 2iP') = \frac{dn^2(P + iP') - k^2 sn^2(P + iP') cn^2(P + iP')}{1 - k^2 sn^4(P + iP')}$$

$$= \frac{k^2 \cdot \frac{1}{k^2} \cdot \frac{k'^2}{k^2}}{1 - \frac{k^2}{k^4}} = \frac{k'^2}{k^2 - 1} = -1 \qquad (6)$$

• **PROBLEM** 24-68

Prove that

a) $\operatorname{sn}(z + 2iP') = \operatorname{sn} z$ (1)

b) $\operatorname{cn}(z + 2P + 2iP') = \operatorname{cn} z$ (2)

c) $\operatorname{dn}(z + 4iP') = \operatorname{dn} z$ (3)

Solution: a)

$$\operatorname{sn}(z + 2iP') = \operatorname{sn}(z - 2P + 2P + 2iP')$$

$$= \frac{\operatorname{sn}(z-2P)\operatorname{cn}(2P+2iP')\operatorname{dn}(2P+2iP') + \operatorname{sn}(2P+2iP')\operatorname{cn}(2-2P)\operatorname{dn}(2-2P)}{1 - k^2 \operatorname{sn}^2(z-2P)\operatorname{sn}^2(2P+2iP')}$$

$$= -\operatorname{sn}(z - 2P) = \operatorname{sn} z \qquad (4)$$

b)
$$\operatorname{cn}(z+2P+2iP') = \frac{\operatorname{cn} z \operatorname{cn}(2P+2iP') - \operatorname{sn} z \operatorname{sn}(2P+2iP') \cdot \ldots}{1 - k^2 \operatorname{sn}^2 z \operatorname{sn}^2(2P+2iP')}$$

$$= \operatorname{cn} z \cdot \operatorname{cn}(2P + 2iP') = \operatorname{cn} z \qquad (5)$$

c) $\operatorname{dn}(z + 4iP') = \operatorname{dn}(z - 4P + 4P + 4iP')$

$$= \frac{\operatorname{dn}(z-4P)\operatorname{dn}(4P+4iP') - k^2 \operatorname{sn}(z-4P)\operatorname{sn}(4P+4iP')\operatorname{cn}(z-4P)\operatorname{cn}(4P+4iP')}{1 - k^2 \operatorname{sn}^2(z-4P)\operatorname{sn}^2(4P+4iP')}$$

$$= \operatorname{dn} z \qquad (6)$$

• **PROBLEM** 24-69

Consider a doubly periodic function f(z) with a pair of primitive periods w and w'. Then a network of points nw + n'w' can be established in the complex plane.

The entire plane is divided into an arrangement of parallelograms (see Fig. 1) whose lattice points are the period-points nw + n'w', where n and n' are integers. Then, the doubly

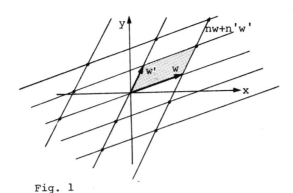

Fig. 1

periodic function assumes the same value or exhibits the same singularity, at congruent points. That is, once we know everything at z_0 we know

$$z_0+w,\ z_0+w',\ z_0+2w+w',\ z_0+2w+2w',\ldots,z_0+5w+7w',\ldots$$

It is sufficient to study the doubly periodic function in the basic period-parallelogram with the vertices

$$0, w, w', w+w'.$$

The first Liouville theorem states that:

Theorem

There exists no non-constant doubly periodic entire function.

Based on the results of the preceding problems establish the period-parallelograms of the following functions

$$\text{sn } z, \quad \text{cn } z, \quad \text{dn } z.$$

Solution:

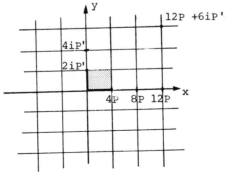

Fig. 2 Period-parallelogram for sn z

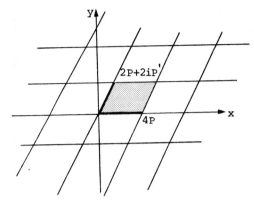

Fig. 3 Period-parallelogram for cn z

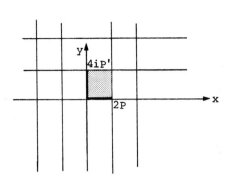

Fig. 4 Period-parallelogram for dn z

ASYMPTOTIC EXPANSIONS

• **PROBLEM 24-70**

Show that the asymptotic expansion of $\Gamma(z + 1)$ is

$$\Gamma(z + 1) \sim \sqrt{2\pi z}\, z^z\, e^{-z}\left[1 + \frac{1}{12z} + \frac{1}{288z^2} + \ldots\right] \quad (1)$$

Solution: Let

$$\ln t - t = -1 - \frac{(t-1)^2}{2} + \frac{(t-1)^3}{3} - \frac{(t-1)^4}{4} + \ldots$$
$$= -1 - v^2 \quad (2)$$

Then

$$v^2 = \frac{(t-1)^2}{2} - \frac{(t-1)^3}{3} + \frac{(t-1)^4}{4} - \ldots \quad (3)$$

and

$$\frac{dt}{dv} = a_0 + a_1 v + a_2 v^2 + \ldots = \sqrt{2} + \frac{\sqrt{2}}{6} v^2 + \frac{\sqrt{2}}{216} v^4 + \ldots \quad (4)$$

Therefore from the formula for the asymptotic expansion we get

$$\Gamma(z+1) \sim \sqrt{\frac{\pi}{z}}\, z^{z+1}\, e^{z(\ln 1 - 1)}\left[\sqrt{2} + \frac{1}{2}\frac{\sqrt{2}}{6}\frac{1}{z} + \frac{1\cdot 3}{2\cdot 2}\frac{\sqrt{2}}{216}\frac{1}{z^2} + \ldots\right] \quad (5)$$

$$= \sqrt{2\pi z}\, z^z\, e^{-z}\left[1 + \frac{1}{12z} + \frac{1}{288z^2} + \cdots\right]$$

For most practical purposes it is enough to use the first term of the expansion

$$\Gamma(z + 1) \sim \sqrt{2\pi z}\, z^z\, e^{-z} \tag{6}$$

● **PROBLEM 24-71**

Estimate the value of 40!

Solution: The Stirling's asymptotic formula for the gamma function is

$$\Gamma(z + 1) \sim \sqrt{2\pi z}\, z^z\, e^{-z} + \cdots \tag{1}$$

Then for positive integers

$$\Gamma(n + 1) = n! \tag{2}$$

and

$$n! \sim \sqrt{2\pi n}\, n^n\, e^{-n} \tag{3}$$

Substituting n = 40 we find

$$\sqrt{2\pi \cdot 40}\, 40^{40}\, e^{-40} \approx 15.8533 \cdot 1.2089 \cdot 10^{64}$$
$$\cdot 4.2484 \cdot 10^{-18} \approx 8.1592 \cdot 10^{47} \tag{4}$$

● **PROBLEM 24-72**

Show that a Bessel function $J_n(z)$ of the first kind has asymptotic expansion

$$J_n(z) \sim \sqrt{\frac{2}{\pi z}} \cos\left(z - \frac{n\pi}{2} - \frac{\pi}{4}\right) \tag{1}$$

Solution: We have

$$J_n(z) = \frac{1}{\pi}\int_0^\pi \cos(nt - z\sin t)\,dt = \operatorname{Re}\left[\frac{1}{\pi}\int_0^\pi e^{-int}e^{iz\sin t}\,dt\right] \tag{2}$$

Then defining

$$F(t) = i\sin t, \text{ we have}$$

$$F'(t) = i\cos t$$

where $F'\left(\frac{\pi}{2}\right) = 0$.

Let $t = \frac{\pi}{2} + u$, then

$$\frac{1}{\pi}\int_{-\frac{\pi}{2}}^{\frac{\pi}{2}} e^{-in(\frac{\pi}{2}+u)}\, e^{izsin(\frac{\pi}{2}+u)}\, du$$

$$= \frac{e^{-in\frac{\pi}{2}}}{\pi} \int_{-\frac{\pi}{2}}^{\frac{\pi}{2}} e^{-inu}\, e^{izcosu}\, du \tag{3}$$

$$= \frac{e^{-in\frac{\pi}{2}}}{\pi} \int_{-\frac{\pi}{2}}^{\frac{\pi}{2}} e^{-inu}\, e^{iz(1-\frac{u^2}{2}+\ldots)}\, du$$

$$= \frac{e^{i(z-n\frac{\pi}{2})}}{\pi} \int_{-\frac{\pi}{2}}^{\frac{\pi}{2}} e^{-inu}\, e^{\frac{-izu^2}{2}+\ldots}\, du$$

Substituting

$$u = \frac{(1-i)v}{\sqrt{z}} \tag{4}$$

we obtain

$$v = \frac{1}{2}(1+i)\sqrt{z}\, u \tag{5}$$

and the last integral becomes

$$\frac{(1-i)e^{i(z-n\frac{\pi}{2})}}{\pi\sqrt{z}} \int_{-\infty}^{\infty} e^{-\frac{(1+i)nv}{\sqrt{z}}}\, e^{-v^2 - \frac{iv^4}{6z} - \ldots}\, dv \tag{6}$$

For large positive values of z

$$\frac{(1-i)e^{i(z-\frac{n\pi}{2})}}{\pi\sqrt{z}} \int_{-\infty}^{\infty} e^{-v^2}\, dv = \frac{(1-i)e^{i(z-\frac{n\pi}{2})}}{\sqrt{\pi z}} \tag{7}$$

Taking the real part, we obtain

$$\frac{1}{\sqrt{\pi z}}\left[\cos\left(z-\frac{n\pi}{2}\right)+\sin\left(z-\frac{n\pi}{2}\right)\right]$$

$$= \sqrt{\frac{2}{\pi z}}\cos\left[z-\frac{n\pi}{2}-\frac{\pi}{4}\right] \tag{8}$$

APPENDIX

TABLE OF TRANSFORMATIONS OF REGIONS

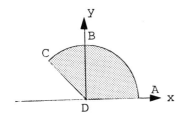
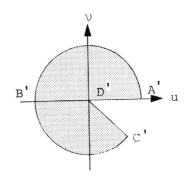

Fig 1. $\omega = z^2$

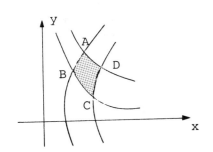
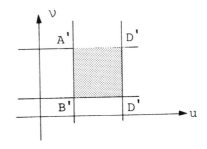

Fig 2. $\omega = z^2$

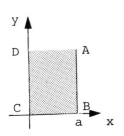
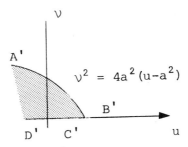

Fig 3. $\omega = z^2$

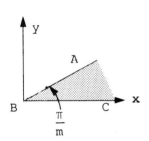 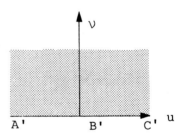

Fig 4. $\omega = z^m$, $m \geq \frac{1}{2}$

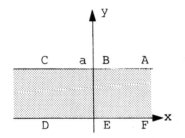 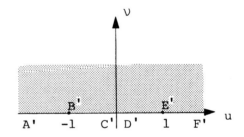

Fig. 5 $\omega = \exp(\pi z/a)$

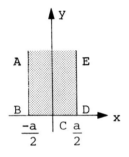 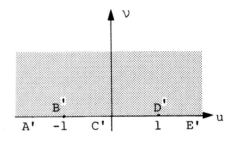

Fig 6. $\omega = \sin \frac{\pi z}{a}$

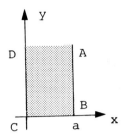 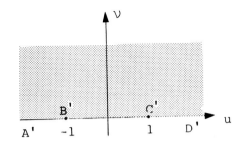

Fig 7. $\omega = \cos \dfrac{\pi z}{a}$

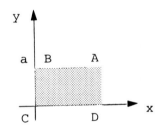 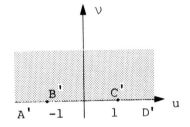

Fig 8. $\omega = \cosh \dfrac{\pi z}{a}$

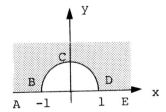 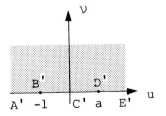

Fig. 9 $\omega = \dfrac{a}{2}(z+\dfrac{1}{z})$

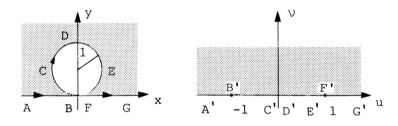

Fig 10 $\omega = \coth(\frac{\pi}{z})$

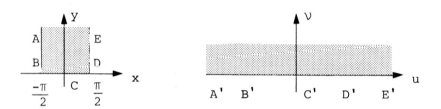

Fig 11 $\omega = \sin z$

Fig 12 $\omega = \sin z$

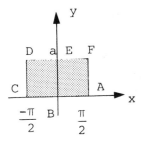 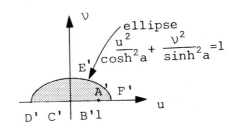

Fig 13 $\omega = \sin z$

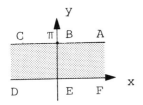 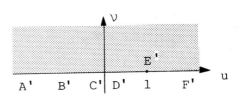

Fig 14 $\omega = e^z$

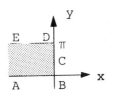 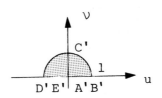

Fig 15 $\omega = e^z$

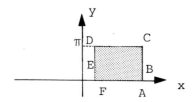 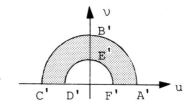

Fig 16 $\omega = e^z$

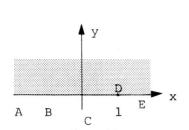 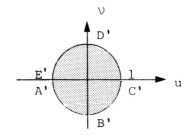

Fig 17 $\omega = \dfrac{i-z}{i+z}$

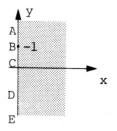 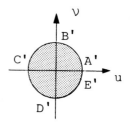

Fig 18 $\omega = \dfrac{z-1}{z+1}$

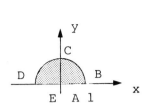 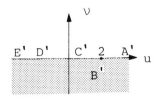

Fig 19 $\omega = z + \dfrac{1}{z}$

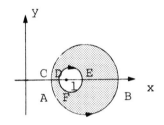 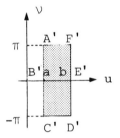

Fig 20 $\omega = \text{Log}\,\dfrac{z+1}{z-1}$

Centers of circles are at $z = \coth a$ and $z = \coth b$, corresponding radii are $r = \operatorname{csch} a$ and $r = \operatorname{csch} b$.

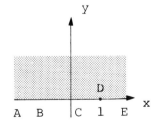 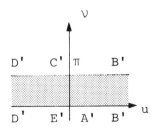

Fig 21 $\omega = \text{Log}\,\dfrac{z-1}{z+1}$

INDEX

Numbers on this page refer to PROBLEM NUMBERS, not page numbers

Abel, Niels, 6-22
Abelian function, 6-22
Abel's theorem, 6-22, 13-13
   application, 13-14
   converse, 13-15
Absolute value, 1-25
Accumulation point, 4-11, 16-36
Airy's stress function, 22-19
Algebra:
   fundamental theorem, 12-14, 21-17
   number, 1-36
   theory, 1-36
Alternating series test, 5-25
Analytic continuation, 15-25, to 15-27, 24-1 to 24-10
Analytic function, 14-27, 16-4, 16-6, 16-11 to 16-14, 16-22, 18-11, 19-6, 19-32 to 19-36, 21-14, 22-1
   at infinity, 8-27
   definition, 8-12, 9-16
   derivative, 8-20
   expansion, 14-1, 14-2, 14-4
   necessary condition, 8-13
   power series, 14-1 to 14-4
   properties, 8-16 to 8-18, 8-21 to 8-26, 9-7, 9-8
   sufficient condition, 8-15
   theorem, 9-16, 21-2
Analytic mappings, 21-8, 21-15, 21-16
Annalen, 1-3
Antecedent, 6-1
Application of:
   Abel's theorem, 13-14
   argument principle, 21-16
   Cauchy-Hadamard theorem, 13-8 to 13-12
   Cauchy-Riemann equation, 8-14

Cauchy's integral formula, 12-6 to 12-11
Cauchy's theorem, 19-16
chain rule, 18-8
complex function, 7-5, 7-9, 7-14
complex numbers, 2-4 to 2-26
Liouville's theorem 21-15
maximum principle, 21-32
Mittog-Leffler theorem, 19-32 to 19-38
Poisson's formula, 21-39 to 21-46
Residue theorem, 17-19 to 17-29, 18-6, 18-7, 19-1, 19-2, 19-18, 19-22, 19-27, 19-32
Rouche's theorem, 21-17, 21-19
Schwarz-Christoffel transformation, 23-2, 23-5 to 23-11, 23-20
Taylor series, 21-11, 21-12
Applications in physics:
   complex potential, 23-25
   electrostatic potential, 23-26
Argand diagram, 2-1
Argand, Jean R., 2-1
Argument principle, 21-16
Associative laws, 1-16
Asymptotic expansions, 24-70, 24-71
   of Bessel function, 24-72
Auxiliary complex plane, 20-25
Axiomatic theory, 1-3
Axis of doublet, 22-37
Ball, 4-5, 4-13
Bernoulli's law, 22-40, 22-45, 22-47
Bernoulli's numbers, 15-17
Bessel functions, 15-16, 24-50, 24-52, 24-53

Numbers on this page refer to PROBLEM NUMBERS, not page numbers

asymptotic expansion, 24-72
generating function, 24-49
of first kind, 24-49
recursion formula, 24-51
Beta function:
   definition, 24-26
   example, 24-27
   functional identities, 24-28 to 24-30
   problems on, 24-31 to 24-33
Biharmonic equation, 22-19
Bilinear transformation, 20-15
Binomial coefficient, 14-18
Blasius theorem, 22-45
Bolzano-Weierstrass theorem, 4-26
Borel Lebesque theorem, 4-31
Boundary:
   natural, 24-1
   point, 4-18
   value problem, 21-43, 21-47, 22-1, 22-4, 22-14
Bounded sequence, 4-23, 5-1, 5-4
Bounded set, 4-5, 4-28
Branch cut, 20-26
Branch lines, 19-7, 19-16
Branch point, 19-7, 20-26
Brouwer fixed-point theorem, 21-19, 21-20

Cantor, George, 1-3
Capacitance, 22-73
Cauchy convergence criterion:
   for complex series, 13-1
   for real series, 13-1
Cauchy-Goursat theorem, 11-14 to 11-16
Cauchy-Hadamard theorem:
   application, 13-8 to 13-12
   definition, 13-7
Cauchy-Riemann equations, 11-22, 11-27, 21-21, 22-11, 22-23, 22-51
   application, 8-14
   necessary condition, 8-14
   polar form, 8-19
   rectangular form, 8-13
Cauchy sequence, 1-3, 4-22, 4-25, 4-30
Cauchy's inequality, 1-29, 1-30, 12-9
Cauchy's integral formula, 12-1, 12-2, 14-3, 17-17
   applications, 12-6 to 12-11
   for derivatives, 12-4, 12-5
   multiply-connected domain, 12-3
   simply-connected domain, 12-3
Cauchy's integral theorem, 11-14, 16-22, 17-16, 17-17
Cauchy's principal value, 19-1
Cauchy's test, 5-34 to 5-36
Cauchy's theorem, 11-15, 11-20 to 11-26, 19-8, 19-16
   application, 19-16
Center of mass, 2-26, 2-27
Chain rule:
   application, 8-8
   definition, 8-7
   generalization, 8-9
Chebyshev polynomials, 3-52
Circle, 2-4
   equation, 2-4, 2-23
   exterior point, 2-25
   inequality, 2-25
   interior point, 2-25
   radius, 2-25
   transformation, 2-23
   unit, 2-4, 2-14, 2-24, 6-25
Circulation, 22-22, 22-31, 22-42
Class, 1-1
Closed polygons, 23-2
Closed sets, 4-13, 4-14, 4-28
Closure:
   of open interval, 4-13
   of a set, 4-16, 4-20
Collinear vectors, 2-7 to 2-9
Commutative laws, 1-16
Compact set, 4-26, 4-28, 4-31
   continuous mapping, 6-12, 6-13
Compact space, 4-30
Complete space, 4-25, 4-30
Complex conjugate, 1-18 to 1-20
   modulus absolute value, 1-21 to 1-30
   polynomials, 1-5, 1-31 to 1-36
   real and imaginary parts, 1-18
Complex constant, 5-17, 5-23
   contraction, 20-1
   expansion, 20-1
   rotation, 20-1
Complex function, 19-3
   applied to limits, 7-5, 7-9, 7-14

Numbers on this page refer to **PROBLEM NUMBERS**, not page numbers

continuity at a point, 7-11, 7-13, 8-2
continuity in a domain, 7-11, 7-12
continuity of absolute value, 7-12
continuity of boundary value, 7-22
derivative at a point, 8-2
derivative, 8-1, 8-6
divergence, 9-5
gradient, 9-4
limits, 7-6, 7-7, 7-10
limits of sequence, 7-21
logarithmic differential, 8-10
multiple-valued, 7-1
properties of derivative, 8-3 to 8-5
single-values, 7-1 to 7-3
test for continuity, 7-15, 7-16
uniform continuity, 7-17 to 7-20
Complex line integrals, 11-1 to 11-8
Complex numbers, 1-5, 1-6
addition and multiplication, 1-7, 1-8
additive neutral element, 1-11 to 1-15
applications in geometry, 1-6, 2-4 to 2-25
applications in physics, 2-26
Bolzano-Weierstrass property, 4-37
diameter, 4-32
diverging sequence, 5-4
dot product, 9-5
equal, 1-6
field, 1-16, 1-17
formulas, 3-33 to 3-35, 3-43, 3-44
fundamental operations, 1-7 to 1-15
geometrical interpretation, 2-1
graphical representation, 3-2
identities, 1-27 to 1-30
imaginary part, 1-18
inequalities, 1-27 to 1-30
isomorphic, 3-54
Lagrange's identity, 1-28
linear algebra, 3-54
matrix representation, 3-54
modulus, 3-5, 3-16
multiplicative neutral element, 1-6

nested intervals, 4-33, 4-34
ordered pairs, 1-17
polar coordinates, 3-1
polynomials, 1-31 to 1-36
product and quotient, 3-4, 3-7
properties of complex space, 4-37
real part, 1-18
representation, 1-5, 2-1 to 2-3
roots, 3-26 to 3-28
roots of unity, 3-31, 3-32
sequence, 4-35, 4-36, 5-8, 5-9, 5-33, 5-34
sequence of intervals, 4-33
series, 5-15, 5-18
set, 1-16
square roots, 3-30
subtraction and division, 1-9 to 1-15
system, 1-6
Complex plane, 2-1, 6-41
auxiliary, 20-25
equation of circle, 2-4
equation of ellipse, 2-4
extended, 6-43, 6-50
motion, 8-29
Complex potential:
applications in physics, 23-25
of flow, 22-23 to 22-26, 22-28, 22-30, 22-32 to 22-40, 22-44
Complex series:
Cauchy convergence criterion, 13-1
Complex variable, 5-17
definition, 7-1
real valued functions, 7-4, 7-5
Complex vector, 2-18
Complex velocity, 22-25, 22-28
Complex zeros, 21-17
Composite function, 14-24
Conformal mappings, 21-5 to 21-7, 22-1, 22-9, 22-44, 23-34
Conformal transformation, 21-4, 21-43, 22-14
Conjugate:
complex, 1-18 to 1-20
Connected set, 4-17, 6-21, 6-28
Connected space, 6-16
Constant:
complex, 5-17, 5-23
dielectric, 22-73
Euler, 1-5
real, 2-23

Numbers on this page refer to **PROBLEM NUMBERS**, not page numbers

Continuous function, 6-6, 21-30
   on a closed bounded set, 7-27, 7-28
   on a compact set, 7-24, 7-25
   on a connected set, 7-23
   on a disjoint compact set, 7-26
Continuous partial derivative, 18-22
Continuous transformation, 20-5
Contours, 19-2, 19-7, 19-9 to 19-11, 19-13, 19-15
   dented, 19-6, 19-7
Contradiction, 6-20
Convergence:
   in a domain, 24-9
   of a sequence, 4-35
Convergent products, 24-37
Convergent sequences, 4-21 to 4-23, 4-35, 5-2, 5-9, 5-11
Convergent series:
   absolute, 13-3
   definition, 5-14
   necessary condition, 5-13, 5-15
   proof of theorem, 5-16
   properties, 13-2, 13-4
Convex set, 6-29
Convolution theorem, 22-17
Covering, K-to-one, 21-10
Critical point, 21-1, 21-10 to 21-13
   theorem, 21-10
Cross product, 3-20, 3-22
Curl of a complex function, 9-5
Curves, 6-24 to 6-28
   arc, 6-27
   compact point, 6-35
   continuous, 6-24, 6-25, 6-28, 6-34 to 6-36
   image, 6-1, 6-2
   Jordan, 6-36, 21-16
   mechanical interpretation, 6-24
   neighborhood, 6-26, 6-31, 6-34
   piecewise smooth, 11-4
   polygonal, 11-6
   rectifiable, 6-27
   smooth, 11-1
   space-filling, 6-24

D'Alembert, Jean-le-Rond, 5-19

D'Alembert's test, 5-35, 5-36
Dedekind, 1-36
Definite integrals, 19-2
Definition of:
   Abel's theorem, 13-13
   analytic function, 8-12, 9-16
   asymptotic expansion, 24-69
   Beta function, 24-26
   Cauchy-Hadamard theorem, 13-7
   chain rule, 8-7
   complex variable, 7-1
   convergent series, 5-14
   elliptic function, 24-61
   exponential function, 10-1
   Gamma function, 24-13
   harmonic function, 9-6, 9-7, 9-16
   hypergeometric function, 24-57
   infinite series, 5-13
   intersection, 4-2
   Legendre polynomial, 15-54
   mapping, 6-4
   series, 5-17, 5-25
   topological space, 6-4
   trigonometric function, 10-4
   union, 4-2
   univalent function, 10-16
   Zeta function, 24-59
Degenerate polygons, 23-5, 23-7 to 23-11, 23-17, 23-20, 23-21
DeMoivre's theorem:
   proof, 3-25
DeMorgan's laws, 4-1
Dented contours, 19-6, 19-7
Derivative:
   continuous partial, 18-22
   inverse function, 10-17, 10-18
   normal, 22-2
Diagonals:
   squared, 2-22
Diameter of a set, 4-32
Dielectric constant, 22-73
Differential equations, 9-13 to 9-15, 24-41 to 24-48
   irregular singular point, 24-42
   ordinary points, 24-42
   power series solution, 24-43 to 24-45
   region of convergence, 24-45
   regular singular points, 24-42
Differential operators, 9-1, 9-2
Dirichlet, 21-27, 21-30, 21-31, 21-32, 21-34, 21-37, 21-38,

Numbers on this page refer to PROBLEM NUMBERS, not page numbers

21-40 to 21-42, 22-1, 22-8,
22-10, 22-71
Discrete metric distance, 4-4
Disk, 2-4, 2-5
Distributive law, 1-16
Divergence of a complex function, 9-5
Diverging series:
application of a theorem, 5-16
Domain, 4-18, 4-19, 6-32, 6-33
convergence, 24-9
finite subfamily, 6-34
Jordan, 6-36
multiply-connected, 11-16
of a function, 15-20, 24-7, 24-8
Dot product, 3-19, 3-20, 3-22
Doubly periodic functions, 24-69
Double pole, 16-24, 17-4 to 17-6, 18-2, 18-18, 19-14, 19-24, 19-26
Double valued function:
branches, 20-28

Einstein, 1-30
Elasticity, 22-19 to 22-21
Electric field, 22-49 to 22-73
complex potential, 22-52
semi-infinite condenser, 22-55
stationary, 22-49
Electrostatic potential:
applications in physics, 23-26
coaxial cylinders, 22-72
infinite cylinder, 22-68
infinite parallel planes, 22-62
long hollow cylinder, 22-58
semi-infinite space, 22-69
Ellipse:
foci, 2-4
major axis, 2-4
Elliptic functions:
definition, 24-61
differentiation, 24-62
identities on, 24-63, 24-65, 24-66, 24-68
periodicity, 24-64
problem on, 24-67
relation with trignometric functions, 24-61
Empty set, 4-14
Entire function, 16-34, 24-41
Equations:
biharmonic, 22-19

Cauchy-Riemann, 11-22, 11-27, 21-21, 22-11, 22-23, 22-51
circle, 2-4, 2-23
ellipse, 2-4
hyperbola, 2-5
Laplace, 9-7, 9-8, 22-1, 22-2
lemniscate, 2-5
line, 7-2
Maxwell's, 22-49
parametric, 11-4
Poisson's, 21-46
polynomial, 12-14, 12-15
roots, 15-23
straight lines, 2-7
Equipotential, 22-25 to 22-28, 22-30, 22-33, 22-51
lines, 22-24, 22-37, 22-38
Equivalence relation, 1-3
Essential singularity, 16-15 to 16-17, 16-23, 16-28, 21-15
Euclidean space, 6-40
Euler, 1-5
constant, 5-23
formula application, 3-6
formula justification, 3-18
Evaluation of integrals, 19-1, 19-3 to 19-5
Expansion:
analytic function, 14-1, 14-2, 14-4
complex constant, 20-1
Lagrange, 15-22, 15-23
Laurent, 16-5, 16-7, 16-8, 16-10 to 16-12, 17-9, 19-6, 19-30, 22-41, 22-44
Maclaurin series, 14-7, 14-8, 14-25, 22-10
series, 19-28
Taylor series, 14-11, 14-12, 16-3, 16-23
Exponential functions:
analytic property, 10-3
definition, 10-1
periodicity, 10-2
properties, 10-1
Extended complex plane, 6-43, 6-50

Family, 1-1
Finite numbers, 5-13
Fixed points, 20-16, 20-17
Flux, 22-22, 22-31
lines, 22-59

Numbers on this page refer to PROBLEM NUMBERS, not page numbers

Formula:
   complex number, 3-33 to 3-35, 3-43, 3-44
   Poisson's, 12-27, 22-10, 22-12
Fourier:
   series, 22-10
   transform, 22-17, 22-20
   transform integral, 18-19
   transform theorem, 18-25, 18-29
Fresnel integral, 19-8
Function:
   Abelian, 6-22
   Airy's stress, 22-19
   analytic, 14-27, 16-4, 16-6, 16-11 to 16-14, 16-22, 18-11, 19-6, 19-32 to 19-36, 21-14, 22-1
   Bessel's, 15-16
   Bessel's generating, 24-49
   complex, 19-3
   complex number, 3-9
   complex variable, 16-25
   composite, 14-24
   continuous, 6-6, 21-30
   continuous on sets, 7-23 to 7-28
   continuously differentiable, 11-9
   domain, 15-20, 24-7, 24-8
   double-valued, 20-26
   doubly periodic, 24-69
   entire, 16-34, 24-41
   even, 14-10, 18-16 to 18-18, 19-5
   exponential, 10-1
   gamma, 24-13
   harmonic, 12-12, 12-28, 21-27 to 21-31, 21-34 to 21-36, 22-19
   hyperbolic, 10-7, 10-8
   hypergeometric, 24-57
   integral, 16-34
   logarithmic, 10-9
   meromorphic, 16-34 to 16-36
   multiple-valued, 19-15, 19-16
   odd, 14-10, 18-16 to 18-18
   principal branch, 20-27
   proper, 19-20
   rational, 16-32, 16-33, 18-1
   real variable, 16-25
   real-valued rational, 18-19
   regular, 12-6
   single-valued, 12-6
   stream, 22-24 to 22-26, 22-34, 22-36
   trigonometric, 10-4
   univalent, 10-16
   zeta, 24-59
Fundamental theorem of algebra, 12-14

Gamma function, 24-14, 24-17 to 24-20, 24-22 to 24-25
   analytic continuation of, 24-16
   definition, 24-13
   infinite products, 24-40
   properties, 24-15, 24-21
   Stirling's asymptotic formula, 24-11
Gauss, 1-5, 1-36
   differential equation, 24-57
   mean value theorem, 12-16
   theorem, 22-56, 22-73
Geometric problems, 3-41, 3-42
   parallelogram, 3-21
   perpendicular lines, 3-11
   quadrilateral, 3-24
   triangle, 3-10, 3-12, 3-13, 3-23
   vectors, 3-3
Geometric series, 14-17, 14-24
Gerhardt, 1-15
Gradient:
   of a complex function, 9-4
   of a real function, 9-2
   operator, 9-2
   property of function, 9-3
Graphical:
   interpretation, 2-3
   sum of numbers, 2-2
Green, G., 11-10
Green's theorem, 11-10 to 11-12, 11-14, 22-50
   application, 21-31
Group, 6-22, 6-23
   Abelian, 6-22

Half-plane, 2-5
   image of, 20-8
Harmonic conjugate, 9-11, 9-12, 21-24 to 21-26, 22-11
Harmonic function, 12-12, 12-28, 21-27, 21-28, 21-29, 21-30, 21-31, 21-34 to 21-36, 22-19
   definition and theorem, 9-16
   definition in polar coordinates, 9-8

Numbers on this page refer to PROBLEM NUMBERS, not page numbers

definition in rectangular coordinates, 9-6, 9-7
 properties, 9-6, 9-9, 9-10
Harmonic series, 5-21
Hausdroff, 1-3
 space, 4-10
Heat flow, 21-32, 22-1 to 22-18,
 complex temperature, 22-7,
  22-11
 isotherms, 22-6
 semi-infinite plate, 22-4
 semi-infinite slab, 22-5
 semi-infinite strip, 22-3
 solid body, 22-2
 steady-state, 22-8, 22-12
 steady-state temperature,
  22-9, 22-10, 22-13 to 22-16
 temperature distribution,
  22-17, 22-18
Homomorphism, 6-7, 6-9, 6-18,
 6-19, 6-25
Horizontal line:
 image of, 20-7
Hyperbola:
 equation, 2-5
 mapping, 20-11, 20-22
Hyperbolic cosine, 14-23
Hyperbolic functions:
 analytic property, 10-7
 definition, 10-7, 10-8
 principal branch, 10-23
 zeros, 10-7
Hypergeometric function, 24-58
 definition, 24-57

i, 1-5
Identity:
 complex numbers, 1-27 to 1-30
 element, 6-22
 elliptic function, 24-63, 24-65
 Lagrange, 1-29
 transformation, 6-22, 6-23
Image, 6-1, 6-2
 curve, 21-9
 traversing, 20-7
Imaginary axis, 2-1
Improper integrals, 18-19, 19-1
Indefinite integrals, 11-27 to
 11-29
Inequality:
 Cauchy's, 1-29, 1-30, 12-9
 circle, 2-25
 complex numbers, 1-27 to 1-30

geometric interpretation, 2-3
Minkowski's, 1-30
triangle, 1-23, 1-30, 2-3, 5-26
Infinite limits, 18-19
Infinite products, 24-34 to 24-39,
 24-41
 for gamma function, 24-40
Infinite series, 5-13
Infinite strip, 20-4,
 image of, 20-9
Integers, 1-1 to 1-4
 positive even, 1-1, 1-2
Integrals:
 complex line, 11-1 to 11-8
 definite, 19-2
 evaluation, 19-1, 19-3 to 19-5
 Fourier transform, 18-19
 Fresnel, 19-8
 function, 16-34
 improper, 18-19, 19-1
 indefinite, 11-27 to 11-29
 independent of path, 11-17
Intersection:
 definition, 4-2
 open sets, 4-9
 sets, 4-14
Integration by parts, 11-29
Interior of intervals, 4-11, 4-12
Intervals, 4-2
Inverse:
 element, 6-22, 6-23
 Fourier transform, 18-19
 image, 6-2
 transformation, 6-22, 6-44
Irrational numbers, 1-1
Irrotational field, 22-23
Isogonal transformation, 21-6
Isolated:
 poles, 16-36
 singularity, 16-2, 16-5, 16-6,
  16-7, 16-12, 16-23
Isometric transformation, 4-3
Isomorphism, 1-6, 1-17

Jacobian of the transformation,
 21-21 to 21-24
Jordan, M., 6-27
Jordan:
 arc, 6-27
 curves, 6-36, 21-16
 domain, 6-36
Joukowski:
 air foils, 21-13

Numbers on this page refer to PROBLEM NUMBERS, not page numbers

theorem, 22-46
transformation, 12-13

K-to-one covering, 21-10
Kummer, 1-36
Kutta-Joukowski theorem, 22-46

Lagrange, 1-36
  expansion, 15-22, 15-23
  identity, 1-29
Lagrange, Joseph, 1-28
Laplace equation, 22-1, 22-2
  polar form, 9-8
  rectangular form, 9-7
Lattice points, 24-69
Laurent expansion, 16-5, 16-7,
  16-8, 16-10 to 16-12, 17-9,
  19-6, 19-30, 22-41, 22-44
Laurent series, 16-9, 16-15,
  16-18, 16-21, 16-26 to 16-28,
  17-1 to 17-3, 17-6, 17-8,
  17-16, 17-18, 17-19, 17-23
Laurent series expansion:
  circle of convergence, 15-1
  domains of convergence, 15-4
  exponential function, 15-3,
  15-13, 15-16
  generalization of Taylor
  series, 15-8
  hyperbolic function, 15-14
  product of polynomials, 15-21
  product of rational functions,
  15-20
  rational function, 15-1, 15-4
  to 15-7, 15-10, 15-11,
  15-15, 15-18, 15-19, 15-25
  Taylor expansion, 15-3
  trigonometric function, 15-9,
  15-12
Laurent theorem, 15-2
Laws:
  associative, 1-16
  Bernoulli's, 22-40, 22-45,
  22-47
  commutative, 1-16
  De Morgan's, 4-1
  distributive, 1-16
Legendre polynomials, 15-24
  definition, 15-54
  problem on, 24-55
  property, 24-54

recursion formula, 24-56
Leibnitz, 1-15
  rule, 12-4, 18-7, 18-21, 18-22
Leipzig, 1-30
Lemniscate, 2-5
  equation, 20-11
  mapping, 20-11
  of Bernoulli, 2-5
L'Hopital's:
  rule, 8-28, 12-19, 14-15,
  17-12, 17-16, 17-25, 19-4
  theorem, 5-24
Limits of:
  complex function, 7-6, 7-7,
  7-10
  sequences, 5-2 to 5-5, 5-9 to
  5-11, 5-29 to 5-34
Lines:
  branch, 19-7, 19-16
  flux, 22-22, 22-31
  polygonal, 6-30 to 6-32
Linear equation, 1-2, 1-3
Linear fractional transformation,
  20-12 to 20-19, 20-23 to 20-25
  circles and lines, 20-13
  mapping planes, 20-14
Linear transformation, 20-1, 20-2
Liouville's theorem, 12-13, 12-14,
  24-69
  application, 21-15
Logarithmic functions, 10-12,
  10-13
  analytic domain, 10-14
  complex constants, 10-12
  definition, 10-9
  principal value, 10-10, 10-11
  property, 10-10, 10-11, 10-15
Logical notation, 4-1
Lower half plane, 18-19
Lower limit of a sequence, 5-29,
  5-30

Maclaurin series, 14-9, 14-16,
  14-24, 17-10
  expansion, 14-7, 14-8, 14-25,
  22-10
  radius of convergence, 14-6,
  14-8
Magnification factor, 21-7, 21-22
Magnitude, 2-2
Mapping, 6-1 to 6-9, 8-7, 8-11
  analytic 21-8, 21-15, 21-16
  bounded, 6-12

Numbers on this page refer to **PROBLEM NUMBERS**, not page numbers

composition, 6-3
conformal, (see conformal mapping)
continuous, 6-4, 6-5, 6-9, 6-19, 6-26, 6-28, 6-42
continuous onto, 6-21
half plane onto semi-infinite strip, 23-21
hyperbola, 20-11, 20-22
inverse, 6-1, 6-3, 6-9, 6-42
lemniscate, 2-5
one-to-one, 6-1, 6-9, 6-20
onto, 6-1, 6-9, 6-20
parabola, 20-20
polygon onto unit circle, 23-12
product, 6-3
real axis onto edges of an infinite strip, 23-17
real axis onto equilateral triangle, 23-14
real axis onto isocles right triangle, 23-18
real axis onto a rectangle, 23-15, 23-16
real axis onto a square, 23-19
real axis onto a triangle, 23-13
Schwarz-Christoffel transformation, 23-1, 23-3
sequential definition, 6-4
topological, 6-7
uniformly continuous, 6-8
unit open disk, 20-23, 20-25
various polygons, 23-22 to 23-24
Maximum modulus theorem, 12-17, 12-18
Maximum principle, 21-32
Maxwell's equation, 22-49
Mean-value property, 21-32
Meromorphic function, 16-34 to 16-36
Metric spaces, 4-3, 4-4, 4-31
Minimum modulus theorem, 12-18
Minkowski, Hermann, 1-30
Minkowski's inequality, 1-30
Mittag-Leffler theorem, 19-35
  applications, 19-32 to 19-38
Mobius, 20-12
Monotonically increasing sequence, 24-34
Morera's theorem, 11-22, 12-8
Multiple point, 6-27
Multiple-valued functions, 19-15, 19-16

branches of, 10-19
branch point, 10-20
principal branch, 10-22
Riemann surface, 10-21
Multiply-connected domains, 11-16
  Cauchy's integral formula, 12-3

Natural boundary, 24-1
Natural number, 1-1
  selection of, 5-7
  sequence of, 5-16
Neighborhood of a point, 4-31, 7-28
Neumann problem, 21-37, 21-41, 21-47, 22-1
Non-critical points, 21-3, 21-7
Non-isolated pole, 16-36
Normal derivative, 22-2
Null sequence, 5-1 to 5-3
  product of, 5-1
  sum of, 5-2
  system of, 5-11
  theorem of, 5-2
Numbers:
  algebraic, 1-36
  arrangement, 5-10
  Bernoulli's, 15-17
  complex, 1-5, 1-6
  finite, 5-13
  irrational, 1-1
  natural, 1-1
  rational, 1-1 to 1-4, 1-33, 1-36
  real, 1-1 to 1-4, 1-14, 1-17, 2-9
  sequence of, 3-53
  transcendental, 1-36
  winding, 21-16

One-to-one correspondence, 1-1, 1-17, 2-1, 5-27, 20-5, 20-12, 20-22 to 20-24, 21-2, 21-8
One-to-one mapping, 6-1, 6-9, 6-20
One-to-one onto, 1-6
Open sets, 4-7 to 4-9, 6-15, 6-19, 6-33
Ordered pairs, 1-17
Orthogonal system, 22-51

Numbers on this page refer to PROBLEM NUMBERS, not page numbers

Parabola, 20-20
Parallelogram, 2-18 to 2-22
   diagonals of, 2-16, 2-21
   theorem of, 2-22
Parametric equation, 11-4
Partial sums, 5-13
Path segment, 6-27
Peano, 6-24
Piecewise smooth curve, 11-4
Point:
   boundary, 4-18
   branch, 19-7, 20-26
   classification, 16-15
   collinear, 2-9
   critical, 21-1, 21-12, 22-2
   fixed, 20-16, 20-17
   image of, 20-1
   lattice, 24-69
   loci of, 2-5
   multiple, 6-27
   neighborhood of, 4-31, 7-28
   non-critical, 21-3, 21-7
   regular, 16-2
   stagnation, 22-43
Poisson's equation, 21-46
Poisson's formula, 12-27, 22-10, 22-12
   application, 21-39 to 21-42
Pole, 16-7 to 16-11, 16-17 to 16-21, 16-34, 18-1, 18-3, 18-4, 18-8
   double, 16-24, 17-4 to 17-6, 18-2, 18-18, 19-14, 19-24, 19-26
   isolated, 16-36
   simple, 16-7, 16-24, 17-4, 17-5, 17-7, 18-6, 18-12, 18-13, 18-19, 18-21, 18-25 to 18-27, 18-29, 18-30, 19-4, 19-5, 19-7, 19-9, 19-11, 19-13, 19-14, 19-18, 19-22, 19-28, 19-33 to 19-36
Polygon:
   close, 23-2
   curve, 11-6
   degenerate, 23-5, 23-7 to 23-11, 23-17, 23-20, 23-21
   line, 6-30 to 6-32
   mapping of, 23-22 to 23-24
   trigonometric identities, 3-41
Polynomial, 21-15
   Chebyshev, 3-52
   complex conjugate, 1-5, 1-31 to 1-36
   equation, 12-14, 12-15

   Legendre, 15-24
   linear equation, 1-2
   prime, 16-33
Power series:
   analytic function, 13-24, 14-1 to 14-4
   circle of convergence, 13-17
   differentiation, 13-25, 13-26
   domain of convergence, 13-22, 13-23
   radius of convergence, 13-5, 13-6, 13-8, 13-9
   ratio test, 13-6
   root test, 13-7 to 13-12
   uniform convergence of, 13-18
   uniformly convergent series, 13-19, 13-20
Preservations of angles, 21-3, 21-4, 21-11
Prime polynomials, 16-33
Primitive periods, 24-69
Principal part, 16-7, 16-27
Product of series, 5-28
Product symbol "Π", 1-35
Proper function, 19-20
Ptolemy's theorem, 3-14

Quadrilateral, 2-18 to 2-20
   bisection of diagonals, 2-19
   vertices of, 2-18, 2-19
Quotient space, 1-3

Raabe, J.L., 5-24
Raabe's test, 5-26
Ratio test, 5-20
Rational function, 16-32, 16-33, 18-1
Rational numbers, 1-1 to 1-4, 1-33, 1-36
Real axis, 1-1
Real numbers, 1-1 to 1-4, 1-14, 1-17, 2-9
   ordered pairs, 1-17
   positive, 2-21
Real parameter, 2-7, 2-8
Rearrangement of a series, 5-27
Rectangles:
   vertices of, 2-24
Rectangular (equilateral) hyperbola, 2-5
Rectangular region:
   image of the, 20-2

Numbers on this page refer to PROBLEM NUMBERS, not page numbers

Reflection principle, 24-12
   Schwarz's, 24-11
Region, 4-18, 4-19
Regular function, 12-6
Regular point, 16-2
Removable singularity, 16-6, 16-8
   to 16-10, 16-16, 16-17, 16-22,
   16-23, 16-27, 19-33, 19-34,
   21-15
Residues, 17-1, 17-2, 17-18,
   18-2 to 18-15, 18-18, 18-21,
   18-25 to 18-31, 19-2, 19-11,
   19-14, 19-19, 19-22, 19-36
   at poles, 17-3 to 17-11
   of function $\frac{p(z)}{q(z)}$, 17-12 to 17-14
   theorem, 12-22, 17-15, 17-17,
   17-18, 18-1
   theorem application, 17-19 to
   17-29, 18-6, 18-7, 19-1, 19-2,
   19-18, 19-22, 19-27, 19-32
Riemann sphere, 6-38, 6-39,
   6-43
   geometry, 6-45 to 6-57
Riemann's mapping theorem,
   21-41
Root test, 5-36
Roots of:
   complex numbers, 3-26 to
   3-28
   equations, 15-23
Rotation, 20-2 to 20-4
Rouche's theorem, 12-24, 12-26
   21-16
   application, 21-17, 21-19

Schwarz-Christoffel transformation:
   application to closed polygons,
   23-2
   application to various regions,
   23-5 to 23-11, 23-20
   mapping of interior of polygon, 23-3
   mapping of polygon onto real axis, 23-1
   property of, 23-4
Schwarz's:
   principle of reflection, 24-11
   theorem, 12-19
Segmental arc, 6-27
Separation of variables, 22-71
Sequence:

bounded, 4-23, 5-1, 5-4
Cauchy, 1-3, 4-22, 4-25,
4-30
complex numbers, 4-33, 4-35,
4-36, 5-8, 5-9, 5-33, 5-34
convergent, 4-21 to 4-23,
4-35, 5-2, 5-9, 5-11
limits of, 5-2 to 5-5, 5-9 to
5-11, 5-29 to 5-34
lower limit, 5-29, 5-30
monotonically increasing,
24-34
natural numbers, 5-16
null, 5-1 to 5-3
of intervals, 14-34
of numbers, 3-53
partial sums, 5-13
positive numbers, 5-7
subsequence, 4-26
upper limit, 5-29 to 5-32
vertical, 5-10
Series:
   absolutely convergent, 5-18,
   5-24, 5-27, 5-28, 5-33, 5-34,
   5-36, 19-30
   algebraic expression, 5-14,
   5-28
   Cauchy product, 14-26
   complex power, 5-17
   convergence of, 5-16, 5-17,
   5-21 to 5-25
   definition of, 5-17
   divergent, 5-16
   expansion of, 19-28
   geometric, 14-17, 14-24
   harmonic, 5-21
   Laurent, 16-9, 16-15, 16-18,
   16-21, 16-26 to 16-28, 17-1 to
   17-3, 17-6, 17-8, 17-16,
   17-18, 17-19, 17-23
   Maclaurin, 14-9, 14-10, 14-24,
   17-10
   product of, 5-28
   rearrangement of, 5-27
   Taylor, 15-1, 15-25, 16-13,
   17-3, 17-16
   tests, 5-20, 5-36
   uniformly convergent, 12-24
Set:
   bounded, 4-5, 4-28
   closed, 4-13, 4-14, 4-28
   closure of, 4-16, 4-20
   compact, 4-26, 4-28, 4-31
   complex numbers, 1-16
   connected, 6-21, 6-28

Numbers on this page refer to **PROBLEM NUMBERS**, not page numbers

convex, 6-29
diameter, 4-32
empty, 4-14
infimum, 4-6
integers, 1-1
intersection, 4-2, 4-14
open, 4-7 to 4-9, 6-15, 6-19, 6-33
path-connected, 6-33
positive even integers, 1-1
positive integers, 1-1
rational numbers, 1-1
real numbers, 1-1
subset, 1-1
supremium, 4-6
union, 4-2, 4-14
Simple pole, 16-7, 16-24, 17-4, 17-5, 17-7, 18-6, 18-12, 18-13, 18-19, 18-21, 18-25 to 18-27, 18-29, 18-30, 19-4, 19-5, 19-7, 19-9, 19-11, 19-13, 19-14, 19-18, 19-22, 19-28, 19-33 to 19-36
Simply-connected domain:
 Cauchy's integral formula, 12-3
Single-valued function, 12-6
Singular point, 11-23, 16-1, 16-2, 16-5, 16-6, 16-17, 16-24, 16-26, 18-2, 19-2
 essential, 16-15, 16-24, 16-34
 removable, 16-34
Singularity, 16-1 to 16-6, 16-20, 16-26, 16-27, 19-9, 19-10, 19-25
 essential, 16-15 to 16-17, 16-23, 16-28, 21-15
 isolated, 16-2, 16-5, 16-6, 16-7, 16-12, 16-23
 of integrands, 18-1 to 18-3, 18-5, 18-6, 18-26, 18-30
 property, 16-13 to 16-21
 removable, 16-6, 16-8 to 16-10, 16-16, 16-17, 16-22, 16-23, 16-27, 19-33, 19-34, 21-15
Smooth curve, 11-1
Solenoidal field, 22-23
Space:
 compact, 4-30
 complete, 4-25, 4-30
 connected, 6-16
 Euclidean, 6-40
 Hausdroff, 1-3
 metric, 4-3, 4-4, 4-31
 quotient, 1-3

rational number, 4-25
topological, 4-10, 6-4
Special functions, 20-25 to 20-28
 double-valued function, 20-26
 principal branch, 20-27
Speed, 22-30, 22-35
Sphere:
 Riemann, 6-38, 6-39, 6-43
 transformation, 6-42
Square:
 vertices of, 19-17, 19-22, 19-28
Stagnation point, 22-43
Steady-state two-dimensional velocity field, 22-22
Stereographic projection, 6-38, 6-43, 6-44, 6-48, 6-49, 6-51
 ordinary, 6-51
Stirling's asymptotic formula, 24-71
Straight line:
 equations of, 2-7
Stream function, 22-24 to 22-26, 22-34, 22-36
Streamlines, 22-24 to 22-28, 22-33, 22-35, 22-38, 22-39
Subsequence, 4-26
Subset, 1-1, 4-5, 4-6
Supremum and infimum of a set, 4-6
System:
 complex numbers, 1-6
 of masses, 2-26, 2-27
 orthogonal, 22-7, 22-51

Tangent:
 to a circle, 20-6, 20-7
 vector, 8-11
Tauber's theorem, 13-16
Taylor expansion, 14-11, 14-12, 16-3, 16-23
Taylor series, 15-1, 15-25, 16-13, 17-3, 17-16
 rational function, 14-17 to 14-22
 theorem, 14-3, 14-4
Taylor series application, 21-11, 21-12
 differential equations, 14-23
 integration, 14-16
 L'Hôpital's rule, 14-15
 to polynomial, 14-14
 truncated series, 14-13

Numbers on this page refer to PROBLEM NUMBERS, not page numbers

Tchebycheff, (see Chebyshev)
Test:
   alternating series, 5-25
   Cauchy's, 5-34 to 5-36
   comparison, 5-19
   continuity of complex function, 7-15, 7-16
   convergence, 5-19 to 5-35
   D'Alembert's, 5-19, 5-35, 5-36
   power series, 13-6 to 13-12
   ratio, 5-19, 5-20
   root, 5-36
   series, 5-20, 5-36
   Weierstrass, 13-21, 14-3
Theorem:
   Abel's, 6-22
   absolute convergence of series, 5-26 to 5-28
   algebra, 12-14, 21-17
   analytic function, 9-16, 21-2
   Blasius, 22-45
   Bolzano-Weierstrass, 4-26
   Borel Lebesque, 4-31
   Brouwer fixed point, 21-19, 21-20
   Casorati-Weierstrass, 16-29, 16-30
   Cauchy, 11-15, 11-20 to 11-26, 19-8, 19-16
   Cauchy-Goursat, 11-14 to 11-16, 17-1
   Cauchy's integral, 11-14, 16-22, 17-16, 17-17
   closed sets, 4-24
   convergent series, 5-16
   convolution, 22-17
   critical point, 21-10
   De Moivre's, 3-25
   Fourier transform, 18-25, 18-29
   Gauss', 22-56, 22-73
   Gauss' mean value, 12-16
   Green's, 11-10 to 11-12, 11-14, 22-50
   harmonic function, 9-16
   integration of series, 18-11
   Joukowski, 22-46
   Kutta-Joukowski, 22-46
   Laurent, 15-2
   L'Hôpital's, 5-24
   Liouville's, 12-13, 12-14, 24-69
   maximum modulus, 12-17, 12-18

   minimum modulus, 12-18
   Mittag-Leffler, 19-35
   Morera's, 11-22, 12-18
   null sequence, 5-2
   parallelogram, 2-22
   Picard's, 16-31
   product of limits, 6-6
   Ptolemy's, 3-14
   residues, 12-22, 17-15, 17-17, 17-18, 18-1
   Riemann's mapping, 21-41
   Rouché's, 12-24, 12-26, 21-16
   Schwarz's, 12-19
   sum of the limits, 6-6
   Tauber's, 13-16
   Taylor series, 15-1, 15-25, 16-13, 17-3, 17-16
   topology, 21-19
   Weierstrass, 13-21, 14-3
Theory:
   algebraic, 1-36
   axiomatic, 1-3
   of aggregates, 1-3
   of convex bodies, 1-30
Topological space, 4-10, 6-4
   neighborhood, 6-4
   properties of, 6-10 to 6-23
Transcendental:
   numbers, 1-36
Transformation:
   bilinear, 20-15
   circle, 2-23
   composition of, 20-1, 20-2
   conformal, 21-4, 21-43, 22-14
   continuous, 20-5
   fractional linear, 20-12 to 20-19, 20-23 to 20-25
   identity, 6-22, 6-23
   inverse, 6-22, 6-44
   isogonal, 21-4
   isometric, 4-3
   Jacobian of, 21-21, 21-23
   Joukowski, 12-13
   linear, 20-1, 20-2
   of regions, 20-26 to 20-28
   sphere, 6-42
   trigonometric, 6-42
Translation, 20-1 to 20-4
Triangle:
   equilateral, 2-14, 2-15
   inequality, 1-23, 1-30, 5-26
   median of, 2-17
   vertices of, 2-17
Trigonometric functions:
   definition, 10-4

Numbers on this page refer to **PROBLEM NUMBERS**, not page numbers

    even and odd property, 10-5
    principal branch of, 10-22
Trigonometric identities, 3-38 to
    3-40, 3-45 to 3-47
    for polygon, 3-41
Trigonometric transformation,
    19-37

Union:
    definition, 4-2
    of closed sets, 4-15
    of sets, 4-2, 4-14, 4-17, 4-29
Unit circle, 2-4, 2-14, 2-24,
    6-25
Unit element, 1-16, 1-17
Unit open disk:
    mapping of, 20-23 to 20-25
Univalent functions:
    definition, 10-16
    maximum domains of, 10-18
Upper bound, 19-3
Upper half plane, 18-19, 18-21,
    18-22, 18-24 to 18-31, 19-14
Upper limit of a sequence, 5-29
    to 5-32

Vector:
    collinear, 2-7 to 2-9
    complex, 2-18, 22-25, 22-28
    non-parallel, 2-16
    notation, 2-2
    position, 2-2, 2-20, 2-21
    velocity, 22-38
Velocity, 22-26, 22-30, 22-34,
    22-35
    potential, 22-24 to 22-26,
    22-36
    vector, 22-38
Vertical sequence, 5-10
Vortex, 22-31

Weierstrass:
    test, 13-21, 14-3
    theorem, 15-19, 19-36, 24-40
Werke, 1-15
Winding number, 21-16
w-plane:
    transforming z-plane into,
    20-1 to 20-27

Zero, 16-2 to 16-4, 16-11, 16-20
    complex, 21-17
    element, 1-16, 1-17
    hyperbolic function, 10-7
    of a function, 10-4
Zeta function, 24-60
    analytic continuation of,
    24-59
Z-plane, 2-1

A particularly useful reference for those in math, science, engineering and other technical fields. Includes the most-often used formulas, tables, transforms, functions, and graphs which are needed as tools in solving problems. The entire field of special functions is also covered. A large amount of scientific data which is often of interest to scientists and engineers has been included.

Available at your local bookstore or order directly from us by sending in coupon below.

RESEARCH & EDUCATION ASSOCIATION
61 Ethel Road W., Piscataway, New Jersey 08854
Phone: (908) 819-8880

☐ Payment enclosed
☐ Visa ☐ Master Card

Charge Card Number

Expiration Date: _____ / _____
　　　　　　　　　　　Mo　　　Yr

Please ship the "Math Handbook" @ $24.95 plus $4.00 for shipping.

Name _____

Address _____

City _____ State _____ Zip _____

"The ESSENTIALS" of Math & Science

Each book in the ESSENTIALS series offers all essential information of the field it covers. It summarizes what every textbook in the particular field must include, and is designed to help students in preparing for exams and doing homework. The ESSENTIALS are excellent supplements to any class text.

The ESSENTIALS are complete, concise, with quick access to needed information, and provide a handy reference source at all times. The ESSENTIALS are prepared with REA's customary concern for high professional quality and student needs.

Available in the following titles:

- Advanced Calculus I & II
- Algebra & Trigonometry I & II
- Anthropology
- Automatic Control Systems / Robotics I & II
- Biology I & II
- Boolean Algebra
- Calculus I, II & III
- Chemistry
- Complex Variables I & II
- Differential Equations I & II
- Electric Circuits I & II
- Electromagnetics I & II
- Electronic Communications I & II
- Electronics I & II
- Finite & Discrete Math
- Fluid Mechanics / Dynamics I & II
- Fourier Analysis
- Geometry I & II
- Group Theory I & II
- Heat Transfer I & II
- LaPlace Transforms
- Linear Algebra
- Math for Engineers I & II
- Mechanics I, II & III
- Modern Algebra
- Numerical Analysis I & II
- Organic Chemistry I & II
- Physical Chemistry I & II
- Physics I & II
- Set Theory
- Statistics I & II
- Strength of Materials & Mechanics of Solids I & II
- Thermodynamics I & II
- Topology
- Transport Phenomena I & II
- Vector Analysis

If you would like more information about any of these books, complete the coupon below and return it to us or go to your local bookstore.

RESEARCH & EDUCATION ASSOCIATION
61 Ethel Road W. • Piscataway, New Jersey 08854
Phone: (908) 819-8880

Please send me more information about your Essentials Books

Name _____

Address _____

City _____ State _____ Zip _____

"The ESSENTIALS" of COMPUTER SCIENCE

Each book in the **Computer Science ESSENTIALS** series offers all essential information of the programming language and/or the subject it covers. It includes every important programming style, principle, concept and statement, and is designed to help students in preparing for exams and doing homework. The **Computer Science ESSENTIALS** are excellent supplements to any class text or course of study.

The **Computer Science ESSENTIALS** are complete and concise, with quick access to needed information. They also provide a handy reference source at all times. The **Computer Science ESSENTIALS** are prepared with REA's customary concern for high professional quality and student needs.

Available Titles Include:
- BASIC
- C Programming Language
- COBOL I
- COBOL II
- Data Structures I
- Data Structures II
- Discrete Stuctures
- FORTRAN
- PASCAL I
- PASCAL II
- PL / 1 Programming Language

If you would like more information about any of these books,
complete the coupon below and return it to us or go to your local bookstore.

RESEARCH & EDUCATION ASSOCIATION
61 Ethel Road W. • Piscataway, New Jersey 08854
Phone: (908) 819-8880

Please send me more information about your Computer Science Essentials Books

Name _____

Address _____

City _____ State _____ Zip _____

"The ESSENTIALS" of ACCOUNTING & BUSINESS

Each book in the Accounting and Business ESSENTIALS series offers all essential information about the subject it covers. It includes every important principle and concept, and is designed to help students in preparing for exams and doing homework. The Accounting and Business ESSENTIALS are excellent supplements to any class text or course of study.

The Accounting and Business ESSENTIALS are complete and concise, with quick access to needed information. They also provide a handy reference source at all times. The Accounting and Business ESSENTIALS are prepared with REA's customary concern for high professional quality and student needs.

Available titles include:

Accounting I & II
Advanced Accounting I & II
Advertising
Auditing
Business Law I & II
Business Statistics I & II
Corporate Taxation
Cost & Managerial Accounting I & II

Financial Management
Income Taxation
Intermediate Accounting I & II
Microeconomics
Macroeconomics I & II
Marketing Principles
Money & Banking I & II

If you would like more information about any of these books, complete the coupon below and return it to us or go to your local bookstore.

RESEARCH & EDUCATION ASSOCIATION
61 Ethel Road W. • Piscataway, New Jersey 08854
Phone: (908) 819-8880

Please send me more information about your Accounting & Business Essentials Books

Name _____

Address _____

City _____ State _____ Zip _____

Available at your local bookstore or order directly from us by sending in coupon below.

RESEARCH & EDUCATION ASSOCIATION
61 Ethel Road W., Piscataway, New Jersey 08854
Phone: (908) 819-8880

☐ Payment enclosed
☐ Visa ☐ Master Card

Charge Card Number

Expiration Date: _____ / _____
                    Mo   Yr

Please ship **"Careers for the '90s"** @ $15.95 plus $4.00 for shipping.

Name _____

Address _____

City _____ State _____ Zip _____

HANDBOOK AND GUIDE FOR COMPARING and SELECTING COMPUTER LANGUAGES

BASIC	PL/1
FORTRAN	APL
PASCAL	ALGOL-60
COBOL	C

- This book is the first of its kind ever produced in computer science.
- It examines and highlights the differences and similarities among the eight most widely used computer languages.
- A practical guide for selecting the most appropriate programming language for any given task.
- Sample programs in all eight languages are written and compared side-by-side. Their merits are analyzed and evaluated.
- Comprehensive glossary of computer terms.

Available at your local bookstore or order directly from us by sending in coupon below.

RESEARCH & EDUCATION ASSOCIATION
61 Ethel Road W., Piscataway, New Jersey 08854
Phone: (908) 819-8880

☐ Payment enclosed
☐ Visa ☐ Master Card

Charge Card Number

☐☐☐☐ ☐☐☐☐ ☐☐☐☐ ☐☐☐☐

Expiration Date: ____ / ____
                    Mo   Yr

Please ship the **"Computer Languages Handbook"** @ $8.95 plus $2.00 for shipping.

Name _____

Address _____

City _____ State _____ Zip _____

The Best Test Preparation for the GRE GRADUATE RECORD EXAMINATION GENERAL TEST

6 full-length exams

Completely Up-to-Date Based on the Current Format of the GRE

- All of the exam sections prepared by test experts in the particular subject fields
- The *ONLY* test preparation book with detailed explanations to every exam question to help you achieve a *TOP SCORE*
- Includes a special section of expert test tips and strategies

PLUS... a **MATH REVIEW** with drills to build math skills for quantitative sections, and a **VOCABULARY LIST** with word drills of the most frequently tested words on the GRE

RESEARCH & EDUCATION ASSOCIATION

Available at your local bookstore or order directly from us by sending in coupon below.

RESEARCH & EDUCATION ASSOCIATION
61 Ethel Road W., Piscataway, New Jersey 08854
Phone: (908) 819-8880

☐ Payment enclosed
☐ Visa   ☐ Master Card

Charge Card Number

Expiration Date: _____ / _____
                    Mo   Yr

Please ship **"GRE General Test"** @ $15.95 plus $4.00 for shipping.

Name _____

Address _____

City _____ State _____ Zip _____

REA's Test Preps
The Best in Test Preparations

The REA "Test Preps" are far more comprehensive than any other test series. They contain more tests with much more extensive explanations than others on the market. Each book provides several complete practice exams, based on the most recent tests given in the particular field. Every type of question likely to be given on the exams is included. Each individual test is followed by a complete answer key. **The answers are accompanied by full and detailed explanations.** By studying each test and the pertinent explanations, students will become well-prepared for the actual exam.

REA *has published 40 Test Preparation volumes in several series. They include:*

Advanced Placement Exams (APs)
Biology
Calculus AB & Calculus BC
Chemistry
Computer Science
English Literature & Composition
European History
Government & Politics
Physics
Psychology
United States History

College Board Achievement Tests (CBATs)
American History
Biology
Chemistry
English Composition

French
German
Literature
Mathematics Level I, II & IIC
Physics
Spanish

Graduate Record Exams (GREs)
Biology
Chemistry
Computer Science
Economics
Engineering
General
History
Literature in English
Mathematics
Physics
Political Science
Psychology

CBEST - California Basic Educational Skills Test
CDL - Commercial Drivers License Exam
ExCET - Exam for Certification of Educators in Texas
FE (EIT) - Fundamentals of Engineering Exam
GED - High School Equivalency Diploma Exam
GMAT - Graduate Management Admission Test
LSAT - Law School Admission Test
MCAT - Medical College Admission Test
NTE - National Teachers Exam
SAT - Scholastic Aptitude Test
TOEFL - Test of English as a Foreign Language

RESEARCH & EDUCATION ASSOCIATION
61 Ethel Road W. • Piscataway, New Jersey 08854
Phone: (908) 819-8880

Please send me more information about your Test Prep Books

Name _____

Address _____

City _____ State _____ Zip _____

REA's **Problem Solvers**

The "PROBLEM SOLVERS" are comprehensive supplemental textbooks designed to save time in finding solutions to problems. Each "PROBLEM SOLVER" is the first of its kind ever produced in its field. It is the product of a massive effort to illustrate almost any imaginable problem in exceptional depth, detail, and clarity. Each problem is worked out in detail with a step-by-step solution, and the problems are arranged in order of complexity from elementary to advanced. Each book is fully indexed for locating problems rapidly.

ADVANCED CALCULUS
ALGEBRA & TRIGONOMETRY
AUTOMATIC CONTROL
   SYSTEMS/ROBOTICS
BIOLOGY
BUSINESS, ACCOUNTING, & FINANCE
CALCULUS
CHEMISTRY
COMPLEX VARIABLES
COMPUTER SCIENCE
DIFFERENTIAL EQUATIONS
ECONOMICS
ELECTRICAL MACHINES
ELECTRIC CIRCUITS
ELECTROMAGNETICS
ELECTRONIC COMMUNICATIONS
ELECTRONICS
FINITE & DISCRETE MATH
FLUID MECHANICS/DYNAMICS
GENETICS
GEOMETRY
HEAT TRANSFER
LINEAR ALGEBRA
MACHINE DESIGN
MATHEMATICS for ENGINEERS
MECHANICS
NUMERICAL ANALYSIS
OPERATIONS RESEARCH
OPTICS
ORGANIC CHEMISTRY
PHYSICAL CHEMISTRY
PHYSICS
PRE-CALCULUS
PSYCHOLOGY
STATISTICS
STRENGTH OF MATERIALS &
   MECHANICS OF SOLIDS
TECHNICAL DESIGN GRAPHICS
THERMODYNAMICS
TOPOLOGY
TRANSPORT PHENOMENA
VECTOR ANALYSIS

If you would like more information about any of these books,
complete the coupon below and return it to us or visit your local bookstore.

RESEARCH & EDUCATION ASSOCIATION
61 Ethel Road W. • Piscataway, New Jersey 08854
Phone: (908) 819-8880

Please send me more information about your Problem Solver Books

Name _____

Address _____

City _____ State _____ Zip _____